Lecture Notes
in Control and Information Sciences 223

Editor: M. Thoma

Springer-Verlag London Ltd.

O. Khatib and J.K. Salisbury (Eds)

Experimental Robotics IV

The 4th International Symposium, Stanford, California, June 30 - July 2, 1995

Springer

Editors

Professor Oussama Khatib
Department of Computer Science, Stanford University, Stanford, CA 94305, USA

Dr J. Kenneth Salisbury
Department of Mechanical Engineering and Artificial Intelligence Lab.
Massachusetts Institute of Technology
Cambridge, MA 02139, USA

Front cover illustration reproduced with the kind permission of Stanford University.

ISBN 978-3-540-76133-4 ISBN 978-3-540-40942-7 (eBook)
DOI 10.1007/978-3-540-40942-7

British Library Cataloguing in Publication Data
Experimental robotics IV : the 4th International Symposium,
Stanford, California, June 30-July 2, 1995. - (Lecture
notes in control and information sciences ; 223)
1.Robotics - Congresses
I.Khatib, Oussama II.Salisbury, J. Kenneth
629.8'92

ISBN 978-3-540-76133-4

Library of Congress Cataloging-in-Publication Data
A catalog record for this book is available from the Library of Congress

Originally published by Springer-Verlag London Limited in 1997

Typesetting: Camera ready by editors

69/3830-543210 Printed on acid-free paper

Preface

Experimental Robotics IV — The Fourth International Symposium was held at Stanford University, California, from June 30 to July 2, 1995. This meeting was the fourth in a series of symposia designed to bring together, in a small group setting, researchers from around the world who are in the forefront of experimental robotics research. The goal of these symposia is to provide a forum for research in robotics that focuses on theories and principles that are validated by experiments.

The experimental robotics symposia are organized every two years in a rotating fashion around North America, Europe, and Asia. The first Symposium, organized by V. Hayward and O. Khatib, was held in in Montréal, Canada in June 1989. The second, organized by R. Chatila and G. Hirzinger, was held in Toulouse, France, in June 1991. The third, organized by T. Yoshikawa and F. Miyazaki, was held in Kyoto, Japan, in October 1993. The proceedings of Experimental Robotics Symposia are published by Springer-Verlag. In addition to the proceedings, these symposia have produced compilations of video segments illustrating the reported research, which are available as Video Proceedings.

The International Program Committee of the Fourth International Symposium on Experimental Robotics was composed of the following individuals:

Oussama Khatib	Stanford University, U.S.A (Co-Chair)
Kenneth Salisbury	MIT, U.S.A (Co-Chair)
Alicia Casals	Universitat Politècnica de Catalunya, Spain
Raja Chatila	LAAS/CNRS, France
John Craig	Adept Technology, Inc., U.S.A.
Paolo Dario	Scuola Superiore S. Anna, Italy
Joris De Schutter	Katholieke Universiteit Leuven, Belgium
Vincent Hayward	McGill University, Canada
Gerhard Hirzinger	DLR, Germany
Fumio Miyazaki	Osaka University, Japan
Yoshihiko Nakamura	Tokyo University, Japan
Jean-Pierre Merlet	INRIA, France
James Trevelyan	University of Western Australia
Tsuneo Yoshikawa	Kyoto University, Japan

The proceedings of the fourth symposium includes the fifty-five contributions that were selected by the International Program Committee and presented during the three days of the meeting. These contributions represent work in the areas of design, perception, control, planning, and robotic applications of research groups from Belgium, Canada, Denmark, England, France, Germany, Italy, Japan, Portugal, Spain, Switzerland, and the United States. In addition, a small group of observers from industry and funding agencies was invited to take part in this meeting.

The fourth symposium began with a keynote address by Brian Carlisle, the Chairman and CEO of Adept Technology, on "Robot Technology for System Integration." Received with interest, the keynote presentation generated numerous questions and comments from the participants.

In addition to the technical sessions, the fourth symposium featured a panel on "Challenges for the Next Decade," with Brian Carlisle and John Craig as moderators. This was the first time at ISER that we organized a session with speakers from industry, and we feel it was a very worthwhile addition. The speakers at this panel were Craig Battles (Boeing), Steve Holland (GM), Vic Scheinman, Tyler Schilling (Schilling), Stan Rosenschein (Teleos), and Antonio Terrible (Tecnomare, Italy). The panel served as a useful vehicle to inspire introspection and interaction among the participants.

As we prepared for this meeting, our foremost objective was to create a collegial atmosphere with ample opportunity for discussion and exchange of the ideas. In view of the feedback we have received, this goal was largely reached. We believe that the credit for this success goes to the authors for the quality of their contributions and to all the participants for their availability and active involvement in the discussions all along the three days of this meeting.

On behalf of the International Program Committee, we would like to thank Dean James Plummer and Professor Bernard Roth for their support and warm welcoming addresses. Also, we would like to express our appreciation and thanks to Adept Technology, Interval Research Corporation, and Lockheed Martin Corporation, for the financial support they extended to the fourth symposium. We are grateful to Jacqui Taylor of MIT for her assistance in the preparation of the final proceedings.

Our special thanks go the staff and students of Stanford University who generously gave of their time to help in the organization of this meeting – many thanks to Arancha Casal, the Symposium Secretary, and to Andreas Baader, Mike Costa, Stan Birch, Ralf Koeppe, Sascha Lewald, Mina Madrigal, Allison Okamura, Francisco Valero, and Stef Sonck.

The international program committee has asked Alicia Casals (Barcelona, Spain) and Anibal Almeida (Coimbra, Portugal) to co-chair the Fifth Symposium to be held in Barcelona, Spain in June 1997.

Oussama Khatib and Kenneth Salisbury
Stanford, California, August 1996

List of Participants

1. Abdou, Sofiane
   INRIA Rhône-Alpes
   46, rue Félix-Viallet
   38031 Grenoble, France

2. Asada, Haruhiko
   Dept. of Mechanical Engineering
   Massachussetts Institute of Technology
   Cambridge, MA 02139

3. Battles, Craig
   The Boeing Co.
   Commercial Airplane Group
   Automation Research
   P.O. Box 3707 M/S 30-RK
   Seattle, WA 98124-2207

4. Becker, Craig
   Robotics Laboratory
   Dept. of Computer Science
   Stanford, CA 94305

5. Bidaud, Philippe
   Laboratoire de Robotique de Paris
   Centre Universitaire de Technologie
   10-12 Avenue de L'Europe
   78140 Velizy, France

6. Bruyninckx, Herman
   Katholieke Universiteit Leuven
   Celestijnenlaan 300B
   B-3030 Heverlee, Belgium

7. Burdick, Joel
   California Institute of Technology
   Department of Mechanical Engineering
   Mail Code 104-44
   Pasadena, CA 91125

8. Burridge, Robert R.
   Department of Electrical Engineering
   and Computer Science
   The University of Michigan
   Ann Arbor, Michigan

9. Cannon, David
   Penn State University
   1761 Princeton Drive
   State College PA, 16803

10. Carlisle, Brian
    Adept Technology, Inc.
    150 Roase Orchard Way
    San Jose, CA 95134

11. Chatila, Raja
    LAAS-CNRS
    7, Av. du Colonel Roche
    31077 Toulouse, France

12. Christiansen, Alan D.
    Dept. of Computer Science
    Tulane University
    New Orleans, LA 70118

13. Chung, Woojin
    Dept. of Mechano-Informatics
    University of Tokyo
    7-3-1 Hongo, Bunkyo-ku
    Tokyo 113, Japan

14. Coste-Manière, Eve
    INRIA Sophia Antipolis
    2004, Route des Lucioles BP93
    06902 Sophia Antipolis, France

15. Craig, John J.
    Silma Incorporated
    1601 Saratoga-Sunnyvale Road
    Cupertino, California 95014

16. Cutkosky, Mark
    Center for Design Research
    Building 02-530, Duena Street

Stanford University
Stanford, CA 94305-4026

17. Dario, Paolo
Scuola Superiore Santa Anna
Via Carducii 40
56127 Pisa, Italy

18. Daviet, Pascal
INRIA, BP 105
78153 Le Chesnay, France

19. De Almeida, Aníbal
Universidade de Coimbra
Departamento de Engenharia Electrotécnica
Largo Marquêz de Pombal
Pombal, 3000 Coimbra, Portugal

20. Donald, Bruce Randall
Robotics and Vision Laboratory
Department of Computer Science
Cornell University
Ithaca, NY 14853

21. Dubowsky Steven
Dept. of Mechanical Engineering
Massachussetts Institute of Technology
Cambridge, MA 02139-4307

22. Eicker, Pat
Sandia National Labs, Dept 1410
1515 Eubank SE.
Albuquerque, NM 87185

23. Elgazzar, Shadia
National Research Council of Canada
Department of Electrical Engineering
Montreal Road, M-50
Ottawa, Ontario, Canada K1A 0R6

24. Espiau, Bernard
INRIA Rhône-Alpes
41, rue Félix-Viallet
38031 Grenoble, France

25. Fraisse, Philippe
LIRMM
Université de Montpellier II
161 Rue Ada
34392, Montpellier, France

26. Garnero, Marie-Agneès
Electricite de France
Groupe Téléopération-Robotique
6, quai Watier - B.P. 49
78401 Chatou, France

27. Gat, Erann Gat
Jet Propulsion Laboratory
4800 Oak Grove Drive
Pasadena, CA 91109

28. Goldberg, Kenneth
Dept. of Computer Science
University of California at Berkeley
Berkeley, CA 94720

29. González-Baños, Hector Hugo
Robotics Laboratory
Dept. of Computer Science
Stanford, CA 94305

30. Ha, Yun-Su
Intelligent Robot Laboratory
University of Tsukuba
1-1-1 Tennodai, Tsukuba
305, Japan

31. Hani, Ahmad Fadzil M.
University Sains Malaysia
USM Perak Branch
31750 Tronoh, Perak, Malaysia

32. Hayward, Vincent
McGill University
3480 University Street
Montreal, Quebec H3A 2A7
Canada

33. Holand, Steven
General Motors Corp.,
NAO Manufacturing Center
30300 Mound Road 1-9
Warren, MI 48090-9040

34. Hollerbach, John M.
Department of Computer Science
University of Utah
Salt Lake City, UT 84112

35. Howe, Robert D.
Division of Applied Sciences
Harvard University
Cambridge, MA 02138

36. Hyde, James
Center for Design Research
Building 02-530, Duena Street
Stanford University
Stanford, CA 94305-4026

37. Inaba, Masayuki
University of Tokyo
Dept. of Mechano-Informatics
7-3-1 Hongo, Bunkyo-ku
Tokyo 113, Japan

38. Inoue, Hirochika
University of Tokyo
Dept. of Mechano-Informatics
7-3-1 Hongo, Bunkyo-ku
Tokyo 113, Japan

39. Kaneko, Makoto
Industrial and Systems Engineering
Hiroshima University
Kagamiyama 1-4-1, Higashi-Hiroshima
Hiroshima 724, Japan

40. Kazerooni, Hami
Dept. of Mechanical Engineering
6189 Etcheverry Hall 1740
University of California at Berkeley
Berkeley, CA 94720-1740

41. Khatib, Oussama
Department of Computer Science
Stanford University
Stanford, California 94305

42. Killough, Steve
Oak Ridge National Laboratory
P.O.Box 2008
Oak Ridge, TN 37831-6364

43. Koeppe, Ralf
Institute for Robotics
Postfach 1116
D-82230 Wessling, Germany

44. Kolarov, Krasimir
Interval Research Corporation
1801 Page Mill Road
Palo Alto, CA 94204

45. Konolige, Kurt
SRI International
333 Ravenswood Avenue
Menlo Park, CA 94025

46. Krotkov, Eric
The Robotics Institute
Carnegie Mellon University
5000 Fobres Avenue
Pittsburgh PA 15213-3890

47. Latombe, Jean-Claude
Robotics Laboratory
Dept. of Computer Science
Stanford, CA 94305

48. Laugier, Christian,
LIFIA-CNRS & INRIA Rhône-Alpes
46, rue Félix-Viallet
38031 Grenoble, France

49. Lawrence, Peter D.
Department of Computer Science
University of British Columbia
Vancouver, B.C. V6T 1Z4, Canada

50. Lenarcic, Jadran
Jozef Stefan Institute
University of Ljubljana, Jamova 39
61111 Ljubljana, Slovenia

51. Martinoli, Alcherio
Microcomputing Laboratory
Swiss Federal Institute of technology
IN-F Ecublens, CH-1015, Lausanne, Switzerland

52. Mavroidis Constantinos
Dept. of Mechanical Engineering
Massachussetts Institute of Technology
Cambridge, MA 02139-4307

53. Menezes, Paulo
Universidade de Coimbra
Largo Marquêz de Pombal
Pombal, 3000 Coimbra, Portugal

54. Merlet, Jean-Pierre
INRIA, Centre de Sophia-Antipolis
2004, Route des Lucioles
06565 Valbonne, France

55. Miyazaki, Fumio
Faculty of Engineering Science
Osaka University
Toyonaka, Osaka, 560, Japan

56. Moritz, Wolfgang
Universitat G. Paderborn
FB 10 - Automatisierungstechnik
Pohlweg 55
D-33098 Paderborn, Germany

57. Morrell, John B.
Artificial Intelligence Laboratory
545 Technology Square
Massachussetts Institute of Technology
Cambridge, MA 02139

58. Nelson, Brad
The Robotics Institute
Carnegie Mellon University
5000 Fobres Avenue
Pittsburgh PA 15213-3890

59. Pai, Dinesh K.
Department of Computer Science
University of British Columbia
Vancouver, B.C. V6T 1Z4, Canada

60. Paquin, Normand
MPB Technologies Inc.
151 Hymus Blvd.
Pointe Claire, PQ, H9R 1E9, Canada

61. Peuch, Alexis
IFREMER, BP 330
83507 Lay Seyne Sur Mer, France

62. Pierrot, Francois
LIRMM
Universite de Montpellier II
161 Rue Ada
34392 Montpellier, France

63. Poirier, Alain
Canadian Space Agency
6767 Route de L'Aeroport
Saint Hubert, Qc J3Y 8Y9, Canada

64. Popović, Milos R.
University of Toronto
5 King's College Road
Toronto, Ontario M5S 1A4, Canada

65. Pratt, Gill A.
Artificial Intelligence Laboratory
545 Technology Square
Massachussetts Institute of Technology
Cambridge, MA 02139

66. Prattichizzo, Domenico
Dipartimento di Sistemi Elettricie
Università di Pisa, Italy

67. López de Mantaras, R
Artificial Intelligence Institute
Campus UAB, 08193 Bellaterra, Spain

68. Ravn, Ole
Institute of Automation
Technical University of Denmark
Building 326-327
DK-2800 Lyngby, Denmark

69. Reboulet, Claude
CERT/DERA
2 Avenue Edouard Belin
31055 Toulouse, France

70. Rizzi, Alfred A.
Department of Electrical Engineering
and Computer Science
The University of Michigan
Ann Arbor, Michigan

71. Rosenschein, Stanley J.
Teleos Research
576, Middlefield Avenue
Palo Alto, CA 94301

72. Roth, Bernard
Dept. Mechanical Engineering
Stanford University
Stanford, CA 94305

73. Rus, Daniela
Dartmouth College
Department of Computer Science
6211 Sudikoff Laboratory
Hanover, NH 03755-3510

74. Russakow, Jeffrey
Aerospace Robotics laboratory
Stanford University
Stanford, CA 94305

75. Salcudean, Tim (S. E.)
Department of Electrical Engineering
University of British Columbia
Vancouver, B.C. V6T 1Z4, Canada

76. Salisbury, J. Kenneth
Artificial Intelligence Laboratory
Massachusetts Institute of Technology
545 Technology Square
Cambridge, MA 02139

77. Schilling, Reuben
Schilling Robotics
1632 Da Vinci Ct
Davis, CA 95616

78. Schütte, Herbert
Universitat G. Paderborn
FB 10 - Automatisierungstechnik
Pohlweg 55
D-33098 Paderborn, Germany

79. Slatkin, Brett A.
California Institute of Technology
Mail Code 104-44
Pasadena CA 91125

80. Slotine, Jean-Jacques E.
Dept. of Mechanical Engineering
Massachussetts Institute of Technology
Cambridge, MA 02139

81. Son, Jae S.
Harvard University
Division of Applied Sciences
Pierce Hall
Cambridge, MA 02138

82. Stevens, Michael
Robotics Research Group
Dept. of Engineering Science
Oxford University
Oxford OX1 3PJ, England

83. Taussig, Robert
Bechtel
50 Beale Street
San Francisco, CA 94105

84. Terribile, A.
Tecnomare Spa
S. Marco 3584, Italy

85. Tomasi, Carlo
Robotics Laboratory
Dept. of Computer Science
Stanford University
Stanford, Ca 94305

86. Tonko, Martin
Inst. Algorithmen & Kognitive Systeme
Universitat Karlsruhe
Postfach 6980
D-76128 Karlsruhe, Germany

87. Tsujita, Katsuyoshi
Dept. of Mechanical Engineering for Computer Controlled Machinery
Osaka University
2-1, Yamada-Oka, Suita City
Osaka 565, Japan

88. Van Vactor, David
Lockheed Martin
3251 Hanover Street 0192-30 - B/250
Palo Alto, CA 94304

89. Verplank, William
Interval Research Corporation
1801 Page Mill Road
Palo Alto, CA 94204

90. Williams, David
Robotics Laboratory
Dept. of Computer Science
Stanford University
Stanford, Ca 94305

91. Williamson, Matthew M.
Artificial Intelligence Laboratory
545 Technology Square
Massachussetts Institute of Technology
Cambridge, MA 02139

92. Yokoi, K.
Robotics Laboratory
Dept. of Computer Science
Stanford University
Stanford, Ca 94305

93. Yoshida, Kazuya
Dept. of Aeronautics and
Space Engineering
Tohoku University
Aoba, Sendai, 980-77, Japan

94. Yoshikawa, Tsuneo
Department of Mechanical Engineering
Kyoto University
Kyoto 606, Japan

Contents

Author Index

Chapter 1

Cooperative Mobile Robots

Cooperative mobile robots are one of several emerging contexts in which multiple agents act in coordination to achieve common manipulation or perception goals. Although multi-finger and multi-arm systems seek similar objectives, cooperative mobile robots intrinsically address larger workspaces and objects while having to face the unique problems of navigation and intra-agent communication.

Martinoli and Mondada describe biologically inspired control of the table-top Khepera robot, utilizing simple cooperative and non-cooperative behaviors. They show that employing more robots to perform a task can either help or hinder the success rate, depending on the task demands and particularly on the interaction between robots.

Böhringer, Brown, Donald, Jennings, and Rus have applied minimalist principals to permit both mobile robots and massively parallel arrays of micro-machined actuators to manipulate objects, large and small. They identify the trade-off between communication and computation resources and have shown how properly distributing resources can significantly reduce the complexity of parallel manipulation tasks.

Alami, Aguilar, Bullata, Fleury, Herrb, Ingrad, M. Khatib, and Robert, introduce a plan-merging algorithm which permits cooperation between mobile robots. By incrementally merging plans into a set of coordinated plans, cooperation is achieved with limited centralized activity. The theory is demonstrated both in simulation and on three Hilare robots.

Amat, Mantàras, and Sierra describe experiments with a group of small autonomous vehicles designed to cooperatively explore unknown environments. The robots individually employ different behaviors to perform the exploration resulting in improved collective information gathering.

Collective and Cooperative Group Behaviours: Biologically Inspired Experiments in Robotics

Alcherio Martinoli
Microcomputing Laboratory, Swiss Federal Institute of Technology
Lausanne, Switzerland
martinoli@di.epfl.ch

Francesco Mondada
Microcomputing Laboratory, Swiss Federal Institute of Technology
Lausanne, Switzerland
mondada@di.epfl.ch

Abstract

This paper describes the implementation of two biologically inspired collective behaviours on a group of Khepera miniature mobile robots. The first experiment is concerned with the gathering and clustering of randomly distributed small cylinders. In the second experiment the group of robots are expected to remove long sticks from holes, requiring a synchronous collaboration between two robots. The results are quantified, analysed and discussed, showing interesting (both positive and negative) aspects of this approach. Furthermore, we compare the results of both experiments with those reported by Deneubourg [1], [2] where similar behaviours are observed in ant colonies.

1. Introduction

In the last years we observe more and more collaborations between biologists and engineers [3]. For instance, common experiences, where biologically inspired control structures are implemented on real robots, allow biologists to understand how living organisms work, and engineers to develop new technologies that can deal with unsolved problems. This interaction between biologists and engineers is bringing new ideas to the traditional computational paradigms of robotics which typically involve several sequential and precise functional processes. First, sensing the environment, then detecting features, then constructing and modifying a world model, reasoning for the task and the world model in order to find some sequence of actions which might lead to success, then executing the action sequence one step at the time while updating the world model and replanning it if necessary at any stage. This is a very time consuming operation and requires a remarkable computational power and basic knowledge.

In the last decade a completely different architectural approach was envisaged: the "subsumption" architecture [4]. The control architecture, inspired from some biological aspects, consists of a small number of simple modules, each capable to sense some limited aspect of the environment and to control part or all of the robot effector systems to achieve some limited tasks. Also in the field of neural networks, the biological inspiration plays an important role in the design of control architectures. At the Microcomputing Laboratory of the Swiss Institute of Technology many efforts has been undertaken to design and realize robots and control structures, based on biological models and to implement autonomous agents.

Let us now talk about the actual state of the research in a particular mutual domain of the biology and of the engineering fields: collective behaviour. Ant colonies are able to collect objects (such as food or dead ants) and place them in particular places. All ants of a given colony place the food at the same place and the carcass in another place. In this way they can collect and store food or carry dead ants to a "cemetery": if a large number of ant corpses or food particles are scattered outside a nest, they will pick them up, carry them for a while, and drop them. Within a short time we can observe that the corpses are being arranged into small clusters and, as time goes on, the number of clusters decreases and their size increases until eventually all the corpses will be in one or two large clusters. The emergence of these clusters has been studied with social insects by Deneubourg [1], who showed that a simple mechanism involving the modulation of the probability of dropping corpses as a function of the local density, was sufficient to generate the observed sequence of the clustering of corpses.

Gaussier and Zrehen [5] carried out an experiment with a group of Khepera robots implementing similar mechanisms with the same property: the probability of dropping corpses was a function of the local density. They mounted a hook behind the robot, which was therefore able, with an appropriate sequence of movements, to grasp and shift small cylindrical objects. Precise rules for the basic behaviours were defined: the perception of the objects and obstacles (Winner-Takes-All neuronal net) as well as dragging and placing objects were preprogrammed in such a way that the global probability of building a cluster was greater than that of destroying it. So, after a few minutes, the first clusters began to appear on the arena (see [5] for more details).

Beckers [6] made the same experiment with robots of approximatively 25 cm in diameter. The collective behaviour was analysed on the basis of the stigmergy principle, which signifies 'incitement to work by the products of the work'. It consists in essentially the production of a certain behaviour in agents as a consequence of the effects produced in the environment by previous actions. The experiment was carried out using 1 to 5 robots in an arena where many pucks of 4 cm in diameter were scattered. The robots were equipped with a frontal surface to push the pucks. A microswitch was installed behind the surface to control the maximal number of pucks which can be pushed at the same time. The robot was able to shift two pucks, but when three pucks were detected by the microswitch, the robot stopped pushing and changed direction. Their results indicated that the optimal density of robots on the arena surface, in order to accomplish the given collective task in a minimal time lapse (relative to the number of robots), was three. According to the authors, the reasons for the presence of this optimum were attributed to the geometry of the clusters and to the constructive and destructive interferences among the agents.

Another kind of experiment was performed by Deneubourg [2] on the collective behaviour of a colony of ants. They rammed some sticks close to the nest and they observed what occurred. After a few minutes, the ants tried to grasp the sticks in order to accumulate building material for their nest, but a single ant was not able to accomplish this task. A few minutes later, the colony solved the problem with a collaborative behaviour among two or more ants.

We can classify the above mentioned experiments into two categories, considering the different tasks involved:

- collective noncooperative behaviour, which does not necessarily need cooperation among the individuals to be solved, i.e., a greater number of agents can only accelerate the work to be accomplished;
- collective cooperative behaviour, which absolutely needs the collaboration of two or more individuals in order to be carried out, because of some physical constraints of a single agent.

This paper aims to describe two experiments on real robots, which can be classed into the two above mentioned categories. The discussion will also focus on the radical difference between the two tasks.

2. Materials and Methods

Khepera is a miniature mobile robot developed to perform "desktop" experiments [7]. Its main characteristic is a diameter of 55 mm. It also has other very interesting possibilities, such as an important processing power (32 bits processor at 16 MHz), autonomy, precise odometry, light and proximity sensors, grasping possibilities, and many other additional features that can be added on a extension bus. In its basic configuration Khepera is equipped with 8 infrared (IR) sensors, 6 on the front and 2 behind its cylindrical structure. On the front these sensors are distributed with a gap of about 14 mm. The wheels are controlled by two DC motors with an incremental encoder (12 pulses per mm of advancement of the robot), and can move in both directions. The simple geometrical shape and the motor layout allow Khepera to negotiate any kind of obstacle or corner. Each robot is extended with a gripper module, which can grasp and carry objects with a maximum diameter of 50 mm (see fig. 2a). A robot of this size is not only optimal for the test of basic features in an individual behaviour, but is also a very interesting platform with which to perform experiments in collective robotics: 20 robots can easily work on $2\,m^2$, which approximately represents an area of 10×20 m for a bigger robot (being for instance 50 cm in diameter).

Our two experiments on collective behaviour use 1 to 5 Kheperas. In both cases, the number of robots involved has been gradually increased and the performance measured and compared. The objects manipulated by the robot in the first experiment have a cylindrical form, with a diameter of 16 mm and a height of 25 mm. In the second experiment, we use longer cylinders (150 mm) with the same diameter as before; they stretch out 50 mm from the ground (see fig. 6b). Both experiments are carried out in a square arena of 80×80 cm. The initial scattering of the objects is arbitrarily predefined. Our experiments are repeated and quantified 3 times and each experiment takes approximately 30 minutes.

It is worth emphasising that in both experiments the robots operate completely autonomously and independently; all sensors, motors and controls are on-board,

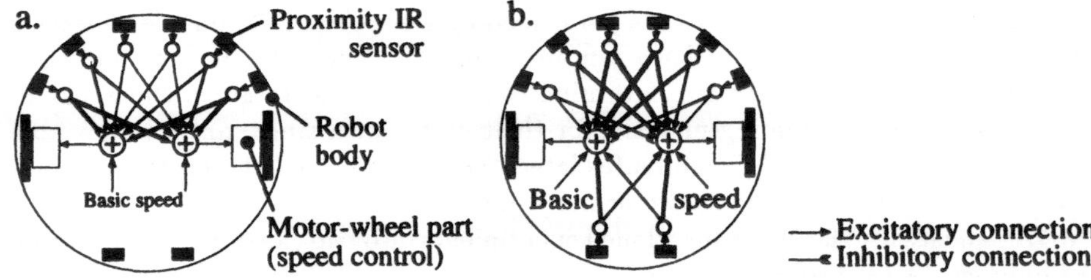

Figure 1. a) Neural connection to implement the search behaviour. These kinds of connections are inspired from the Braitenberg vehicle 3c [6]. The IR proximity sensors have an activation proportional to the proximity of an obstacle. The thickness of the connection represents the value of the weight of the connection. b) Neural connection that implements obstacle avoidance.

and there is not explicit communication (IR or radio link) with other robots or with the experimenters. The only possible interactions among robots are the reciprocal avoidance of collisions and an indirect form of messages, which arise from the modifications of the environment (i.e., for instance the cluster geometry).

2.1. First Experiment

In this first experiment, the task performed is object clustering. The object searching behaviour is based on a Braitenberg vehicle as described in figure 1a. The real object recognition is made when the front proximity sensors exceed a given activity threshold (in most cases this activity corresponds to a distance between the robot and the object of about 20 mm): the recognition algorithm differentiates between the proximity values of the two central and two lateral sensors. Two cases can occur: if all the four sensors are very active, this indicates to Khepera that there is a large obstacle in front of it and has to avoid it, independently from its structure (wall, another robot, or an array of objects); in this case the control algorithm commutes his behaviour to obstacle avoidance (implemented as described in figure 1b) and stays in this mode until none of the proximity sensors notice any obstacle. On the other hand, if the two lateral proximity sensors are not saturated, it indicates to Khepera that there is a small object in front of it; if Khepera is not carrying an object, it moves slightly backwards and grasps the object with the gripper; if Khepera is carrying an object, it moves further backwards and drops the carried object close to the new object found; then, in both cases, it turns about 180 degrees and begins a new search. With the control algorithm mentioned above, the typical form of a cluster will be more or less an array of objects.

2.2. Second Experiment

In the second experiment the task is completely different. The robot group has to remove long cylinders from holes in the ground. This is only possible if two robots collaborate together at the right moment (see fig. 6c). The individual behaviour is exactly the same as in the first experiment. However, due to the different global task, the robots do not need to transport the objects but only remove them from the ground. For this reason we have added a timeout to the grasping behaviour, to give a maximal time lapse for removing the stick, in case that an other robot does not collaborate.

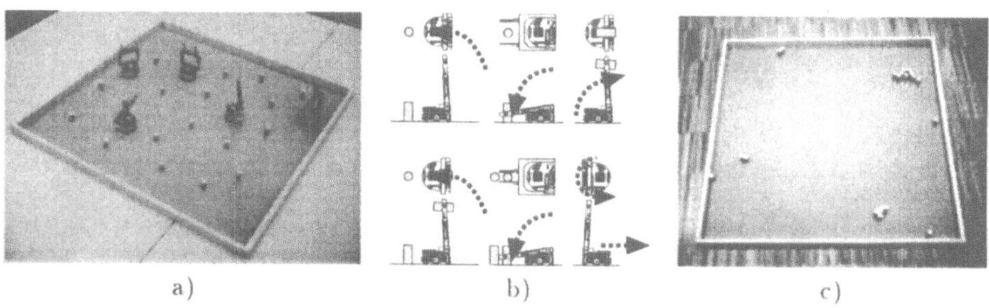

Figure 2. a) Setup for the first experiment. b) Behaviour of the robots. c) Environment after 30 minutes.

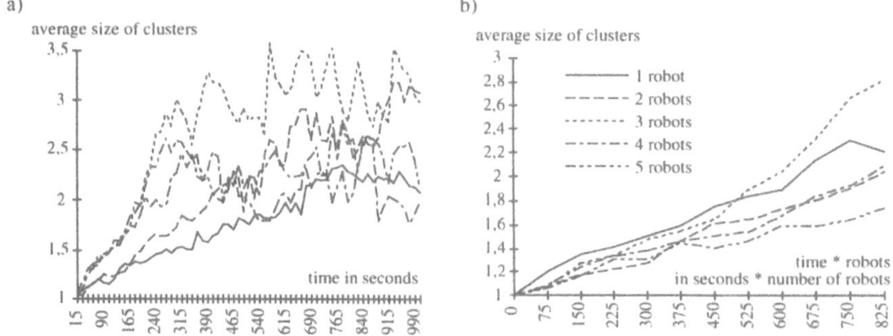

Figure 3. a) Absolute performance of the robot groups. b) Relative performance of the robot groups.

3. Results and Discussion

3.1. First Experiment

Figure 2 shows the setup employed for the first experiment and one of the resulting environments after 30 minutes.

In figure 3, the performances of the robot groups are presented. The number of robots involved in the experiment varies from 1 to 5. It is necessary to outline two skills of the absolute performance (i.e. versus time, fig. 3a): first, the cluster building process can be speeded up by increasing the number of robots on the arena; second, after a rapid rise in the first phase of the experiment (almost 10 minutes), the performance indexes reach a saturated zone, i.e. the mean size of clusters does not grow any more. This means that the "cluster system" presents a dynamical equilibrium point. Figure 3b gives indications on the relative performance (i.e. versus time * robots) of the robot colony: 30 minutes of work using a single robot have to be compared with 15 minutes of work using two robots, with 10 minutes of work using three robots, and so on. Notice that the performances of all groups of robots with two or more agents are worse than the performance of the single robot. The only exception is represented by the performance of the three robots shown on the right part of the graphic. However, we are convinced that this result is a fortuitous case and we do not consider the group of three robots as a density optimum in order to solve the given task. Observing carefully the

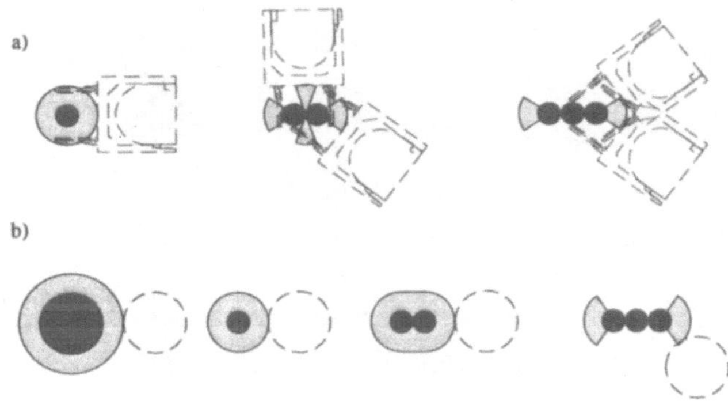

Figure 4. Geometrical representation of access areas for Khepera in order to destroy (a) a cluster of one object, a clusters of two objects and a clusters of three objects. b) Access areas for Khepera in order to build a group of one object from nothing (wrong detection of another robot), to build a group of two objects from one object, a group of three objects from a group of two and a group of four objects from a group of three.

experiment, we notice that a major density of robots on the arena causes only an increasing rate of destructive interferences (e.g., avoidance of the other robots, false perception of the object to grasp), as demonstrated in the first and more reliable part of the experiment. Therefore, the results of this experiment contribute to the proof that a collective but noncooperative task can not always be achieved with better performances (in this case the judgement criterion was the relative speed) increasing the density of robots.

Let us analyse in more detail the process rules of the first experiment. Figure 4 presents a geometrical explanation of the building-destroying mechanism. We draw the access area in order to perform one action or the other. We consider the cluster and the robot gripper geometry. The cluster building area is always greater than the cluster destroying area. The size of these areas is directly proportional to a sort of mean global probability; hence, we can assert that the building process is more probable than the destroying one and the result is, as illustrated by figure 3a, a final dynamical equilibrium.

In figure 5, the normalised results of figure 4 are presented. The probability of building a two objects cluster from two single object clusters (i.e., destroying them) is very high. On the other hand the probability to build greater clusters is inversely proportional to their size. In effect, it is a question of conditional probability: an 8 objects cluster can only be built if a 7 objects cluster previously exists.

3.2. Second Experiment

The setup of the second experiment is depicted in figure 6. The event showed in figure 6c has occurred almost every 20 minutes. How we mentioned above, this experiment implements a collective cooperative task, this means that the stick can not be removed by a single Khepera. Because of the very low rate of collaboration, it is difficult to formulate any consideration concerning the optimal density of robots on the arena. Furthermore, there are more destructive interferences during this

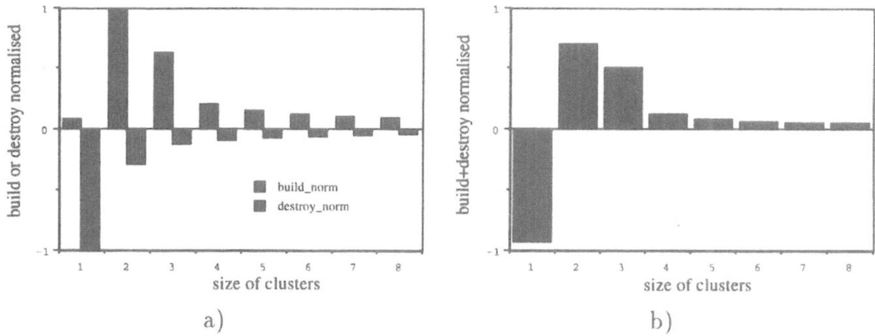

Figure 5. a) Histogram of the probability to build or destroy a cluster with a given number of objects. b) Sum of both probabilities for a given cluster size.

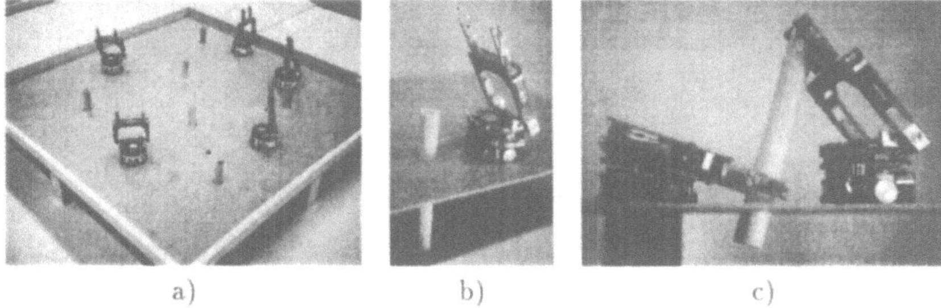

Figure 6. a) Setup for the second experiment. b) Detail of the setup: Khepera in front of a stick. c) Collaboration between two robots in order to achieve the task.

experiment than during the first one, because there are fewer objects to grasp, more robots are moving towards the sticks and therefore more collisions occur.

3.3. Remarks

Some drawbacks in the basic behaviour of robots emerge from both experiments. Often it occurs that the perception algorithm does not operate correctly: a robot is not always differentiated from an object. Very often a robot drops an object in front of another robot and the latter grasps the object (object exchange). For the same reason, the robots try to grasp each other and it often occurs that they become tangled for a few seconds. The default opening width of the gripper is too large in order to grasp only a single object: sometimes the robot grasps two objects at once but it can not carry them and therefore fall.

4. Conclusion

Our first experiment shows that only an absolute time speed-up can always be reached. The performance expressed in "work achieved" versus "time * robots" usually decreases with an increasing number of robots. This results from the fact that the task achieved in this case is purely collective and noncooperative. The fact that several robots are working together does not help to achieve the task. On the contrary, the collective aspect introduces many destructive interactions, such as collisions between robots, errors of detection due to moving obstacles, conflicting situations, etc. Probably the single robot performance, which represents the top limit in this type of task and with this kind of performance measurement, can only be improved by groups of robots if some kind of differentiation of the individual behaviour is introduced.

The results of the second experiment show that a good interaction between the robots and the environment, also with a very simple behaviour, can give rise to a collective task where a real cooperation is present.

These two experiments show interesting characteristics of collective performances where the individual behaviour is very simple and the same on every robot of the group. The interactions between many robots and between the robots and the environment play a crucial role in the performances of the group. A very simple modification of these relationships, as illustrated in the two experiments, can radically modify the performances and the behaviour of the group. Both types of working organisations (cooperative and noncooperative) can be very useful in robotics to extend the capacity of a single robot. To exploit this group structure in new design methodologies and apply it to real useful applications, we still need to better understand the basic mechanisms of collective work.

5. Acknowledgements

We would like to thank Edo Franzi and André Guignard for the important work in the design of Khepera, and Paolo Ienne for the review of this paper. Alcherio Martinoli and Francesco Mondada have been partially supported by the Swiss National Research Foundation (project FN and PNR23).

References

[1] J. C. Deneubourg, S. Goss, N. Franks, A. Sendova, A. Franks, C. Detrin, and L. Chatier. The dynamics of collective sorting: Robot-like ant and ant-like robot. In J. A. Mayer and S. W. Wilson, editors, *Simulation of Adaptive Behavior: From Animals to Animats*, pages 356–365. MIT Press, 1991.

[2] J.C. Deneubourg, 1994. Personal Communication.

[3] J. C. Deneubourg, P. S. Clip, and S. S. Camazine. Ants, buses and robots self-organization of transportation systems. In P. Gaussier and J-D. Nicoud, editors, *Proceedings of the conference From Perception to Action.* IEEE Press, Los Alamitos, CA, 1994.

[4] R. A. Brooks. A robust layered control system for a mobile robot. *IEEE Robotics and Automation*, RA-2:14–23, March 1986.

[5] P. Gaussier and S. Zrehen. A constructivist approach for autonomous agents. In Thalmann D. & N., editor, *Artificial Life in Virtual Reality*, pages 97–113. John Wiley and Sons, London, 1994.

[6] R. Beckers, O.E. Holland, and J.L. Deneubourg. From local actions to global tasks: Stigmergy and collective robotics. In R. Brooks and P. Maes, editors, *Proceedings of the Fourth Workshop on Artificial Life*, Boston, MA, 1994. MIT Press.

[7] F. Mondada, E. Franzi, and P. Ienne. Mobile robot miniaturization: A tool for investigation in control algorithms. In *Proceedings of the Third International Symposium on Experimental Robotics*, Kyoto, Japan, 1993.

Distributed Robotic Manipulation: Experiments in Minimalism

Karl Böhringer, Russell Brown,
Bruce Donald, Jim Jennings
Cornell University
Ithaca, NY 14853, USA
brd@cs.cornell.edu

Daniela Rus
Dartmouth College
Hanover, NH 03755, USA
rus@cs.dartmouth.edu

Abstract

Minimalism pursues the following agenda: For a given robotics task, find the minimal configuration of resources required to solve the task. Thus, minimalism attempts to reduce the resource signature for a task, in the same way that (say) Stealth technology decreases the radar signature of an aircraft. Minimalism is interesting because doing task A without resource B proves that B is somehow inessential to the information structure of the task. We will present experimental demonstrations and show how they relate to our theoretical proofs of minimalist systems.

In robotics, minimalism has become increasingly influential. Marc Raibert showed that walking and running machines could be built without static stability. Erdmann and Mason showed how to do dextrous manipulation without sensing. Tad McGeer built a biped, kneed walker without sensors, computers, or actuators. Rod Brooks has developed online algorithms that rely less extensively on planning and world-models. Canny and Goldberg have demonstrated robot systems of minimal complexity. We have taken a minimalist approach to distributed manipulation. First, we describe how we built distributed systems in which a team of mobots cooperate in manipulation tasks without explicit communication.[1] Second, we are now building arrays of micromanipulators to perform sensorless micromanipulation. We describe how well our experimental designs worked, and how our manipulation experiments mirrored the theory.

This paper describes research done in the Robotics and Vision Laboratory at Cornell University. Support for our robotics research was provided in part by the National Science Foundation under grants No. IRI-8802390, IRI-9000532, IRI-9201699,

[1] No RF or IR messages are sent between the robots.

by a Presidential Young Investigator award to Bruce Donald, by an NSF/ARPA S.G.E.R. in MEMS, and in part by the Air Force Office of Sponsored Research, the Mathematical Sciences Institute, Intel Corporation, and AT&T Bell Laboratories.

1. Introduction

This paper describes our experience in building distributed systems of robots that perform manipulation tasks. We have worked at both the macroscopic and the microscopic scale. First, we describe a team of small autonomous mobile robots that cooperate to move large objects (such as couches). The robots run SPMD[2] and MPMD[2] manipulation protocols with no explicit communication. We developed these protocols by distributing offline, sequential algorithms requiring geometric models and planning. The resulting parallel protocols are more on-line, have reduced dependence on *a priori* geometric models, and are typically robust (resistant to uncertainty in control, sensing, and initial conditions).

Next, we discuss our work on sensorless manipulation using massively parallel arrays of microfabricated actuators. The single-crystal silicon fabrication process opens the door to building monolithic microelectromechanical systems (MEMS) with microactuators and control circuitry integrated on the same chip. Our actuators are servoed to uniquely orient (up to symmetries) an object lying on top, and require no sensing. We can also program the array as a sensorless geometric filter—to sort parts based on shape or size.

We developed both the macroscopic and the microscopic systems by distributing and parallelizing sequential manipulation algorithms with global control, to obtain distributed algorithms running on independent physical agents. Our MEMS control algorithms for micromanipulation are SPMD; for the macroscopic (furniture-moving) task, we describe implementations and experiments with both SPMD and MPMD control.

We have implemented and extensively tested our macroscopic distributed manipulation strategies. We have built MEMS prototypes, and we are now fabricating and testing our biggest micro-array yet (the entire wafer is tiled with 7000 microactuators). Our macroscopic algorithms use no direct communication between the agents, but do employ sensing. Our microscopic algorithms are sensorless, but require a small amount of local communication to initialize and synchronize. Our theory predicts a trade-off between communication and sensing when we parallelize a manipulation strategy. We will discuss experiments we have performed to experimentally observe and validate these trade-offs.

2. Reorienting Large Objects with Autonomous Mobile Robots

We are interested in large-scale manipulation of objects by small mobile robots. In Sections 2 and 3 of this paper, the manipulated objects have comparable size and dynamic complexity to the robots. Objects used in our experiments are up to 6 times the robot's diameter in length, and up to twice the mass of one of the robots. Repositioning and reorientation of these objects may be possible only through active cooperation of a team of mobile robots; for other objects, employing multiple robots may yield performance or other benefits, such as ease of programming.

[2] SPMD (MPMD) = *S̲ingle (M̲ultiple) P̲rogram, M̲ultiple D̲ata.*

Consider the task whose goal is to change the orientation of a large object by a given amount. This is called the *reorientation* task. We have described and analyzed in detail the reorientation task in [16]. Figure 1 depicts one robot reorienting a large object. A robot can generate a rotation by applying a force that is displaced from the center of friction. This property relates the dynamics and the geometry or reorientations [12] and it can be used to effect continuous reorientations with mobile robots. The idea is to compliantly apply a sliding force on the face of the object[3]. We call this action a *pushing-tracking* step. When the end of the face is reached, the robot may turn around to reacquire contact and repeat the pushing-tracking. A robot that has gone past the end of a face effectively losing contact with the object has *broken contact* with the object. A robot whose maximum applied force (defined by a threshold) does not change the pose of the object has encountered an *impediment.*

One robot may effect any desired reorientation by repeated pushing-tracking steps if it can apply a large enough force, but it may require a large workspace area for the rotation. We are interested in strategies for reorienting objects *in place.* This can be done by a team of k robots[4]. (See Figures 1(b), 2, and 3.) The k robots can simultaneously constrain the motion of the object and execute *pushing-tracking* operations. We can measure the degree of parallelism in the reorientation algorithm by counting how many of the robots are *active*, *i.e.*, that execute push-tracking motions, and how many of the robots are *stationary*, *i.e.*, that constrain the motion of the object by staying fixed in place. We show how to select the active and stationary robots and how to decide on role-switching over the course of an algorithm.

Figure 1. (a): Reorientation by one robot executing pushing-tracking. (b) A system of two robots reorienting a couch. The robot motions are shown in an object-fixed frame. Each robot executes a pushing-tracking motion. Robots recognize when they reach the end of a face by breaking contact, and execute a spinning motion to turn around and reacquire contact.

We now present four different but "equivalent" reorientation protocols that have different degrees of parallelism, synchrony, and resource utilization. Our notion of "equivalence" comes from looking at the task as a dynamical system. Consider the configuration space C of the manipulated couch. We call two manipulation protocols *equivalent* if the protocols have the same forward-attracting compact limit set in C [7] (p. 284) and [6]. All of our reorientation protocols rely on the ability of robots to execute *push-tracking* motions.

[3]This strategy can be implemented by a force that forms an acute angle on the contacting edge. This is similar to hybrid control [14] which would be used for a dexterous hand [17].

[4]For an excellent survey of cooperative mobile robotics see [4].

2.1. An Off-line, Global Control Protocol

The off-line global control strategy denoted by GLOBAL-OFFLINE requires three robots and is described and analyzed in detail in [17]. The algorithm consists of a sequence of pushing-tracking steps, where one robot is active (pushing) and two robots are stationary at all times (see Figure 2). A global controller sends and receives control signals to the robots. Under the assumption that the robots are already in contact with the object, the algorithm can be summarized as follows:

Active robot:	Stationary robots:
1. push-track until position or force termination	1. sense for relative motion (*slip*)
2. signal the global controller	2. if no slip, signal the global controller
3. become stationary	3. when signaled, become active

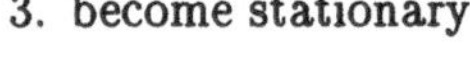

Figure 2. A two-step pushing-tracking sequence of snapshots for reorienting an object with three robots. The right sequence denotes three snapshots of the first step. The left sequence denotes three snapshots of the second step. The black circles denote stationary robots. The white circles denote active robots.

A planner [17] takes as input the geometric model of the object we wish to manipulate. It outputs the initial contact locations for the robots and the order in which the robots become active, which is called the push-tracking *schedule*. The termination of each each step is characterized by "jamming" the object between the three robots and thus can be detected by an active slip sensor. The planner guarantees that *no robot will break contact* in a multi-step execution.

In the *setup* phase, the robots make contact with the object in the areas generated by the planner, using motion plans generated by a trajectory planner. The output of the planner is also used by the global controller to signal the robots when to become active.

A system of robots executing this algorithm requires the following *skills* of each robot:

- `(goto` ⟨*position*⟩`)` to contact the object during the setup phase in the areas computed by the planner. `(goto` ⟨*position*⟩`)` can be implemented by using a trajectory generator and a localization system, or even dead-reckoning [3].

- `(push-track)` to control the pushing-tracking motion of each active robot. This entails exerting a force in the normal direction to the face of the object while commanding a velocity in the tangential direction. (See Figures 1 and 2.)

- `(active-slip?)` used by the stationary robots to detect slip at their points of contact.

2.2. Summary of Two "Intermediate" Protocols

Due to space limitations, we omit the development and discussion of two of our reorientation strategies, and instead summarize them briefly.[5] They may be considered "intermediate" in the sense that they represent successive transformations of the Off-line, Global Control Protocol described above. A final transformation yields the On-line, Uniform, Asynchronous Protocol described below in Section 2.3.

- **An Off-line, Local Control Protocol:**

  A variant of the Off-line, Global Control Protocol can be derived for three (or more) independent robots. This system of autonomous robots cooperates to complete a reorientation and does not have a global controller. Instead, the robots use communication (IR or RF) to synchronize their actions, which are performed according to the same push-tracking *schedule* as in the previous protocol.

- **An On-line, Synchronous Protocol:**

  The two previous protocols require a planner. We now ask: do reorientation protocols depend crucially on planning? We present in the full text of this report another version of the algorithm that does not use a geometric model for the object being manipulated and does not necessitate a planner. It is denoted SYNCH-ONLINE. Without a reorientation planner, we note that (1) the initial setup phase is randomized, and (2) there is now no guarantee that all of the robots maintain contact at all times. While maintaining contact is important for fine manipulation within a dexterous hand, it is not necessary for the task of large-scale manipulation with mobile robots. Therefore, we are willing to give up maintaining contact in favor of an online algorithm without a planner and without a world model.

2.3. An On-line, Uniform, Asynchronous Protocol

Figure 3. Two mobile robots cooperating to reorient a couch: a snapshot taken from a couch reorientation experiment.

The three previous protocols require explicit communication. We now ask: how essential is the explicit communication for reorientation protocols? We present a

[5]The full text of this report, including the discussion of the omitted protocols may be found online in `ftp://flamingo.Stanford.edu/pub/brd/iser-95.ps.gz`.

protocol (which we denote by ASYNCH-ONLINE) that is *uniform* (SPMD), in that the robots execute the same program asynchronously and in parallel and there is no explicit communication between the robots.

For this protocol, two robots suffice. All the robots are active all the time. Assuming that the robots are in contact with the object, the following algorithm achieves the reorientation.

Active robots (all):
1. push-track until contact break or impediment
2. if contact break then spin
3. if impediment then graze

The intuition is that the robots try to maintain the push-tracking state. When the end of the face is reached, a *spinning* motion is employed to reacquire contact. *Spin* is a motion that is executed when a robot has gone past the end of the face – the robot turns around and reacquires contact by traveling around a circular trajectory (see Figure 1(b)). Alternatively, if an impediment is encountered, the robot executes a guarded move near-parallel to, but towards the face of the object, effectively *grazing* the object. *Graze* terminates when the robot recontacts the object or when it detects that it has traveled past the end of the face. Hence, graze terminates in (I) reacquiring the object at a contact with a longer moment arm, or (II) missing altogether. (I) is detected with a guarded move. (II) is detected using a sonar-based wall/corner detection algorithm of [10]. When (II) occurs, the robot executes `(spin)`.

In the setup phase the robots contact the object using the same procedure as in protocol SYNCH-ONLINE. The robots may reach the object at different times. As soon as a robot makes contact with the object, it proceeds with executing a push-tracking motion. The following skills are required of the robots:

- `(guarded-move` ⟨*direction*⟩ `)`, `(blocked?)`, `(end-of-face?)`, and `(push-track)`,
- `(spin)` Each robot must be able to reacquire contact when the end of the face is reached.
- `(graze)` Each robot must be able to translate near-parallel to the face of an object to find a new contacting point. Graze uses sonar.

We have implemented a system to move furniture in our lab that includes the asynchronous online reorientation protocol. Our experiments suggest that the system is very robust.

3. MPMD Manipulation

Section 2 presents an implemented and tested SPMD manipulation protocol. We now present another implemented and tested strategy, an MPMD protocol called the *Pusher/Steerer system*. A detailed description and analysis of this sytem is given in [2, 3].

Despite conventional wisdom regarding the complexities of programming a multi-robot system, a key feature of the Pusher/Steerer system is its ease of use – the actual robot code is simple and elegant, and yet there remains great flexibility in methods of path specification. We present the following properties of the Pusher/Steerer system:

- The code that implements the Pusher/Steerer strategy is simple because it relies on the "natural" kinematic and dynamic interactions between the robots and the manipulated object to achieve the goal.
- The Pusher/Steerer system is easily adapted to employ either an offline path planner, or an online navigation system in which the path to the goal is not known *a priori*. This is possible because of the decoupling of the steering and pushing components of the strategy.
- Finally, an information invariant analysis (see [7]) of the Pusher/Steerer strategy reveals several formal properties, which we may express informally here:
  1. The Pusher/Steerer system is exactly a redistribution of the same resources (computation, state, sensors, etc.) of a comparable single-robot manipulation system.
  2. There is no explicit communication between the Pusher and the Steerer.
  3. The addition of a clock and some state to each robot (the Pusher and the Steerer) increases the power of the system significantly. With clocks, the Pusher and Steerer may exchange roles in an online fashion, and thus execute complex paths, such as "parallel parking" maneuvers.

3.1. Details of Pusher/Steerer

In this protocol, the robots take on the role either of the Pusher, in which:

- Torque-controlled translations push the object in front of the robot,
- the robot follows the object by continually turning to align its front bumpers with the rear face of the object (the rotational and translational motions here are decoupled and occur in parallel), and
- the robot does *not* know the path that the object is supposed to follow.

or the Steerer, in which:

- The robot knows a path that it is supposed to follow,
- the robot is translationally compliant (controlling only the heading of its wheels), and
- the robot moves forward as a result of being pushed by the object (which is itself being pushed by the Pusher).

The manipulated object sits between the robots. We use *no direct communication* between the two robots, but employ only *indirect* communication through the mechanics of the robots-and-box system. One advantage is that the robots can trade roles, allowing such maneuvers as the "back-and-fill" that automobile drivers use for turning cars around on narrow roads. Objects of varying size, mass distribution, and surface friction may be manipulated by our system over a wide range of paths. Figure 4 shows a drawing of two robots moving a rectangular object through a circular arc.

Our analyses of the mechanics of Pusher/Steerer protocols for translational manipulation only, circular-arc following, and more general trajectory-following are omitted here. We will summarize our analyses of the Pusher/Steerer system with respect to information invariants, however.

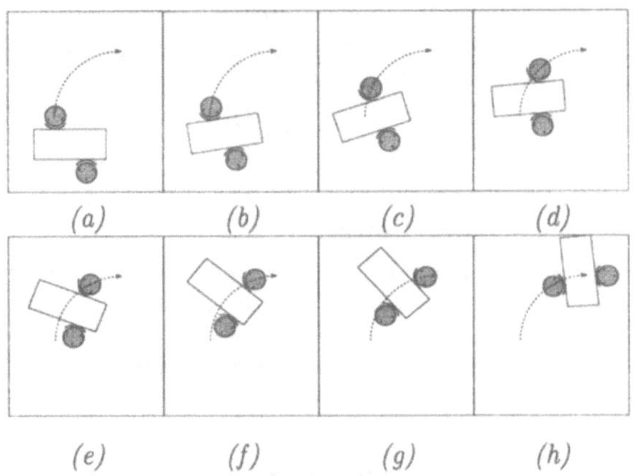

Figure 4. This series of figures depict a box being guided through a 90 degree arc by a steering robot (in front, following the arc), and a pushing robot. The box begins with its front and rear faces approximately perpendicular to the path. In parts *(b)* and *(c)*, the box rotates in the wrong direction, due to poor initial placement of the Pusher relative to the Steerer. By part *(d)*, the Pusher, with no model of the box or the path and with no communication, has compensated for the poor initial configuration. By part *(h)*, the box has traversed the arc and rotated until its front and rear faces are approximately perpendicular to the path.

For the purpose of illuminating the information and resource requirements of our manipulation system, we turn to the framework of information invariants [7], which defines formal reductions between protocols. We say "Protocol B reduces to Protocol A" when we can use the resources (state, communication, computation, sensors, effectors) of Protocol A to build Protocol B. In writing a reduction, we list explicitly any new resources (including communication) added to Protocol A to assemble Protocol B. We use reductions to make formal comparisons between different protocols that achieve the same task; for instance, we can calculate equivalence between protocols. While a presentation of the formal reductions between single-robot pushing and our 2-robot Pusher/Steerer system is beyond the scope of this report, we will pause to discuss those results informally.

In [7], Donald claims that the spatial distribution of resources has a critical effect on the capabilities of a system. The Pusher/Steerer system validates that claim. Consider a *single-robot* manipulation algorithm such as, e.g., [11]. As implemented on the Cornell mobile robots, the execution system consists of the following skills:

- a *pushing* primitive, `(prim-push)` (given a heading direction, each robot has the ability to compliantly exert a pushing force);
- `(align)` (the robot actively aligns its heading to the object face using the relative angle between the robot and the object, which our robots can measure directly using a ring of contact-sensors[10]);
- a *steering* primitive, `(steer)`; and
- *a priori* path information.

The Pusher's (entire) control system (`Pusher`) is obtained by the *parallel* composition of (`align`) and (`prim-push`). The skill (`push-track`) used in Section 2 can be shown to be the *sequential* composition of (`align`) and (`prim-push`). In the language of information invariants [7], this implies the equation[6]

$$(\texttt{push-track}) =_0 (\texttt{prim-push}) + (\texttt{align}). \tag{1}$$

Similarly, Brown [3] shows that

$$(\texttt{Pusher}) =_0 (\texttt{prim-push}) + (\texttt{align}), \tag{2}$$

and hence, from Equation(1) it follows that

$$(\texttt{push-track}) =_0 (\texttt{Pusher}). \tag{3}$$

We can synthesize the Pusher/Steerer system by redistributing items from the list above into *two separate physical locations.* The *pushing* and *alignment* primitives become the Pusher, and the *steering* primitive and path information comprise the Steerer. Clearly we have added a second robot to the system. But did we actually add resources, or just move them around? The Steerer gets the rotation subsystem from our single-robot strategy; the Pusher gets the translation subsystem. The Steerer gets the path information; the Pusher gets the alignment subsystem (a relative orientation sensor and rotation capability). So we did add a resource! Both the Pusher and Steerer need to rotate.

The Pusher/Steerer system consists of a redistribution of the resources of the single-robot manipulation system described above, plus one rotate motor. Yet, if we substitute a slightly different robot for the Pusher, we find that we do not need to add the rotate motor at all. The Cornell mobile robot CAMEL has a flat bumper instead of a semi-circular one, and when CAMEL is the Pusher, the *alignment* resource is not needed. CAMEL passively maintains correct alignment with the object face due to rotational compliance and the mechanics of line-contact or "blade" pushing. Thus, the Pusher is *rotationally compliant,* but controls translations, while the Steerer is *translationally compliant,* but controls rotations. There is an explicit tradeoff between the choice of robot bumper geometry and the need for an active *alignment* primitive.[7]

In summary, if we choose an appropriate pushing robot, we can build the Pusher/Steerer system by redistributing exactly the resources that would be used in a single-robot manipulation system. There is thus a de facto equivalence, in terms of resource usage, between the two strategies: Pusher/Steerer and single-robot manipulation. It appears, however, that the Pusher/Steerer system admits a larger class of executable paths; moreover, the system may have other advantages which are less easy to quantify. The benefits of Pusher/Steerer do not derive from an addition of resources, but rather from the spatial redistribution of existing resources.[8]

---

[6] $A =_0 B$ when $A \leq_0 B$ and $B \leq_0 A$.

[7] These tradeoffs are precisely quantified in the information invariants theory.

[8] It should be noted that the combined internal state of the Pusher and Steerer is no greater than that of the single-robot manipulator described above, and that no extra computation nor communication is needed.

Table 1. Degrees of arc traversed at given turn radius for several boxes. w is the box dimension between the contact faces; ℓ is the box dimension between the non-contact faces. The values presented are averaged over 5 runs.

	Turning Radius (mm)			
Box ($w \times \ell \times m$)	1000	1500	2000	2500
51cm × 58cm × 3Kg	1000	1050	1220	1440
35cm × 23cm × 2Kg	225	234	528	618
33cm × 58cm × 4Kg	153	342	475	656

3.2. MPMD Manipulation Experiments

We have performed over one hundred constrained manipulation experiments using the Pusher/Steerer protocol running on several pairs of Cornell mobile robots. In these experiments, boxes and similar objects of varying size, mass, mass distribution, and material properties were manipulated along complex paths up to 50 feet in length. On the basis of these experiments, described in [3], we have observed the system to be quite robust in practice. Additional experiments using online navigation methods (human guidance in one case, and visual landmark recognition in another) have demonstrated the flexibility of the system.

One set of experimental tests is summarized here. The task is circular arc following, as an endurance test: how far around a circle, on average, could TOMMY and LILY carry each of a set of test objects? We ran the protocols at each of a number of turning radii on each of several boxes 5 times, and present here (table 1) the average arc distance traversed before the Steerer loses control (breaks contact with the object). The maximum distance traversed for any test is 1440 degrees (four complete circumferences).

There are two main lessons we have learned from our experiments with distributed reorientation and with the Pusher/Steerer system.

1. *Information invariants theory indicates that we should be able to distribute a manipulation task across multiple robots with essentially no additional resource cost. The Pusher/Steerer system is an example.*

2. *A mechanics analysis of large-scale manipulation indicates that distributing a manipulation task across multiple robots using a Pusher/Steerer model will allow robots with limited control and sensing to perform that manipulation task in a manner equivalent to a single-mobot system with much more sophisticated control and sensing.*

4. Manipulation with microelectromechanical actuator arrays

Next, we discuss our work on sensorless manipulation using massively parallel arrays of microfabricated actuators [1]. The single-crystal silicon fabrication process opens the door to building monolithic microelectromechanical systems (MEMS) with microactuators and control circuitry integrated on the same chip. Our actuators are servoed to uniquely orient (up to symmetries) an object lying on top, and require

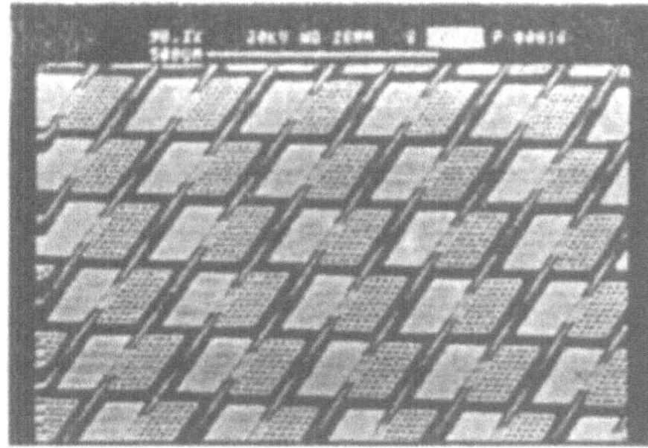

Figure 5. A prototype M-CHIP fabricated in 1995. A large unidirectional actuator array (scanning electron microscopy). Each actuator is $180 \times 240\,\mu m^2$ in size. Detail from a $1\,in^2$ array with more than 11,000 actuators.

Figure 6. Released asymmetric actuator for the M-CHIP (scanning electron microscopy). Left: Dense grid ($10\,\mu m$ spacing) with aluminum electrode underneath. Right: Grid with $5\,\mu m$ high poles.

no sensing. We can also program the array as a sensorless geometric filter—to sort parts based on shape or size.

4.1. Device Fabrication and Properties

In recent years much progress has been made in microelectromechanical systems (MEMS). They consist of structures in the micrometer range which are usually fabricated with VLSI technology. Unlike conventional circuits, MEMS devices have an electrical *and* a mechanical component, i.e. moving parts that can be controlled or monitored with electrical voltage or current.

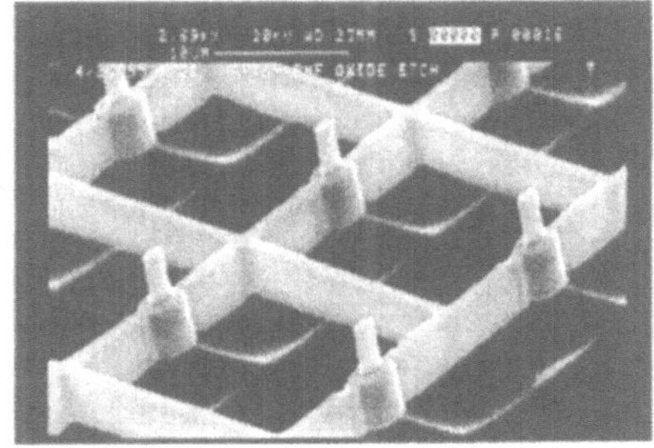

Figure 7. Released M-CHIP actuator (detail) consisting of single-crystal silicon with $5\,\mu m$ high tips (described in [1]).

Fabrication of our devices consists of a sequence of depositions and etches (called SCREAM process, for Single Crystal Reactive Etching and Metallization [13]) which define and partially release the structures from the silicon substrate. Thus, the devices consist of a single crystal silicon core, which is usually covered with aluminum, isolated by a thin film of dielectric silicon oxide. Typically these devices resemble grid or truss structures, because only beams up to a few μm wide (but up to $1\,mm$ long) can be released with the SCREAM process. The fabrication process is compatible with conventional VLSI. Control logic can be integrated on the same chip or even within the silicon structures. Figures 5 through 7 show such actuators at different magnification levels. Each of them consists of a grid structure suspended on a torsional beam. Electrostatic forces cause the device to rotate out of plane by several degrees. When applying an AC voltage the actuator oscillates, with resonance in the kHz range. The design of the grid is asymmetric, with tips only on one side of the grid (see Figure 7). This ensures that when the actuator is in contact with an object, it will generate a lateral force.

Recently we have fabricated arrays with up to 7000 individual actuators for massively parallel manipulation. There is a huge potential of applications. Such MEMS actuator arrays can be used as bulk-fabricated (cheap), ultra-thin transport mechanisms e.g. for paper in copy machines or printers. At the other end of the spectrum, recent advances have brought within reach arrays equipped with tips that can probe and move single atoms [18]. Such devices, employed in a massively parallel fashion, will yield tremendous data storage capacities.

The MEMS array that we present here is designed for "medium size" applications in which objects in the millimeter range are moved, e.g. for an automatic stage of a microscope, or for the assembly of small parts.

4.2. Part Positioning and Orienting

We want to use arrays of up to hundreds or thousands of microactuators to manipulate and orient parts in the millimeter to centimeter range. Each individual actuator is approximately $200\,\mu m \times 300\,\mu m$ in size and can generate a force of up to $10\,\mu N$. Clearly this is a strongly distributed manipulation task, as it cannot be achieved with an individual actuator. However, by distributing the task among cooperating actuators the joint force is sufficient (the weight of paper per actuator area is approximately $60\,nN$, more than two orders of magnitude less than the force generated by an actuator).

Let us consider the task in which a flat polygonal part $\mathcal{P}$ has to be oriented to a specific angle on the array, starting from an arbitrary initial configuration. A global control strategy could act as follows:
(1) Determine the current position of $\mathcal{P}$.
(2) If $\mathcal{P}$ is in the goal configuration, stop.
(3) Compute a motion that brings $\mathcal{P}$ closer towards the goal.
(4) For each actuator, compute the force necessary to induce this motion, and tell the actuator to generate this force.
(5) Repeat.

This strategy is rather complex, and requires individual communication with each actuator, as well as sensing. In the following we show how a simpler and more effective distributed manipulation strategy can perform the same task without sensing, and with the use of only very limited communication resources.

Suppose the array generates a "squeeze" pattern in which all actuators push perpendicularly towards a straight line l. A polygonal part $\mathcal{P}$ on the array will experience a force that causes translation towards the line l. When l intersects $\mathcal{P}$, it also experiences a torque. This can be modeled as forces acting on the respective centroids of $\mathcal{P}$ on either side of l (see Figure 8). These forces are proportional to the surface area of each section of $\mathcal{P}$.

In earlier work we have shown [1] that every polygonal part has a small finite number of stable equilibrium configurations in such an actuator pattern. In equilibrium the forces and moments balance out, such that l becomes a bisector of the part, and the line connecting the centroids is perpendicular to l (see Figure 9).

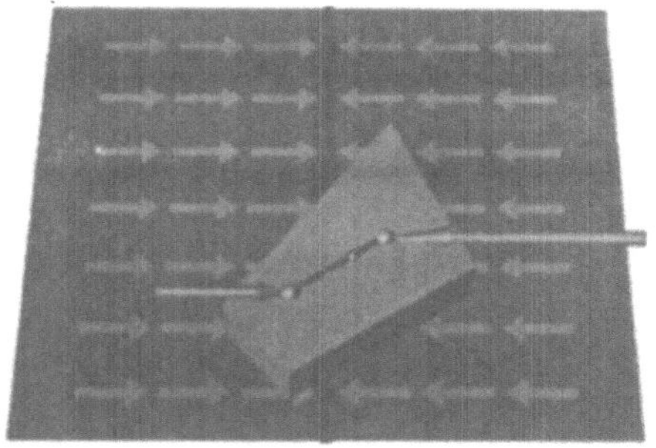

Figure 8. Part in a force field generated by an actuator array. The resulting forces for the left and the right section of the part are shown acting at the respective centers of mass: the part experiences a translational force and a moment.

Figure 9. Part in equilibrium: The resulting forces for the left and the right section of the part are of equal magnitude and opposite direction, and the resulting moment is zero.

All possible equilibrium configurations can be predetermined from the geometry of the part. For now let us assume that the part $\mathcal{P}$ has only one equilibrium. $\mathcal{P}$ will reach this equilibrium when put on an array with a squeeze pattern. Thus, to orient $\mathcal{P}$ up to symmetry it is sufficient to generate a squeeze pattern with the center line l at the appropriate place.

Note that this strategy does not require any sensing, and that it is not necessary to have individual communication with each actuator. If each actuator knows its relative place in the array, a single broadcast of the location of l is sufficient for each actuator to determine the direction of pushing. We see that by distributing computing resources and state information we can significantly reduce the amount of communication necessary. This SPMD approach simplifies both the control strategy (software) as well as the communication circuitry (hardware) for MEMS actuator arrays.

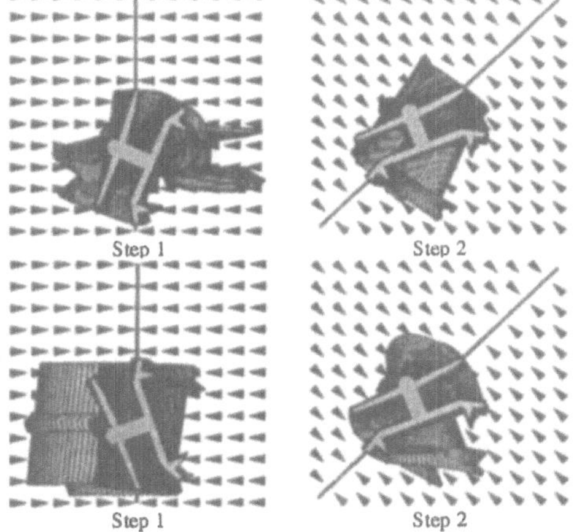

Figure 10. Sensorless parts alignment using force vector fields: Parts reach unique poses after two subsequent squeezes.

4.3. Multi-step part alignment

In the previous section we made the assumption that the part $\mathcal{P}$ has only one stable equilibrium. For parts that have multiple equilibria we can still use the same basic idea to align parts up to symmetry. We employ a multi-step strategy in which we successively reduce the number of possible configurations in which $\mathcal{P}$ can be. This is achieved by a sequence of squeeze patterns at specific angles. As Goldberg [8] has shown for the related problem of aligning parts with a robot gripper, there always exist efficient multi-step strategies. We have extended his results for general manipulation in force fields [1].

As an example consider the traces of a two-step strategy in Figure 10. The ratchet-shaped part is put on the array at two different random initial configurations. After two squeeze steps, the part ends up in the same orientation.

4.4. Summary

We have presented bulk-fabricated MEMS actuator arrays to perform massively parallel manipulation tasks, and described efficient control strategies for part manipulation. We discovered a trade-off between communication and computation resources and have shown how distributing resources can significantly reduce the complexity of parallel manipulation tasks.

Acknowledgements

We thank Jean-Claude Latombe for his great hospitality during our stay at the Stanford Robotics Laboratory.

References

[1] Böhringer, K., Donald, B., Mihailovich, R., and MacDonald, N., Sensorless manipulation using massively parallel microfabricated actuator arrays, In *Proc. IEEE Int. Conf. on Robotics and Automation*, pages 826–833, San Diego, CA, May 1994.

[2] Brown, R., and Jennings, J., Manipulation by a pusher/steerer, In *Proceedings of Intelligent Robot Systems*, Pittsburgh, PA, August 1995.

[3] Brown, B., *Algorithms for Mobile Robot Localization and Building Flexible, Robust, Easy to Use Mobile Robots*, PhD thesis, Cornell University, Ithaca, NY, 1995.

[4] Cao, Y., Fukunaga, A., Kahng, A., and Meng, F, Cooperative mobile robots: Antecedents and directions, Technical Report, UCLA Department of Computer Science, 1995.

[5] Chandler, D., Atom by atom, *The Boston Globe*, May 5:37ff, 1995.

[6] Donald, B., Jennings, J., and Rus, D., Information invariants for distributed manipulation, *The First Workshop on the Algorithmic Foundations of Robotics*, eds. K. Goldberg, D. Halperin, J.-C. Latombe, and R. Wilson, A. K. Peters, pages 431–459, 1994.

[7] Donald, B., Information invariants in robotics, *Artificial Intelligence*, 72:217–304, 1995.

[8] Goldberg, K., Orienting polygonal parts without sensing, *Algorithmica*, 10(2/3/4):201–225, August/September/October 1993.

[9] Jennings, J., *Distributed Manipulation with Mobile Robots*, PhD thesis, Cornell University, Ithaca, NY, (forthcoming, January 1996).

[10] Jennings, J., and Rus, D., Active model acquisition for near-sensorless manipulation with mobile robots, In *IASTED International Conference on Robotics and Manufacturing*, pages 179–184, Oxford, England, September 1993.

[11] Lynch, K., and Mason, M., Stable pushing: Mechanics, controllability, and planning, In *Proceedings of the 1994 Workshop on the Algorithmic Foundations of Robotics*, San Francisco, CA, 1994.

[12] Mason, M., Manipulator grasping and pushing operations, *International Journal of Robotics Research*, 5(3):53–71, 1995.

[13] Mihailovich, R., Zhang, Z., Shaw, K., and MacDonald, N., Single-crystal silicon torsional resonators, In *Proc. IEEE Workshop on Micro Electro Mechanical Systems*, pages 155–160, Fort Lauderdale, FL, February 1993.

[14] Raibert, M., and Craig, J., Hybrid position/force control of manipulators, *Journal of Dynamic Systems, Measurement, and Control*, 102, 1981.

[15] Rees, J., and Donald, B., Program mobile robots in scheme, In *Proc. of the 1992 IEEE International Conference on Robotics and Automation*, Nice, France, 1992.

[16] Rus, D., Donald, B., and Jennings, J., Moving furniture with mobile robots, In *Proceedings of Intelligent Robot Systems*, Pittsburgh, PA, August 1995.

[17] Rus, D., *Fine motion planning for dexterous manipulation*, PhD thesis, Cornell University, Ithaca, NY, August 1992.

[18] Xu, Y., Miller, S., and MacDonald, N., Microelectromechanical scanning tunneling microscope, *Bulletion of the American Physical Society*, 40(1):63, 1995.

A General Framework for Multi-Robot Cooperation and its Implementation on a Set of Three Hilare Robots

R. Alami, L. Aguilar, H. Bullata, S. Fleury, M. Herrb,
F. Ingrand, M. Khatib, F. Robert
LAAS / CNRS
7, Avenue du Colonel Roche
31077 Toulouse - France

Abstract

We present a general concept for the control of a large fleet of autonomous mobile robots which has been developed, implemented and validated through various experiments.

The robots demonstrate advanced autonomous features including non-holonomic motion planning, environment modeling, sensor-based obstacle avoidance, and decentralized cooperation schemes at mission and trajectory level.

1. Introduction

We present an implemented system which allows to run a fleet of autonomous mobile robots in a route network or an in-door environment with a very limited centralized activity. The robots are endowed with all the necessary ingredients for planning and executing navigation missions expressed at a very high level, as well as for multi-robot cooperation.

The cooperation scheme is based on a generic paradigm called *Plan-Merging Paradigm*, where robots incrementally merge their plans into a set of already co-ordinated plans. This is done through exchange of information, between robots, about their current state and their future plans (§3).

The robot control architecture is derived from the generic architecture developed at LAAS. We present its instantiation in the case of a set of cooperating mobile robots (§4).

The last section (§5) reports on the implementation and tests of the complete system. Numerous tests on various environments have been performed. Two examples will be presented in the following: (i) a simulation which runs ten robots in an in-door environment, and (ii) a real robot demonstration involving three Hilare robots.

2. Related Work

Various methods have been proposed which deal with multi-robot systems. Besides, the term "cooperation" has been used in several contexts with different meanings. We will not consider here cooperation schemes at servo level nor contributions dealing with "intelligent groups" of simple robots.

A thorough analysis of the literature is still to be made; we simply mention here some representative contributions which involve an effective cooperation (at plan or program level) between several robots.
Several approaches have been proposed, such as generation of trajectories without collision (e.g. [6, 22]), traffic rules [12, 15], dynamic adaptation of trajectories [11], negotiation for dynamic task allocation [17, 4], and synchronization by programming [18, 25, 24].
Concerning more particularly multi-robot motion planning methods, numerous contributions have been made which are generally based on a central planner, specially designed to cope with the intrinsic complexity of the problem [21, 8, 19]. While these methods are not complete or cannot be used for a "large" number of robots (more than 3), recent techniques based on randomized search in the Global Configuration Space [20], allow most often to obtain a solution in a reasonable time, even though, in the worst case, they fall into the unavoidable problem complexity.
Most of these contributions are based on a pre-defined set of situations or on task specific planners. We claim that our Plan-Merging Paradigm is a generic framework which can be applied in different contexts, using different planners (action planners as well as motion planners). It has some clean properties (and clear limitations) which should allow, depending on the application context, to provide a coherent behavior of the global system without having to encode explicitly all situations that may be encountered. Another advantage of our method is that it allows, most of the time, to solve a conflict without using a full multi-robot planner and even without stopping the execution of the other robots.

3. A Fleet of Autonomous Mobile Robots

The problem consists in devising a system which allows to run a large fleet of autonomous mobile robots in a route network (resp. in-doors) composed of lanes (corridors), crossings and open areas (rooms).
Typical applications of this problem are load transshipment as dealt with in the MARTHA project[1] which focused on the development of a large fleet of autonomous mobile robots for the transportation of containers in harbors, airports and marshaling yards environments.
The *Plan-Merging Paradigm* we propose is well suited to such applications as it allows to deal with a great number of robots, treating conflicts locally (whenever it is possible) and in a decentralized manner while maintaining a global coherence of the system. Indeed, it limits the role of the central system to the assignment of tasks and routes to the robots (without specifying any trajectory or any synchronization between robots) taking only into account global traffic constraints, for a better management of the robot fleet ([23]).

3.1. Mission Specification

The system is composed of a Central Station and a set of autonomous mobile robots. As mentioned above, the environment is a route network composed of entities like lanes, crossings and open areas. Basically, the robots navigate through an oriented graph of cells. Lanes and crossings are composed of a set of connected cells, while areas consist of only one cell.

[1]MARTHA: European ESPRIT Project No 6668. "Multiple Autonomous Robots for Transport and Handling Applications"

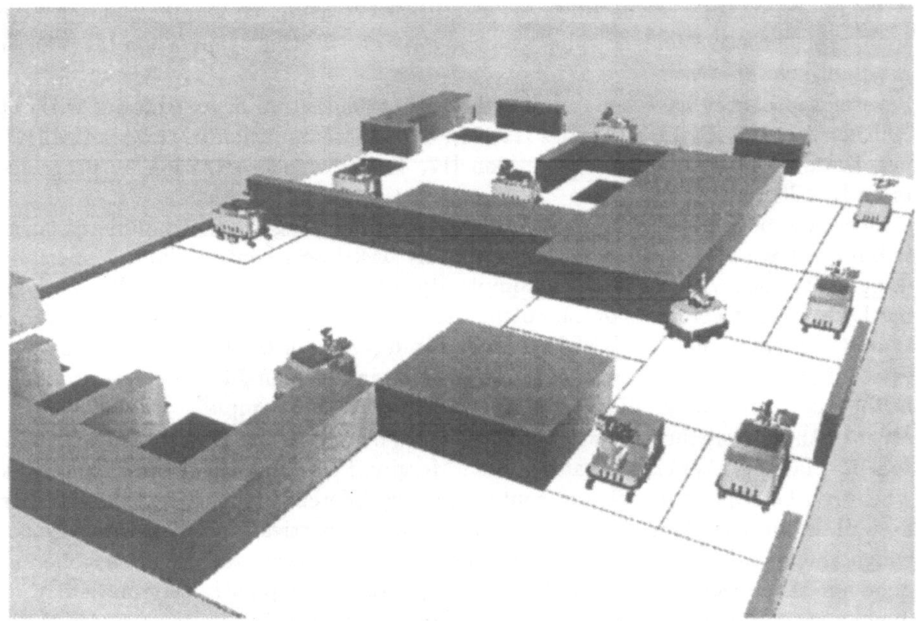

Figure 1. An environment model with 10 simulated robots

Thus, the Environment Model, which is provided to each robot, contains a hierarchical description of the site topology and geometry:

1. A network describing the connections of areas and crossings by oriented lanes. This is the only information used by the Central Station to elaborate routes.
2. A lower level topological description (cell level). The graph of cells is oriented, in order to provide a nominal (but not exclusive) direction for lanes and crossings use. However, cells adjacency is also provided in order to allow robots to use complementary spatial resources when necessary.
3. The geometry of cells (polygonal regions) and of permanent obstacles.
4. Additional information concerning landmarks (for re-localization), station descriptions for docking and load handling actions.

Figure 1 illustrates a typical environment model.

In the developed system, the central station is in charge of producing the high level plans. The produced plans take into account the topological model of the environment as well as the availability of such or such robot. However, it does not further specify robot trajectories or the sequence of robots going through a crossing (these decisions are left to the robots locally concerned), nor does it forces the robot to remain on the specified lanes (in case it needs to move away from an unexpected obstacle).

3.2. A Plan-Merging Protocol for Multi-Robot Navigation

The cooperation scheme we use is based on a general paradigm, called *Plan-Merging Paradigm* [2], where robots incrementally merge their plans into a set of already coordinated plans. This is done through exchange of information about their current state and their already planned actions.

For the case of a number of mobile robots in a route network environment, we have devised a specific *Plan-Merging Protocol* based on spatial resource allocation (see [3]). It is an instance of the general protocol described in [2], but in this context, the *Plan-Merging Operation* (PMO) is done for a limited list of resources: a set of cells which will be traversed during the plan to merge. Due to place limitations, we will not describe in more detail this protocol. A full description may be found in [2]. One of the most interesting attributes of this particular protocol is that it allows several PMOs to be performed simultaneously if they involve disjunctive resource sets. This proved to be particularly efficient for the cooperative navigation of a large number of mobile robots inducing several local conflicts at the same time.

▷ **Plan-Merging for cell occupation:** In most situations, robot navigation and the associated Plan-Merging Operation are performed by trying to maintain each cell of the environment occupied by at most one robot. This allows the robots to plan their trajectories independently, to compute the set of cells they will cross and to perform Plan-Merging at cell allocation level.

We have chosen an allocation strategy which makes the robots allocate one cell ahead when they move along lanes, while for crossing, they must allocate all the cells necessary to traverse and leave it. This is done in order not to constrain unnecessarily the other robots.

▷ **When reasoning about cells is not sufficient:** While, most of the time, the robots may restrict their cooperation to cells allocation, there are situations where this is not enough. This happens when they have to cross non-structured areas (rooms) or when an unexpected obstacle, encountered in a lane or in a crossing, forces a set of robots to maneuver simultaneously in a set of cells. In such situations, a more detailed cooperation (using the same protocol but a different planner: the motion planner) takes place allowing the robots to coordinate their actions at trajectory level.

Thus, we have a hierarchy of PMOs:

1. first, at the cell level, based on resource (cells) allocation;
2. then, depending on the context, at trajectory level: motion planning in a set of common cells determined by the first level.

This hierarchy authorizes a "light" cooperation, when possible, and a more detailed one, when necessary.

4. The Robot Control System

For each robot, the Robot Control System (RCS) architecture is derived from the generic control architecture for autonomous mobile robots developed at LAAS [5, 1, 9]. The control architecture of an autonomous robot must include both some decision-making processes, embedded in the Decisional Level, and real-time execution processes which are gathered in the Functional Level.

The Decisional Level which includes the refinement of the missions and a coherent control of the actions, calls for centralized knowledges and decisions, whereas the Functional Level, the reactive part of the architecture, calls for a distribution of the low level functions (*e.g.* the sensor/actuator control loops).

4.1. The Decisional Level

From an architecture point of view, the decisional level is organized into three independent layers running in parallel: a mission layer, a coordination layer and a execution layer (Figure 2). At each layer planning aspects and execution control aspects are independent and run separately to satisfy different response time constraints [1].

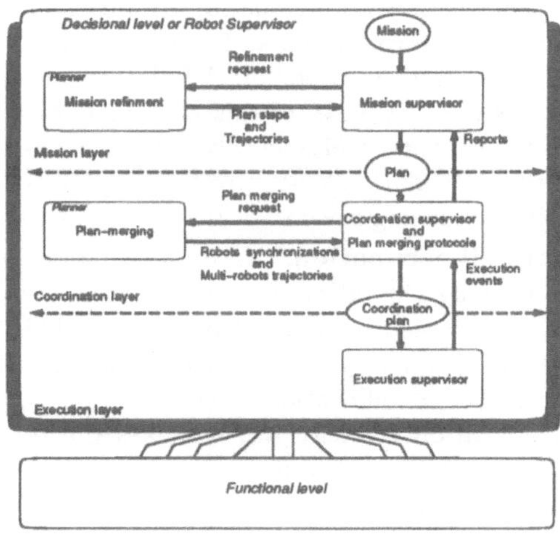

Figure 2. The robot supervisor architecture

⊳ **The mission layer** deals with mission refinement, mission control and interactions with the Central System. The mission (see a typical mission in Figure 3) are first refined as if the robot was alone. Refinement consists in planning all trajectories[2] according to the specified actions (go to a station, dock,...). The plan is annotated with cell entry and exit monitoring operations which will be used to maintain the execution state of the robot and to synchronize its actions with other robots.
Figure 4 shows a refined plan corresponding to the refinement of the first action of the mission in Figure 3 (the robot being at the end of the lane 0).

```
(mission (.
    (action 1 (goto (station 1))
        (using (lane 10)))
    (action 2 (dock))
    (action 3 (putdown))
    (action 4 (undock))
    (action 5 (goto (station 3))
        (using (lane 12) (lane 8)))
    (action 6 (dock))
    (action 7 (pick-up (container 5)))
    (action 8 (undock))
    (action 5 (goto (end-lane 0))
        (using (lane 9) (lane 0))) .))
```

Figure 3. Example of mission sent by the Central Station

```
(plan (.
    (plan-step 1 (report (begin-action 1)))
    (plan-step 2 (monitor (entry (cell 4))))
    (plan-step 3 (monitor (entry (cell 5))))
    (plan-step 4 (monitor (exit (cell 14))))
    (plan-step 5 (monitor (exit (cell 4))))
    (plan-step 6 (exec-traj 0))
    (plan-step 7 (exec-traj 1))
    (plan-step 8 (exec-traj 2))
    (plan-step 9 (monitor (entry (cell 0))))
    (plan-step 10 (monitor (exit (cell 5))))
    (plan-step 11 (monitor (entry (station 1))))
    (plan-step 12 (exec-traj 3))
    (plan-step 13 (exec-traj 4))
    (plan-step 14 (exec-traj 5))
    (plan-step 15 (report (end-action 1))) .))
```

Figure 4. Example of refined mission

⊳ **The coordination layer** is involved in plan merging operations, interactions with others robots (plans or events exchange), the control of the plan and the production of the coordination plan. Before executing its plan, a robot must validate

[2]the route can be given by the Central Station or computed by the robot itself.

it in the multi-robot context. This is done through incremental plan merging. The plan merging protocol allows the coordination supervisor to incrementally build a new plan called the coordination plan. This plan specifies all trajectories and actions to be executed, but also all the events to be monitored and sent to another or awaited from another robot. Note that this plan is also the one exchanged between robots during plan merging operations.
Figure 5 presents a coordination plan example corresponding to the 8 first plan steps of the refined mission given above.

```
(coordination-plan
  (.(exec-plan 1 (report (begin-action 1)))
    (exec-plan 2 (wait-exec-event robot-3 9))
    (exec-plan 3 (wait-exec-event robot-7 48))
    (exec-plan 4 (monitor (entry (cell 4))))
    (exec-plan 5 (monitor (entry (cell 5))))
    (exec-plan 6 (monitor (exit (cell 14))))
    (exec-plan 7 (monitor (exit (cell 4))))
    (exec-plan 8 (exec-traj 0))
    (exec-plan 9 (exec-traj 1))
    (exec-plan 10 (exec-traj 2)) .))
```

Figure 5. Example of coordination plan

▷ **The execution layer** is in charge of the interpretation and the execution of the coordination plan. As a result, it is responsible of most interactions with the functional level. The coordination plan is executed while taking into account synchronization between robots. When plan merging is done only at the "cell allocation level", entry or exit of cells are monitored in order to produce exchanged execution events. If "trajectory level" plan merging has been necessary, then curvilinear abscissa along trajectories are monitored for execution synchronization purposes.

4.2. The Functional Level

The functional level implements all the basic capabilities of the robot in sensing, acting and computing. These functionalities are grouped according to data or resource sharing, and integrated into *modules*.
Beside real time capabilities to ensure closed-loop control and reactivity, this level fulfills several conditions towards the others layers: bounded response time to request, observability and programmability. To guaranty these properties and to allow a coherent and homogeneous integration of a diversity of modules in the system, a generic canvas for a module has been specified. This has allowed to elaborate a "module specification language" and to implement an automatic module generator called G^{en}$_{o}$M (see [9]).
Roughly, the structure of a module is composed of a control level and an execution level. The control capacities are related to the asynchronous handling of requests and replies and the control of the functions of the execution level.
At the execution level, the implemented functions (i.e. embedded algorithms) are classified in four categories according to their execution modalities (i.e. starting and ending conditions, periodicity, ...), and consequently the way to control them : *servers, filters, servo-processes, monitors.* All these functions are interruptible by the module controller.
Figure 6 shows the functional level, for the presented application, including 8 modules, their client/server relations and 4 exported data structures (called "posters"). The Robot Supervisor is a client of all the modules of the functional level. It manages

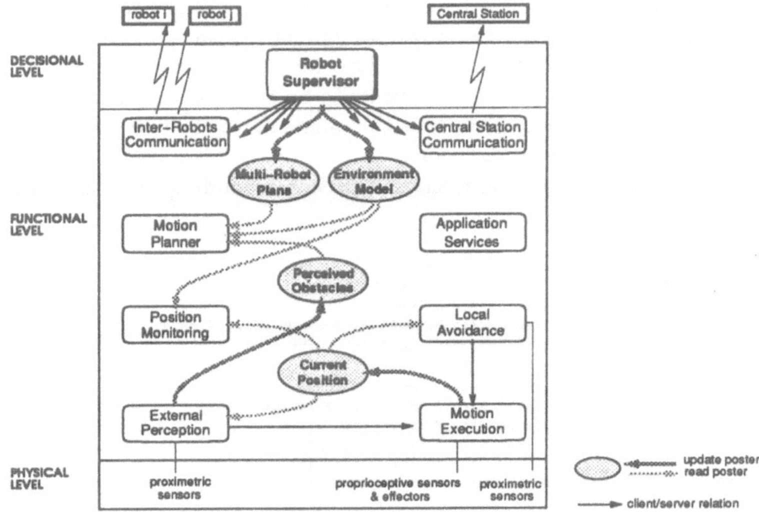

Figure 6. Architecture and interactions of the functional level

itself the poster named **ENVIRONMENT** which contains the topological and geometrical model of the environment (cf §3.1).

The modules currently implemented are:

- The Motion Planner module is composed of a Topological Planner, a Geometrical Planner and Multi-robot Scheduler. It is used in order to compute not only feasible trajectories but also synchronization events between different robot trajectories.
- The Motion Execution module [10] embeds all robot motion primitives.
- The Local Obstacle Avoidance module [16] allows to execute (non-holonomous) trajectories while avoiding, when possible, unknown obstacles. Figure 7 illustrates an example of its capabilities.
- The External Perception module [7] performs landmark-based (wall and fixed furniture) localization (see Figure 8) and allows to build a model of the local obstacles. This feature is used to update the world model after the detection of a situation which requires a drastic change of a planned trajectory (when local avoidance fails).
- The Position Monitoring module provides a variety of position based monitors for detecting the entry of exit of regions for synchronization purposes, as established after a Plan-merging operation.
- The External Communication modules: two independent systems which handle the communication with the Central Station and the direct communication between robots. This last one allows also to broadcast messages to all robots.

5. Implementation of Various Testbeds

We have developed a complete robot control system which includes all the features described above. The robot supervisor is coded using a procedural reasoning system: C-PRS [14, 13]. The target implementation runs on-board a multi-processor VME rack, under the VxWorks real-time system.

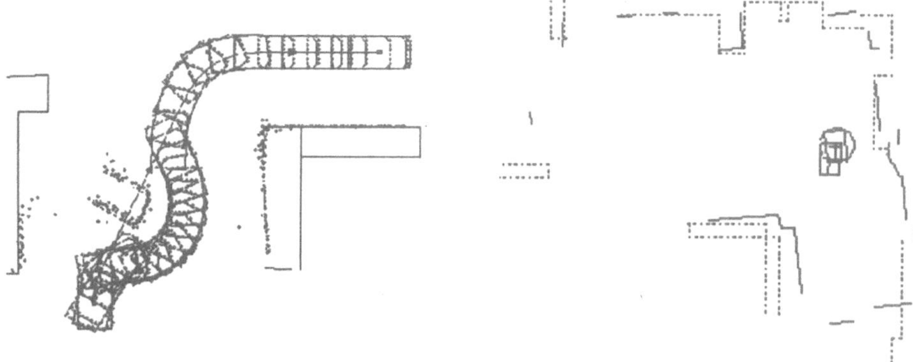

Figure 7. An example of an obstacle avoidance

Figure 8. An example of re-localization based on walls and furniture

▷ **Simulated environments:** For testing purposes, and to demonstrate the co-ordination scheme for a large number of robots, we have built an emulation of the communication and multi-tasking primitives of the real-time system, that allows us to run all elements of a robot (the supervisor and all functional modules) on a Unix workstation as a set of Unix processes.
The motion execution is simulated at the lowest level by sending the elementary motion controls to a robot simulator. Radio communications between robots and with the central station are simulated on the Ethernet network.
Moreover, virtual obstacles can be given to the external perception module. Their presence on the robot path will be detected on-line and signaled to the supervisor.
A 3-D graphic server has been built in order to visualize the motions and the load/unload operations performed by all the robots in their environment (Figures 1,11). It receives position updating from the motion execution modules of all the robots, each running on its own workstation.
Experiments have been run successfully on fifteen workstations, each workstation running a complete robot simulator. Figure 1 is a snapshot of a typical run: 10 robots in an in-door environment.

▷ **An experiment using three real robots:** We have also tested the implementation and run several demonstrations on three real mobile robots (Figure 9).

Figure 10 illustrates the hardware used in the experimental testbed.
The Hilare robots have been equipped with ultra-sonic sensors (for obstacle avoidance), a laser range finder (for re-localization and environment modeling), and radio modems. The three robots are all different, but they perform the same missions with the same interface.
These robots have performed numerous "Martha-like" missions in a very constrained in-door environment.
Besides a behavior similar to the simulations, they demonstrated integrated capabilities of re-localization, environment modeling, real-time obstacle avoidance, ..., as well as capabilities to recover from errors in a multi-robot context and in a coherent manner.
The 3D graphic system can also be used as a server which draws periodically the

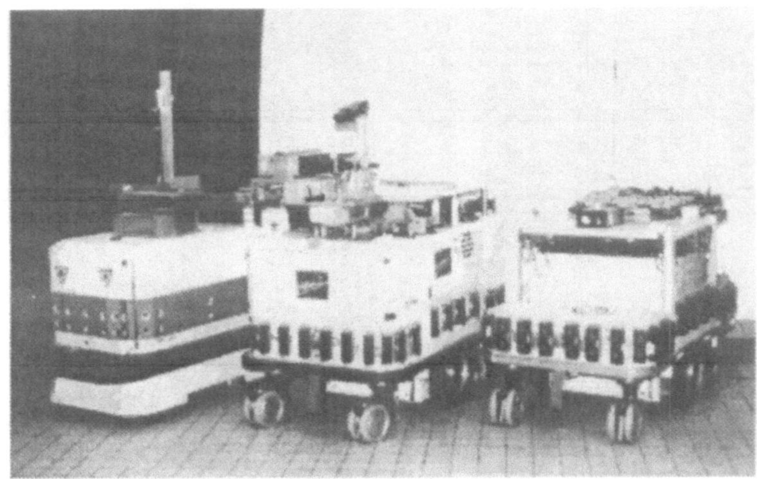

Figure 9. The 3 Hilare robots

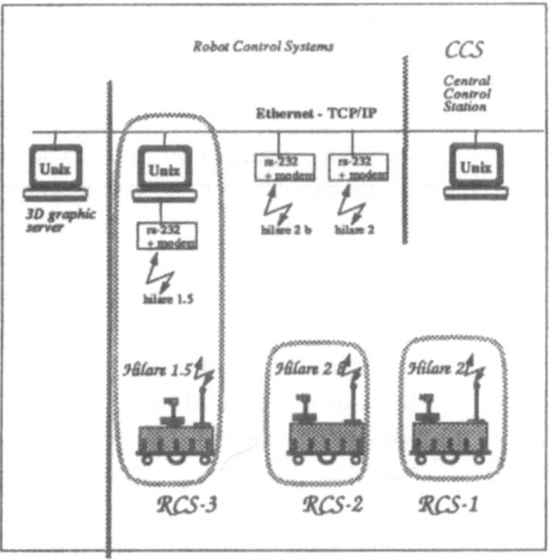

Figure 10. The experimental testbed

positions of the real robots in the model as well the result of the various actions they perform (landmark-based localization, obstacle modeling, pick-up and put-down actions...). An example of such use is represented in Figure 12.

6. Illustrative Runs

We shall now illustrate the plan-merging paradigm and its capabilities with some sequences from our experimentation with simulated robots in a route network and real robots in an indoor environment. The first example presents a PMO at a crossing with simulated robots, the second, a PMO in an open area with real robots.

6.1. An Example of PMOs at a Crossing (Simulation)

The example is presented by Figure 11.

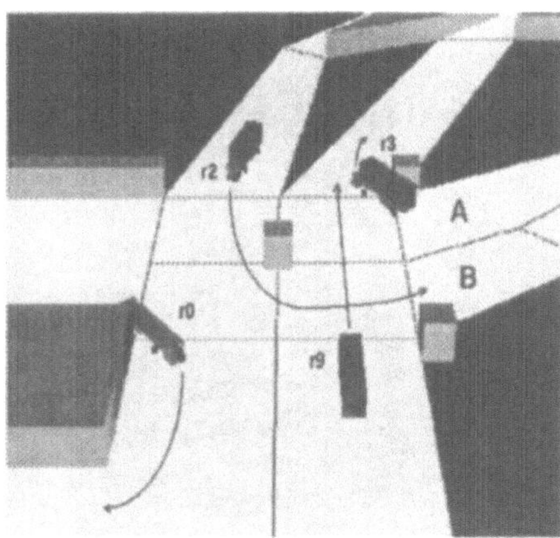

Crossing: step 1

Step 1: Robots $r3$ and $r0$ have disjunctive list of resources: they perform PMOs in parallel and decide to traverse the crossing without any synchronization. The PMO of $r9$ failed: it has to wait for $r3$ to plan to leave the cell beyond the crossing. On the other hand, $r2$ going to **B** can merge its plan. **Step 2:** $r2$ traverses the crossing after a synchronization with $r0$. $r9$ has received the awaited "planning event" from $r3$ and succeeds its PMO. **Step 3:** $r2$ frees the crossing. On the reception of the synchronization events from $r3$ and $r2$, $r9$ traverses on its turn.

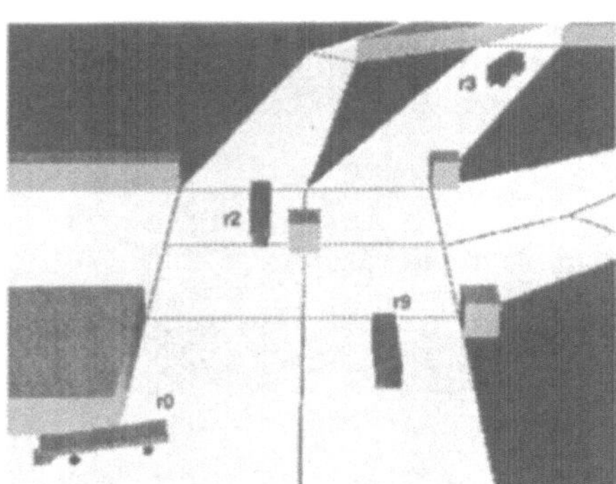

Crossing: step 2

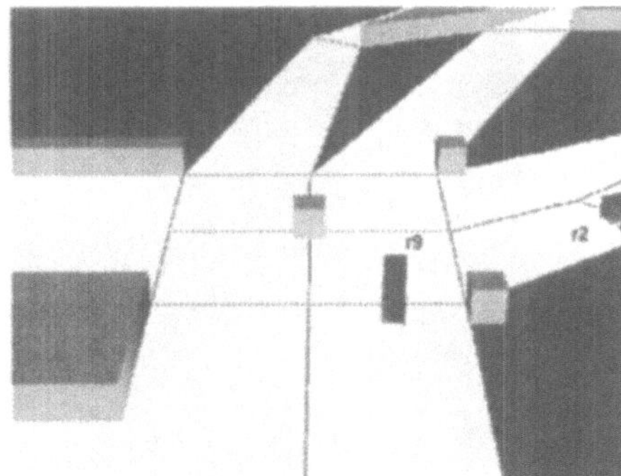

Crossing: step 3

Figure 11. **Coordination at a crossing**

▷ **Quantitative Results:** As an example, we give here below some measures from an experiment where 10 simulated robots execute each one mission similar to the one presented above (1 km long). The average message length is 100 bytes (without any optimization).

The experiment lasted around 15 minutes for 41000 bytes exchanged. A very simple optimization (shortening the id exchanged) would result in a 20000 bytes exchanged for the whole mission. These results show that the bandwidth required by the PMO protocol is compatible with the low baud rate provided by the inter-robot communication system. 587 messages were exchanged:

. 225 broadcasts for plan merging,
. 87 plans exchanged,
. 95 messages for execution synchronization.

During all these missions several types of conflicts have been encountered and solved: 8 conflicts for simultaneous PMO for common resources, 24 messages to re-synchronize PMOs and 52 messages to update the deadlock detection graph.

Depending on the length and difficulty of their missions, individual robots have produced from 20 to 90 messages.

6.2. An Example of PMOs in an Open Area: 3 Real Robots in the LAAS Robotics Room.

Figure 12. The three Hilare in the robotics room (3D view point)

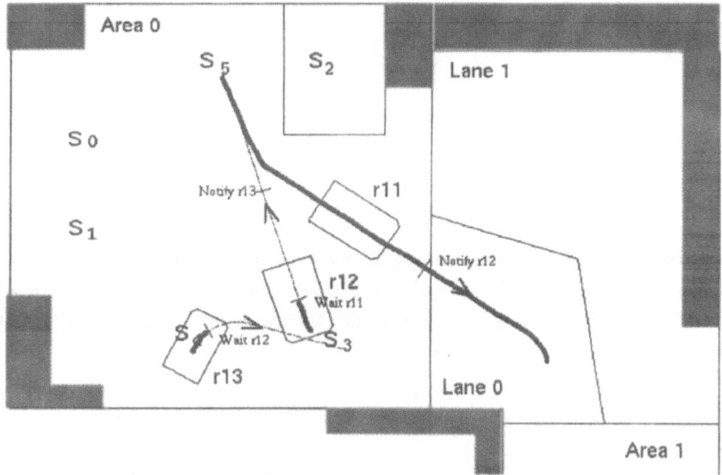

Figure 13. The three Hilare in the robotics room (2D view point)

As mentioned earlier, open areas are not exclusive resources. In our example (see Figures 13 and 12 which represent the same situation in 3D and 2D), $r11$ (Hilare 2) goes from station $S5$ to *Lane* 0, $r12$ (Hilare 2bis) moves backward from station $S3$ to station $S5$, and $r13$ (Hilare 1.5) goes from station $S4$ to station $S3$.

After synchronizing at the PMO level, the three robots performed theirs PMOs, at the "trajectory level", in the following order: $r11$, $r12$ and $r13$.

In short, $r13$ waits for $r12$ to move away, and $r12$ waits until $r11$ clears its path. If we analyze the situation more carefully, one can see that $r12$ has moved until position *Wait_r11*, at which point, it waits until $r11$ reaches the point *Notify_r12* and then notifies $r12$ of this event. As a consequence, $r12$ will proceed to station $S5$ and will notify $r13$ (upon reaching point *Notify_r13*), which is waiting at point *Wait_r12*, that it can now proceed to station $S3$.

▷ **Quantitative Results:** We here below some measures from an experiment where the 3 Hilare robots execute each navigation missions during 2 hours! Each robot traveled about 600 meters in a 70 square meters environment. 1381 messages were exchanged:
. 473 broadcasts for plan merging,
. 533 plans exchanged,
. 125 messages for execution synchronization (at trajectory level),
. 7 messages for execution synchronization (at cell level).
During all these missions several types of conflicts have been encountered and solved:
. 18 conflicts for simultaneous PMO for common resources,
. 72 messages to re-synchronize the PMO,
. 71 messages to update the deadlock detection graph.
For each robot, 3000 requests were issued by the Robot Supervisor to the various modules.

6.3. Discussion

To conclude on these two examples presented above, one can note the following interesting properties of the PMO:

- Planning and execution are done in parallel.
- Several robots may use a crossing simultaneously ($r0$ and $r3$).
- The example exhibits the two types of synchronization: synchronization based on *planning events* ($r9$ and $r3$ in the crossing example) and synchronization based on *execution events* ($r11$, $r12$ and $r13$ in the area example).
- Each robot produces and merges its plans iteratively, and the global plan for the use of the crossing is incrementally built through several PMOs performed by various robots.
- It is not a first arrived first served execution. In the crossing example $r9$ arrived second, was blocked by $r3$, but did not block the crossing and let $r2$ enter the crossing before.

7. Conclusion

The system described in this paper paper presents many original contributions to the field of research on autonomous mobile robot. To our knowledge, it is is the first time such a large fleet of autonomous robot is put together to execute high level missions given by a central station.

Our experimentation using large number of emulated robots has shown the feasibility and the embarkability of our solution.

The Plan-Merging Paradigm we propose has the following properties;

1. It "fills the gap" between centralized, very high level planning and distributed execution by a set of autonomous robots in a dynamic environment.
2. It makes possible for each robot to produce a coordination plan which is compatible with all plans executed by other robots.

3. No system is required to maintain the global state and the global plan permanently. Instead, each robot updates it from time to time by executing a PMO.
4. The PMO is safe, because it is robust to plan execution failures and allows to detect deadlocks.
5. The Plan-Merging paradigm is particularly efficient when there are uncertainties concerning actions durations; it allows to take advantage of the available information about the current context and its short term evolution.
6. Incremental and local plan merging of short-term plans do not, in most cases, over-constrain the other vehicles. This operation can be performed with anticipation and is fast enough to be done in parallel with execution. Indeed, experiments confirm that the robots do not stop unnecessarily.

The current implementation have shown that the Plan-Merging Paradigm applied to navigation works and allows for far more than fifteen robots to cooperate. In fact, considering the locality of the conflict resolution, i.e. the ability of robots in a group to coordinate their plans and actions without disturbing the rest of the fleet, one can easily see that this protocol can scale to a much larger number of robots (hundreds). This protocol allowed us to make a large number of autonomous robots behave coherently and efficiently without creating a burden on the central system activity.

References

[1] R. Alami, R. Chatila, and B. Espiau. Designing an intelligent control architecture for autonomous robots. *International Conference on Advance Robotics.* ICAR'93, Tokyo (Japan), November 1993.

[2] R. Alami, F. Robert, F. F. Ingrand, and S. Suzuki. A paradigm for plan-merging and its use for multi-robot cooperation. *IEEE International Conference on Systems, Man, and Cybernetics, San Antonio, Texas (USA)*, 1994.

[3] R. Alami, F. Robert, F. F. Ingrand, and S. Suzuki. Multi-robot cooperation through incremental plan-merging. *IEEE ICRA'95*, Nagoya (Japan), May 1995.

[4] H. Asama, K. Ozaki, et al. Negotiation between multiple mobile robots and an environment manager. *ICAR'91*, Pisa (Italy), June 1991.

[5] R. Chatila, R. Alami, B. Degallaix, and H. Laruelle. Integrated planning and execution control of autonomous robot actions. *IEEE International Conference on Robotics and Automation*, Nice, (France), May 1992.

[6] H. Chu, H.A. EiMaraghy. Real-time multi-robot path planner based on a heuristic approach. *IEEE ICRA'92*, Nice (France), May 1992.

[7] M. Devy and H. Bullata. Landmark-based vs Feature-based Localization of a Mobile Robot in a Structured Environment. *International Conference on Advanced Robotics (ICAR)*, Barcelona (Spain), 1995.

[8] Michael Erdmann and T. Lozano-Perez. On multiple moving objects. *IEEE International Conference on Robotics and Automation*, San Francisco (USA), April 1986.

[9] S. Fleury, M. Herrb, and R. Chatila. Design of a modular architecture for autonomous robot. *IEEE International Conference on Robotics and Automation*, San Diego, California, (USA), 1994.

[10] S. Fleury, P. Soueres, J.P. Laumond, and R. Chatila. Primitives for smoothing mobile robot trajectories. *IEEE International Conference on Robotics and Automation*, Atlanta (USA), May 1993.

[11] T. Fraichard and C. Laugier. Path-velocity decomposition revisited and applied to dynamic trajectory planning. *IEEE Transactions on Robotics and Automation*, pages 40–45, 1993.

[12] D.D. Grossman. Traffic control of multiple robot vehicles. *IEEE Journal of Robotics and Automation*, 4(5):491–497, Oct. 1988.

[13] F. F. Ingrand, R. Chatila, R. Alami, and F. Robert. Embedded control of autonomous robots using procedural reasoning. *International Conference on Advanced Robotics (ICAR)*, Barcelona (Spain), 1995.

[14] F. F. Ingrand, M. P. Georgeff, and A. S. Rao. An Architecture for Real-Time Reasoning and System Control. *IEEE Expert, Knowledge-Based Diagnosis in Process Engineering*, 7(6):34–44, December 1992.

[15] S. Kato, S. Nishiyama, J. Takeno. Coordinating mobile robots by applying traffic rules. *IEEE IROS'92*, Raleigh (USA), July 1992.

[16] M. Khatib and R. Chatila. An extended potential field approach for mobile robot sensor-based motions. *Intelligent Autonomous Systems (IAS'4), Karlsruhe (Germany)*, Eds *E. Rembold, R. Dillmann, L.O. Hertzberger, T. Kanade*, IOS Press, pages 490–496, 1995.

[17] C. Le Pape. A combination of centralized and distributed methods for multi-agent planning and scheduling. *IEEE ICRA'90*, Cincinnati (USA), May 1990.

[18] F.R. Noreils. Integrating multi-robot coordination in a mobile-robot control system. *IEEE IROS'90*, Tsuchiura (Japan), July 1990.

[19] P.A. O'Donnell and T. Lozano Perez. Deadlock-Free and Collision-Free Coordination of Two Robot Manipulators. *IEEE Transactions on Robotics and Automation*, 1989.

[20] P. Svestka and M.H. Overmars. Coordinated Motion Planning for Multiple Car-Like Robots using Probabilistic Roadmaps. *IEEE ICRA'95*, Nagoya (Japan), May 1995.

[21] Pierre Tournassoud. A strategy for obstacle avoidance and its application to multi-robot systems. *IEEE International Conference on Robotics and Automation*, San Francisco, CA, April1986.

[22] T.Tsubouchi, S.Arimoto. Behavior of a mobile robot navigated by an "iterated forecast and planning" scheme in the presence of multiple moving obstacles. *IEEE ICRA'94*, San Diego (USA), 1994.

[23] T. Vidal, M. Ghallab, R. Alami Incremental mission allocation to a large team of robots. *IEEE ICRA'96*, Minneapolis (USA), April 1996.

[24] J. Wang. On sign-board based inter-robot communication in distributed robotic systems. *IEEE ICRA'94*, San Diego (USA), 1994.

[25] S. Yuta, S. Premvuti. Coordination autonomous and centralized decision making to achieve cooperative behaviors between multiple mobile robots. *IEEE IROS'92*, Raleigh (USA), July 1992.

Acknowledgments: This work was partially supported by the MARTHA (ESPRIT III) Project, the CNRS, the Région Midi-Pyrénées and the ECLA ROCOMI Project.

This project is the fruit of a very intensive collaboration between numerous researchers. We would like to acknowledge the help and the effective involvement of S. Suzuki, J. Perret, T. Siméon, M. Devy, G. Bauzil, C. Lemaire, B. Dacre-Wright, C. Dousson, P. Gaborit.

Cooperative Autonomous Low-cost Robots for exploring Unknown Environments

Josep Amat
Automatic Control Department, UPC
Barcelona, Catalonia (Spain)
amat@esaii.upc.es

Ramon López de Màntaras, Carles Sierra
IIIA - Artificial Intelligence Research Institute, CSIC
Bellaterra, Catalonia (Spain)
{mantaras, sierra}@iiia.csic.es

Abstract

In this paper we present the results obtained with a troup of small autonomous vehicles designed to cooperatively explore unknown environments. In order to improve the covering of the explored zone the vehicles show different behaviours. The host that controls the troup of vehicles generates the most plausible map of the environment from the information obtained by the different components of the troup, which at the end of their mission return back. To perform the map generation a two-step algorithm, fusion and completion, based on fuzzy techniques, is presented.

1. Introduction

With the aim to explore an environment that is unknown but easily passable, a system constituted by a set of low cost, small autonomous vehicles has been developed. These vehicles follow the already classical line of insect robots [1] [2] [3] [4] [5] [6]. The goal of these autonomous vehicles is to obtain partial information about the environment during their exploration runs and afterwards to supply this information to a Master Autonomous Robot that in turn will be able to compute the most plausible map. With this information, the Master should be able to perform a given mission within structured environments. Using this Master-Multislave strategy to generate a model of the environment, we expect to achieve a better efficiency and a safer procedure than that which would be obtained based only on the Master perception capabilities.

The behaviour of these small autonomous vehicles has been programmed in an individual basis for obtaining a behaviour similar -to some degree- to that of ants in two aspects. First, in order to increase the coverage of the environment, the vehicles have a partially random moving behaviour; and second, the vehicles cooperate in the process of environment information gathering by transfering each other the perceived environment when they meet. Sharing data in this way, allows the Master to get the information not only from the vehicles that successfully return after an exploratory run, but also from those that cannot return, provided that they have encountered vehicles that have safely returned.

The sensing capability of each small vehicle, to build its own partial map, comes from two kind of sensors: IR proximity sensors for environment data acquisition, and a relatively accurate odometric system for the estimation of the vehicle position during its run.

The next section of this paper describes the structure and the behaviour of the vehicles. Section 3 describes the fuzzy logic-based algorithms that we have developped in order to compute the most plausible map based on the partial maps perceived by the succesfully returning vehicles. Section 4 describes the results obtained to date and in section 5 we briefly point to some future work.

2. Structure of each mobile vehicle

Each vehicle has been designed with the aim of being small and cheap. It must have a high autonomy and be endowed with a computer, with which they can get and memorize, with the highest resolution, the observed portion of the environment. All these requirements have lead to a compromise solution consisting on small vehicles with three wheels. Two of them are steering wheels having independent motors.

The vehicles environment perception system and the communications with the host or with the other vehicles are based on IR impulse modulated sensors. Since some of the monirobots may not return to deliver their map to the Master robot, the following communication process is established: when two of them meet along their exploration run, they back-up from one to the other all the information they have acquired so far from the environment. In this way, when a vehicle reaches the host, it delivers both, the information acquired during its own run as well as the information obtained from other vehicles that it has encountered. This communication process allows to get all the information of a non-returning minirobot that had been transferred to a returning one.

2.1. Mechanical characteristics

The vehicle is 21 cm. long and 15 cm. wide (see Fig. 1). The 5cm. driving wheels allow the vehicle to save some small obstacles such as carpets or electrical wires. With the motors utilised, the vehicles can reach a speed up to 0.6 m/sec., and since the battery has a one hour autonomy at full regime, each vehicle can do a run of about 2000 m. long.

2.2. Sensing Capability

The vehicle is equipped with the following sensing elements:

- Impulse generators at each wheel for odometry.
- Five I.R. proximity sensors for obstacles detection.
- A proximity sensor to detect terrain horizontal discontinuities.
- Safety microswitches to detect collisions.
- One omnidirectional IR Emitter/Receiver sensor to detect other vehicles and to transmit data.

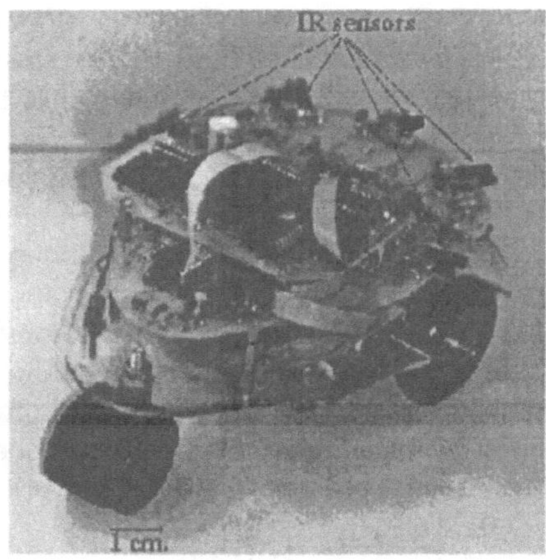

Figure 1. Autonomous Mini-Robot structure

- One IR Emitter with a sector scope of 90° to generate a priority signal (right hand preference).

The obtained odometric precision produces a 2x20 cm. uncertainty ellipse after a 10 meters run without direction changes. In order to use the data of this particular odometer, to get at each instant the vehicle position and also the environment information, a quality factor has been experimentally elaborated. This quality factor depends on the distance traveled L and on the number of turns N done. This quality factor is:

$$FC = \frac{1}{(1 - \frac{L}{K_l})(1 - \frac{N}{K_n})}$$

K_l and K_n are constant parameters that control the rate of decay of the quality factor.

When the vehicle detects and follows a wall, the corresponding information is incorporated into memory together with its associated quality factor. The host will generate a map of the environment by integration of the data supplied by the different vehicles in an efficient way (see Section 3).

2.3. Navigation Strategy

The navigation system incorporated to each vehicle has a random behaviour: The vehicle does a ±45° or ±90° turn, either randomly or when it detects an obstacle.

The random turns are done according to three significantly different probabilities: $P_1 > P_2 > P_3$, modelling three different behaviours:

- P_1 = vehicle with an "anxious" behaviour.
- P_2 = vehicle with "normal" behaviour.
- P_3 = vehicle with "routine" behaviour.

When the vehicle finds a frontal obstacle, the turn can be done to the right or to the left based also on a probability value P_4. The vehicles having a probability

$P_4 < 0.5$ will show a tendency to turn right more often than to turn left, whilst the vehicles having a probability $P_4 > 0.5$ will behave inversely.

Consequently, the different vehicles of the exploration troup will not show an identical behaviour. They can behave in six different ways corresponding to the different combinations of behaviours and turning tendencies. By using a troup of vehivles showing different behaviours, we expect to obtain a better covering of an environment than with vehicles showing the same behaviour.

Once a vehicle has run a length L_{P_i}, depending on the probability P_1, P_2 or P_3, where $L_{P_1} < L_{P_2} < L_{P_3}$, it starts its way back towards the host.

The algorithm used to guide the vehicle back towards the starting point generates a trajectory following an improvement of the travelled path contained in memory (in order to assure a path possibly free of obstacles). This improvement consists in eliminating loops.

2.4. Control System

The control unit in each vehicle has been designed having in mind that the hardware had to be as simple as possible but, on the other hand, it had to allow achieving a behaviour sufficiently smart in order to navigate efficiently. Furthermore the vehicle had to be based on a hardware flexible enough to allow for experimentation of navigation and control strategies. These requirements have resulted in a design which contains three different functional modules: The *navigation module* generates the trajectory to be followed; the *steering module* controls the motors in order to follow the generated trajectory; and the *perception module* acquires information of the environment by means of IR sensors. However, it is possible to replace this module by other modules adapted to different types of sensors.

The computer used to implement the navigation control unit is a 80C186 with a 1MB RAM to store the data of the perceived portion of the environment. The environment is discretized with a resolution of 4 cm which means that each vehicle can store maps of up to 40x40 m.

The steering control module operates with a much higher resolution since each encoder corresponds to a displacement of only 2 mm. and it is implemented on a 80C552

3. Environment map generation

The goal of map generation is to obtain the most plausible position of walls and obstacles. The information about their position conveyed by the different vehicles is imprecise. Furthermore, vehicles can detect portions of walls or obstacles with different degrees of imprecision. The main problem is to decide whether several detected portions, represented by imprecise segments, belong to the same wall or obstacle or not. This decision depends on the relative position of the segments as well as on their imprecision. The relative position of the segments can be represented by their euclidean distance. The distance is compared to a threshold to decide whether or not the segments represent different portions of the same wall or obstable. If two segments represent the same wall or obstacle, a segment fusion procedure is applied to produce a single segment. This process of segment fusion is followed by a completion process in which hypothesis are made with respect to non observed regions. The completion process is performed by means of hypothetical reasoning based on declarative heuristic knowledge about the structured environments in which vehicles evolve.

The map generation algorithm consists of two steps. The first one is the fusion of the map perceived by each vehicle, taking into account that the same vehicle can observe more than one portion of the same wall. The second step consists in a global fusion and completion of the maps perceived by several troup members. The fusion function is the same in both steps. However, since not all troup members may return home to give the information to the master and since, on the other hand, we want to have at any time the most plausible map based on the information obtained so far, it is necessary to develop an incremental approach to the map generation algorithm. Any time a vehicle returns home the algorithm is executed to update the map. The overall algorithm is as follows:

Function Map_generation(NewAntsMap, CurrentMap) =
Begin
NewCurrentMap = Fusion(CurrentMap ∪ Fusion(NewAntsMap, α, β), α, β);
Return Completion(NewCurrentMap)
End

where α and β are decision threshold parameters that will be explained later. In section 3.1 we give details of the segment fusion procedure and in 3.2 we outline the completion process.

3.1. Segment fusion

Let us start by defining the basic object of our approach which is the imprecise segment.

An *imprecise segment* S is a function that gives for any cartesian coordinates (x, y) the degree of possibility of the coordinates as being part of a wall. That is $S : \Re \times \Re \rightarrow [0, 1]$. This function can be seen as a possibility distribution [7] in the sense of fuzzy logic [8] and is determined by a segment that, for simplicity, we assume has a precise length, plus the imprecision measure of its position with respect to both axis. Furthermore, for each segment we keep the list of coordinates of the singular points that have been detected (i.e. corners and gaps in the real world).

Given the orthogonality of the structured environment, we have only two possible orientations for a segment: vertical or horizontal. That is $S = ((x, y_1), (x, y_2), e, p_1, ..., p_n)$ or $S = ((x_1, y), (x_2, y), e, p_1, ..., p_n)$. Where e is the imprecision and the p_i are the singular points. For simplicity, in the sequel we drop the list of singular points from the notation of segments.

For example, an imprecise horizontal segment $S = ((x_1, y), (x_2, y), e)$ means that there is a portion of a wall with coordinates $((a, b)(c, b))$ such that $y - e < b < y + e$ and $x_1 + e > a > x_1 - e$ and $x_2 - e < c < x_2 + e$. Similarly for the orthogonal case. Figure 2 shows an example of an imprecise horizontal segment.

As we have said, the main problem is to decide whether several imprecise segments represent portions of the same wall or obstacle or not. This decision depends on the relative position of the imprecise segments, like for example the case shown in figure 3.

$$s = ((x, y_1), (x, y_2), e), s' = ((x', y_1'), (x', y_2'), e')$$

This relative position is represented by two parameters Δ_x and Δ_y that represent the minimum distance with respect to the two axis:

$$\Delta_x(s, s') = |x - x'|$$

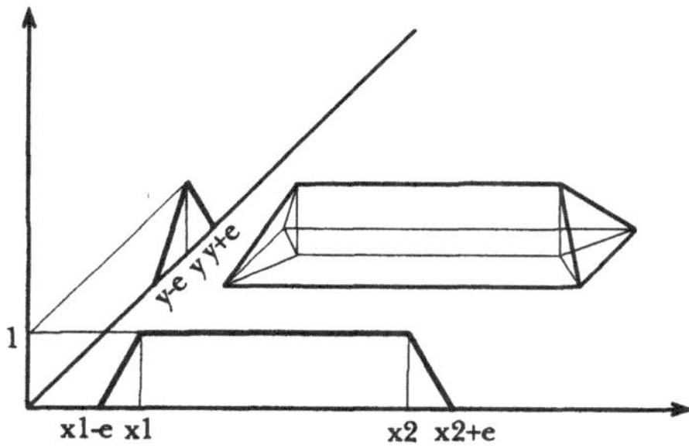

Figure 2. Horizontal Imprecise segment

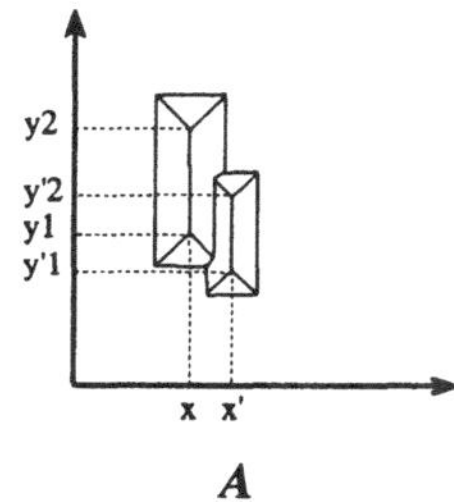

Figure 3. Example of two overlapping imprecise segments

$$\Delta_y(s,s') = \begin{cases} 0 & \text{if } (y_1 - y_2') * (y_2 - y_1') < 0 \\ min(|y_1 - y_2'|, |y_2 - y_1'|) & \text{otherwise} \end{cases}$$

In the Δ_y expression, the first part stands for the case in which both segments overlap. In that case both terms in the product have a different sign.

Given a map M, represented as a list of segments ordered by imprecision, the "Fusion" function sketched below computes a new map M' whose difference with M is that some segments have been fused.

To do that, for each segment s, starting with the most precise, we compute its distance to the segments in M' with respect to both axis. Then, with those segments in M such that both distances are below a threshold we build the list of candidates for merging with s (the $MayMerge$ list).

Next, the function *Select_Min_Dist* selects for fusion the segment s' in M' with the lowest average distance with respect to both axis. Finally, we merge s and s' with the Merging fuction and we replace s' in M' and s in M by the resulting merged segment s". Furthermore, the singular points in M and M' are updated in order to keep the singular points of the merged segments properly connected with other segments meeting at these singular points. Obviously, in the case where the $MayMerge$ list for s is empty we simply add s to M'. This process is iterated while $M' \neq M$.

Function Fusion(M, α, β) =
{M is a Map represented as a list of segments ordered by
imprecision, α and β are decision thresholds}
M' : Map
Begin
M' = M;

```
    M = nil;
    While M' ≠ M do
        M = M';
        M' = nil;
        Foreach s in M
            MayMerge = nil;
            Foreach s' in M'
                If Δx(s, s') < α  and  Δy(s, s') < β Then MayMerge = MayMerge + s'
                endif
            Endforeach
            If MayMerge = nil Then M' = M' + s
            Else  s" = Select_Min_Dist(MayMerge, s);
                  M' = M' - s" + Merging(s", s)
                  If singular(s") or singular(s) Then
                      M = update_sing_points(M,s",s);
                      M'= update_sing_points(M',s",s)
                  Endif
            Endif
        Endforeach
    Endwhile;
    Return M'
End
```

The function $Select\_Min\_Dist(MayMerge, s)$ selects the segment s', in the list $MayMerge$, that has the minimum average distance to s with respect to both axes, that is, $\Delta_x(s, s') + \Delta_y(s, s')/2$.

The function $Merging(s", s)$ checks whether the orientation of the imprecise segments to merge is vertical or horizontal and uses the appropriate function: $Vert\_Merging$ or $Hor\_Merging$.

The vertical merging function is sketched next (the horizontal one is similar):

```
Function Vert_Merging (s, s') =
    {s = ((x, y1), (x, y2), e, p1...pn), s' = ((x', y1'), (x', y2'), e', p1'...pm')}
    Begin
        y1" = min(y1, y1');
        y2" = max(y2, y2');
        e" = e * e'/(e + e');
        dist = abs(x - x');
        x" = e * Dist/(e + e');
        For i = 1 to n do pi" = (x", projy(pi));
        For i = 1 to m do p"i+n = (x", projy(pi'));
        Return ((x", y1"), (x", y2"), e", p1"...p"n+m)
    End
```

The coordinates $y_1^"$ and $y_2^"$ are obvious. We can easily see that x" is the center of gravity of the masses $\frac{1}{e}$ and $\frac{1}{e'}$ of s and s', the x coordinate of $p_i^"$ is x" and the y coordinate $p_i^"$ is the y coordinate of p_i, and e" is the degree of imprecision of the merged segments computed as a function of e and e' in such a way that the mass of the merged segments $\frac{1}{e^"}$ is the sum of the masses of the components of the merge $\frac{1}{e} + \frac{1}{e'}$, however alternative functions [9], like for example $\frac{1}{e^"} = max(\frac{1}{e}, \frac{1}{e'})$, are also possible. Which function to use can only be decided experimentally. Figure 4 shows an example of vertical segments fusion.

Figure 5 graphically shows the merging result (B), along the x axis, for two vertical segments (A) with no singular points

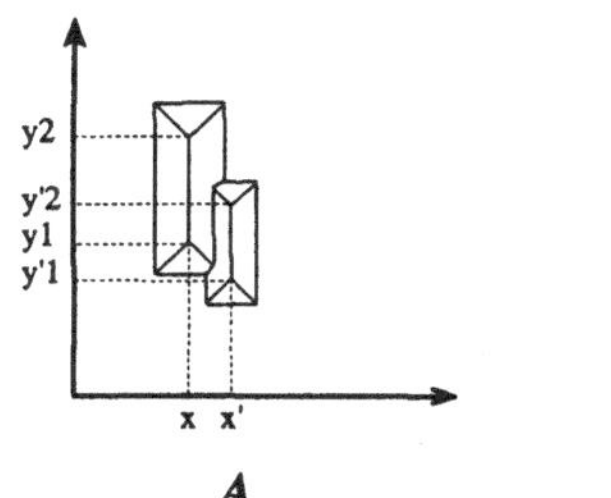

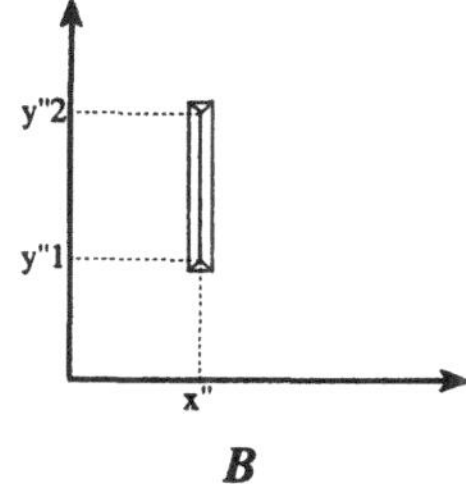

Figure 4. Vertical segments in A are fused into the segment in B

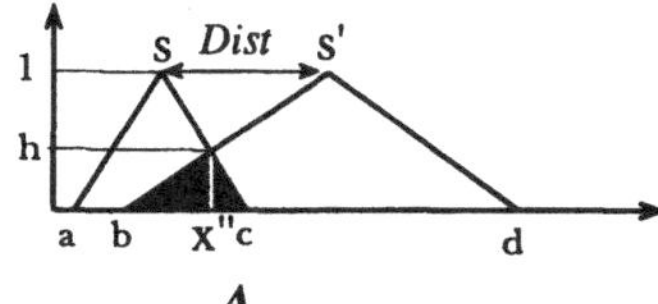

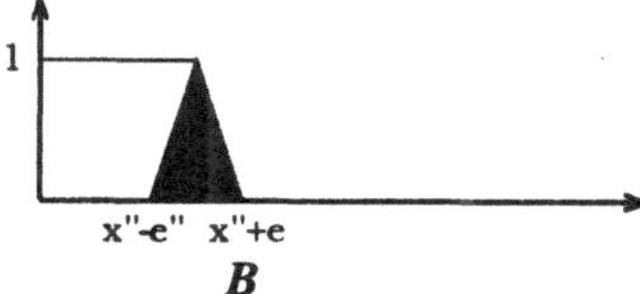

Figure 5. Merging of two vertical segment projections along the x axis

3.2. Global environment map completion

The fusion procedure is an initial step that only merges segments on a local basis using the geometrical properties of the segments. To improve the map, we perform a completion process based on a global view of the environment, including the map that each vehicle has from the other vehicles that it has encountered during its exploration, taking into account knowledge about the characteristics of the environment and knowing the paths followed by the vehicle. We adopt an heuristic knowledge-based approach. One example of such knowledge is expressed by the following heuristic rule:

If there exists a segment s_1 **and**
there exists a segment s_2 **and**
s_1 and s_2 are parallel **and**
there are no singular points in the closest extremes of s_1 and s_2 **and**
$\Delta_x(s_1, s_2) < \alpha$ **and**
$k \times \beta > \Delta_y(s_1, s_2) > \beta$ **and**
there is no vehicle path crossing the gap
Then Assume($Vert\_Merging(s_1, s_2)$)

A set of fifteen rules covering many different situations has been identified.

4. Results

A first prototype of an autonomous vehicle has been physically built and tested. Comparing the environment map with obstacles obtained through IR sensing in real operating conditions with the real map of the environment allowed us to evaluate the accumulation of error due to the odometer and then to determine the values of the parameters K_l and K_n (see 2.2). This allowed us to compute the quality factor that is memorized for all segments.

Figure 6 shows different coverings of the environment obtained by computer simulation of two different behaviours (anxious and routine). We can observe that the percentage of covered environment is very high with a relatively short running time. This is true for a major part of the experimented environments.

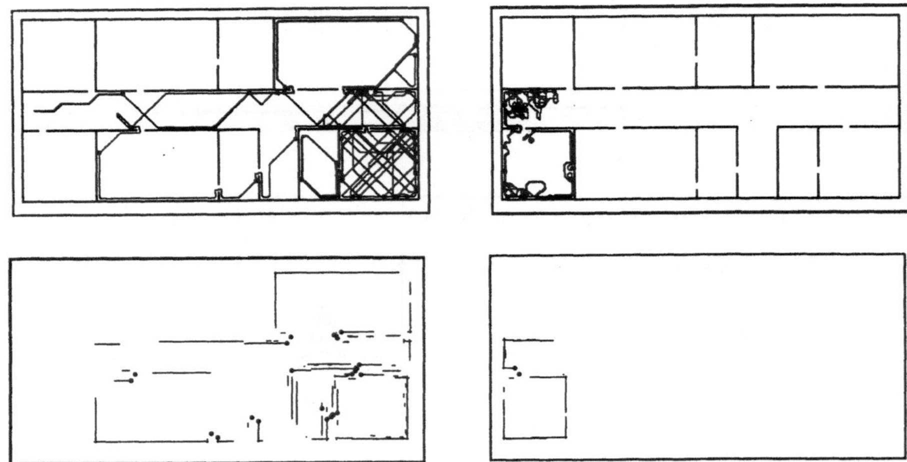

Figure 6. Examples of paths followed and maps generated with routine and anxious behaviours

Figure 7 shows the results obtained after applying the fusion process to a set of maps, including those of figure 6.

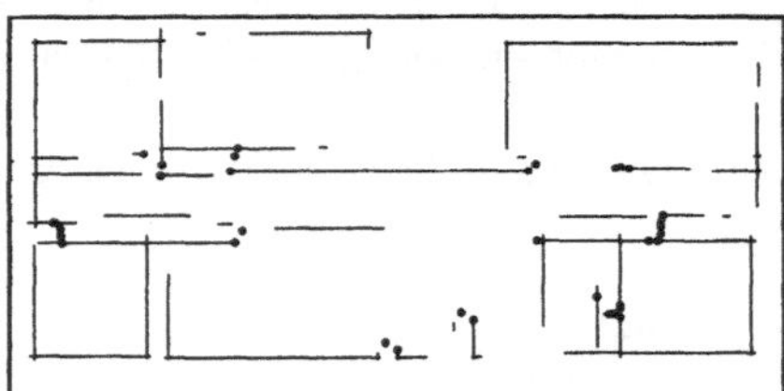

Figure 7. Fusion process result

5. Future work

The results obtained with this first prototype are encouraging enough to undertake the construction of a whole troup consisting of 15 vehicles. This troup will allow us to perform exhaustive experimentation with different number of vehicles and with different behaviour combinations. Concerning map generation, future work will be focused on refining fusion and implementing the completion procedure. Another interesting future work will be to set up the conditions under which it may be advisable to send further ants to explore those portions of the environment that remain incomplete.

6. Acknowledgements

We acknowledge the contributions of Francesc Esteva in the discussions concerning the fuzzy logic approach to map generation as well as the contributions of Gil

Arbós, Maite López, Albert Sánchez and Angel Toribio who have worked hard in the implementation and experimentation of the prototype.

References

[1] R. Alami, R. Chatila, B. Espiau, Designing an Intelligent Control Architecture for Autonomous Robots, ICAR'93, Tokyo, 1993.

[2] R. A. Brooks, A Robust Layered Control System for a Mobile Robot, IEEE Journal of Robotics and Automation, RA-2, pp. 14-23, 1986.

[3] R. A. Brooks, Intelligence Without Reason, Proc. of IJCAI'91, pp. 569-595, 1991.

[4] R. A. Brooks, L. A. Stein, Building Brains for Bodies, Memo 1439, MIT, AI Lab., Cambridge, Massachussets, 1993.

[5] B. R. Donald, J. Jennings, D. Rus, Information invariants for cooperating Autonomous Mobile Robots, in Robotics Research, the Sixth International Symposium, pp. 29-48, Pittsburgh, USA, 1993.

[6] B. Jouvencel, J.E. Symphor, The variable Modelling of Mobile Robot Environments, IEEE IROS'91, Osaka, Japan, 1991.

[7] R. López de Màntaras, Aproximate Reasoning Models, Ellis Horwood Series in Artificial Intelligence, GB, 1990.

[8] L. A. Zadeh, Fuzzy sets as a basis for a theory of possibility, Fuzzy Sets and Systems, 1, 3-28, 1978.

[9] D. Dubois, J-L. Koning, Social choice axioms for fuzzy set aggregation, Fuzzy Sets and Systems, 43, 257-274, 1991.

Acknowledgment. [illegible] Angel [illegible] who have worked [illegible] the implementation and experimentation of the [illegible].

References

[1] [illegible]

[2] [illegible] Journal [illegible]

[3] [illegible]

[4] [illegible]

[5] [illegible] International [illegible]

[6] [illegible]

[7] [illegible]

[8] [illegible]

[9] [illegible]

Chapter 2

Dextrous Manipulation

Dexterity is a term in robotics that has come to suggest many ideals. Initially dexterity was used to describe the qualities (or hoped-for qualities) of articulated hands. As our abilities to control mechanisms and sense task state have improved, it has recently come to suggest more general capability. In general, dexterity suggests a degree of facility or deftness in manipulating objects. Robotics researchers seek to achieve dexterity in a range of domains ranging from micro-manipulators, to hands, to fleets of large cooperating vehicles. Achieving effective dexterity requires advances in mechanism design and control, in sensing and perception, and in the "cognitive" elements that tie these together.

Hyde, Tremblay, and Cutkosky discuss the development of a control frame-work for articulated end-effectors which decomposes tasks into specific phases detectable by specific events. Their paper focuses on a phase-manager which detects phases and mediates transitions between phase-appropriate control laws and reports on experiments demonstrating the utility of the approach.

Burridge, Rizzi, and Koditschek look at the problem of obstacle avoidance during the performance of tasks in dynamic environments. They define the goal of dynamical safety which addresses the need to avoid workspace limits and obstacles during the performance of tasks in which intermittent contact is central (such as juggling and hopping).

Horaud, Dornaika, Bard, and Laugier address the problem of using a single camera to guide the approach of an end-effector for grasping an object. Their technique relies on finding the image Jacobian, the differential transform between robot joint motion and motion in the camera's frame of reference.

Prattichizzo, Salisbury, and Bicchi develop a method for assessing the quality of grasp robustness. The algorithm is useful in the grasp planning stage and in the grasp execution stage. Their paper reports on experiments in which grasp quality is measured in real-time from a sensor equipped end-effector.

Son and Howe analyze the problem of modulating the stiffness of an object grasped by an articulated hand. They show how the effective stiffness is limited by the performance characteristics of the hand, the location of the center of stiffness, and the geometry of the grasp configuration.

An Object-Oriented Framework for Event-Driven Dextrous Manipulation

James M. Hyde
Sarcos Research Corporation
Salt Lake City, UT 84108
j.hyde@sarcos.com

Marc R. Tremblay
Virtual Technologies, Inc.
Palo Alto, CA 94305
tremblay@virtex.com

Mark R. Cutkosky
Center for Design Research
Stanford University
Stanford, California 94305-2232
cutkosky@cdr.stanford.edu

Abstract

Multi-fingered robotic end-effectors have not yet made significant inroads into practical applications, partly due to the complexity of dextrous manipulation tasks. This paper develops an approach for assembling tasks from relatively simple phases which are punctuated by discrete events, signaling the transfer of operation to the next phase in a sequence. We examine the constraints active during phases, and develop methods for conducting smooth transitions between phases. Techniques for robust event detection in the presence of disturbances are also described. Experimental data is shown in support of the phase transition and event detection methods.

1. Introduction

A dextrous manipulation task can be viewed as a sequence of control phases punctuated by events. For example, as the fingers of a hand close upon an object they are driven using position control, but when manipulating an object they are driven to maintain control of internal forces. In this example, the sensation of contact is an event that signals the transition from one control phase to the next.

In manipulation, many events are associated with changing contact conditions and attendant changes in the kinematic and dynamic equations that describe the behavior of the fingers and object. The changing dynamics and kinematic constraints require different control laws and/or controller gains during each phase.

In the physiology literature, it has been observed that when humans grasp and manipulate objects they employ a sequence of responses triggered by events detected using a combination of haptic sensors [Joh91]. One motivation of the present work is to emulate the human ability to shift smoothly between responses.

The basic concept of a phase/event approach to dextrous manipulation and a discussion of the tactile sensors used to detect events have been given elsewhere [Cutkosky and Hyde 1993]. In this paper we focus on a framework developed for defining and executing phases and transitions and for detecting events.

In the present work, a task phase includes a specific control law, along with explicit force/motion constraints and provisions for force or motion trajectory specifications for the duration of the phase. The transition from one phase to the next is triggered by events which can be expected or unexpected. A phase-manager selects an appropriate subsequent phase based on the detected event. Using phase-based control, a complex manipulation task can be decomposed into comparatively simple elements, making the manipulation process tractable.

The concept of decomposing a task into discrete units is hardly new. However, there are some special considerations in the context of dextrous manipulation:

- Events are primarily associated with changes in contact status and are detected with (comparatively noisy) haptic sensors.
- Time constants are short; for example, if incipient slippage is detected, corrective action must be taken in milliseconds. Consequently the control framework must strike a balance between flexibility and computational efficiency.

- It is essential to maintain smooth control across the transitions associated with changes in the kinematic structure of the grasp. Rough transitions will disturb the sensors that are being relied on to detect events.

Our goal is a general purpose programming environment for dextrous manipulation which accounts for these considerations and permits flexible, robust designation and execution of manipulation tasks.

Our control framework is implemented as an object-oriented system with object classes for phases and events. We describe the framework and its motivating design decisions. We provide results of simple manipulation experiments conducted using the framework and discuss what they have taught us, motivating future extensions of the system.

2. Related Work

Although multifingered robotic hands have been available for over a decade, their use has been largely confined to a few research laboratories and their application in practical tasks under autonomous control is virtually nonexistent. One reason for the slow progress in putting such hands to work is the complexity of dextrous manipulation from the standpoints of kinematics, dynamics, sensing, control, and planning.

As Table 1 indicates, the requirements for successful manipulation can be broadly divided into three levels. At the highest level, one is concerned with task planning, grasp choice, etc. At this level, events such as acquiring or releasing a part are treated as symbolic entities that occur instantaneously and shift the hand/tool system from one state to the next. Considerable research has been conducted on discrete event systems for robotic manipulation and many of the techniques including fuzzy Petri Nets [Cao93] and discrete event systems [Sob92, Kat94, McC93] are applicable to dextrous manipulation.

<table>
<tr><td>High-level
(symbolic events, states)</td><td colspan="2">task planning, grasp choice,
discrete event systems,
Petri nets, etc.</td></tr>
<tr><td>Mid-level</td><td>phase, transition
control</td><td>event detection</td></tr>
<tr><td>Low-level
(co-op control, trajectory
specification)</td><td colspan="2">operational space dynamics,
object impedance control,
kinematics, forces, etc.</td></tr>
</table>

Table 1:High-, mid- and low-level requirements for dextrous manipulation

At a lower level, dextrous hands pose formidable challenges in terms of coordinating the motions and forces applied by fingers grasping and manipulating an object. The literature in this area is extensive and includes important developments in kinematics, dynamics and control. Our own framework particularly builds on the operational space dynamics formulation of Khatib [Kha87], the representations of internal forces developed by Nagai and Yoshikawa [Nag93], the object impedance control formulation of Schneider [Sch89], and the MDL force/motion trajectory language of Brockett [Bro88].

While there is an extensive literature concerning the high-level and low-level issues in table 1, the middle level, at which the framework presented in this paper resides, has received comparatively little attention. Some important exceptions include the work of Schneider [Sch89], Brock [Brk93], and Brockett [Bro94].

Schneider combined his object-impedance controller for cooperating arms with state tables of events and associated control actions. The finite state programming method employed branching and looping to improve task execution robustness. Brock introduced task segments described by a context, in which a particular action was executed in the pursuit of a goal. Brockett extended the theoretical framework of [Bro88] to address hybrid systems composed of continuous dynamics and discrete events.

Other relevant work includes techniques for smoothly and stably switching between motion and force control. Several techniques from the literature are reviewed by Hyde and Cutkosky [Hyd93] and compared with a method based on input command shaping.

The success of decomposing a task into phases also depends critically on the ability to detect events that will trigger the transition from one phase to the next. Eberman and Salisbury [Ebe94] used fingertip force/torque sensor information to label some simple events through a combination of signal processing and sequential hypothesis testing. In the present work we use a variety of dynamic tactile sensors which have been demonstrated in previous work [How90, Son94, Tre93] to be especially suited for registering contact events in dextrous manipulation.

3. Phase/Event Control Framework

Our control framework spans the gap between high and low level control noted in Table 1 by establishing a set of building blocks (phases and events) that are used to assemble manipulation tasks. Phases control the operation during a task segment in which a particular set of constraints is active, and events signal the shifts from one phase to the next. These building blocks communicate with a low level data manager to obtain sensor information and output commands to actuators.

Chains of Phases and Events

All phases store a list of events that might terminate that phase, along with the corresponding next phase. This structure of event/next-phase pairs allows rapid shifts between phases as events fire, and establishes a chain of phases to describe the task. The chains may contain branches and loops to promote robustness.

There are two types of phases: manager and action phases. Manager phases are responsible for activating the proper phase in a chain, and may govern multiple parallel sub-chains – useful when the fingers of a manipulator may be operating either independently or cooperatively. For example, three fingers might be cooperatively manipulating an object while a fourth finger executes an independent free motion to obtain a better grasping site. Chains may contain other manager phases, defining a hierarchical chain structure.

The action phases are responsible for actually controlling the fingers and/or grasped object. We have defined a class hierarchy of action phases corresponding to the different kinds of control laws needed:

- finger(s) independent versus cooperatively manipulating an object
- finger(s) or object in free motion versus in contact with, and constrained by, the environment
- finger(s) or object in stationary versus sliding contact

Phases are activated in response to events, which are implemented here as objects that store a confidence variable and methods for updating that variable. When an event confidence reaches a certain level, the event "fires" and the chain manager activates the next phase in the sequence. Note that with this interpretation events are instantaneous; the occurrence of an event means that the framework has committed to a particular event. This commitment logically takes the system from one state to the next.

Figure 1 shows events, manager and action phases for a simple grasp-lift-replace task involving a two-fingered hand. The entire task is governed by a single manager phase Top, which controls the task sequence: Acquire, Lift, Replace, Retract. We note that the Acquire phase is also a manager phase, controlling the execution of two parallel chains in which the left and right fingers independently approach the object, make contact and exert a prescribed contact force. If the fingers fail to make contact with the object within a certain time, a timeout event occurs, signalling the transition to a different chain designed to handle this problem.

Other branches are established by the multiple events that might terminate the Lift and Replace phases. If the object is stripped from the grasp during Lift, for instance, the detection of this event triggers a loop back to the start of the Acquire phase. Ultimately, a sequence of phases

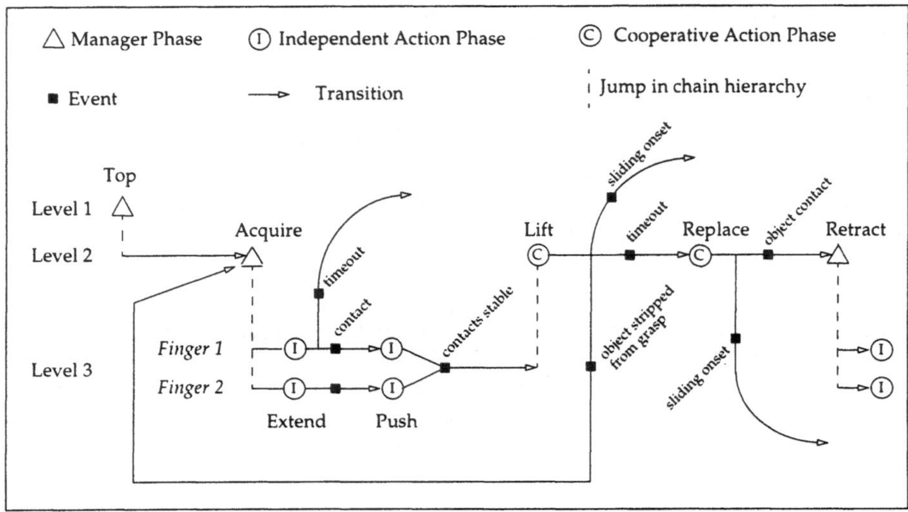

Figure 1: The phase/event structure for an Acquire, Lift, Replace task in which two fingers alternate between independent and cooperative phases in response to detected events. Arcs denote transitions into phase chains not shown here.

and sub-chains with branches and loops may become complicated. The literature on task planning and analyses of reachability, convergence, etc. will be applicable in these situations.

The notation in Figure 1 approximately matches the standard notation of discrete event systems. Action phases, denoted by circles, correspond to states (or "places" in Petri net notation); events are box labels that prompt the system to shift from one state to the next. Note that the term "transition" in the Petri net literature corresponds to what we would call an "event."

Transitions and Event Confidence

We define transitions as the beginning and ending sections of phases that are used to either prepare for an impending event or ramp up operation in the phase following an event. Figure 2 presents an example of a transition that would occur between the *Extend* and *Push* phases of Figure 1. The top plot in Figure 2 shows the evolving confidences of the events tracked in the pre-contact phase. When one or more of those confidences reaches a "preparatory threshold," an "alert action" is triggered, slowing the fingertip to a constant velocity as it approaches the object. More generally, alert actions are automatic adjustments in trajectory and/or control gains taken to facilitate transitions to subsequent phases. Because we do not yet know which event will actually occur, we can only take preparations within the context of the current phase and its control law and constraints. Eventually, the finger contacts the object, causing the corresponding event confidence to breach a "commitment threshold," triggering the switch to the contact phase. Once we have committed to a particular event and subsequent phase it is only logical to initiate control actions and event detection computations within the context of the new phase.

Returning to Figure 2, after the phase shift occurs, the "startup action" for the contact phase halts the controller setpoint inside the object, causing an increase in the interaction force exerted between the fingertip and object. The event confidences are reset, and the set of scanned events in the new phase may differ from those in the previous phase. The new events are monitored and their confidences will evolve to trigger additional alert actions and phase shifts.

Action Phases & Constraints

The Free Motion and Contact phases noted in Figure 2 are examples of Action phases. Action phases, as mentioned above, are responsible for specifying the laws that control the behavior of a

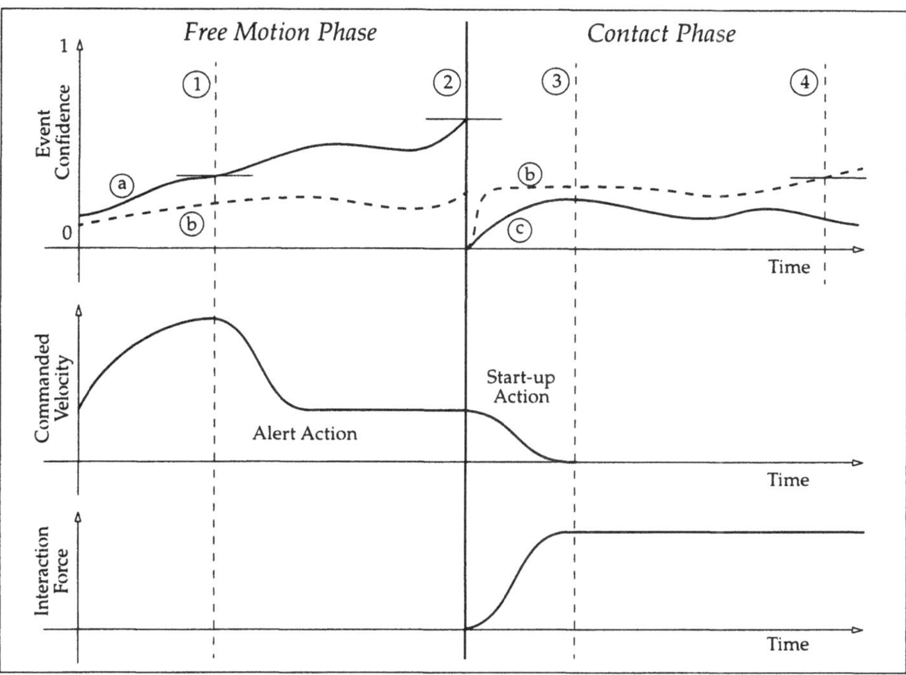

Figure 2: Event Confidences, Alert Action, and Startup Action for a transition prompted by a contact event. The Free Motion phase monitors the confidence levels of events a and b. At time 1, event a's confidence exceeds a preparatory threshold, prompting an alert action which slows the approach to a constant velocity. At time 2, event a crosses a commitment threshold, shifting execution to a Contact phase and resetting the event confidences. The new phase starts by halting the setpoint inside the object, causing the interaction force to increase. The startup action ceases at time 3. In the Contact phase, events b and c are monitored. Event b eventually reaches a preparatory threshold at time 4, causing another alert action (not shown).

fingertip or grasped object. This responsibility includes specifying trajectories of setpoints for the phase control laws; following the approach of Brockett [Bro94] it is understood that trajectories may include both forces and positions with variable control gains.

Action phases also explicitly store the constraints acting on the system. Constraints may be natural, reflecting the holonomic or non-holonomic constraints imposed by environment surfaces, or user-specified, denoting artificial constraints created to ensure the proper execution of a phase. For instance, to avoid actuator saturation leading to non-linear behavior, constraints can be imposed on commanded setpoint accelerations or exerted forces. Natural and user-specified constraints are stored explicitly within each phase, rather than embedded implicitly into phase control laws. In this way a modest repertoire of basic phase types can be customized to meet task-specific constraints regarding contact kinematics, friction, etc.

Figure 3 provides a breakdown of the constraints active during a fingertip sliding phase. In the figure, a fingertip slides along a constraining surface under impedance control. The gap between the setpoint and actual fingertip position establishes an impedance force, f_{imp},which is subject to user-specified soft and hard constraints. In addition, the contact prevents further motion into the surface and defines a constraint on the normal and tangential interaction force components. These are natural constraints.

If the fingertip slides far enough, the manipulator links will encounter a workspace obstacle. Further setpoint motion will cause the impedance force to violate its soft constraint, prompting

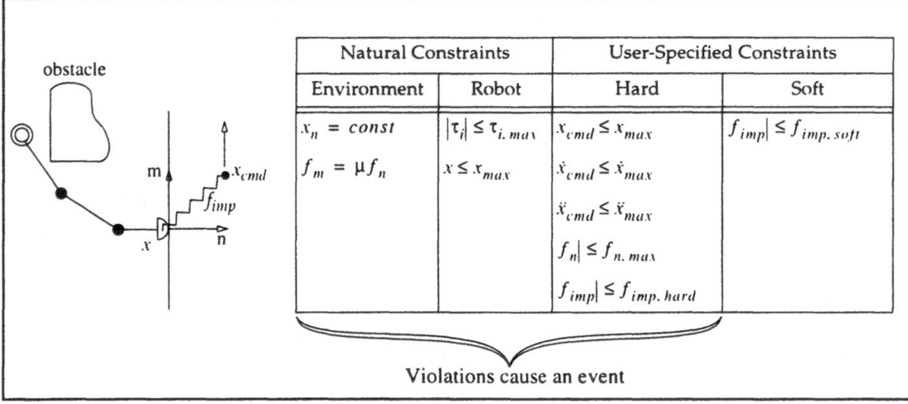

Natural Constraints		User-Specified Constraints	
Environment	Robot	Hard	Soft
$x_n = const$	$\lvert\tau_i\rvert \le \tau_{i,max}$	$x_{cmd} \le x_{max}$	$\lvert f_{imp}\rvert \le f_{imp,soft}$
$f_m = \mu f_n$	$x \le x_{max}$	$\dot{x}_{cmd} \le \dot{x}_{max}$	
		$\ddot{x}_{cmd} \le \ddot{x}_{max}$	
		$\lvert f_n\rvert \le f_{n,max}$	
		$\lvert f_{imp}\rvert \le f_{imp,hard}$	

Figure 3: Constraint categorization for a Sliding phase using impedance control. The controller setpoint is moving in the "m" direction. If the links contact the obstacle, further setpoint motion may cause a violation of the impedance force soft or hard constraints. A soft constraint violation prompts the phase to take action to back away from the constraint to avoid actuator saturation. Violation of the hard constraints or natural constraints triggers a shift to a different phase.

reactive behavior from the phase to avoid actuator saturation. If the impedance force exceeds its hard constraint, or if any of the other constraints in the first three columns of Figure 3 are violated, events fire, triggering a shift to a subsequent phase.

4. Event Detection

Events have been described thus far as objects used to signal the shifts between phases. The reliable detection of events is a rich problem in itself. Events have built-in functions for updating their confidences on the basis of a combination of sensor information and context.

In our framework, the individual fingertips are equipped with multiple sensors, each of which can provide several types of information. For example, tip position, velocity and acceleration can be computed from joint sensor data and short-time energy can be obtained from skin acceleration sensor data. We refer to each of these types of information as sensor-based features. When contextual information about the robot's behavior is included as a feature (desired velocity or force, for example) we refer to these features as context-based features.

For each phase in a manipulation task, if one identifies the possible events and the features required to detect them, one can construct a feature space for the phase. Let us define a feature F as a set of discrete real numbers f corresponding to all possible values for that feature and let Φ be the current phase of a manipulation task. Now let us define an n-dimensional Euclidean feature space corresponding to the cartesian product of the family of sets, or observed features, denoted by:

$$F^n_{\Phi} = F_1 \times F_2 \times \ldots \times F_n$$

At any given moment during the phase, there will be an n-tuple $(f_1, f_2, \ldots, f_n)$ which corresponds to the current feature values.

Within this feature space, each event $e^{q_i}_{i\Phi}$ will occupy a set of regions ξ corresponding to a q_i - dimensional subspace:

$$e^{q_i}_{i(\Phi)} = \left\{ \bigcup_{j=1}^{k} \xi_{ij} \,\middle|\, \xi_{ij} \subseteq e^{q_i}_{i(\Phi)} \text{ for all } i \in \aleph_p \right\}$$

where k corresponds to the number of regions ξ associated with each event e. If we define an n-tuple in the n-dimensional feature space of a given phase as

$$f_{\Phi}^{n} = (f_1, f_2, \ldots, f_n)$$

It then follows that the condition

$$f_{\Phi}^{n} \in E_{\Phi}^{p}$$

must be satisfied for an event to have *possibly* occurred.

The decision as to whether an event has occurred is made using confidence distribution functions (c.d.f.'s), or $\psi(f)$ defined as:

$$\psi(f) : F \rightarrow [0, 1]$$

where 0 means that from that feature's point of view, the event could not possibly have occurred and a value of 1 means that from that feature's point of view, all requirements have been met and the event certainly *could* have occurred. The functions are similar to the membership functions in fuzzy set theory.

Once confidence values have been obtained for each feature, they need to be combined in order to get overall confidence values for each event. By observing these continuously varying confidences, the framework can make decisions relating to event occurrence. Whereas the sensor-based features are typically uncertain and noisy, the context-based features represent current knowledge and serve mainly to rule out certain events. Therefore, the overall confidence for an event consists of the weighted sum of the confidences of the sensor-based features multiplied by confidence values for each of the context based features. Thus a single context-based feature can significantly influence the overall confidence value.

Let the overall confidence Ψ that an event ε is occurring at sampling period k be defined as

$${}^{\varepsilon}\Psi_k = \left(\prod_{i=1}^{m} {}^{\varepsilon}_{i}\psi(\tilde{f}_i)\right) \times \left(\sum_{j=1}^{n} \omega_j \; {}^{\varepsilon}_{j}\psi(\hat{f}_j)\right)$$

where m corresponds to the number of context-based features, n corresponds to the number of sensor-based features, ω_j is the weight assigned to a sensor-based confidence values and ψ is the c.d.f. of a given feature. Note that the same feature will have different c.d.f.'s depending on which event is being considered. As shown in Figure 2, the relationship of the event confidences to established preparatory and commitment thresholds defines when the framework commits to an event and shifts operation from one phase to the next.

5. Experiments and Results

We conducted a series of tests of the event detection and phase transition schemes described in the preceding sections. For our experiments, we chose a "cooperative motion" phase in which two fingers manipulate a grasped object. In this phase, we consider four types of events that cause sensor excitation thereby triggering reactions. These events are: object/world contact, link collision, mechanical disturbance and object slip. Four sensors were used for the experiments: position, force, skin acceleration and stress-rate sensors. The sensor-based features included maximum short-time energies of force and skin acceleration sensors as well as position and force errors. The context-based features included commanded object acceleration. Details on the sensors and features are provided in [Tre95].

The overall confidence value for each monitored event was computed for a variety of runs conducted over a range of speeds from 1 to 10 cm/s and with hard and soft objects. In all test cases, the scheme quickly identified the correct event. Figure 4 shows the results for 4 typical runs where the overall confidence for each event is displayed as a function of time and the title of the plot indicates the actual event that occurred. For all plots, the event occurs at t=50ms and one can see that in every case the correct event is detected quickly.

Figure 5 shows the results for a typical phase transition. In this case the event is an object/world contact, as in the upper left plot of Figure 4. At approximately 1.4 seconds, the contact event passed the preparatory threshold, prompting a constant velocity approach. At 2 seconds the

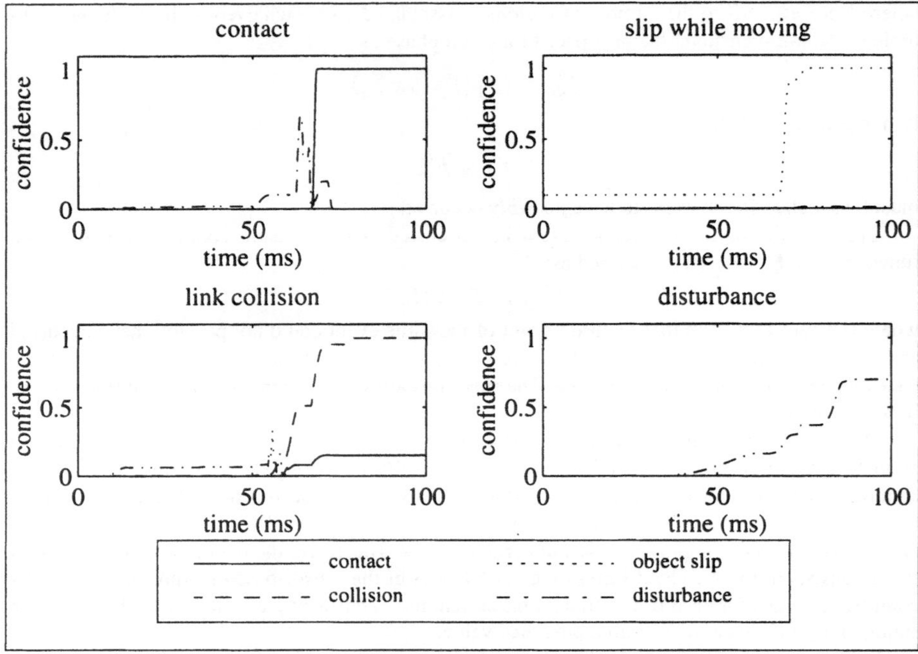

Figure 4: Data from four event detection experiments.

object/world contact event occurred, shifting execution to a cooperative manipulation Contact phase. The object/world interaction force increases during the start-up action to its desired level of 2 N. The transition is executed quickly without imparting significant vibration into the system. Details on this and other transitions are given in [Hyd95].

6. Conclusion

Our preliminary experiments conducted on a system of two fingers have demonstrated the efficacy of this architecture. The underlying object orientation of the phases and chains permits simple extensibility and rearrangement of the chains. The provision of explicit constraints, preparatory and startup actions, and possible terminating events for each phase allows a modest set of basic phase types to be applied to many specific tasks. The data manager, phase, transition, and event objects combine into a general programming environment for dextrous manipulation tasks. Using this environment, we can design and execute manipulation tasks with flexibility and robustness.

7. Acknowledgments

This paper describes work performed at the Stanford University Dextrous Manipulation Laboratory. Funding for this research was provided in part by the Office of Naval Research under the University Research Initiative contract #N-00014-90-J-4014-P01.

8. References

[Brk93] Brock, D. L., "A sensor based strategy for automatic robotic grasping," Massachusetts Institute of Technology Ph.D. Thesis, Department of Mechanical Engineering, 1993.

[Bro88] Brockett, R. W., "On the computer control of movement," IEEE Int'l. Conf. on Robotics and Automation, 1988.

[Bro94] Brockett, R. W., "Dynamical systems and their associated automata, in Systems and Networks, Mathematical Theory and Applications, Academie Verlag, Berlin, 1994.

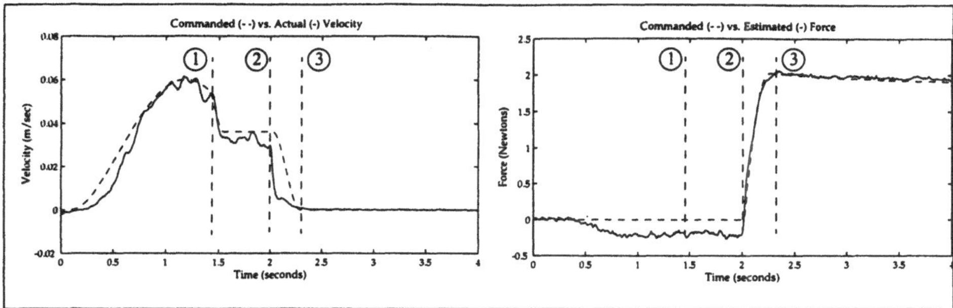

Figure 5: Data from a free motion -> contact transition experiment where a grasped object is brought into contact with an environment surface. Compare with Figure 2. At time 1, an alert action is triggered, causing the object to adopt a constant velocity approach. The object/world contact event is detected at time 2, and the startup action for the contact phase ceases at time 3.

[Cao93] Cao, T., and Sanderson, A. C., "A fuzzy petri net approach to reasoning about uncertainty in robotic systems," IEEE Int'l. Conf. on Robotics and Automation, 1993.

[Cut93] Cutkosky, M. R., and Hyde, J. M., "Manipulation control with dynamic tactile sensing", 6th International Symposium on Robotics Research, Hidden Valley, Pennsylvania, 1993.

[Ebe94] Eberman, B., and Salisbury, J. K., "Application of change detection to dynamic contact sensing," Int'l Journal of Robotics Research, v. 13, n. 5, pp. 369-394, 1994.

[Hog85] Hogan, N., "Impedance control: an approach to manipulation: parts I, II, and III," ASME Journal of Dynamic Systems, Measurement, and Control, v. 107, pp. 1-24, 1985.

[How90] Howe, R. D., "Dynamic tactile sensing," Ph.D. Thesis, Stanford University, October 1990.

[Hyd93] Hyde, J. M., and Cutkosky, M. R., "Contact transition control: an experimental study," IEEE Int'l. Conf. on Robotics and Automation, 1993.

[Hyd95] Hyde, J. M., "A phase management framework for event-driven dextrous manipulation," Stanford University Ph.D. thesis, 1995.

[Joh91] Johansson, R. S., and Westling, G., "Afferent signals during manipulative tasks in man," In Franzen, O., Westman, J. (eds.): Information processing in the somatosensory system: Proceedings of an International Seminar at the Wenner-Gran Center, Macmillan, New York, 1991.

[Kat94] Katayama, Y., Nanjo, Y., and Shimokura, K., "Event-driven motion-module switching mechanism for robot motion control: concept and experiment," ASME Journal on Dynamic Systems and Control, v. 55, n. 1, 1994.

[Kha87] "Unified approach for motion and force control of robot manipulators: the operational space formulation," IEEE Journal of Robotics and Automation, v. 3, n. 1, pp. 43-53, 1987.

[McC93] McCarragher, B. J., and Asada, H., "A discrete event approach to the control of robotic assembly tasks," IEEE Int'l. Conf. on Robotics and Automation, 1993.

[Nag93] Nagai, K., and Yoshikawa, T., "Dynamic manipulation/grasping control of multi-fingered robot hands," IEEE Int'l. Conf. on Robotics and Automation, 1993.

[Sch89] Schneider, S. A., "Experiments in the dynamic and strategic control of cooperating manipulators," Stanford University Ph.D. thesis, 1989.

[Sob92] Sobh, T. M., and Bajcsy, R., "Autonomous observation under uncertainty," IEEE Int'l. Conf. on Robotics and Automation, 1992.

[Son94] Son, J. S., Monteverde, E. A., and Howe, R. D., "A tactile sensor for localizing transient events in manipulation," IEEE Int'l. Conf. on Robotics and Automation, 1994.

[Tre93] Tremblay, M. R., and Cutkosky, M. R., "Estimating friction using incipient slip sensing during a manipulation task," IEEE Int'l. Conf. on Robotics and Automation, 1993.

[Tre95] Tremblay, M. R., "Using multiple sensors and contextual information to detect events during a manipulation task," Stanford University Ph.D. thesis, 1995.

Toward Obstacle Avoidance in Intermittent Dynamical Environments

Robert R. Burridge, Alfred A. Rizzi, and Daniel E. Koditschek

Department of Electrical Engineering and Computer Science
The University of Michigan, Ann Arbor, Michigan, USA

Abstract

In this paper we discuss a robotic task requiring *dynamical safety* in the face of an intermittent environment. We define and offer examples of this notion. We then construct a dynamically safe composite controller from dynamically safe constituents, and present empirical evidence of its effectiveness.

1. Introduction

This paper and a companion paper [1] develop an approach to building up complexes of controllers from simpler constituents. These ideas, first set out in [2], arise from our previous experience with dynamically dexterous robot tasks. The composite controllers we develop result in more varied and capable closed loop robot behavior than can be accomplished by any one of the constituent controllers acting in isolation. The companion paper [1] presents a method for extending the domain of attraction of a goal set via an appropriate composition procedure. This paper introduces the notion of *dynamical safety* for systems with only intermittent contact between robot and environment. In other words, we seek to add obstacle avoidance capability to robot controllers that can already successfully manipulate objects that must be periodically released into a dynamical environment along the way toward their goal. Although all the discussion in the companion paper is limited to the dynamical setting, we believe that our framework for controller composition is useful over a much wider class of robot task domains. In contrast, the novelty of the obstacle avoidance problem considered here is peculiar to the intermittent setting.

To introduce the new notion of a *safe* control strategy we find it useful to add to the formal definition (i) a specific extended example along with (ii) some empirical results reflecting our efforts to implement the example on our three degree of freedom Bühgler robot [3].

1.1. Intermittent Dynamically Dexterous Tasks

We are interested in constructing machines that can interact with an environment possessing significant dynamics, and that can accomplish tasks affording only intermittent influence over that environment. As in the past, we focus on robot juggling (ball bouncing) tasks as good representatives of this larger class which we believe to include such capabilities as catching, throwing, and hopping. Our past work has resulted in a family of machines [3, 4] that exhibit excellent dexterity in a narrow domain, as well as an immature but growing body of theory to explain how [5].

This paper extends the work reported in [1]. In particular, we wish to develop methods by which we can guarantee that constituent feedback based dynamical manipulation strategies are *safe* with respect to certain generalized obstacles. We show that these ideas will afford the automatic construction of complex combinations which are themselves also guaranteed to be *safe* with respect to the same obstacles.

1.2. Dynamical Obstacles

Our robot, the three degree of freedom Bühgler [6], is equipped with a (roughly) one meter long flat paddle and a stereo vision system. The task we wish to explore – the "batted pick and place" – consists of a ball being thrown into the workspace without warning with the requirement that the robot capture it (prevent it from escaping the workspace), bat it as necessary, then bring it to rest at a pre-specified location on the paddle.

The edge of the workspace defined by the end of the paddle provides a natural obstacle since once the ball passes that boundary, it will never return. To avoid this obstacle, we need to make certain not only that the ball isn't currently *at* the boundary, but also that future interactions between the robot and the ball will not cause it to *reach* the boundary before we can hit it again. This observation motivates our definition of *safety* in Section 2.

1.3. Robot Control Architecture

All of the experimental work described in this paper has been implemented on the Bühgler robot [6, 3] as elaborated in [1]. This machine senses ball positions using 2 CCD cameras located above and outside the workspace, and senses impacts using a microphone attached to the paddle. The raw data is filtered and integrated by an observer, producing a "continuous" stream of estimates of the ball's position and velocity, as reported in [6][1]. The estimate of the ball's state is fed through a nonlinear transformation, or "mirror law", $M(b)$, to arrive at a desired reference trajectory for the robot, which is in turn passed to a smoothing "follow-through" generator. The resulting reference trajectory is tracked by a high performance adaptive inverse dynamics controller [7]. The signal flow through this sequence of processing steps from camera image plane all the way down to the joint

[1] Although space limitations prevent an exhaustive list of modifications to the original system, we will note that both the window manager and dynamical observers have been significantly modified to allow balls to be thrown into the workspace rather than carefully presented, as had been the case heretofore.

level torque commands amounts to a feedback control policy that we shall refer to as Φ. Different choices of observers or mirror laws give rise to a characteristic robot response and in this sense there are many such Φ that might be brought to bear on any particular problem.

Throughout the work presented here, we presume that the state estimates emerging from the observer are correct (i.e. correctly converged to the true ball state), and that the robot will accurately track whatever reference trajectory we command. In fact we only require that the robot be "on track" whenever an impact occurs. Thus far our laboratory experiences have consistently supported these assumptions.

1.4. Sequential Composition of Controllers

As described in [1], our current program of research focuses on the use of multiple control strategies and switching schemes to carry out tasks that are impossible with a single continuous controller [8, 9].

For example, to carry out the "batted pick and place" task described in the companion paper we employ three control strategies – juggling (Φ_J), catching (Φ_C), and palming (Φ_P). Indeed, we foresee the need for several *deployments* of each of these strategies, each with different set points and gains. In order to organize these various strategies and deployments toward achieving the overall goal, we use a version of the backchaining algorithm originally described in [10] and extended in [1].

This algorithm uses knowledge of limit sets – sets of points, $\mathcal{G}_\Phi$, in the phase space of the ball, to which a particular control policy, Φ, tends to bring the ball – and domains of attraction – sets of points, $\mathcal{D}_\Phi$, from which Φ is guaranteed to deliver the ball to the associated limit set, $\mathcal{G}_\Phi$ – in order to partition the ball's state space into regions where individual controllers should be active. The resulting partition of the state space provides the basis for a switching mechanism between the controllers which is guaranteed to drive the system to the goal. It turns out that if the component controllers are "safe" with respect to an obstacle, then the composite controller has the same property. While local estimates of the domains may be achieved by standard linear analysis, we wish to work with the larger global domain for which analytical estimates may not be available. This motivates the experimental determination of $\mathcal{D}_\Phi$ presented in the sequel.

2. Safe Behavior

2.1. Setting

Let $b \in \mathcal{B} \approx \mathrm{IR}^6$ be the full state of the ball (position and velocity) in Cartesian coordinates. Let $r \in \mathcal{R} \approx \mathrm{IR}^6$ be the state of the robot in joint space.

The ball in flight will be modeled by Newtonian dynamics with gravity pointing down along the z-axis. Due to the simplicity of the ball flight dynamics, we can derive a closed form expression for the ball position at time t in the future as a function of present state, denoted

$$b(t) = F^t(b). \tag{1}$$

When the ball and paddle collide we use the standard restitution model of collisions (see Synge and Griffith [11] for a discussion of restitution models). In short, we assume that only the ball's velocity component normal to the paddle is affected, while neither the tangential component nor the velocity of the robot is altered by impact. Unless the ball and robot are in continuous contact, it is natural to divide the trajectory of the ball into epochs of time punctuated by collisions. The kth epoch starts with the ball in state b_k, and ends immediately after the next impact, in state b_{k+1}.

Since the robot has no effect on the ball except at contact, we will ignore from now on the trajectory of the actuator system, and only consider the state of the robot at an impact event. Given that the robot strategies used here are feedback strategies (entirely based on the state of the ball), it follows that we can compute a return map of the form

$$b_{k+1} := f_\Phi(b_k) \tag{2}$$

for a particular robot strategy Φ.

We can now make formal the notions of limit sets and domains of attraction introduced in Section 1.4. Suppose there is an attracting set, $\mathcal{G}$, arising from iteration of f_Φ. We call $\mathcal{G}$ the goal of Φ, and the domain of attraction of Φ to $\mathcal{G}$ is given by

$$\mathcal{D}_\Phi(\mathcal{G}) = \{b \in \mathcal{B} | f_\Phi{}^\infty(b) \in \mathcal{G}\}. \tag{3}$$

2.2. Definition

To introduce the notion of safety we must first consider what an obstacle is. Let the set of all ball states which are unacceptable be denoted by $\mathcal{O}$. This set includes the positions where the ball is in a physical obstacle, as well as all states where the ball is beyond the cylinder of the workspace, or too fast to be viewed by, or outside the field of view of the current vision system.

The time until a ball, b, will hit this obstacle set is given by

$$\tau_\mathcal{O}(b) = \min t \ : \ F^t(b) \in \mathcal{O}, \tag{4}$$

while the time to next impact for a particular controller, Φ, is given by $\tau_\Phi(b)$.

We define a controller Φ to be *safe* if there can be found a set $\mathcal{D}_{\Phi S} \subseteq \mathcal{D}_\Phi$ such that

$$\mathcal{D}_{\Phi S} \subseteq \{b \in \mathcal{D}_\Phi | \tau_\Phi(b) < \tau_\mathcal{O}(b)\} \tag{5}$$

and

$$f_\Phi(\mathcal{D}_{\Phi S}) \subseteq \mathcal{D}_{\Phi S}. \tag{6}$$

Assuming that the robot is acting according to Φ, then the first criterion requires that the domain not contain any ball states from which the trajectory of the ball will lead to the obstacle before the next contact. The second criterion – invariance to impacts – implies that once in the safe domain the robot will never knock the ball out (again assuming that the robot continues in Φ). Together, these criteria insure that once a ball is in $\mathcal{D}_{\Phi S}$, it will neither leave that set nor hit an obstacle as long as the robot remains in Φ.

2.3. Two Examples with Formal Demonstration

For the remainder of this paper we will only consider purely configuration space obstacles which are vertically aligned generalized cylinders. That is to say that the obstacle set $\mathcal{O}$ can be completely characterized by its projection onto the x-y plane of the ball's phase space.

2.3.1. The Horizontally Regulated Juggle

Consider the juggling strategy, Φ_J, used in [6] and [1]. It is easily demonstrated that any ball which lies at the horizontal set-point with zero horizontal velocity is in fact "trapped" in a positive-invariant submanifold of the ball's state space [5]. If we consider this two dimensional set of ball states as $\mathcal{D}_{\Phi_J S}$ it follows that this juggling strategy will be safe provided the horizontal set-point lies outside the obstacle set. First we note that since this set is invariant under the controller Φ_J it follows that (6) is trivially satisfied. Satisfaction of (5) follows from the fact that $\tau_{\mathcal{O}}$ is ∞ for any ball in $\mathcal{D}_{\Phi_J S}$ since, by presumption, no ball in this set has any horizontal velocity and thus can not reach $\mathcal{O}$ (since $\mathcal{O}$ is limited to being a vertically aligned cylinder).

2.4. Local Safety

We have empirically verified that the final equilibrium-point for the juggling strategy mentioned above is asymptotically stable. From this fact it follows that there exists a positive definite matrix, P, such that the set defined by

$$\mathcal{D}_{\Phi L}(\epsilon) = \left\{ b \in \mathcal{B} \mid \|f_\Phi(\bar{b})\|^2{}_P < \|\bar{b}\|^2{}_P < \epsilon \right\} \tag{7}$$

is nonempty for some $\epsilon > 0$ (here $\bar{b}$ stands for $b - \mathcal{G}_\Phi$). Furthermore there must exist an ϵ^* such that $\mathcal{D}_{\Phi L}(\epsilon^*) \subseteq \mathcal{D}_\Phi(\mathcal{G})$ and $\mathcal{D}_{\Phi L}(\epsilon^*) \cap \mathcal{O} = \emptyset$. This set is by design a positive invariant set under f_Φ and thus (6) is immediately satisfied. Verification of (5) however is more difficult. First we note that no impact between the robot and machine can occur within the obstacle set if we start the ball in $\mathcal{D}_{\Phi L}(\epsilon^*)$ – this is a direct result of the invariance of this set and the lack of an intersection between it and $\mathcal{O}$. However in order to conclude that the overall system is *safe* we must also ensure that during the ball's flight (between impacts) the boundary of the obstacle is not crossed. From our simple Newtonian model for flight we know that the projection of a ball's flight onto the x-y plane will be a straight line, and as a result of the convexity of the set $\mathcal{D}_{\Phi L}(\epsilon^*)$ we can conclude that that the entire flight of a ball remains "over" the projection of this set onto the x-y plane and thus can never intersect the boundary of the obstacle set.

3. Experimental Determination of Safe Domains

In order to determine the domain of attraction for our juggling controller, a "typical" set point, p_0, was chosen and the juggle controller was activated. Two series of experiments were conducted, as outlined below.

3.1. Notation

For the purposes of this section it is convenient to consider apex points rather than the full ball state. Essentially, an apex point has all ball states except $\dot{z}$. Furthermore, we will express the horizontal component of these apex points in polar coordinates, rather than cartesian. Thus we have ϕ and r and their velocities instead of x and y and their velocities. In apex space, our nominal set point is $p_0 = (-0.3, 0.55, 4.5, 0, 0)$, where the first value is the polar angular position, the second is the radial distance from the origin, the third is the vertical energy, and the last two are the velocities $\dot{\phi}$ and $\dot{r}$ (which will always be zero at juggle set points).

During ball flight, the apex point should remain constant, thus each epoch can be represented by a single point in apex space.

3.2. Experiments

In the first set of experiments, a small region around the fixed point, $\mathcal{N}_{p_0}$, was chosen based on the steady-state performance of the controller. Every time the apex point entered $\mathcal{N}_{p_0}$, the robot "kicked" the ball in a random direction and then attempted to recover by juggling. If it succeeded in regulating the ball again, then all of the points after the "kick" were considered "good". If the juggler lost the ball after batting it at least once, then all the apex points between the perturbation and the loss were considered "bad".

In the second set of experiments, the robot regulated the ball to a different set point then switched on the original juggle controller. If it successfully negotiated the switch, "good" points were generated and the second set point was moved further away and the process repeated. If not, "bad" points were generated.

All "good" apex points from both experiments were combined into a single set, and all "bad" apex points into another.

3.3. Presentation

Due to the nature of the experiments, as well as very fast vertical regulation, the data set is sparse for apex points with z outside a roughly 10cm band. Furthermore, examination of all the data reveals that the height of the apex had no significant effect, thus we are able to ignore z and project all the data onto the four-dimensional horizontal state space, $(\phi, r, \dot{\phi}, \dot{r})$.

The velocity space, $(\dot{\phi}, \dot{r})$, was divided into twenty-five bins, or sectors, and the sets of good and bad points were divided among them. For each velocity sector, the corresponding positions, (ϕ, r), were plotted. In figure 3 we show one "spine" of these plots, with $\dot{r}$ near zero, and $\dot{\phi}$ varying from negative to positive as we progress down the page. The center plot shows the positions for horizontal velocities nearest zero.

In figure 2 we show the plots of r against $\dot{r}$ and ϕ against $\dot{\phi}$ for all the recorded apex points, with superimposed ellipses approximating the "safe" regions of each.

3.4. Discussion

In figure 2, as $\dot{\phi}$ goes negative, we see fewer bad points on the right, and those on the left begin to creep in. There are also more failures toward the end of the paddle. As $\dot{\phi}$ goes positive, we see the opposite effect, as the bad points on the right creep in toward the set point.

The bad points appearing within each of the ellipses in figure 2 represent points outside of the other ellipse. Thus if all the bad points outside the r-$\dot{r}$ ellipse were removed from the ϕ-$\dot{\phi}$ plot, none would remain inside the ϕ-$\dot{\phi}$ ellipse.

4. Conclusion

4.1. Combining Safe Controllers

Having chosen a "safe" ellipsoid approximating the safe domain for the juggle controller, we next chose several such controllers with different set points to cover a large portion of the zero-velocity configuration space. Using the partition method described in [1] we combined them together to create a larger, more robust controller whose domain was the union of all the individual domains. Figure 1 shows a representative run of such a controller when the ball is introduced outside the domain of the goal controller.

4.2. Future Work

The methods employed to generate data to determine the boundaries of the domain for juggling do not generate a rich enough set. We are currently looking at other methods, including simulation and analysis, to help determine the juggling domain.

Nonetheless, the conservative approximation we were able to derive enabled us to successfully build a safe complex controller that successfully negotiated ball states that would have been lost by a single juggle algorithm.

References

[1] Robert R. Burridge, Alfred A. Rizzi, and Daniel E. Koditschek. Toward a dynamical pick and place. In *IROS*, pages 2:292–297, August 1995.

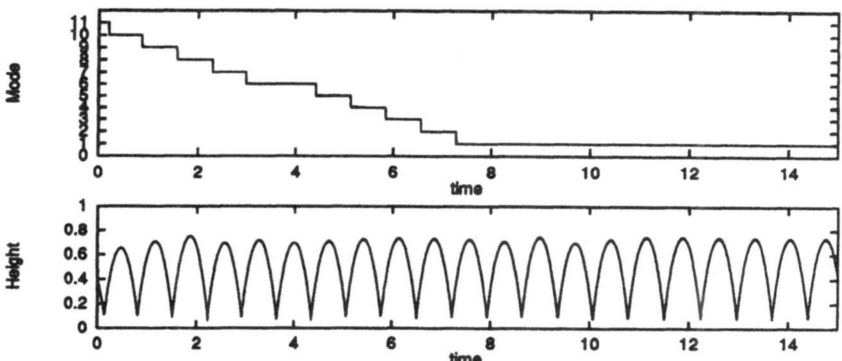

Figure 1. The current control mode indexed against time with the height of the ball plotted as a reference.

[2] R. R. Burridge, A. A. Rizzi, and D. E. Koditschek. Dynamical pick and place. Technical Report CSE-TR-235-95, University of Michigan, Ann Arbor, MI, 48105, April 1995.

[3] Alfred A. Rizzi, Louis L. Whitcomb, and D. E. Koditschek. Distributed real-time control of a spatial robot juggler. *IEEE Computer*, 25(5), May 1992.

[4] Martin Bühler. *Robotic Tasks with Intermittent Dynamics.* PhD thesis, Yale University, New Haven, CT, May 1990.

[5] A. A. Rizzi and D. E. Koditschek. Further progress in robot juggling: Solvable mirror laws. In *Int. Conf. Rob. and Aut.*, pages 2935–2940, 1994.

[6] Alfred A. Rizzi. *Dexterous Robot Manipulation.* PhD thesis, Yale University, 1994.

[7] Louis L. Whitcomb, Alfred Rizzi, and Daniel E. Koditschek. Comparative experiments with a new adaptive controller for robot arms. *IEEE Transactions on Robotics and Automation*, 9(1):59–70, 1993.

[8] Roger W. Brockett. Asymptotic stability and feedback stabilization. In Roger W. Brockett, Richard S. Millman, and Hector J. Sussman, editors, *Differential Geometric Control Theory*, chapter 3, pages 181–191. Birkhäuser, 1983.

[9] Daniel E. Koditschek. Assembly: Another source of nonholonomy in robotics. In *Proc. American Control Conference*, pages 1627–1632, Boston, MA, June 1991. American Society of Control Engineers.

[10] Tomás Lozano-Perez, Matthew T. Mason, and Russell H. Taylor. Automatic synthesis of fine-motion strategies for robots. *The International Journal of Robotics Research*, 3(1):3–23, 1984.

[11] J. L. Synge and B. A. Griffith. *Principles of Mechanics.* McGraw Hill, London, 1959.

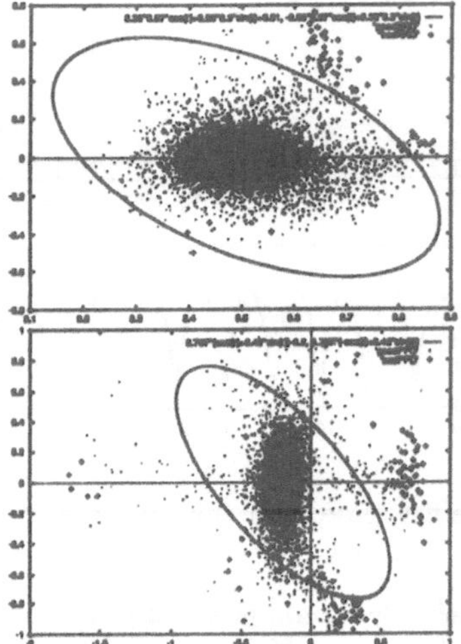

Figure 2. r plotted against $\dot{r}$, and ϕ plotted against $\dot{\phi}$ for all the data points, with superimposed ellipses approximating the safe domains.

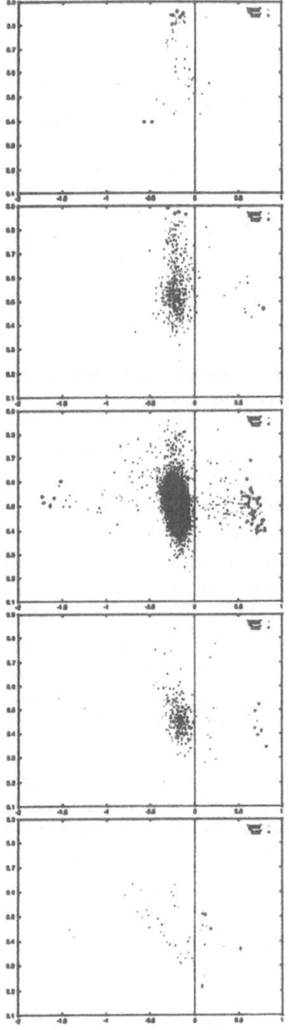

Figure 3. In all plots, $-0.1 < \dot{r} < 0.1$. As the plots proceed down the page, $\dot{\phi}$ moves between the intervals bounded by -1.0, -0.5, -0.2, 0.2, 0.5, and 1.0.

Integrating Grasp Planning and Visual Servoing for Automatic Grasping*

Radu Horaud, Fadi Dornaika, Christian Bard, and Christian Laugier

GRAVIR–IMAG & INRIA Rhône-Alpes
655, avenue de l'Europe
38330 Monbonnot Saint Martin, FRANCE
Radu.Horaud@imag.fr

Abstract

In this paper we describe a method for aligning a robot gripper — or any other end effector — with an object. An example of such a gripper/object alignment is grasping. The task consists of, first computing an alignment condition, and second servoing the robot such that it moves and reaches the desired position. A single camera is used to provide the visual feedback necessary to estimate the location of the object to be grasped, to determine the gripper/object alignment condition, and to dynamically control the robot's motion. The original contributions of this paper are the following. Since the camera is not mounted onto the robot it is crucial to express the alignment condition such that it does not depend on the intrinsic and extrinsic camera parameters. Therefore we developp a method for expressing the alignment condition (the relative location of the gripper with respect to the object) such that it is *projective invariant*, i.e., it is view invariant and it does not require a calibrated camera. The central issue of any image-based servoing method is the estimation of the image Jacobian. This Jacobian relates the 3-D velocity field of a moving object to the image velocity field. In the past, the exact estimation of this Jacobian has been avoided because of the lack of a fast and robust method to estimate the pose of a 3-D object with respect to a camera. We discuss the advantage of using an exact image Jacobian with respect to the dynamic behaviour of the servoing process. From an experimental point of view, we describe a grasping experiment involving image-based object localization, grasp planning, and visual servoing.

1. Introduction and background

One of the most common operations in Robotics is grasping. Although the importance of grasping has been recognized for many years now, there are only a few grasping systems that can operate in a complex environment. This is mainly due to

*The work described herein has been supported by the European ESPRIT-III programme through the SECOND project (Esprit-BRA No. 6769).

the difficulty to execute precise robot hand motions in the presence of various perturbations: the robot's kinematic is known only partially, unpredictable obstacles may be located in the neighbourhood of the object to be grasped, and the location of the object to be grasped may not be known in advance. Moreover, the task of grasp planning itself is a complex one because the planner has to analyse a large number of situations.

One of our research goals is to demonstrate the advantage of integrating planning, sensing, and control in a number of robot tasks. Among these tasks, we have chosen grasping because of its importance in a number of applications: nuclear instrument manipulation, material handling, automated assembly, space robotics, etc.

Our approach to perform automatic grasping is as follows. A grasp plan is built off line [1]. This plan takes advantage of a priori knowledge such as the robot direct and inverse kinematics, the shape and the location of the object to be grasped, the shape of the gripper, etc. The output of this off line grasp planning process is the selection of a grasp, more precisely, the rigid transformation that maps the object's reference frame onto the gripper's reference frame.

Given such a grasp the problem to be solved on line is to control the robot's motion such that the final gripper/object alignment is as close as possible to the previously computed rigid transformation. In this paper we describe a *visual servoing* method for controlling the robot's motion (and hence, the gripper's motion) such that final goal position just described is properly reached. For that purpose, a camera is fixed in such a position and orientation that it sees both the gripper in some initial position and the object to be grasped.

The object-to-robot relationship is not known, the robot's kinematics is not perfectly known, and camera calibration is an almost impossible task; therefore, the robot's motion must be guided by some *relative rather than absolute* sensor measurements. To summarize, visually guided grasping proceeds in the following steps:

1. Locate the object to be grasped in some camera centered frame,

2. Predict the final position of the gripper in the image,

3. Detect the current gripper position in the image,

4. Attempt to align the gripper and the object by minimizing the discrepancy between the current gripper position in the image and its final image position (image-based servoing).

The problem of determining the position and orientation (location) of an object with respect to a camera is a classical and well studied one in computer vision. Recently we came up with an algorithm [10] which computes object location in 2 milliseconds. Because the gripper-to-object transformation to be reached is known in advance, one is able to combine these two relationships (gripper-to-object and object-to-camera) and to project gripper features onto the image. These image predictions amount to a goal position to be reached by the gripper in order to align with the object. The task of moving the robot such that it's gripper reaches this goal position is carried out by a visual servoing algorithm. Many such algorithms have been proposed in the literature [4], [9], [13], [8], [5], [3], [2], [15]. Among them we found that the image-based visual servoing method developed by Espiau, Chaumette and Rives [4] is a very powerful one. With respect to the method cited above [4], we modified it such that (i) we were able to deal with a camera that was not rigidly attached to the robot being controlled and (ii) we showed that dynamic estimation of the hand-to-camera relationship improved the behaviour of the visual servoing algorithm [12].

The remainder of this paper is organized as follows. In section 2 we show how to represent an alignment between two objects in 3-D projective space. The alignment condition thus derived is projective-invariant in the sense that it can be used in conjunction with an un-calibrated camera (nor the intrinsic neither the extrinsic camera parameters are known) as a goal position for visual servoing. The method is inspired by recent work in computer vision and has some similarities with [7]. The major difference is that, with our approach, one camera is sufficient for carrying out the servoing while with the approach proposed in [7] two cameras are necessary. In section 3 we overview the visual servoing method which is an extension of the method described in [4]. Finally section 4 describes an experiment performed with a visually guided robot.

2. Projective-invariant alignment

Whenever one wants to align two objects (an object and a gripper, a peg and a hole, etc.) it is necessary to characterize somehow the relationship between these two objects. In Robotics this relationship is traditionally represented by some Euclidean transformation. In this section we show how to characterize alignments, like the ones just mentioned, such that the characterization is non Euclidean and hence is view-invariant. We will be considering alignment conditions associated with an un-calibrated camera and with the classical pin-hole model associated with the camera.

First we consider 6 points onto the 3-D object to be grasped, e.g., Figure 1. The approach can be easily generalized to any number of points provided that there are at least 6 points and that 5 among these points are in general position (i.e., no 4 among them are coplanar). This object has a Cartesian frame associated with it and the 3-D coordinates of the 6 (or more) object points are known in this frame – the object frame. We also consider a robot gripper and a number of points onto this gripper. The coordinates of these gripper points (or features) are known in some Cartesian gripper centered frame – the gripper frame.

The object-to-gripper Euclidean alignment condition states that the required 3-D rigid displacement D (a 4×4 homogeneous matrix) between the gripper frame and the object frame is known. We show now how to represent this alignment condition in the 3-D projective space rather than in the 3-D Euclidean space. We claim that this projective representation of the alignment condition is projective invariant because it is straightforward to map it onto any image *without calibrating the camera.*

Second we consider a 4×4 invertible matrix P that describes the mapping between the object Cartesian frame and the 3-D projective space. Matrix P is defined up to a scale factor and can be easily computed as follows. Let $\mathbf{A}_1$ through $\mathbf{A}_5$ be 4-vectors (i.e., homogeneous coordinates in the object frame) associated with 5 object points in general position and let $\mathbf{A}_6$ be the 4-vector associated with the sixth object point.

We denote by $\mathbf{A}_i^p$ the 4-vectors associated with the coordinates of the same

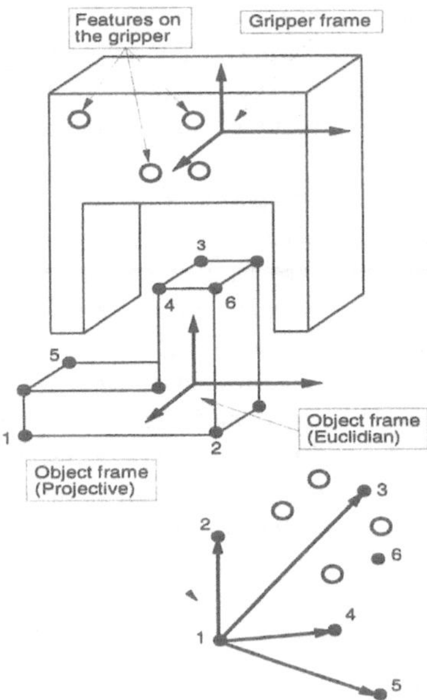

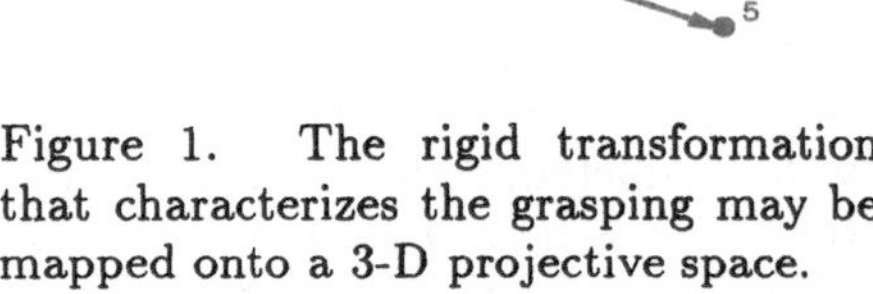

Figure 1. The rigid transformation that characterizes the grasping may be mapped onto a 3-D projective space.

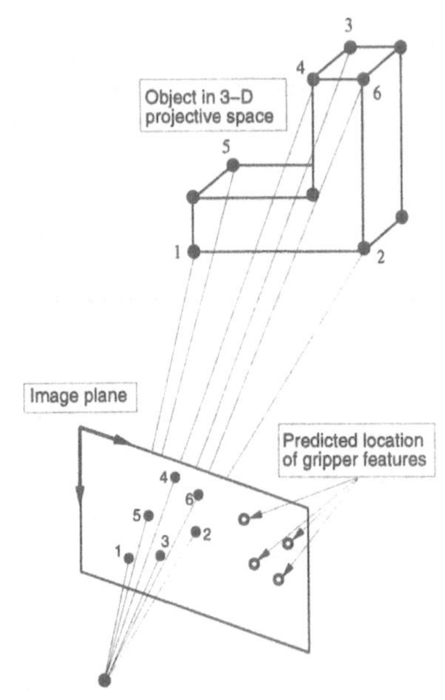

Figure 2. Object and gripper points may be mapped from the 3-D projective space onto an image, provided that at least 6 image-to-object point correspondences are known.

points in projective space. In this projective space let the first 5 points form a projective basis and hence have the following canonical coordinates associated with them:

$$(0\,0\,0\,1)\ \ (1\,0\,0\,1)\ \ (0\,1\,0\,1)\ \ (0\,0\,1\,1)\ \ (1\,1\,1\,1)$$

Every projective mapping:

$$\mathbf{A}_i^p = \lambda\ P\ \mathbf{A}_i \tag{1}$$

provides 3 linear constraints for the entries of P. Therefore, the 16 entries of P may be determined using 5 such point mappings (15 linear constraints) and an additional constraint. One possibility is to use the following constraint:

$$\sum_{i,j=1}^{4} p_{ij} = 1$$

Matrix P can therefore be determined using first 5 object points and their canonical projective coordinates. Once P is determined one can map the sixth object point (or any other point) onto the projective space spanned by the projective basis just defined.

Moreover, let $\mathbf{B}_j$ be a 4-vector associated with a gripper point. One can easily determine the 4-vector $\mathbf{B}_j^p$ which represents its projective coordinates in the canonical projective basis (D is the rigid transformation from gripper frame to object frame):

$$\mathbf{B}_j^p = \lambda\ P\ D\ \mathbf{B}_j$$

The sets of 4-vectors $\mathbf{A}_1^p$ through $\mathbf{A}_6^p$ and $\mathbf{B}_j^p$ ($j \geq 4$) constitute the projective invariant representation of the alignment condition. The reason for which there must be a minimum of 4 gripper points is because 4 object-to-image point matches constrain the position and orientation of the object with respect to the camera.

Finally, we consider a camera that "sees" the object to be grasped in an unknown position and orientation – neither the camera-to-object nor the camera-to-robot relationships are known, e.g., Figure 2. Let $\mathbf{a}_1$ through $\mathbf{a}_6$ be the images of the 6 object points. As already mentioned, the imaging process is a projective mapping from 3-D to 2-D and let M be a 3×4 matrix describing this mapping:

$$\mathbf{a}_i = M\ \mathbf{A}_i^p \tag{2}$$

with $\mathbf{a}_i = (su_i\ sv_i\ s)^T$ where u_i and v_i are the image coordinates of $\mathbf{a}_i$ and s is a scale factor. It is well known that at least 6 pairs of 3-D/2-D point correspondences are needed to completely define M. Therefore, the 3-D projective coordinates of the 6 object points together with their 2-D image coordinates completely define the matrix M.

Once the mapping M is determined it is straightforward to predict image coordinates for the gripper points by projecting their view invariant coordinates onto the image:

$$\mathbf{b}_j = M\ \mathbf{B}_j^p \tag{3}$$

It is important to stress the fact that the estimation of the projective mapping M outlined above does not amount for camera calibration. Indeed, matrix M maps points from a 3-D projective space onto the image and ***not*** from a 3-D Euclidean space onto the image. Therefore, the alignment condition as defined herein has no metrics associated with it.

In practice, the computations above are decomposed into two steps. The first step (estimation of $\mathbf{A}_1^p$ through $\mathbf{A}_6^p$ and $\mathbf{B}_j^p$) is performed off-line and the second step (estimation of M and of $\mathbf{b}_j$) is performed on-line. Notice that the numerical computations associated with the estimation of M are particularly simple because of the use of the canonical coordinates for 5 among the 6 object points.

The major difficulty associated with the approach described in this section is to find 6 matches between points onto the object to be grasped and image points. This matching problem amounts to object recognition from a single view and is a current research issue in its own right.

3. Visual servoing

In this section we consider a camera that observes a moving robot gripper. First we determine the image Jacobian associated with such a configuration. Second we define a visual servoing process that allows the camera to control the robot motion such that the gripper reaches a previously computed image position – one way to compute such an image position is, for example, to use the view-invariant alignment condition.

Let, as before, $\mathbf{B}_j$ be a 3-D point onto the robot gripper and let x_j, y_j, and z_j be its coordinates in the camera centered Cartesian frame. The projection of this point onto the image has as coordinates:

$$u_j = \alpha_u \frac{x_j}{z_j} + u_c \tag{4}$$

$$v_j = \alpha_v \frac{y_j}{z_j} + v_c \tag{5}$$

where α_u, α_v, u_c, and v_c are the well known intrinsic camera parameters associated with a pin-hole model.

Let's assume that the gripper moves in the space and that the translational velocity of the gripper frame origin is $\mathbf{V}_O$ and the angular velocity associated with the gripper frame is Ω_O. Let us denote by $\mathbf{T}_O$ this 3-D velocity screw (a 6-vector) expressed in the gripper frame and by $\mathbf{T}_c$ the 3-D velocity screw associated with the same gripper motion but expressed in the camera frame. The relationship between these velocity screws is:

$$\begin{pmatrix} \mathbf{V}_c \\ \Omega_c \end{pmatrix} = \begin{pmatrix} \mathbf{R} & -\mathbf{R}S(-\mathbf{R}^T\mathbf{t}) \\ 0 & \mathbf{R} \end{pmatrix} \begin{pmatrix} \mathbf{V}_O \\ \Omega_O \end{pmatrix} \tag{6}$$

In this equation $\mathbf{R}$ and $\mathbf{t}$ are the rotation matrix and translation vector associated with the gripper-to-camera rigid transformation and $S(\mathbf{a})$ is the skew-symmetric matrix associated with a 3-vector $\mathbf{a}$. By computing the time derivatives of u and v in equations (4) and (5), we obtain the relationship between the velocity of a gripper point $\mathbf{B}_j$ and its image velocity:

$$\begin{pmatrix} \dot{u}_j \\ \dot{v}_j \end{pmatrix} = \begin{pmatrix} \alpha_u/z_j & 0 & -\alpha_u x_j/z_j^2 \\ 0 & \alpha_v/z_j & -\alpha_v y_j/z_j^2 \end{pmatrix} \begin{pmatrix} \dot{x}_j \\ \dot{y}_j \\ \dot{z}_j \end{pmatrix}$$

If one applies the well known law of rigid motion: $\mathbf{V}_{B_j} = \mathbf{V}_C + \Omega_C \times \overrightarrow{CB_j}$ relating the velocity of a point $\mathbf{V}_{B_j} = \begin{pmatrix} \dot{x}_j & \dot{y}_j & \dot{z}_j \end{pmatrix}^T$ and the translational and angular velocities $\mathbf{V}_C$ and Ω_C, and by combining with eq. (6), then it is straightforward to obtain:

$$\underbrace{\begin{pmatrix} \dot{u}_j \\ \dot{v}_j \end{pmatrix}}_{2\times 1} = J(\mathbf{B}_j) \underbrace{\begin{pmatrix} \mathbf{V}_O \\ \Omega_O \end{pmatrix}}_{6\times 1} \tag{7}$$

Since the goal of visual servoing is to execute robot motions, eq. (7) must be somehow inverted. This inversion supposes that (i) one is able to measure 2-D velocities of image points, (ii) at least three points are available such that one obtains at least 6 equations if no free motion is wished, (iii) the rank of the linear system of equations thus obtained is equal to 6, and (iv) than some reasonable model of the image Jacobian J is available. In what follows we will derive the robot control law and make clear the items mentioned above.

As it has been already explained, we consider 3-D points ($\mathbf{B}_j$) onto the robot gripper together with their projections onto the image ($\mathbf{b}_j$). Let $\mathbf{s}$ be the image vector formed with the coordinates of all the points $\mathbf{b}_j$. For n points, the vector $\mathbf{s}$ has $2n$ components:

$$\mathbf{s} = (u_1\ v_1 \ldots u_j\ v_j \ldots u_n\ v_n)^T$$

We denote by $\mathbf{s}^*$ the vector of image point positions in the final (goal) position. This goal position may correspond, for example, to an alignment condition for grasping (section 2) or to any other goal position that one wants to reach.

Therefore, the task consists of moving the robot such that the Euclidean distance between the current position $\mathbf{s}$ and the goal position $\mathbf{s}^\star$ is minimized. Hence, one may constrain the image velocity of each point being considered to be proportional to the difference vector separating the current position from the goal position. This ideal desired behaviour writes as:

$$\dot{\mathbf{s}} = g\,(\mathbf{s}^\star - \mathbf{s}) \tag{8}$$

where g is a positive scalar that controls the convergence rate of the visual servoing.

It is now possible to combine eq. (7) with eq. (8) and we obtain:

$$J \begin{pmatrix} \mathbf{V}_O \\ \Omega_O \end{pmatrix} = g\,(\mathbf{s}^\star - \mathbf{s}) \tag{9}$$

With:

$$J = \begin{pmatrix} J(\mathbf{B}_1) \\ \vdots \\ J(\mathbf{B}_n) \end{pmatrix}$$

In these equations J is a $2n \times 6$ matrix that depends on the following parameters: α_u and α_v which are the horizontal and vertical scale factors associated with the camera, x_j, y_j, and z_j are the coordinates of the gripper points expressed in the camera frame, and $\mathbf{R}$ and $\mathbf{t}$ are the rotation and translation that map the gripper centered coordinate frame onto the camera centered coordinate frame. Notice that the latter parameters are time-varying because the camera is not rigidly attached to the robot.

It is therefore desirable to estimate the values of x_j, y_j, z_j and of $\mathbf{R}$ and $\mathbf{t}$ any time one needs to compute the matrix J. This can be done by a *pose* computation method. Pose computation is a classical problem in computer vision and photogrammetry and many closed-form and/or numerical solutions have been proposed in the past (see [14] for a review). Nevertheless, past solutions to the object pose computation problem are not entirely satisfactory. This is the main reason for which the current solution used in visual servoing consists of considering that the pose parameters do not vary too much over time and hence J has constant values for its coefficients [4]. Details of a real-time pose computation method that we recently developed can be found elsewhere [10].

Consequently, the control velocity screw may be computed as:

$$\begin{pmatrix} \mathbf{V}_O \\ \Omega_O \end{pmatrix} = g\left(J^T J\right)^{-1} J^T (\mathbf{s}^\star - \mathbf{s}) \tag{10}$$

The visual servoing algorithm can now be summarized as follows:

1. Grab an image of the robot gripper.
2. Detect image points that correspond to some pre-specified gripper points.
3. Match these current image point positions with the goal positions.
   If the current position is close enough to the goal position *then* stop.
   Else go to the next step.
4. Compute the pose of the gripper with respect to the camera.

5. Compute the matrix J as well as its pseudo-inverse.

6. Compute the velocity screw associated with the gripper (eq. (10) and update the current velocity screw with the newly computed values.

7. Go to step 1.

There are a number of advantages associated with steps 4. and 5. of this algorithm. First, an exact estimation of J guarantees an exponential decrease of the image error, i.e., $\|\mathbf{s}^\star - \mathbf{s}\|$. Second, the task of tracking the feature points in the image is more robust because one can predict more precisely the locations of these points. With an approximate Jacobian there is a risk that these feature points run out of the field of view of the camera because the transient image trajectory of the robot is unpredictable. Third, the time necessary for reaching the final position is optimized.

To conclude this section we compare the dynamic behaviour of the algorithm in the case of an approximated (constant) Jacobian with the case of a exact (continuously updated) Jacobian. Let $J^\dagger$ be the pseudo-inverse of J and let $\mathbf{T}$ be a 6-vector representing the velocity screw. Eq. (10) can be written as:

$$\mathbf{T}_O = gJ^\dagger(\mathbf{s}^\star - \mathbf{s})$$

In practice we use an estimation of J, $\hat{J}$, and the previous formula becomes:

$$\mathbf{T}_O = g\hat{J}^\dagger(\mathbf{s}^\star - \mathbf{s})$$

The time derivative of the image error $\mathbf{e} = (\mathbf{s}^\star - \mathbf{s})$ is:

$$\begin{aligned} \dot{\mathbf{e}} &= -\dot{\mathbf{s}} \\ &= -J\mathbf{T}_O \\ &= -gJ\hat{J}^\dagger\mathbf{e} \end{aligned}$$

The time derivative of the module of the image error vector $\mathbf{e}$ is:

$$\begin{aligned} \frac{d}{dt}\left(\frac{1}{2}\|\mathbf{e}\|^2\right) &= \mathbf{e}^T\dot{\mathbf{e}} \\ &= -g\mathbf{e}^T J\hat{J}^\dagger\mathbf{e} \end{aligned}$$

It is clear now that:

- If $\hat{J}$ is a good estimation such that $\hat{J} \approx J$ we have an exponential convergence of the algorithm because the matrix $J\hat{J}^\dagger$ is a positive matrix.
- If $\hat{J}$ is constant over time the positivity of $J\hat{J}^\dagger$ is not guaranteed for any robot position.

4. Experiments

As already mentioned, on-line grasping is performed in two steps: alignment and servoing. The task of alignment is to determine an image configuration associated with the final grasp configuration to be reached by the robot. The task of servoing is to actually move the robot from an arbitrary position to its final position.

In practice, alignment consists in the following steps:

- *Image processing*: Edges are extracted from the row image and these edges are further segmented into straight lines. The endpoints of these lines form junctions (Figure 3 top-right). A network of lines and junctions is thus built. This network is then treated as a graph which is further split into connected components. Finally, the connected component that is most likely to represent a polyhedral object is selected (Figure 3 middle-right).

- *Object model prediction*: A 3-D wire frame representation of the object is available. This wire frame is projected onto an image using approximate intrinsic and extrinsic camera parameters. The choice of these parameters may be done interactively such that the image of the model is as closed to the real image of the object as possible. The view of the model thus obtained is a network of lines and junctions.

- *Image-to-model matching*: In order to determine the 3-D to 2-D projective mapping M described in section 2 at least 6 point matches are necessary. This is done by directly matching the network (lines and junctions) associated with the image of the model against the network associated with the actual image of the object. Notice that, because of noise corruption and because the camera parameters are not exactly the same, this network-to-network matching is not a trivial task. Figure 3 (middle) shows a set of 9 junction matches that were obtained using the method described in [6].

- *Prediction of final gripper position*: The final alignment step is to be able to predict the position of the gripper in the image. The gripper has 4 white marks on it. The geometry of these marks — their x, y, and z coordinates in gripper frame — was determined off-line using a hand/eye calibration technique [11]. It is therefore possible to map the 3-D Euclidean coordinates of these marks onto the 3-D projective space and then to project them onto the image. The projected marks thus obtained are shown on Figure 3 bottom-left.

The image marks thus determined constitute the goal position to be reached by the gripper. Any other more "natural" marks could be used in theory. In practice, the white marks greatly simplify the low-level image processing associated with the servoing task. More formally, the image coordinates of the centroids of the marks are the components of the position vector $\mathbf{s}^*$ — see section 3.

The object grasping with visual feedback method that we just outlined is illustrated on Figure 3. This figure shows the image of the object to be grasped together with the robot gripper (top-left). This image is segmented into edges and these edges are described in terms of lines and vertices (top-right). A matching algorithm establishes vertex-to-vertex assignments between the image and a wire-frame description of the object (middle-left and middle-right). A projective invariant characterization of the grasp allows to predict image locations for the gripper marks (bottom-left). Finally, the robot gripper is servoed such that the actual image positions of the gripper marks are aligned with their predictions (bottom-right).

5. Discussion

In this paper we described a method for aligning a robot gripper — or any other end effector — with an object. An example of such a gripper/object alignment is grasping. The task consists of, first computing an alignment condition, and second servoing the robot such that it moves and reaches the desired position. A single camera is used to provide the visual feedback necessary to estimate the location of the object to be grasped, to determine the gripper/object alignment condition, and to dynamically control the robot's motion.

The original contributions of this paper are the following. Since the camera is not mounted onto the robot it is crucial to express the alignment condition such that it does not depend on the intrinsic and extrinsic camera parameters. Therefore we developed a method for expressing the alignment condition (the relative location of the gripper with respect to the object) such that it is *projective invariant*, i.e., it is view invariant and it does not require a calibrated camera. The central issue of any image-based servoing method is the estimation of the image Jacobian. This Jacobian relates the 3-D velocity field of a moving object to the image velocity. In the past, the exact estimation of this Jacobian has been avoided because of the lack of a fast and robust method to estimate the pose of a 3-D object with respect to a camera.

From an experimental point of view, we showed the interest of exact versus approximate Jacobian estimation. It is important to stress the fact that, in all the visual tasks that have been described above, the camera is either not calibrated or poorly calibrated. The only intrinsic parameter whose value is accurately needed, is the ratio between the horizontal and vertical scale factors. This ratio is known to be provided by the camera manufacturer with great accuracy.

The use of visual feedback for object grasping and for other alignment tasks is a promising method and an active research topic because its use allows for various disturbances and because it does not require robot-to-world calibration. An alternative to the use of vision is force feedback. However, vision and force are complimentary because force is effective only if the robot end effector touches something. Coordination of vision and force is a promising research topic.

References

[1] C. Bard, C. Bellier, J. Troccaz, C. Laugier, B. Triggs, and G. Vercelli. Achieving dextrous grasping by integrating planning and vision based sensing. *International Journal of Robotics Research*, 14, 1995. To appear.

[2] P. I. Corke. Video-rate robot visual servoing. In K. Hashimoto, editor, *Visual Servoing*, pages 257–283. World Scientific, 1993.

[3] P. I. Corke. Visual control of robot manipulators – a review. In K. Hashimoto, editor, *Visual Servoing*, pages 1–32. World Scientific, 1993.

[4] B. Espiau, F. Chaumette, and P. Rives. A new approach to visual servoing in robotics. *IEEE Transactions on Robotics and Automation*, 8(3):313–326, June 1992.

[5] J. T. Feddema, C. S. G. Lee, and O. R. Mitchell. Feature-based visual servoing of robotic systems. In K. Hashimoto, editor, *Visual Servoing*, pages 105–138. World Scientific, 1993.

[6] P. Gros. Matching and clustering: two steps towards automatic model generation in computer vision. In *Proceedings of the AAAI Fall Symposium Series: Machine Learning in Computer Vision: What, Why, and How?, Raleigh, North Carolina, USA*, pages 40–44, October 1993.

[7] G. D. Hager. Real-time feature tracking and projective invariance as a basis for hand-eye coordination. In *Proceedings of the 1994 IEEE Computer Society Conference on Computer Vision and Pattern Recognition*, pages 533–539, Seattle, Washington, June 1994.

[8] G. D. Hager, G. Grunwald, and G. Hirzinger. Feature-based visual servoing and its application to telerobotics. In *Proceedings of the IEEE/RSJ/GI International Conference on Intelligent Robots and Systems*, volume 1, pages 164–171, September 1994.

[9] K. Hashimoto, T. Kimoto, T. Ebine, and H. Kimura. Manipulator control with image-based visual servo. In *Proceedings of the 1991 IEEE International Conference on Robotics and Automation*, volume 3, pages 2267–2272, Sacramento, California, April 1991.

[10] R. Horaud, S. Christy, F. Dornaika, and B. Lamiroy. Object pose: Links between paraperspective and perspective. In *Proceedings Fifth International Conference on Computer Vision*, pages 426–433, Cambridge, Mass., June 1995. IEEE Computer Society Press, Los Alamitos, Ca.

[11] R. Horaud and F. Dornaika. Hand-eye calibration. *International Journal of Robotics Research*, 14(3):195–210, June 1995.

[12] R. Horaud, F. Dornaika, C. Bard, and B. Espiau. Visually guided object grasping. Technical report, INRIA, March 1995. Submitted to *IEEE Trans. on Robotics & Automation.*

[13] N. Maru, H. Kase, S. Yamada, A. Nishikawa, and F. Miyazaki. Manipulator control by visual servoing with the stereo vision. In *Proceedings of the 1993 IEEE/RSJ International Conference on Intelligent Robots and Systems*, volume 3, pages 1866–1870, Yokohama, Japan, July 1993.

[14] T. Q. Phong, R. Horaud, A. Yassine, and D. T. Pham. Object pose from 2-D to 3-D point and line correspondences. *International Journal of Computer Vision*, 15(3):225–243, July 1995.

[15] R. Sharma and S. Hutchinson. On the observability of robot motion under active camera control. In *Proceedings of the 1994 IEEE International Conference on Robotics and Automation*, volume 1, pages 162–167, San Diego, California, May 1994.

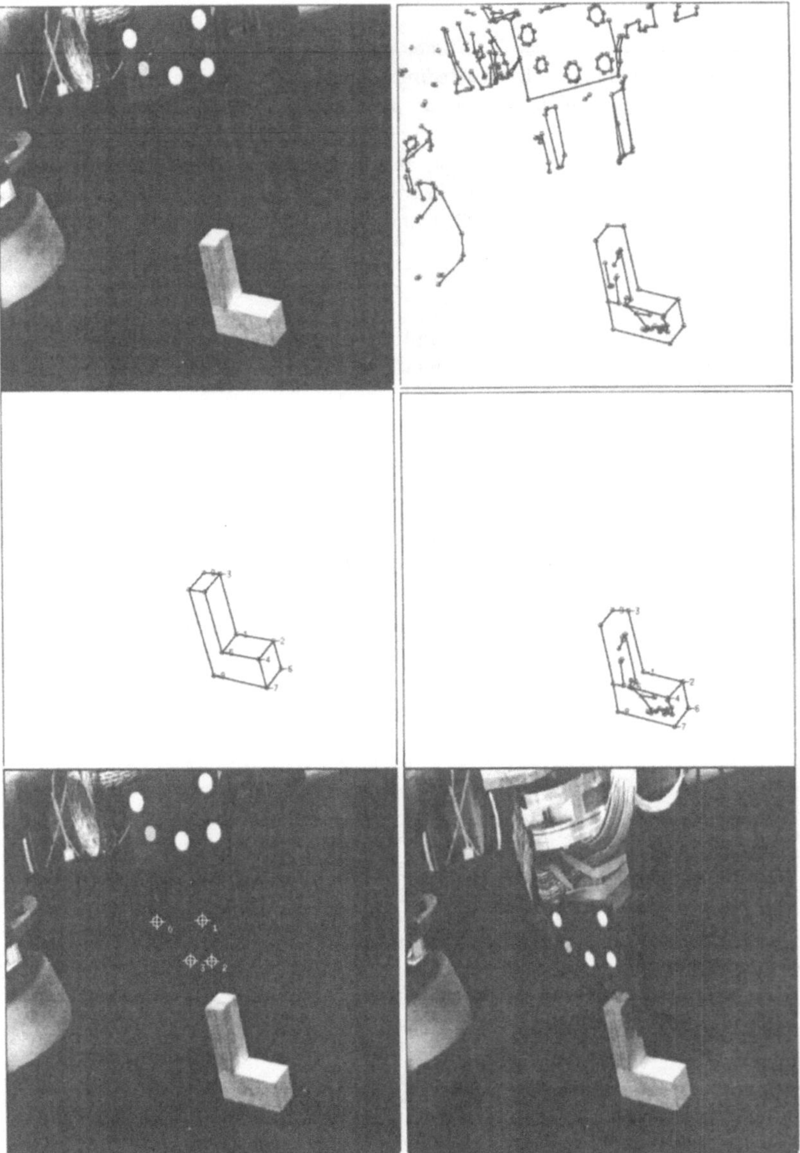

Figure 3. An example of applying the visually guided grasping method (see text). These images (top-left and bottom) are those grabbed by the camera performing both the alignment and servoing tasks.

Contact and Grasp Robustness Measures: Analysis and Experiments

Domenico Prattichizzo [*†]
John Kenneth Salisbury, Jr.[*]
Antonio Bicchi[†]

[*] Artificial Intelligence Laboratory, Massachusetts Institute of Technology, USA
[†] Dipartimento di Sistemi Elettrici e Automazione, Università di Pisa, Italia
[†] Centro "E.Piaggio", Università di Pisa, Italia

Abstract

In this paper we discuss some aspects related to the practical assessment of the quality of a grasp by a robotic hand on objects of unknown shape, based on sensorial feedback from tactile and force sensors on the hand. We briefly discuss the concept of contact and grasp robustness, pointing out that the former is an easily computable but overconservative sufficient condition for the latter. Some experimental results on a simple gripper, the so-called "Instrumented Talon", are reported as an illustration.

1. Introduction

This paper presents procedures for the assessment of the quality of grasps by robotic hands. The interest of having a good measure of the quality of the grasp is twofold: during planning of a manipulation sequence, it allows the optimization of the positioning of the hand with respect to the object to be grasped, and the grasping forces; during the execution of a grasping task, such measure can be used as a performance index according to which local optimization techniques can be used in order to react, at least sub-optimally, to external disturbances and modeling errors.

In the literature, there is a wide interest in the problem of planning good grasps. In [2] the quality criteria of grasp are based on the minimization of the sum of the maximum finger force (L_∞ metric) and of the total finger force (L_1 metric). In [4] the goodness of a grasp is defined in the space of object wrenches and is given as the radius of the largest closed ball, centered in the origin of the space, contained in the set of all the possible wrenches that can be resisted by applying at most unit forces at contacts. An optimality criterion depending on the specific task to be executed has been addressed by Li and Sastry in [5]. In [10] the approach to the analysis of the grasp quality consists of looking at the distance from the vector of contact forces to the nearest contact constraint, suggesting that the farther is the worst-case finger force from violation of a constraint, the better the grasp is. This approach is very intuitive and has been widely used in literature. Naturally, the choice of internal forces by the controller affects the grasp quality and one is led to consider, for each grasping configuration, a quality measure related to the best force distribution that an optimizing grasp force controller can possibly achieve.

In this paper, we build upon previous contributions by analyzing more closely two aspects that influence the concept of "good" grasp. Our analysis is performed in a quasi–static setting. Its first peculiarity is related with the fact that enveloping (alias "power", or "whole–hand") grasping is explicitly considered. In such style of grasping, not only the fingertips, but also the inner parts of the gripper are exploited in order to achieve a more robust hold on the object. This fact implies that contact constraints on the object may be imposed by members of the robotic hand which only enjoy limited mobility and are, therefore, not able to exert arbitrary forces at the contact at will. Secondly, it is observed that in most practical grasps, the set of contact constraints is redundant, in the sense that violation of some of them (slipping or detaching the contact) may well not imply mobilization of the object in the grasp. We therefore suggest that "contact robustness" measure is distinguished from "grasp robustness" measure, where the former is related to distances from the violation of any contact constraint, while the latter is concerned with actually overcoming the immobilization constraint of the object. It thus turns out that contact robustness is an easily computable but overconservative sufficient condition for grasp robustness, which is the property of actual concern in grasping.

Experimental activity on a testbed comprised of a simple enveloping gripper, the "Instrumented Talon", developed at the MIT AI Lab is finally described.

2. Contact Model

When the manipulation system is modeled by rigid-bodies, the i–th contact imposes that some components of the relative velocity between the surfaces are zeros. Mathematically, this can be written as $\mathbf{H}_i\,({}^h\dot{\mathbf{c}}_i - {}^o\dot{\mathbf{c}}_i) = \mathbf{0}$, where $\mathbf{H}_i$ is a constant selection matrix depending on the physical model assumed for the i–th contact (cf. [12]) and ${}^h\mathbf{c}_i$, ${}^o\mathbf{c}_i$ are vectors locally describing the posture of reference frames attached to the surface of the hand and of the object, respectively. For the sake of simplicity, in this paper we only focus on hard–finger contact models. Small displacements of the contact frames can be expressed as a linear function of small displacements of the object $\delta\mathbf{u}$ and of the joints $\delta\mathbf{q}$, respectively, as $\delta^o\mathbf{c}_i = \tilde{\mathbf{G}}_i^T\,\delta\mathbf{u}$ and $\delta^h\mathbf{c}_i = \tilde{\mathbf{J}}_i\,\delta\mathbf{q}$. In juxtaposed vectorial notation, one has that rigid–body constraints can be summarized by the equation $\mathbf{H}(\tilde{\mathbf{J}}\delta\mathbf{q} - \tilde{\mathbf{G}}^T\delta\mathbf{u}) = 0$. The matrix $\mathbf{G} = \tilde{\mathbf{G}}\mathbf{H}^T$ is usually referred to as the "grasp matrix" (or "grip transform") while the matrix $\mathbf{J} = \mathbf{H}\tilde{\mathbf{J}}$ is called "hand jacobian".

As mentioned in the introduction, in this paper we allow for general grasping conditions, including enveloping grasps that exploit kinematically defective links to contact and constrain the object. Kinematic defectivity reflects in the fact that the hand jacobian is not full row rank. It has been shown in previous work of the authors [8] that, in enveloping grasping, the rigid body model in general is not adequate to describe unambiguously the system and in particular its force distribution problem. Moreover the rigid body model does not emphasize the dynamic of contact force control loops. Accordingly, a more accurate model describing how elastic energy can be stored in the system is necessary. We consider a simplified model of elasticity in the system, i.e. at each contact i we introduce a set of lumped "virtual springs" with characteristic stiffness $\mathbf{K}_{i_s}$. This allows us to describe small displacements of the contact force $\mathbf{t}_i$ from its equilibrium configuration as $\mathbf{t}_i = \mathbf{K}_{i_s}\mathbf{H}_i(\delta^h\mathbf{c}_i - \delta^o\mathbf{c}_i)$. Juxtaposing the n contact force vectors $\mathbf{t}_i$ in a single vector $\mathbf{t}$, one obtain: $\delta\mathbf{t} = \mathbf{K}_s(\mathbf{J}\delta\mathbf{q} - \mathbf{G}^T\delta\mathbf{u})$, where $\mathbf{K}_s = \mathrm{diag}(\mathbf{K}_{1_s}, \ldots, \mathbf{K}_{n_s})$.

Due to the unisense nature of contact forces and to friction, in order to avoid slippage and detachment of contacts, contact forces must satisfy unilateral constraints and the Coulomb's friction law. Letting t_{ik} ($k = x, y, z$) be the component of the contact force $\mathbf{t}_i$ along the k–axis of the i–th contact frame (henceforth L_i) fixed to the object and chosen with the z–axis normal to the contact tangent plane, such constraints are written as

$$\text{a) } t_{i_z} \geq 0, \qquad \text{b) } \sqrt{t_{i_x}^2 + t_{i_y}^2} \leq \mu_i t_{i_z}, \tag{1}$$

where μ_i is the static coefficient of friction at the i-th contact.

3. Contact and Grasp Robustness

Suppose that a robotic hand grasps an object by means of n contacts and its configuration is of static equilibrium with balance equations: $\tau = \mathbf{J}^T\mathbf{t}$ and $\mathbf{w} = -\mathbf{G}\mathbf{t}$, where τ is the vector of joint torques and $\mathbf{w} = \left[\mathbf{f}^T, \mathbf{m}^T\right]^T$ is the external (disturbing) object wrench. We introduce the following hypotheses (cf. [8]):
H1: the subspace of *under–actuated* object displacements $\mathsf{ker}(\mathbf{G}^T)$ is empty;
H2: the manipulation system is asymptotically stabilized in the equilibrium point by a joint–position feedback controller with steady state gain $\mathbf{K}_p$;
H3: contact points do not change by rolling (this assumption is reasonable whenever the relative curvature is large).
Let us consider the vector

$$\mathbf{d}(\mathbf{t}) = \left[(d_{1_c}, d_{1_f}), \ldots, (d_{n_c}, d_{n_f})\right]^T, \tag{2}$$

where d_{i_c} and d_{i_f} are the distances of $\mathbf{t}_i$ from the tangent plane to the object surface at the contact i and from the friction cone, respectively. The vector $\mathbf{d}(\mathbf{t})$ indicates how far the grasp is from violating contact constraints (1) and plays a fundamental role in the evaluation of the grasp quality. For instance, Kerr and Roth [10] base the quality measure of the grasp on the minimum component of the vector $\mathbf{d}(\mathbf{t})$.

3.1. Contact Robustness

In $\mathbb{R}^{3n}$, the inequality $\|\delta\mathbf{t}\| \leq \|\mathbf{d}(\mathbf{t})\|_\infty$ describes a sphere centered in the equilibrium contact force and provides a sufficient condition on the maximum euclidean norm of contact force perturbations $\delta\mathbf{t}$ in order to avoid slippage and detachment at all contacts. In order to assess the contact robustness of a grasp, the limitation of $\|\mathbf{d}(\mathbf{t})\|_\infty$ expressed in the contact force space, needs to be reflected in the space of external disturbances acting on the object. We will denote such disturbances as $\delta\mathbf{w}$ (referring to departures from the equilibrium condition). In the quasi–static setting chosen for this paper, the map from contact forces to object disturbance wrenches is $\delta\mathbf{w} = -\mathbf{G}\delta\mathbf{t}$ (via the principle of virtual work). However what is needed to assess contact robustness is the inverse of such map, namely the force distribution map from $\delta\mathbf{w}$ to $\delta\mathbf{t}$. As discussed e.g. in [7] we have, under (**H2**), that

$$\delta\mathbf{t} = -\mathbf{G}_K^R\, \delta\mathbf{w}; \qquad \text{with} \quad \mathbf{G}_K^R = \mathbf{K}\mathbf{G}^T(\mathbf{G}\mathbf{K}\mathbf{G}^T)^{-1}, \tag{3}$$

where $\mathbf{K} = (\mathbf{K}_s^{-1} + \mathbf{J}\mathbf{K}_p^{-1}\mathbf{J}^T)^{-1}$ is the composite grasp stiffness matrix (cf. [1]).

In order to make our following arguments independent from measurement units, we assume that the wrench vector $\delta\mathbf{w}$ is scaled with respect to the nominal value

$\delta\mathbf{w}_n$ of expected external disturbance wrenches in the task under consideration, such that $\delta\mathbf{w}$ is adimensional.

By using the inverse map (3), we can relate the limitation $\|\mathbf{d}(\mathbf{t})\|_\infty$ on contact forces $\delta\mathbf{t}$ with a limitation on the external disturbances $\delta\mathbf{w}$: the relationship $\delta\mathbf{t}^T\delta\mathbf{t} = \delta\mathbf{w}^T\mathbf{G}_K^{R\,T}\mathbf{G}_K^R\delta\mathbf{w} \leq \|\mathbf{d}(\mathbf{t})\|_\infty^2$ describes an ellipsoid in the wrench space centered in zero and with principal axes $2\|\mathbf{d}(\mathbf{t})\|_\infty/\sigma_k(\mathbf{G}_K^R)$. Under the assumed hypotheses, the inscribed sphere with radius $\|\mathbf{d}(\mathbf{t})\|_\infty/\sigma_{max}(\mathbf{G}_K^R)$ represents a limit for the euclidean norm of $\delta\mathbf{w}$, ensuring that all contact constraints hold, notwithstanding the wrench disturbance. In other words, under the hypotheses **H1–H3** and in quasi-static conditions, a given grasp is able to resist any disturbance wrench $\delta\mathbf{w}$ without violating constraints (1), that is to say without detachment and slippage at any contact point, provided that

$$\|\delta\mathbf{w}\| \leq \frac{\|\mathbf{d}(\mathbf{t})\|_\infty}{\sigma_{max}(\mathbf{G}_K^R)}, \tag{4}$$

where $\sigma_{max}(\mathbf{G}_K^R)$ is the maximum singular value of the $\mathbf{K}$-weighted right-inverse $\mathbf{G}_K^R$, (3). Accordingly, the right-hand side term of (4) is defined as the **measure of contact robustness.** Note that condition $\mathsf{ker}(\mathbf{G}^T) = \emptyset$ (**H1**) guarantees that the inscribed sphere is 6-dimensional. A similar measure of contact robustness has been studied for non-defective manipulators in [11]. Our contribution here consists in pointing out the role of the complete stiffness matrix $\mathbf{K}$ in evaluating contact robustness for enveloping grasps. Furthermore, as observed in [5], the proposed measure is only a partial information on the grasp, since two ellipsoids may share the maximal inscribed sphere, though having different shapes.

An important aspect of grasping is that the contact force distribution can be modified by acting on internal forces, i.e. those self-balanced contact forces belonging to the nullspace of the grasp matrix $\mathbf{G}$. In the sequel we investigate how contact robustness may be improved by a redistribution of contact forces. In enveloping grasps the kinematic defectivity of the gripping mechanism may prevent the actual controllability of all the internal forces so that the best policy for redistributing contact forces in the grasp must be confined to modifying only some of the forces in $\mathsf{ker}(\mathbf{G})$. According to [7] and [8], the subspace of internal forces that are asymptotically reproducible (hence, quasi-statically controllable) is given by

$$\mathcal{F}_a = \mathsf{ker}(\mathbf{G}) \cap (\mathrm{im}(\mathbf{KJ}) + \mathrm{im}(\mathbf{KG}^T)) = \mathrm{im}(\mathbf{I} - \mathbf{G}_K^R\mathbf{G})\mathbf{KJ}. \tag{5}$$

Letting $\mathbf{E}$ denote a matrix whose columns span $\mathcal{F}_a$, the general solution to the balance equation $\mathbf{w} = -\mathbf{Gt}$ is given by $\mathbf{t} = -\mathbf{G}_K^R\mathbf{w} + \mathbf{Ey}$ where $\mathbf{y}$ parameterizes controllable internal forces in the basis $\mathbf{E}$.

It is an easy matter to show that $\|\mathbf{d}(\mathbf{t})\|_\infty$ can be infinitely improved by squeezing harder the object (i.e. by increasing the internal contact force). Therefore, in order to compare different grasps, upper bounds on the intensities of contact forces must be considered:

$$\|\mathbf{t}_i\| \leq f_{i,max} \qquad \text{with} \qquad f_{i,max} > 0. \tag{6}$$

Note that the inclusion of such upper bounds in grasp controller becomes necessary whenever we deal with limitations on actuator torques (i.e. for power expenditure) and fragility of the object and/or tactile sensors.

The following definition of contact robustness is based on previous arguments and consists in a maximization of the measure (4) on the subspace of reproducible internal forces: $\mathcal{F}_a = \text{range}(\mathbf{E})$. By including in the distance vector $\mathbf{d}(\mathbf{t})$ new components, $d_{i,max} = f_{i,max} - \|\mathbf{t}_i\|$ (for all contacts), the **measure of potential contact robustness** is now defined as

$$\text{PCR} = \max_{\mathbf{y}} \frac{\|\mathbf{d}(\mathbf{G}_K^R \mathbf{w} + \mathbf{E}\mathbf{y})\|_\infty}{\sigma_{max}(\mathbf{G}_K^R)}. \tag{7}$$

Let $\hat{\mathbf{y}}$ be the maximizing vector for the given upper bound $\mathbf{f}_{max}$ on $\|\mathbf{t}_i\|$. Any disturbance wrench $\delta\mathbf{w}$ such that $\|\delta\mathbf{w}\| \leq \text{PCR}$ can be resisted without any contact slippage or detachment, provided that the (potential) internal force $\mathbf{E}\hat{\mathbf{y}}$ is actuated.

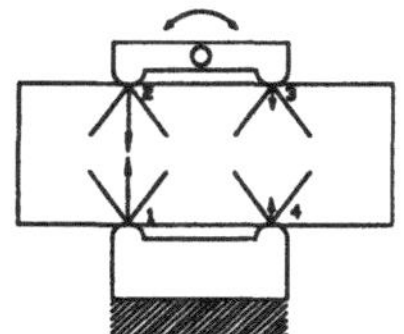

Figure 1: Four contacts grasp.

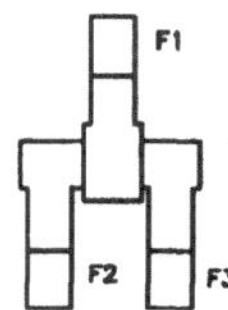

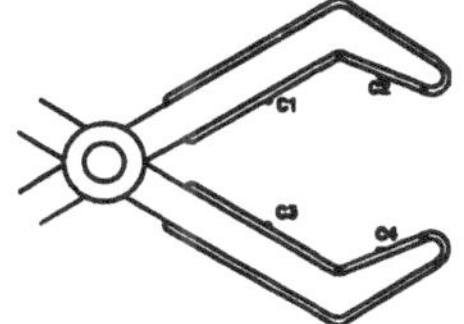

Figure 2: Instrumented talon.

3.2. Grasp Robustness

To illustrate the concept of grasp robustness, consider the planar grasp of fig. 1. Because of the contact forces at contacts 1 and 2, the grasp is intuitively firm and robust, although the minimum distance $\|\mathbf{d}(\mathbf{t})\|_\infty = d_{3f} = d_{4f} \approx 0$ and consequently the measure of contact robustness (4) is nearly zero. To obtain a less conservative estimate of how large an external disturbances can actually be resisted by the grasp, the fact that some of the contact constraints may be violated should be allowed, provided that a sufficient set of unviolated constraints remain to ensure immobilization of the object. We explicitly note that local slippage or contact detachment are possible because of the elasticity of bodies in contact, similarly to the theory of incipient slippage in classical contact mechanics. On the other hand, such elasticity is considered as lumped in virtual springs interposed at the contacts so that bodies still move as rigid bodies in space. The local details of friction and elasticity at the contacts may have large influence on the phenomena occurring in grasping under the above conditions, and they render an exact treatment very complex. In the following, we consider some simplifying assumptions that will allow a safe estimate of grasp robustness, with a degree of conservativeness however inferior to that of contact robustness estimates.

Our method is based on a set of simplified assumptions on the structure of the stiffness matrix $\mathbf{K}_{i_s}$ at the i–th contact that, consequently to the action of a disturbance $\delta\mathbf{w}$, can be in different contact states:
i) when, at the i–th contact, both constraints (1) are fulfilled, the corresponding stiffness matrix (in the local contact frame L_i fixed to the object at the contact i) is assumed diagonal and definite positive, ${}^{L_i}\mathbf{K}_{i_s} = \text{diag}(K_{i_{tx}}, K_{i_{ty}}, K_{i_n})$;
ii) if the friction constraint (1–b) is violated, stiffnesses in the tangent plane are set to zero, i.e. ${}^{L_i}\mathbf{K}_{i_s} = \text{diag}(0, 0, K_{i_n})$;
iii) if the contact condition (1–a) is violated at one contact point, the contact stiffness

at that point is assumed to be null: ${}^{L_i}\mathbf{K}_{i_s} = \mathbf{0}$.
Clearly, such assumptions conservatively disregard the fact that locally slipping contacts continue to contribute to the force balance.

For a given grasp consisting of n contact points, let $\mathcal{C}$ be the set of all possible combinations of the three contact states above (the cardinality of $\mathcal{C}$ is 3^n). For each grasp configuration $\mathcal{C}_j$ in $\mathcal{C}$, consider the global stiffness matrix $\mathbf{K}(\mathcal{C}_j) = (\mathbf{K}_s^{-1} + \mathbf{J}\mathbf{K}_p^{-1}\mathbf{J}^T)^{-1}$, where $\mathbf{K}_s = \mathrm{diag}(\mathbf{K}_{1_s}, \ldots, \mathbf{K}_{n_s})$ and local stiffness matrices are defined according to the state of the corresponding contact in $\mathcal{C}_j$. Remember that $\mathbf{K}_{i_s} = {}^{L_i}\mathbf{R}^{T\,L_i}\mathbf{K}_{i_s}{}^{L_i}\mathbf{R}$, where ${}^{L_i}\mathbf{R}$ is the rotation matrix of the contact frame L_i. Finally, the **measure of potential grasp robustness** (PGR) can be defined as

$$\max_{\mathcal{C}_j} \max_{\mathbf{y}} \quad \frac{\|\mathbf{d}(\mathbf{G}^R_{K(\mathcal{C}_j)}\mathbf{G}\mathbf{t} + \mathbf{E}(\mathcal{C}_j)\mathbf{y})\|_\infty}{\sigma_{max}(\mathbf{G}^R_{K(\mathcal{C}_j)})}, \tag{8}$$

$$\text{subject to} \quad \ker(\mathbf{K}(\mathcal{C}_j)\mathbf{G}^T) = \emptyset, \tag{9}$$

where $\mathbf{G}^R_{K(\mathcal{C}_j)}$ and $\mathbf{E}(\mathcal{C}_j)$ are the weighted pseudoinverse (3) and the basis matrix of asymptotically reproducible internal forces (5), respectively, evaluated with $\mathbf{K}(\mathcal{C}_j)$ modified as described above. Note that condition (9) implies that candidate grasp configurations $\mathcal{C}_j$'s need only be considered among those that can actually immobilize the object, thus effectively reducing the dimension of the set to be searched for the highest contact robustness.

In a few words, such a measure is similar to measure (7) but here we take into account the fact that some of the contact constraints (as friction constraints of contacts 3 and 4 in fig. 1) may be violated without that the grasp of the object fails. The overconservativeness of contact robustness (7) with respect to grasp robustness (8) will be illustrated by the following experiment.

4. Experiment

Grasp analysis tools discussed in section 3 have been employed in an experimental testbed consisting of a simple one–degree-of–freedom gripper or "Instrumented Talon" developed, in part, through collaboration with Harvard University for use on the M.I.T. Whole–Arm Manipulator [13]. The talon (fig.2) has three fingers, each of them equipped with four tactile–sensitive piezoelectric pads [3], and strain–gage based force sensors at the base of the fingers. In its present version, the instrumented talon is only able to sense forces in the finger plane.

The instrumented talon shares its computational resources with the robotic system it is a part of. The complete computational architecture consists of five Motorola 68040 single board computers working in parallel on a real–time software environment. At present, sensory data from the talon are acquired through an interface board and two fiber optic lines to a 68040 VME. The bulk of grasp analysis computations are carried over by a second 68040 board.

In a first experiment, the talon was used to grasp a 1 Kg parallelepiped (see fig. 3). For this experiment, the stiffness of the finger position controller was set very high. The composite grasp stiffness matrix $\mathbf{K}$ results with $\mathbf{K}_p \to \infty$ as $\mathbf{K} \approx \mathbf{K}_s$. Moreover, due to the homogeneity of materials used, we assume that the stiffness matrix has a very simple form: $\mathbf{K} = k\mathbf{I}$, where k is scalar. It can be easily shown that k does not influence the grasp analysis of previous sections.

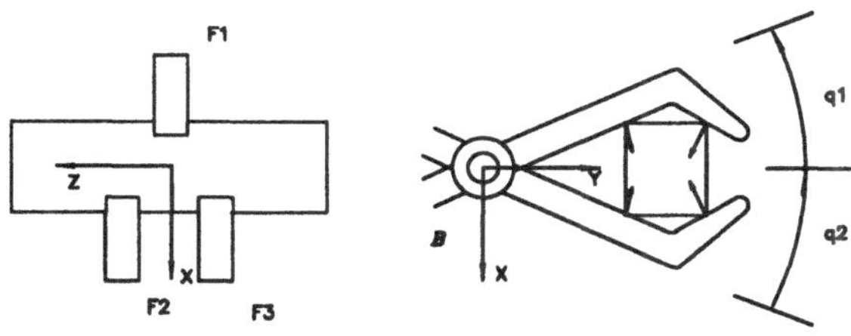

Figure 3: Talon grasping.

In fact, under the condition $\mathbf{K} = k\mathbf{I}$, the weighted pseudoinverse (3) together with the basis matrix of (5) do not depend upon k, and consequently measures (4, 7 and 8) do not depend upon k as well. The friction coefficient for the considered contact conditions is ca. 0.78. For the grasp under consideration (fig. 3), joint angles are $q_1 = 35$ deg and $q_2 = -35$ deg, and the constraint distance vector (2) is $\mathbf{d}(\mathbf{t}) = [(0.1, 0.002)\ (0.47, 0.2)\ (2, 0.1)\ (2.8, 1.5)\ (2, 0.12)\ (2.8, 1.5)]^T$ (in N), showing that the contact c_1 is closest to slippage. In this case, the measure of contact robustness (4) is $\|\mathbf{d}(\mathbf{t})\|_\infty / \sigma_{max}(\mathbf{G}_K^R) = 0.002/26.2 = 0.000076$, where $\delta\mathbf{w}$ is scaled with respect to $[(1,1,1)\mathrm{N}, (1,1,1)\mathrm{Nm}]$. In order to evaluate how much the grasp can be improved by modifying internal forces, it is necessary to determine the subspace of asymptotically reproducible internal forces. Note that, in this example, while $\dim \mathsf{ker}(\mathbf{G}) = 12$, we have that $\dim \mathcal{F}_a = 1$. In particular, contact forces in $\mathcal{F}_a$ are depicted in fig. 3.

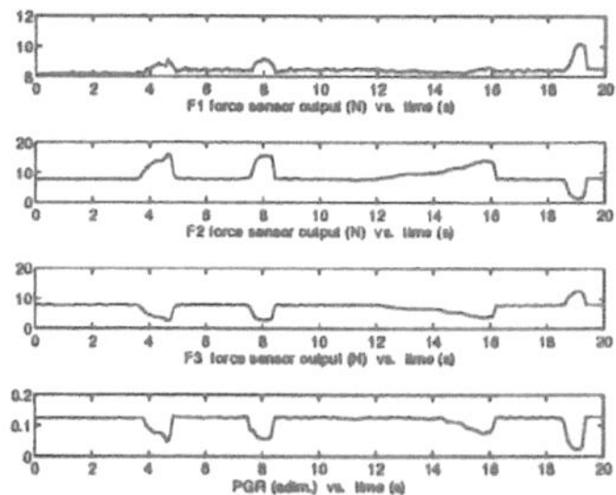

Figure 4: Force sensor outputs and estimated PGR corresponding to application of four external disturbance on the grasped object.

The algorithm described in [9] for real time evaluation of the potential contact robustness measure (7) (having set $f_{i,max} = 20$N) was implemented on a devoted 68040 processor, yielding a rate of 10 KHz, while sensor measurements were processed at a faster rate to estimate the external wrench $\mathbf{w}$. The potential contact robustness (PCR) for this grasp (obtained through averaging because of sensor noise) was $\overline{\mathrm{PCR}} = 0.002$, thus showing the beneficial effect of increasing internal forces.

Finally, we consider the measure of potential grasp robustness (PGR). For the equilibrium grasp configuration of fig. 3, the (averaged) value of $\overline{\mathrm{PGR}} = 0.1594$ was obtained, corresponding to the grasp state when the innermost contacts on the three fingers are supposed to locally slip. Note that $\overline{\mathrm{PGR}} \approx 80\overline{\mathrm{PCR}}$. The measure of potential grasp robustness has been evaluated in real time while unknown disturbances were manually applied to the object grasped in two different configurations. Results are reported in fig. 4, showing the decrease of the PGR corresponding to the increase of the disturbing action on the object.

5. Conclusions

We have investigated the robustness of robotic grasping with respect to external disturbances in a more general framework than previous works on the same topic. Namely, we considered the case when the gripper is kinematically defective (as happens in simple grippers and in whole–arm mechanisms), and we underscored the difference between contact robustness and grasp robustness. Some preliminary experimental results have been presented, indicating the viability of the proposed tools for real–time implementation of optimizing force policies in grasping.

Acknowledgements

The authors would like to gratefully acknowledge financial support for this work from the following sources: ESPRIT grants CO/032/94/TS, WG 8474 "LEGRO"; The National Aeronautics and Space Administration under NASA Contract No: 959774; and the Office of Naval Research University Research Initiative Program under ONR contract N00014-86-K-0685.

References

[1] M.R. Cutkosky, and I. Kao, "Computing and controlling the compliance of a robotic hand," *IEEE Trans. on Robotics Automat.*. vol. 5, no. 2, pp. 151–165, Apr. 1989.

[2] C. Ferrari and J. Canny, "Planning optimal grasps," in *Proc. IEEE Int. Conf. Robotics and Automat.*, May 1992, pp. 2290–2295.

[3] D.H. Howe and M. R. Cutkosky, "Sensing skin acceleration for slip and texture perception" in *Proc. IEEE Int. Conf. Robotics and Automat.*, 1989, pp. 145–150.

[4] D.G. Kirkpatrick, B. Mishra, and C.K. Yap, "Quantitative Steinitz's theorems with applications to multifingered grasping," in *Proc.* 20th *ACM Symp. on Theory of Computing*, May 1990, pp. 341–351.

[5] Z. Li and S. Sastry, "Task oriented optimal grasping by multifingered robot hands,"*IEEE Trans. Robotics Automat.*, vol. 4, np 1, pp. 32–44, Feb. 1988.

[6] J. Kerr and B. Roth, "Analysis of multifingered hands," *Int. J. Robotics Res.*, vol. 4, no. 4, Winter 1986.

[7] A. Bicchi, "Force distribution in multiple whole–limb manipulation," in *Proc. IEEE Int. Conf. Robotics and Automat.*, 1993.

[8] A. Bicchi and D. Prattichizzo, "A standard form for the dynamics of general manipulation systems," in *Proc. IEEE Int. Conf. Robotics and Automat.*, 1995.

[9] A. Bicchi, "On the closure properties of robotic grasping," *Int. J. Robotics Res.*, vol. 14, no. 4, 1995.

[10] J. Kerr, and B. Roth, " Analysis of multifingered hands," *Int. J. Robotics Res.*, vol. 4, no. 4, pp. 3–17, 1986.

[11] Y. Nakamura, K. Nagai, and T. Yoshikawa, "Dynamics and stability in coordination of multiple robotic systems," *Int. J. of Robotics Res.*, vol. 8, no. 2, pp. 44–61, Apr. 1989.

[12] J. K. Salisbury, "Kinematic and force analysis of articulated hands," *Ph.D. thesis*, Stanford University, Dept. of Mechanical Engineering, 1982.

[13] J. K. Salisbury, "Whole-arm manipulation" in *Proc. of the* 4th *Int. Symp. of Robotics Res.,* Santa Cruz, CA. MIT Press, 1987.

Performance Limits and Stiffness Control of Multifingered Hands

Jae S. Son and Robert D. Howe
Division of Applied Sciences
Harvard University
Cambridge, MA 02138 USA
jae@hrl.harvard.edu, howe@das.harvard.edu

Abstract

This paper investigates the fundamental performance limitations of robot hands in terms of their ability to modulate the stiffness and the center of compliance (CC) of a grasped object. We consider the operation of a Cartesian object stiffness controller in two stages. First, object position is determined from joint measurements. Sliding, rolling, and uncertainty in the initial grasp pose can produce errors in the object location calculated from these joint angles. Since the controller produces restoring forces proportional to the calculated object position, these errors can lead to inaccuracy in the commanded forces. Errors in object position estimation can also cause difficulties in the second stage of the controller operation, force generation, which we analyze in terms of the grasp kinematics errors. In practical terms, the lower limit to attainable stiffness may be set by friction in the robot hand, which increases "effective stiffness" until the frictional forces are exceeded. One upper bound to object stiffness is finger tip compliance. Maximum stiffness is also related to the geometry of the grasp configuration: due to controller stability limits, the maximum usable stiffness decreases as the CC is moved away from the fingers, and increases as the object width increases. These results are experimentally confirmed in a precision assembly task with a two-fingered robot hand.

1. Introduction

Multifingered robot hands have the potential to handle a wide variety of objects with great dexterity. Unfortunately, the robot hands constructed to date have not fulfilled this promise. Among the problems are limitations in mechanical design which result in friction and backlash, difficulties in coordination of the many degrees of freedom, and inadequate contact sensing. As a result, little progress has been made in implementing real tasks. The few examples of experimental task execution that have appeared (e.g. [1]) do not provide an understanding of the basic issues.

In this paper we investigate the fundamental performance limitations of robot hands in terms of their ability to modulate the impedance of a grasped object. Hogan [2] argues that impedance is a fundamental way to characterize mechanical interactions, and several studies have demonstrated that impedance control permits effective robotic execution of certain tasks [3,4,5]. To this end, a number of algorithms have been proposed for controlling a robot hand to achieve the desired impedance of a grasped object [6,7,8,9]. For simplicity, in this paper we limit consideration to stiffness, the static component of mechanical impedance that relates displacement of the object to the restoring force. We consider this modulation of stiffness in the context of a particular class of tasks: precision assembly, as typified by close-tolerance peg-in-hole insertion. Whitney [4] has provided a thorough analysis of the mechanical interactions involved in the insertion process, and we investigate the requirements for robots to minimize forces in the insertion process by analyzing theoretical limits and experimental execution of such tasks.

We begin by reviewing Cartesian stiffness control for multifingered hands, and describe its implementation on the planar two-fingered hand used in the experiments reported here. Next, we consider the factors that affect the ability to control object stiffness. Because stiffness control works by measuring object displacement and generating proportional forces, controller function can be separated into position sensing and force generation processes. Position sensing is limited by the ability to determine the contact location, due to uncertainty in initial grasp or because the object may roll or slide against the finger tips. Force generation is also limited by this uncertainty, and by disturbance torques such as friction at the joints of the fingers.

We then measure and analyze the limits to generating arbitrary object stiffness. The minimum stiffness is limited by friction and other disturbance torques. The maximum stiffness is determined by finger tip compliance, and by the the geometry of the grasp configuration, in particular the ratio of the distance between the finger tips to the distance between the finger tips and the desired center of compliance (CC). We show that this ratio acts as a gain on the finger tip controller which can adversely affect stability. Finally, the ability to control object stiffness and CC location is tested by experimental execution of a close-tolerance peg insertion task.

2. Cartesian Object Stiffness Control

From the many proposed schemes for coordinating the control of multifingered hands, we have selected Salisbury's Cartesian object stiffness controller [6] for its lucidity and immediate applicability to the peg insertion task. Figure 1 shows the concept behind this controller: the aim is to make the object behave as if rotational and translational springs are attached from ground to a center of compliance. The location of the CC and the stiffness about it can be arbitrarily specified in Cartesian coordinates. If the stiffness matrix is diagonal, the object stiffness is decoupled at the CC, so forces acting through the CC cause only translation, while pure moments cause only rotations about the CC.

To facilitate the following analysis, we now briefly review the development of the Cartesian object stiffness controller; see [6] for details. The controller measures the difference between the desired and actual CC locations, and calculates a restoring force on the object by multiplying by the stiffness gain,

$$f_{obj} = K_{obj} \delta x_{obj} \tag{1}$$

where f_{obj} is the net force on the object (which is augmented to include grasp force and object weight), K_{obj} is the desired stiffness matrix at the CC, and δx_{obj} is the difference between the desired and actual CC locations. This object force is mapped

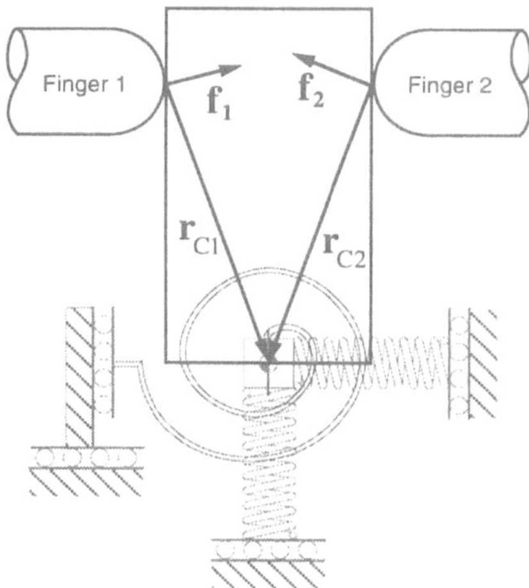

Figure 1. *The Cartesian object stiffness controller allows arbitrary specification of the Center of Compliance (CC) location and the stiffness about that CC.*

to the finger tips with transpose of the grasp matrix G^T,

$$f_{tip} = G^T f_{obj}. \tag{2}$$

The finger tip forces are then mapped to joint torques τ with the transpose of the Jacobian matrix J^T, yielding

$$\tau = J^T f_{tip} = J^T G^T K_{obj} \delta x_{obj}. \tag{3}$$

To determine the object position error, we start with measured joint angles q, then calculate the manipulator forward kinematics $f(q)$, and finally the finger tip to object grasp kinematics $g(q)$

$$x_{obj} = g(f(q)). \tag{4}$$

The position error δx_{obj} is the difference between this actual position and the desired object position $x_{obj,desired}$. Combining the above, the motor torque is computed from the object stiffness, finger and grasp kinematics, and position error as

$$\tau = J^T G^T K_{obj}[x_{obj,desired} - g(f(q))]. \tag{5}$$

2.1. Determining Object Location

We can separate the function of the controller into two components: the determination of the object position x_{obj} from joint measurements, and the generation of the appropriate restoring forces f_{obj} through the finger tips. We start by considering object location determination, as in equation (4). Determination of finger tip position from joint angles using the kinematic model of the manipulator $f(q)$ is straightforward and accurate for current rigid-link manipulators with good joint encoder resolution. However, going from the finger tip location to the contact location and then the object location, as represented by the grasp kinematics $g(f(q))$, is less certain. The object's pose may not be well-determined in the initial grasp, and the

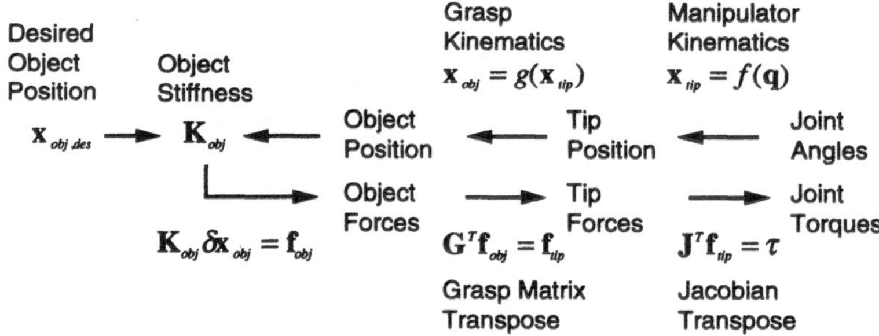

Figure 2. *In Cartesian object stiffness control, object position sensing and joint torque generation are coupled through the object stiffness matrix.*

object can roll and slide against the compliant finger tips as it is manipulated. This uncertainty can be a major source of error in precision manipulation tasks.

It should be noted that the controller cannot compensate for these kinematic errors by increasing object stiffness, but force errors can be reduced with higher gains. If the controller erroneously calculates that the desired position has been obtained because of a kinematic error, no compensating forces will be generated. Solutions to this problem include modeling the finger-object interaction [9,10], and using tactile sensing to determine the contact location [11,12,14]. For this paper, we minimize this problem through the use of finger tips which provide near ideal point contact. The limitations of this approach are that the object orientation between the fingers can be uncertain within the friction cone, and grasp stability is reduced. Use of tactile array sensors mounted on larger radius finger tips can provide better object orientation information, in addition to improved grasp stability [14].

2.2. Limits to Force Generation

The second component of the stiffness controller involves generating desired forces on the object. The force that is actually generated in response to object displacements may contain several sources of error. We can explicitly enumerate these errors by combining equations (2) and (3) above to find

$$f_{obj} = G^{-T} f_{tip} = G_{act}^{-T} J_{act}^{-T} \tau_{act} \tag{6}$$

where the subscript "*act*" refers to the actual physical situation, as opposed to the calculated quantities based on joint angle measurements. We can describe the actual torque, τ_{act}, as composed of the desired component, τ_{calc}, calculated from (3) above, plus a disturbance term, $\tau_{disturb}$, due to joint-level friction, motor torque ripple, D/A quantization, amplifier noise, and other shortcomings of the system. Combining (3) with (6) we find

$$\begin{aligned} f_{obj} &= G_{act}^{-T} J_{act}^{-T} (\tau_{disturb} + \tau_{desired}) \\ &= G_{act}^{-T} J_{act}^{-T} \tau_{disturb} + G_{act}^{-T} J_{act}^{-T} J_{calc}^{T} G_{calc}^{T} K_{obj} \delta x_{obj}. \end{aligned} \tag{7}$$

If there are no kinematic errors, then $J_{act} = J_{calc}$ and $G_{act} = G_{calc}$, and the joint torques will be accurately mapped to the object force. Errors, particularly in contact location, will degrade the mapping, and incorrect object forces will result. The magnitudes of these errors will vary greatly with the details of the grasp and joint configuration; explicit values may be estimated with the above relation.

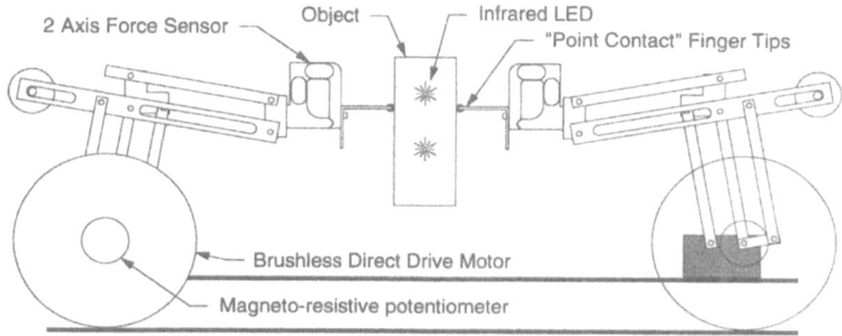

Figure 3. *Two-fingered planar hand grasping a block containing LEDs for optical tracking of object position.*

The most serious source of disturbance torques with present robotic hands is often friction, which is a particular problem in this context because it is highly nonlinear and difficult to model, especially for complex tendon drive systems used in many robot hands. Other sources of torque such as motor torque ripple and amplifier gain variation often can be corrected. In addition, joint torques or finger tip forces may be directly measured by sensors at the appropriate locations in the manipulator, and a controller can compensate, at least partially, for these disturbance torques.

3. Experiment Hardware

3.1. Planar Hand

To experimentally investigate the limits to stiffness control, we implemented the Cartesian object stiffness controller on the two-fingered planar robot hand shown in Figure 3. Each finger has 2 degrees of freedom, and the workspace is oriented in the vertical plane. In contrast to most robot hands, this manipulator trades off large workspace and many degrees of freedom for a clean mechanical design with high bandwidth and good dynamic range. The manipulator uses brushless DC motors in a parallel link, direct drive configuration to minimize backlash, friction, and moving mass. A contactless magneto-resistive potentiometer is used for determining joint angle position, and the joint velocities are electronically generated from the position signal using a differentiating amplifier. Further details of the manipulator design can be found in [13].

3.2. Experimental Setup

For these experiments, our manipulator was fitted with thin, rigid finger tips to provide approximate point contacts with the grasped objects. The pointed finger tips were formed by 25.4 mm long aluminum brackets 1.5 mm thick. To increase friction at the contact surface, the contact points were wrapped with a 0.4 mm thick silicone rubber tube with a radius of 2 mm. We modeled the resulting contact as a point without considering the finger tip radius or compliance.

The objects used in the experiments were a rectangular block and a machinist's clearance inspection peg, both fitted with a pair of infrared LEDs. An optical tracking device measured the x and y location of each LED, from which object position and orientation were readily calculated. Electrical noise resulted in 0.1 mm of peak-to-peak position uncertainty, but multiple samples could be averaged to reduce this error. To test stiffness controller performance, the object was perturbed by a probe attached to a two-axis force sensor, and the applied forces along with the resulting object displacements were measured using a separate data acquisition computer.

4. Stiffness Limits

Limits to the range of attainable object stiffness can have a profound effect on manipulation task. In assembly tasks, low apparent stiffness of the grasped part can minimize contact forces. Analysis of the mechanics of peg-in-hole insertion shows that minimal forces are generated if the object CC is placed just below the tip of the peg, and the stiffness in lateral and rotational directions are kept to a minimum [4]. At the other extreme, high stiffness provides good position control, despite external disturbances. In this section we are concerned with finding the factors with determine the minimum and maximum object stiffness attainable with robot hands.

4.1. Minimum Object Stiffness Measurements

The first set of experiments involved measuring the object stiffness and comparing it to the desired stiffness. The purpose of the measurements was to observe the magnitudes of the stiffness errors, and observe the effects of friction on object stiffness. We commanded a low object stiffness and measured the actual stiffness with two levels of friction: the intrinsic friction in the manipulator mechanism, and, for comparison, a higher friction level produced by wedging wooden splints between the motor shaft and the support frame.

The manipulator was commanded to pick up an object and hold it in space. Forces were applied to the object in the y (vertical) direction through the CC using the probe described above, and object displacement was recorded with the optical tracker. The probe was moved down until 1 N was measured on the force sensor, then back up again until 1 N was reached in the other direction, followed by moving down again until the force sensor reading went back to zero. Each force/displacement data point was taken after moving the probe and allowing the forces to stabilize to ensure quasi-static stiffness measurements.

The force/displacement measurements are shown in Figure 4. The object exhibits substantial hysteresis due to the friction associated with the manipulator joints. Without adding friction, the frictional force at the tip is approximately 0.1 N, while the wooden splints in the joints increased this friction to 0.8 N. Frictional force was calculated as half the difference between the two force values at positions in the center of the hysteresis loop.

Since the friction prevents motion at the joints until the frictional force is exceeded, the addition of friction effectively increases the stiffness. This high stiffness is observed, rather than a vertical line, because of compliance in the manipulator finger tips, joints, and linkages. The desired stiffness was set to 200 N/m; with and without added friction, the measured stiffness beyond the friction dead bands were 320 N/m and 240 N/m respectively. The stiffness associated with the dead band for the former case is on average 2,400 N/m. As shown in the peg insertion experiments below, this high friction results in much higher insertion forces.

For small object forces, the lowest obtainable stiffness is limited by the frictional force. Although the desired stiffness of the object can be set to zero, any disturbances which are smaller than the friction force will not generate a restoring force from the controller. Instead, the restoring force will be generated by the finger tip and the manipulator structure. With respect to the peg insertion task, the minimum usable

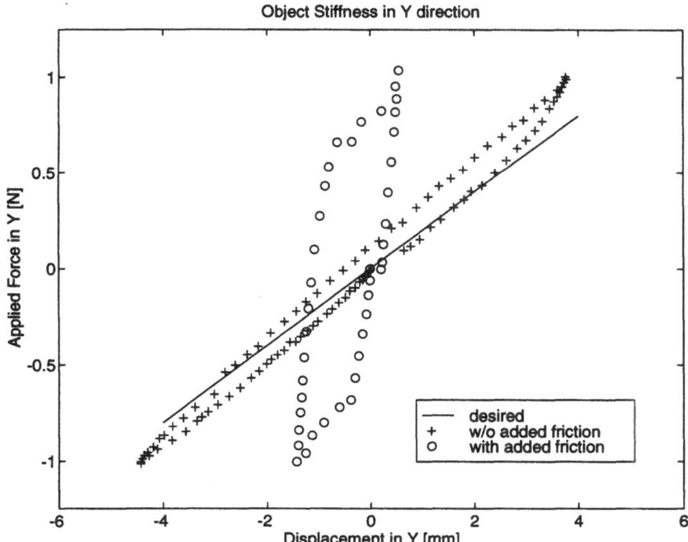

Figure 4. *Desired and measured object stiffness in the y direction showing that friction forces must be exceeded before the desired stiffness behavior can be obtained.*

stiffness is set by the chamfer width, δx_{req}. Assuming that the peg's location is accurately known, then the stiffness must be large enough that the controller will generate a force larger than the effective tip friction to move the peg if it is not within the chamfer distance, or

$$k_x > \frac{f_{friction}}{\delta x_{req}}. \tag{8}$$

Once the peg has crossed the chamfer, the commanded stiffness can decrease, but the frictional force will still bound the attainable stiffness.

4.2. Maximum Stiffness Gain

The upper limit to stiffness was determined by several factors, particularly the finger tip compliance and the geometry of the finger tip-grasped object system. To demonstrate that one upper bound on stiffness is set by the finger tips, the actual stiffness of the object was measured at high stiffness gains. The stiffness measurements indicate that the actual stiffness was much lower than the desired stiffness. For the y axis, the stiffness gain could be set as high as 20,000 N/m and the manipulator remained stable. However, the highest stiffness measured was 6,300 N/m for the case with friction. This is due to the compliance of the rubber-covered finger tips. Although this finger tip compliance would saturate at higher forces, the manipulator could not generate the large forces required to see this phenomena in our experiments. In general, soft finger tips can be advantageous, as they have high friction, and their passive compliance can increase grasp stability. However, these advantages must be weighed against the limits on obtainable maximum stiffness and precise positioning.

Next, we consider the stiffness limits due to CC location. Here the maximum

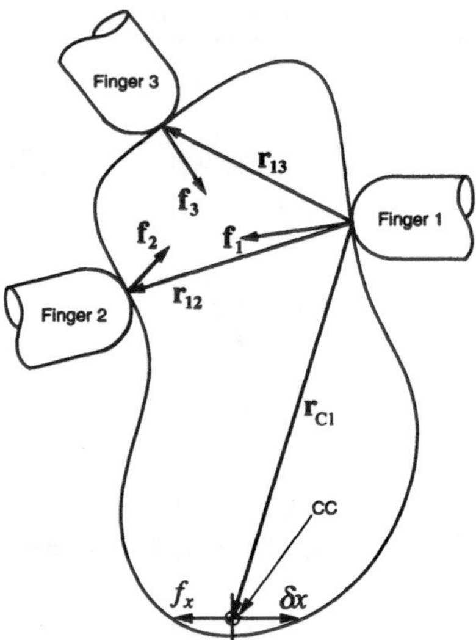

Figure 5. *Because of the large lever arm associated with the distance to the CC in comparison to the distances between the finger tips, small disturbance at the CC requires large finger tip forces to balance the resulting torque.*

stiffness depends on the ratio of the distance from the finger tips to the CC and the distance between the finger tips. In other words, for a given grasp configuration, as the distance to the desired CC increases, the maximum attainable stiffness decreases. The situation is illustrated in Figure 5, where an external disturbance displaces the object's CC laterally by δx. The object does not rotate, so if we assume a diagonal stiffness matrix, a pure restoring force f_x should be generated, with no net object torque. The controller must generate finger tip forces to provide this restoring force, and these forces must cancel to avoid creating a torque.

In general, for quasi-static equilibrium the following equation must be satisfied

$$r_{c1} \times k_x \delta x = \sum_{i}^{n} r_{1i} \times f_i \tag{9}$$

where r_{c1} is the position vector from finger 1 to the CC, k_x is the translational object stiffness at the CC in the direction of the disturbance, δx is the position displacement vector, r_{1i} is the position vector from the chosen finger 1 to the another finger i, f_i is the force vector exerted by finger i, and n is the number of fingers grasping the object. From equation (9) we can see that the effective finger tip gain, defined as the force generated at the finger tips f_i for a given object displacement δx, is proportional to k_x and r_{ci}, and inversely proportional to r_{1i}. Because finger tip gain is inevitably bounded by stability considerations, this means that the stiffness (k_x) is limited as well, and this limit must vary with CC location (r_{ci}) and object width (r_{1i}).

To experimentally verify the variation of maximum stiffness gain as a function of r_{c1} and r_{1i}, the manipulator was commanded to hold objects with various widths, and the stiffness, k_x or k_y, was increased until the system became unstable for a range

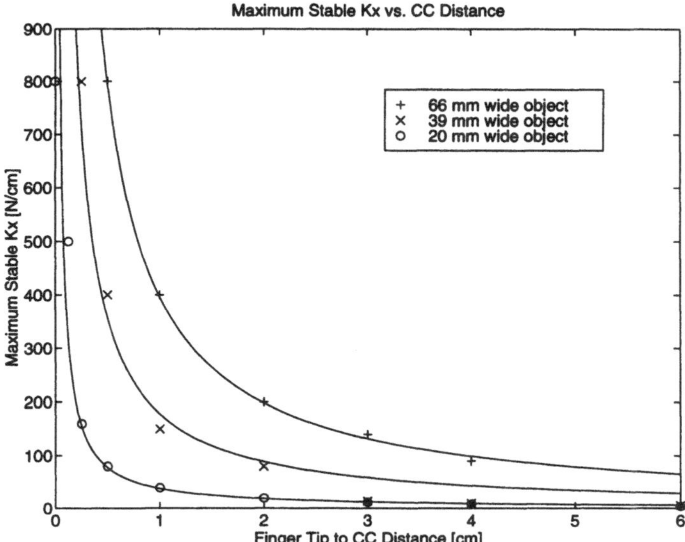

Figure 6. *Maximum stable stiffness k_x as a function of finger tip to CC distance r_{c1}. Solid curves are $k_x r_{c1}$ = constant; symbols are experimental measurements of the maximum stable stiffness for three object widths.*

of r_{c1} values. Although small high frequency vibrations were allowed, the criteria used for determining stability of a gain setting was that the measured finger tip velocities fell below 1 mm per second when the object was held steady. In addition, any disturbance to the peg had to damp out quickly. The maximum gains which met these criteria were recorded. Afterwards, the CC was moved to a new location and the procedure was repeated. Figure 6 shows the maximum k_x when r_{c1} is varied in y direction. Below the maximum stiffness obtainable with the CC between the fingers, the product of k_x and r_{c1} is nearly constant for a particular object width.

This relationship breaks down for large r_{c1} due to other limitations in the controller, such as the noise associated with object position values and unmodeled higher-order dynamics. For example, object position and velocity error in the x direction is now a function of translational error plus the rotational error multiplied by the lever arm r_{c1}.

Similar analysis for the rotational stiffness requirement shows that

$$k_\theta \delta\theta = \sum_i^n r_{1i} \times f_i \tag{10}$$

where k_θ is the rotational stiffness vector and $\delta\theta$ is the rotational displacement vector. In this case, only the distance between the finger tips amplifies the stiffness gains.

5. Peg Insertion

As an example of the implications of these limits in task execution, we now briefly consider the role of stiffness in peg insertion, a prototypical precision assembly task. Whitney [4] has provided a thorough analysis of the mechanics of peg insertion.

This analysis shows that the insertion force experienced by the peg increases with a number of factors, including lateral stiffness of the peg, position error, friction coefficient, and the distance between point of support and the contact point with the hole. This analysis explains the success of the remote center of compliance device (RCC) for close-tolerance assembly tasks. The RCC is a passive mechanism that helps minimize insertion forces by placing the CC near the tip of the peg while reducing the lateral and rotational stiffness. In theory, the Cartesian stiffness controller's ability to arbitrarily place the CC and modulate the stiffness about that CC should enable a robot hand to emulate RCC performance.

To demonstrate the utility of controlling the CC and the object stiffness, we used the two-fingered hand to perform the peg insertion task while varying both the CC and the friction at the joints. Since the manipulator motions are constrained in the plane, we used a slot oriented out of the plane instead of a round hole. This slot is assembled from machinist's parallels, which provides high accuracy and uniformity of the slot geometry. We used a precision go/no-go clearance inspection shaft mounted in a flat-sided handle as the peg. The peg diameter was 12.70 mm, and the slot was slightly undersize, resulting in a slight friction fit of the peg into the slot; this tight fit emphasized the need for good control of stiffness to minimize insertion forces. Finger tip friction forces measured with and without added friction were 0.1 and 0.4 N respectively for these peg insertion trials. Since the chamfer width was only 0.5 mm for the 12.7 mm diameter peg, the gain required for the case with added friction was relatively high to obtain the positional accuracy required to start the peg into the slot. The gains used on all trials were $k_x = 4.0$, $k_y = 10$ N/cm, $k_\theta = 100$ N-cm/rad.

The manipulator grasped the peg while it rested in the slot. After making contact with both fingers, the actual object position was recorded as the desired object position in the x direction. After lifting the peg out of the hole, the manipulator moved the peg 5 mm to one direction then back to the slot before attempting to insert the peg. The purpose of this motion was to stay consistently on one side of the friction dead band. Object position and tip forces were recorded at 100 Hz during the insertion task.

Without the added friction, insertion forces were minimal, whether the CC was located at the tip of the peg or at between the fingers. Even when the slot was tilted over 4.6 deg from vertical, the insertion was successful for both cases with minimal forces. It appears that since the stiffness in the x direction was so low, the location of the CC did not have a large effect on performance.

Figure 7 shows typical trials with and without friction with the CC located at the tip of the peg. Adding friction to the joints increased the insertion force by an order of magnitude. Part of the increase in force is due to an increase in positional error for the case with added friction, but this increase was only three times as large compared to the case without friction added. Therefore, the increase in insertion force is due to the higher effective stiffness of the object, which is created by the additional friction at the joints.

6. Conclusions

These results from the peg insertion task demonstrate the importance of stiffness modulation in precision manipulation tasks. In our analysis of the limitations on the ability to modulate stiffness, we have considered the controller's operation in

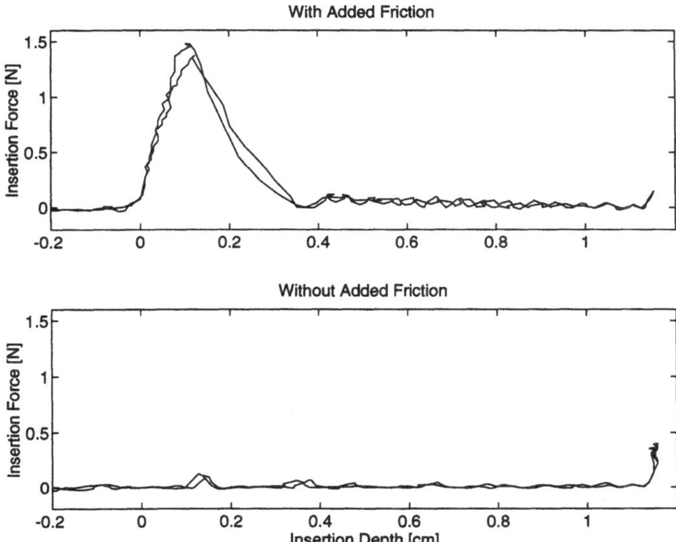

Figure 7. *Peg insertion force with and without additional friction at the joints. The CC was located at the tip of the peg in both cases.*

two stages. First, object position is determined from joint measurements. Sliding, rolling, and uncertainty in the initial grasp pose can produce errors in the object location calculated from these joint angles. Since the controller produces restoring forces proportional to the calculated object position error, this can can lead to errors in the commanded forces.

Errors in object position can also cause errors in the second stage of the controller operation, force generation. If the object location is incorrect, then the mapping between finger tip forces and object forces will be incorrect. Equation (7) relates the actual object force generated by the manipulator to errors in the grasp kinematics. Potential solutions to these problems include modeling the dynamics at the finger-object contact, and use of tactile sensing to determine the actual object pose.

In practical terms, the lower limit to attainable stiffness may be set by friction in the robot hand, which increases "effective stiffness" until the frictional forces are exceeded. Since friction is often an extremely non-linear process, compensation can be difficult. One limit on the upper bound to object stiffness is finger tip compliance, but this compliance is often important for reliable manipulation. Maximum stiffness is also related to the geometry of the grasp configuration: due to controller stability limits, the maximum usable stiffness decreases as the CC is moved away from the fingers, and increases as the object width increases.

Acknowledgments

The authors would like to thank Mark Cutkosky of Stanford University for instructive and insightful discussions during his sabbatical at Harvard. This work was supported by the Office of Naval Research under ONR Grant No. N00014-92-J-1814. Support for the first author was provided by a Doctoral Fellowship from the GM Hughes Electronics Company.

References

[1] Starr, G., Experiments in Assembly Using a Dextrous Hand, *IEEE Transactions on Robotics and Automation*, 6(3):342-347, June 1990.

[2] Hogan, N., Impedance control: An approach to manipulation, Parts 1, 2, and 3, *Transactions of the ASME Journal of Dynamic Systems, Measurement, and Control*, 107:124, March 1985.

[3] Hanafusa, H. and Asada, H., A robot hand with elastic fingers and its application to assembly process. In Brady et al., editors, *Robot Motion: Planning and Control*, Cambridge, MA, MIT Press, 1983.

[4] Whitney, D., Quasi-static assembly of compliantly supported rigid parts, *ASME Journal of Dynamic Systems, Measurement, and Control*, 104:65-77, March 1982.

[5] Kazerooni, H., Direct-drive active compliant end effector (active RCC), *IEEE Journal of Robotics and Automation*, 4(3):324-333, June 1988.

[6] Salisbury, J. K., Active stiffness control of a manipulator in Cartesian coordinates, *Proceedings of the 19th IEEE Conference on Decision and Control*, Albuquerque, Dec. 10-12, 1980, pages 95-100. Also reprinted in M. T. Mason and J. K. Salisbury, editors, *Robot Hands and the Mechanics of Manipulation*, MIT Press, Cambridge, 1985, pages 95-108.

[7] Khatib, O., A unified approach for motion and force control of robot manipulators: the operational space formulation, *IEEE Journal of Robotics and Automation*, 3(1):43-53, February 1987.

[8] Cutkosky, M. and Kao, I., Computing and controlling compliance of a robotic hand, *IEEE Transactions on Robotics and Automation*, 5:151-165, April 1989.

[9] Cole, A., Hsu, P. and Sastry, S., Dynamic regrasping by coordinated control of sliding for a multifingered hand, *Proceedings of 1989 IEEE International Conference on Robotics and Automation*, Scottsdale, AZ, May 1989, pages 781-786.

[10] Montana, D., Tactile Sensing and the Kinematics of Contact, Ph.D. thesis, Division of Applied Sciences, Harvard University, August 1986.

[11] Maekawa, H., Komoriya, K. and Tanie, K., Manipulation of an unknown object by multifingered hands with rolling contact using tactile feedback, *Proceedings of the 1992 IEEE/RSJ International Conference on Intelligent Robots and Systems, IROS '92*, Raleigh, NC, July 7-10, 1992, pages 1877-1882.

[12] Son, J., Cutkosky, M., and Howe, R., A Comparison of Object Localization by Contact Sensors, *Proceedings of the 1995 IEEE/RSJ International Conference on Intelligent Robots and Systems, IROS '95*, Pittsburgh, August 1995.

[13] Howe, R., A force reflecting teleoperated hand system for the study of tactile sensing in precision manipulation, *Proceedings of 1992 IEEE International Conference on Robotics and Automation*, Nice, France, May 1992, pages 1321-1326.

[14] Son, J. and Howe, R., Tactile sensing and stiffness control with multifingered hands, Technical Report, Harvard University Robotics Laboratory, September 1995.

Chapter 3

Autonomy via Vision

As a sensory modality, vision is of extreme importance in developing autonomous robots in that it can provide feedback to guide both manipulation and locomotion actions. The papers in this chapter reflect increased interest in using machine vision techniques in tightly coupled, real-time coordination with robotic mechanisms.

Inaba, Kagami, and Inoue describe a system in which the visual processing elements of the robot are located off-board, providing a convenient and modular experimental environment. This permits use of computational units physically too large to be carried on the robot. They report on results with a fast matching function implemented with a correlation chip that permits image flow and stereo matching techniques, which are applied to a variety of manipulation and mobility demonstrations.

Colombo, Allotta, and Dario describe visual servoing in which the camera-object model are locally linearized. This allows affine transformations to be used to deduce the differential mapping between camera motion and task goals. Experiments with an end-effector mounted camera demonstrate the robustness of the approach for vision-based navigation, active exploration, and human-robot interaction.

Nelson and Khosla propose a visual servoing framework that uses 3-D environment models to describe the task. These models are augmented by sensor mappings that represent the visual sensors used by the system and are used to drive an image-based, visually servoed agent to execute a task. Experiments with a single camera used to guide a 3-D insertion are discribed, and the extension to multiple cameras and other sensory modalities is discussed.

Hong and Slotine describe an active vision system and discuss the algorithms used to track and catch moving objects. Image segmentation is achieved via a color-keyed approach, permitting camera tracking of the target and rapid stereo localization and target path estimation. The paper reports on how successive estimates of the catch point are made, and describes the catching of balls and other more dynamically complex objects.

Andersen, Henriksen, and Ravn address the problem of using visual servoing techniques to guide the docking of non-holonomic vehicles. The difficulties imposed by docking precision and non-holonomic planning complexity are discussed. Experiments using a variety of visual and navigational algorithms on a test-bed vehicle are reported.

Real-time Vision plus Remote-Brained Design Opens a New World for Experimental Robotics

Masayuki Inaba, Satoshi Kagami, Hirochika Inoue
Department of Mechano-Informatics
The University of Tokyo
7-3-1 Hongo, Bunkyo-ku, Tokyo, JAPAN

Abstract

We present our approach for experimental robotics based on real-time tracking vision and remote-brained design. A robot with remote-brained design does not bring its own brain within the body but leaves the brain in the mother environment. The robot talks with it by radio links. The brain is raised in the mother environment inherited over generations. The key idea of the remote-brained approach is that of interfacing intelligent software systems with real robot bodies through wireless technology. In this framework the robot system can have a powerful vision system in the brain environment. We have applied this approach toward the creation of vision-based dynamic and intelligent behaviors in various robot configurations. In this paper we introduce our robot vision system and the remote-brained approach and describe visual processes for vision-based behaviors with remote-brained robots.

1. Introduction

This paper presents a new experimental environment for robotics research. The key concept of the system is the remote-brained approach. Another key technology which the system employs is a real-time visual tracking system. The combined use of the tracking vision system with remote brained robots provides us with a very friendly, flexible, and open environment for experimental robotics.

Visual information plays a very important role for robot-environment interaction. Provided with visual sensing, the repertoire of robotic behavior becomes very rich and fruitful. For this purpose we developed very fast vision system. The system is implemented as a transputer-based vision system augmented with a high speed correlation processor chip. Using the chip, we have developed a correlation based tracking vision system. This system is implemented as a multi-processor configuration to greatly enhance performance. The basic functionality provided is (1) real-time tracking of moving objects, (2) depth map generation, (3) real-time optical flow calculation, and so on.

The key concept of the system design is the "Remote-Brained Approach". Generally, a robot consists of a brain and a body. The robot brain is a computer, and the robot body is a machine which behaves in the real world. The typical design for a robot is a stand-alone, self-contained robot in which all the parts including computer and power source are built in. Another design connects a robot with its computer and power source by a wire harness. In the former case, both computing and physical power are limited by payload capacity. In the latter case, the wire harness often places severe limitations on motion and dynamic performance. We took a new approach, the remote-brained approach.

In this approach, a robot body consisting of mobility, manipulator and vision is designed as a wireless robot configuration. The robot does not carry its computer, instead it is connected with the powerful vision system and computer by radio link. Thus, this approach enables very compact design of a robot body which actually behaves in real world. The combined use of real-time vision and remote-brained robots provides us with very friendly and enjoyable environments for robotic experiments. Our students enjoy this environment and extend it themselves, while developing their imagination and creativity. In this paper, several experiments on the system are presented and the merits and limitation of the approach are discussed. From our experience on this approach, we can say that the approach is quite suitable not only for software study of robots but also for the study of the dynamic behavior of the whole body.

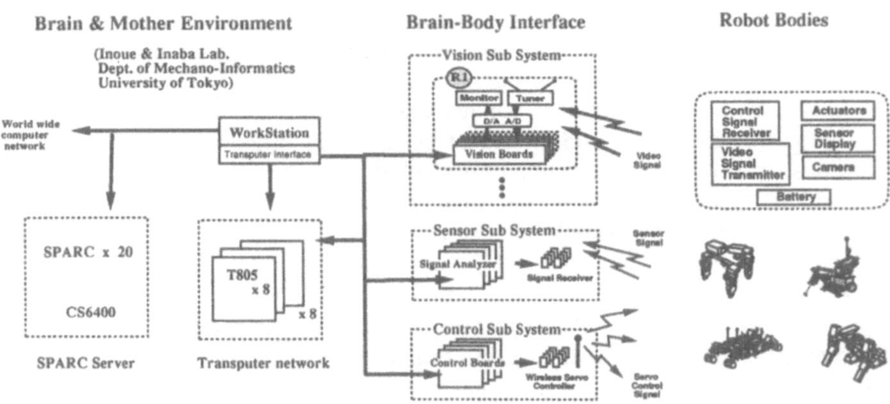

Figure 1. Hardware Configuration of the Remote-Brained System

2. The Remote-Brained Design

The remote-brained robot does not carry its own brain within the body. It leaves the brain in the mother environment and communicates with it by radio links. This allows us to build a robot with a free body and a heavy brain. The connection link between the body and the brain defines the interface between software and hardware. Bodies are designed to suit each research project and task. This approach enables us to advance in performing research with a variety of real robot systems[1].

1. **Cooperative Research Tools and Environment**

   In order to build an environment for real robot experiments, we need to work on mechanisms, on control interface, and on software. Until everything has been integrated, we cannot perform any experiments on robotic behaviors. This is one of the things that make the work time-consuming. However, the remote-brained approach can help; it partitions the work on mechanism, on interface, and on software. This approach provides a cooperative environment where each expert can concentrate on his own role and share tools with others.

2. **Building a Robot with Powerful Brain and Lightweight Body**

   A major advantage of remote-brained robots is that the robot can have a large and heavy brain based on super parallel computers. Although hardware technology for vision has advanced and produced powerful compact vision systems, the size of the hardware is still large.

   Another advantage of remote-brained approach is that the robot bodies can be lightweight. This opens up the possibility of working with legged mobile robots. As with animals, if a robot has 4 limbs it can walk. We are focusing on vision-based adaptive behaviors of 4-limbed robots, mechanical animals, experimenting in a field as yet not much studied.

3. **Evolving Mother Environment**

   The brain software is raised in the mother environment inherited over generations. The brain and the mother environment can be shared with newly designed robots. A developer using the environment can concentrate on the functional design of a brain. For robots where the brain is raised in a mother environment, it can benefit directly from the mother's 'evolution', meaning that the software gains power easily when the mother is upgraded to a more powerful computer.

In the remote-brained approach the design and the performance of the interface between brain and body is the key. Our current implementation adopts a fully remotely brained approach, which means the body has no computer onboard. Figure 1 shows the hardware configuration of the remote-brained system which consists of brain base, robot body and brain-body interface. The current system consists of three blocks of vision subsystems and the motion control system. A block can receive video signals from cameras on robot bodies. Two of the vision subsystems are parallel sets each consisting of eight vision boards.

A body just has a receiver for motion instruction signals and a transmitter for sensor signals. The sensor information is transmitted from a video transmitter. It is possible to transmit other sensor information such as touch and servo error through the video transmitter by integrating the signals into a video image[2]. The actuator is a geared module which includes a servo circuit and receives a position reference value from the motion receiver. The motion control subsystem can handle up to 184 actuators through 23 wave bands and send the reference values to all the actuators every 20msec.

3. The Real-time Robot Vision

Vision for mechanical animals requires powerful functions in dynamic visual processing. Our approach is based on a visual tracking function in order to continuously observe the moving scene. We have developed a tracking vision board using a correlation chip[3]. The vision board consists of a transputer augmented with a special LSI chip(MEP : Motion Estimation Processor) which performs local image block matching. Figure 2 shows the hardware configuration of our robot vision system.

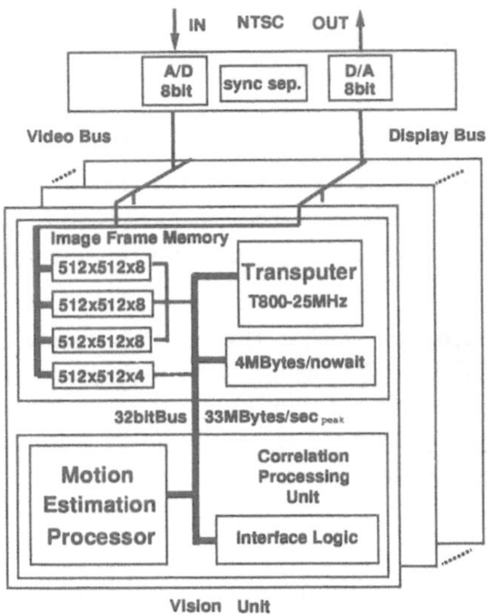

Figure 2. Hardware Organization of the Robot Vision System and Overview of the Vision Unit

The inputs to the processor MEP are an image as a reference block and an image for a search window. The size of the reference block is up to 16 by 16 pixels. The size of the search window depends on the size of the reference block; it is usually up to 32 by 32 pixels so that it can include 16 * 16 possible matches. The processor calculates 256 values of SAD (sum of absolute difference) between the reference block and 256 blocks in the search window and also finds the best matching block, that is, the one which has the minimum SAD value.

1. **Rotation and Scalable Block Matching by Software Window**

   In our current implementation, the main processor (a Transputer) sends the reference block and the search window images to the processor MEP from the input frame memory. Thus we can select the pixels for the reference block and the search window from the frame memory in software, which provides us with a flexible spatial sampling rate for the block.

   With this configuration, we can implement several kinds of template matching such as scalable and rotatable matching by software program. One calculation to find the best-matching block takes about 1.5 msec. We can thus perform about 20 block matchings in the interval between video samples (33.3msec).

2. **Motion Detection and Tracking**

   When the vision system tracks an image, it applies the block matching process with the reference block stored in the initial frame (time $t = t_0$) and the search window selected to be centered around the last tracking position from the current frame ($t = t_i$). This allows vision to follow the initial template image using this local searching method.

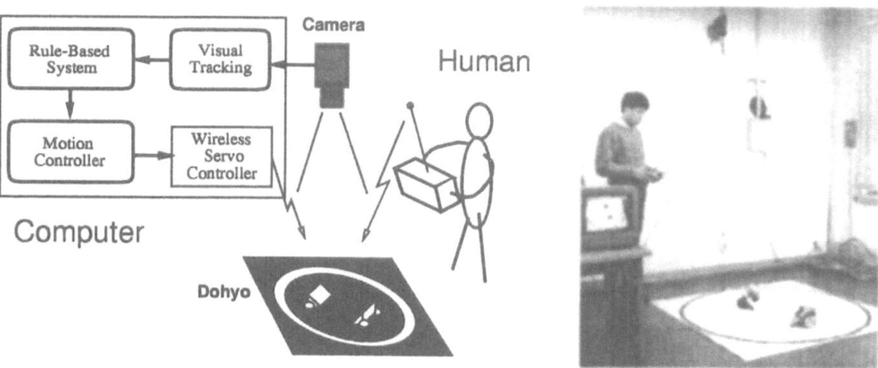

Figure 3. Overview of Human-Computer Sumo Wrestling Environment

When the vision system detects flow vectors in an image, it samples both the reference block ($t = t_i$) and the search window images($t = t_{i+1}$) continuously. In our system, we can use a thinned 256 × 256 image computed by the transputer to reduce the influence of image sampling error. As one visual processing unit can calculate 32 flow vectors, eight processors system can perform 256 optical flow generation at video-rate.

4. Vision-Based Human-Computer Sumo Wrestling

In our vision system, a transputer vision board with a MEP can perform about 20 block matchings at video frame rate. We have built an environment where the robot can play sumo games in real time with a human controlled robot as shown in Figure 3. As the human has experience playing games, he can improve the tactics and the game strategies. Then the developer of the computer-controlled robot has to match improvement to win games. This competitive situation requires advanced tools for developing software for real-time reactive behaviors based on real-time vision[5] which can track the rotating robot using rotated image templates.

As we adopted a remote-brained approach, it is possible for a human to interact through robot bodies via remote-control devices. This method, robot development motivated by the task of playing games with humans, stimulates us (especially students) to improve the mother environment for the robot brain.

5. Vision-Based Mobile Manipulation in Rope Handling

Handling flexible objects requires a robot where sensing and action are tightly coordinated. We have developed two sets of mobile hand-eye robots. Each robot has a camera, an 3DOF arm with a gripper and two wheels mounted on a base. The robot is very compact, with the size of the base being 20cm by 20cm. A team of two of these robots can succeed at rope handling[4].

The remote-brained approach affects the design of the arms. The gripper is small enough that vision can observe the target object without occlusion by the gripper. As the wires making up the gripping effector are flexible, the picked object may change orientation. In general, such a gripper is not considered to be an appropriate robot hand, because the robot can't estimate the location of the grasped object. However, a vision-based robot can tolerate such flexibility because vision allows it

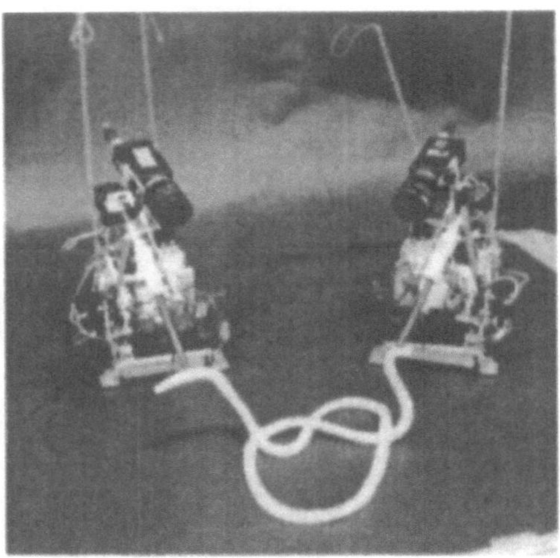

Figure 4. The hand-eye mobile robots handling a flexible rope

to know the actual location of the object, not just the gripper location. Thus, an approach based on intensive powerful vision allow us to design simpler robot bodies.

6. Vision-Equipped Mechanical Animals

We have developed several 4-limbed robots such as quadruped robots and apelike robots. We call them mechanical animals and use them to investigate sensor-based behaviors.

The main electric components of the body of a mechanical animal are joint servo actuators, control signal receiver, battery for actuators, a camera, a video transmitter, and the battery for the vision systems. There is no computer onboard. A servo actuator includes a geared motor and analog servo circuit in the box. The control signal to each servo module is position reference. The torque of servo modules available cover 2Kgcm - 14Kgcm with the speed about 300deg/sec.

The control signal transmitted on radio link encodes eight reference values. In our remote-brained environment, as we can transmit 35 wave bands simultaneously, so all of the mechanical animals can move at the same time. The body structure is designed and simulated in the mother environment, and in particular, the kinematic model of the body is described in an object-oriented lisp, Euslisp[6] which has enabled us to describe the geometric solid model and window interface for behavior design.

7. Vision-Based Balancing in Apelike Robot

Figure 5 shows an experiment in balancing. In this experiment, a human tilts the ground board on which the robot is standing. The robot vision tracks the scene in the front view. It remembers the vertical orientation of an object as the reference for visual tracking and generates several rotated images of the reference image. The rotational visual tracker[5] can track the image at video rate. If the vision tracks the reference object using the rotated images, it can measure the body rotation. In order to keep the body balance, the robot feedback controls its body rotation to

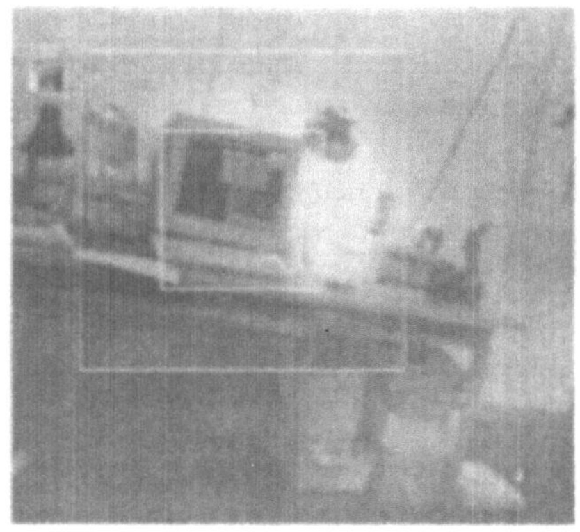

Figure 5. Keeping Balance, as Done by Humanlike Robot

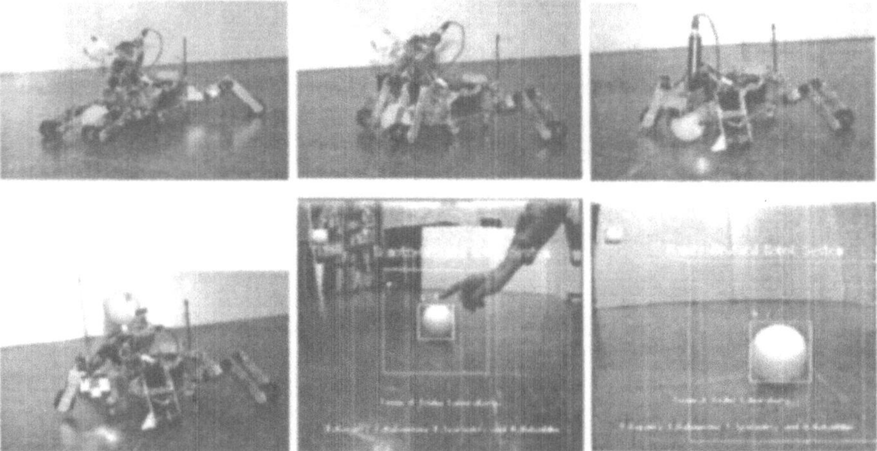

Figure 6. Holding a Ball and Images in Tracking a Ball

control the center of the body gravity.

8. Vision-Based Action Selection in Quadruped Robot

A quadruped robot is designed to interact with human by catching a ball. It can chase a ball and grab it. Figure 6 shows a sequence of motions and the image which the robot sees. Vision is the only sensor in this interaction. It detects the target ball by template matching, analyzes the relationship between its body and the target ball, and verifies the picking up action.

The robot can show several vision-based adaptive behaviors. When the robot chases a moving ball, it controls the stride of both sides of legs to change its walking speed and direction. In order to keep tracking the ball in the middle of the view, the robot continuously controls the neck orientation in parallel with walking motion. When the target detection fails, a looking-for-action is performed and the robot tries to find the target by block matching by moving its head around.

When the robot tracks the target, it keeps watching the target and measures the distance to the target. If the distance to a static target does not change when walking, it means some obstacles are in the way. If the walking direction does not point to the goal, it means that one of the legs is probably being hindered thus avoid and climb actions emerge to give adaptive capability in chasing the target.

9. Vision-Based Swing Control

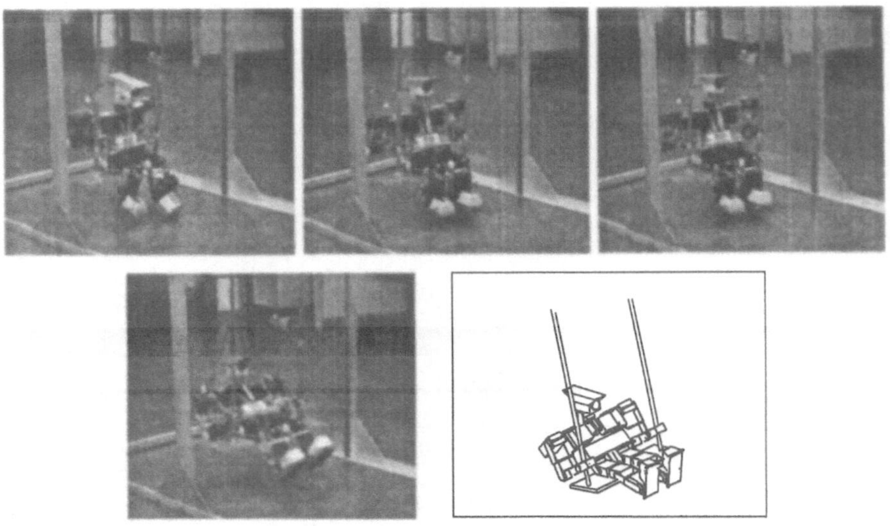

Figure 7. Apelike Robot Swinging

When a robot sits on a swing and swings, it requires a fast visual tracker and a free body. The remote-brained design allows us to make an apelike robot perform vision-based motion control on a swing. Figure 7 shows the experiment. The robot vision can measure the frequency and the phase of the swing using optical flow generation. In order to accelerate swinging, the robot controls the height of the center of gravity of its body using the visual measurement. Although a human can write a control program based on the theory of parametric oscillation, we have succeeded automatic generation of swing motions on a neural-net based controller where the weights of the neural network are generated by a genetic algorithm. This is a first step to automatic action acquisition for mechanical animals. As the GA method requires long time trials, we developed a simulation system for the swing world. The simulation program is written in Euslisp and simulates not only the physical motion of the robot and the swing but also visual processing with time-base filters. Generally, it is hard to generate accurate optical flows. We adopted a Butterworth-filter to get the motion velocity of the robot body from the optical flows. The visual process with a time-based filter is indispensable to get stable motion control in dynamic actions.

10. Concluding Remarks

This paper has introduced our approach to experimental robotics. The wireless connection between a brain and a body allows us to do a variety of vision-based experiments. Software for vision-based behaviors can be trained by playing games with humans; this and other software can be kept in the mother environment and inherited by the next generation of robots. Further, this approach allows us to design new types of robot body, such as the apelike body.

The key feature in our vision system is the fast matching function implemented with motion estimation LSI. It allows us to implement important visual functions for robot vision: moving object detection, object segmentation based on optical flow generation, object tracking and shape reconstruction in three dimension using stereo matching. The vision system provides the mechanical robot bodies with dynamic and adaptive capabilities in their behaviors.

The research environment based on tracking vision and the remote-brained design allows us to progress in vision-based experimental robotics. In our laboratory it has enabled the development of a new research environment, better suited to robotics and real-world AI.

References

[1] Masayuki Inaba. Remote-Brained Robotics: Interfacing AI with Real World Behaviors. In *Robotics Research: The Sixth International Symposium*, pp. 335–344. International Foundation for Robotics Research, 1993.

[2] M. Inaba, S. Kagami, K. Sakaki, F. Kanehiro, and H. Inoue. Vision-Based Multisensor Integration in Remote-Brained Robots. In *1994 IEEE International Conference on Multisensor Fusion and Integration for Intelligent Systems*, pp. 747–754, 1994.

[3] H. Inoue, T. Tachikawa, and M. Inaba. Robot vision system with a correlation chip for real-time tracking, optical flow and depth map generation. In *Proceedings of the 1992 IEEE International Conference on Robotics and Automation*, pp. 1621–1626, 1992.

[4] M. Inaba, T. Kamada, and H. Inoue. Rope Handling by Mobile Hand-Eye Robots. In *Proceedings of the International Conference on Advanced Robotics ICAR '93*, pp. 121–126, 1993.

[5] Masayuki Inaba, Satoshi Kagami, and Hirochika Inoue. Real time vision-based control in sumo playing robot. In *Proceedings of the 1993 JSME International Conference on Advanced Mechatronics*, pp. 854–859, 1993.

[6] Toshihiro Matsui and Masayuki Inaba. EusLisp: An Object-Based Implementation of Lisp. *Journal of Information Processing*, Vol. Vol. 13, No. 3, pp. 327–338, 1990.

Experimental Validation of an Active Visual Control Scheme Based on a Reduced Set of Image Parameters

Carlo Colombo   Benedetto Allotta   Paolo Dario
ARTS Lab
Scuola Superiore Sant'Anna
Via Carducci 40
56127 Pisa, ITALY
columbus@shamash.sssup.it

Abstract

An experimental validation of some visual control schemes for the relative positioning of a robot camera with respect to an object is presented. The schemes are based on the bi-dimensional appearance of objects in the image plane. The proposed approach allows the design of both reactive control strategies, such as fixation of a moving object, and/or active control strategies, such as the positioning of the camera in a desired location with respect to a moving object.

1. Introduction

As available computing power has increased, it has become possible to build and experiment vision systems that operate continuously. Much work has been done in the last few years on the design of specific architectures for the control of camera motion in the visual environment, or *visual servoing* [1]. A new approach to visual servoing has been proposed in [2], in which the visual loop is closed at the image level instead than in space, with significant improvements in terms of decreased sensitivity to camera calibration and kinematic modeling uncertainties. Active gaze control based on the learning of *visual appearance* has been described in [3]. In [4], an appearance-based visual servoing framework for relative positioning of a robot with respect to an object has been presented, and a constraint on the fixation of the centroid of the object was introduced.

In this paper, the visual servoing strategy presented in [4] is improved by eliminating the constraint of centroid fixation. Experimental results are shown to demonstrate the feasibility of the proposed strategy. A linear model of camera-object interaction, which dramatically simplifies visual analysis, is introduced. The linear model allows one to describe rigid motions (translations and rotations) and other affine transformations (stretching, scaling) of (close-to-) planar objects in the image plane by a set of only 6 parameters. The desired camera-object relative position and orientation is mapped to a desired object shape in the image plane. Based on the current state of the image, a desired deformation of the object shape is generated based on a control algorithm that contains both feedback and feedforward information. The current implementation relies on active contour analysis for tracking object shape. Polynomial planning of object contour evolution is used to generate desired trajectories for the object in the image plane. Based on the linear differential mapping existing between the relative motion of the camera with respect to the object and image curve transformations, velocity commands for the camera are then generated in order to fulfill the task.

Experiments with an eye-in-hand configuration are shown, which demonstrate the robustness of the approach, and its feasibility for applications in the fields of visual navigation, active exploration and perception, and man-robot interaction.

The paper is organized as follows: in Sect. 2, the proposed control strategy is described as well as the camera-object interaction model; in Sect. 3, the experimental setup and the results experiments are presented. Conclusions and directions for future work are then provided in Sect. 4.

2. Formulations

2.1. Control strategy

Changes in viewpoint determined by relative motion of camera and object influence the raw image data according to the nature of both camera projection and object shape. A suitable set of n image differential parameters can be chosen in order to describe changes in the image caused by a velocity twist of the camera with respect to the object. The mapping between $n \times 6$ *interaction matrix* $\mathcal{L}$, first defined in [2], maps the 3D relative velocity twist of camera and object to 2D changes of visual appearance:

$$\boldsymbol{p} = \mathcal{L}(\boldsymbol{V}_c - \boldsymbol{V}_o), \tag{1}$$

where $\boldsymbol{V}_c = [V_{cx}\ V_{cy}\ V_{cz}\ \Omega_{cx}\ \Omega_{cy}\ \Omega_z]^{\mathrm{T}}$ is the velocity twist of the camera and $\boldsymbol{V}_o = [V_{ox}\ V_{oy}\ V_{oz}\ \Omega_{ox}\ \Omega_{oy}\ \Omega_z]^{\mathrm{T}}$ is the velocity twist of the object.

A relative positioning task is described in the image plane by the desired evolution of the object shape and centroid position toward goal ones. If $\boldsymbol{p}$ is the chosen set of differential parameters, the task can be represented in a synthetic way by a trajectory $\boldsymbol{p}^{\mathrm{des}}(t)$.

Once the set $\boldsymbol{p}$ has been introduced and the structure of $\mathcal{L}$ has been identified, we propose the following control strategy which makes use of both feedforward and feedback information.

$$\boldsymbol{V}_c^{\mathrm{des}} = \underbrace{\widehat{\boldsymbol{V}}_o + \mathcal{L}^+ (\boldsymbol{p}^{\mathrm{des}}}_{feedforward} + \underbrace{k_p \widehat{e}}_{feedback}), \tag{2}$$

where $\boldsymbol{V}_c^{\mathrm{des}}$ is the required motion of the camera, $\widehat{\boldsymbol{V}}_o$ is an estimate of object motion, $\boldsymbol{p}^{\mathrm{des}}$ is the current desired set of image parameters, and $\widehat{e}$ is an error signal derived from a suitable comparison between the current object appearance and the desired one.

2.2. Camera-object interaction

We refer to a pinhole camera with fixed focal length f and optical axis Z, as shown in Fig. 1 (*left*). Projecting a point of the object's visible surface of camera-centered coordinates $(X, Y, Z(X, Y))$ onto an image point (x, y) yields:

$$[x\ y]^{\mathrm{T}} = \frac{f}{Z(X,Y)}[X\ Y]^{\mathrm{T}}. \tag{3}$$

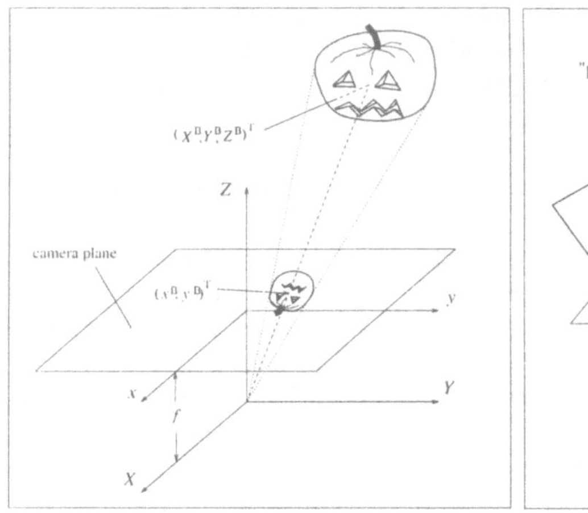

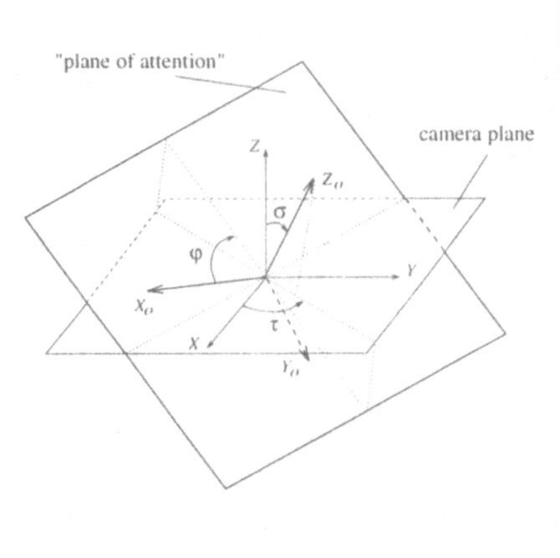

Figure 1. The geometry of camera-object interaction. *Left*: camera projection. *Right*: definition of extrinsic parameters. The camera-centered frame has been translated for convenience in the object-centered frame's origin.

Suppose now the camera to have a *narrow field of view in the sense of vanishing squares*, i.e. that the transversal dimensions of the sensor are small enough with respect to focal length to assume that:

$$(x/f)^2 \approx 0; \quad (y/f)^2 \approx 0, \tag{4}$$

thus constraining also by eq. (3), in order for the object to be visible, its transverse dimensions to be small with respect to its depth[1]. Specifically, assume the depth function $Z(X,Y)$ to be sufficiently smooth for its quadratic and higher order terms in the development around (X^B, Y^B) to be negligible with respect to Z^B, where the superscript 'B' indicates a convenient point in space—say, the centroid of the visible surface. The visible surface itself can then be approximated by a *planar surface*, referred to as *plane of attention*, whose equation is:

$$Z(X,Y) = pX + qY + c. \tag{5}$$

By combining eqs. (3) and (5) and defining $z(x,y)$ by

$$z(fX/Z, fY/Z) = Z, \tag{6}$$

we obtain:

$$z(x,y) = \frac{c}{1 - p\frac{x}{f} - q\frac{y}{f}}, \tag{7}$$

where the terms p and q are the components of the gradient of the plane of attention, and c is the Z-coordinate of the intercept between the optical axis and the plane of attention. Eq. (7) expresses object depth directly in terms of image coordinates. Notice that $c = Z^B - pX^B - qY^B$ is the plane intercept with the optical axis, and p and q, the X- and Y-components of the depth gradient at (X^B, Y^B).

[1] In the case of a wide field of view, the approximation above holds approximately true in the case of an object almost centered in the visual field, and sufficiently far from the camera plane.

As a straightforward consequence of the above, *the mapping between any two object views* $\{(x_1, y_1)\}$ *and* $\{(x_2, y_2)\}$ *is affine*, and it holds:

$$[x_2 - x_2^B \; y_2 - y_2^B]^T = \mathcal{A}_{12} \, [x_1 - x_1^B \; y_1 - y_1^B]^T, \tag{8}$$

where $\mathcal{A}_{12} = \mathcal{T}_2 \mathcal{T}_1^{-1}$.

Eq. (8) can be used for planning a contour evolution: starting from the initial contour, a smooth trajectory of the affine transformation matrix $A(t)$ can be devised in order to describe a possible desired evolution of the contour toward a goal one. A contour evolution at time t can be described by the following set of differential parameters:

$$\boldsymbol{p} = [u_C \; v_C \; \text{div} \; \text{curl} \; \text{def}_1 \; \text{def}_2]^T, \tag{9}$$

where u_C and v_C are the X- and Y-components of the centroid velocity, and the other differential parameters describe the first-order changes of shape for the contour [5].

If affine transformations hold, the interaction matrix $\mathcal{L}$, seen in Sect. 2 is the following:

$$\mathcal{L} = \begin{bmatrix} \frac{-f}{z} & 0 & \frac{x_C}{c} & 0 & -f & y_C \\ 0 & \frac{-f}{z} & \frac{y_C}{c} & f & 0 & -x_C \\ \frac{p}{c} & \frac{q}{c} & \frac{2}{c} & 0 & 0 & 0 \\ \frac{-q}{c} & \frac{p}{c} & 0 & 0 & 0 & -2 \\ \frac{p}{c} & \frac{-q}{c} & 0 & 0 & 0 & 0 \\ \frac{q}{c} & \frac{p}{c} & 0 & 0 & 0 & 0 \end{bmatrix}, \tag{10}$$

where x_C and y_C are the coordinates of the object centroid, f is the focal distance, z is given by eq. (7) and c is the distance of the camera from the object plane. Notice that $\mathcal{L}$ is singular if p and q are both zero (i.e. the focal axis is orthogonal to the object plane). For this reason we have assumed that the condition that the task can be fullfilled without passing through singular configurations of $\mathcal{L}$.

3. Experiments

3.1. Experimental setup

The hardware system consists of a PUMA 560 manipulator with MARK III controller equipped with a wrist-mounted camera and a PC equipped with a frame grabber and a graphic accelerator. The PC features a 80486-66 Mhz processor. The MARK III controller runs VAL II programs and communicate with the PC via the ALTER real-time protocol using an RS232 serial interface. The ALTER protocol allows to modify the cartesian setpoint of the robot arm every 28 ms. Due to the burden of calculation of tracking algorithms, a multirate real-time control has been implemented. New velocity setpoints are generated by the PC with a sampling rate $T_2 = N \, T_1$ where $T_1 = 28$ ms is the sampling rate of the ALTER protocol and N is an integer, depending on the number of control points of B-spline contour. By using 10 points for the B-spline contour, the resulting computation time is less than 100 ms. Hence, the sampling time T_2 has been chosen equal to $4\,T_1$. A "fast" communication process maintains the handshake with the MARK III controller sending the most recent velocity twist setpoint generated by the high level "slow" process, available in a mailbox. The proposed approach is well suited for objects whose visible surface can be approximated with a planar one. So far, planar objects have been chosen for experiments.

3.2. Task to be performed

The task chosen for experiment consists in positioning the camera with respect to an object at rest. At power up, A n-point snake is attached to the object contour. The snake will track the object contour during the relative motion. Then the robot moves the camera without any feedback and an estimate of 3D parameters, namely p, q and c, is performed. After a reasonable estimate of p, q, and c is available, the positioning task can start. The goal contour can either be known in advance or be obtained from the initial one performing successive affine transformations. A mouse-driven interface has been implemented in order to modify the initial contour and synthesize the goal one. The initial contour and the goal one are used to generate smooth contour trajectories. Cubic polinomials are used in the current implementation.

3.3. Control schemes

Experiments have been performed to compare two slightly different control schemes described in Sect. 2. The first one (scheme 1) is a simple servo that make use of planning. The current desired contour is matched via least squares against the current one in order to obtain an error signal. The error signal is then tranformed to desired image parameters and finally mapped to a desired velocity twist of the camera. The second one (scheme 2) combines feedforward and feedback information. In addition to the servo term, a feedforward term based on planned contour evolution is added to the control. In both cases, no on-line estimation of 3D parameters (p, q, c) is performed. As a result, the interaction matrix $\mathcal{L}$ is not updated correctly.

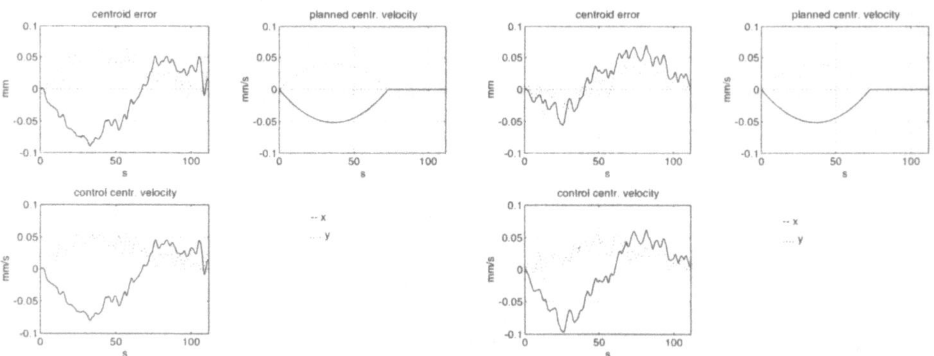

Figure 2. Comparison between scheme 1 (feedback only, *left*) and scheme 2 (feed-forward plus feedback, *right*): centroid.

3.4. Results

Experimental results concerning scheme 1 and scheme 2 are reported. Despite the fact that the interaction matrix $\mathcal{L}$, is only roughly computed (no on-line estimation of 3D parameters is performed), both the control schemes seem to be effective. The same IIR digital filters have been used for smooting sensory data. The gains of the filters as well as those of the feedback control term where tuned experimentally and were the same in the 2 experiments. In Fig. 2, the centroid error is shown as well as the planned centroid velocity in the image plane and the velocity resulting from the control algorithm. In Fig. 3, the translational and angular velocities of the

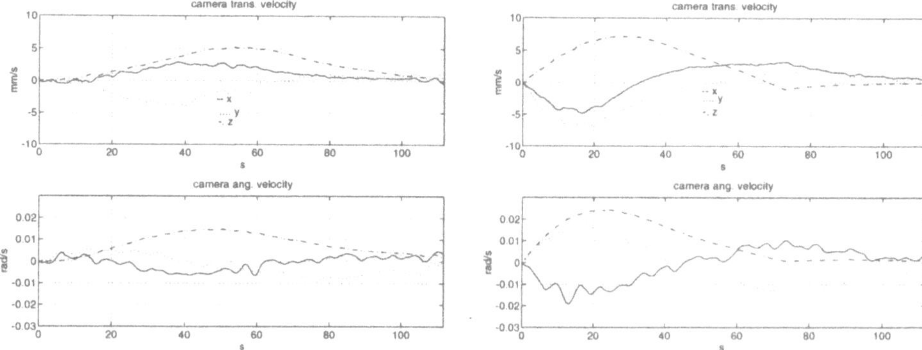

Figure 3. Comparison between scheme 1 (feedback only, *left*) and scheme 2 (feedforward plus feedback, *right*): velocities.

camera are shown. As one can see, by using feedforward information in addition to feedback (scheme 2), at the end of the planning phase (75 s), the 'z' components of translational and rotational velocity are almost zero. This because inaccuracies in the estimate of 3D parameters (p, q, c) have a great influence only on the mapping between image centroid velocity and differential invariants on one side and V_x, V_y, Ω_x and Ω_y components of camera cartesian velocity twist. In Fig. 4, the planned

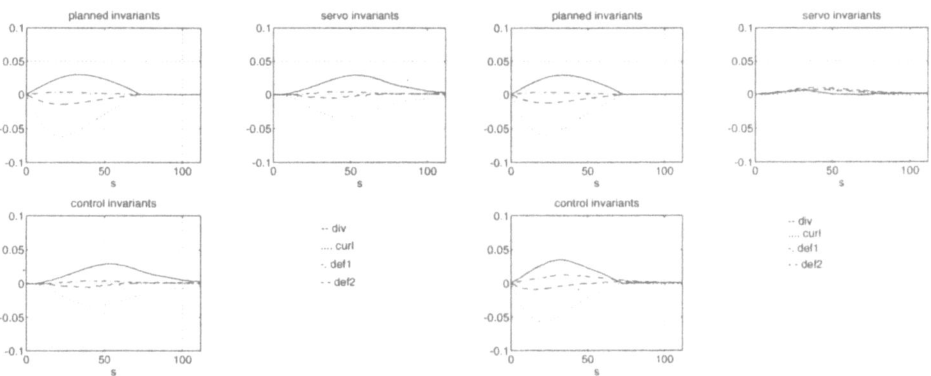

Figure 4. Comparison between scheme 1 (feedback only, *left*) and scheme 2 (feedforward plus feedback, *right*): invariants.

differential invariants are shown as well as those generated by the feedback (servo invariants) and those used to produce camera cartesian velocity twist (control invariants). Notice that the effect of feedforward control is twofold. On the one hand, it significantly reduces the job of the feedback. On the other, it makes system performance more sensitive to inaccuracies in the estimation of the interaction matrix.

4. Conclusions and future work

The affine visual servoing framework has been improved by relaxing the constraint on the fixation of the object centroid. Experiments have been performed in order to validate the effectiveness of the approach. Results have shown the importance of adding feedforward information to feedback in visual servoing tasks even with a non perfect knowledge of the interaction matrix due to the fact that 3D parameters are not updated on-line. A main drawback of the approach is the presence of singularities of the interaction matrix whenever both p and q are zero, i.e. the object surface is orthogonal to the optical axis. We suggest that using a technique such as the damped least square inverse used for the inverse differential kinematic control of robot manipulators might solve the problem. Although we have not yet verified this conjecture, in our experiments we have used a rough estimate of matrix $\mathcal{L}$ and the control schemes work in a satisfactory way. Future work will concern the addition of on-line estimation of 3D parameters (p, q, c) to improve the effectiveness of both feedback and feedforward control. Further effort will be devoted to overcome the problem of singularities in the interaction matrix.

Acknowledgements

The authors warmly thank Prof. Bernard Espiau for useful comments on an earlier draft of this work, and Mr. Giacomo Borlizzi for his precious help during the implementation phase.

References

[1] L.E. Weiss, A.C. Sanderson, and C.P. Neuman. Dynamic sensor-based control of robots with visual feedback. *IEEE Journal of Robotics and Automation*, 3(5):404–417, 1987.

[2] B. Espiau, F. Chaumette, and P. Rives. A new approach to visual servoing in robotics. *IEEE Transactions on Robotics and Automation*, 8(3):313–326, 1992.

[3] S.K. Nayar, H. Murase, and S.A. Nene. Learning, positioning and tracking visual appearance. In *Proceedings of the 1994 IEEE International Conference on Robotics and Automation, San Diego, California, 1994*, 1994.

[4] C. Colombo, B. Allotta, and P. Dario. Affine visual servoing: a framework for relative positioning with a robot. In *Proceedings of the 1995 IEEE International Conference on Robotics and Automation, Nagoya, Japan, 1995*, pages 464–471, 1995.

[5] J.J. Koenderink and A.J. van Doorn. Invariant properties of the motion parallax field due to the movement of rigid bodies relative to an observer. *Optica Acta*, 22:773–791, 1975.

Task Oriented Model-Driven Visually Servoed Agents

Bradley J. Nelson
Department of Mechanical Engineering
University of Illinois at Chicago
Chicago, Illinois USA
bnelson@uic.edu

Pradeep K. Khosla
Department of Electrical and Computer Engineering
Carnegie Mellon University
Pittsburgh, Pennsylvania USA
pkk@cmu.edu

Abstract

Most proposed visual servoing control strategies use image-based visual servoing techniques in which the visual servoing control loop is closed using 2D image plane coordinates rather than 3D world coordinates. These strategies do not allow for explicit representations of the 3D world. We propose a visual servoing framework that uses 3D environment models to describe the task. These models are augmented by sensor mappings that represent the visual sensors used by the system and are used to drive an image-based visually servoed agent to execute the task in the real world. This framework allows for the use of 3D reasoning within the environment model, while taking advantage of the benefits 2D image-based visual servoing provides. In this paper, we describe our proposed framework and present experimental results which show that the framework can be used to successfully perform visually servoed manipulation tasks.

1. Introduction

Visual servoing is a robust technique for guiding a manipulator through an uncertain environment using poorly calibrated camera-lens-manipulator systems. The technique employs a differential view of the world in which only small scene changes between image frames are assumed and compensated for by any of a number of proposed control techniques.

Most proposed visual servoing control strategies use image-based techniques[1]. In its basic form, image-based visual servoing is a purely reactive manipulator motion strategy. The objective of the control system is simply to make some set of measured feature states defined in the camera's sensor space match some set of desired states defined in the same space by moving the manipulator. While this strategy is relatively easy to implement because it avoids the need to perform an explicit inverse perspective mapping, manipulation tasks exist in the real 3D world and must often be described to motion planners in terms of this world. Thus, in order for the execution of a robotic task to benefit from the advantages visual servoing provides, a 3D representation of the task must exist. This representation must also be capable of providing appropriate reference inputs to the visual servoing control loop.

In this paper, we describe and experimentally verify a framework for visually servoed manipulation that uses an expectation-based approach to task execution. The task is represented by the actions and interactions of dynamic 3D geometric models of objects in the environment. Each model is augmented by sensor mappings representing the visual sensors used to provide feedback for task execution. We refer to these augmented environment models as *object schemas*, because they simultaneously represent the task and provide an understanding of how the environment is perceived. The mappings are used to predict how the sensors will view the environment model while the given task is being executed. This provides reference inputs for visual servoing controllers which then guide execution of the task. The visual servoing systems we call visually servoed *port-based agents*, because these systems provide a level of system autonomy beyond that normally associated with simple feedback controllers.

We begin by discussing our task representation in terms of object schemas. Next, a visually servoed port-based agent is presented, followed by the overall system framework that combines object schemas, port-based agents, and the real world. Finally, we present experimental results that demonstrate the framework operating within a dynamically varying, imprecisely calibrated environment.

2. A Task Representation Using Object Schemas

A key component of our system framework is the structure and content of the task representation. In order to provide the system with the capability to reason about manipulation strategies for objects, the internal representation must contain three dimensional geometric knowledge of the objects being manipulated. The internal representation must also be capable of being correlated with visual information in order to direct the visually servoed manipulator to perform the task. This implies that a visual representation of the task must also exist internally.

We have defined a representation for objects to be manipulated that we call an *object schema*. Our object schema definition includes a texture mapped geometric environment model of an object augmented by sensor mappings that describe how the object will be perceived by the actual system sensors. Forward projective sensor mappings and their associated Jacobian matrix for each visual sensor that exists in the system provides this representation. For a single camera these forward mappings are simply perspective projection mappings and are of the form

$$x_S = \frac{fX_C}{s_x Z_C} \qquad y_S = \frac{fY_C}{s_y Z_C} \tag{1}$$

where x_S and y_S are the projected image coordinates of a point on the object in the internal representation located with respect to the camera frame at (X_C, Y_C, Z_C), f is the focal length of the camera lens, and s_x and s_y are the pixel dimensions of the CCD array. The Jacobian of this mapping is of the form

$$\delta\mathbf{x}_S = \mathbf{J}(\phi)\delta\mathbf{X}_T \tag{2}$$

where $\delta\mathbf{x}_S$ is an infinitesimal displacement vector in sensor space and $\delta\mathbf{X}_T$ is an infinitesimal displacement vector in task space. $\mathbf{J}(\phi)$ is the Jacobian matrix and is a function of the extrinsic and intrinsic parameters of the visual sensor as well as the number of features used for tracking the object and their locations on the image plane.

For the experimental results to be presented, an orthogonal stereo pair is used. Figure 1 shows the coordinate frame definitions for this type of camera-lens configuration. If the axes are aligned as shown in the figure, the Jacobian mapping from task space to sensor space for a single feature can be written as

$$\mathbf{J} = \begin{bmatrix} \frac{f}{s_x Z_{Cl}} & 0 & -\frac{x_{Sl}}{Z_{Cl}} & -\frac{x_{Sl} Y_T}{Z_{Cl}} & \left[\frac{f Z_T}{s_x Z_{Cl}} + \frac{x_{Sl} X_T}{Z_{Cl}}\right] & -\frac{f Y_T}{s_x Z_{Cl}} \\ 0 & \frac{f}{s_y Z_{Cl}} & -\frac{y_{sl}}{Z_{Cl}} & -\left[\frac{f Z_T}{s_y Z_{Cl}} + \frac{y_{Sl} Y_T}{Z_{Cl}}\right] & \frac{y_{Sl} X_T}{Z_{Cl}} & \frac{f X_T}{s_y Z_{Cl}} \\ \frac{x_{Sr}}{Z_{Cr}} & 0 & \frac{f}{s_x Z_{Cr}} & \frac{f Y_T}{s_x Z_{Cr}} & \frac{x_{Sr} Z_T}{Z_{Cr}} - \frac{f X_T}{s_x Z_{Cr}} & \frac{x_{Sr} Y_T}{Z_{Cr}} \\ \frac{y_{Sr}}{Z_{Cr}} & \frac{f}{s_y Z_{Cr}} & 0 & -\frac{f Z_T}{s_y Z_{Cr}} & \frac{y_{Sr} Z_T}{Z_{Cr}} & \frac{y_{Sr} Y_T}{Z_{Cr}} - \frac{f X_T}{s_y Z_{Cr}} \end{bmatrix} \quad (3)$$

In (3), we assume the camera-lens parameters are identical for both cameras. The other terms in (3) correspond to Figure 1. The tracking of features will be described in Section 3.2.

Our previous work in defining the sensor placement measure *resolvability* [2] uses the Jacobian mapping to compare various visual sensor systems from a control standpoint in terms of the accuracy of control. This measure quantifies the ability of various monocular and stereo camera configurations to resolve the position and orientation of visually servoed objects. As discussed in [2], the measure is easily extendable to other visual sensors including multi-baseline stereo and laser rangefinders. A key component of this measure, the Jacobian mapping from task space to sensor space, is also a critical component of our visual servoing control strategy.

Figure 2 shows the resolvability of two different stereo camera configurations. A sensitivity analysis of resolvability with respect to the variable extrinsic and intrinsic camera-lens parameters shows how camera-lens systems can be reconfigured in order to improve the precision with which objects can be observed and manipulated [3].

Figure 3 shows a diagram of an object schema, including the current internal pose of the geometric model; its desired pose (for controllable schemas) as determined by a meta-schema (an assembly planner or supervisor); and sensor mappings used to direct variable sensor parameters as well as update the current internal pose of the schema based on actual sensor feedback. Note that the pseudoinverse of the Jacobian mapping used for determining sensor resolvability is also used to "servo" the geometric model in order to reduce errors between the current internal visual representation of the object and the actual visual representation.

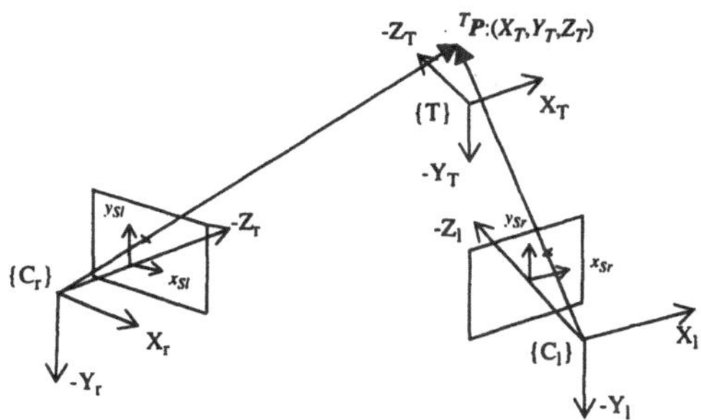

Figure 1. Task frame-camera frame definitions.

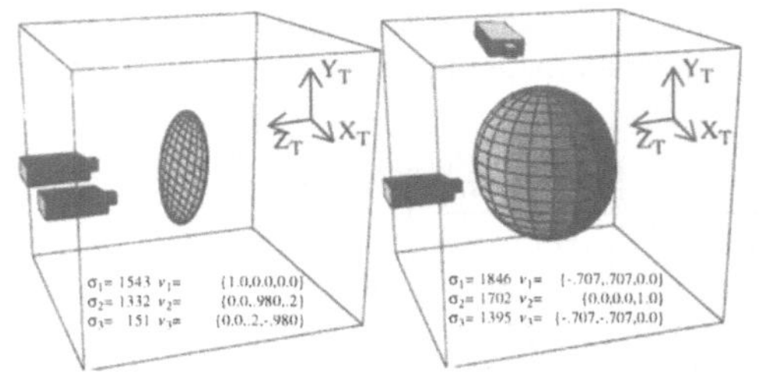

Figure 2. Resolvability ellipsoids for two different stereo configurations. The singular-values of $\mathbf{J}$ and the eigenvectors of $\mathbf{J}^T\mathbf{J}$ are given for each configuration.

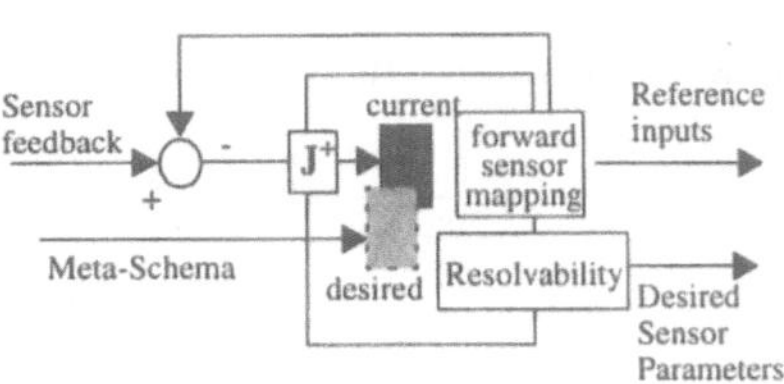

Figure 3. An object schema diagram.

3. Visually Servoed Port-Based Agents

The desired visual representation of a task is derived from schemas. This desired visual representation provides reference inputs to a visual servoing system. The visual servoing system we consider to be an agent that continuously accepts reference commands from object schemas and attempts to achieve the desired visual representation of the task based on these inputs.

3.1. Controller Formulation

The controlled active vision paradigm [4] is used to derive a control strategy for the visually servoed port-based agent. A state equation for the visual servoing system can be derived by discretizing (2) and writing this discretized equation as

$$\mathbf{x}(k+1) = \mathbf{x}(k) + T\mathbf{J}(k)\mathbf{u}(k) \qquad (4)$$

where $\mathbf{x}(k) \in R^{2M}$ and is a vector of feature states, T is the sampling period of the vision system, $\mathbf{u}(k) = \left[\dot{X}_T\ \dot{Y}_T\ \dot{Z}_T\ \omega_{X_T}\ \omega_{Y_T}\ \omega_{Z_T}\right]$ is the commanded manipulator end-effector velocity, and M is the number of features being tracked. Using optimal control techniques, an objective function that places a cost on feature error and control energy of the form

$$F(k+1) = [\mathbf{x}(k+1) - \mathbf{x}_D(k+1)]^T \mathbf{Q}\,[\mathbf{x}(k+1) - \mathbf{x}_D(k+1)] + \mathbf{u}^T(k)\mathbf{L}\mathbf{u}(k) \qquad (5)$$

can be minimized at each time instant to obtain the control law

$$\mathbf{u}(k) = -(T^2\mathbf{J}^T(k)\mathbf{Q}\mathbf{J}(k) + \mathbf{L})^{-1}\, T\mathbf{J}^T(k)\mathbf{Q}\,[\mathbf{x}(k) - \mathbf{x}_D(k+1)] \qquad (6)$$

The vector $\mathbf{x}_D(k+1)$ represents the desired feature state, i.e. the desired visual representation of the scene, at the next time instant. $\mathbf{Q}$ and $\mathbf{L}$ are weighting matrices and allow the user to place a varying emphasis on the feature error and the control input. Extensions to this control strategy and guidelines for choosing the matrices $\mathbf{Q}$ and $\mathbf{L}$ can be found in [5]. New sensor configurations can be quickly and easily added to the system by substituting the correct $\mathbf{J}$ and by adjusting $\mathbf{Q}$ and $\mathbf{L}$ accordingly.

3.2. Visual Tracking of Features

In this section the determination of the feature state $\mathbf{x}(k)$ is described. The measurement of the motion of the features on the image plane must be done continuously and quickly. Our method for measuring this motion is based on optical flow techniques and is a modification of the method proposed in [6]. This technique is known as Sum-of-Squared-Differences (SSD) optical flow, and is based on the assumption that the intensities around a feature point remain constant as that point moves across the image plane. A more complete description of the algorithm and its original implementation can be found in [5].

By integrating texture mapped environment models with our visual tracker, we are able to use context-based vision strategies to take advantage of a large amount of previously unavailable information. We use texture-mapped models in two ways: 1. during initial feature selection, and 2. during feature tracking.

Our original feature selector was based on a technique proposed in [7]. This feature selector assumes that the best features to track are features which have the strongest omni-directional gradients, for example well defined corners or small circles. One problem we have encountered during feature selection is that exterior features are often selected as the best features because exterior boundaries often exhibit strong gradients. Tracking algorithms are more prone to failure if feature templates are located near the exterior boundaries of objects due to background "noise." Because texture mapped models inherently maintain the correct segmentation of objects, this segmentation is used to avoid choosing features which lie near the exterior edges of the object.

During feature tracking, environment models are also used to aid in tracking. Although the feature template used for tracking is obtained from actual image data rather than from the texture mapped model, correlations are periodically performed with the current feature template on the model in order to determine if a particular feature template is unable to successfully track an object because of scale changes or occlusions. If tracking performance decreases, a new feature is selected using the most recent image.

4. A Framework of Schemas and Agents

Our system framework of object schemas and port-based agents for robotic manipulation draws many of its concepts from past work in the development of expectation-based and verification approaches for guiding mobile robots. In the section we discuss these past approaches and present representations of our system framework.

4.1. Expectation/Verification Approaches

An expectation-based approach to scene understanding was first explicitly proposed by Dickmanns [8]. His work was originally concerned with guiding autonomous mobile systems in rapidly changing environments, particularly autonomous vehicles and aircraft. Roth and Jain propose a "verification-based" approach to navigation in the world [9]. A key point of both the expectation and verification-based approaches is that strong internal models of the recent world state are maintained. This significantly reduces the number of hypotheses that must be considered when determining the current state of the world based on sensory data. Neisser's view of the human "perceptual cycle" [10], as Jain points out [11], is similar in many ways to a verification or expectation-based approach. Figure 4 shows a modified representation of Neisser's "perceptual cycle." This figure illustrates our view of the relationship between the schemas of the world, the real world, and where the visually servoed agents exist within this scheme. The counter-clockwise flow of information represents the cyclical nature of the system; sensory data updates schemas, which in turn provide reference inputs to the visually servoed agents, which provides sensory data obtained from the real world to the environment model and the schemas existing within the environment model. This cycle illustrates the interaction between perception of the world, actions taken within this world, and plans made about the world.

Our proposed integration of visually servoed agents into systems capable of performing manipulation tasks in the real world uses this cycle to clearly delineate the flow of information from object schemas to visually servoed port-based agents. By clearly separating the system in this way, different types of sensor-based manipulation strategies can

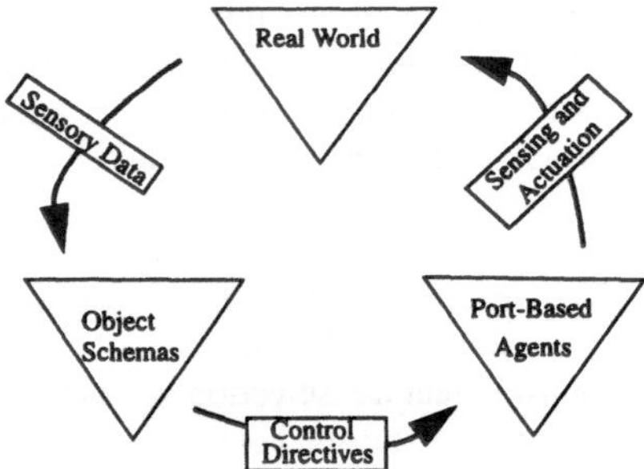

Figure 4. A modified "perceptual cycle" for visually servoed manipulators.

be more easily integrated into the system, because the internal representation of the task becomes clearly separate from the hardware and low-level behavior-based control systems that perform actions and sensing in the real world. The aim is toward plug-and-play types of actuation/sensor components.

4.2. System Framework

Figure 5 shows a block diagram structure of the expectation-based visually servoed system. The upper loop represents a servoed schema. The lower loop represents a single visually servoed agent. The goal of the system is to maintain a temporal coherence between schemas in the environment model and objects in the real world using visual feedback and visual servoing. We assume that if this coherence is maintained the task is being properly executed, within the limits of sensor resolution.

One can also view the block diagram presented in Figure 5 as the closed-loop system shown in Figure 6. Within the figure, $\mathbf{\Phi}$ represents the forward sensor mapping given in (1), C is the visual servoing control law given in (6), $\mathbf{P}$ is the visually servoed arm, $\mathbf{V}$ is the vision system that determines the feature state $\mathbf{x}(k)$, $\mathbf{X}(k)$ is the 3D model state, and $\mathbf{X}_D(k+1)$ is the desired state of the world as determined by "goal achieving functions." In this view, the "goal achieving functions" are provided by a supervisor or some type of motion planner. A supervisor uses a graphical representation of the internal schema representation, rather than live video imagery, to guide his or her actions. This is similar to the remote teleoperative techniques described in [12] and [13].

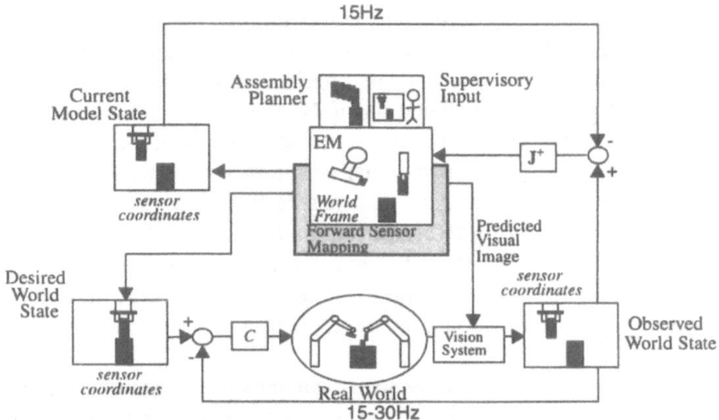

Figure 5. The block-diagram representation of the framework for visually servoed manipulation.

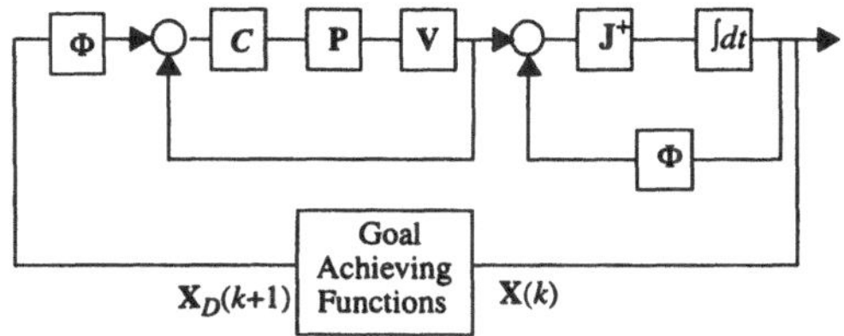

Figure 6. A closed-loop view of the system framework.

5. Hardware Implementation

The schema-agent framework has been implemented on a robotic assembly system consisting of three Puma 560's called the Troikabot. The Pumas are controlled using the Chimera 3.0 reconfigurable real-time operating system[14]. An Adept robot is also used for providing accurate target motion for experimental purposes.

A Datacube Maxtower Vision System calculates the optical flow of the features using the SSD algorithm discussed in Section 3.2. An image can be grabbed and displacements for up to five 16x16 features in the scene can be determined at 30Hz. Stereo system implementations result in half the sampling frequency because only a single digitizer exists on the Datacube. For the stereo system with which experimental results were obtained, visual servoing is performed at 15Hz while tracking ten features. The vision system VME communicates with the robot controller VME using BIT3 VME-to-VME adapters.

The environment model exists within a Silicon Graphics Indigo2 Extreme. The robot simulation package Telegrip™ is used for modeling and texture mapping objects. Unix sockets and the Chimera Enet protocol are used to communicate between the Silicon Graphics machine and the robot controller and vision system. The rate of communication between the Silicon Graphics and the robot controller is approximately 15Hz.

6. Experimental Results

In order to illustrate how our proposed framework operates, an example of a servoed schema is shown, followed by an example of an insertion task guided by a supervisor within the environment model. Figure 7 shows images of the real scene and the visual internal representation of the same scene from a single camera. The fingers of a gripper mounted at the manipulator flange can be seen, as well as a texture-mapped rotor-stator motor pair. An Adept robot provides periodic motion to the surface on which the rotor-stator pair rest. The servoed schema technique treats the motion as a disturbance and updates the position of the model accordingly within the internal representation. Figure 8

Figure 7. A live image of the environment is shown to the left and the corresponding visual representation of the same scene is to the right.

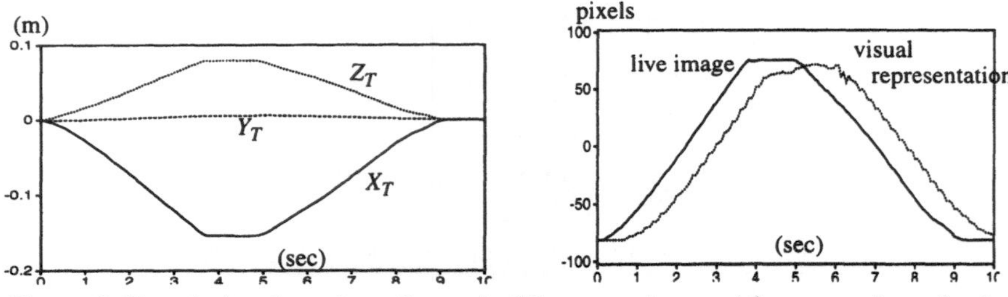

Figure 8. Translational motion of rotor in 3D versus time and feature trajectories in the live image and the visual representation of the image.

shows the 3D translational motion of the object and motion of a feature in the live image and the position of the same feature in the visual representation. Coordinate axes correspond to those shown in Figure 1. Maximum object speeds of approximately 6cm/s were induced and successfully tracked. As can be seen in the right plot, a significant time delay is introduced due to the use of an ethernet link as a communication mechanism between the environment model that exists on the Silicon Graphics machine and the controller VME running Chimera. The link introduces a delay of approximately 0.6sec into the servoed modeling feedback loop. The two plots show that relatively fast tracking speeds can be achieved, but latency causes significant error between the magnitude of feature errors on the image plane for the live image and the corresponding expected feature error obtained from the internal visual representation of the task.

In Figure 9, feature trajectories are shown in which a supervisor guides the rotor into the stator after the rotor is grasped by a manipulator. The supervisor guides the rotor by clicking on its geometric model with a mouse and dragging the model in a plane within the visual representation of the environment. Visual servoing is used to maintain a temporal coherence between the environment model and the real world. Two feature states are shown. The solid line corresponds to the feature locations within the visual representation. The dashed line corresponds to the same feature in the live image. Again, the effects of latency due to the ethernet link are evident, but relatively fast tracking speeds are still obtained. The main limitation of the latency inherent in the ethernet link is that the speed at which moving objects can be grasped is reduced. We are currently investigating techniques for reducing this latency.

Finally, in Figure 10 feature templates from the rotor are shown. The left template is the selected feature, and the right template is the corresponding feature matched on the texture mapped environment model that contains information concerning the location of the feature on the object and possible occlusions.

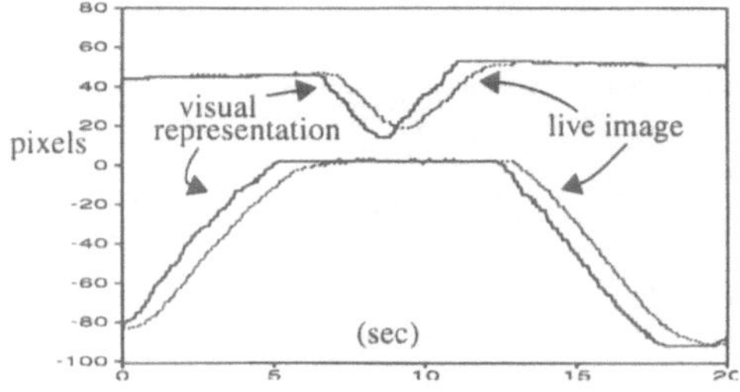

Figure 9. Feature trajectories during an insertion.

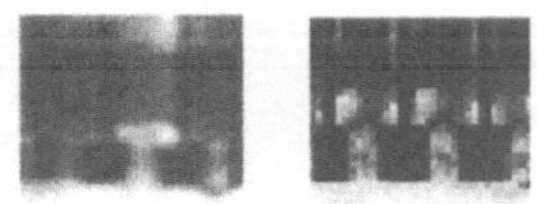

Figure 10. Example templates derived from a live image on the left and from the visual representation on the right.

7. Conclusion

A framework and task representation for integrating visually servoed manipulation strategies with sensor-based robotic systems has been described and initial experimental results presented. Our goal is to develop a framework that will allow the use of plug-and-play types of sensor-based manipulation components. A key component of the framework is the use of an expectation-based approach in which augmented texture mapped geometric models called object schemas are used to describe and reason about the environment. Object schemas are continuously updated based on sensor feedback to ensure that the current state of the environment model agrees with current sensory input. Visually servoed port-based agents are used to ensure that desired actions described within the environment model are carried out in the real world. Resolvability provides a shared ontology among various visual sensor configurations and allows the quick and easy incorporation of different visual sensor configurations into the servoed schema and visual servo control loops. Ongoing work aims to demonstrate system extendability by integrating other visual sensors into the system as well as feedback from other types of sensors such as force.

Acknowledgments

This research was supported in part by the U.S. Army Research Office through Grant Number DAAL03-91-G-0272 and by Sandia National Laboratories through Contract Number AC-3752D.

References

[1] L.E. Weiss, "Dynamic visual servo control of robots: an adaptive image-based approach," Ph.D Thesis CMU-RI-TR-84-16, Pittsburgh, PA:The Robotics Institute Carnegie Mellon University, 1984.

[2] B.J. Nelson and P.K. Khosla, "The Resolvability Ellipsoid for Visual Servoing," in *Proc. IEEE Conf. on Computer Vision and Pattern Recognition (CVPR94)*, pp. 829-832, 1994.

[3] B. Nelson and P.K. Khosla, "Integrating Sensor Placement and Visual Tracking Strategies," in *Experimental Robotics III: The Third International Symposium, Kyoto, October 28-30, 1993*, eds. T. Yoshikawa and F. Miyazaki, Springer-Verlag, London, pp. 169-181, 1994.

[4] N.P. Papanikolopoulos, Khosla, P.K. and Kanade, T., "Adaptive robotic visual tracking," *Proc. of the American Control Conference*. Evanston, IL.:American Automation Control Council, pp. 962-967, 1991.

[5] N.P. Papanikolopoulos, B. Nelson, and P.K. Khosla, "Full 3-d tracking using the controlled active vision paradigm," *Proc. 1992 IEEE Int. Symp. on Intelligent Control (ISIC-92)*, pp. 267-274, 1992.

[6] P. Anandan, Measuring visual motion from image sequences. Tech. Rept. COINS-TR-87-21, Amherst, Mass.:University of Massachusetts COINS Department, 1987.

[7] C. Tomasi and T. Kanade, "Detection and tracking of point features," Tech. Rept. CMU-CS-91-132. Pittsburgh:Carnegie Mellon University School of Computer Science, 1991.

[8] E.D. Dickmanns, "Expectation-based Dynamic Scene Understanding," in *Active Vision*, eds. A. Blake and A. Yuille, 303-335, The MIT Press, Cambridge, 1992.

[9] Y. Roth and R. Jain, "Verification versus Discovery in Vision-Based Systems," Technical Report CSE-TR-110-91, The University of Michigan, 1991.

[10] U. Neisser, *Cognition and Reality*, W.H. Freeman and Co., New York, 1976.

[11] R. Jain, "Environment Models and Information Assimilation," Technical Report RJ 6866(65692), IBM-Yorktown Heights, 1989.

[12] C. Fagerer, D. Dickmanns, and E.D. Dickmanns, "Visual Grasping with Long Delay Time of a Free Floating Object in Orbit," *Autonomous Robots*, 1(1):53-68, 1994.

[13] G. Hirzinger, "ROTEX-the first space robot technology experiment," *Experimental Robotics III: The Third Int. Symp., Kyoto, Japan, Oct. 28-30, 1993*, eds. T. Yoshikawa and F. Miyazaki, Springer-Verlag, pp.579-598, 1994.

[14] D.B. Stewart, D.E. Schmitz, and P.K. Khosla, "The Chimera II real-time operating system for advanced sensor-based control systems," *IEEE Trans. Sys., Man Cyber.* 22(6):1282-1295., 1992.

Experiments in Hand-Eye Coordination Using Active Vision

Won Hong
Massachusetts Institute of Technology
Cambridge, MA, USA
jesse@ai.mit.edu

Jean-Jacques E. Slotine
Massachusetts Institute of Technology
Cambridge, MA, USA
jjs@mit.edu

Abstract

Robot hand-eye coordination has recently enjoyed much attention. Previous research at MIT has examined combining vision and manipulation applied to the task of tracking and catching tossed balls in controlled environments. Building upon the foundations of this past research, this paper presents work which incorporates a new *active* vision system which requires a minimally controlled environment, and implements methods for object tracking, robot/camera calibration, and new catching algorithms. Experimental results for real time catching of free-flying spherical balls are presented. The system was tested on under-hand tosses from random locations approximately 1.5-2.5 meters distant from the base of the arm. The best performance results were found to be 70-80% success for similar tosses.

1. Introduction

This paper presents work on the application of a manipulator arm (the Whole Arm Manipulator (WAM) [1] [2] [3]) and an active vision system (the Fast Eye Gimbals (FEGs) [4]) to the task of real time catching of free flying objects. The purpose of this research is to investigate the challenges involved in achieving successful catching and also to lay the foundations for further study into hand-eye coordination and robot learning with this system.

Although there is little current research in the area of robotic catching, there is a wealth of research in the general area of applying manipulators and vision systems to accomplishing dynamic tasks. Good examples of hand-eye coordination systems are the robot ping pong players [5] [6] [7]. These have exceptional vision systems combined with fast manipulators, but are constrained to work in heavily controlled environments. Previous work has also studied robot juggling [8] [9].

As for active vision, a good survey and definition of future research directions has come from the NSF Active Vision Workshop, 1991 [10]. More recently, there have been some portable vision systems designed to study topics in active vision [11] [12] [13]. In contrast with these systems, our system uses two independent two degree of freedom actuators in combination to achieve stereo vision. Our system also utilizes a very simplified approach to vision processing to achieve fast 60 Hz information, with the disadvantage that it is not as general purpose as the vision processing on these other systems.

Past research at our site has examined robotic catching. Successful catching results were first presented using the WAM combined with two stationary black and white cameras [14]. Adaptive visual tracking algorithms were later added which provided cleaner vision information and improved catching reliability [15] [16].

Our current system uses the WAM once again, combined with new vision hardware. The new vision system is comprised of two color CCD cameras, each mounted on two DOF gimbals. Vision processing is accomplished through fast color keyed blob detection. This method of vision processing simplifies a number of complex tasks such as target detection, feature extraction, and target/background separation. As the system becomes more mature and requires more detailed information, we will incorporate general purpose vision processing hardware to perform concurrent processing tasks.

Additional changes to the system consist of a new "hand" and new algorithms. A three DOF end effector replaces the previous pneumatic on/off gripper. Algorithms for incorporation of the active vision system have been developed. And methods for catch point determination and manipulator path generation have been implemented.

The catching task to which we apply the current system consists of a human participant tossing a specially colored ball at random locations 1.5-2.5 meters distant from the base of the WAM. The ball is tossed under-hand and the time of travel from the tosser's hand to the time of interception is approximately 0.5 seconds. All computations are done in real time during the duration of the toss.

The remainder of this paper is organized as follows. Section 2 describes the vision system and algorithms for object tracking and cross calibration. Section 3 presents the object path prediction, catch point determination, and WAM path generation techniques used in catching. Some experimental results are presented in Section 4. Section 5 presents some recent results on extending our system to use wavelet network based path prediction. Section 6 offers brief closing remarks.

2. Vision System

We begin by presenting a description of the new active vision system and its vision processing hardware. Basic methods for utilizing the vision system to locate and track objects are described, followed by a method for cross calibration between the manipulator and vision system.

2.1. Description

The vision system is comprised of a pair of two DOF gimbals (FEGs) mounted to a ceiling rafter (Figure 1). The FEGs have $\pm$ 90 degree and $\pm$ 45 degree ranges of motion on the tilt and pan axes respectively. The camera lenses have approximately a $\pm$ 5 degree field of view for higher resolution, better focusing of attention, and less distortion. The FEGs are controlled using PD, PID, and adaptive/feedforward [18] controllers depending upon the nature of the trajectory they are commanded.

The camera output from the FEGs are directed to simple blob detector vision boards [19]. The vision boards initially convert the color camera image into a binary

Figure 1. A picture of the FEGs mounted to a ceiling rafter.

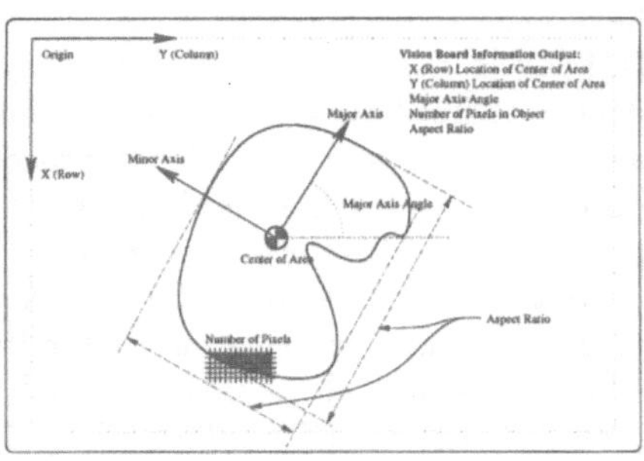

Figure 2. Image showing the types of output from the blob detector boards.

image through comparison with a color histogram (which is re-trainable). Subsequent processing on the binary image yields the center of area, number of pixels, major axis angle, and aspect ratio of the largest "blob" in the current image (Figure 2). The boards output information at 60 Hz by using the interlaced NTSC output from the cameras. Using this low-cost simplified approach to vision allows for fast information extraction and requires a minimally controlled environment, with simple marking of targets of interest.

2.2. Object Location and Tracking

Due to the small field of view of the lenses, the FEGs must execute a search to locate objects of interest. Once an object is seen by both cameras, the coordinates of the object can be determined by solving the triangulation problem shown in Figure 3. The cameras are separated by a baseline distance of $2D$ (0.8255 meters in our case). The origin of the camera coordinate frame is located at the midpoint of the baseline. The location of the object in the camera coordinate frame is then

$$x = \frac{D(\tan\theta_2 + \tan\theta_4)}{\tan\theta_4 - \tan\theta_2} \quad y = \frac{2D\cos\theta_2\cos\theta_4}{\sin(\theta_2 - \theta_4)}\cos\theta_1 \quad z = \frac{2D\cos\theta_2\cos\theta_4}{\sin(\theta_2 - \theta_4)}\sin\theta_1 \quad (1)$$

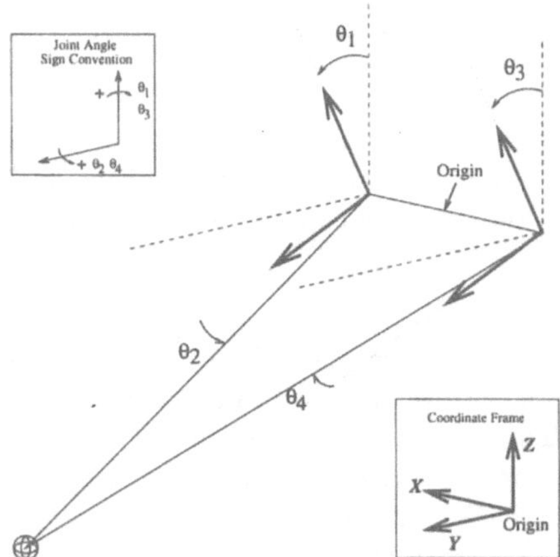

Figure 3. Triangulation is used to determine the coordinates of the object.

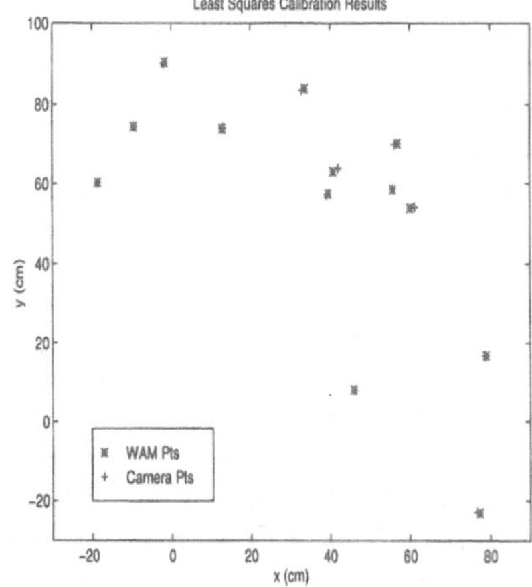

Figure 4. Results from least squares cross calibration.

In addition to locating static objects, the cameras are capable of tracking objects moving up to 1.3 - 1.4 radians/second. The tracking is accomplished through state estimation and path prediction. The information which we receive from the vision boards is delayed by approximately 0.350 seconds due to image acquisition and processing. The tracking algorithm estimates the velocity and acceleration of the object at this previous time using time-varying first order filters and weighted sums based on prediction error. These estimates are then used to integrate over the time delay assuming constant acceleration to obtain predictions for the current position and velocity of the object. These values are then fed to the FEG adaptive/feedforward controller [17].

It should be noted that there is a trade-off between control and vision performance. The large focal length of the lenses yields higher resolution, lower distortion, and focuses attention better. But it limits the fastest tracking capabilities of the system based on control and actuator performance and necessitates the use of search routines to initially locate objects.

2.3. Cross Calibration

The cameras and the manipulator each have their own coordinate frames. In order to relate the two, a transformation between the coordinates frames is required. We simplify the calibration problem here by taking advantage of the physical placement of the cameras and the WAM. The z axes for both coordinate frames are parallel. This leaves one rotational degree of freedom and a translation to be determined.

A ball is placed at the end of the WAM and a number of data points are collected as seen by the cameras and as measured by the WAM. The translation is found by determining the vector which overlaps the centroids of the two sets of data. The angle of rotation is found by using a least squares approach [20]. Referencing each data point relative to its centroid, the problem becomes a minimization of

$$\sum_{i=1}^{n}(\delta x_i^2 + \delta y_i^2) \quad \text{where} \quad \begin{bmatrix} \delta x_i \\ \delta y_i \end{bmatrix} = \begin{bmatrix} x_i' \\ y_i' \end{bmatrix} - \begin{bmatrix} \cos\theta & \sin\theta \\ -\sin\theta & \cos\theta \end{bmatrix} \begin{bmatrix} x_i \\ y_i \end{bmatrix} \tag{2}$$

The resulting angle of rotation for the rotation matrix is then given by

$$\tan\theta = \frac{\sum \mathbf{r}_i' \times \mathbf{r}_i \cdot \hat{\mathbf{z}}}{\sum \mathbf{r}_i' \cdot \mathbf{r}_i} \tag{3}$$

where $\mathbf{r}_i = (x_i, y_i, 0)^T$ corresponds to data points from the camera and $\mathbf{r}_i' = (x_i', y_i', 0)^T$ corresponds to data points from the WAM.

Figure 4 shows the results from a calibration run. The sum of the square error was 6.55 cm^2 over 13 points, with an average error of 0.7 cm. These errors result from small inaccuracies in mounting of the cameras and also from partial obstruction of the ball or from lighting effects on color matching by the vision boards.

3. Catching Algorithms

We now present methods which are used specifically for catching. These methods rely upon the object tracking and cross calibration discussed in the previous section. The flow of this section follows the sequence of events which occur during catching. First, we present toss triggering and object path prediction methods. Next, we present the methods for catch point selection. And finally, we present our method for generating an intercepting path for the WAM.

3.1. Toss Triggering and Object Path Prediction

Detection of the start of the ball toss is required to begin the catching process. We use a radial distance trigger to determine when the ball has been tossed. The trigger is set to be a constant radial distance from the lowest point which the ball has reached prior to triggering. This attempts to compensate for differences in back-swings among various people.

Once the toss has been triggered, data storage, fitting and prediction begins. The future path of the tossed object is predicted using recursive least squares techniques [16] assuming a parabolic model for the trajectory. The algorithm is recursive, therefore the computation required for each new data point is independent of the number of data points already collected. Similarly, each data point is weighted equally, the last having as much effect as the first. The least squares fit yields parabolic constants which allow us to compute the future path of the object.

3.2. Catch Time/Point Determination

Using the predicted parabolic constants, a satisfactory catch time/point must be determined. The process begins with an initial prospective catch point as the closest point on the object path to the base of the WAM. By using this point rather than the point closest to the end of the WAM we allow for greater room for the arm to accelerate. This initial point is then checked against the workspace constraints shown in Figure 5 and Figure 6 and modified as necessary. The minimum radius and the minimum z height constraints are offset from the physical limits in order to allow for sufficient room for post-catching deceleration. If the catch point cannot be made to satisfy the constraints, or the catch time moves below 0.3 seconds, then the ball is deemed uncatchable.

The catch time/point is updated with each change in predicted parabolic constants (approximately 60 Hz). As a result, the catch time/point vary as the toss progresses. But once the command to close the hand has been given, the catch time must become fixed to assure that the ball and hand are coincident at the catch time. Therefore, for the 0.33 seconds required to close the hand, the catch time is held constant.

3.3. WAM Path Generation

Once a satisfactory catch point is determined, the WAM attempts to intercept and match position and velocity with the object. Third order polynomials in x, y, and

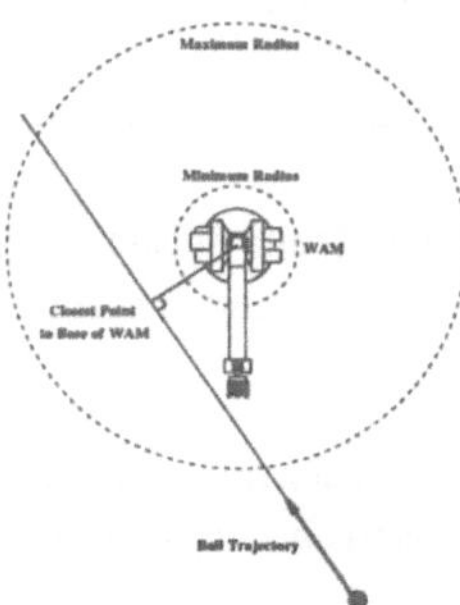

Figure 5. Safety constraints for the WAM based on radial distance.

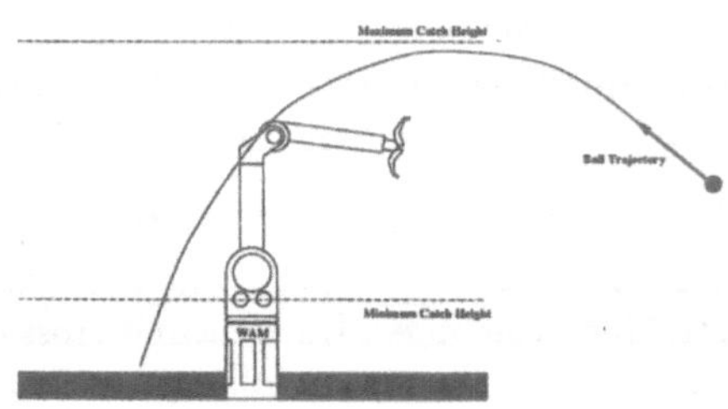

Figure 6. Safety constraints for the WAM based on z height.

z are used to generate an intercepting path for the WAM [15].

$$\mathbf{p}(t) = \frac{1}{6}\mathbf{jp}t^3 + \frac{1}{2}\mathbf{ap}t^2 + \mathbf{vp}t + \mathbf{pp} \tag{4}$$

The constants in the polynomial path are determined each servo loop from knowledge of the current position and velocity of the WAM ($\mathbf{p}(t_1)$ and $\mathbf{v}(t_1)$) and the prospective catch point position and velocity ($\mathbf{p}(t_2)$ and $\mathbf{v}(t_2)$).

$$\mathbf{jp} = \frac{-6}{(t_2-t_1)^2}\left\{\frac{2}{(t_2-t_1)}(\mathbf{p}(t_2)-\mathbf{p}(t_1)) - \alpha\mathbf{v}(t_2) - \mathbf{v}(t_1)\right\} \tag{5}$$

$$\mathbf{ap} = \frac{2}{t_2-t_1}\left\{-\frac{1}{6}\mathbf{jp}(t_2^2+t_1t_2-2t_1^2) + \frac{1}{t_2-t_1}(\mathbf{p}(t_2)-\mathbf{p}(t_1)-\mathbf{v}(t_1)\right\} \tag{6}$$

$$\mathbf{vp} = \mathbf{v}(t_1) - \mathbf{ap}t_1 - \frac{1}{2}\mathbf{jp}t_1^2 \tag{7}$$

$$\mathbf{pp} = \mathbf{p}(t_1) - \mathbf{vp}t_1 - \frac{1}{2}\mathbf{ap}t_1^2 - \frac{1}{6}\mathbf{jp}t_1^3 \tag{8}$$

The α term in the first equation is added to compensate for instances when the polynomial path approaches the outer workspace limit of the WAM. The initial value of α is 1.0. If the path exceeds the outer limit, α is reduced, thereby scaling down the velocity to be matched, which reduces the "out-swing" of the generated path. This resulting polynomial path is then fed to the WAM controller which utilizes adaptive control to achieve high performance tracking [21].

After the WAM has reached the catch point, the WAM matches trajectories with the object for a short duration of time and then begins deceleration along the previous path of the object. The path matching and deceleration ensure smooth graceful catching and allow for a greater tolerance for timing errors.

In addition to path generation for the end point of the WAM, the forearm of the end-effector is oriented such that the fingers close in a direction which is perpendicular to the path of the object. This prevents the ball from impacting the outside of the fingers as they attempt to close.

4. Experimental Results

4.1. Experimental Setup

The system is controlled via a multiprocessor VMEbus system. Figure 7 shows the elements of our system and the boards which comprise our VMEbus. Our system contains five 68040 boards, a dual DSP board, and additional I/O boards. The vision boards are separate from the VMEbus and communicate via serial lines at 38400 bps.

Figure 8 shows the configuration used in catching. The FEGs are mounted on a ceiling rafter approximately 2.1 meters from the ground and 1 meter behind the WAM. At full extension, with the current end-effector, the WAM is approximately 1.05 meters in length from the center of the shoulder joint.

4.2. Catching Results

The following experimental results were obtained without special calibration or tuning required. Although one time training of the vision boards and FEG/WAM cross calibration are necessary, these calibrations need not be repeated each time the system is run.

Figures 9, 10, and 11 show results from a successful catch. Figure 9 shows plots of x, y, and z versus time. The solid line is the ball data, the dash-dot line is the

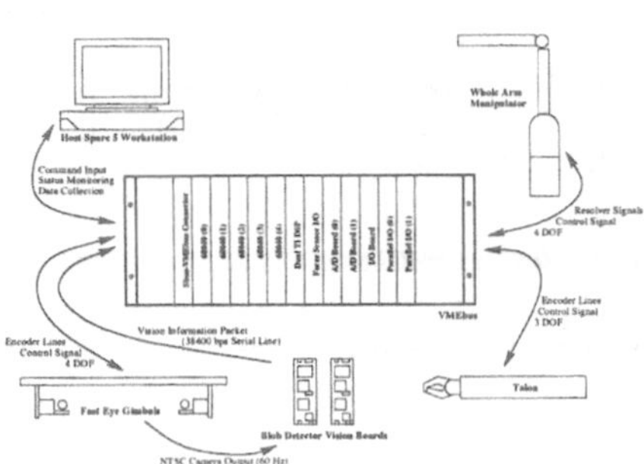

Figure 7. Overall system and connections to the VMEbus.

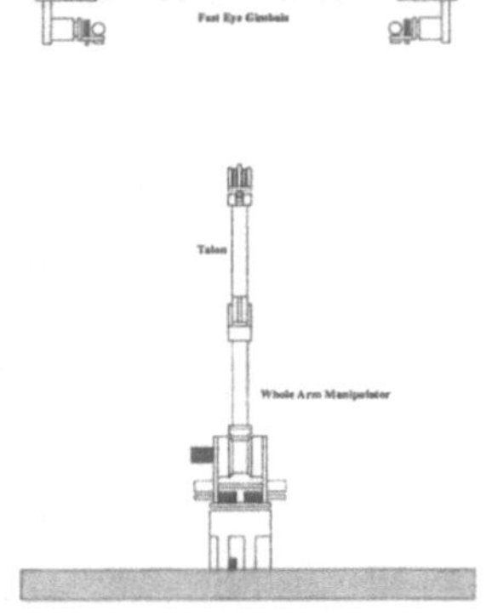

Figure 8. The system configuration for catching.

recursive least squares fit, the dashed line is the desired path for the WAM hand, and the dotted line is the actual path of the WAM. At the catch time (0.482 seconds) and for a short duration after, all four lines can be seen to overlap as a result of the intercepting/matching trajectory of the WAM. Figure 10 shows the evolution of the prospective catch time/point versus the time of the toss. The catch time is fixed once the close command has been given. The lower plot shows how the catch point converges as the toss progresses. Figure 11 shows a sequence of images for a sample catching run.

Testing was done to measure the repeatability of the system. Tosses were made to the same general area of the WAM's workspace, with all of the final catch points lying within a two foot cubic region of space. The success rate was found to be roughly 70% - 80%. One testing run showed 53 successful catches from a set of 75 attempts, with a sequence of 14 successful catches.

In failed attempts, the ball impacted the end effector, exhibiting errors of approximately 1.25 cm. In the best case, the hand can afford approximately 1.25 cm position error and less than 0.005 seconds timing error and still succeed in grasping the object. The majority of failures can be attributed to noisy data from the vision system which results in more error in the prediction. The noisy data results from

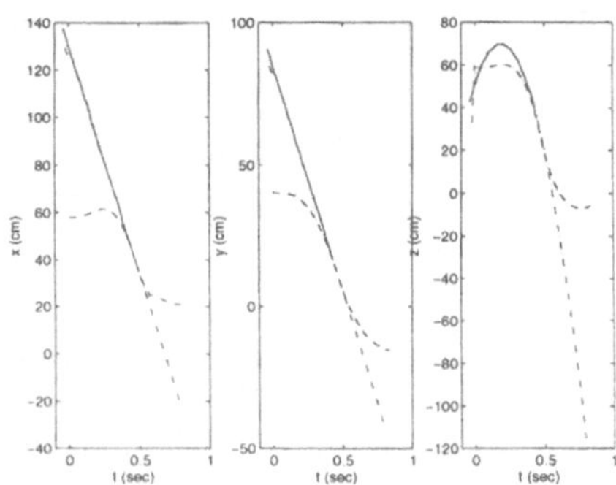

Figure 9. Plots of xyz vs. time.

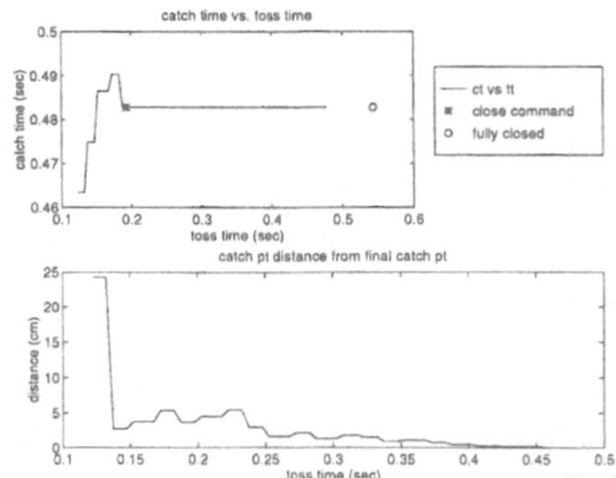

Figure 10. Plots of catch time/point versus toss time.

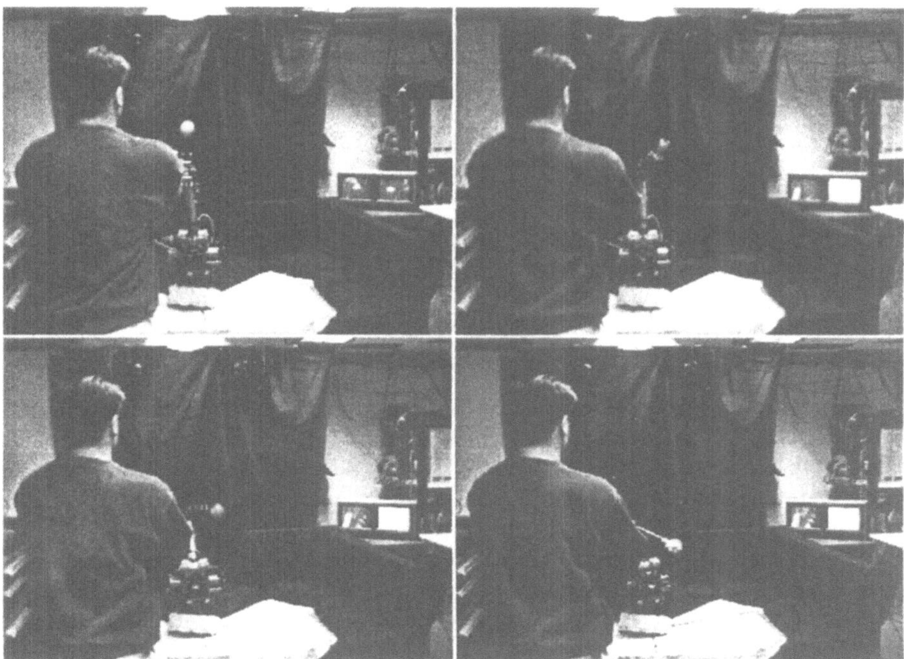

Figure 11. A sequence of images for a sample catching run.

inaccurate compensation for time delays and adverse lighting effects. It is difficult to accurately compensate for the latency in the vision system since it is not constant. Each vision board outputs information once it has finished computation on a frame. Depending upon the number of pixels, blobs, and edges in the image, the timing of completion will vary. In addition, non-uniform lighting effects also contribute significantly to the noise. As the ball travels towards the WAM, it passes under lighting fixtures in the room and over different backgrounds. In the bright regions, the light reflecting off the surface of the ball cause white spots in the image which do not register as the appropriate color from training. In dark regions, there are more shadows and again less of the ball is seen.

5. Current Research

In addition to successful catching, this work has laid the foundations for further research with this system for coordination tasks. Current research is being done to examine catching of objects with different aerodynamic characteristics. Objects of interest are light-weight foam balls, paper airplanes, and other additional items with non-parabolic trajectories. The current model based least squares prediction methods for the tossed object are replaced by wavelet network based prediction methods [22] [23].

Interfacing this with the catching algorithms used here have yielded good preliminary results. Figure 12 shows 3D plots of successful catches using the network based path prediction method. The figure shows results for a regular ball, a light sponge ball, and a paper airplane. The circles show the actual path of the object, the asterisks show a calculated path with an assumption of only gravitational

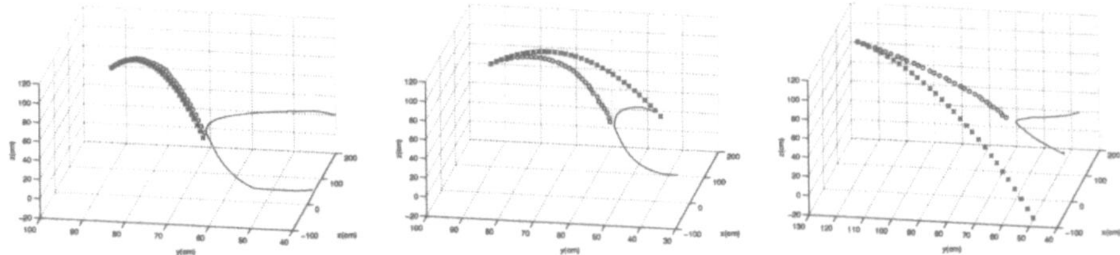

Figure 12. Object and WAM paths for catching a regular heavy ball, a light sponge ball, and a paper airplane (left to right).

acceleration, and the solid line shows the path of the WAM hand.

6. Conclusion

This paper has presented some of the methods and results for our experiments in real-time robotic catching of free flying objects. We have used model based prediction methods to predict the path of a tossed ball combined with third order polynomials to generate an intercepting path for the WAM. Success rates of up to 70% - 80% have been obtained.

Future work will attempt to incorporate concurrent general purpose vision processing for more complex vision tasks, attempt catching of more difficult objects such as rotating cylinders, and incorporate learning to increase system performance over repeated trials.

Acknowledgments

The authors would like to thank Ken Salisbury for stimulating discussions and for leading the development of the experimental hardware. This research also greatly benefited from interactions with Günter Niemeyer, Ichiro Watanabe, Akhil Madhani, and Daniel Theobald. This paper describes research done at the Nonlinear Systems Laboratory and the Artificial Intelligence Laboratory at the Massachusetts Institute of Technology. This research was supported in part by NASA/JPL Contract No: 959774, "Vision and Touch Guided Grasping of Stationary and Moving Objects", by grants from Fujitsu and Furukawa Electric, and by the Starr Foundation. Development of the WAM was sponsored in part by the Office of Naval Research, University Initiative Program Grant N00014-92-J-1814. Development of the FEG camera actuators was sponsored in part by a grant from the Sloan Foundation.

References

[1] J. K. Salisbury, *Whole Arm Manipulation*, Proc. 4th Int. Symposium on Robotics Research, Santa Cruz, CA, August, 1987.

[2] J. K. Salisbury, W. T. Townsend, B. S. Eberman, D. M. DiPietro, *Preliminary Design of a Whole-Arm Manipulation System (WAM)*, Proc. 1988 IEEE Int. Conf. on Robotics and Automation, Philadelphia, PA, April 1988.

[3] W. T. Townsend, *The Effect of Transmission Design on Force-Controlled Manipulator Performance*, PhD Thesis, Department of Mechanical Engineering, MIT, April 1988. (See also MIT AI Lab Technical Report 1054).

[4] N. Swarup, *Design and Control of a Two-Axis Gimbal System for Use in Active Vision*, S.B. Thesis, Department of Mechanical Engineering, MIT, Cambridge, MA, 1993.

[5] J. Billingsley, *Robot ping pong*, Practical Computing, May 1983.

[6] H. Fässler, H. A. Beyer, and J. Wen, *A robot ping pong player: optimized mechanics, high performance 3D vision, and intelligent sensor control*, Robotersysteme 6, Springer-Verlag, pp. 161-170, 1990.

[7] R. L. Andersson, *A robot ping-pong player: Experiment in real time control*, MIT Press, Cambridge, MA, 1987.

[8] E. Aboaf, S. Drucker, and C. Atkeson, *Task-level robot learning: juggling a tennis ball more accurately*, Proc. IEEE Int. Conf. on Robotics and Automation, pp. 1290-1295, Scottsdale, AZ, May 1989.

[9] A. A. Rizzi and D. E. Koditschek, *Further progress in robot juggling: Solvable mirror laws*, Int. Conf. on Robotics and Automation, pp. 2935-2940, 1994.

[10] *Promising Directions in Active Vision*, NSF Active Vision Workshop, University of Chicago, August 5-7, 1991, in Int. Journal of Computer Vision, Vol. 11:2, pp. 109-126, 1993.

[11] J. C. Fiala, R. Lumia, K. J. Roberts, and A. J. Wavering, National Institute of Standards and Technology, *TRICLOPs: A Tool for Studying Active Vision*, Int. Journal of Computer Vision, Vol. 12:2/3, pp. 231-250, 1994.

[12] A. J. Wavering and R. Lumia, *Predictive Visual Tracking*, SPIE Volume 2056, Intelligent Robots and Computer Vision XII, 1993.

[13] H. K. Nishihara, H. J. Thomas, *Real-Time Tracking of People Using Stereo and Motion*, IS&T/SPIE Symposium on Electronic Imaging: Science & Technology, San Jose, California, February 6-10, 1994.

[14] B. M. Hove and J.-J. E. Slotine, *Experiments in Robotic Catching*, Proc. of the 1991 American Control Conf. Vol. 1, Boston, MA, pp. 380-385, June 1991.

[15] H. Kimura, N. Mukai, and J.-J. E. Slotine, *Adaptive Visual Tracking and Gaussian Network Algorithms for Robotic Catching*, DSC-Vol. 43, Advances in Robust and Nonlinear Control Systems, Winter Annual Meeting of the ASME, Anaheim, CA, pp. 67-74, November 1992.

[16] H. Kimura, *Adaptive Visual Tracking Algorithms for 3-D Robotic Catching*, M.S. Thesis, Department of Mechanical Engineering, MIT, Cambridge, MA, August 1992.

[17] W. Hong, *Robotic Catching and Manipulation Using Active Vision*, M.S. Thesis, Department of Mechanical Engineering, MIT, Cambridge, MA, August 1995.

[18] J.-J. E. Slotine and W. Li, *Applied Nonlinear Control*, Prentice Hall, Englewood Cliffs, New Jersey, 1991.

[19] A. Wright, *A high speed low-latency portable vision sensing system*, SPIE, September 1993.

[20] B. K. P. Horn, *Robot Vision*, MIT Press, Cambridge, MA, 1986.

[21] G. Niemeyer and J.-J. E. Slotine, *Performance in Adaptive Manipulator Control*, December 1988, Int. Journal of Robotics Research, 10(2).

[22] M. Cannon and J.-J.E. Slotine, *Space Frequency Localized Basis Function Networks for Nonlinear System Estimation and Control*, Neurocomputing, 9(3), 1995.

[23] I. Watanabe and J.-J.E. Slotine, *Stable real-time prediction of the trajectories of light objects in air using wavelet networks*, MIT-NSL 100195, 1995.

Visual Positioning and Docking of Non-holonomic Vehicles

Nils A. Andersen, Lars Henriksen and Ole Ravn

Institute of Automation, Technical University of Denmark,
Building 326, DK-2800 Lyngby, Denmark

or@iau.dtu.dk

Abstract

The paper describes the benefits and drawbacks of different strategies for docking and positioning of autonomous vehicles based on visual feedback. Three algorithms are described in detail, and extensive experiments with an implementation of the algorithms on our test-bed Autonomous Guided Vehicle are documented.

1. Introduction

Mobile platforms play a central role in automation. An important factor regarding systems including mobile platforms is the positioning of the vehicles. Here focus is put on end precision, speed, need for sensory information, and need for physical space to perform the manoeuvres. Especially precision of the positioning is interesting as it enables manipulators to grab an object for further processing.

The precision is a result of both sensory processing, actuators and the control law used. In this paper focus is put on three distinct control strategies. The vehicle is non-holonomic which adds to the complexity of the control. The sensing is established through vision because of the combination of flexibility and precision.

The three algorithms are implemented and tested for a docking situation. This is done for three reasons:

- The docking is interesting because it requires significant precision,
- docking enables restoration of energy resources and contact to high bandwidth communication and
- finally docking can yield a physical fix.

This enables the vehicle to act through dead reckoning in an extended area. Energy restoration and communications greatly enhances system autonomy.

2. Positioning of non-holonomic Vehicles

It is well known that accurate positioning of non-holonomic vehicles such as a typical AGV with two driven wheels and a castor wheel is difficult. Samson shows that there exists no stabilizing C^1 feedback control for such a vehicle [Samson and Ait-Abderrahim, 1991]. These control problems have led some researchers to advocate holonomic vehicle designs but as these designs are more complex and therefore more expensive it may be disadvantageous to use them unless the application really need the full manoeuvrability.

During the last years several methods for positioning of non-holonomic vehicles have been proposed. A comparative study of three different methods based on ideas reported in the literature is presented.

- Non-linear feedback from the distance to goal, the angle between vehicle and direction to goal, and the angle between direction of goal and desired vehicle direction. [Aicardi et al., 1995]
- Multi-objective control using weighting-functions. The three objectives are: going to the line defined by the goal and the desired direction, aligning the vehicle with the desired direction and stopping the vehicle at the right position. [Steer and Larcombe, 1991]
- Feedback tracking of a reference cart moving to the goal. [Samson and Ait-Abderrahim, 1991]

2.1. Central properties of the navigation algorithms

Below are outlined the main ideas behind for the three navigation algorithms for non-holonomic vehicles. An summary is given in Table 1. See the references for more detailed information and equations.

2.1.1 [Aicardi et al., 1995]

[Aicardi et al., 1995] proposes a set of control equations for linear and angular speed references which can be proved to be Lyapunov stable when not considering vehicle dynamics. The effect of including vehicle dynamics has not been described but instability has not been a problem with the vehicle used for the tests. The position of the vehicle converges exponentially, why the goal of the navigation must be set in front of the desired physical goal if the goal is to be reached in finite time. The algorithm converges smoothly near the goal but can be corrupted by noise in the position and orientation estimates. The algorithm is tuned with three parameters. This is relatively easy especially as formulas for the interrelation between the parameters are given. If these are fulfilled it is guarantee that the vehicle approaches the target frame by asymptotically proceeding along the rectilinear path aligned with the vehicle itself. An approximate value of the maximum of the speed of approach can be specified explicitly.

2.1.2 [Steer and Larcombe, 1991]

The navigation by [Steer and Larcombe, 1991] is performed by a fuzzy controller using three fuzzy sets with corresponding control equations for generation of linear and angular speed references:

- 'Seek-back-goal', is active when far from the goal. It leads the vehicle to towards the back-goal which is a point located at some distance behind the goal.
- 'Seek-to-goal-line' is active near the goal and places the vehicle on the line going through the goal, having the goal orientation.
- 'Align-with-goal-line' is also active near the goal and aligns the vehicle with the goal orientation.

The different control equations are rather simple and may easily be improved (in the implementation tested here a few inconveniences have been removed to yield stability). The method is more a concept than a set of equations. This has both advantages and drawbacks. One drawback is that there are generally a large number of parameters to tune making it difficult and time consuming to tune intuitively even when using simulations. Tuning experiments have shown that it can be quite difficult to obtain a set of parameters that fit a large number of starting positions and headings. On the other hand the flexibility of the basic fuzzy concept that the algorithm is based upon is intuitively easy to understand and hence to apply to fairly different applications as for instance object avoidance. The linear speed is constant except for change of sign and can be specified implicitly. Near the goal additional control equations are needed as the main equations are not designed for smooth convergence near the goal.

2.1.3 [Samson and Ait-Abderrahim, 1991]

The navigation method designed by [Samson and Ait-Abderrahim, 1991] controls a point in front of the vehicle to reach the corresponding point on a reference cart. It is shown that the reference cart needs to be moving to guarantee stability. Hence the feedback is a trajectory tracking. In the current navigation experiments the reference cart traverses along a straight line leading to the fixed goal used by the other two navigation algorithms. The velocity of the reference cart is constant until a certain distance behind the goal where the velocity is ramped down to a low speed until the goal is reached. The stability when adding vehicle dynamic is not described. Simulations show however that for the current vehicle stability does not seem to be a problem. For tuning of the algorithm there are four parameters which are intuitively closely coupled to the behaviour of the AGV and they are relatively easy to tune. The maximum of speed of approach can be difficult to control.

	[Aicardi et al]	**[Steer & Larcombe]**	**[Samson et al]**
Concept	Exponential convergence	Fuzzy	Trajectory tracking
Parameters, tuning of	3, easy	7, quite difficult	5, easy
Inconveniences	Slow convergence	Need additional control near goal. Quite difficult to tune	Goal must be moving to ensure stability
Advantages	Good performance near goal in noiseless case	Adaptable to a wide set of problems	Quite fast
Computational load	Low 15 flops 4 math	High 62 flops 7 math	Medium 26 flops 4 math

Table 1. Some central features of the navigation algorithms. (math includes trigonometric and square root functions).

3. Simulations

The initial implementation and testing of the navigation algorithms were performed in an AGV simulator developed in MATLAB/Simulink at IAU. The parameters of the different algorithms were tuned to the desired performance and tests not possible in real-life application such as the noiseless case were performed. It is to be kept in mind that the parameters of the algorithms may not be optimal why the full potential may not be revealed. The conditions for the navigation: A speed of approach was approx. 0.2-0.4 m/s. The target position of the AGV was (x,y)=(0, 0.55) m with a heading of 0 degrees.

3.1. Performance with known position and orientation

The parameters of the three algorithms have been tuned so the algorithms works in a large area. This is done in order to provoke the algorithms to reveal their possible weaknesses. One difficulty is the starting point. The AGV is placed well off the docking line and with a heading of zero degrees i.e not heading towards the dock. Also the AGV is given a relatively short distance for manoeuvring before the goal is reached. The AGV position and orientation is in this way (x,y,θ) = (-0.15 m, 1.7 m, 0 deg.). Results from these simulations are shown in table 2. Results from simulations with an even more difficult case are also shown in table 2. Here the AGV is started from (x,y,θ) = (-0.25 m, 1.70 m, 0 deg.) necessitating more vigorous control.

The position and orientation of the AGV in the real-life experiments is extracted using a camera mounted on the AGV viewing two guidemarks placed a little behind the goal. Under the difficult starting position a larger angle between the optical axis of the camera and line-of-sight to the guide marks occur. When using a fixed camera this will put the visual contact with the guide marks and thereby the on-line estimation of position and orientation at risk. The parameters of the navigation algorithms have been tuned to comply with a visual constraint of 22 degrees between optical axis and line-of-sight which is the critical angle of the camera. If the pan/tilt unit is active or the estimations of position and orientation is obtained in other ways for example through dead-reckoning the algorithms can be tuned to reach the desired position and orientation or trajectory at shorter distance than shown in the simulations presented here.

From Table 2 it is seen that the goal is reached with quite different precision.

- [Aicardi et al., 1995] is indisputably the most accurate in both position and heading while also consuming the most time.
- [Samson and Ait-Abderrahim, 1991] is also quite accurate with a precision of a few mm's and degrees with a relatively low time consumption.
- [Steer & Larcombe] is also quite fast but only able to place the vehicle with a few cm's precision.

It is also seen that the end precision of the algorithms by [Aicardi et al., 1995] and [Samson and Ait-Abderrahim, 1991] is only affected a little by the increased difficulty of starting in (x,y,θ) = (-0.25 m, 1.70 m, 0 deg.). Moreover they are able to retain the visual contact with the guidemarks. The error of the end heading on [Samson and Ait-Abderrahim, 1991] is doubled though, but still small. The performance of [Steer and Larcombe, 1991] is somewhat worse. End precision in x is halfed and the visual contact is not maintained.

	[Aicardi et al.]	**[Steer & Larcombe]**	**[Samson et al.]**
End precision ($\Delta x, \Delta y, \Delta\theta$) [mm, mm, °].	[0.01,-0.4,-0.02] {0.01, 0.0,-0.01}	[18.2, -7.3, 0.94] {38.5, -7.7,1.65}	[1.89, -5.8, 0.26] {1.65, -5.7, 0.49}
Time consumption: [s]	[18.16] {18.32}	[11.0] {11.0}	[8.24] {8.24}
$\left(\int v, \int \lvert\omega\rvert\right)$ m, degrees	[1.163, 26.48] {1.187, 43.34}	[1.176, 34.57] {1.212, 57.61}	[1.169, 27.62] {1.19, 43.0}
$max(v, \lvert\omega\rvert)$ m/s, degrees/s	[0.212, 9.29] {0.2116, 14.83}	[0.2, 8.72] {0.2, 14.12}	[0.289, 14.63] {0.313, 20.1}

Table 2. Simulation results for starting points: []: (x, y, θ) =(-0.15 m, 1.7 m, 0 deg.)
{}: (x, y, θ) =(-0.25 m, 1.7 m, 0 deg.)

3.2. Vision estimation of position and orientation

For the estimation of the position and the orientation of the AGV from video images a number of different algorithms have been developed. Some of these are exact solutions to the problem but they have shown to be quite sensitive to uncertainties in the measurement of the physical location of the guidemarks. Therefore we have developed a set of approximative formulas designed for the insensitivity to these uncertainties. The introduction of approximations do however introduce errors in the position and orientation estimates but it has been possible to keep these below a few cm and degrees in the working range of the AGV. The vision algorithms have also been designed to comply with uncertainty on the pixel locations of the image of the guide marks.

3.3. Performance with vision estimation

In order to assess the influence of both the approximations in the vision routines and noise on pixel locations a number of simulations have been performed. These show that the approximations in the vision routines do not affect the navigation to a very large extend, typically below 0.5 mm and 0.5 degrees at end location. Figure 1 shows the simulated track of the AGV in this case.

Also all three algorithms seems relatively insensitive to noise corruption of the pixel locations. Both the trajectories and the end precision are affected only little during the ride. When regarding the control effort shown on figure 2 it shows that the linear and angular speed references are affected at different stages of the navigation by the noise. The algorithm by [Aicardi et al., 1995] is most sensitive at the end of the navigation while the algorithms by [Samson and Ait-Abderrahim, 1991] and [Steer and Larcombe, 1991] are sensitive in the beginning. The statistical results from 50 simulations are shown in table 3.

Mean and standard deviation	**[Aicardi et al.]** [μ], (σ)	**[Steer & Larcombe]** [μ], (σ)	**[Samson et al.]** [μ], (σ)
End precision: $(\Delta x, \Delta y, \Delta\theta)$ [mm, mm, °]	0.018, 0.13 -0.17, 0.087 0.032,0.89	14.68, 9.09 -7.95, 0.70 0.82, 0.20	1.35, 1.05 -6.85, 0.73 0.17, 0.094
Time consumption [seconds]	20.0, 0.11	11.1, 0.23	8.95, 0.21
$\left(\int v, \int \lvert\omega\rvert \right)$ [m, °]	1.16, 0.026 30.2, 5.7	1.17, 0.027 34.8, 6.08	1.17, 0.028 28.1, 5.24
$max\,(v, \lvert\omega\rvert)$ [m/s, °/s]	0.212, 0.0048 9.12, 2.21	0.20, 0.00 8.99, 1.53	0.291, 0.0086 14.7, 2.55

Table 3. Simulations for 50 starting points randomly distributed: $(x,y,\theta) = ([-0.2; -0.1]$ m, $[1.65; 1.75]$ m, $[-3; 3]$ °). Position is estimated and pixel noise is 0.5 pixel.

3.4. Behaviour near the goal

The three navigation algorithms have quite different behaviour near the goal. In the case where the estimates of position and orientation are not corrupted by noise the method by [Aicardi et al., 1995] converges smoothly to reach the goal. As convergence is exponential the goal is not reached in finite time. In the case of noise corrupted position and ori-

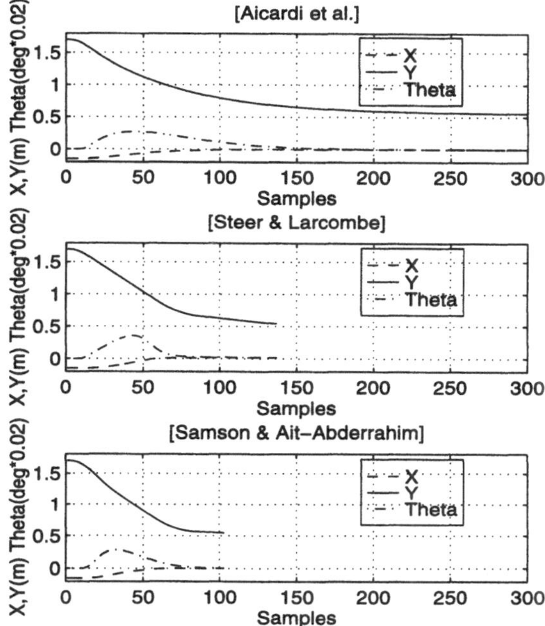

Figure 1. (x, y, θ) simulated for the three algorithms without noise.

entation the performance near the goal change. The control signals are relatively unaffected in the beginning of the navigation but grows corrupted as the goal is approached. Especially the control signal for ω is affected as it is seen in figure 2 and figure 3.

The reason for this is that the vehicle need to turn vigorously if a position measurement corrupted to be a little to the side of the vehicle when the distance to the goal is low necessitating a large turning angle. If the navigation is not aborted a few cm before the goal is reached the vehicle becomes unstable. This inconvenience can be reduced by navigating towards a goal behind the real goal and stopping when the real goal is reached.

The algorithm supplied by [Steer and Larcombe, 1991] is not designed to position the vehicle at the goal but only to lead it towards the goal. Therefore additional capturing algorithms are needed. Such algorithms have not been implemented. An approximative positioning can be performed by ramping the linear and rotational velocities to zero when the goal is reached.

As the basic principle of [Samson and Ait-Abderrahim, 1991] is trajectory tracking the method as proposed is not capable of positioning the vehicle at the goal. An approximated positioning can be performed as for [Steer and Larcombe, 1991]

4. Experiments

Extensive experiments have been carried out in the laboratory verifying and comparing the algorithms described above. The test bed autonomous vehicle is described in greater detail in [Ravn and Andersen, 1993]. The scenario examined is the AGV docking a charging station from a distance of approximately 1.5 m using a camera and visual gui-

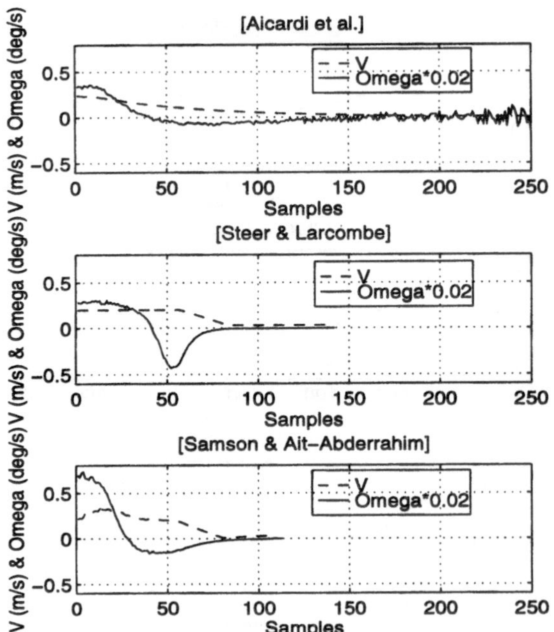

Figure 2. Simulated linear and rotational vel. ref.. Starting point: [-0.15 m 1.7 m 0 deg.]

demarks to estimate the current position of the AGV relative to the charging station is described earlier.

Internal data in the AGV is logged including the linear and rotational velocity references. In figure 3 results from simulations (without noise) and experiments for the algorithms are compared in order to demonstrate the validity of the simulator.

Note the unstable behaviour of especially the rotational velocity for the algorithm of [Aicardi et al., 1995] near the goal in the presence of noise. Note also the exponential approach to the goal similar to the simulations in figure 3. This is very similar to what was observed using the simulator. Based on several experiments like the one shown on figure 3 a good agreement between the simulator and the real world can be established.

Another camera mounted in the ceiling is observing the position and orientation of the AGV based on the measurement of two lamps on the AGV. This external measurement is used for the evaluation of the accuracy of the final position of the AGV. In figure 4 the logging of position and orientation the AGV from this camera is shown. The camera is calibrated so the orientation and position measures are correct at the goal.

The measurement is done every 40 ms, the time between each frame in the plot is 0.4 s. The determination of the position of the LEDs is done with subpixel accuracy.

The externally measured position and orientation for the three algorithms are plotted in figure 5. The figure shows good agreement with simulated data shown on figure 1.

In Table 4 results for several runs with the same nominal starting point are shown, this should be compared to the results in Table 3.

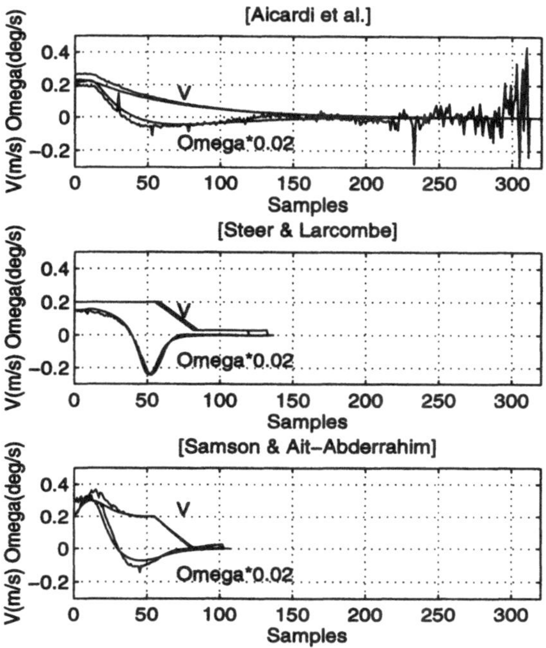

Figure 3. Experimental linear and rotational velocity references. Starting point: [-0.15 m 1.7 m 0 deg.]

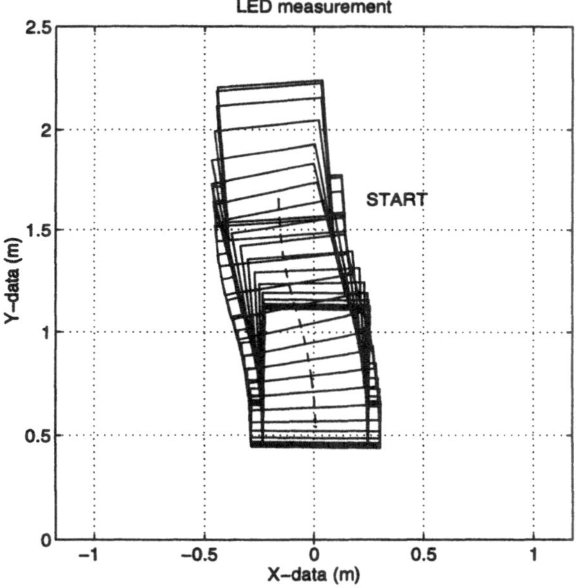

Figure 4. Experimental measurement of the AGV position and orientation using the external camera measurements. Samson & Ait-Abderrahim.

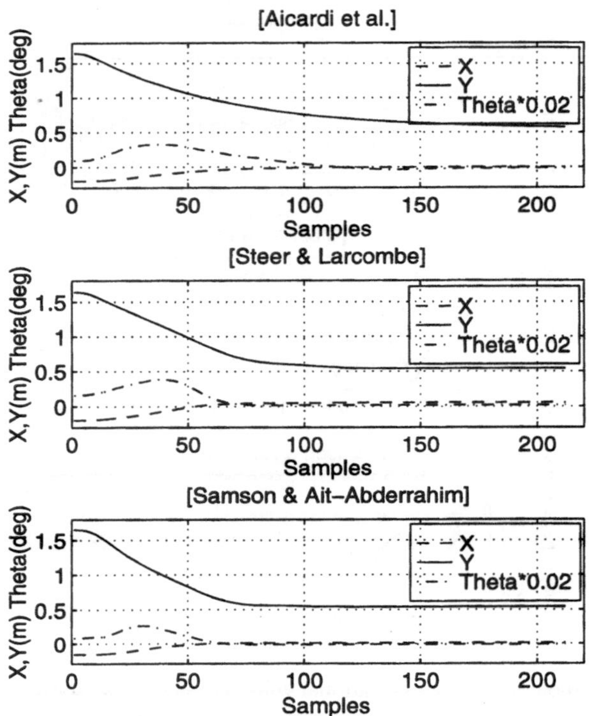

Figure 5. Experimental measurement of the AGV position and orientation using the external camera.
Nominal starting point: [-0.15 m 1.7 m 0 deg.]

Mean and standard deviation $[\mu]$, (σ)	**[Aicardi et al.]** $[\mu]$, (σ)	**[Steer & Larcombe]** $[\mu]$, (σ)	**[Samson & Ait-Abderrahim]** $[\mu]$, (σ)
End precision: $(\Delta x, \Delta y, \Delta\theta)$ mm and deg	3.2, 4.25 -0.14, 3.7 -2.00, 0.88	54.1, 3.00 -8.8, 1.27 1.27, 0.20	14.0, 3.5 -8.8, 2.6 -0.93, 0.33
$\left(\int v, \int \lvert\omega\rvert \right)$ m, degrees	1.15, 0.030 59.9, 5.6	1.17, 0.014 55.4, 3.55	1.18, 0.023 49.8, 7.0
$max\,(v, \lvert\omega\rvert)$ m/s, deg/s	0.268, 0.037 14.0, 1.9	0.235, 0.0057 13.6, 2.04	0.338, 0.0090 17.7, 2.54

Table 4. Experimental results for several runs and the same nominal starting point.

For the experiments and simulations as a whole the following remarks are in place:

- [Aicardi et al., 1995] is the most accurate in both position and heading while consuming the most time. Near goal noise sensitivity problems.
- [Steer & Larcombe] is complex to tune because of many parameters and no good directions on the tuning, quite fast but only able to place the vehicle with a few cm's precision.

- [Samson and Ait-Abderrahim, 1991] is quite accurate with a precision of about 10 mm and with a low time consumption.

The algorithms all have very limited support for parameter tuning making real life implementation time consuming. Near goal modifications of the algorithms are needed in all cases to make them work in a practical setting.

5. Conclusion

The aim of this work has been to explore and compare different algorithms for positioning and docking of non-holonomic autonomous vehicles equipped with a vision system. The procedures developed has been successfully tested and compared on the test bed autonomous vehicle.

The simulator has been shown to give good results and the sensitivity of the performance of the algorithms to noisy measurements and other changes from a non-ideal state has been shown to be small.

Using these algorithms is seems possible to position a non-holonomic vehicle within a few cm's and degrees or better. This makes a good case for the use of these vehicles except in very demanding applications.

6. References

Aicardi, M., Casalino, G., Bicchi, A., and Balestrino, A. (1995). Closed loop steering of unicycle-like vehicles via lyapunov techniques. *IEEE Robotics and Automation Magazine*, 2(1).

Buccolini, S., Loggini, R., and Santilli, C. (1994). Closed loop steering of mobile robot: Design implementation of lyaponov based techniques. In *Proceeding of TELEMAN Student Research Projects Congress*, pages 115–121.

Ravn, O. and Andersen, N. A. (1993). A test bed for experiments with intelligent vehicles. In *Proceedings of the First Workshop on Intelligent Autonomous Vehicles*, Southampton, England.

Ravn, O., Andersen, N. A., and Sørensen, A. T. (1993). Auto-calibrations in automation systems using vision. In *Proceedigs of the Third International Symposium on Experimental Robotics.*

Samson, C. and Ait-Abderrahim, K. (1991). Feedback control of a nonholonomic wheeled cart in cartesian space. In *Proceedings of the IEEE International Conference on Robotics and Automation*, pages 1136–1141.

Steer, B. and Larcombe, M. (1991). A goal seeking and obstacle avoiding algorithm for autonomous mobile robots. In *Proceedings of the IEEE International Conference on Robotics and Automation*, pages 1518–1528.

Chapter 4

Human Augmentation

Machines have long been used to extend human capability. Vehicles and computers are among the many devices used to expand our abilities. Applying robotic technology to human extension via the combination of motive power and computer mediated control allows us to venture into domains otherwise inaccessible to humans. Robots with humans in their control loop offer us the possibility to explore and work in environments distant in location or scale, or inaccessible due to hazard.

Becker, González-Baños, Latombe, and Tomasi present a mobile robotic system which is guided by high-level commands from a human operator. Using a variety of visual guidance techniques, the system allows the operator to select targets for inspection and navigation with the robot taking responsibility for completing the indicated task.

Slatkin, Burdick, and Grundfest describe a novel robotic endoscope for gastrointestinal diagnosis and therapy. The device is designed to propel itself through tubes by gates composed of various sequences of expansion and elongation. The paper also reports on encouraging preliminary experiments with this device.

Kazerooni describes the design of a series of exo-skeletal human extenders. By sensing the motion and force the human exerts on the inside of the device, power flows to large actuators on the extender to provide multiplication of the human strength. The performance characteristics of several laboratory prototypes are discussed.

Lawrence, Salcudean, and Sepehri address the problem of joystick control of a remote manipulator in which force on the joystick results in resolved-rate control of the arm. Utilizing a magnetically-levitated joystick, the paper shows how controlling the joystick's stiffness as a function of end-point force contributes to stable and effective velocity control of the arm. Successful application to the control of a large hydraulic excavator is described.

An Intelligent Observer*

C. Becker H. González-Baños J.C. Latombe C. Tomasi

CS Robotics Laboratory
Stanford University, CA 94305
{cdb, hhg, latombe, tomasi}@flamingo.stanford.edu

Abstract

This paper describes an integrated mobile robotic system dubbed the *intelligent observer* (IO). The IO is a mobile robot which moves through an environment (such as an office building or a factory) while autonomously observing moving targets selected by a human operator. The robot carries one or more cameras which allow it to track objects while at the same time sensing its own location. It interacts with a human user who issues task-level commands, such as indicating a target to track by clicking in a camera image. The user could be located far away from the observer itself, communicating with the robot over a network. As the IO performs its tasks, the system provides real-time visual feedback to the user. We have implemented a prototype of the IO which integrates basic versions of five major components: landmark detection, target tracking, motion planning, motion control, and user interface. We have performed initial experiments using this prototype, which demonstrate the successful integration of these components and the utility of the overall system.

1. Introduction

This paper describes the concept, design, and initial implementation of a system we call the *intelligent observer* (IO). Our goal for the IO is to develop a system that provides a human user with intuitive, high-level control over a mobile robot which autonomously plans and executes motions to visually track a moving target (see Figure 1.a). The user sends commands, such as "follow the next moving object which enters the view", and receives real-time feedback, such as a graphical display of the positions of the observer and target overlaid on a map of the environment.

Although the IO can be seen as an extension to a traditional teleoperation system, there are several major differences. First, it responds to high-level commands which are specified at the task level. There is no need for a "virtual joystick" or any other such control. Removing such low-level responsibilities from the human user provides many benefits, such as reducing the likelihood of human error and allowing the user to focus attention on more important, higher-level issues.

*This research was funded by ARPA grant N00014-94-1-0721-P01 (ONR) and by an NSF/ARPA contract for ANVIL through the University of Pennsylvania. C. Becker is supported in part by an NSF Graduate Fellowship.

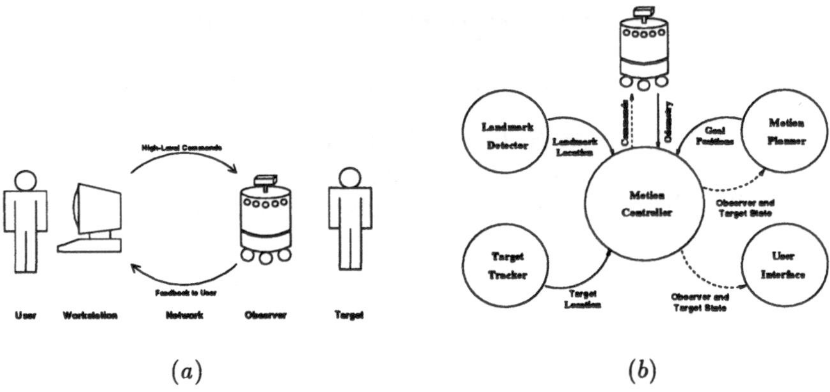

Figure 1. (*a*) *Interaction between a user and the intelligent observer;* (*b*) *The components of the IO system.*

Second, it uses its internal representation to provide a more flexible feedback mechanism than would be possible by simply displaying the image seen by the observer's cameras. The IO can fuse information from various sensors and, using geometric information about the environment, reconstruct a view of the observed scene. This view could be a simple two-dimensional, "top-down" view or a more realistic three-dimensional view rendered from an arbitrary viewpoint.

The IO project brings together concepts and algorithms from computer vision, motion planning, and computer graphics in order to create a robust, useful, and integrated system. One important aspect of the system is that vision, often viewed merely as a mechanism for aiding robot navigation, itself becomes the central objective of the system.

There are a number of possible applications for the IO system. Consider an engineer who must remotely monitor operations on a factory floor in order to analyze the performance of an assembly line. The IO would provide a convenient and automatic telepresence which could, for example, automatically track a part as it moves through the assembly sequence. The engineer need not focus on the mundane task of moving the observer, but may instead concentrate on the assembly line itself.

As another example, consider a security surveillance system for a large building. Typically a human operator is presented with a large number of views from various cameras positioned throughout the building. To be sure of catching any unusual activity, the user must cycle regularly through many views, possibly watching several views simultaneously. Here, the IO would alert the operator of motion in any camera view, and also plan new views as required to keep any moving objects in view. In this case the mobile robot is replaced by a large number of cameras, each of which could be either fixed or able to pan and tilt. The basic problem, however, is the same: the IO must still choose the best view of the target. This is accomplished by either moving the current camera or switching to a different camera.

The rest of this paper is organized as follows. In Section 2 we present the overall design of the IO and the roles of its different components. In Section 3 we describe an initial implementation of each component of the IO. In Section 4 we describe our initial experiments with the IO. Finally, in Section 5 we draw conclusions from our work and discuss possible directions for further research.

2. Overall Design

The complete IO consists of five major modules: landmark detection, target tracking, motion planning, user interface, and motion control. The relationships between these modules and the actual robot are shown in Figure 1.b. Each module is described below.

Landmark Detection As the observer moves around in its environment it must keep track of its current position. Our approach to this problem involves placing artificial landmarks throughout the environment. Many researchers have studied the use of landmarks in robot navigation; for examples see [1, 3, 5, 6, 7].

In our system, the positions of the landmarks are provided as part of a map which is provided to the IO. Each landmark induces a *landmark region* such that the landmark is visible to the robot whenever it moves within that region. The robot localizes itself by visually detecting landmarks and determining its position relative to them. Since the success of the robot depends on this self-localization, the vision algorithms used to detect the landmarks must be fast, accurate, and robust.

Target Tracking The central task of the IO is to observe moving objects, or targets. There are two main requirements: first, that the IO recognizes when a new target enters its field of view; and second, that the IO is capable of tracking the desired target as it moves. Since the robot must respond to the movement of objects, all tracking must happen in real time. In general, target motions may be almost totally unconstrained and the targets themselves may by nonrigid and of unknown shape; humans are a good example. In order to handle such targets, we can apply real-time versions of tracking algorithms such as those described in [4]. Currently, however, we use a simplified approach as detailed in Section 3.

Motion Planning Traditional motion planning problems involve finding a collision-free path from an initial region I to a goal region G in the presence of obstacles whose geometry is known in advance. The assumption of a static world allows the entire path to be precomputed in an off-line fashion. A robot can then follow the path without fear of collision.

In the context of the IO, the planning problem is quite different. The goal is no longer a fixed location; instead, our goal is to remain in view of a moving target at each point in time. Since the target motion is not known in advance, we must employ an on-line algorithm. In order to maintain a view of the target we must avoid occlusions due to obstacles and, as in the traditional case, we must also avoid collisions with obstacles. Any given obstacle may obstruct the robot's view, or its motion, or both. For example, a glass wall obstucts motion but not visibility, while a table may obstruct both, depending on its height.

We define an *on-line view planning problem* which must be solved by the IO as follows. The planner is given as inputs the position R_t (at time t) of the observer, the position T_t of the target, the region $\mathcal{R}_{t+1}$ of reachable locations of the observer at time $t+1$, and a function $\mathcal{T}_{t+1}$ which gives the probability of the target moving to any position T_{t+1}. As output, the planner produces a new observer position $R_{t+1} \in \mathcal{R}_{t+1}$ which maximizes the probability of maintaining an unobstructed view of the target for all possible T_{t+1}.

The problem outlined above is "local" in the sense that it only considers the probability one step in advance. A less local version would attempt to maximize the probabilty over some window of time, instead of only at time $t+1$.

User Interface One simple way of providing feedback to the user of the IO is to display live video from the robot's camera. There are, however, several drawbacks to this approach. First, it makes no use of the higher-level representation kept by the IO. Second, it limits the viewpoint to that of the robot's camera. Third, we would like to avoid "information overload" in cases where there are multiple cameras and/or observers. Finally, full-motion video requires high transmission bandwidth; this problem is especially important when the user is far away from the observer.

Instead, we have chosen to use a different approach. The user is presented with a synthetic reconstruction of the environment in which the IO is operating. This reconstruction can be either a two-dimensional overhead view or a three-dimensional view from an arbitrary vantage point. This module relies on an appropriate geometric model of the environment as well as information about the positions of the observer and the target.

Assuming that the environment is largely static, only a small amount of new information is required to update a scene, alleviating the bandwidth problem. This also allows us to fuse input from multiple cameras into a single, unified view.

Motion Control The motion controller takes the role of the top-level supervisor in the IO system (see Figure 1.b). It coordinates communication with the other components and produces appropriate low-level commands to control the physical robot. In addition, it periodically updates the current estimate of the robot's position based on feedback from the landmark detector and the odometric information from the robot. When appropriate, it also requests a new goal point from the motion planner. Finally, is sends updated information to the user interface.

The IO system is made up of a number of different processes, each of which has its own characteristic cycle time. Because of this, the motion controller must communicate asynchronously with the other processes, using new information as it becomes available. The exception is communication with the path planner, which follows a transaction-based model: the controller requests new a goal once it has achieved the previous one.

3. Implementation

This section describes our initial implementation of each of the five components of the IO. Each component has been implemented as a separate Unix process. The communication between processes uses standard TCP/IP protocols, making it possible to run them on different machines to increase performance. In fact, during our experiments we ran the landmark detector, motion controller, and user interface on a Sun SPARCstation 20, the target tracker on an SGI Indigo^2, and the motion planner on a DEC Alpha/AXP. The implementation of each process is detailed below.

Landmark Detection As discussed earlier, we rely on artifical landmarks to localize the robot. Our landmarks, shown in Figure 2.a, are placed on the ceiling at known positions throughout the robot's workspace. Each landmark consists of a black square with a 4×4 pattern of smaller squares inside of it. The detection algorithm first identifies edge pixels in the image (using a variant of the algorithm presented in [2]), and then looks for edge chains that are consistent with the boundary of a square. When such a chain is found, the corners are detected and then lines are fit to the pixels which make up each edge. The slopes of these lines yield the orientation of the landmark; their intersection points locate the landmark's corners. Once the landmark is localized, the positions of the inner squares are computed and their intensities are read from the image using bilinear interpolation. These intensities are grouped into "black" and "white" subgroups to determine the 16 binary values they represent. Four of these values are used to disambiguate the landmark's orientation, and the others encode the landmark's unique ID.

The landmark detector uses 320 × 240 grayscale images as input. Each frame requires approximately 0.5 seconds to process; this includes detecting, localizing, and identifying the landmark. The algorithm is very accurate: the translational error has zero mean and a standard deviation of 0.75 inches, while the rotational error is also zero mean and has a standard deviation of 0.5 degrees.

Target Tracking This component is used by the observer to detect and track a moving target. Currently the target consists of a number of black, vertical bars on a white cylindrical "hat" which sits on top of the target robot (see Figure 2.c). The tracking algorithm detects these bars in each image and, given the camera parameters and the physical size of the bars, computes the target's location relative to the camera.

The tracking system operates at approximately 5 frames per second using 160 × 120 grayscale images as input. For each frame, it determines the angle θ and distance r to the center of the target. Typically the target can be detected at distances ranging from 2 feet to 10 feet from the camera. Experimentation shows that θ is accurate to within ± 1 degree, and r is accurate to within ± 4 inches.

Motion Planning Our implemented solution to the on-line view planning problem is deliberately simple so that planning time is reasonable. The planner is given a polygonal map of the workspace, as well as bounds v_R and v_T on the maximum velocities of the observer and target, respectively. These velocities are given in terms of distance per planning cycle.

Suppose a goal position is requested at time t. The planner is given the positions R_t and T_t of the observer and target. We first find $\mathcal{R}_{t+1}$ by constructing a disk of radius v_R centered at R_t, and intersecting it with the free space in the environment. Likewise, we take $\mathcal{T}_{t+1}$ to be the intersection of the free space with a disk of radius v_T centered at T_t. We assume that all points within this region are equally likely.

To compute a new goal position R_{t+1}, we use the following approach. First we uniformly sample $\mathcal{T}_{t+1}$, producing m possible positions of the target. For each position we then compute the resulting visibility polygon V_i ($i = 1, \ldots, m$). We then sample $\mathcal{R}_{t+1}$, yielding n possible goals for the observer. From these samples we select the one which falls within the maximum number of visibility regions V_i. Ties can be broken in any number of ways; currently we choose the goal which results in the smallest motion of the observer. Assuming that we sample densely enough, this approach approximately maximizes the probabilty of maintaining a view of the target at time $t + 1$.

In the experiments we have performed, a single application of this planning technique requires between 0.2 and 0.8 seconds depending on the configuration of the observer and target relative to the obstacles in the environment.

User Interface In our initial system we have adopted a simple, two-dimensional user interface as shown in Figure 2.b. Physical obstacles are shown in black and configuration space obstacles are dark gray. The positions of the observer (R) and target (T) are updated in real time. In addition, the visibility region of the observer is shown in light gray.

Motion Control As mentioned before, this module not only controls the motion of the robot, but it also coordinates communication between all of the other modules. In a sense it is the top-level supervisory process for the whole system.

The most basic task of the motion controller is to issue velocity commands to the robot. This is accomplished using a tight feedback loop which runs at approximately 10 cycles per second. At each cycle, the loop reads odometry information from the robot and computes a velocity command to move the robot toward the current goal position. The control system measures and compensates for the communication delays present in the system. When new information is available from the landmark detector, the controller uses this to update the current position of the robot; this is done to compensate for errors which would build up over time if only odometric information were used.

The landmark detector returns the position of the robot relative to a landmark at the time s that a particular image is captured by the camera. The information is not received by the motion controller until some later time t. Due to latency in our digitizing hardware and to the image processing time, the difference between s and t can be as long as two seconds. To compensate for this delay, the controller keeps a history of the robot's position over the past several seconds. An updated position estimate at time t is computed by adding the difference between the stored positions at times t and s to the position sensed at time s using landmarks.

The error signal for steering and translation is based on the (x, y) goal points received from the motion planner. An error in both cartesian directions is computed by using the current position estimate of the observer. The steering and translation errors are then calculated in such way that the required steering is minimized. Both of these errors are then provided to the control law.

A totally separate control loop is used to position the camera in response to feedback from the target tracking system. The camera is mounted on a turret which can rotate independently of the robot's drive wheels. At each cycle, the turret control loop obtains the (r, θ) coordinates of the target, where r is the distance from the observer to the target and θ is the angle between the camera's optical axis and the line between the observer and the target. To account for delays, the actual target position is estimated using the last reading from the target tracker, an estimate of the target velocity, and the measured delay between actually acquiring an image of the target and completing the processing of that image. The estimated target position is the error signal used by the turret control law.

The control laws for the turret, steering, and translation are all similar in structure. These are simple P-controllers with compensation for the time-delay found inherent in communication with the robot. This time delay is present both when new odometric information is requested and when new velocity commands are issued. The control laws attempt to compensate for this lag based on estimates of previous delays. This compensation ideally performs a linearization by feedback, and the whole system is reduced to three second-order systems once the loops are closed. The simplification is valid as long as the feedback gain is not excessive.

All three P-controllers have exactly the same gain. In our system we attempted to minimize both time-response and oscillations. In theory, the ideal gain is the one that makes the two closed-loop poles at each loop identical. In practice the ideal value is 80-92% of the theoretical one, so the system was effectively reduced to three second-order systems under the imposed operating conditions.

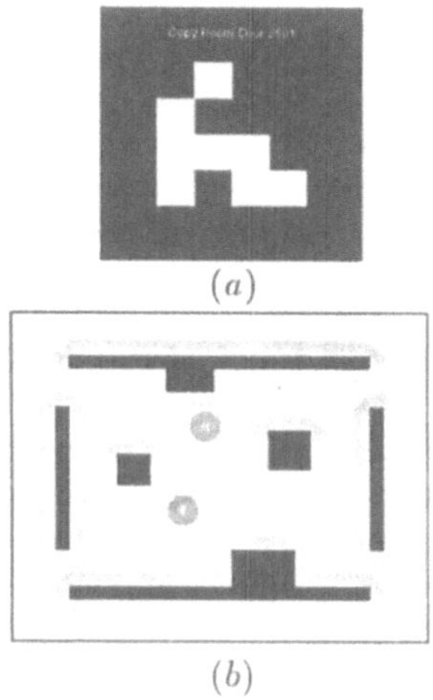

(a)

(b)

(c)

Figure 2. (*a*) *A sample ceiling landmark;* (*b*) *The view presented to the user;* (*c*) *A view of the IO tracking a second robot.*

4. Experimental Setup

This section describes our initial experiments with the IO system. Our experimentation has two goals: first, to validate the utility and robustness of the chosen algorithms for each of the various system components; and second, to demonstrate the feasibility of integrating all of the components into a unified system.

Our experiments took place in our laboratory, a fairly typical office environment. As obstacles we used desks, chairs, and large cardboard boxes. A map, as required by the motion planner, was constructed by approximating the obstacles by polygons; this map is shown in Figure 2.a.

The observer itself is a NOMAD-200 robot with an onboard Pentium-based computer. It is equipped with an upward-pointing camera for landmark detection and a forward-pointing camera for target tracking. Both cameras are mounted rigidly to the robot's turret, which can rotate independently of its drive wheels. This allows the turret (along with the tracking camera) to rotate based on the motions of the target without affecting the motion of the robot.

As a target, we used a second NOMAD-200 equipped with a special "hat" required by our simplified tracking algorithm, as described above. The target was moved under joystick control by a human operator. Figure 2.c shows a view of both the target and the observer.

As stated above, the purpose of our experiments was to verify each of the components in the IO, as well as to show that they could be successfully integrated. Although we have been able to quantify the performance of some of the components, it is too early to quantitatively describe the performance of the whole system. Qualitatively, however, the system was successful: the observer was consistently able to visually track the target and keep it in view in the presence of obstacles.

The most important conclusion of our experimentation is that, even though each of the components which makes up the system may periodically fail, the overall system is robust in the face of those failures. For example, the target tracker or landmark detector may periodically given an inaccurate result, but these errors are tolerable since they are random variables with zero mean and a relatively small variance. In addition, the sampling rate is high enough relative to the reaction time of the control system that sporadic sensing failures do not significantly alter the long-term performance of the system.

5. Conclusion

In this paper we have described the concept and high-level design of the intelligent observer. We have also described our initial implementation of the components of the system as well as early experiments with the system as a whole. Up to this point, our goal has been to design the overall system and to develop simple versions of all of the system components. Our continuing work focuses on two types of extensions: first, those which increase the generality and robustness of current components; and second, those which add new functionality to the concept of the overall IO system.

In terms of improving upon the current components, we have the following goals. First, we are working on ways to remove the discretization in our planning implementation by developing new visibility algorithms more directly related to the view planning problem. Second, we are working on ways to allow the planner to look farther ahead in time when it considers possible target moves. Third, we are working to implement a more general target tracking mechanism which removes the need for a special visual cue on the target.

In terms of adding functionality to the IO system, we are working on several extensions. First, we plan to add a component which automatically builds a 3D model of the observer's environment. The model will be generated using a laser rangefinder (for geometry) and a camera (for color). This model would then be used to reconstruct a view for the user. Second, we plan to extend the view planner to deal with multiple observers which can cooperate to track a single target. Finally, we will also consider the case of multiple targets and the problem of dynamically assigning observers to these targets.

References

[1] C. Becker, J. Salas, K. Tokusei, and J.C. Latombe. Reliable navigation using landmarks. In *Proc. IEEE Int'l Conference on Robotics and Automation*, 1995.

[2] J.F. Canny. A computational approach to edge detection. *IEEE Transactions on PAMI*, 8(6):679–698, 1986.

[3] S. Hutchinson. Exploiting visual constraints in robot motion planning. In *Proc. IEEE Int'l Conference on Robotics and Automation*, pages 1722–1727, 1991.

[4] D.P. Huttenlocher, J.J. Noh, and W.J. Rucklidge. Tracking non-rigid objects in complex scenes. Technical Report Tech. Rep. 92-1320, Cornell University Department of Computer Science, 1992.

[5] D.J. Kriegman, E.Triendl, and T.O. Binford. Stereo vision and navigation in buildings for mobile robots. *IEEE Transactions on Robotics and Automation*, 5(6):792–803, 1989.

[6] A. Lazanas and J.C. Latombe. Landmark-based robot navigation. *Algorithmica*, 13:472–501, 1995.

[7] T.S. Levitt, D.T. Lawton, D.M. Chelberg, and P.C. Nelson. Qualitative navigation. In *Proc. DARPA Image Understanding Workshop*, pages 447–465, 1987.

The Development of a Robotic Endoscope

A. Brett Slatkin   Joel Burdick
Department of Mechanical Engineering
California Institute of Technology
Pasadena, California USA
brett, jwb@robby.caltech.edu

Warren Grundfest, M.D.
Cedars-Sinai Medical Center
Los Angeles, California USA

Abstract

This paper describes the development of a prototype robotic endoscope for gastrointestinal diagnosis and therapy. The goal of this device is to access, in a minimally invasive fashion, the portions of the small intestine that cannot be accessed by conventional endoscopes. This paper describes the macroscopic design and function of the device, and the results of preliminary experiments that validate the concept.

1 Introduction and Motivation

This paper describes initial progress in the development of a *gastrointestinal robot* or *robotic endoscope* for traversing the human gastrointestinal system. Such a device could provide minimally invasive access to large sections of the gastrointestinal system which are currently accessible only through invasive methods.

One of the biggest trends in medical surgery in the 1990's is the shift to minimally invasive alternatives from traditional open practice [4]. Arthroscopic knee surgery, colonoscopic polypectomy, and laparoscopic gall bladder removal are widely recognized examples of this trend. Minimally invasive surgery typically involves the use of slender surgical instruments inserted into body cavities through naturally occuring or surgically produced orifices in order to reduce the amount of peripheral cutting required to reach the site of diseased or injured tissue. This can translate to dramatic improvements in patient care at greatly diminished health care costs. These improvements include, but are not limited to, reduced patient recovery times and reduced postoperative discomfort. There can be tremendous fiscal incentives as well for the adoption of these techniques. Approximately 21,000,000 surgeries are performed each year in the United States [4]. It is estimated that 8,000,000 of these surgeries can potentially be performed in a minimally invasive manner; however, only about 1,000,000 surgeries are currently so performed annually due in part to limitations in minimally invasive surgical technology. The complete adoption of minimally invasive techniques could be expected to save up to $\sim$ \$28,000,000,000 annually in hospital residency costs and lost patient wages [4]. One of the biggest technological impediments to increasing usage of minimally invasive approaches is lack of minimally invasive access to interior body cavities. While the focus of our current development effort is a device for minimally invasive access to the gastrointestinal system, we hope that successful deployment of this robot will pave the way for other applications of robotic technology to minimally invasive medicine.

Conventional endoscopes provide minimally invasive visualization of interior lumens and cavities, such as the stomach, colon, urinary tract and respiratory tract. Flexible endoscopes are comprised mainly of fiber optic bundles, or distal CCD chip cameras, for transmitting an optical image. Their fiber optic bundles can also transmit laser beams that cut, cauterize, or vaporize tissue. Typically, larger diameter endoscopes contain devices for steering their distal tips permit the deployment of simple surgical instruments for manipulation and dissection of tissue.

Gastrointestinal endoscopy represents the diagnosis and treatment of diseases of the alimentary canal by the use of flexible endoscopes. *Gastroscopes* are used to visualize the inner surfaces of the stomach and *colonoscopes* provide visualization of the large intestine, or colon. But these two ends of the alimentary canal represent only thirty percent of the length of the gastrointestinal tract. The remaining seventy percent, also known as the small intestine, cannot be reached without abdominal incisions using current endoscopic technology. The latest estimates from the Journal of Clinical Gastroenterology (1992) indicate that diseases of the small intestine, such as Crohn's disease, ulcerative colitis, and intestinal blockage, affict ~430,000 people in the United States. At present, the diagnosis of small intestine diseases is done either by assessing the "external" symptoms, often with radiological substantiation (e.g., via barium ingestion), or by invasive surgical exploration. And, in many instances, the current therapeutic intervention of these disease processes requires very invasive surgery, typically with a prolonged recovery period.

This paper reports our ongoing efforts to develop a robotic endoscope that can directly access and visualize, in a minimally invasive manner, the entire gastrointestinal tract. The long term goals of these efforts are to produce a surgeon guided robot that can semiautonomously locomote in the small bowel to perform medical diagnostic procedures. Such a device could eliminate or minimize the need for highly invasive diagnosis. A longer term goal is to supply therapeutic intervention when possible. For example, the robot may assist in the removal of blockages, or to accurately deliver drugs which may help in the non-invasive treatment of Crohn's disease. The required first step is to develop a machine that can dependably travel through the small intestine. Unless this technical challenge is overcome, none of the other medical goals can be reached. The focus of this paper is on the robotic locomotion mechanism.

2 Relation to Previous Work

Endoscopy is generally considered to be a mature medical technology with many commercial devices available for the diagnosis and treatment of disease processes found within various lumens of the human body. To use an existing endoscope, the endoscopic surgeon holds the proximal end and moves the distal end by pushing, pulling and twisting the device from outside of the body. Larger diameter endoscopes also allow limited active bending at the distal end, but none provide sophisticated control of their shape. It is desired that the endoscope slide easily inside of the tunnel-like lumen, but often this is not the case. Many lumens in human physiology are curved along their length; thus, endoscopes designed to traverse them must be sufficiently flexible to lateral bending. Unfortunately, these devices are advanced in the lumen by pushing from behind, which requires them to be sufficiently stiff to prevent buckling. These are two contradictory requirements, and, thus, flexible endoscopes are design compromises. Buckling of the flexible endoscope is a main cause of patient discomfort and is potentially injurious to fragile, diseased, surrounding tissues. It also limits the maximal depth of endoscopic penetration because it is

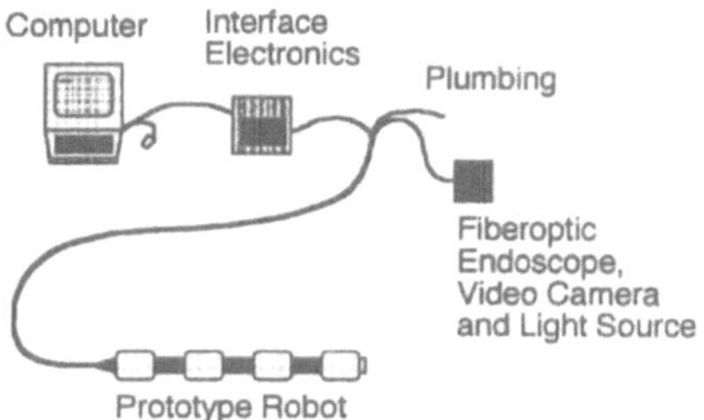

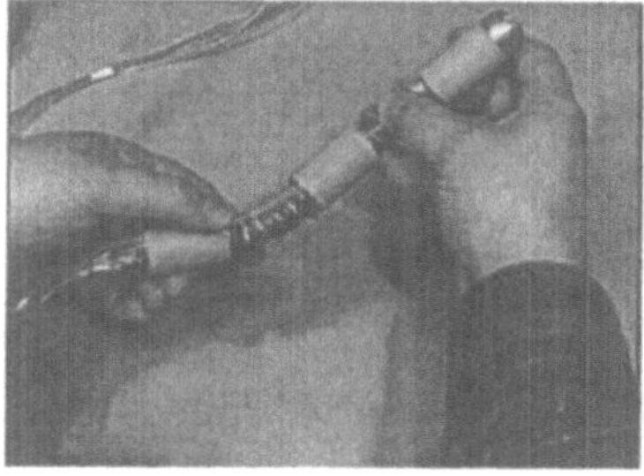

Figure 1: (a) Schematic Overview of the Robotic Endoscope System and (b) Photograph of one Robotic Endoscope Prototype

more likely to occur when the unsupported length of the endoscope increases inside the body. Thus, endoscopic accessibility is limited, and great skill is required of endoscopists to position such devices deep within the body.

Many have recognized that improvements in endoscopy could be effected by introducing actively controlled electromechanical articulation along the endoscope's length. For example, Sturges et. al. [10] have investigated the use of articulated bead-chain mechanisms for endoscope design. Fukuda and coworkers have likewise engineered prototype catheters which can actively bend along their length [2]. Ikuta and collaborators have developed hyper-redundant robotic endoscope prototypes [8, 7]. But in all of these cases the devices are advanced into the body by forces produced at their proximal ends, which are located outside of the patient. This type of actively articulated endoscope design inherently limits its overall length and hence its ultimate reach into the body.

As described below, our design is reminiscent of "pipe crawling" robots that have previously been investigated for inspection of buried pipes and channels. Fukuda and coworkers have developed self-propelled robotic systems for inspecting small and medium size pipes[3], and Shishido et. al. [9] have been granted a United States patent for such an invention [9]. While the topology of the human intestine is analogous to a pipe, there are many significant differences which prevent a simple adaptation of prior pipe crawling robot principles to our problem. First of all, the diameter of the human intestine can vary by a factor of four over its length. Conventional pipe crawling designs do not handle such variations. And, in addition, the intestine is highly flexible, fragile, and slippery. Thus, the traction mechanisms used in many prior pipe crawling devices would likely cause significant injury to the intestinal lining (provided they could produce sufficient traction at all).

It was with consideration of these concerns that the authors have conceived and patented the robotic endoscope described herein [6, 5].

3 The Robot Endoscope

The following sections of this paper describe a class of mechanisms and locomotion modalities which may be incorporated into a robotic endoscope that can propel itself in a flexible lumen for the purposes of medical diagnosis and therapeutic intervention. After providing a detailed presentation of the locomotion concepts for endoscopic propulsion, this section will describe the mechanical components and electronic systems that implement these concepts in our prototype.

3.1 System Overview

Figure 1 shows a schematic diagram of the prototype endoscopic system and a photograph of one of the prototypes that have been developed to date. There is an

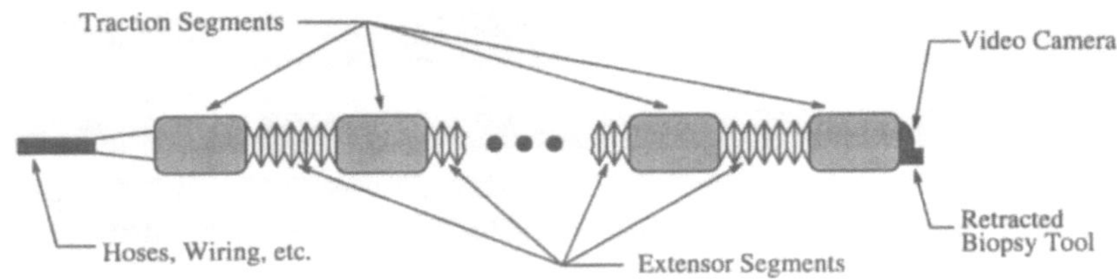

Figure 2: Schematic Diagram of the Robotic Endoscope

endoscopic robot with a trailing cable which consists of: electrical wiring for control signals; tubing to connect the pneumatic actuators to high and low pressure sources; and an optical fiber bundle for illumination and imaging of the area in front of the robot. The optical fiber bundle consists of a 1.4 millimeter diameter flexible endoscope inserted through a channel placed within the robot. The electric wiring carries currents to solenoid valves located within the robot. These currents are sourced by interface electronics which interpret signals from an external computer. Fluid power actuation for these devices was chosen because conventional endoscopic procedures require carbon dioxide gas, saline solution, and partial vacuum for insufflation, irrigation, and suction of gastrointestinal lumens. Hence, it is convenient and efficient to use these fluids as power sources for locomotion. Additionally, since large amounts of pneumatic or hydraulic mechanical power can be controlled by small electric signals, this approach minimizes the danger of electric shock to the patient. Another safety measure of these designs is that the working pressures of the actuators are kept intentionally small. In our prototypes, the high pressure source is typically maintained at 12 psig, while the low pressure is nominally -14.2 psig (vacuum).

Referring to Figure 2, our endoscopes appear outwardly to be electromechanical analogs of segmented worms, such as earthworms. In order to propel itself, this robotic endoscope employs mechanisms along its length which can be described as "grippers" and "extensors." The primary purpose of the grippers, or traction devices, is to provide traction against the lumen wall by expanding radially outward. The extensors provide extensibility between the grippers, i.e. they cause the mechanism to locally expand or contract in length. Locomotion is the process of generating net displacement of the robotic endoscope inside a flexible lumen by specific sequences of gripping and stretching actions. Such sequences are commanded by an external computer, and, thus, changes in the selection and control of mechanical movements can be easily accomplished in software. Furthermore, these machines are not limited by their mechanical design to locomote by a particular gripper and extensor sequence. Practical experience dictates that robust locomotion of the machine through lumens which exhibit changing geometric or material characteristics along their length (e.g. varying cross-sectional diameter) require repeated changes in gaits, or maneuvering sequences.

3.2 Methods for Locomotion

A *gait* is a distinct cycle of changes in the state of the component gripper and extensor segments that leads to a unit of displacement, which we term a *stride*. The length of the displacement is termed the *stride length*. Repetition of a gait leads to net displacement, or *locomotion*. Depending upon the number and arrangement of traction and extension mechanisms, a given device will typically be able to implement more than one gait.

In Figure 3, two gaits are shown for endoscopes that are comprised of three gripper segments. In these figures, the *phase*, or *state*, of the machine is shown for sequential moments of time. Between the given phases, the states of expansion of

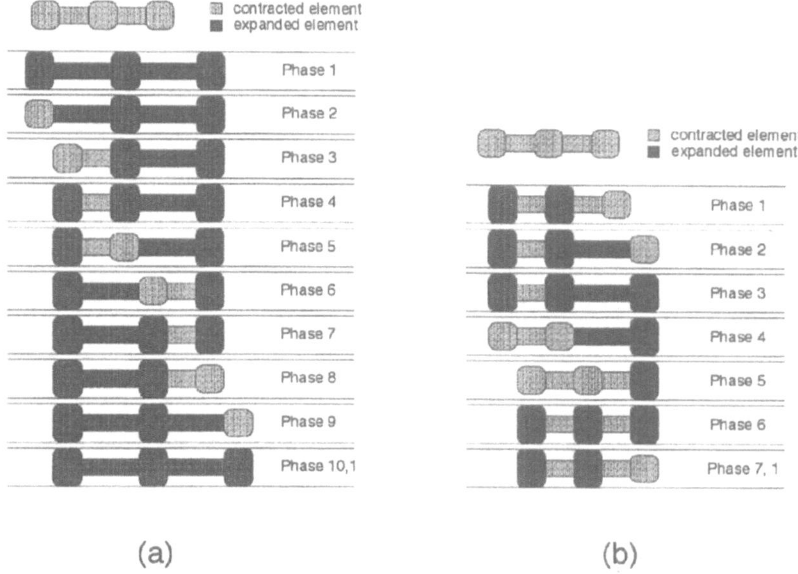

Figure 3: Two Locomotion Gaits for a 3 Gripper, 2 Extensor Mechanism

individual gripper and extensor segments change. Since subsequent phases of the locomotion gaits are portrayed in these figures as vertically displaced lumens, the stride length of each gait shown can be deduced. While the lumen is represented as a straight tube, the inch-worm type locomotion schemes will work if the lumen is curved and has reasonable variations in its cross-sectional shape and diameter. This is possible because the extensor devices are engineered to exhibit passive compliance to lateral bending loads.

3.3 Gaits for a 3-Gripper/2-Extensor Mechanism

Let us now consider means by which a device consisting of 3 grippers and 2 extensors (like the robot of Figure 1(b)) can locomote. This version can effect at least ten distinct gaits. This paper will discuss two of these gaits which point out important characteristics of our device.

The first gait for this robot that we will characterize is elucidated in Figure 3(a). In Phase 1, all of the grippers and extensors are in their expanded states. The rear gripper is retracted in Phase 2. Next, the rear extensor is retracted (either partially, or fully) in Phase 3. Subsequently, the rear gripper is expanded to make contact with the lumen (Phase 4). In Phase 5, the middle gripper is retracted. Next, the forward extensor is retracted while the rear extensor is expanded (Phase 6). Then, the middle gripper is expanded to make contact with the lumen. In Phases 8 and 9, the forward gripper is retracted, and then the forward extensor is extended. Finally, in Phase 10, the forward gripper is again expanded. At this point the mechanism has returned to its original state, but moved forward by one stride length. This cycle can be repeated for continual forward movement, or it can be reversed for rearward motion.

From the explanation of this gait sequence, it can be seen that contact points with the lumen advance along the lumen wall in a "wave-like" pattern. Hence, we can describe this motion sequence as a "traveling wave gait" in much the same manner as those described in [1]. It should be recounted that lateral compliance of the extensors will enable the mechanism to move in curved lumens as well. Furthermore, it is clear that at any instant, only relatively short segments of the device are being pushed forward through the lumen. This overcomes the buckling problems inherent in conventional endoscope designs.

Figure 3(b) shows a second gait example for a 3-gripper/2- extensor device. It should be noted that in all phases of this gate, the rear extensor is retracted. The same exact sequence of maneuvers could also be used if the rear extensor were extended. In other words, this gate makes no use of the rear extensor. Thus, if the rear extensor were to become inoperable during use, this gait could be used to enable endoscope locomotion even in the event of the failure of the rear extensor. Similarly, a gait exists which will be robust to a failure of the front extensor. And, thus, because the endoscope can switch between gaits, this design is robust to a single extensor failure. In addition, one can also derive a gait which will be robust to the failure of one of the grippers (assuming that the gripper fails in its retracted state).

In general, it is advantageous for an endoscope implementation to be capable of a large number of gaits. This in turn implies that highly adaptable endoscopes will have many gripper and extensor segments. As shown above, some gaits typically have more grippers in contact with the lumen. These gaits tend to be more stable, though the progress of the device through the lumen tends to be slower. A slower but more robust gait would be useful when the robot moves from a region of the lumen which produces strong traction into one which does not. Further, gaits with more grippers in contact can in some instances generate greater forward forces, which might be useful for unblocking intestinal blockages. Conversely, it may be desirable to select a faster gait when the robot moves into a region of strong traction from one of weaker traction. In addition, there exist gaits which are robust to the failure of particular components, as illustrated in Figure 3(b). The ability to switch between gaits as the situation dictates is a key feature of this device.

3.4 Component Designs

This section describes the mechanisms which implement the actions described above.

Traction Segment Designs: The action of a gripper or traction segment is to grasp the inside surface of the encompassing lumen by locally increasing its effective diameter. It is important to note that these mechanisms require special features in order to guarantee that this medical robot is safe and efficacious. With regard to the gripper segments, their design should allow the machine to gently grasp the inner linings of the lumen so as to minimize any potential injury to the possibly fragile tissues. Although many different mechanisms can be conceived which will produce this effect, the prototypes utilize pneumatically inflated toroidal balloons to grip the lumen. Solenoid valves which control the flow in and out of these balloons are located within each gripper segment. For scale, the gripper segments of the robot shown in Figure 1(b) are 0.875 inches in diameter and 1.500 inches in length.

Extensor Segment Designs: The extensors provide the local axial expansion and contraction required for inch-worm type locomotion. Since lumens of human physiology are often substantially curved along their length, the endoscope must be able to bend laterally. Our recent prototypes employ a modified bellows to provide extension of over fifty percent of the contracted length of 1.35 inches. Internal stiffening elements provide the appropriate bending and longitudinal stiffnesses. This modified bellows structure allows the extensor segments to expand preferentially in the axial direction under internal pneumatic inflation while simultaneously acting to provide the desired lateral bending compliance. In the present prototypes, onboard solenoid valves are used to control the flow of fluid in and out of the extensor segments.

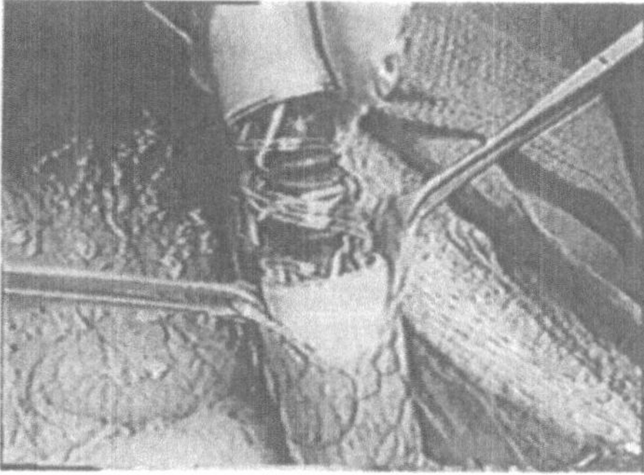

Figure 4: (a) Approach to a Simulated Polyp, and (b) Insertion of a Prototype Robot into the Small Intestine of a Pig

4 Experiments and Results

To date, we have developed several 2-gripper/1-extensor and 3-gripper/2-extensor endoscope prototypes (such as the one in Figure 1(b). This section describes the results of early experiments used to test the locomotion concepts and mechanical designs of these endoscopic robots. These experiments include tests of the components and systems in fairly rigid urethane tubing as well as swine intestines.

4.1 Experiments in Urethane Tubing

The first goal of these experiments is to verify the reliability of the locomotion and and structural integrity of these prototype devices. The intestines through which they are intended to travel are extremely flexible, slippery, and curved with relatively small radii of curvature. Although the urethane tubing is quite rigid, by lubricating its interior surface and bending it along its length, the ability of our robots to locomote in slippery and curved environments can be tested. The use of this laboratory apparatus continues to provide proof of concept for the hardware and software development of these devices. The prototype of Figure 1(b) houses a conventional fiberoptic endoscope for the illumination and imaging of the robot's environment. Figure 4(a) is an image taken from within the tube as the robotic endoscope was driven under remote control toward a simulated polyp.

The software for this prototype can command the robot to locomote in any of eight gaiting sequences as selected by the user, including ones which provide reliable locomotion with failed actuators. This was especially valuable when one of the extensor actuators failed during an experiment. While the robot was still located within the tube, the user commanded the robot to move according to a gait which does not involve that extensor, and the test continued successfully. It is expected that subsequent versions of these robotic endoscopes will consist of many more segments than this prototype which will result in machines that can endure multiple component failures and still be safely removed from a patient.

4.2 Animal Experiments

In addition to the "clean" testing of these robotic devices, modest animal studies have also been undertaken. It is accepted that certain animal models represent reasonable equivalents to human physiology. In particular, the intestines of an adult pig strongly resemble those of a juvenile human in size and mechanical properties. And the intestines of a juvenile human are typically smaller and considerably more flexible than those of an adult human. As such, the intestines of a pig are a conservative model with which to test an actively propelled endoscopic robot since it is considered to be more difficult to move through a smaller and more flexible lumen than through a larger, rigid one.

These in vivo experiments took place in a surgical laboratory with the animal under general anesthesia as was required to keep the appropriate tissue alive. The first in vivo experiment was a simple dissection of intestinal tissue for analysis of its mechanical properties. The results of this test indicated that the development of a machine that could locomote within this environment would be quite difficult; the flexibility and lubricity of this lumen would probably prove to be problematic. However, a subsequent experiment to test the available traction of a balloon type gripper device indicated that very substantial traction was indeed possible, and modest studies of robotic locomotion within swine intestines commenced. Figure 4 (b) shows a prototype robot as it was inserted through an incision into the small intestine of a pig. Ultimately, these devices will be launched from an endoscopic delivery system from either the stomach or the colon, but, to date, they have been introduced into the small intestines through surgical incisions. Since the outer surface of the small intestines has been visible during these experiments, it was possible to observe the movements of the prototype inside the lumen. Although this machine could indeed move through a portion of the small intestine, it was clear that further development of more advanced prototypes is required to support extensive in vivo experiments.

5 Conclusions and Future Work

There is a clear need for endoscopic devices that can access the human small bowel in a minimally invasive fashion. In this paper we described our preliminary efforts to develop such a system. Our experiments have shown that the locomotion concepts work extremely well for machines moving within relatively rigid tubes and appear to apply to the flexible and slippery small intestines as well. These efforts represent an encouraging first step toward robotic endoscopy.

Our ongoing work is focused on the design and fabrication of a new modular endoscope robot for continuing in vivo locomotion experiments. Our short term goal is to reduce the robot's size as much as is practical. To achieve this reduction in size, we are currently developing a new generation of miniature pneumatic valves, since the size of the on-board valves represents the limiting constraint on overall endoscope size. The newest valves are roughly sixty percent of the size of those used in the previous prototypes, and, thus, the robot prototype under development utilizes gripper segments that are 0.700 inches in diameter, 1.000 inch in length.

In Section 4.1, it was noted that the urethane tubing experimental apparatus cannot precisely match the environment within the intestines due to its structural rigidity. But, within the body, there exist other physiological lumens which are considerably stiffer and, therefore, would be well modelled by the urethane tubing. Unfortunately, these tubes in the body are significantly smaller in diameter than the intestines, and, thus, an endoscopic robot designed to traverse them must likewise be considerably smaller than those intended for the intestines. Therefore, it is a future goal of these researchers to build yet smaller versions for these applications.

References

[1] G.S. Chirikjian and J.W. Burdick. The kinematics of hyper-redundant locomotion. *IEEE Trans. on Robotics and Automation*, to appear, 1994.

[2] T. Fukuda, S. Guo, K. Kosuge, F. Arai, M. Negoro, and K. Nakabayashi. Micro active catheter system with multi degrees of freedom. In *Proc. IEEE Int. Conf. on Robotics and Automation*, San Diego, May 1994.

[3] T. Fukuda, H. Hosokai, and M. Uemura. Rubber gas actuator driven by hydrogen storage alloy for in-pipe inspection mobile robot with flexible structure. In

Proc. IEEE Int. Conf. on Robotics and Automation, pages 1847–1852, Scottsdale, AZ, 1989.

[4] The Wilkerson Group. New developments in medical technology. 1991.

[5] W.S. Grundfest, J.W. Burdick, and A.B. Slatkin. Robotic endoscopy. *U.S. Patent pending.*

[6] W.S. Grundfest, J.W. Burdick, and A.B. Slatkin. Robotic endoscopy. *U.S. Patent No. 5337732*, August 16 1994.

[7] K. Ikuta, M. Nokata, and S. Aritomi. Hyper-redundant active endoscope for minimum invasive surgery. In *Proc. First Int. Symp. on Medical Robotics and Computer Assisted Surgery*, Pittsburg, PA, 1994.

[8] K. Ikuta, M. Tsukamoto, and S. Hirose. Shape memory alloy servo acuator system with electric resistance feedback and application for active endoscope. In *Proc. IEEE Int. Conf. on Robotics and Automation*, pages 427–430, Tokyo, 1988.

[9] Y. Shishido, H. Adachi, H. Hibino, T. Yamamoto H. Miyanaga, S. Takayama, Y. Ueda, Y. Aoki, and S. Yamaguchi. Pipe-inspecting apparatus having a self propelled unit. *U.S. Patent No. 5090259*, March 25 1986.

[10] R.H. Sturges and S. Laowattana. A flexible, tendon-controlled device for endoscopy. In *Proc. IEEE Int. Conf. on Robotics and Automation*, Sacramento, CA, 1991.

The Extender Technology: An Example of Human-Machine Interaction via the Transfer of Power and Information Signals

H. Kazerooni

Mechanical Engineering Department
University of California, Berkeley, CA 94720 USA
E-Mail: kazerooni@euler.berkeley.edu

Abstract

A human's ability to perform physical tasks is limited by physical strength, not by intelligence. We define "extenders" as a class of robot manipulators worn by humans to augment human mechanical strength, while the wearer's intellect remains the central control system for manipulating the extender. Our research objective is to determine the ground rules for the design and control of robotic systems worn by humans through the design, construction, and control of several prototype experimental direct-drive/non-direct-drive multi-degree-of-freedom hydraulic/electric extenders. The design of extenders is different from the design of conventional robots because the extender interfaces with the human on a physical level. Two sets of force sensors measure the forces imposed on the extender by the human and by the environment (i.e., the load). The extender's compliances in response to such contact forces were designed by selecting appropriate force compensators. This paper gives a summary of some of the selected research efforts related to Extender Technology, carried out during 80's. The references, at the end of this article, give detailed description of the research efforts.

1. Introduction

This article presents an overview of a new human-integrated material handling technology being developed at the University of California, Berkeley. This material handling equipment is a robotic system worn by humans to increase human mechanical ability, while the human's intellect serves as the central intelligent control system for manipulating the load. These robots are called extenders due to a feature which distinguishes them from autonomous robots: they extend human strength while in physical contact with a human.

The human becomes a part of the extender, and "feels" a force that is related to the load carried by the extender. Figure 1 shows an example of an extender [1]. Some major applications for extenders include loading and unloading of missiles on aircraft; maneuvering of cargo in shipyards, foundries, and mines; or any application which requires precise and complex movement of heavy objects.

The goal of our research at the University of California, Berkeley, is to determine the ground rules for a control system which lets us arbitrarily specify a relationship between the human force and the load force. In a simple case, the force the human feels is equal to a scaled-down version of the load force: for example, for every 100 pounds of load, the human feels 5 pounds while the extender supports 95 pounds. In another example, if the object being manipulated is a pneumatic jackhammer, we may want to both filter and decrease the jackhammer forces: then, the human feels only the low-frequency, scaled-down components of the forces that the extender experiences. Note that force reflection occurs naturally in the extender, so the human arm feels a scaled-down version of the actual forces on the extender without a separate set of actuators.

Three elements contribute to the dynamics and control of this material handling system: the human operator, an extender to lift the load, and the load being maneuvered. The extender is in physical contact with both the human and the load, but the load and the human have no physical contact with each other. Figure 2 symbolically depicts the communication patterns between the human, extender, and load. With respect to Figure 2, the following statements characterize the fundamental features of the extender system.

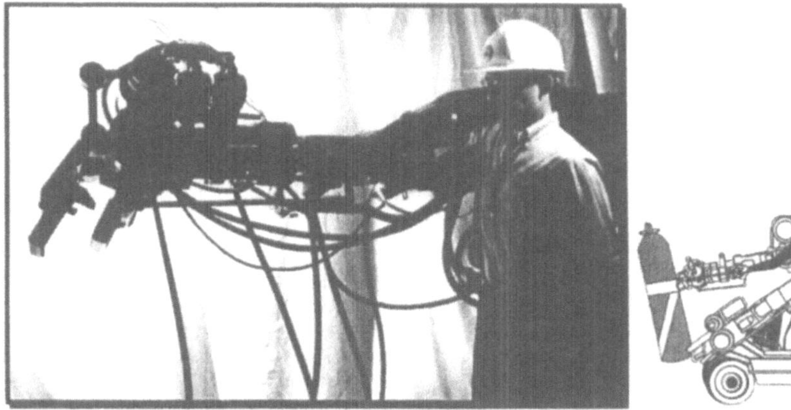

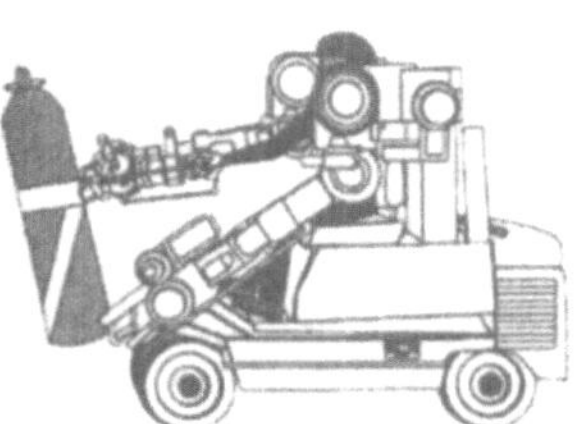

Figure 1: Experimental Six-Degree-of-Freedom Hydraulic Extender designed for loading and unloading aircrafts.

1) The extender is a powered machine and consists of: 1) hardware (electromechanical or hydraulic), and 2) a computer for information processing and control.
2) The load position is the same as the extender endpoint position. The human arm position is related kinematically to the extender position.
3) The extender motion is subject to forces from the human and from the load. These forces create two paths for power transfer to the extender: one from the human and one from the load. No other forces from other sources are imposed on the extender.
4) Forces between the human and the extender and forces between the load and the extender are measured and processed to maneuver the extender properly. These measured signals create two paths of information transfer to the extender: one from the human and one from the load. No other external information signals from other sources (such as joysticks, pushbuttons or keyboards) are used to drive the extender.

The fourth characteristic emphasizes the fact that the human does not *drive* the extender via external signals. Instead, the human moves his/her hands naturally when maneuvering an object. Considering the above, human-machine interaction can be categorized into three types:

1) <u>Human-machine interaction via the transfer of power</u>
In this category, the machine is not powered and therefore cannot accept information signals (commands) from the human. A hand-operated carjack is an example of this type of machine; to lift a car, one imposes forces whose power is conserved by a transfer of all of that power to the car. This category of human-machine interaction includes screw-drivers, hammers, and all similar unpowered tools which do not accept information signals but interact with humans or objects through power transfer.

2) <u>Human-machine interaction via the transfer of information</u>
In this category, the machine is powered and therefore can accept command signals. An electric can opener is a machine which accepts command signals. No power is transferred between the can opener and the human; the machine function depends only on the command signals from the human [2].

3) <u>Human-machine interaction via the transfer of both power and information signals</u>
In this category, the machine is powered and therefore can accept command signals from the human. In addition, the structure of the machine is such that it also accepts power from the human. Extenders fall into this category. Their motions are the result not only of the information signals (commands), but also of the interaction force with the human [3].

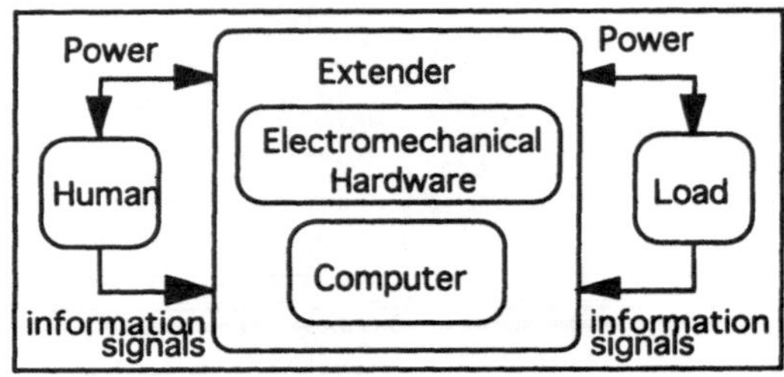

Figure 2: The extender motion is a function of the forces from the load and the human, in addition to the command signal from the computer.

Our research focuses on the dynamics and control of machines belonging to the third category of interaction involving the transfer of both information signals and power. The information signals sent to the extender computer must be compatible with the power transfer to the extender hardware. The objective of our research effort is to determine the rules for the control of robotic systems worn by humans through the design, construction, and control of a prototype experimental extender.

2. Work at UC Berkeley

It is important to note that previous systems (described in [4, 5, 6, 7, 8, 9, 10 and 11]) operated based on the master-slave concept [12], rather than on the direct physical contact between human and manipulator inherent in the extender concept. Unlike the Hardiman and other man-amplifiers, the extender is not a master-slave system (i.e. it does not consist of two overlapping exoskeletons.) There is no joystick or other device for information transfer. Instead, the human operator's commands to the extender are taken directly from the interaction force between the human and the extender. This interaction force also helps the extender manipulate objects physically. In other words, information signals and power transfer simultaneously between the human and the extender. The load forces naturally oppose the extender motion. The controller developed for the extender translates this interaction force signal into a motion command for the extender. This allows the human to initiate tracking commands to the extender in a very direct way. The concept of transfer of power and information signals is also valid for the load and extender. The load forces are measured directly from the interface between the load and the extender and processed by the controller to develop electronic compliancy [13, 14, 15, 16, and 17] in response to load forces. In other words, information signals and power transfer simultaneously between the load and the extender [18]. Several prototype experimental extenders were designed and built to help clarify the design issues and verify the control theories for various payloads and maneuvering speeds.

2.1 One-Degree-of-Freedom Extender

To study the feasibility of human force amplification via hydraulic actuators, a one-degree-of-freedom extender was built (Figure 3 and 4). This experimental extender consists of an inner tube and an outer tube. The human arm, wrapped in a cylinder of rubber for a snug fit, is located in the inner tube. A rotary hydraulic actuator, mounted on a solid platform, powers the outer tube of the extender. A piezoelectric load cell, placed between the two tubes, measures the interaction force between the extender and the human arm. Another piezoelectric load cell, placed between the outer tube and the load, measures the interaction force between the extender and the load. Other sensing devices include a tachometer and encoder (with corresponding counter) to measure the angular speed and orientation. A microcomputer is used for data acquisition and control. We developed a stabilizing control algorithm which creates any arbitrary force amplification and filtering. This study led to understanding the nature of extender instability resulting from human-machine interaction [11].

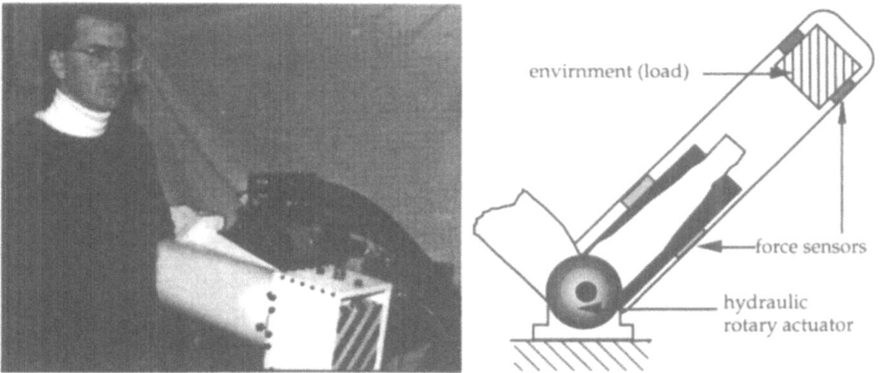

Figure 3: A one-degree-of-freedom hydraulic extender during an unconstrained maneuver.

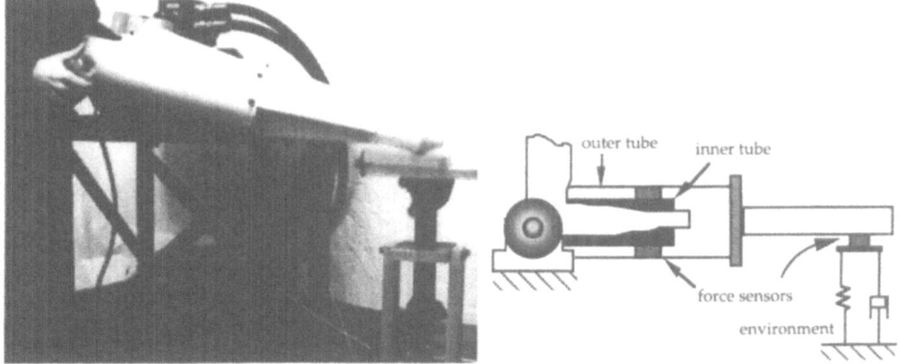

Figure 4: A one-degree-of-freedom hydraulic extender during a constrained maneuver.

2.2 Two-Degree-of-Freedom Direct-Drive Electric Extender

To rapidly maneuver light loads (weighing less than 50 lb), the bandwidth of the extender's actuators must be wider than the human largest maneuvering bandwidh. The dynamic behavior of the extender system at high speeds is non-linear. To develop and test nonlinear control algorithms, a direct-drive, electrically-powered extender was built (Figure 5). The direct connection of the motors to the links (without any transmission systems) produces highly nonlinear behavior in the extender. This extender has two degrees of freedom corresponding to a shoulder and an elbow. Two motors are located at the same height as the average human shoulder. Force sensors are located at the human-extender and extender-load interfaces. A third degree of freedom may be added: either rotation about a vertical axis or roll about a horizontal fore-aft axis. Figure 5 shows the seven-bar-linkage mechanism used for our prototype laboratory system. Force sensors are mounted at the human-machine and machine-environment interfaces. Motor 2 rotates link 4 causing the main arm (link 6) to move up and down via a four-bar linkage (links 4, 5, 6, and 3 as the ground link). In another four-bar linkage (links 1, 2, 3, and 7), motor 1 rotates link 1 causing the follower link (link 3) and the main arm (link 6) to move in and out. Both motors 1 and 2 are connected to bracket 7 which is mounted on a platform at the same height as the human shoulder.

A gripper is mounted on link 6 where the operator force, is measured along two directions. When the human holds onto the gripper, his/her upper arm parallels link 3 and his lower arm parallels link 6. The materials and the dimensions of the device components are chosen to preserve the structural-dynamic integrity of the haptic interface. Each link is machined as one solid piece rather than as an assembly of smaller parts. Link 1, 3, 6 are made of high strength 7075 aluminum alloy to reduce the weight of the haptic interface device. Link 2, 4, 5 are made of steel to fit in limited design space. This study led to understanding the nonlinear stability analysis and the trade-offs between stability and performance for extenders with nonlinear behavior [19].

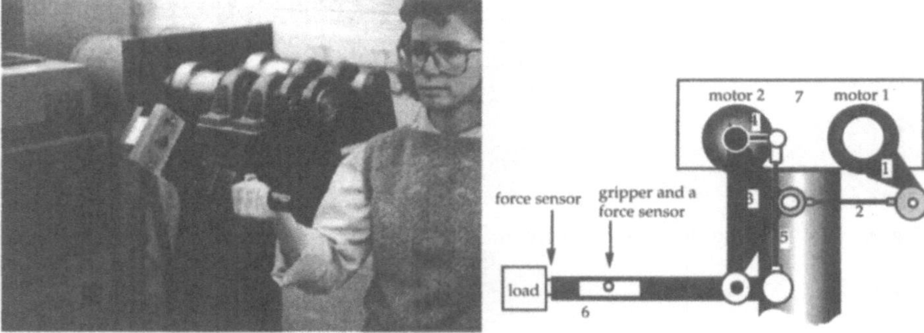

Figure 5: A two-degree-of-freedom direct-drive experimental extender to verify the feasibility of using direct-drive actuators for high-speed maneuvers of loads weighing less than 50 lb.

2.3 A Two-Degree-of-Freedom Non-Direct-Drive System to Measure Human Arm Dynamics

We have learned from our research work that human arm dynamic behavior is the dynamic element which varies the most in human-machine systems, both from person to person and also within one person [20, 21, 22, 23].

To understand significant variations in human arm dynamics when the human wears an extender, a computer-driven XY table was designed and built for measuring the human arm dynamics in horizontal maneuvers. A piezoelectric force sensor between the handle and the table measures the human's force along two orthogonal directions. Two high-resolution encoders on the motors measure the table's motion to within two microns [18]. In a set of experiments where the table was maneuvered by a computer, the operator tried to move her hand and follow the table so that zero contact force was created between her hand and the table. Since the human arm cannot keep up with the high frequency motion of the table when trying to create zero contact forces, large contact forces would consequently be expected at high frequencies. Based on several experiments, analysis of the power spectral density of the table position and the contact forces resulted in a human arm impedance along two orthogonal horizontal directions.

2.4 Extender Walking Machine

Figure 7 shows one of our experimental walking machines which helped us learn how to control a machine that must stand on its own. This experimental machine has two links which are powered relative to each other by a DC motor. The DC motor is housed in the lower link and it powers joint 2 via a cable and speed reducer at joint 2. Joint 1 at ground level is not powered: a motor at joint 1 would require an prohibitively large and lengthy foot similar to a snow ski. We have developed an actuation mechanism and a control technique which stabilize the mass on top of the second link without the use of any gyro. The control of this *under actuated* dynamic system is the very first and very fundamental issue in the design and control of walking machines for the extender. Our results will be published in the IEEE International Conference on Robotics and Automation, 1994.

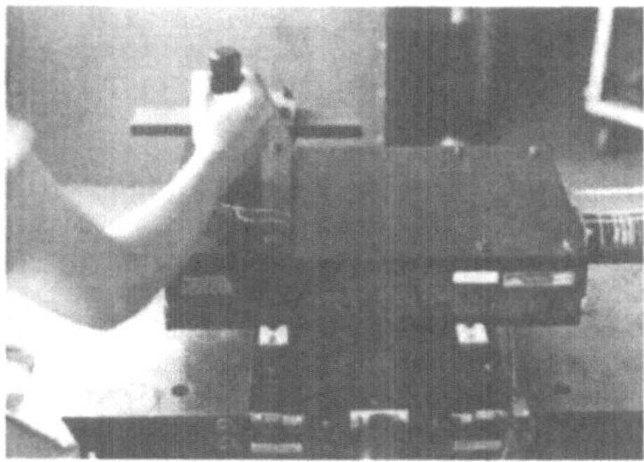

Figure 6: To understand significant variations in human arm dynamics when the human wears an extender, a computer-driven XY table was designed and built for measuring the human arm dynamics in horizontal maneuvers

2.5 Industrial Extender

A six-degree-of-freedom hydraulic extender (Figure 1) was designed and built for manipulating heavy objects [1]. The extender's hand linkage performs the grasping function while the arm mechanism executes the load manipulations. The arm mechanism (shown in Figure 8) consists of a forearm and an upper arm and has three degrees of freedom. The rotational axes of the extender arm are designed to coincide with those of the human arm joints. Both the upper arm and the forearm are planar four-bar linkages. The upper arm is driven by actuator A2, while the forearm is driven by actuator A3. Both actuators A2 and A3 are located on a bracket which rotates in response to the rotation of actuator A1. Actuator A1 is anchored to a base which is attached to the ground so the extender is not mobile. The arm uses four-bar linkages because: 1) the reaction torque of A2 does not act directly on A3, and vice versa, 2) the weight of one motor is not a load on the other, and 3) both actuators may be placed outside of the human viewing area. Figure 8 also shows a three-direction piezoelectric force sensor that measures the first three components of human force.

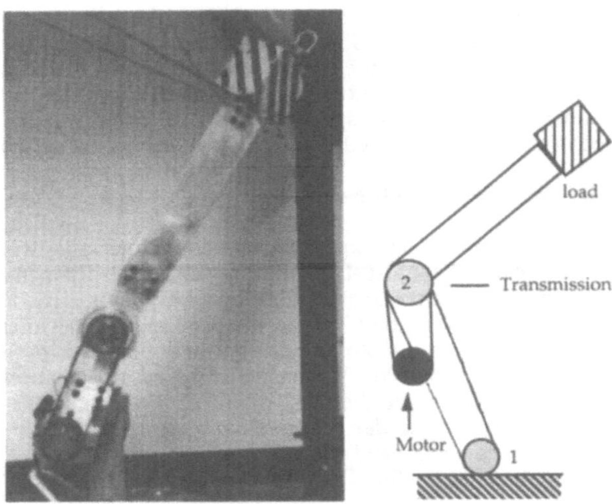

Figure 7: An experimental walking machine consisting of two links. Only joint 2 is powered. Joint 1 is not powered and sits on the ground. The goal of our research has been to stabilize this under actuated system. The actuator at Joint 2 must be controlled in such a way that the center of mass of the entire system passes through Joint 1.

The extender hand mechanism is shown in Figure 9; it has three degrees of freedom. Due to its complexity, the interface mechanism between the extender hand and the operator hand is not shown. The need to minimize operator fatigue requires that the extender hand be easily integrated with human hand postures. Actuator A5 drives the axis of wrist flexion and extension. For tasks that require secure grasping, humans apply opposing forces around the object. Opposing forces in the extender hand are created by motion of the thumb relative to the wrist via actuator A6. A5 and A6 are both located on a bracket connected to actuator A4 which twists the whole hand mechanism. The human hand is located in a glove which has force sensors mounted at the interfaces between the hand and the glove for measuring the second component of the human hand force. A thin elastic material is used at these contact points for the comfort of the human hand. The human operator can adjust the tightness of this glove simply by moving his/her elbow position horizontally. Several force sensors, not shown in the figure, are also mounted at the gripper to measure the load force in six directions. The materials and the dimensions of the extender components are chosen to preserve the structural-dynamic integrity of the extender. Each link is machined as one solid piece rather than as an assembly of smaller parts. The links are made of high strength 7075 aluminum alloy to reduce the weight of the extender.

This industrial extender is capable of lifting of objects up to 500 lb when the supply pressure is set at 3000 psi. Since the high frequency maneuvers of 500 lb load is rather unsafe, the experimental analysis on the extender dynamic behavior was carried out at low level of force amplification. In order to observe the system dynamics within the extender bandwidth, in particular the extender instability, the supply pressure was decreased to 800 psi and low force amplification ratios were chosen for analysis. This allows us to maneuver the extender within 2 Hz. Force amplifications of 7 times in the vertical direction and 5 times in the horizontal direction was prescribed on the controller. The human operator maneuvers the extender irregularly (i.e., randomly).

Figure 10 shows the FFT of the ratio of the load force to the human force along the horizontal direction where the load force is more than human force by a factor of 5. It can be seen that this ratio is preserved only within the extender bandwidth. Figure 11 shows the load force versus the human force along the horizontal direction where the slope of -5 represents the force amplification by a factor of 5. Figures 12 and 13 are similar to Figures 10 and 11, except that they show the system performance in the vertical direction. Figure 12 shows the FFT of the ratio of the load force to human force along the vertical direction where the amplification of 7 is preserved within the extender bandwidth. Figure 13 shows the load force versus the human force along the vertical direction where the slope of -7 represents the force amplification by a factor of 7.

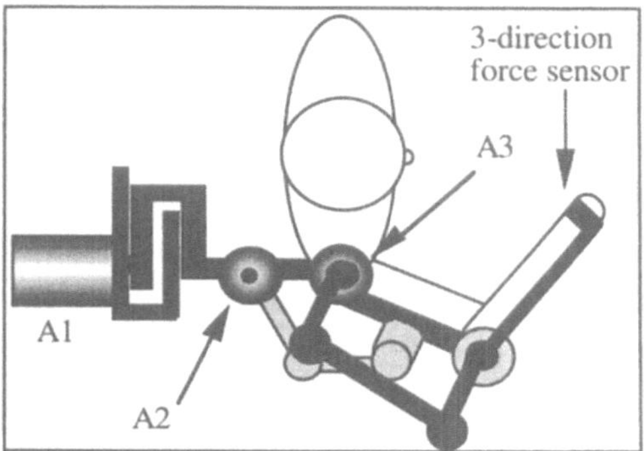

Figure 8: The extender arm has three degrees of freedom. The upper arm is driven by actuator A2, while the forearm is driven by actuator A3. Both actuators A2 and A3 are located on a bracket which rotates in response to the rotation of actuator A1.

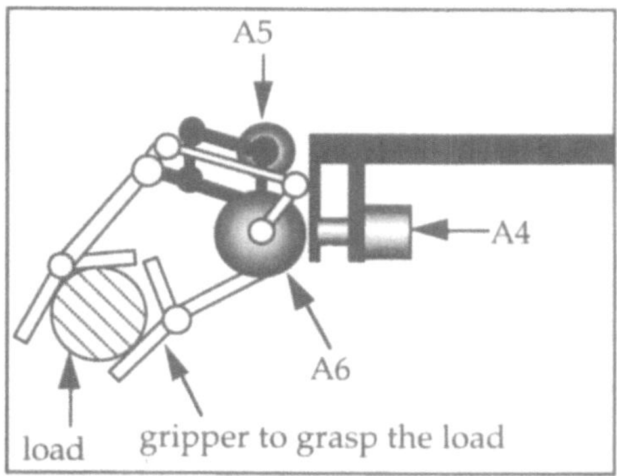

Figure 9: The extender hand has three degrees of freedom. Actuator A4 twists the whole arm while actuators A5 drives the axis of wrist felxion and extension. Actuator A6 opens and closes the gripper. Due to its complexity, the interface mechanism between the extender hand and the operator hand is not shown here.

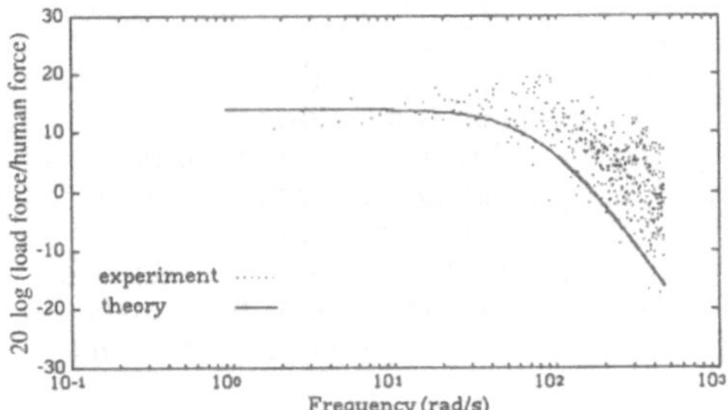

Figure 10: Theoretical and experimental force amplification ratio along the horizontal direction.

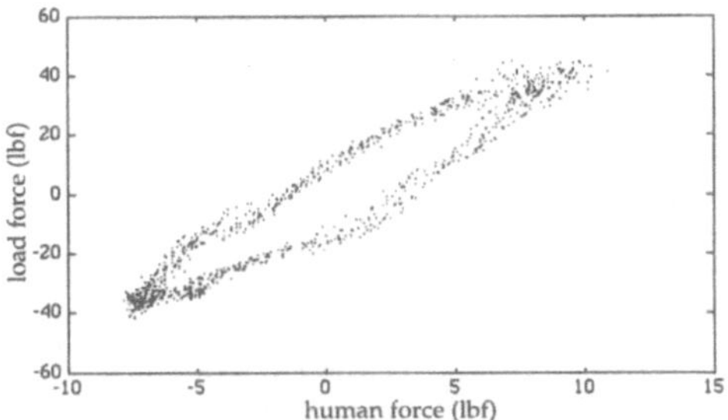

Figure 11: Load force versus human force along the horizontal direction. Slope is approximately 5.

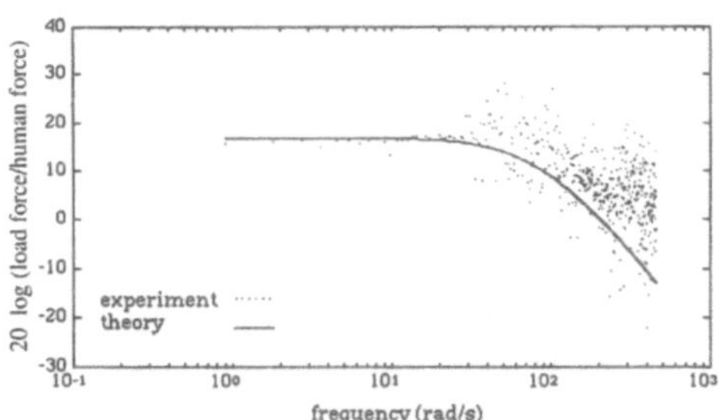

Figure 12: Theoretical and experimental force amplification ratio along the vertical direction.

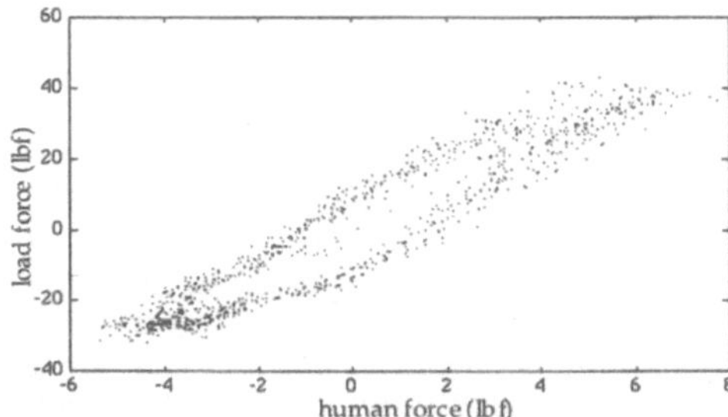

Figure 13: Load force versus human force along the vertical direction. Slope is approximately 7.

3. Concluding Remarks

This paper describes a summary of some of the selected projects during 80's. We are now in the process of design and fabrication of a whole suit: arms, hands, legs, ankles and torso for some specific applications. We have very little time for publication at this time. Hopefully when we finish the project and we have some time, we will publish some of our results. Contact H. Kazerooni at University of California for more information about our recent projects.

4. References

[1] Kazerooni, H., and Guo, J., "Human Extenders," ASME Journal of Dynamic Systems, Measurements, and Control, Vol. 115, No. 2(B), June 1993.

[2] Sheridan, T. B., Ferrell, W. R., Man-Machine Systems: Information, Control, and Decision, Model of Human Performance, MIT Press, 1974.

[3] Kazerooni, H., "Human-Robot Interaction via the Transfer of Power and Information Signals," IEEE Transactions on Systems and Cybernetics, Vol. 20, No. 2, March 1990.

[4] Clark, D. C. et al., "Exploratory Investigation of the Man-Amplifier Concept", U.S. Air Force AMRL-TDR-62-89, AD-390070, August 1962.

[5] GE Company, "Exoskeleton Prototype Project, Final Report on Phase I", Report S-67-1011, Schenectady, NY, 66.

[6] GE Company, "Hardiman I Prototype Project, Special Interim Study", Report S-68-1060, Schnectady, NY, 1968.

[7] Groshaw, P. F., "Hardiman I Arm Test, Hardiman I Prototype", Report S-70-1019, GE Company, Schenectady, NY, 1969.

[8] Makinson, B. J., "Research and Development Prototype for Machine Augmentation of Human Strength and Endurance, Hardiman I Project", Report S-71-1056, General Electric Company, Schenectady, NY, 1971.

[9] Mizen, N. J., "Preliminary Design for the Shoulders and Arms of a Powered, Exoskeletal Structure", Cornell Aeronautical Laboratory Report VO-1692-V-4, 1965.

[10] Mosher, R. S., "Force Reflecting Electrohydraulic Servomanipulator", Electro-Technology, pp. 138, Dec. 60.

[11] Mosher, R. S., " Handyman to Hardiman", SAE Report 670088.

[12] Kazerooni, H., Tsay, T. I., and Hollerbach, K., "A Controller Design Framework for Telerobotic Systems," IEEE Transactions on Control Systems Technology, Vol. 1, No. 1, March 1993.

[13] Kazerooni, H., Sheridan, T. B. and Houpt P. K., "Fundamentals of Robust Compliant Motion for Manipulators," IEEE Journal of Robotics and Automation, Vol. 2, No. 2, June 1986.

[14] Kazerooni, H., "On the Contact Instability of Robots When Constrained by Rigid Environments," IEEE Transactions on Automatic Control, Vol. 35, No. 6, June 1990.

[15] Kazerooni, H., "On the Robot Compliant Motion Control," ASME Journal of Dynamic Systems, Measurements, and Control, Vol. 111, No. 3, September 1989.

[16] Kazerooni, H., Waibel, B. J., and Kim, S., "Theory and Experiments on Robot Compliant Motion Control," ASME Journal of Dynamic Systems, Measurements, and Control, Vol. 112, No. 3, September 1990.

[17] Kazerooni, H., and Waibel, B. J., "On the Stability of the Constrained Robotic Maneuvers in the Presence of Modeling Uncertainties," IEEE Transactions on Robotics and Automation, Vol. 7 No. 1. February 1991.

[18] Kazerooni, H., and Mahoney, S. L., "Dynamics and Control of Robotic Systems Worn By Humans," ASME Journal of Dynamic Systems, Measurements, and Control, Vol. 113, No. 3, pp. 379-387, September 1991.

[19] Kazerooni, H., and Her, M. G., "The Dynamics and Control of a Haptic Interface Device," IEEE Transactions on Robotics and Automation, Vol. 10, No. 4, August 1994.

[20] Cooke, J. D., "Dependence of human arm movements in limb mechanical properties", Brain Res., Volume 165, pp: 366-369, 1979.

[21] Cooke, J. D., "The Organization of Simple, Skilled Movements", Tutorials in Motor Behavior, edited by G. Stelmach and J. Requin, Amsterdam,: Elsevier, 1980.

[22] Houk, J. C., "Neural control of muscle length and tension", in: Motor control, ed. V. B. Brooks. Bethesda, MD, American Physiological Society Handbook of Physiology.

[23] Stein, R. B., “What muscles variables does the nervous system control in limb movements?”, J. of the behavioral and brain sciences, 1982, Volume 5, pp 535-577.

Coordinated and Force-Feedback Control of Hydraulic Excavators

P. D. Lawrence, S. E. Salcudean, N. Sepehri
D. Chan, S. Bachmann, N. Parker
M. Zhu and R. Frenette
Department of Electrical Engineering
University of British Columbia
Vancouver, Canada
peterl@ee.ubc.ca

Abstract

The human interface of a Caterpillar 325FB feller-buncher was modified to allow the operator to use (i) a 5-DOF joystick, and (ii) a 6-DOF magnetically-levitated joystick with stiffness feedback. While the operator commanded the velocity of the endpoint, an onboard computer system managed total system power, solved the inverse kinematics, servoed the joint actuators, and controlled the magnetically-levitated joystick.
It was found that there were significant benefits to single joystick endpoint velocity control including smoothness of motion, less damage to product (trees), and ease of operation. Controlling joystick stiffness as a function of endpoint force, was found to be both a stable and effective form of feedback for a system where joystick position maps to endpoint velocity.
Two different hydraulic systems were implemented and evaluated. The first used valve control, as in a standard excavator. The second used hydrostatic control, by variable displacement pumps, and was found to lead to lower power consumption and higher operating speeds.

1. Introduction

The human interface for controlling hydraulic excavators has not changed significantly over many years. Traditionally, two spring-centered joysticks, each with two degrees of freedom (see Figure 1), have been used by operators to control the joint rates of a 4-DOF excavator arm in a joint velocity control scheme. Machine operators have to implicitly learn the inverse Jacobian of machines with different kinematic parameters - making use of vision to estimate the elements of the Jacobian at each moment. Since no single joystick mapping standard has emerged in the industry, operators may have to adapt to different mappings between joystick axes and arm joint axes when they change machines. In addition, the standard joysticks do not provide any force or tactile feedback from the environment.
Although the difficulty of machine operation is most evident in new operators, some

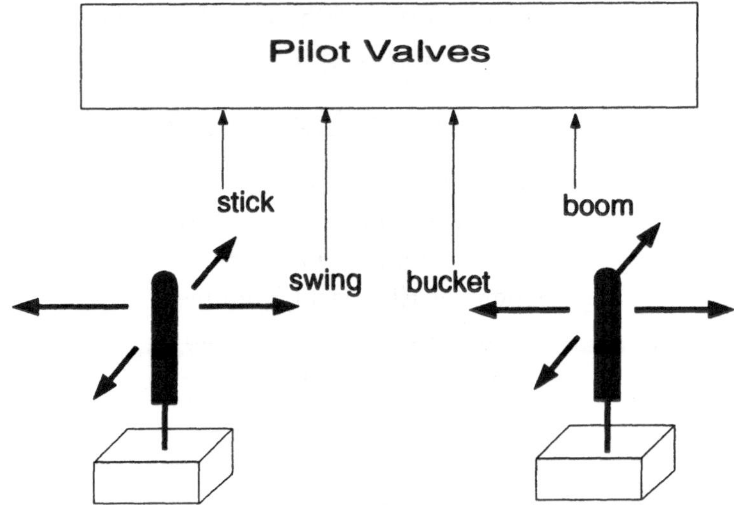

Figure 1. Standard Hand Controls

well-trained operators are much better than others in their ability to carry out workspace tasks using jointspace controls. A number of tasks such as digging deep trenches or leveling require the operator to work blindly or rely on secondary information such as engine noise or cab level to control bucket forces. These difficulties are even more pronounced when remote operation of machines is required, such as in the handling and removal of hazardous waste.

Our previous work and the work reported here was motivated by a possible reduction in learning time, reduction in adaptation time to different machines, reduction in fatigue, enhancement of safety through remote operation, and improvement in productivity and uniformity of production between operators with different motor coordination skills.

Specifically, the objective of the project has been to study the benefits of endpoint velocity control (EVC) with and without force feedback (FFB) to the operator.

Early equipment for telerobotics used spatially-corresponding master/slave manipulators to achieve endpoint position control for "hot-cell" manipulation [1]. Bilateral force-feedback was either mechanically or electrically incorporated in these systems. With the availability of high-speed computation, non-spatially-corresponding masters were later used with bilateral force-feedback to control manipulators [2].

As the ratio of the length of the slave arm to the length of the master arm increases, it becomes difficult to accurately control the slave using endpoint position control. An alternative is to use endpoint velocity control for which the joint rates can be found using the manipulator inverse Jacobian [3]. Small hand controls for this purpose that are similar in size to the joint velocity hand controls shown in Figure 1 have been proposed for space manipulator control [4].

A project was initiated at the University of British Columbia (UBC) in 1985 to examine new control interfaces for hydraulic machines used in forest harvesting. For ease in communicating the concept, endpoint velocity control in an excavator-based forestry machine was termed "coordinated motion control". The results of early

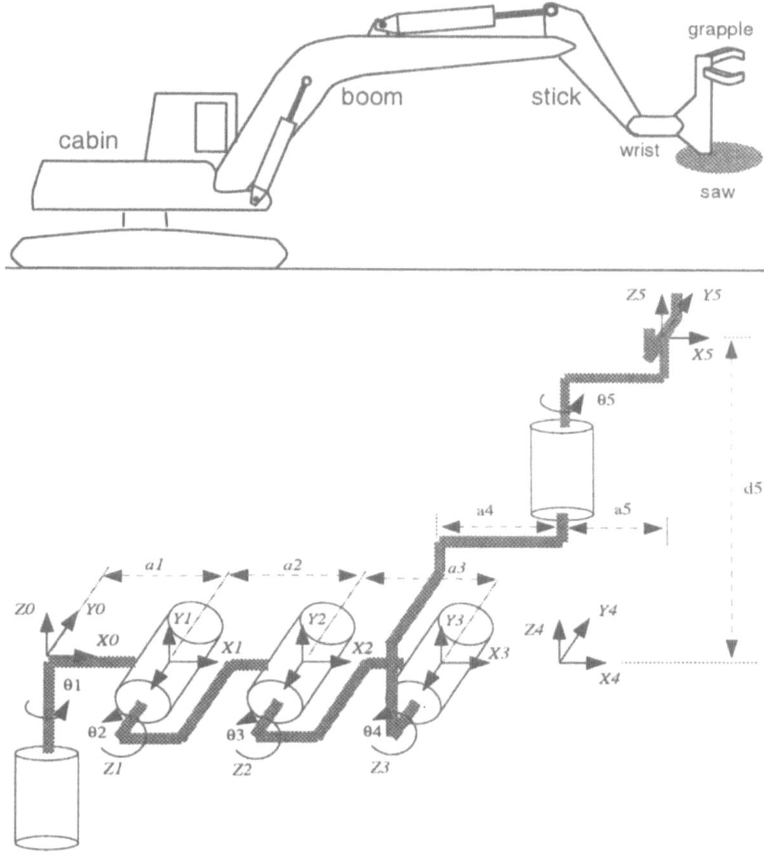

Figure 2. Feller-Buncher Schematic and Kinematic Configuration

experiments were reported by Wallersteiner *et al.* for excavator-based machines [5] in which the operator rotates with the machine. A Swedish project studied truck-based log-loaders in which the operator works in a fixed world reference frame [6]. Also at about the same time, there was a proposal for telerobotic excavation [7]. The UBC project has addressed issues of:
(i) implementation of coordinated control on large hydraulic machines, including the design of joysticks, machine instrumentation and hydraulic system modifications,
(ii) evaluation of various coordinated motion schemes and joysticks from a human factors point of view, and
(iii) implementation of force-feedback to the operator by using an active 6-DOF joystick.

A number of machine simulators were developed ranging in complexity from simple kinematic models to full dynamic models [8] including the machine hydraulic system and actuators. The graphics interface to the simulators was implemented on workstations displaying the operator's view from the cab. Force feedback computed for simple dynamic interaction such as contact with the environment or payload inertial force was presented to the operator using an active 6-DOF joystick [9].

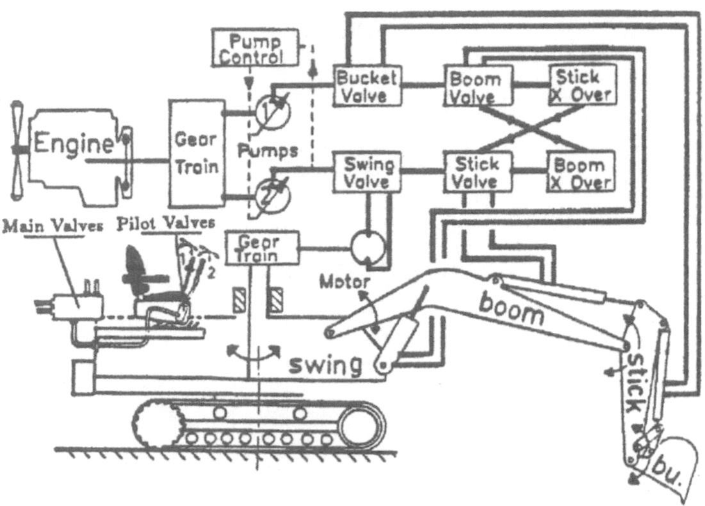

Figure 3. Excavator Hydraulics Circuit

Extensive experiments were carried out with a CAT215B machine configured as an excavator, as a log-loader, and with a CAT325FB feller-buncher. This led to a month-long field demonstration of endpoint velocity control in tree harvesting and to the control of a machine using a single 6-DOF force-feedback joystick to control end-effector linear motion, angular motion, forces and torques.

This paper concentrates on the experiments carried out on the Caterpillar 325FB feller-buncher and includes i) a comparison of the two hydraulic systems experimented with, the first one employing electrohydraulic pilot valves which push the spools of the main hydraulic valves, the second using variable displacement pumps instead of main valves for each actuating cylinder, (ii) a discussion of bilateral (force-feedback) control modalities and their effect on stability and performance.

2. Machine and Control System

An excavator has four major degrees of freedom - a cab rotation with respect to the base, a boom rotation (proximal arm segment), a stick rotation (distal arm segment), and a bucket curl (i.e. pitch). Different implements (end-effectors) can replace the bucket and these implements can have several degrees of freedom such as a grapple which can rotate (yaw direction), and open/close; or a feller-buncher head. A feller-buncher head, used for felling and accumulating trees, can pitch, roll, grasp (with two pairs of arms called "accumulator arms" and "grab arms"), and cut using an approximately 1m diameter circular saw (see Figure 2).

The Caterpillar 325 excavator used in these studies was modified by Balderson Inc. and Finning Inc. into a feller-buncher designated as a CAT325FB. The joints of a basic excavator are controlled by a set of main hydraulic valves which divert the hydraulic oil flow from the pumps into each of the single-ended cylinders that move the joints. The hydraulic system is shown in Figure 3. The simplest conversion to computer control of each joint is to replace the manually-operated pilot valves shown in Figure 3 with electrohydraulically-operated pilot valves, and add joint an-

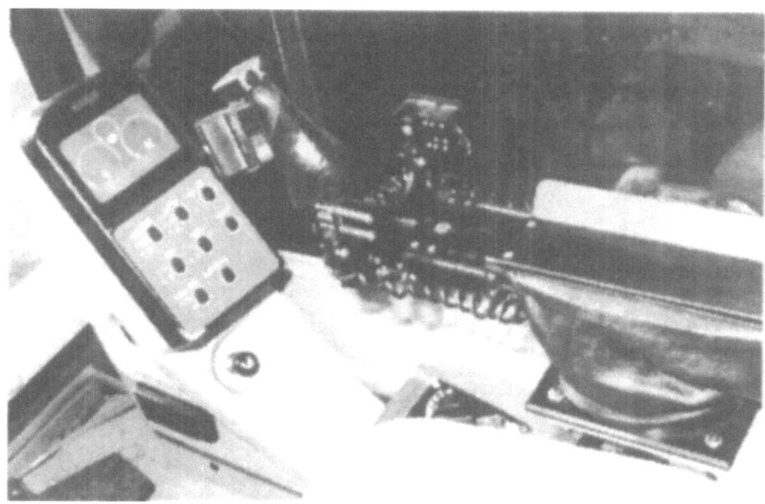

Figure 4. Feller-Buncher Hand Controller

gle sensors to each joint of the machine.
A VME-Bus based machine computer system was assembled and consists of a UNIX^{TM} host, a VxWorks^{TM} host and active joystick controller, a Transputer-based board controlling the machine in resolved-rate mode and various input/output boards. Major variables such as joint angles, system pressures, and valve control signals were monitored in real-time using Stethescope^{TM}. Feasible desired endpoint velocities are computed based on machine hydraulic constraints. A PD controller then uses a form of compensation for load described in [10].

3. Coordinated Motion Control

A 5-DOF hand controller (see Figure 4) for endpoint velocity control was designed that provided preferential motion (using slides) in the direction of cutting the tree (i.e. in the forward direction as the saw in the head cuts through the tree).

Two different hydraulic systems were investigated. The first, called "Valve Control", utilized electrohydraulic pilot valves which applied pressures to each end of the main valve spools. A description of this system and its performance is given in [10].
The second system called "Pump Control" was designed to replaced the original hydraulic system with a set of 8 variable displacement pumps in a hydrostatic drive configuration. In this arrangement, each main hydraulic actuator (tracks, cab rotation, boom, stick, feller-buncher head pitch and saw motor) is driven by a single variable displacement pump. The swash-plate angle on the pump is electrohydraulically controlled to vary the flow to the actuator. Each side of the actuator is connected to the corresponding side of the pump. A "hot-oil valve" and charge system are used to compensate for flows resulting from the difference in piston areas [11]. The less demanding functions (accumulator arms, grab-arms, and head roll) were valve-controlled. Figure 5 shows the outside cluster of 4 pumps symmetrically located on the face of the engine transmission. Beneath this layer is another layer of 4 pumps driven in tandem by the transmission.

Figure 5. Hydrostatic Drive System

The potential benefits of using pump control are that (i) there are no main valves to generate heat, and (ii) at each moment in time, excess energy from one function can be absorbed by another function. As an example, the potential energy released from a tree falling in the "pitch" direction while in the feller-buncher head's grasp, would be recovered by the saw pump through the transmission. A similar case can be made for combined boom and stick motions.

4. Rate Control with Stiffness Feedback

The benefits of force-feedback in telemanipulation are well known. In addition to faster completion time of assembly tasks [12], force feedback can also reduce stresses in the manipulator and the materials being handled (this is particularly important if the machine does not have six degrees of freedom). Additional information such as power consumption or hazardous operation can also be presented in an intuitive way as force-feedback information.

The use of force-feedback in the teleoperated control of excavators was considered before in [13], where a kinematically equivalent master controlling the excavator position and providing joint-level force feedback was proposed. The approach presented here addresses the problem of providing force-feedback in rate mode and used a 6-DOF magnetically levitated (maglev) hand controller designed and built at UBC [14] (motion range ± 5 mm, $\pm 6°$, maximum continuous force 20 N).

When force-feedback is used in rate mode, it can be shown that if the end-point force is scaled and returned to the operator directly, the teleoperator cannot be transparent. A mass load would be felt as a viscous force, a damper as a spring, etc.. Although perfect transparency is still achievable when force-feedback is used in rate mode, it requires integration of hand forces and differentiation of environment forces, and has quite limited stability robustness [15, 9, 16].

An alternative approach has been presented in [15, 17], will be referred to as "stiffness feedback", and is shown schematically in Figure 6. In Figure 6 the master and operator hand impedances are incorporated in Z_m, and the slave, having impedance Z_s, is assumed to be controlled in a stable manner by a compensator C_s

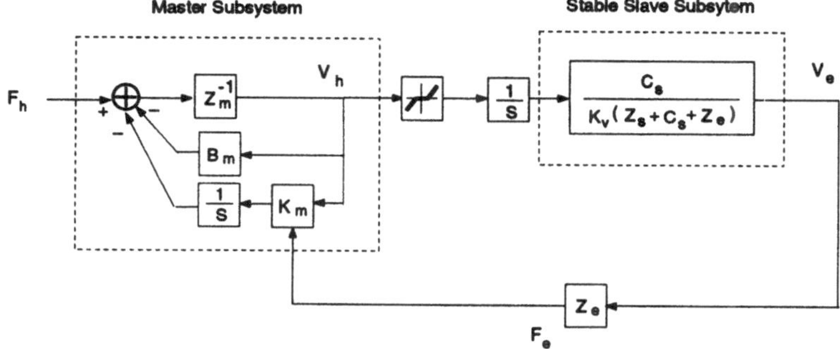

Figure 6. Schematic Diagram of Rate Control with Stiffness Feedback

against an environment impedance Z_e. Instead of direct force feedback, the stiffness K_m of the master is modulated by the sensed environment force according to the rule,

$$K_m \quad = \quad \text{sat}[K_{nom} + f_e K_r \text{sgn}(x_h)] \tag{1}$$

which is illustrated in Figure 7. The saturation function in (1) limits the joystick stiffness to $K_m \in [K_{min}, K_{max}]$ ($K_m > 0$), while the the sign function is needed to avoid having the slave "stuck" in a stiff environment because the master is centered by a high stiffness value.

Within the linear area of the stiffness adjustment rule, the operator experiences an environment-dependent force equal to $K_r|x_h|f_e$, which is essentially direct force feedback modulated by the joystick displacement. This suggests that the "transparency" of stiffness control should be good, at least locally. Stability proofs for stiffness control have been obtained so far only under very limiting assumptions on the operator impedance and either slow-varying assumptions on f_e, or large joystick damping B_m (see Figure 6) [9]. However, simulations and experimental results have been very good [15, 9].

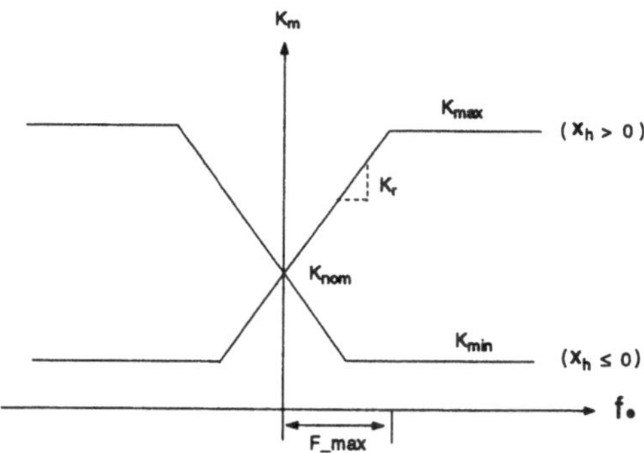

Figure 7. Stiffness Adjustment Rule

5. End-Point Force Estimation

The accurate measurement and/or estimation of end-point forces for excavator-type machines is a difficult problem. The machine arms cannot be easily modified to accept multi-degree-of-freedom force-torque sensors mounted at the end-effectors, and, even if they did, large shock loading would make this option of dubious reliability [15]. Instead, it is better to measure cylinder forces via load-cells or differential pressure sensors. The cylinder pressures were measured by transducers mounted close to the hydraulic pumps. The vector of joint torques τ of the machine arm is computed from the cylinder forces by solving simple two-bar and four-bar linkage problems [15, 9]. The end-point wrench vector f_e can be computed from the arm dynamic equations and a Jacobian transformation as follows:

$$\begin{aligned} J^T(q)f_e &= \tau - D(q)\ddot{q} - C(q,\dot{q})\dot{q} - g(q) \qquad (2) \\ &= \tau - D(q)\ddot{q} - C(q,\dot{q})\dot{q} - Y_g(q)p_g \end{aligned}$$

where q, J, D, C and g are the joint variables, arm Jacobian, mass matrix, centrifugal and Coriolis terms and the gravity force, respectively, and the gravitational terms have been re-written in a "linear-in-parameters" form [18]. Since neither the excavator nor the feller-buncher have six degrees of freedom, the Jacobian in (2) is not invertible. Either a pseudo-inverse (smallest norm satisfying (2) or setting components of f_e to zero can be used to solve this equation.

Due to the significant noise component in the velocity and acceleration signals $\dot{q}$ and $\ddot{q}$, and due to the uncertainty in the link inertial parameters, only the gravitational terms of the arm rigid body dynamics were used in the above computation. It is essential that the gravity terms be computed because, without gravity compensation, machine operation with an active joystick would be both tiring and hazardous. The parameter vector p_g was identified from $\tau - Y_g(q)p_g = 0$ ($Y_g(q) \in \mathbb{R}^{5\times 7}$) by driving the free CAT325FB arm very slowly to random locations and collecting the cylinder forces and joint variables, then using a recursive least-squares algorithm [15, 9]. With better joint-angle sensors, the complete forward dynamics could be used in the above equation. If the arm parameters are not known exactly, the identification procedure could be repeated, although there would be significantly more parameters to be identified.

6. Coordinated Control Experiments

The most objective comparison of coordinated control *vs* standard joint control would be to have two identical machines and have an operator use each machine on identical terrain and identical trees. This type of study had been previously carried out on a log loader (see [19]) but was not deemed to be feasible for a feller-buncher since the same trees cannot be cut twice. Instead, extended demonstrations were set up, to which machine operators and other representatives of forest companies, harvesting contractors, suppliers and machine manufacturers were invited to come and operate the feller-buncher.

For valve control, the machine was delivered to a forest harvesting site near Kelowna B.C. for a period of one week. For pump control, the feller-buncher was moved to a different forest harvesting test site near Kelowna B.C. It was also evaluated for one week at a second site near Quesnel B.C.

Operator Reaction

Each visitor to any of the sites observed a demonstration, operated the machine for a period of time and then responded to a questionnaire.

It was observed [20] that novices could be reasonably comfortable in controlling the machine after several hours compared to much longer times of at least several days

to a week for the standard machine controls. A major problem with long learning times is the potential damage to the machine during the learning period.
Experienced operators observed that the machine was smoother in operation and somewhat faster - especially for the pump-controlled version. This was hard to quantify since operators use different manufacturer's machines and there was no direct comparison between machines on the site.
Because of the large diameter of the saw and the necessity of perfectly coordinating the joints to produce a horizontal motion during cutting, it is not uncommon that the butt of the tree will be damaged (split) during felling using standard controls. It was observed that that the incidence of butt damage was much reduced with coordinated control.

Power and Speed Comparison

During the feller-buncher field trials, estimates of head speed and power consumed by boom, stick and head pitch were computed. The bottom trace in Figure 8 shows a typical recording of head speed during valve controlled operation. The top trace shows the power consumed by the three functions. Figure 9 shows the corresponding recordings during a typical pump controlled motion. It can be seen from these curves that higher speeds (conservatively 10-15%) can be achieved at about 30% lower power. The net benefit would be that machines using pump control would require less cooling, could consume less fuel, and would have a longer working life. There were two disadvantages of pump control that were evident. The first is that the hydraulic system is more complex and consequently would cost more than a conventional system. This could be reduced somewhat by improved manifold design. The second is a small inherent drift in cylinder position that has to be actively controlled using the joint sensors (in valve control, centering the valve stops cylinder motion). This has safety implications and requires reliable joint or cylinder displacement sensors.

7. Stiffness Feedback Experiments

Force feedback control was also implemented. The UBC maglev hand-controller was mounted inside the machine cab. End-point forces were computed according to (2), in the cab-attached frame, assuming that the cab yaw torque component is negligible (this is equivalent to using a 5×5 Jacobian in (2)). Direct force feedback in rate mode was implemented and found to be unsatisfactory, since stability could only be maintained for very low force-scaling back to the master. The stiffness control scheme illustrated in Figure 6 was implemented along each axis of the hand-controller, except yaw, with the following parameters (see Figure 2):

Axis	K_{min}	K_{max}	B_m	Deadband
	N/mm	N/mm	N/mm/s	mm
X_0	3.00	8.00	0.06	1.5
Y_0	2.00	4.00	0.06	1.5
Z_0	3.00	9.00	0.06	1.5
	Nm/rad	Nm/rad	Nm/rad/s	mrad
X_0-rot.	0.30	0.80	0.04	35.0
Y_0-rot.	0.30	0.80	0.04	35.0

Thresholds for the computed machine forces were also implemented in order to avoid feeding back machine cylinder stiction forces, as well as pressure oscillations due to the pumps, hose elasticity, etc.. These thresholds were in the range of 6,000 N. End-effector forces as high of 55,000 N were encountered.

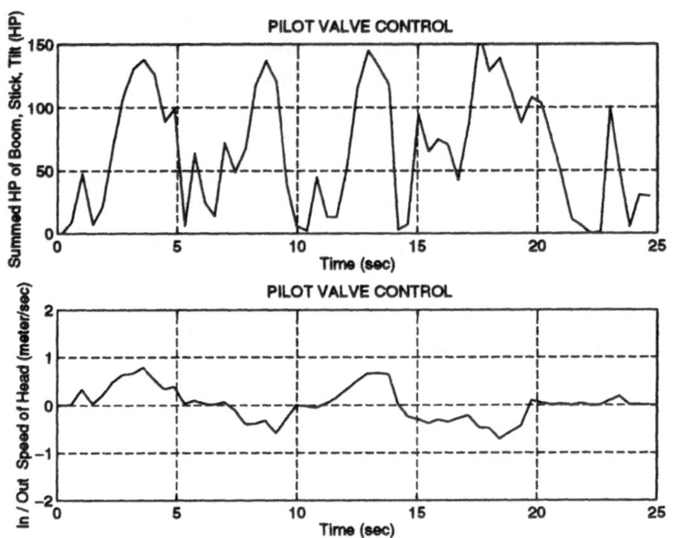

Figure 8. Machine Power and Head Speed with Valve Control

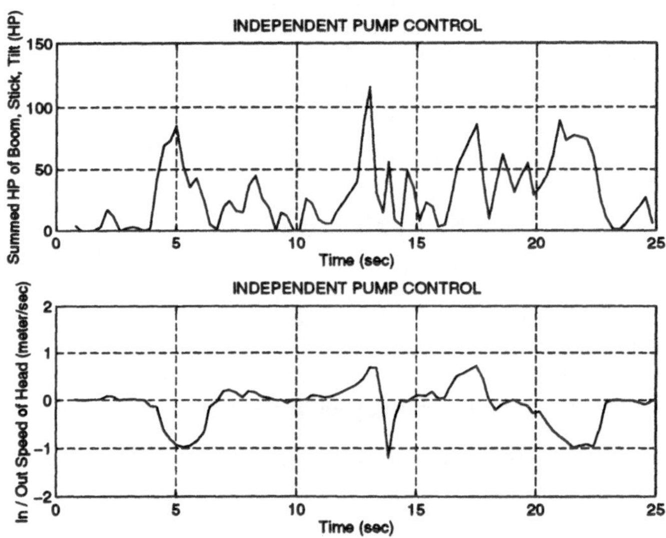

Figure 9. Machine Power and Head Speed with Pump Control

The overall haptic feedback provided by the above scheme was excellent. Although a thorough set of measurements could not be completed due to the difficulty of working with such a large machine, the ability to apply controlled forces was demonstrated. For example, Figure 10, displays the result of an experiment requiring the operator to apply an end-effector force against the ground as displayed on the screen of a lap-top computer. By contrast, if no stiffness feedback is employed, the task becomes impossible to perform (Figure 11). Finally, Figure 12 shows the radial force encountered during tree-cutting.

These experiments have shown that stiffness feedback presents a viable way of controlling a large excavator-based machine in a stable fashion with sufficient haptic feedback to enable the application of controlled forces. Further work is necessary to improve the sensed cylinder forces, to quantify the degree of "transparency" provided by the stiffness feedback, and to demonstrate that real tasks can be completed faster when force information is displayed to the operator.

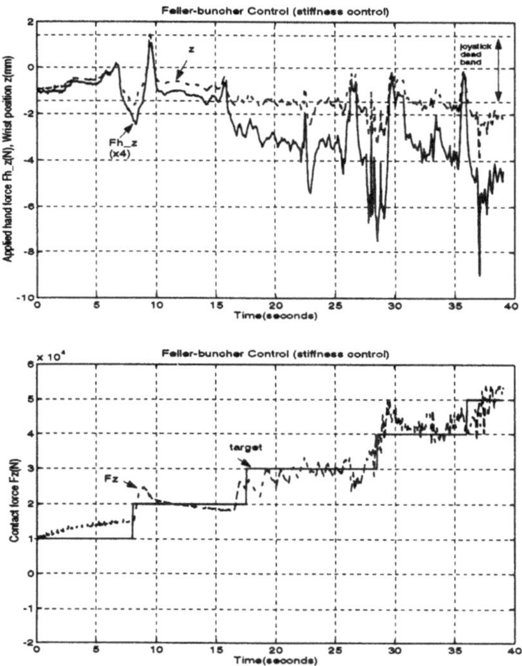

Figure 10. Controlled Force Application using Stiffness Feedback

8. Conclusions

Coordinated motion control in either pump or valve controlled systems was found by experienced operators to provide smoother machine motion and less damage to the tree stems during cutting. It was also found that novice operators were able to operate the machine more readily than with the standard controls.
It was found that, although somewhat more complex, a pump controlled machine is both faster and consumes lower power than a valve controlled machine.
Stiffness control was found to be a stable and effective method of controlling force application in endpoint velocity controlled systems.
As a result of these experiments, we feel that there is a considerable potential for improvement in the human interfaces in excavator type machines.

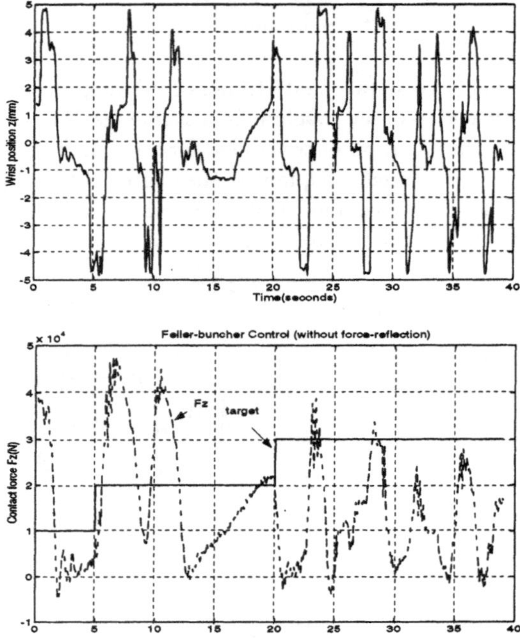

Figure 11. Attempts at Controlled Force Application without Force Feedback

9. Acknowledgements

We would like to acknowledge the following organizations and persons who greatly contributed to the success of this project. RSI Research of Sidney B.C. integrated the pump controlled hydraulic system (Derek Foster), and the joint angle sensors into the machine. The late George Olorenshaw of Vancouver designed the hand controller used on the feller buncher. Project management was carried out by Rudi Wagner and, arrangements for the field tests and operator interviews were carried out by Marv Clark of the Forest Engineering Research Institute of Canada.

Support for the project was derived from the Advanced Research Institute of B.C., Industry Canada, Western Diversification and 6 member companies of FERIC.

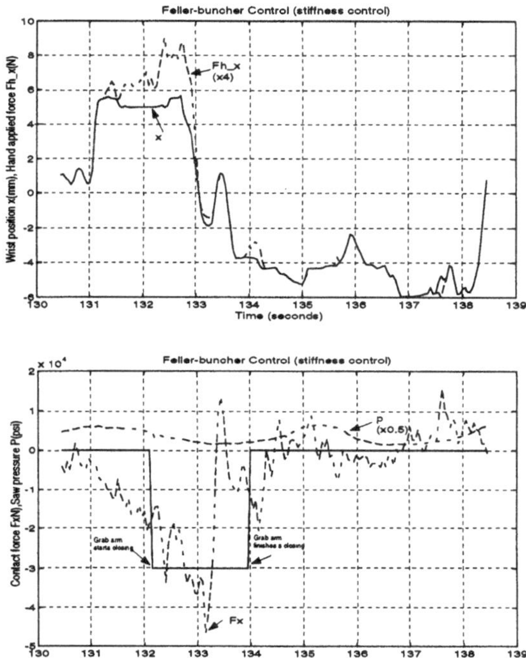

Figure 12. Hand-Controller and Feller-Buncher Radial Displacements and Forces During Tree-Cutting Process

References

[1] Edwin G. Johnsen and William R. Corliss. *Human Factors Applications in Teleoperator Design and Operation.* John Wiley and Sons, New York, New York, 1971.

[2] B.Hannaford, L.Wood, B.Guggisberg, D.McAffee, and H.Zak. Performance evaluation of a six-axis generalized force-reflecting teleoperator. Technical Report 89-18, California Institute of TechnologyPasadena, 1989, Pasadena, California, Jun. 1989.

[3] Daniel E. Whitney. Resolved motion rate control of manipulators and human prostheses. *IEEE Transactions on Man-Machine Systems*, MMS-10(2):47–53, Jun. 1969.

[4] John M. O'Hara. Telerobotic control of a dextrous manipulator using master and six-dof hand controllers for space assembly and servicing tasks. In *Proceedings of the Human Factors Society-31st Annual Meeting*, pages 791–795. Human Factors Society, 1987.

[5] U. Wallersteiner, P. Stager, and P.D. Lawrence. A human factors evaluation of teleoperator hand controls. In *International Symposium on Teleoperation and Control*, pages 291–296, Bristol, England, Jul. 1988.

[6] Bjorn Lofgren. Kranspetsstyrning. Technical Report Meddelande Nr 18, Skogsarbeten, Box 1184, 16422 Kista, Sweden, 1989.

[7] M. Ostoja-Starzewski and M. Skibniewski. A master-slave manipulator for excavation and construction tasks. *Robotics and Autonomous Systems*, 4:333–337, 1989.

[8] Darrell Wong, Peter D. Lawrence, and Ahmed Ibrahim. Efficient computation of the mass matrix for tree structured systems using orthogonal complements. In *Canadian Conference on Electrical and Computer Engineering*, pages 342–346, Montreal,

Canada, Sep. 1989.

[9] M. Zhu. Master-slave force-reflecting resolved motion control of hydraulic mobile machines. Master's thesis, University of British Columbia, December 1994.

[10] N. Sepehri, P.D. Lawrence, F. Sassani, and R. Frenette. Resolved-mode teleoperated control of heavy-duty hydraulic machines. *ASME Journal of Dynamic Systems, Measurement and Control*, 116(2):232–240, Jun. 1994.

[11] Alan J. Hewitt. Hydraulic circuit flow control. U.S. Patent, allowed Jan 12, 1994, 1994. U.S. filing Jan 21, 1993.

[12] Dragan Stokic, Miomir Vukobratovic, and Dragan Hristic. Implementation of force feedback in manipulation robots. *International Journal of Robotics Research*, 5(1):66–76, Spring 1986.

[13] M. Ostoja-Starzewski and M. Skibniewski. A Master-Slave Manipulator for Excavation and Construction Tasks. *Robotics and Autonomous Systems*, 4:333–337, 1989.

[14] S.E. Salcudean and N.M. Wong. Coarse-fine motion coordination and control of a teleoperation system with magnetically levitated master and wrist. In *Third International Symposium on Experimental Robotics*, Kyoto, Japan, Oct 28-30, 1993.

[15] N. Parker. Application of force feedback to heavy duty hydraulic machines. Master's thesis, University of British Columbia, October 1992.

[16] M. Zhu and S. Salcudean. Achieving transparency for teleoperator systems under position and rate control. In *Proceedings of IROS'95*, Pittsburgh, Philadelphia, August 5-9 1995.

[17] N.R. Parker, S.E. Salcudean, and P.D. Lawrence. Application of Force Feedback to Heavy Duty Hydraulic Machines. In *Proceedings of the IEEE International Conference on Robotics and Automation*, pages 375–381, Atlanta, USA, May 2-6, 1993.

[18] C.H. An, C.G. Atkeson, and J.M. Hollerbach. *Model-Based Control of a Robot Manipulator*. MIT Press, Cambridge, MA, 1988.

[19] U. Wallersteiner, P. Lawrence, and B. Sauder. A human factors evaluation of two different machine control systems for log loaders. *Ergonomics*, 36(8):927–934, Aug 1993.

[20] Marvin Clark. Coordinated motion of a feller-buncher. Technical Report TN-223, Forest Engineering Research Institute of Canada, Vancouver, B.C., Canada, 1994.

Chapter 5

Perception

Perception implies a level of understanding of sensory information higher than just the measurement of physical quantities. Typically, perception techniques rely on the matching to, or creation of, underlying models of objects. The choice of the correct underlying model has a significant effect on the efficiency and precision of perception. This chapter explores these issues for a number of different sensory modalities available to robots.

Tomasi, Zhang, and Redkey develop a real-time system for reconstructing shape and motion of objects in the robot environment from a time sequence of images. Experiments performed in a 2-dimensional environment (planar motion) demonstrate the feasibility of their approach and identify sensitivity to noise and calibration error.

Krotkov, Klatzky, and Zumel describe an approach for discriminating materials by analyzing the acoustic response to a mechanical impulse. In particular, they show that the relationship between response decay rate and material type is not an invariant with frequency as previous studies had suggested. Their experiments indicate that the dependence of decay rate on frequency may well encode material type in a shape invariant manner, which has yet to be identified.

Tonko, Schafer, Gengenbach, and Nagel describe an approach for visual tracking of known polyhedral objects in real-time. They employ optical flow techniques to estimate the object's velocity. This information is used to augment a feature-based pose estimator. The reported results underline the synergy and reduced cycle time obtained by their multi-level approach.

Kaneko, Kanayama, and Tsuji report on the development of techniques which permit an active antenna (mechanical probe) to 'detect contact location along its length via active palpation. They utilize the variation in beam stiffness with contact point to infer contact location. They describe how tangential friction at the contact effects precision, and report on experiments which demonstrate a method for compensating for these errors.

Menezes, Dias, Araújo, and De Almeida describe a stochastic model for sonar sensors and their activity that enables the detection of obstacles. The techniques are applied to the navigation of a mobile platform around obstacles by using the perceived information to modify potential fields in the navigation module.

Experiments With a Real-Time Structure-From-Motion System

Carlo Tomasi John Zhang
David Redkey
Computer Science Department, Stanford, USA
tomasi,zhang,dredkey@flamingo.stanford.edu

Abstract

We present a real-time system for the reconstruction of shape and motion from a sequence of single-scanline images. With this system, experiments on reconstruction can be run effortlessly and make it possible to explore the delicate sensitivity aspects of the problem. We identify three singular values of a certain matrix of image measurements as the key elements for a sensitivity analysis. Our experiments suggest that reconstruction is feasible with sufficient accuracy at least for navigation.

1. Introduction

A vision system that can provide scene shape and camera motion information reliably and in real time would be of great usefulness, for instance, in the control of manipulators and in robot navigation. While the literature on the subject is vast, no system is known that works reliably and in real time. In [6], we proposed a new formulation of this problem that faces directly the poor conditioning of shape reconstruction. In this formulation, the computation is expressed in terms of well-observable parameters only, using redundant data, and paying close attention to the numerical issues. A setup that allows running experiments with little effort is crucial for understanding the problem of shape and motion reconstruction from image sequences. In fact, the problem itself, and not just this or that implementation, is inherently sensitive to noise. Even a good algorithm will fail if the input data are not very good. Subpixel amounts of positional uncertainty in the measured image feature coordinates can defeat the most sophisticated algorithm. Camera miscalibration can have even worse effects, since it introduces systematic errors. Thus, attention to the implementation details is part of a successful system at least as much as attention to the mathematical and numerical aspects of the computation.

In summary, the following three elements are essential for a good shape and motion reconstruction system. First, the problem must be carefully formulated so that only well observable quantities are made part of the required solution. Second, the image measurements must be good enough to bring the input of the reconstruction algorithm to within the "basin of attraction" of a solution. Third, close attention must be paid to the numerical aspects of the computation, so that the particular implementation of the algorithm does not add failure modes of its own.

This research was supported by the NSF under contracts IRI-9496205 and IRI-9509149.

Our reconstruction system is for flatland, where the camera moves in a plane and images are single scanlines. A two-dimensional version of shape reconstruction, besides being an interesting intermediate step towards a three-dimensional system, is useful in its own right. For instance, indoor robots often travel on a smooth and level surface, so the camera scanline that shares a horizontal plane with the camera's optical center satisfies the flatland assumption. One or more separate vision systems, can then reconstruct one horizontal slice of the environment each. In this paper, we describe a real-time implementation of this system. This implementation required reinventing the computation completely with respect to [6]. Two linear stages, solved incrementally as new images are acquired, provide an approximate solution which is then fed to an iterative refinement procedure.

In the following section, we summarize our formulation of the reconstruction problem (section 2) and its solution (section 3). We then discuss the basic numerical issues (section 4) and implementation problems (section 5). In section 6, we show our experimental results, and in the conclusion (section 7) we point out the main problems to be overcome for a satisfactory solution.

2. Formulation

Suppose that the camera and the world points all live in a two-dimensional world. Point 0 serves as the origin of the global reference system. For every frame $f = 1, \ldots, F$ the camera records the tangents t_{fp} of the angles formed by the projection ray of the feature points $1, \ldots, P$ with that of feature point 0, so with $P+1$ feature points there are P tangents per frame. The tangent t_{fp} can be found by simple geometry to be (see also [6])

$$t_{fp} = \frac{u_f z_p - w_f x_p}{1 - u_f x_p - w_f z_p} \tag{1}$$

where (x_p, z_p) is the position of feature number p in the world and

$$\mathbf{k}_f = \begin{bmatrix} u_f \\ w_f \end{bmatrix} = \mathbf{m}_f / |\mathbf{m}_f|^2 \tag{2}$$

is the vector obtained by reflecting the camera coordinates $\mathbf{m}_f$ across the unit circle. This reflection is introduced to make equation (1) bilinear in motion and shape.

The FP measurements $t_{11}, \ldots, t_{FP}$ can be collected into an $F \times P$ matrix T. Each row represents one snapshot, and each column is the evolution of one tangent over time. If the reflected camera coordinates u_f, w_f and the shape coordinates x_p, z_p are collected in a $F \times 2$ reflected-motion matrix and a $2 \times P$ shape matrix, equation (1) can be rewritten in matrix form for the entire sequence as follows:

$$T = \pi(K, S) \tag{3}$$

where the projection function π operates on the f-th row of K and on the p-th column of S to produce entry t_{fp} of T according to equation (1).

3. Solution Procedure

The matrix equation (3) is now solved for shape S and reflected motion K in a series of steps, each of which either solves a linear system or takes a ratio of scalars. Initial estimates of shape are refined and the new camera coordinates are computed every time a new image becomes available. We first list the steps of the procedure, and we then show how each of them works.

1. Find shape $\hat{S}$ up to an affine transformation for each quadruple of points with subscripts $(0,1,2,p)$ where p ranges from 3 to P. Points $0,1,2$ establish a common affine reference system for all the quadruples.

2. Compute Euclidean shape S by determining a 2×2 matrix A such that

$$S = A\hat{S}\ . \tag{4}$$

3. Compute the matrix K of reflected camera positions from equation (3).

4. Determine the matrix M of camera positions by reflecting the rows of K back across the unit circle through the inverse of transformation (2),

$$\mathbf{m}_f = \mathbf{k}_f/|\mathbf{k}_f|^2\ . \tag{5}$$

We now show that all these steps involve solving a linear system or computing ratios of scalars.

1. If the scalar projection equation (1) is repeated three times for points $1,2,p$, the reflected camera coordinates u_f, w_f can be eliminated to yield the following homogeneous linear equation (see [6] for details)

$$\begin{aligned} a_1^{(p)}t_{f1}(t_{fq}-t_{fr}) + a_2^{(p)}t_{fr}(t_{fq}-t_{f1}) + a_3^{(p)}t_{f1}(1+t_{fq}t_{fr}) \\ +a_4^{(p)}t_{fq}(1+t_{f1}t_{fr}) + a_5^{(p)}t_{fr}(1+t_{f1}t_{fq}) = 0\ . \end{aligned}$$

where

$$\begin{aligned} a_1^{(p)} &= -x_p(x_1-x_2) - z_p(z_1-z_2)\ , \quad a_2^{(p)} = -x_1(x_2-x_p) - z_1(z_2-z_p)\ , \\ a_3^{(p)} &= -x_2z_p + z_2x_p\ , \quad a_4^{(p)} = x_1z_p - z_1x_p\ , \quad a_5^{(p)} = -x_1z_2 + z_1x_2\ . \end{aligned} \tag{6}$$

Writing this equation once for every frame $f = 1,\ldots,F$ yields an $F\times 5$ homogeneous linear system in $a_1^{(p)},\ldots,a_5^{(p)}$:

$$T\mathbf{a}^{(p)} = 0\ . \tag{7}$$

The solution can only be determined up to a multiplicative constant, that is, only $\lambda\mathbf{a}^{(p)}$ can be determined. An $F\times 5$ system of the form (7) must be solved for every point $p = 3,\ldots,P$.

The affine coordinates of point p are then (see [6]) $\hat{x}^p = -a_5^{(p)}/a_7^{(p)}$ and $\hat{z}^p = -a_6^{(p)}/a_7^{(p)}$. These coordinates can be collected into a $2\times P$ affine-shape matrix

$$\hat{S} = \begin{bmatrix} 1 & 0 & \hat{x}_3 & \cdots & \hat{x}_P \\ 0 & 1 & \hat{z}_3 & \cdots & \hat{z}_P \end{bmatrix}\ . \tag{8}$$

2. The affine coordinates in $\hat{S}$ differ from the Euclidean coordinates by a 2×2 transformation A (see equation (4)). In order to compute A, we notice that the final solution is determined up to a similarity. Therefore we can assume without loss of generality that

$$A = \begin{bmatrix} 1 & a \\ 0 & b \end{bmatrix} \tag{9}$$

which amounts to keeping the origin at point 0 and the unit point of the x axis at point 1. Then, equations (4), (8), and the last four equations (6) yield

$$\lambda a_2^{(p)} = \hat{x}_p + a(\hat{z}_p - 1) \, , \quad \lambda a_3^{(p)} = -b\hat{x}_p \, , \quad \lambda a_4^{(p)} = b\hat{z}_p \, , \quad \lambda a_5^{(p)} = -b \, .$$

Using the first of these equations to eliminate λ gives

$$\begin{bmatrix} a_3^{(p)}(1-\hat{z}_p) & a_2^{(p)}\hat{x}_p \\ a_4^{(p)}(1-\hat{z}_p) & a_2^{(p)}\hat{z}_p \\ a_5^{(p)}(1-\hat{z}_p) & -a_2^{(p)} \end{bmatrix} \begin{bmatrix} a \\ b \end{bmatrix} = \begin{bmatrix} a_3^{(p)}\hat{x}_p \\ a_4^{(p)}\hat{x}_p \\ a_5^{(p)}\hat{x}_p \end{bmatrix} .$$

A triple of equations like these can be written for every $p = 3, \ldots, P$, so that the two unknown entries a, b of A are found as the solution of an overdetermined system of $3(P-2)$ equations,

$$C \begin{bmatrix} a \\ b \end{bmatrix} = \mathbf{c} \, . \tag{10}$$

3. Once Euclidean shape S is known, the matrix projection equation (3) is linear in the reflected camera coordinates K. Each of the rows of K,

$$\mathbf{k}_f = (u_f, w_f)^T \, ,$$

can be found as the solution to the overdetermined $P \times 2$ system

$$\begin{bmatrix} z_1 + t_{f1}x_1 & x_1 + t_{f1}z_1 \\ \vdots & \vdots \\ z_P + t_{fP}x_P & x_P + t_{fP}z_P \end{bmatrix} \mathbf{k}_f^T = \begin{bmatrix} t_{f1} \\ \vdots \\ t_{fP} \end{bmatrix} . \tag{11}$$

4. The camera motion matrix M can be found by the reflection (5).

5. Shape and motion computed from the previous steps are suboptimal because shape is computed one quadruple at a time, rather than from all the points at once, and because system (10) uses only four out of the five equations (6) to avoid nonlinear equations. Therefore, the resulting shape and motion are used as starting values for an iterative procedure that refines shape and motion by repeatedly solving the original projection equation (3). Since this equation is bilinear in motion and shape, the iterative procedure interleaves solutions to linear systems.

4. Numerical Aspects

To summarize, we need to solve the following linear systems:

1. $P-2$ versions of the $F \times 5$ system (7), one for each quadruple of points $0, 1, 2, p$ for $p = 3, \ldots, P$. This is a homogeneous system. The number of its rows grows with time, and the solution improves in quality. This system is solved incrementally to avoid unbounded storage and computation time.

2. The $3(P-2) \times 2$ system (10) for the computation of the two unknown entries a, b of the matrix A (equation (9)). The size of this system is fixed, but its solution must be recomputed afresh with every frame because its entries depend on the improving affine shape estimates $\hat{x}_p, \hat{z}_p$.

3. The $P \times 2$ system (11) that computes reflected camera coordinates u_f, w_f. Also this system is fixed size and isrecomputed for every frame.
4. P systems of size $F \times 2$ for the refinement of shape and F systems of size $P \times 2$ for the refinement of motion through equation (3). Both the size of the shape systems and the number of the motion systems grow with the number of frames. To keep storage and computation time fixed, we only remember a fixed-size set of past frames for this computation. The choice of these frames is discussed in section 5.

All linear systems can be easily updated if features disappear because of occlusions or other reasons or if features are added. However, disappearance of any of the reference features causes reconstruction to stop. We are working on overcoming this limitation. We have essentially two different types of linear systems: the growing, homogeneous, $F \times 5$ systems (7), and several fixed systems of size $m \times 2$ with different, usually large m. Both systems are first converted to square (5×5 or 2×2), upper-triangular systems by incorporating one row at a time into the R matrix of a QR decomposition. These square, triangular systems are then solved by backsubstitution. For the homogeneous system, one of the diagonal elements of R is zero in the absence of noise. With noise, we assume that the element that must be zeroed is the smallest diagonal element of R.

5. Implementation

In our experiments, we used a Pulnix CCD camera with a 6.6×8.8 mm sensor and a high quality Schneider Cinegon 1.8/4.8mm lens. Because of the wide field of view (105 degrees along the diagonal), distortion in unavoidable. We calibrated it away by the procedure described in [1]. Frames are acquired by a Digital J300 frame grabber that interfaces directly with the TurboChannel bus of a Digital Alpha 600 workstation. Features are tracked by a one-dimensional version of the system described in [4] at a rate of about one feature per frame per millisecond. Feature selection at this point requires user interaction and the camera is required to remain still until the selection is completed.

Although the tracker updates image feature coordinates at every frame, shape computation waits until the changes in these coordinates are large enough to warrant incorporating a new set of input data. To check for this event, we first monitor the RMS displacement of all the features in the scanline. Once this measure has exceeded one pixel, a new row of the matrix T of angle tangents is computed, and its RMS variation with respect to the previous row used for reconstruction is checked against another threshold (0.005 radians in our implementation). Only when this threshold is exceeded is the most recent frame passed to the reconstruction algorithm. Consequently, the more expensive part of the computation is performed only rather occasionally for a slowly moving camera. In summary, all the frames produced by the camera are tracked, but only a few of them, called *significant frames*, are used for reconstruction.

For the iterative refinement stage described in section 3 a set of 2^k *key frames* is stored, where k is a fixed number (3 in our implementation). These key frames are spread as uniformly as possible over the past tracking history by a replacement policy that after tracking about 2^K significant frames remembers one significant frame every 2^{K-k}.

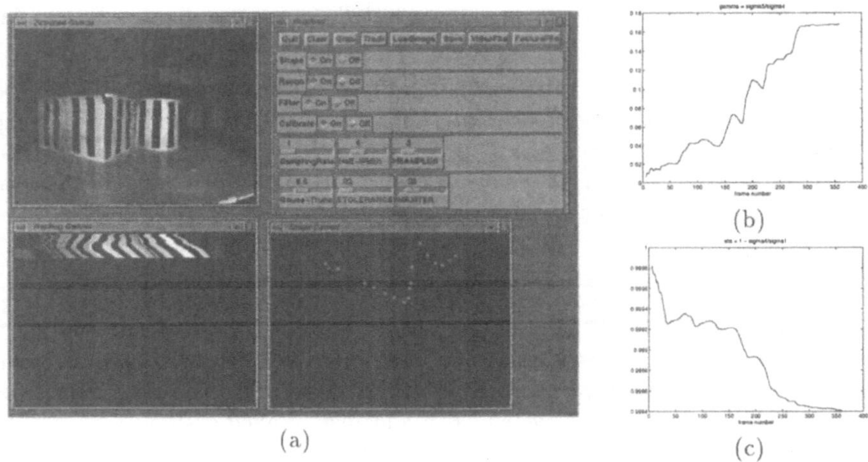

(a) (b) (c)

Figure 1. (a) The user interface. (b) Noise factor γ and (c) sensitivity η as a function of "significant frame" number for a forward moving camera.

6. Experiments

In our experiments, either the camera or the object is moved by sliding it on a table. Figure 1 (a) shows the program interface. The upper left window shows the scene with the selected features superimposed. All features are taken from the central scanline. The lower-left window stacks the significant frames on top of each other, and the lower-right window displays shape and camera position estimates. The menu in the upper right window allows setting various system parameters.

Sensitivity to noise is the dominant issue in the reconstruction problem. For meaningful results, the field of view of the camera must be wide enough [3], [2], motion must be sufficiently extended around the object [5], and image measurements must be both sufficiently accurate (low bias) and precise (low standard deviation).

From a numerical standpoint, since the homogeneous $F\times 5$ system (7) is expected to have exactly one affine shape solution with perfect data and nontrivial motion, its matrix T should be of rank 4. Of its singular values $\sigma_1 \geq \ldots \geq \sigma_5$, σ_4 should be nonzero and σ_5 should be zero. In reality, noise increases σ_5, and closeness to degenerate shape or motion decreases σ_4. For instance, if the camera does not move, all the rows of T are ideally equal, and T is rank 1: $\sigma_2 = \ldots = \sigma_5 = 0$. Also, when the camera's field of view approaches zero width (telephoto lens) the rank of T tends to 3: $\sigma_4 = \sigma_5 = 0$ [6]. Yet reconstruction under perspective requires a stable, substantial gap between the last two singular values, leading to a low *noise factor*

$$\gamma = \frac{\sigma_5}{\sigma_4} \ll 1$$

and a good conditioning of the rank-4 part of the system, leading to a low *sensitivity factor*

$$\eta = 1 - \frac{\sigma_4}{\sigma_1} \ll 1$$

(the condition number is usually defined as σ_1/σ_4; we prefer η because it remains between 0 and 1; $\eta = 0$ is ideal). These two conditions can be optimized by proper setup (η) and good image measurements (γ). Figures 1 (b) and (c) show the two parameters γ and η for a typical point quadruple seen by a forward-moving camera. The object is initially 25 cm away, and the four points span about 15 degrees of the field of view. These plots show the difficulty of the problem. While noise is relatively under control ($\gamma < 0.2$), the sensitivity factor η is dangerously close to 1 throughout. Furthermore, the sensitivity η declines very slowly when more frames are added (the camera moved forward by about 8 cm during the 360 frames), and the noise factor γ increases. The increase of the noise factor γ may be counterintuitive at first, but is due to the fact that new frames often add more noise than new shape information.

7. Conclusion

Our system for reconstructing shape and motion from image sequences in real time makes it possible for us to run many experiments with little effort, and at the same time forces us to consider the real difficulties of the problem.

Our experiments suggest that reconstruction is indeed possible with sufficient accuracy at least for navigation and as a guidance to manipulation. Sensitivity to image noise is by far the dominant problem in reconstruction, and can be understood by looking at the singular values of the matrix T that collects the tangents of the angles between projection rays. For good results, the field of view must be wide enough and the camera must move by a sufficiently large amount; image measurements must be accurate (good calibration) and precise (low noise); the formulation of the problem must be in terms of a minimal number of parameters (camera and feature positions, but no camera rotation); and the algorithms must be numerically sound.

The extension of reconstruction to three dimensions is mathematically far from straightforward, and the computation requires more time or resources than in two dimensions. However, the sensitivity of the problem should, if anything, improve, because the ratio of unknowns to measurements is reduced by a factor of 3/4 from approximately 2(P+F)/PF to 3(P+F)/2PF, where P is the number of points and F is the number of frames. In conclusion, the results are auspicious and encourage us to continue our investigation of solutions of the reconstruction problem.

References

[1] M. Fleck. Shape and wide-angle image. Tech. Rep. 04, U. of Iowa, 1994.

[2] B. K. P. Horn and E. J. Weldon Jr. Direct methods for recovering motion. *IJCV*, 2:51–76, 1988.

[3] J. J. Koenderink and A. J. van Doorn. Facts on optic flow. *Biol. Cyb.*, 56:247–255, 1987.

[4] J. Shi and C. Tomasi. Good features to track. *CVPR*, 593–600, 1994.

[5] C. Tomasi and T. Kanade. Shape and motion from image streams under orthography – a factorization method. *IJCV*, 9(2), 137–154, 1992.

[6] C. Tomasi. Pictures and trails: a new framework for the computation of shape and motion from perspective image sequences. *CVPR*, 913–918, 1994.

Robotic Perception of Material: Experiments with Shape-Invariant Acoustic Measures of Material Type

Eric Krotkov Roberta Klatzky Nina Zumel

Robotics Institute Psychology Department Robotics Institute
Carnegie Mellon University
5000 Forbes Avenue
Pittsburgh, Pennsylvania 15213

Abstract

We present an active approach for discriminating different materials by impulsively contacting (hitting) them, and sensing and interpreting the resulting sounds. In theory, the angle of internal friction is diagnostic of material, but invariant over object shape. In our experiments, we observe that the angle of internal friction is not invariant over the frequency of the sound spectrum for which it is estimated. Hence, samples of different shapes, which exhibit power concentration at different frequencies, exhibit different values of the angle of internal friction. However, the results suggest that shape-invariance may be encoded in the functional form of the relation between the angle of internal friction and frequency.

1. Introduction

Consider a metal rod and a wooden rod of the same length. When you strike the metal rod with your knuckle, it rings; when you strike the wooden rod, it produces a much shorter "thud" sound. This difference in the sound despite the same excitation is due to the difference in the way that the materials vibrate, which in turn is due to stress/strain properties. The rods sound different because they have fundamentally different material properties, so sound waves travel through them quite differently.

Now consider two metal rods that are identical except that one is twice as long as the other. Given the same excitation, the shorter rod will "ring" at a higher frequency than the longer rod. The rods sound different because the waves travel different distances inside them.

How can these differences in the way things sound, one due to material and one due to shape, be resolved? Or, in other words, *What acoustic information is diagnostic of material, but invariant over object shape?*

This fundamental question, and the example that led up to it, concern the sensory modality of audition. And that will be the central topic of this paper. However, we are developing a general approach, applicable to all sensing modalities. Before plunging into the central topic of the paper, we first describe the more general approach.

By definition, a material property is independent of the size and shape of a particular sample. Although there are visual cues to material properties (for example, surface luminance is a cue to the coefficient of friction), reliable determination of the material composition of an unknown object generally requires contact with it. Humans who wish to determine material properties show stereotypical patterns of manual exploration; they press, poke, tap, heft, squeeze, shake, rub, and strike, according to the type of information desired [3]. We are developing a robotic approach analogous to these patterns of human behavior.

In our approach, materials are disambiguated by actively contacting and probing them and by sensing the resulting forces, displacements, and sounds. One can visualize this capability by imagining a game of non-verbal "Twenty Questions," in which one player is the robot and the other player is any object placed in the robot workspace. The robot probes (presses, pokes, taps, etc) the object, in effect asking questions about the object stiffness, density, and other material properties. At the end of the game the robot announces its decision about the material composition of the object.

The capability to perceive material has many potential applications. In general, knowledge of material properties and classes can improve performance of many tasks involving physical interaction. In many real-world scenarios, such knowledge is not given in advance; instead, it must be determined at a worksite or in the field, without jigs or fixtures. Specific applications include grasping, non-destructive evaluation and inspection, reasoning about functionality, handling hazardous waste, recycling, excavating, and traversing natural terrain.

In this paper, we describe a shape-invariant measure of material type, derived through acoustic sensing, and present results of experiments that confirm theoretical predications that the measure is diagnostic of material type. We begin by briefly referring to related research. We present the theoretical framework for the shape-invariant acoustic measure in Section 3. Next, we describe our novel approach to estimating material type in Section 4, and we present experimental results in Section 5. We conclude the paper with a critical discussion of progress to date.

2. Related Research

Compared to the perception of shape or position, perception of material properties is a field in its early infancy.

The artificial intelligence, robotics, civil engineering, mechanical engineering, and materials literature documents two families of techniques to estimate mechanical and mass properties, one employing non-contact sensing, the other employing contact sensing. We explore this literature in detail elsewhere [2].

3. Theory

Wildes and Richards [4] have advanced a theoretical approach to recovering the material type of an object from the sound generated when it is struck. They restrict their attention to anelastic solids, and study the modulus of compliance as the key to understanding the vibration of the struck object. Following classical analysis they relate the modulus of compliance to the angle of internal friction. This is a shape-invariant property of a given material.

They propose two methods for determining the angle of internal friction of an unknown sample: one that impulsively excites the sample and then measures the acoustic decay rate; another that periodically (say, sinusoidally) excites the sample and then identifies the bandwidth of the acoustic signals. They did not experimentally verify either method, although they did cite supporting evidence from earlier empirical studies [1].

Let us consider the decay rate approach first, because experimentally it is simpler to provide an impulsive excitation than a periodic one. In this method, the angle of internal friction ϕ is determined by the time t_e it takes the amplitude of vibration to decay to $\frac{1}{e}$ of its original value after the material sample is struck. According to Wildes and Richards,

$$\tan\phi = \frac{1}{\pi f t_e}, \tag{1}$$

where f is the observed frequency associated with the amplitude. Thus, the problem of determining the angle of internal friction ϕ reduces to determining t_e.

Let θ be a retention parameter representing the proportion of the amplitude present at time t_i that is still present at t_{i+1}. For an exponential process, θ is constant for all i, and the amplitude $A(t)$ at time t is then given by

$$A(t) = A_0\theta^t ,$$

where A_0 is the initial amplitude.

If the amplitude has decayed to a proportion of $\frac{1}{e}$ of the initial value, then $A(t) = \frac{A_0}{e}$. Note that at this point, $t = t_e$, by definition. So, $\frac{A_0}{e} = A_0\theta^{t_e}$. Canceling A_0, taking the natural logarithm of both sides, and rearranging leads to

$$t_e = -\frac{1}{\log\theta} . \tag{2}$$

Thus, the problem of determining t_e reduces to determining $\log\theta$.

Assuming an exponential decay process,

$$A(t) = A_0\theta^t = A_0 e^{-t\log\theta} .$$

Taking logarithms yields $\log A(t) = \log A_0 - t\log\theta$. Thus, the plot of log amplitude against time will be linear with slope equal to $\log\theta$. So, we can determine $\log\theta$ by finding the envelope of the signal waveform as a one-dimensional curve, plotting this curve on a logarithmic scale (this will be linear), and identifying the slope of the plotted line.

Summarizing, in theory we can determine $\log\theta$ from the original waveform, and then determine t_e using (2), and finally determine $\tan\phi$ from (1).

4. Approach

We analyze the discrete digital signal $x[n]$ in four main steps: (1) Compute the signal spectrogram; (2) Determine where the contact transient ends, and where the "signal" begins; (3) Find bands of concentrated signal energy; (4) For each band, determine the angle of internal friction.

4.1. Spectrogram

The spectrogram of a signal describes the distribution of the signal energy in the time-frequency plane. The spectrogram is a popular representation in fields such as speech recognition and acoustic analysis. Formally, the spectrogram $S[l,k]$ is the squared modulus of $X[l,k]$, where

$$X[l,k] = \sum_{n=-\infty}^{+\infty} x[n]g[n-l]e^{-j\frac{2\pi k}{n}l}$$

is the discrete-time Fourier transform of a windowed version $x[n]g[n-l]$ of the original signal $x[n]$.

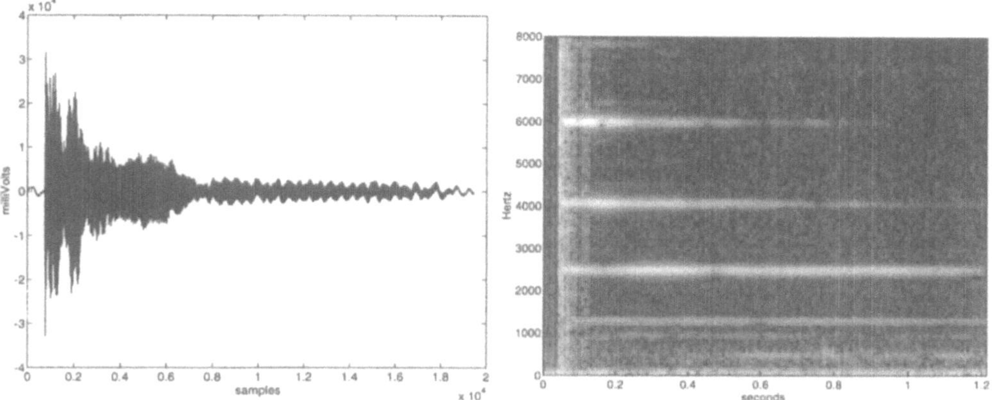

Figure 1. Microphone output after striking aluminum rod (left), spectrogram (right)

We compute the spectrogram by performing the following steps: (1) Split the given signal into $N_{overlap}$ overlapping segments; (2) For each segment, establish a Hanning window of size N_{FFT}; (3) Compute the Fourier transform of each windowed segment.

As an example, Figure 1 shows the original signal recorded at f_{sample} = 16 kHz after striking an aluminum rod. The figure also shows the spectrogram of that signal, computed with $N_{FFT} = 256$ samples and $N_{overlap} = N_{FFT}/2$. The energy is concentrated in four main bands, at approximately 1200, 2500, 4000, and 6000 Hz. Note that these bands are not harmonics of a common fundamental.

4.2. Transient

Due to the impulsive contact, the early part of the signal contains energy at all frequencies. This transient effect, which sounds like a click, does not convey meaningful modal information, so we desire to exclude this segment of the signal from analysis.

For this, we examine correlations between the spectrogram magnitudes at adjacent temporal windows, computed across frequencies. During the click, the spectrogram magnitudes are highly correlated from instant to instant. Immediately after the click, as energy begins to concentrate in relatively narrow bands, the spectrogram frequencies with high magnitudes are not the same as those in the click. This causes the correlation between adjacent windows to dip. Well after the click, the spectrogram magnitudes are again highly correlated. Based on these observations, we determine where the signal begins by identifying when the correlation coefficients rise from the dip caused by the transition from the click to the residual excitation of the rod.

Let $S[t]$ be a vector of length $N_{FFT}/2$ representing the spectrogram magnitudes at time t. In effect, this is a column in the spectrogram shown in Figure 1. For all pairs of time-adjacent vectors, we compute the correlation coefficient $\rho_t = corr(S[t], S[t+1])$. We search for the time at which ρ_t takes on its globally minimum value. We then search for the time, following that, at which ρ_t takes on a locally maximum value. We treat this as the end of the transient. Figure 2 illustrates the time selected as the start of the signal.

4.3. Bands

Once we have discovered when the signal starts, we then identify those frequency bands with a significant concentration of energy. We accomplish this in five steps: (1) Find initial location of bands; (2) Refine initial location estimates; (3) Filter out weak bands; (4) For

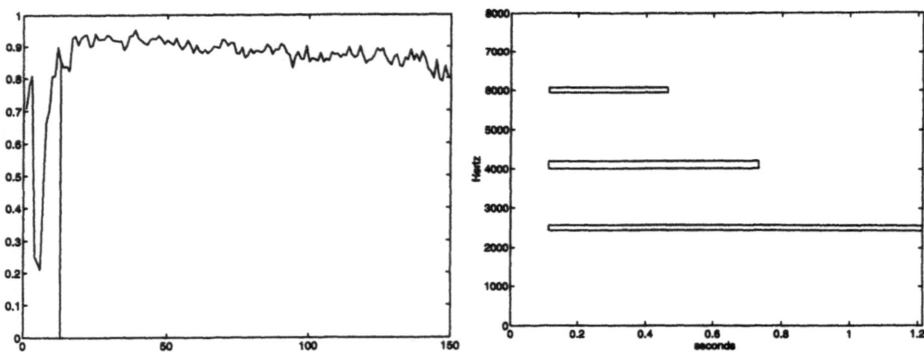

Figure 2. Correlation versus time for data in Figure 1 (left), spectral bands (right)

each remaining band, determine when signal starts and ends; (5) For each remaining signal in a band, determine the frequency with greatest power.

First, to find the initial location of bands, we begin with a thresholding operation: For each time step, we mark as background those spectrogram magnitudes that contribute little to the total power at that time. In the current implementation, the threshold value is set to be the power we would expect if the energy was distributed uniformly across all frequencies. Next, we perform a region growing operation and connected component analysis on the surviving magnitudes. This results in a coarse estimate of the band locations.

Second, to improve this coarse estimate, we seek the smallest sub-bands that contain a given fraction (currently, 99 percent) of the within-band power. This produces bands that are significantly more focussed.

Third, we eliminate those bands which contain significantly less power than the others. In the current implementation, we delete all bands that contribute less than 1 percent of the total power.

Fourth, for each remaining band, we identify when the signal starts and ends. The start is the point of maximum power, and the end is the point at which the power has declined to 0.2 percent of the maximum power.

Finally, for each band, we determine the frequency which contains the most power. We store this value to be used in the computation of $\tan\phi$.

Figure 2 illustrates the result of these five processing steps applied to the spectrogram in Figure 1. The algorithm identified three bands. The band at 1000 Hz was discarded because it contained less than one percent of the total power.

4.4. Angle of Internal Friction

For each band computed in the previous stage, we fit a line to the within-band log power. At the same time, we compute the goodness (r value) of the linear fit and the length of the line. We filter out those lines with $r < 0.866$ (the fit accounts for 75 percent of the variance) and with length less than 10 time steps. The slope of each line determines the $\log\theta$ term in (2). Now it is possible to determine $\tan\phi$ for each frequency band by substituting (2) into (1).

Figure 3 illustrates the total power associated with the three bands (band 1 is at 6000 Hz, band 2 is at 4000 Hz, and band 3 is 2500 Hz). It also shows the background, defined as the sum of all spectrogram magnitudes that did not pass the threshold test described in

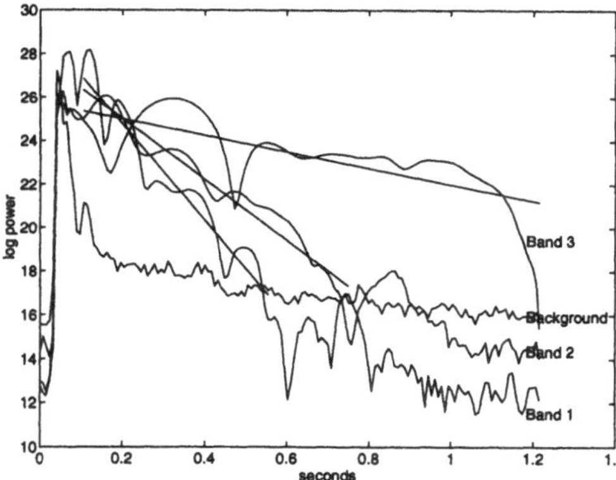

Figure 3. Linear fits to log power within bands

the previous section. In addition, the figure shows the lines fit to the three power curves. Bands 1 and 2 are reasonably linear (r values above 0.95, that is, the fit accounts for more than 90 percent of the variance), and band 3 is not (r value of 0.70).

5. Experiments

To assess the validity of the decay rate approach to identifying the angle of internal friction, we produced thin rods of wood, brass, aluminum, glass, and plastic. For each material, we produced two rods, one of length $L = 15$ cm and one of length $2L$.

We suspended each rod by string from above. We struck the rods with "found" objects, including a soldering pencil stand (selected because its vibrations damped out more rapidly than did the other objects we tried) and the plastic handle of a screwdriver. We used an electret condenser microphone, and fed the signal to an analog/digital converter installed on a Macintosh workstation, operating at sampling rate of $f_{sample} = 22$ kHz.

Figure 4 plots the observed data relating $\pi \tan \phi$ and the frequency of greatest power (the π was omitted from the figure labels). Each point represents a single band; a given trial may provide from 1–4 points. We fit quadratic functions $\pi \tan \phi = a_2 f^2 + a_1 f + a_0$ to the data points for each material, finding the a_i coefficients minimizing the squared error. The fitting procedure combined data from short and long samples, with the exception of brass, for which the single frequency available for the short rod appears anomalous with respect to the function obtained with the long rod.

In summary, the salient features of these graphs are the following:

(1) The variability associated with a single frequency tends to be reasonably small, relative to the variability across frequencies, suggesting some stability to the estimated $\tan \phi$ at a frequency.

(2) The $\tan \phi$ values exhibit clear variation across frequencies. Hence different samples, which produce bands at different frequencies, will give different distributions of $\tan \phi$.

(3) With the exception of brass, it appears that the relation between $\tan \phi$ and frequency can feasibly be fit by a single quadratic, with r^2 ranging from 0.35 to 0.84 in the present data.

(4) Whether the quadratic is concave upward or downward varies with the material.

Table 1 tabulates the parameters of the quadratic. The R^2 term represents the variance

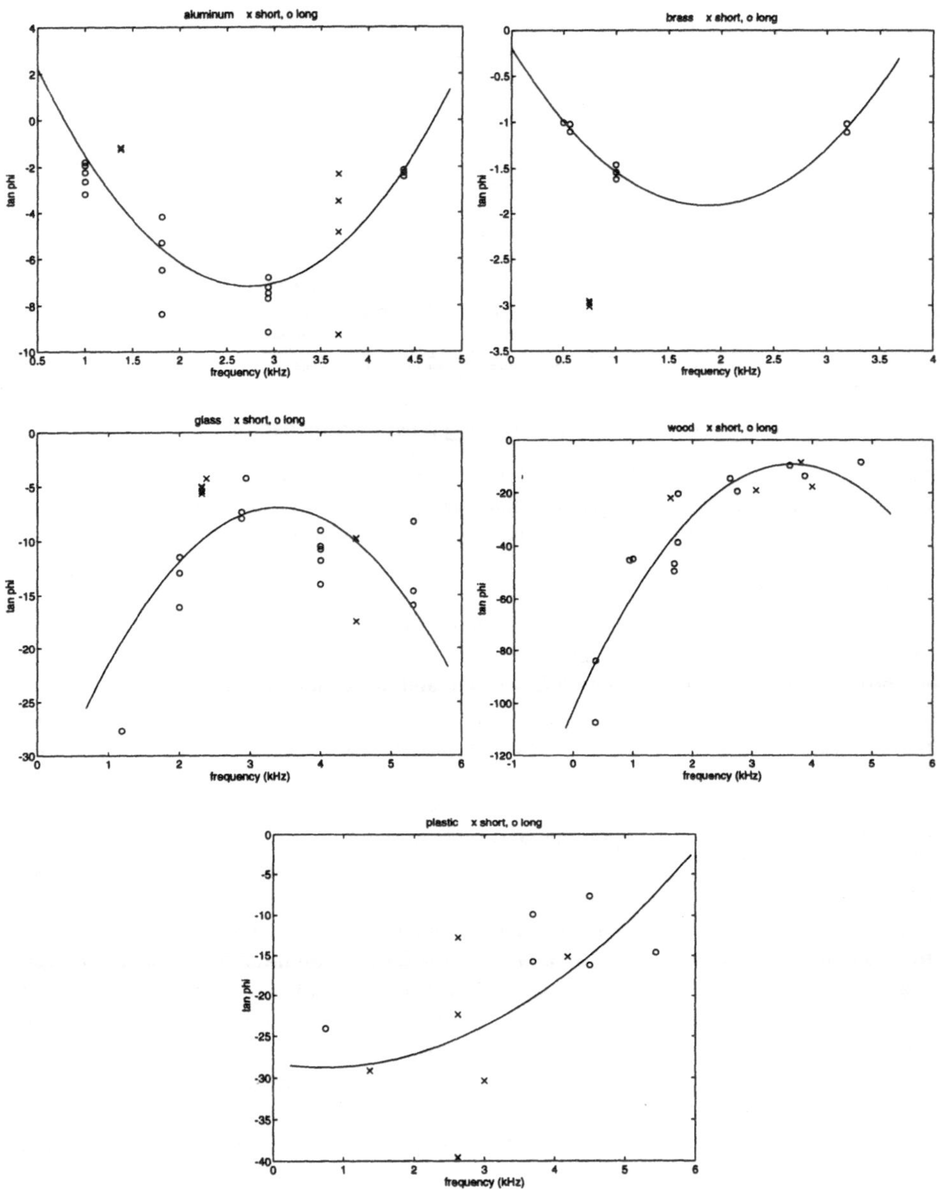

Figure 4. Experimental results

accounted for by the quadratic function, which we interpret as the goodness of the fit.

6. Discussion

On the basis of the theory presented in Section 3, we expected the value of $\tan \phi$ to vary with material type and to remain constant despite changes in the length of the material sample. We find that the angle of internal friction is not invariant over the frequency of

Material	a_2	a_1	a_0	r^2
Aluminum	1.88	-10.29	6.92	0.61
Brass (long)	0.49	-1.84	-0.19	0.96
Glass	-2.54	17.24	-36.16	0.38
Wood	-6.97	51.09	-102.76	0.84
Plastic	0.97	-1.44	-28.19	0.35

Table 1. Parameters of quadratic function $\pi \tan \phi = a_2 f^2 + a_1 f + a_0$

the sound spectrum for which it is estimated. We are surprised to observe the frequency dependence of the $\tan \phi$ values, since the theory does not predict it. This frequency dependence suggests that the shape invariance may be encoded in the functional form of the relation between $\tan \phi$ and frequency. Further, the surprising results indicate the need for a new theory.

Our experience during the course of the experiments has yielded a number of methodological insights that will influence future investigations. The first observation is that the experimental method for suspending the sample is more important than we first believed. In particular, the motion of the sample after striking must be taken into account or otherwise "matched." The second observation is that a higher sampling rate is needed to acquire reliable measurements from non-metals. Finally, the spectrogram appears to be a very powerful representation that is well-suited to the discrimination task at hand.

We hope to apply the insights gained by this investigation in at least two fields of application. In the field of non-destructive evaluation, we will develop new materials testing procedures based on quantitative measurement of the angle of internal friction. In the field of virtual reality, we will develop low-bandwidth representations of real-world sounds that can be used to create multimodal events.

A great deal of work remains before such applications are feasible. Future work will concentrate on implementing these methodological insights in a new experimental setup including more repeatable striking mechanisms and faster sampling devices. Future work will also expand the inquiry to encompass more thorough study of shape effects, beginning with variable length rods and extending to plates, solids, and irregularly formed objects.

In the more distant future, alternative measures and approaches need to be pursued before the new field of perception of material emerges from its early infancy, and realizes its potential for revolutionizing robotic interaction with the real world.

Acknowledgments

We thank Richard Stern for his significant contributions to the work reported here.

References

[1] A. Gemant and W. Jackson. The measurement of internal friction in some solid materials. *Philosophical Magazine*, 157:960–983, 1937.

[2] E. Krotkov. Perception of Material Properties by Robotic Probing: Preliminary Investigations. In *Proc. IJCAI*, pages 88–94, Montreal, August 1995.

[3] S. Lederman and R. Klatzky. Hand movements: A window into haptic object recognition. *Cognitive Psychology*, 19:342–368, 1987.

[4] R. Wildes and W. Richards. Recovering material properties from sound. In W. Richards, editor, *Natural Computation*, pages 357–363. MIT Press, Cambridge, Massachusetts, 1988.

Multi-Level 3D-Tracking of Objects Integrating Velocity Estimation based on Optical Flow and Kalman-Filtering

M. Tonko[†], K. Schäfer[†], V. Gengenbach[‡], H.-H. Nagel[†‡]

[†] Institut für Algorithmen und Kognitive Systeme, Universität Karlsruhe (TH)
Postfach 6980, D–76128 Karlsruhe, Germany
and
[‡] Fraunhofer-Institut für Informations- und Datenverarbeitung
Fraunhoferstr. 1, D–76131 Karlsruhe, Germany

Email: tonko@ira.uka.de, kschaefe@ira.uka.de, gen@ira.uka.de, hhn@iitb.fhg.de

Abstract This contribution describes an extensively tested approach to three-dimensional tracking of a known polyhedral object at two levels of visual robot control, a position-based tracking level and a velocity-based tracking level. At the higher position-based level, an iterated extended Kalman-Filter is used to track a workpiece at a control-cycle of about two hundred milliseconds. This slower cycle operates on top of the video-rate tracking-loop of the velocity-based tracking level. While features such as vertices, edges, and ellipses extracted from gray-value images are used to adjust the 3D pose-estimate at the higher level of control, optical flow is used to initialize the velocity parameters of the upper level and to locally update poses at the lower level.

1. Introduction

Vision-guided disassembly has been hindered by the fact that vision-systems were not capable to operate at video-rate in the presence of complex tasks like tracking a known solid workpiece in order to control a robot manipulator during a dismantling task. As state-of-the-art systems like *MiniVISTA*[1] are now available (see [2]), which are able to perform the computation of the first and second derivatives of the spatio-temporal gray-value function as well as image evaluation tasks such as contour point detection in video-rate, efficient vision-guided servoing methods can be applied to employ disassembly units that are able to cope with the combination of the following boundary conditions:

[1] *MiniVISTA* is an acronym for Miniaturized Visual Inspection System for Technical Applications.

- During disassembly, workpiece positions and points of manipulation may not be known with the same precision as during assembly.
- Parts may slip during disassembly in an unexpected and unpredictable manner.
- The workspace needs to be larger with the consequence that the calibration problems for the camera-robot configuration become more severe.
- Manipulator control during the approach based on machine vision may be more efficient than tactile search which can be employed only after the tool has contacted the workpiece's surface.

Researchers have focused on image-based servoing methods [1, 3, 4, 8], where parameters of extracted image features are used as reference and feedback parameters of the control loop. Admittedly, such methods are very fast during control, but they require much time to compute the sequence of reference parameters necessary to guide the robot during a dismantling task. Furthermore, if a dismantling task requires the application of several power tools with different geometries, the computation of the sequence of reference parameters must be repeated for each tool. In contrast to this, position-based servoing methods [5] use relative positions, e.g. the pose of a workpiece relative to the actuator-camera mounted on the robot manipulator, as control parameters. Naturally, during control, position-based systems require more time to compute the feedback parameters than image-based systems. As computing power is steadily increasing, this seems to be no substantial drawback.

Even though position-based systems operating at video-rate are available, certain problems still remain. Consider a moving object that does not allow for the consecutive estimation of its position and orientation in video-rate, because it moves simply to fast w.r.t. the cycle rate of the control loop. In this case, the estimation of the three-dimensional motion parameters of the object in addition to the estimation of its pose will help to solve this problem.

2. System Overview

Basically, our system can be divided into two parts, the vision system *MiniVISTA* and the robot with its associated controller (see figure 1). *MiniVISTA* is a modular real-time vision system, which is capable to operate at video-rate. Features of *MiniVISTA* are image aquisition, iconic evaluation of image data and, simultaneously, the computation of the first- and second-order derivatives of the spatio-temporal gray-value function (see [2]). Two digital signal processors (DSP1, DSP2) are used to compute vertices, edges, and ellipses by means of the iconic contour point information. DSP3 uses the extracted features to update and predict the object's pose at the position-based tracking level. Six Transputers (T) are employed to compute the optical flow (T1–T4) based on the derivatives of the spatio-temporal gray-value function and to cluster (T5, T6) the optical flow field. T7 determines the object's velocity using the clustered optical flow field. One can say that the DSPs realize the position-based tracker (PT) and the transputers realize the velocity-based tracker (VT).

The robot manipulator is equipped with a CCD-camera mounted on the endeffector, and with automatically attachable and detachable power tools like screwdriver and gripper. Every 80ms, the robot controller submits the true pose of the TCP

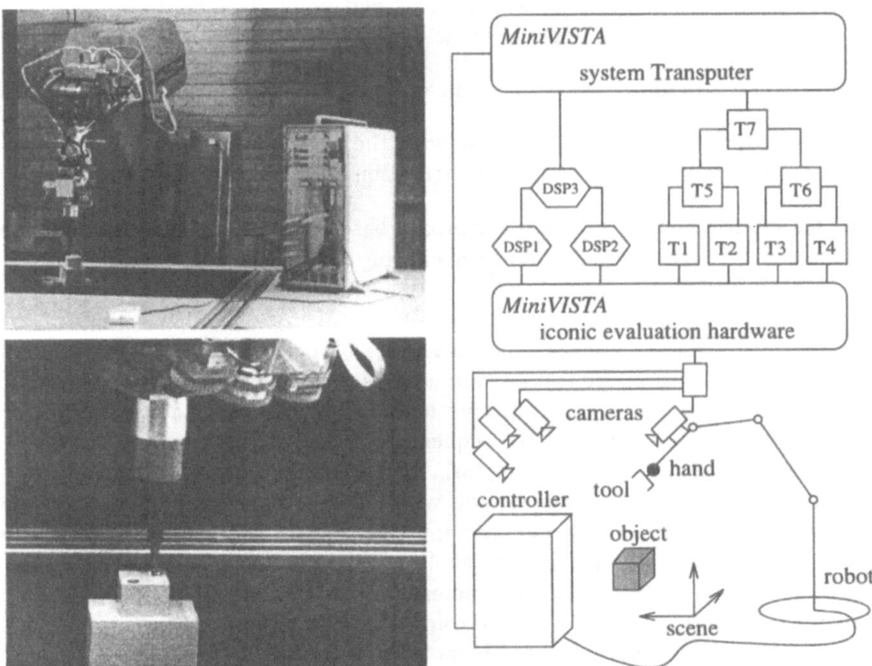

Figure 1. Overview of the system: *MiniVISTA* and robot (upper left), loosening of a screw (lower left), system schematic and network topology (right).

and accepts a new pose to move to. The interpolation of the true pose and the new pose faciliates the estimation of the robot's position at each instant of time.

Three stationary CCD-cameras residing atop the scene are used to monitor the workspace in order to spot new workpieces. The classification of a workpiece and the computation of its initial pose in scene coordinates relies on a stereo triangulation and verification scheme.

Prior to disassembly, several actions take place:

- The initial pose of the workpiece, a homogeneous transformation relating object frame o to scene frame s is determined during workpiece classification using a polyhedral CAD-model.

- The cameras are internally and externally calibrated, i.e. the homogeneous transformations $\mathcal{T}_{hc}$ relating camera to hand frame, $\mathcal{T}_{ic}$ relating camera to image frame, and $\mathcal{T}_{rs}$ relating scene frame to robot base are known. The subscripts h, c, i, r denote the robot hand frame, the endeffector camera frame, its image plane, and the robot base, respectively.

- All available power tools are calibrated. Calibration is done with respect to the hand frame h in order to convert camera-relative poses to tool-relative poses at runtime.

3. Position-Based Tracking

The PT is realized as an iterated extended Kalman-Filter (IEKF), which estimates the pose of the object relative to the scene. In order to meet the boundary conditions (see Section 1), no restrictions concerning the position and orientation of the workpiece are imposed. This results in a six-dimensional filter state consisting of three angular and three translational parameters describing an unconstrained rotation matrix and an unconstrained translation vector that relate the object frame o to the scene frame s.

The filter state is updated by associating extracted vertices, edges, and ellipses to those projections of model-primitives which pass a hidden-surface test. During update, the most likely (MAP) pose of the object is determined by a Levenberg-Marquardt minimization technique, which is a combination of gradient descent and Newton Iteration.

The updated filterstate is predicted by applying a modified Rodrigues' formula to the filter state,[2] i.e. a closed-form solution to the differential motion constraint equation (MCE)

$$\dot{\boldsymbol{x}}_o = \boldsymbol{\omega}_o \times \boldsymbol{x}_o + \boldsymbol{v}_o, \tag{1}$$

assuming constant angular and translational velocities $\boldsymbol{\omega}_o$ and $\boldsymbol{v}_o$. Equation (1) describes the velocity $\dot{\boldsymbol{x}}_o$ of a point $\boldsymbol{x}_o$ in space. As the application requires the declaration of angular and translational velocity parameters of the object, these are determined by the VT. In the absense of the VT, they are set prior to disassembly and are assumed to be constant throughout the entire process.

As we update the object's pose using vertices, edges, and ellipses, the model of a workpiece is constrained to consist only of elements that project to those primitives. Of course, this also constrains the geometry of a real workpiece itself.

4. Velocity-Based Tracking

In order to track objects that move at considerably higher speeds than the PT is able to tolerate, we have realized, in addition and parallel to the higher level, this lower level of control of the robot manipulator that aims at the tracking of a moving object at video-rate, while the higher level is still updating the current pose for increased accuracy. Every time the PT comes up with a predicted pose of the object, the VT is re-initialized using this pose. It updates this pose by integrating the consecutively estimated three-dimensional velocities of the object over time. Additionally, the estimated angular and translational velocities of the object are saved in a FIFO-queue. This ensures that the estimated VT-velocities can also be applied to predict the filter state between successive PT-updates.

4.1. Problem Statement and Analytic Solution

Given the optical flow field of the image of the object, the object's pose estimated employing the PT, and all necessary homogeneous transformations at instant t of image aquisition, we want to compute the object's velocities relative to its own frame of reference in order to track the object while it is moving faster than the PT is able to tolerate. We assume that the solid object is moving with constant angular and translational velocities relative to the camera frame and relative to the object frame

[2] Modified in the sense that it fits our representation of the filter state.

within a short period of time.[3] Also, the knowledge of the hand-relative velocities of the TCP is prerequisite.

Let $\boldsymbol{x}_{c_j} = (x_{c_j},\ y_{c_j},\ z_{c_j})^T$ denote the j-th vertex of the workpiece model with respect to the camera frame c. A homogenized vector $\bar{\boldsymbol{x}}_{c_j}$ of $\boldsymbol{x}_{c_j}$ is determined using

$$\bar{\boldsymbol{x}}_{c_j} = \mathcal{T}_{co} \cdot \bar{\boldsymbol{x}}_{o_j} = \mathcal{T}_{hc}^{-1}\mathcal{T}_{rh}^{-1}\mathcal{T}_{rs}\mathcal{T}_{so} \cdot \bar{\boldsymbol{x}}_{o_j}. \tag{2}$$

As mentioned above, the transformations $\mathcal{T}_{hc}$ and $\mathcal{T}_{rs}$ are obtained by hand-eye calibration. The actual robot pose $\mathcal{T}_{rh}$ at the instant of image aquisition is calculated by interpolating robot poses that have been interchanged during the last robot-*MiniVISTA* handshake. Matrix $\mathcal{T}_{so}$ equals the last pose estimate of the VT. If velocity-based tracking has just begun, this is also equal to the IEKF state prior to PT-update.

Defining the internal calibration, i.e. the homogeneous transformation relating the camera frame c to the image frame i, as

$$\mathcal{T}_{ic} = \begin{pmatrix} s_x f & s_{xy} f & c_u & 0 \\ 0 & s_y f & c_v & 0 \\ 0 & 0 & 1 & 0 \end{pmatrix}, \tag{3}$$

where s_x, s_y, s_{xy} are scaling factors, f is the focal length and $(c_u,\ c_v)^T$ denotes the principal point of the image plane, the projection $\boldsymbol{x}_{i_j}$ of $\boldsymbol{x}_{c_j}$ in the image plane i can be obtained by dehomogenizing $\bar{\boldsymbol{x}}_{i_j} = \mathcal{T}_{ic} \cdot \bar{\boldsymbol{x}}_{c_j}$.

Supposing that the actuator and its endeffector-camera are moving, we introduce a sequence c of moving camera frames for the purpose of treating the problem as a differential one. If $\boldsymbol{\omega}_{c(t)}, \boldsymbol{v}_{c(t)}$ denote the angular and translational velocities of the solid object at the instant t, it moves unconstrained relative to the camera frame $c(t)$ according to the MCE

$$\dot{\boldsymbol{x}}_{c(t)_j} = \boldsymbol{\omega}_{c(t)} \times \boldsymbol{x}_{c(t)_j} + \boldsymbol{v}_{c(t)} \tag{4}$$

Supposing that we know the angular and translational velocities $\boldsymbol{\mu}_{c(t)}$ and $\boldsymbol{\nu}_{c(t)}$ of $c(t)$, we describe the observable motion of moving object vertices $\boldsymbol{x}_{c(t)_j}$ by

$$\dot{\boldsymbol{x}}_{c(t)_j} = \left(\boldsymbol{\omega}_{c(t)} - \boldsymbol{\mu}_{c(t)}\right) \times \boldsymbol{x}_{c(t)_j} + \left(\boldsymbol{v}_{c(t)} - \boldsymbol{\nu}_{c(t)}\right). \tag{5}$$

This is motivated by the following facts:

- In case of a stationary camera, i.e. $\boldsymbol{\mu}_{c(t)} = \boldsymbol{\nu}_{c(t)} = \mathbf{0}$, equation (5) must specialize to equation (4).

- If camera and object move with exactly the same velocities, they maintain their relative positions. Therefore, if $\boldsymbol{\omega}_{c(t)} = \boldsymbol{\mu}_{c(t)}$ and $\boldsymbol{v}_{c(t)} = \boldsymbol{\nu}_{c(t)}$, then $\dot{\boldsymbol{x}}_{c(t)_j} = \mathbf{0}$.

Since we are interested in estimating the object's velocities based on the evaluation of optical flow, we assume that the gradient vector $\dot{\boldsymbol{x}}_{i_j}$ of the projection

[3]This assumption allows the application of linear system theory to determine the object's velocity relative to the camera and its motion relative to the object frame.

$\boldsymbol{x}_{i_j}$ of the j-th vertex $\boldsymbol{x}_{c_j}$ is equal to the optical flow vector $\boldsymbol{u}_j$ of $\boldsymbol{x}_{i_j}$. Utilizing equation (3), we calculate

$$\begin{aligned} \boldsymbol{u}_j &= \dot{\boldsymbol{x}}_{i_j} && (6)\\ &= \underbrace{\frac{1}{z_{c_j}} \begin{pmatrix} s_x f & s_{xy} f & -(x_{i_j} - c_u) \\ 0 & s_y f & -(y_{i_j} - c_v) \end{pmatrix}}_{\Delta} \dot{\boldsymbol{x}}_{c_j} . && (7) \end{aligned}$$

Substituting $\dot{\boldsymbol{x}}_{c_j}$ in equation (7) with the right hand side of equation (5) yields

$$\boldsymbol{u}_j = \frac{1}{z_{c_j}} \left(\Delta\ \boldsymbol{v}_{c(t)} - \Delta\ [\boldsymbol{x}_{c_j}]_\times\ \boldsymbol{\omega}_{c(t)} - \Delta\ \dot{\boldsymbol{y}}_{c_j} \right), \tag{8}$$

where $\dot{\boldsymbol{y}}_{c_j} = \boldsymbol{\mu}_c \times \boldsymbol{x}_{c_j} + \boldsymbol{\nu}_c$ and matrix $[\boldsymbol{x}_{c_j}]_\times$ is defined in terms of $[\boldsymbol{x}]_\times \cdot \boldsymbol{y} = \boldsymbol{x} \times \boldsymbol{y}$. Here, $\times$ denotes the cross product.

Having at least $n \geq 3$ linear independent visible vertices $\boldsymbol{x}_{c_j}$, $j = 1, \ldots, n$ with projections $\boldsymbol{x}_{i_j}$ and associated optical flow vectors $\boldsymbol{u}_j$, the angular and translational velocities $\boldsymbol{\omega}_{c(t)}$ and $\boldsymbol{v}_{c(t)}$ can be determined solving

$$\begin{pmatrix} \Delta & -\Delta\ [\boldsymbol{x}_{c_1}]_\times \\ \vdots & \vdots \\ \Delta & -\Delta\ [\boldsymbol{x}_{c_n}]_\times \end{pmatrix} \begin{pmatrix} \boldsymbol{v}_{c(t)} \\ \boldsymbol{\omega}_{c(t)} \end{pmatrix} = \begin{pmatrix} z_{c_1}\ \boldsymbol{u}_1 + \Delta\ \dot{\boldsymbol{y}}_{c_1} \\ \vdots \\ z_{c_n}\ \boldsymbol{u}_n + \Delta\ \dot{\boldsymbol{y}}_{c_n} \end{pmatrix}. \tag{9}$$

Equation (9) can be solved applying a householder technique or similar methods. Special care has to be taken to achieve numerical stability.

Next, the velocities $\boldsymbol{\omega}_{c(t)}$ and $\boldsymbol{v}_{c(t)}$ have to be expressed with respect to the object frame o. Rewriting the inverse of equation (2) in order to separate the rotational and translational components one obtains

$$\boldsymbol{x}_{o_j} = \mathrm{R}_{oc(t)} \boldsymbol{x}_{c(t)_j} + \boldsymbol{T}_{oc(t)}. \tag{10}$$

The substitution of equation (4) into the temporal derivation of equation (10), namely

$$\dot{\boldsymbol{x}}_{o_j} = \frac{\partial \mathrm{R}_{oc(t)}}{\partial t} \boldsymbol{x}_{c(t)_j} + \mathrm{R}_{oc(t)} \frac{\partial \boldsymbol{x}_{c(t)_j}}{\partial t} + \frac{\partial \boldsymbol{T}_{oc(t)}}{\partial t}, \tag{11}$$

results in

$$\begin{aligned} \dot{\boldsymbol{x}}_{o_j} = {} & \left(\mathrm{R}_{oc(t)} \boldsymbol{\omega}_{c(t)}\right) \times \boldsymbol{x}_{o_j} + \left[\mathrm{R}_{oc(t)} \boldsymbol{v}_{c(t)} - \left(\mathrm{R}_{oc(t)} \boldsymbol{\omega}_{c(t)}\right) \times \boldsymbol{T}_{oc(t)}\right] + \\ & \frac{\partial \mathrm{R}_{oc(t)}}{\partial t}\ \mathrm{R}_{co}\ \boldsymbol{x}_{o_j} + \frac{\partial \boldsymbol{T}_{oc(t)}}{\partial t}. \end{aligned} \tag{12}$$

Having $\boldsymbol{\mu}_c$ and $\boldsymbol{\nu}_c$ assumed to be constant, we get

$$\begin{aligned} \frac{\partial \mathrm{R}_{oc(t)}}{\partial t} &= \frac{\partial \mathrm{R}_{oc} \cdot \mathrm{R}_{cc(t)}}{\partial t} = \mathrm{R}_{oc}\ [\boldsymbol{\mu}_c]_\times \quad \text{and} && (13)\\ \frac{\partial \boldsymbol{T}_{oc(t)}}{\partial t} &= \frac{\partial \mathrm{R}_{oc} \cdot \boldsymbol{T}_{cc(t)} + \boldsymbol{T}_{oc}}{\partial t} = \mathrm{R}_{oc}\ \boldsymbol{\nu}_c. && (14) \end{aligned}$$

As we are interested in the description of the object's velocities $\boldsymbol{\omega}_o$ and $\boldsymbol{v}_o$ according to MCE (1), we only have to compare equations (12) and (1). We get

$$\begin{aligned} \boldsymbol{\omega}_o &= \mathrm{R}_{oc} \left(\boldsymbol{\omega}_{c(t)} + \boldsymbol{\mu}_c\right) \quad \text{and} && (15)\\ \boldsymbol{v}_o &= \mathrm{R}_{oc} \left(\boldsymbol{v}_{c(t)} + \boldsymbol{\nu}_c + \left(\boldsymbol{\omega}_{c(t)} + \boldsymbol{\mu}_c\right) \times \boldsymbol{T}_{oc}\right). && (16) \end{aligned}$$

The velocities ω_o and v_o can be used to predict the object's pose at the VT-level. Prediction is carried out by integration, calculating

$$\mathcal{T}_{so}(t+\Delta t) = \mathcal{T}_{so}(t) \cdot \begin{pmatrix} Q\omega_o{}^T & Qv_o \\ 0\,0\,0 & 1 \end{pmatrix}^{-1}. \tag{17}$$

The matrices $Q\omega_o(\Delta t)$ and $Qv_o(\Delta t)$ are part of the solution of equation (1) assuming constant angular and translational velocities and can be obtained by integration of the MCE (1) over the cycle interval Δt

$$Q\omega_o = I_{3\times3} + \frac{\sin(\|\omega_o\| \cdot \Delta t)}{\|\omega_o\|}[\omega_o]_\times + \frac{1-\cos(\|\omega_o\| \cdot \Delta t)}{\|\omega_o\|^2}[\omega_o]_\times^2 \tag{18}$$

$$Qv_o = \Delta t \cdot I_{3\times3} + \frac{1-\cos(\|\omega_o\| \cdot \Delta t)}{\|\omega_o\|^2}[\omega_o]_\times + \frac{\|\omega_o\| \cdot \Delta t - \sin(\|\omega_o\| \cdot \Delta t)}{\|\omega_o\|^3}[\omega_o]_\times^2, \tag{19}$$

where Δt denotes the cycle time of the VT, here 40ms. A derivation of these equations can be found in [2].

Utilizing equations (15) and (16), which transform velocities between frames fixed relative to each other if $\mu_c = \nu_c = \mathbf{0}$, the velocities μ_c and ν_c are determined using the known velocities μ_h and ν_h of the TCP.

4.2. Computation of Optical Flow

In order to compute optical flow in video-rate, a fast algorithm has to be employed that should yield precise results. The optical flow field calculated needs not to be dense, since we want to relate optical flow vectors with projections of certain vertices of the workpiece.

To meet these conditions, we have employed a method suggested by [6, 7] for the following reasons:

- The estimation of the gradient of the spatio-temporal gray-value function is more reliable than the estimation of higher-order derivatives.
- The suggested computation of optical flow is a very efficient method, because it depends only on the gradient of the gray-value function.

Let $g(\boldsymbol{x}_{i_j}, t)$ be the gray-value at position $\boldsymbol{x}_{i_j}$ at time instant t and let $(g_x,\ g_y,\ g_t)(\boldsymbol{x}_{i_j}, t)$ be its associated gradient. Based on the optical flow constraint equation (OFCE)

$$g_x(\boldsymbol{x}_{i_j}, t) \cdot u_j + g_y(\boldsymbol{x}_{i_j}, t) \cdot v_j + g_t(\boldsymbol{x}_{i_j}, t) = 0, \tag{20}$$

where $\boldsymbol{u}_j = (u_j,\ v_j)^T$ denotes the optical flow at position $\boldsymbol{x}_{i_j}$ and time instant t, and the assumption of locally constant optical flow in some neighborhood $\mathcal{N}_j = \{\boldsymbol{y}_i^{(k)} | k = 1, \ldots, n\}$ of $\boldsymbol{x}_{i_j}$, we determine $\boldsymbol{u}_j$ solving

$$\begin{pmatrix} g_x(\boldsymbol{y}_i^{(0)}) & g_y(\boldsymbol{y}_i^{(0)}) \\ \vdots & \vdots \\ g_x(\boldsymbol{y}_i^{(n)}) & g_y(\boldsymbol{y}_i^{(n)}) \end{pmatrix} \cdot \begin{pmatrix} u_j \\ v_j \end{pmatrix} = - \begin{pmatrix} g_t(\boldsymbol{y}_i^{(0)}) \\ \vdots \\ g_t(\boldsymbol{y}_i^{(n)}) \end{pmatrix}. \tag{21}$$

For simplicity, the argument t of the partial derivations is omitted in equation (21).

In order to reduce the computation of optical flow vectors to only those positions, where the underlying image provides sufficient information, *MiniVISTA* is equipped with a module that tests for each position in the image whether there is sufficient variation of the gradient of the gray-value function g or not. This test ensures a unique solution of the OFCE.

4.3. Clustering and Data Association

Consider a sparse gradient matrix of an image, as it is computed by *MiniVISTA* employing the method described in Subsection 4.2. In order to determine the neighborhoods $\mathcal{N}_j$ of the projections $\boldsymbol{x}_{i_j}$ of all visible vertices $\boldsymbol{x}_{c_j}$, we cluster this gradient matrix using a 3×3-neighborhood-mask, where only the off-diagonal positions are valid and checked. Therefore, each neighborhood is an equivalence class of the transitive closure of the 3×3-neighborhood relation. For each cluster, an optical flow vector is computed. Finally, the clusters are associated to the projections of visible model-vertices by means of their euclidian distance in the image plane.

5. Experiments

The following experiment shows the feasibility and the accuracy of our approach. Initially, the robot is moved to an initial pose and the scene-relative pose of a workpiece consisting of two solid aluminum blocks, which are attached two each other by two screws, is computed. The job of the robot is to maintain its object-relative pose, while the object is moved around. As we want to compare the system

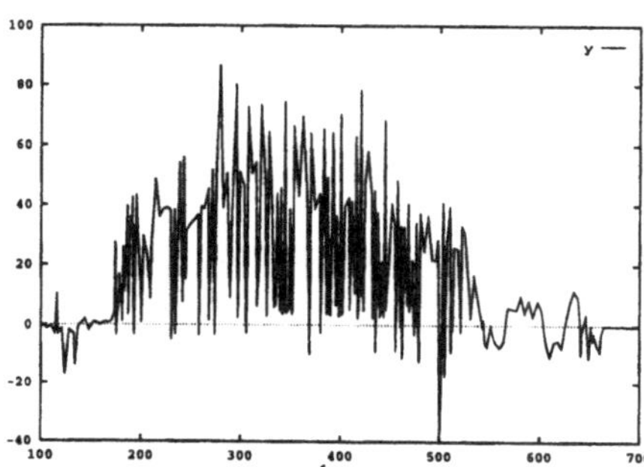

Figure 2. Measured dragging speed in mm/s of the object.

performance with known ideal conditions, we have taken precautions that the object is only be moved linearly along the y-axis of its frame of reference. So, during the experiment, we expect no object-relative rotational velocity at all as well as negligible x- and z-components of any object-relative translational velocity measured while tracking is enabled. Figure 2 shows the y-component of the translational velocity, i.e. the measured dragging speed in millimeters per second (mm/s). The results are shown from the 100th to the 700th frame while PT-/VT-tracking is enabled, i.e. a period of $600 \cdot 40ms = 24s$ is covered. Figure 3 shows the measured x- and z-components of the translational object velocity, which are given in mm/s and should be zero all the time. Figure 4 shows the components of the rotational velocity of the moving object, which is moving purely translational. The unit used is deg/s. Clearly, the ability of tracking extremely fast moving objects is bounded by the

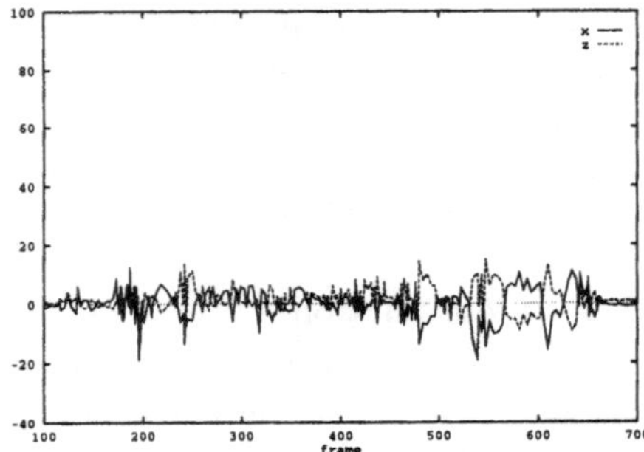

Figure 3. Measured x- and z-components of the translational velocity in mm/s. They should be zero.

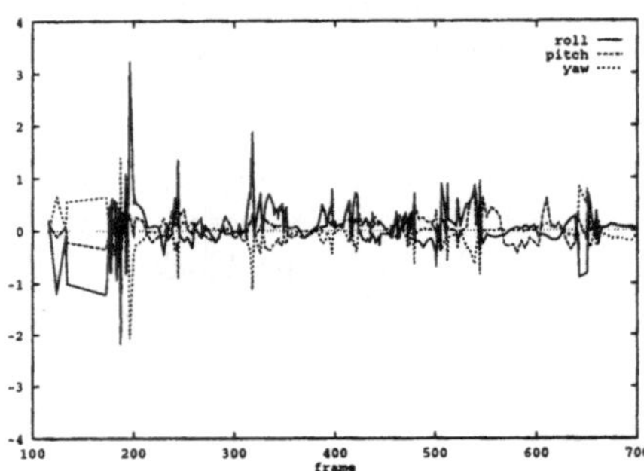

Figure 4. Measured components of the rotational velocity in deg/s. All components should be zero.

convolution kernel size of available hardware convolvers.

6. Conclusion

We have presented a method that significantly improves the tracking performance of position-based systems and, in particular, of our disassembly system. Clearly, by employing a lower level of visual-servo control, the reduction of the critical cycle time could be achieved. In fact, compared with image-based systems, the VT compensates the main disadvantage of position-based systems, namely, the reduced cycle-rate introduced by its computational requirements.

But it is not only the reduced cycle-time that contributes to the improved performance of our system. Indeed, the measurements of the object's dynamic behavior relative to its own frame of reference improves the performance of visual servo control more than position-based tracking of the object's pose at reduced cycle-time or an indirect description like it is produced by an image-based system, e.g. tracking simply the centroid of the projection of an object.

The computation of the object's velocities relies on the source of information that is best suited for the job, namely optical flow. We believe that this leads to more accurate results than an estimation soly based on extracted image features, in particular, because there is no data association problem.

Acknowledgements: This work has been supported in part by a grant of the Deutsche Forschungsgemeinschaft within the Sonderforschungsbereich 314 'Künstliche Intelligenz – Wissensbasierte Systeme'.

References

[1] Feddema, J.T., Lee, C.S.G., Mitchell, O.R., *Model-Based Visual Feedback Control for a Hand-Eye Coordinated Robotic System*, IEEE Computer **25**:8 (1992) 21–31.

[2] Gengenbach, V., *Einsatz von Rückkopplungen in der Bildauswertung bei einem Hand-Auge-System zur automatischen Demontage*, Dissertation, Institut für Algorithmen und Kognitive Systeme, Universität Karlsruhe (TH), 1994. Published in Dissertationen zur Künstlichen Intelligenz (DISKI) **72**, infix-Verlag, Sankt Augustin/Germany, 1994.

[3] Hashimoto, K., Kimoto, T., Ebine, T., Kimura, H., *Manipulator Control with Image-Based Visual Servo*, in T.C. Hsia (ed.), Proc. IEEE Int. Conf. on Robotics and Automation, Sacramento/CA, April 9-11, 1991, IEEE Computer Society Press, Los Alamitos/CA, 1991, pp. 2267–2271.

[4] Jang, W., Bien, Z., *Feature-Based Visual Servoing of an Eye-In-Hand Robot with Improved Tracking Performance*, in T.C. Hsia (ed.), Proc. IEEE Int. Conf. on Robotics and Automation, Sacramento/CA, April 9-11, 1991, IEEE Computer Society Press, Los Alamitos/CA, 1991, pp. 2254–2260.

[5] Koivo, A.J., Houshangi, N., *Real-Time Vision Feedback for Servoing Robotic Manipulator with Self-Tuning Controller*, IEEE Trans. on Systems, Man, and Cybernetics **SMC-21**:1 (1991) 134–141.

[6] Nagel, H.-H., *Displacement Vectors Derived from Second-Order Intensity Variations*, Computer Vision, Graphics, and Image Processing **21** (1983) 85–117.

[7] Nagel, H.-H., *Analyse und Interpretation von Bildfolgen*, Informatik-Spektrum **8** (1985) 178–200, 312–327.

[8] Papanikolopoulos, N.P., Khosla, P.K., Kanade, T., *Visual Tracking of a Moving Target by a Camera Mounted on a Robot: A Combination of Control and Vision*, IEEE Trans. on Robotics and Automation **RA-9**:1 (1993) 14–34.

Experimental Approach on Artificial Active Antenna

Makoto Kaneko Naoki Kanayama Toshio Tsuji
Industrial and Systems Engineering, Hiroshima University
Higashi-Hiroshima 739, JAPAN
kaneko@huis.hiroshima-u.ac.jp

Abstract

This paper discusses an artificial Active Antenna that can detect the contact location between an insensitive flexible beam and a 3D environment through the measurement of the rotational compliance of the beam in contact with the environment. The lateral slip, which possibly occurs for the 3D Active Antenna, overestimates the rotational compliance, and as a result, brings a large sensing error for the localizing contact point. The goal of this paper is to find the contact point under such conditions. In the first step, we push the antenna to the environment. If a lateral slip is confirmed, the pushing direction is changed continuously until we finally avoid the development of any lateral slip. We explore how to detect a lateral slip and how to determine the new pushing direction to avoid it. We experimentally verify an algorithm which can search the contact distance, even under the appearance of a lateral slip during the first step.

1. Introduction

Active Antenna is a new sensing system enabling us to detect the contact location through the measurement of the rotational compliance of an insensitive antenna in contact with an environment. In our former work, we have shown that for a planar type Active Antenna, the contact location is a function of the rotational compliance alone, and that one active motion is necessary and sufficient for localizing the contact point irrespective to friction at the point of contact, if the straight beam is utilized [1], [2]. A big advantage of Active Antenna is that a contact point is obtained through a surprisingly simple active motion, while sophisticated active motions should be prepared for most of contact sensing to avoid a large interaction force between sensor and environment. This is because the flexibility of the antenna successfully relaxes the contact force by itself, even under a large positional error.

In the 2D Active Antenna, we implicitly assumed that the antenna never makes a slip in the lateral direction, while a longitudinal slip inevitably occurs to satisfy the geometrical relationship between the antenna shapes before and after a bending deformation. The lateral slip (see Fig.1), which is the inherent characteristic for 3D Active Antenna [3], strongly depends on the direction of pushing, the friction at the point of contact, and the normal surface of the environment where the antenna makes contact. Generally, such a lateral slip overestimates the rotational compliance of the antenna which is in contact with the environment, and as a result, deteriorates the sensing accuracy directly, while the longitudinal slip brings a minor error only. In this paper, we discuss an algorithm which can detect the contact distance with a sufficient accuracy, even under the appearance of a lateral slip during the first step.

2. Related Works

A simple flexible beam sensor can take the form of a short length of spring piano wire or hypodermic tubing anchored at the end. When the free end touches an external object, the wire bends. This can be sensed by a piezoelectric element or by a simple switch [4]. A more elaborate sensor is described by Wang and Will [5]. Long antennae-like whisker sensors were mounted on the SRI mobile robot, Shakey [6], and on Rodney Brook's six-legged robot insects [7]. Hirose, et. al. discussed the utilization of whisker sensors in legged robots [8]. The sensor system is composed of an electrode and a whisker whose end is fixed at the base. This sensor unit has been arranged in an array around each foot of the legged robot, Titan III, so that it can monitor the separation between each foot and the ground to allow deceleration of the foot before contact. This sensor is also conveniently used to confirm which part of the foot is in contact with the ground. Similarly shaped whiskers have been considered for legs of the Ohio State University active suspension vehicle [9]. Russell has developed a sensor array [10] by mounting whisker sensors on a mobile robot, and succeeded in reconstructing the shape of a convex object followed by the whisker. In his work, it is assumed that the whisker tip is always in contact with the environment, and that when the whisker contacts the environment except for the tip, it is assigned to a failure mode. The major difference between previous works [4]-[10] and ours is that the Active Antenna enables us to localize a contact point between the beam and the environment, while previous works do not.

3. Basic Structure of Active Antenna and Main Assumptions

3.1. Basic structure

Figure 1 shows an overview of the 3D Active Antenna and its coordinate system, where Σ_B (or upper script "B") and Σ_s (or upper script "s") denote the base coordinate system and the sensor coordinate system, respectively. The 3D Active Antenna is composed of an insensitive flexible beam, two actuators to move the beam in 3D space, two position sensors to measure the angular displacements ϕ_1 and ϕ_2, and a two-axis moment sensor to detect moments around both x_s and z_s axes. The moment sensor is designed so that each sensing axis can intersect with the center of rotation (the origin of the sensor coordinate system). With this sensor design, we have a common moment arm y_0 for each moment axis, as shown in Fig.2. Now, let us consider the reaction force f_c. This force can be decomposed into an axial force and a non-axial force components, respectively. Since the antenna is assumed to have a sufficiently large stiffness in the longitudinal direction, we neglect the effect due to the axial force component. So, we consider the effect due to the non axial force component f_s only. We also define the sensing plane Π, with the plane spanned by two unit vectors whose directions coincide with x_s and y_s. The design orientation taken for the two-axis moment sensor enables us to evaluate the direction of the contact force projected on the plane Π from the outputs of the moment sensor, since f_s is parallel to the sensing plane Π.

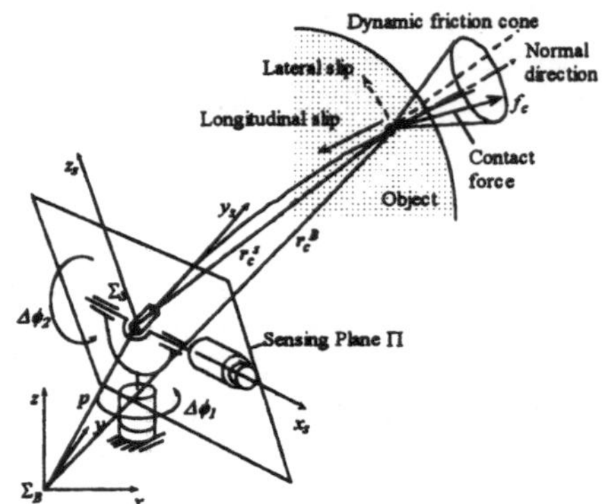

Figure 1. Basic structure and its coordinate system of 3D Active Antenna.

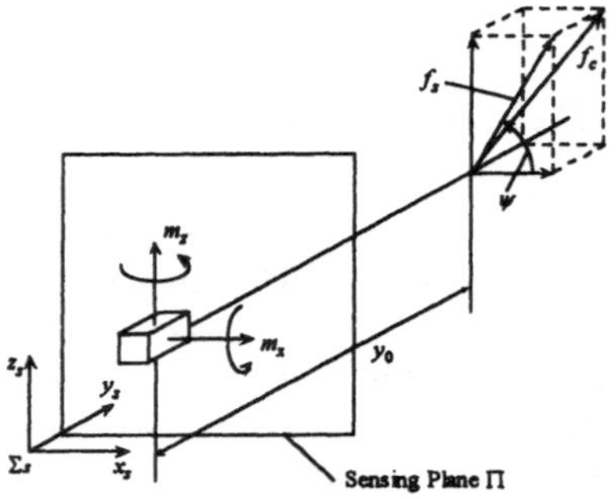

Figure 2. Sensing plane Π and contact force decomposition.

3.2. Main assumptions

For simplifying our discussions, we set the following assumptions:

Assumption 1: The antenna has equal compliance in a plane perpendicular to the longitudinal axis.

Assumption 2: The antenna deformation is small enough to ensure that we can apply a linear approximation.

Assumption 3: Antenna has a sufficiently large stiffness for the axial direction to enable us to neglect the longitudinal elongation of the antenna.

Assumption 4: The compliance of the environment is sufficiently small compared with that of the antenna.

Assumption 5: Before applying an active motion, the antenna is already in contact with an environment with zero force.

4. Working Principle

4.1. Without lateral slip

In our former works [1], [2], we have shown that if there is no lateral slip, the contact distance y_0 is given by the following equation:

$$y_0 = kc, \tag{1}$$

where k is a constant and c denotes the rotational compliance of the antenna in contact with an environment. For a beam whose cross sectional area is circular, $k = 3EI$, where E and I are Young's modulus and the second moment of cross section of the antenna, respectively. c can be computed by the torque increase and the pushing angle after the antenna makes contact with an environment. Figure 3 shows the experimental result obtained under no lateral slip, where the real line denotes the theoretical result and the circles denote experimental data for five trials. It can be seen from Fig.3 that the agreement between theoretical and experimental results is fairly good and the repeatability of experiments is also fine.

4.2. Lateral slip and its effect on contact sensing

The antenna is extremely stiff in the longitudinal direction and is a little elongated along the direction of an axial force (Assumption 3), while it easily deforms for a

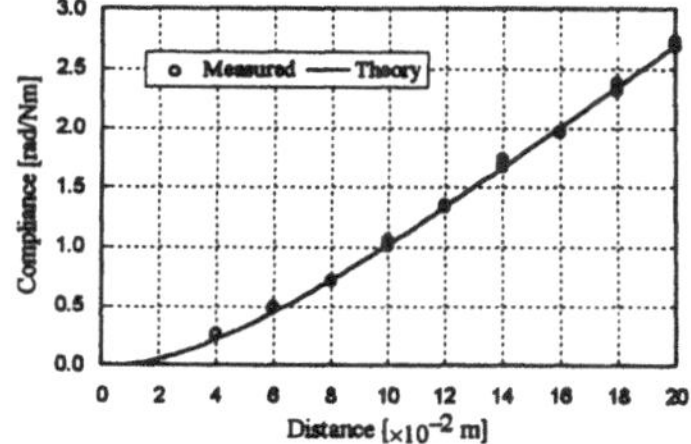

Figure 3. Calibration test without any lateral slip.

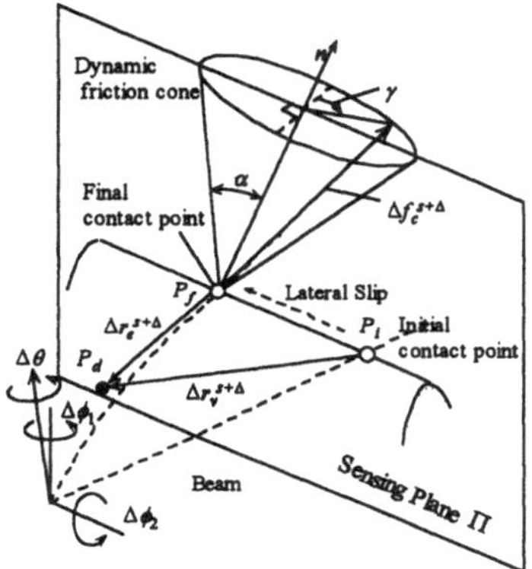

Figure 4. A general view of 3D Active Antenna when a lateral slip occurs.

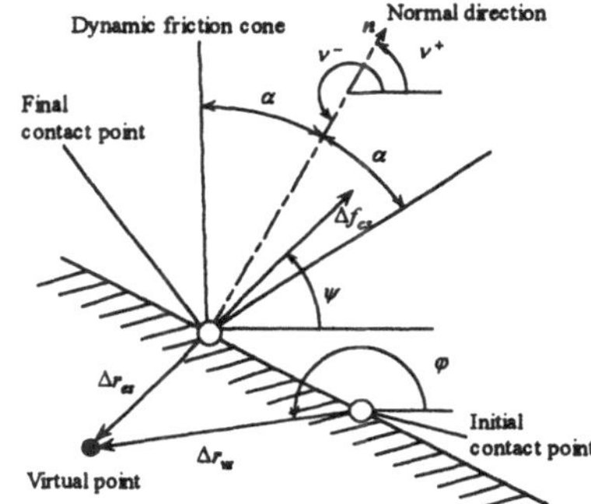

Figure 5. Notations on the sensing plane Π.

bending moment. When the angular displacements, $\Delta\phi_1$ and $\Delta\phi_2$, are imparted to the antenna, it deforms while keeping contact as shown in Fig.1. During this pushing motion, it continuously makes a longitudinal slip on the point of contact to the environment. This allows us to consider only the case of the dynamic friction cone during the whole active motion.

Figure 4 shows the relationship among the virtual displacement vector $\Delta\boldsymbol{r}_v^{s+\Delta}$, the effective displacement vector $\Delta\boldsymbol{r}_e^{s+\Delta}$ and the dynamic friction cone after a pushing motion, where α is the angle of the dynamic friction cone. Since we assume the uniform compliance by Assumption 1, the contact force always appears in the opposite direction against the effective displacement vector $\Delta\boldsymbol{r}_e^{s+\Delta}$. This means that the contact point moves so that $\Delta\boldsymbol{r}_e^{s+\Delta}$ and $\Delta\boldsymbol{f}_c^{s+\Delta}$ are colinear each other. It should be noted that the force projected on the sensing plane can exist not only on the boundary but also within the dynamic friction cone, while the contact force should always exist on the boundary of dynamic friction cone. Since $\|\Delta\boldsymbol{r}_v^{s+\Delta}\| > \|\Delta\boldsymbol{r}_e^{s+\Delta}\|$ while under a lateral slip, $\Delta\boldsymbol{f}_c^{s+\Delta}(\|\Delta\boldsymbol{r}_v^{s+\Delta}\|) > \Delta\boldsymbol{f}_c^{s+\Delta}(\|\Delta\boldsymbol{r}_e^{s+\Delta}\|)$, which means that the contact force under no lateral slip is larger than that under one for the same $\Delta\phi_1$ and $\Delta\phi_2$. Thus, the rotational compliance is generally overestimated under a lateral slip. As a result, the sensing accuracy will be deteriorated.

Figure 5 shows several physical parameters on the sensing plane Π, where all components are projected on the sensing plane Π, and ν^+, ν^-, φ and ψ denote outer and inner normal directions of the environment's surface, the direction of $\Delta\boldsymbol{f}_{cs}$, and the pushing direction with respect to the sensor coordinate system, respectively.

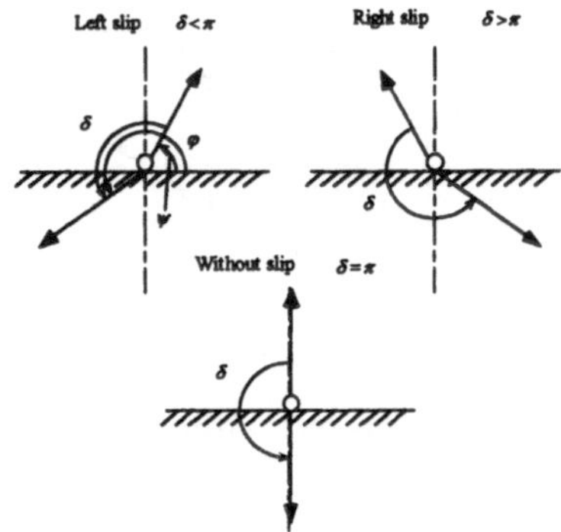

Figure 6. Three patterns of the contact states.

4.3. Judgement of contact states

There are two physical parameters obtained by the sensors.

(1) The direction of contact force projected on Π, ψ:
The moment sensor can involve the following information.

$$m_x = y_0 \|\Delta \boldsymbol{f}_{cs}\| \sin\psi \tag{2}$$
$$m_z = -y_0 \|\Delta \boldsymbol{f}_{cs}\| \cos\psi \tag{3}$$

By dividing each side, we obtain,

$$\tan\psi = -\frac{m_x}{m_z} \tag{4}$$

(2) The pushing direction φ:
Since each actuator has a position sensor, the pushing direction can also be measured. φ can be regarded as an input for this sensor system, and then it provides ψ as an output. Based on these informations, the sensing algorithm should be planned.

There are three patterns for judging the contact states, as shown in Fig.6, where δ is the angle expressing the difference between ψ and φ. If δ is less than π, it means that the antenna made a slip in the left direction. If δ is greater than π, it means that the antenna made a slip in the right direction. If δ is equal to π, it means that there was no slip during the pushing motion.

4.4. Sensing strategy

The main goal is to find the pushing direction avoiding any lateral slip, because once we find such a direction, the problem reduces to that of a planar Active Antenna [2], [3]. Figure 7 explains the sensing strategy eventually leading to the normal direction of environment.

1) The 1st trial:
For the first trial, the antenna is pushed toward an arbitrary direction $\varphi^{(1)}$. Let $\psi^{(1)}$ be the direction of contact force resultantly obtained by the first pushing motion. Since $\delta < \pi$ in this particular example (Fig.7), the sensor system can recognize that the antenna made a slip in the left direction.

2) The 2nd trial:
The second pushing direction is chosen so that the slip may appear in the opposite

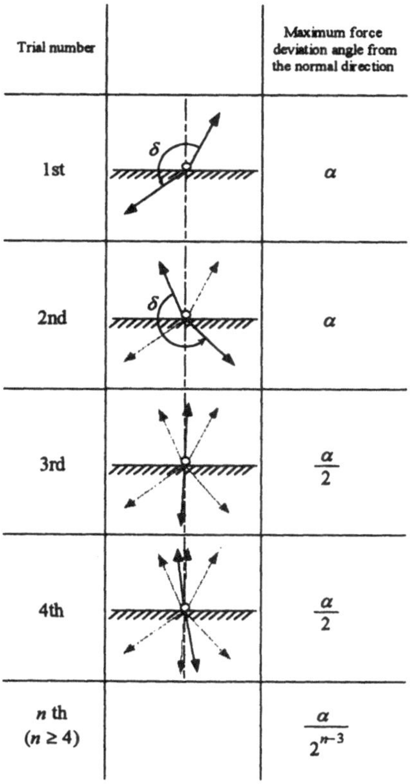

Trial number		Maximum force deviation angle from the normal direction
1st		α
2nd		α
3rd		$\frac{\alpha}{2}$
4th		$\frac{\alpha}{2}$
n th ($n \geq 4$)		$\frac{\alpha}{2^{n-3}}$

Figure 7. Maximum force deviation angle from the normal direction for each trial.

direction against the first one. This can be done by choosing the pushing direction in the following,

$$\varphi^{(2)} = \varphi^{(1)} - \mathrm{sign}(\delta^{(1)} - \pi) \cdot \frac{\pi}{2}, \tag{5}$$

where $\delta^{(1)} = \varphi^{(1)} - \psi^{(1)}$. When we choose the second pushing direction according to eq.(5), $\psi^{(1)}$ and $\psi^{(2)}$ always lie in two different regions, namely, one is the right half plane with respect to the outer normal vector, and the other is the left one. Since the contact force projected on Π can never be away from the boundary of the dynamic friction cone projected on Π, the following condition holds.

$$\nu^{+} - \frac{\alpha}{2} < \frac{\psi^{(1)} + \psi^{(2)}}{2} < \nu^{+} + \frac{\alpha}{2} \tag{6}$$

Equation (6) implies that the angle between ν^{+} and the arithmetic mean of $\psi^{(1)}$ and $\psi^{(2)}$ is less than $\alpha/2$.

3) The 3rd trial:

The third pushing direction is chosen so that the extended line may divide $\psi^{(1)}$ and $\psi^{(2)}$ equally, namely,

$$\varphi^{(3)} = (\psi^{(1)} + \psi^{(2)})/2 + \pi \tag{7}$$

From ineq.(6) and eq.(7), the difference between $\psi^{(3)}$ and ν^{+} is at most $\alpha/2$.

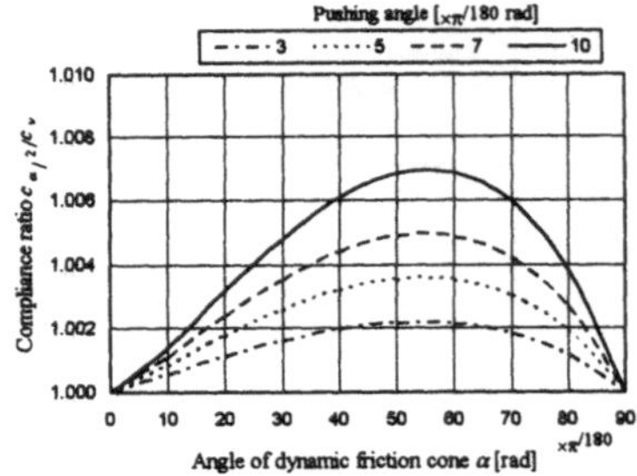

Figure 8. Compliance ratio when pushing to $\alpha/2$ from normal direction.

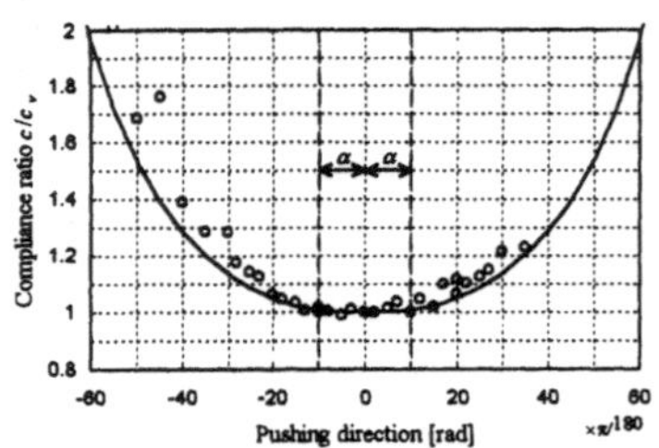

Figure 9. The relationship between c/c_ν and φ ($\alpha = \pi/18$, $\Delta\theta = \pi/36$).

4) The 4th trial:
Picking up $\psi^{(3)}$ and either $\psi^{(1)}$ or $\psi^{(2)}$, we choose the fourth pushing direction in the following.

$$\varphi^{(4)} = (\psi^{(3)} + \psi^{(1|2)})/2 + \pi \tag{8}$$

where $(a|b) \equiv \{x|\min(|\delta^{(x)} - \pi|), x = a, b\}$ By introducing this function, we can always choose the trial whose lateral slip is smaller than the other.

In general, for the n-th trial ($n \geq 4$), we can easily show that the following inequality exists.

$$|\psi^{(n)} - \nu^+| < \frac{\alpha}{2^{n-3}} \quad , n \geq 4 \tag{9}$$

Since $\lim_{n\to\infty} \{\alpha/2^{n-3}\} = 0$, $\lim_{n\to\infty} \{\psi^{(n)}\} = \nu^+$, which means that the direction of contact force finally reaches the outer normal direction, namely, the pushing direction coincides with the inner normal direction. Thus, we can find the normal direction of the environment. Once we can detect the direction, the contact distance can be obtained through the pushing motion along the direction detected, because it is guaranteed that no lateral slip occurs. Thus, the convergence of the algorithm can be ensured if we impart infinite active motions to the antenna, while it is not efficient.

4.5. When can we stop the trial?

A natural question that comes up is when we can stop the trial. In other words, how many active motions are really necessary for keeping a sufficient accuracy. Figure 8 shows the compliance ratio $c_{\alpha/2}/c_\nu$ when the pushing motion is imparted in the direction of $\alpha/2$ from ν^-, where c_ν and $c_{\alpha/2}$ denote the compliances under the pushing in ν^- and under the pushing in $\alpha/2$ from ν^-, respectively. From this figure, it can be found that if the pushing direction is selected within $\alpha/2$ from ν^-, it is ensured that the theoretical sensing error can be suppressed less than 1% irrespective to dynamic friction cone. By combining the results in Fig.7 and 8, we can say that three active motions are sufficient for keeping theoretical sensing error less than 1%.

Figure 9 shows the relationship between the pushing direction φ and compliance ratio between pushing to ν^- and pushing to φ, where the real line denotes the analytical result and the circles show the experimental results, and the angle of the dynamic friction is approximately 10 degrees. Experimental results exhibit almost unity for the pushing angle between $-\alpha/2$ and $\alpha/2$, which supports our analytical result.

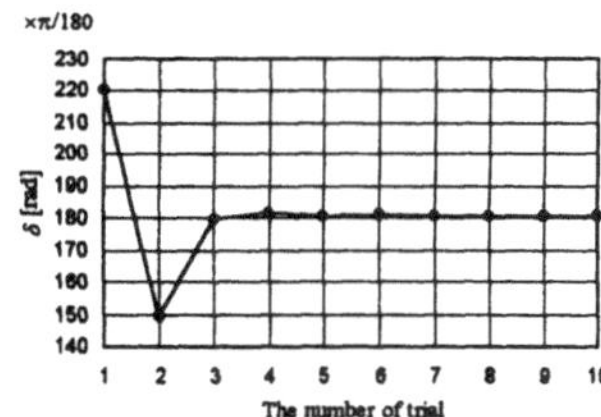

(a) δ versus the number of trial.

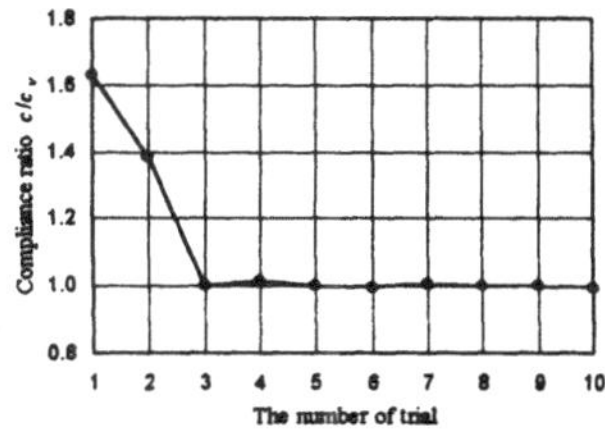

(b) c/c_ν versus the number of trial.

Figure 10. Convergence process for δ and c/c_ν.

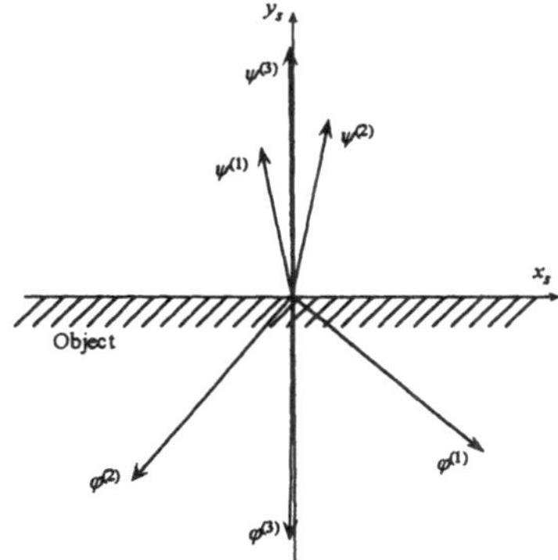

Figure 11. The relationship between $\varphi^{(i)}$ and $\psi^{(i)}$.

4.6. Convergence process by experiment

Figure 10 (a) and (b) show the process of convergence of δ and c/c_ν with respect to the number of trial, respectively. It should be noted that both δ and c/c_ν converges very quickly as the number of trial increases. With three times trials, both δ and c/c_ν almost reach their equilibrium points. When c/c_ν is equal to unity, we can obtain the contact distance very accurately.

Figure 11 shows the relationship between i-th pushing direction and the direction of i-th contact force. For the first pushing, the antenna makes a lateral slip toward the left side, and for the second one, it again results in failure in avoiding a lateral slip and as a result it moves toward the right side. The third pushing direction computed by eq.(7) provides nearly the inner normal direction, and, therefore, the direction of contact force is also very close to the outer normal direction.

5. Future Works

In this work, we did not consider the effect of the curvature of the environment where the antenna makes contact. The curvature will more or less affect on the sensing accuracy. In our future works, we will address this problem in some details.

Finally, we would like to express our sincere thanks to Dr. Ivan Godler, Harmonic Drive Systems Co. Ltd. for his support in designing the hardware of 3D Active Antenna.

References

[1] Kaneko, M: Active Antenna, *Proc. of the 1994 IEEE Int. Conf. on Robotics and*

Automation, pp2665–2671, 1994.

[2] Kaneko, M., N. Ueno, and T. Tsuji: Active Antenna (Basic Working Principle), *Proc. of the 1994 IEEE Int. Conf. on Intelligent Robotics and Systems*, pp1744–1750, 1994.

[3] Kaneko, M., N. Kanayama, and T. Tsuji: 3D Active Antenna for Contact Sensing, *Proc. of the 1995 IEEE Int. Conf. on Robotics and Automation*, pp1113–1119, 1995.

[4] Russell, R. A.: Closing the sensor-computer-robot control loop, *Robotics Age*, April, pp15–20, 1984.

[5] Wang, S. S. M., and P. M. Will: Sensors for computer controlled mechanical assembly, *The Industrial Robot*, March, pp9–18, 1978.

[6] McKerrow, P.: Introduction to Robotics, *Addison-Wesley*, 1990.

[7] Brooks, R. A.: A robot that walks; Emergent behaviors from a carefully evolved network, *Neural Computation*, vol.1, pp253–262, 1989.

[8] Hirose, S., et. al.: Titan III: A quadruped walking vehicle, *Proc. of the Second Int. Symp. on Robotics Research*, MIT Pres.

[9] Schiebel, E. N., H. R. Busby, K. J. Waldron: Design of a mechanical proximity sensor, *Robotica*, vol.4, pp221–ss7, 1986.

[10] Russell, R. A.: Using tactile whiskers to measure surface contours, *Proc. of the 1992 IEEE Int. Conf. on Robotics and Automation*, pp1295–1300, 1992.

Low Cost Sensor Based Obstacle Detection and Description

Experiments with Mobile Robots using Grid Representation*

Paulo Menezes, Jorge Dias, Helder Araújo, Aníbal de Almeida
Instituto de Sistemas e Robótica
Departamento de Engenharia Electrotécnica
Largo Marquês de Pombal, 3000 COIMBRA, PORTUGAL
{paulo,jorge,helder,adealmeida}@isr.uc.pt

Abstract

This article describes an algorithm for detection of obstacles, through the use of low cost sonar sensors, for the navigation of mobile robots. The use of a grid representation for the cartesian space, associated to probabilistic operations on it, allowed the implementation of a fast, simple and robust algorithm for obstacle detection. This algorithm runs always that the mobile robot navigates, and allows to overcome obstacles that appear unexpectedly in the robot's path. The information used to represent the obstacles is based in the information provided by sonar sensors. A stochastic model for the sonar sensors and their *activity*, associated to the platform motion, are explored by the algorithm. Some experimental results obtained through the use this algorithm are described at the end of the article.

1. Introduction

The operation of a mobile robot in a unstructured environment is only possible, if the robot has ability to cope with changes in the environment. These changes, which include the presence of new obstacles, can be detected using sensorial systems.

In most cases a trajectory planner, using some *a priori* description of the robot's workspace, generates feasible paths to execute the desired missions. Using sensorial information, it is possible to improve the mobile robot capabilities, allowing a real-time reaction to avoid the collision and/or to re-plan the trajectory as proposed by Chatila [2].

Normally sonar sensors are chosen to acquire information around the mobile robot. Although sonar sensors present several problems which are related with the limited information that can be obtained from raw data. The choice was based mainly on their low cost, the ease of their use and the relatively low computational power required.

*This research was partially sponsored by NATO's Scientific Affairs Division in the framework of the Science for Stability Program

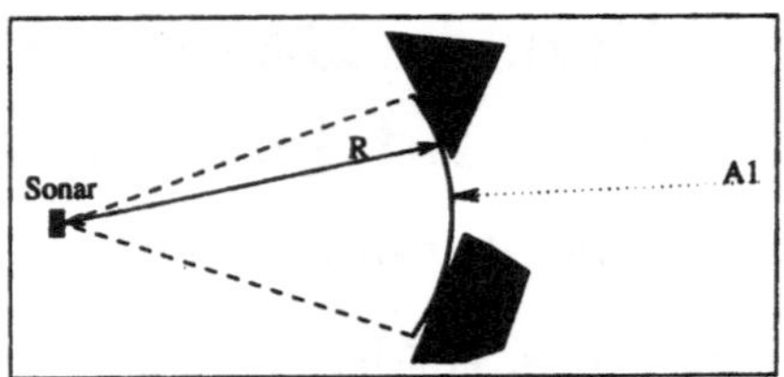

Figure 1. Both targets T1 and T2 produce the same range reading.

2. Problem formulation

The aim of this work is to allow the collision free navigation of a mobile robot in a semi-structured environment through the use of sensorial information. The inputs available are the *range data* provided by the ultrasonic range sensors, the robot's *position* and *orientation*, and the *direction of motion*. Range data and the localization of the robot are used to detect the presence of obstacles and their possible localization.

The direction of motion is used to explore the sensors' activity. Robot's activity serves two purposes: the first is to determine which region is of concern, because we don't need to care about obstacles located in regions where we are not going through, and using this information we can decide which set of sensors should be scanned. This has implication on the time processing since scanning more sensors is time costly both in data acquisition and also in data processing. The second reason is that, for a given location, multiple readings performed by a static sensor don't bring any new information about the extent of the obstacle, although they could be averaged to remove sensor noise.

2.1. Uncertainty in sonar range measures

In the majority of the mobile robots that use sonar sensors, the range information is obtained from the time-of-flight (t.o.f.). However from these measures, the only information that can be obtained is the distance to the closest point of the obstacle that reflected the wave, back to the sensor. This is not sufficient to characterize the obstacle, whose presence was detected. The wide opening angle presented by most sonar sensors introduces another uncertainty factor: to which direction corresponds the measure? Figure 1 illustrates a situation where two possible obstacles could produce the same range reading. From a range value we can only say that there is a region in which every point is a possible location for the detected obstacle. Considering the cartesian space and ignoring some errors that appear in range values, this region takes the form of one arc centered in the sensor, with radius equal to the range reading and angular value equal to twice the sensor opening.

It is also clear that any target tangent to the arc $A1$ would produce the same result. By this reason this arc is commonly referred as a Region of Constant Depth (RCD)[3], which is nothing more than a set of points equally candidates to be considered as the one that has produced the echo.

Calibration experiments performed in our lab have shown that sonar sensors can be approximated as ideal sensors whose outputs are corrupted with gaussian noise of zero mean and standard deviation given by the following expression

$$\sigma_D(z) = 0.0052 \times z + 0.002[m] \tag{1}$$

where z is the distance between the sensor and the target in meters [5].

Using this model we can write the following expression that gives the probability of obtaining a sonar reading r given a target at a distance z.

$$p(r|z) = \frac{1}{\sqrt{2\pi}\sigma_D} exp\left(\frac{-(r-z)^2}{2\sigma_D^2}\right) \tag{2}$$

2.2. Representation

As we have shown above, to obtain a description of the obstacles in terms of extents and orientation, these have to be observed from multiple *points of view*, due to the poor information that can be extracted from each observation. Here arises a problem, since our main goal is navigation we can not move the robot around an obstacle to obtain its full description. We, as long as possible, have to acquire information while the robot is moving. But to do so a suitable representation has to be used, allowing the fusion of information obtained in different robot's positions. The choice was to use a discretization of the operational space by dividing it into cells, each one representing a small area. For each cell, a value is assigned that represents the knowledge level regarding its occupancy. This value is updated, each time the area it represents, is scanned. In our implementation, a cell size of $10 \times 10cm$ was used as a compromise between the smallest detectable object and computational load. Notice that decreasing these dimensions has direct implications on the amount of information that must be processed and is represented by this grid structure. Algorithms operating on these type of structures have a general trend for $O(n^2)$ computational complexity and consequently in the increase of the processing time.

2.3. Obstacle Detection

Using the stochastic sensor model shown in equation (2), and using Bayes' formula we can obtain an expression for the existence of an obstacle (O), given a range measure, as suggested by Elfes [4]

$$p(O|r) = \frac{p(r|o)p(o)}{p(r|o)p(o) + p(r|u)p(u)} \tag{3}$$

In this equation the letter o means occupied and u means unoccupied and could be used to estimate each cell state.

Since the RCD length grows with distance and, with it the uncertainty about the target location, only range measures below 3 meters are used. This has the advantage of reducing the errors produced by multiple reflections and also reduce processing time, because the number of cells that have to be updated is smaller.

Substituting these values in expression (1), we can see that the standard deviation of range errors are ~ 6 times smaller than the cell size we chose. From this it can be seen that the sensor model can be substituted by a step model like the one shown in figure 2 in order to reduce computation time.

So, for a range value r, there is a set of cells that enclose all possible target locations that could produce this value. This set also contains the arc above referred to as RCD (figure 3) and can be viewed as its discretization.

For short range values, the RCD is contained on a single cell, the meaning being that it contains surely the point that produced the echo. In these cases it is assigned

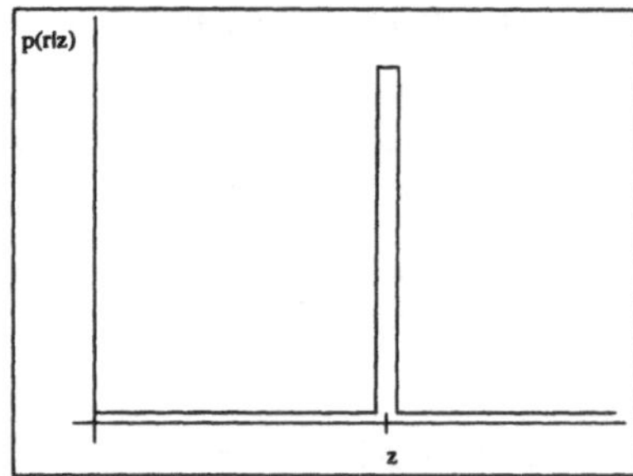

Figure 2. Approximation of the sonar measures by a step model

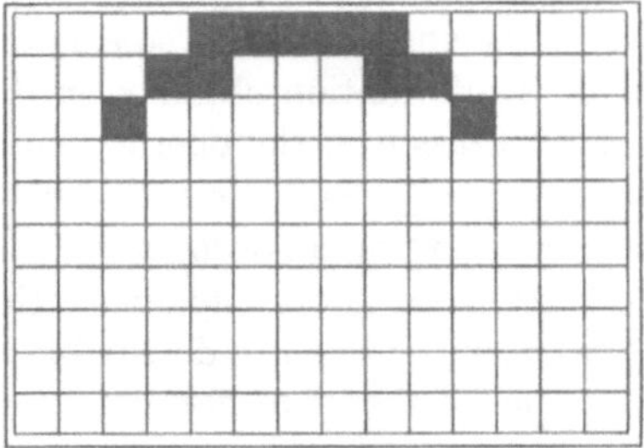

Figure 3. A discrete RCD

a probability of 100% of occupancy. For larger range values the RCD spans over several cells, and then this certainty must be shared by all of them. So each one is updated by $1/n$, where n is the number of cells that contain the RCD.

2.4. Updating the grid

Once a sensor measurement is obtained, several steps are needed; obtaining the robot's position and orientation at the point where the measurement was taken, the determination of the sensor world coordinates is given by:

$$^{W}\vec{\mathbf{x}} = ^{W}\mathbf{T}_{S_i} * ^{S_i}\vec{\mathbf{x}} \tag{4}$$

where $^{W}\vec{\mathbf{x}}$ are the sensor world coordinates, $\mathbf{T}$ the transformation matrix, and $^{S_i}\vec{\mathbf{x}}$ the sensor coordinates related with the robot frame.

For a sensor with an opening angle θ and a measurement r, the cells that must be updated are those that fall inside the cone defined by the two line segments and the arc, as shown in figure 4. The cells that contain the RCD will get their value increased, and the ones between the sensor and the RCD get it decreased because if there exists an obstacle the range value would be smaller. The letter E signifies that the probability of occupancy in that cell will be decreased and the opposite will be made for the letter O.

2.5. Exploring Sensor Activity

The information about the localization and/or dimensions of an obstacle using sonar sensors can not be obtained from a single observation. Even when using multiple

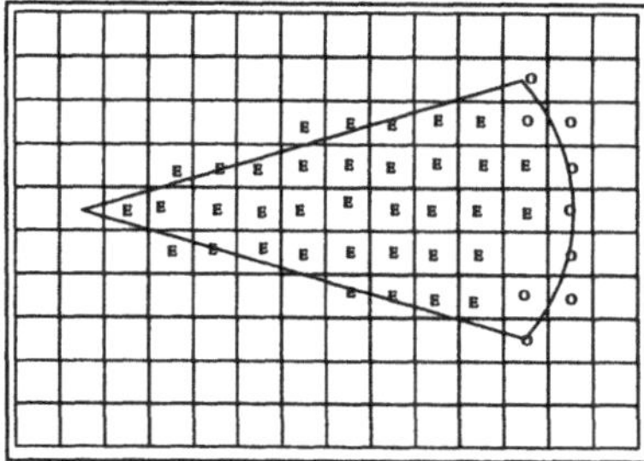

Figure 4. Cells whose probabilities are affected. The letter E signifies that the probability of occupancy in that cell will be decreased. For the letter O the probability will be increased.

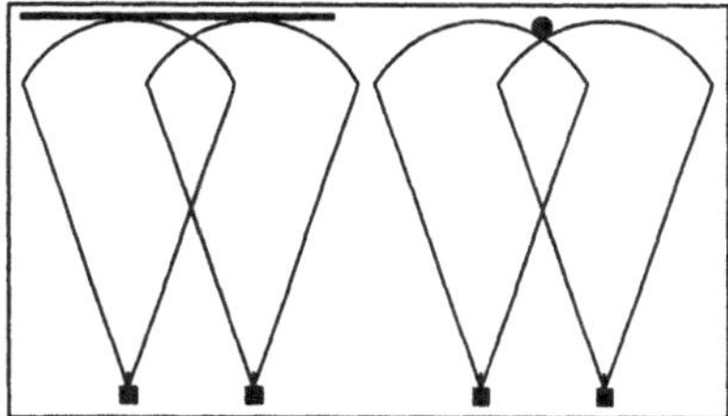

Figure 5. Using two sensors it is not possible to differentiate the kind of target that produced the range values

sensor configurations, the characterization of one obstacle is not possible due to the low spatial sensor resolution.

From figure 5 we can conclude that even using two sensors it is not possible to differentiate between two types of targets because in both situations the range measurements would give the same the values. In fact, although in each scan the grid values are usually slightly changed, after performing a certain number of readings from the same position, all the cells that contain the RCD will reach the maximum level. To overcome this problem, measures are taken only when the robot's position changes.

The information about the direction of motion is also important because, even though the robot has a belt of 24 sensors, only the sensors directed to the regions of interest are scanned allowing once again to reduce the processing time.

2.6. Potential Grid

A trajectory finder module was implemented to generate an alternative path after the detection of an obstacle. This module is based in the potential field proposed by Khatib [6]. Potential field based methods have been used to overcome the detected obstacles with success, the only drawback is being that the repulsive forces have to be calculated at every instant. Using the above described approach to detect the obstacles it becomes a time consuming process to calculate the overall repulsive force from the contribution each of the occupied cells.

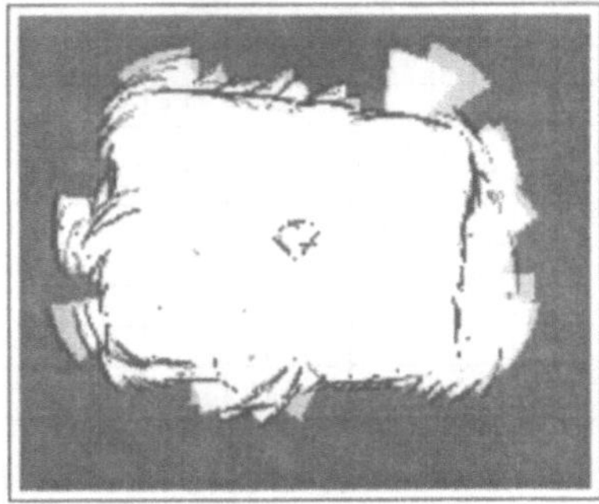

Figure 6. Results obtained in a small room having a vacuum cleaner in the middle

Figure 7. Potential grid generated from simulated sonar scans

However using the measure of the potential field along the path and for each crossed cell, the task for trajectory finder module will be much simpler. This potential grid is constructed from the occupancy grid, by adding to the cell structure another field representing the potential value. This potential grid is calculated from the occupancy grid using the method proposed by Plumer [8], where for each grid cell the corresponding local potential value ϕ_{ij} is calculated based on the value of ϕ of the 8 neighbor cells.

Figure 8. Potential Field obtained for different obstacle configurations

3. Results

Tests have shown good results using the approach described above. Given an *a priori* map of the laboratory whith some boxes randomly placed in the room, the robot was able to detect their presence and, using the artificial potential field, avoid the collision and reach the goal position. Success was also obtained when using obstacles without planar faces, like baskets and even people.

Figure 6 shows the results of the obstacle detection module after performing some movements in our laboratory, with a vacuum cleaner in the center.

Currently, we are integrating the above described methods to generate the potential grid simultaneously with obstacle detection, in order to allow a faster operation of the robot in the presence of new obstacles. In figure 7 are shown the results obtained simulating the robot and the environment. Figure 8 shows the potential field obtained in a room with three different configurations of the obstacles where dark areas represent low potential values and conversely white areas high potential values. It can be seen that there are marked areas outside the walls, this is due to the occurence of spurious reflections.

4. Final Remark

The implemented methods cope with changes in the environment, allowing the operation of a mobile platform in a human environment where the existent objects are subject to change. For example, if an object has been moved, the corresponding cells that were marked as occupied, will be marked as empty and those that correspond to the new position of the box will be marked as occupied. The online generation of a potential grid, allows the faster operation of the obstacle avoidance system.

References

[1] P. Menezes, H. Araújo, J.Dias, M.I. Ribeiro, "Obstacle Detection in Mobile Robots Using Sonar Data", Associação Portuguesa de Controlo Automático Int. Conf. - Controlo 94, I.S.T., Lisbon, September 14-16, 1994.

[2] R. Chatila, R. Alami, B. Degallaix, H. Laruelle, "Integrated Planning and Execution Control of Autonomous Robot Actions", Proc. IEEE Int. Conference on Robotics & Automation, Nice, May 1992, pp. 2689-2696.

[3] J.J. Leonard, H.F. Durrant-Whyte, Directed Sonar Sensing for Mobile Robot Navigation, Kluwer Academic Press, 1992.

[4] A. Elfes," Sonar-Based Real-World Mapping and Navigation", IEEE Journal of Robotics and Automation, vol. RA-3, No. 3, June 1987.

[5] F. Moita, A. Feijao, U. Nunes, A. de Almeida, "Modeling and Calibration of an Ultrasonic Ranging System of a Mobile Robot", Associação Portuguesa de Controlo Automático Int. Conf. - Controlo 94, I.S.T., Lisbon, September 14-16, 1994.

[6] O. Khatib, "Real-time obstacle avoidance for manipulators and mobile robots", Proc. IEEE Int. Conference on Robotics & Automation, 1985, pp. 500-505.

[7] B. Barshan and R. Kuc, "Differentiating Sonar Reflections from Corners and Planes by Employing an Intelligent Sensor", IEEE Trans. on Pattern Analysis and Machine Intelligence, Vol. 12 No. 6 June 1990, pp. 560-569.

[8] E. Plumer, "Neural Network Structure for Navigation Using Potential Fields",IEEE 1992.

Chapter 6

Modeling and Design

Central to good robot performance is the correct design of the mechanical system and the accurate modeling of the resulting plant. The papers in this chapter address the design of mechanisms for improved force control, and present techniques for accurate modeling and calibration of existing mechanisms.

Hayward and Cruz-Hernández analyze the effect of various design decisions on the performance of tendon transmissions. They address tradeoffs in choosing reduction ratio and transmission stiffness and discuss their impact on the bandwidth and dynamic range of a 2N transmission that employs a force sensor to achieve better closed-loop response

Pratt, Williamson, Dillworth, Pratt, and Wright describe an approach to design force-controllable systems, in which the transmission is intentionally compliant. By placing appropriate sensors in the system, the elongation of the transmission can be directly controlled to achieve robust and good bandwidth force control.

Morrell and Salisbury report on a dual-actuator concept, in which a "macro" actuator is used to provide the major component of force output, and a "micro" actuator is used to provide high-bandwidth corrective forces to the output. The resulting system is shown to exhibit high-dynamic range and high bandwidth.

Schütte and Moritz present a formalism for the analysis of the dynamics of multi-body tree-structured systems in which discrete and distributed masses and elasticities are taken into consideration. The approach is applied to a six-degree-of-freedom robot and used to identify 56 parameters which characterize its dynamics.

Hollerbach and Nahvi employ a total least squares approach which acknowledges input and output noise to calibrate robotic systems. By considering the statistics of commanded joint values and achieved end-point values they are able to demonstrate improved calibration over traditional output-only least squares methods.

Parameter Sensitivity Analysis for Design and Control of Tendon Transmissions

Vincent Hayward and Juan Manuel Cruz-Hernández

McGill University
Center for Intelligent Machines
3480 University Street
Montréal, Québec, Canada, H3A 2A7

Abstract

We apply sensitivity analysis to the design and control of a tendon transmission. With this approach, some prefered values for the system parameters and a feedback compensator can be proposed. The controller has the special characteristic of being designed based on a linear plant using a robust loopshaping technique, yet it compensates also for the nonlinear behavior of the plant, while exhibiting good disturbance rejection and robustness. Experimental results using a test bench are discussed.

1. Introduction

There is freedom in the design of tendon transmissions. The question arises of how to choose the design parameters to improve performance. To answer this, we select performance objectives which are relevant to a haptic device[1] [7]— and look at the sensitivity of the parameters with respect to the performance objectives. These include extending the frequency response to the widest range possible, as well as reducing friction and inertia as experienced from the load side of the transmission.

The design of the compensator would be straightforward if a linear model could be used. Unfortunately, a transmission exhibits friction, so precise control requires the compensation of non-linear friction effects. We will develop a scheme which can be tuned for a wide class of systems and which neither rely on a detailed knowledge of the non-linear behavior of friction, nor requires measurement of velocity.

The purpose of the tendon drive is to transmit mechanical signals from a remote location so that the actuators can be mechanically grounded. Because mechanical signals are transmitted by taking advantage of the cohesive forces in a material, large amounts of energy can be transmitted by small amounts of material. This is why cable and tendon transmissions have been a technique of choice for the implementation of teleoperators, hand controllers and now haptic devices for almost five decades, [6, 10, 2].

The transmission is of type 2N [8], with two actuators per channel. This type of transmission minimizes the average tension, while reducing stresses in the supporting structure and idler pulleys. It results in lower friction, simplified assembly and

[1]A haptic device may be viewed as a high fidelity force reflecting hand controller

tuning, as well as higher reliability and work life since the tendons are stressed only during the transients. In addition, friction is reduced dynamically.

Friction in tendon drives tends to grow linearly with tension. Remark first that if we look at friction as noise on mechanical signals, we see that it will grow with the intensity of the force signal. Consider now the function of a haptic device which is to display signals to the hand of an operator. Displacements, forces and other mechanical sensations, obey the same laws as other sensations, following a Fechnerian scale expressed, for example, by a Weber fraction $\Delta I/I$, where I is the intensity of the stimulus. This means that the sensitivity to changes in the signal (noise here) decreases with the signal intensity, in other terms, the relative sensitivity is constant. The mechanical signal-to-noise ratio in a transmission of type 2N can be made roughly constant across its dynamics range instead of decreasing with the signal intensity as in a conventional transmission.

Analogously to class B electronic power amplifiers, each motor is driven by a half-wave signal, as illustrated in Figure 1. Practice has shown that the switching

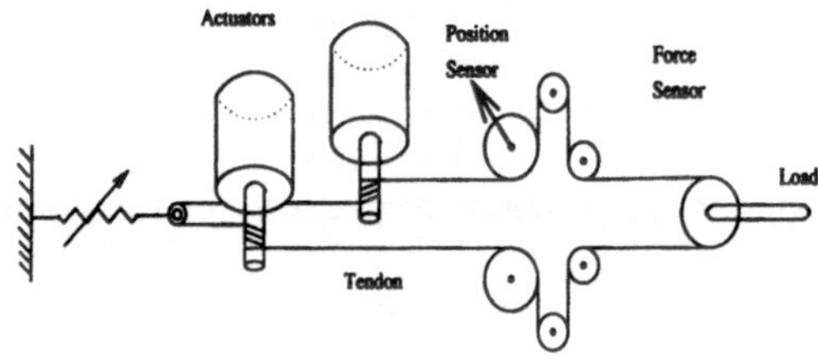

Figure 1. Transmission structure.

nature of the signal did not cause significant distortion, provided that the stiffness of the proximal portion of the transmission is sufficiently high to prevent excessive amounts of stored elastic energy. The generation of the actuator signal was simply accomplished using clamping diodes on the path of a single current amplifier shared by the two actuators. The current amplifier (linear amplifier) effectively inverts by feedback the electrical transfer function (roughly an RL circuit) of the actuators and insures that current, and therefore torque, tracks precisely the input control signal accross a bandwidth much larger than the mechanical bandwidth of the drive.

Displacement and force are measured directly on the tendon path via optical sensors developed in our laboratory. Both rely on differential measurements of infrared light intensities sensed by PIN diodes. This type of sensor has the usual benefits of optical sensing techniques — that of absence of contact, low noise, immunity to environmental conditions and EMF perturbations.

2. Model

Figure 2 illustrates an engineering model of the plant. It includes the inertia of the motor I_M, linear damping lumped into damper B, r the pulley ratio between a capstan and a driven pulley, k_1 the elasticity of the proximal section of the transmission, k_2 the elasticity of the distal part, I_c the inertia of the driven manipuladum, and Z_H an arbitrary impedance representing the load, an operator's hand for example. We call

$$k_e = (k_1 k_2)/(k_1 + k_2), \tag{1}$$

a factor expressing the degree of "co-location" of the force measurement.

The relevant signals are F_r the force generated on the tendon by the actuator, F_n a disturbance force signal representing friction in the motor, F_d another disturbance signal representing the friction in the transmission, and the two signals sensed: X_t, the tendon displacement, and F_m the differential tendon force, both measured at some intermediate location between the actuator and the load. Looking at the

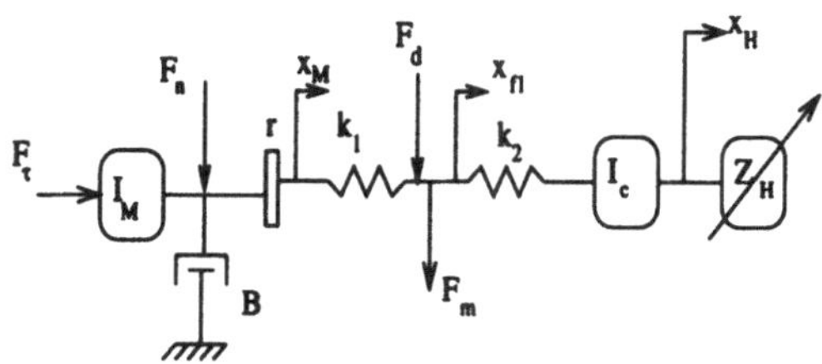

Figure 2. Model of the plant.

transmission of force from the actuator to the load, the transfer function can be worked out for both the actuator force and the actuator friction signal. For the rest of this paper, the variable load will be simplified to a single elasticity k_z, which can be viewed as a worst case as far as stability is concerned:

$$\frac{F_m}{F_r} = \frac{F_m}{F_n} = \frac{N(s)}{D(s)} = \frac{k_e I_c s^2 + k_z k_e}{r^2 I_M I_c s^4 + r^2 I_c B s^3 + (k_e Ic + r^2 I_M (k_z + k_e))s^2 + r^2 B (k_e + k_z)s + k_z k_e} \tag{2}$$

The disturbance friction signal due to the transmission and seen by the sensor is:

$$\frac{F_m}{F_d} = \frac{r^2 I_M I_c}{D(s)} \tag{3}$$

3. Analysis

3.1. Effect of r

A first remark is in order: the numerator of the transfer function (3) is highly sensitive to r and independent from frequency. Increasing r is equivalent to increasing the apparent inertia of the actuator as far as the transmission is concerned. The counter-intuitive result here is: the higher the actuator inertia, or equivalently the higher the pulley (or gear) ratio is, the better the friction disturbance rejection is. This can be understood from another perceptive, considering that for a given desired output signal and a given disturbance, a more inert actuator (or higher ratio) will demand a larger input signal, improving the signal-to-noise ratio. The trade-off is as follows: if the ratio is made too high, the actuator might saturate. The other downside of increasing the ratio is the increase of apparent inertia when the transmission is backdriven. However, since force feedback is applied, the apparent inertia is divided by the loop gain. Thus, an optimal design will result from the the highest possible ratio that will not saturate the actuator and which will permit the highest loop gain possible for a given desired phase margin, since r appears in the denominator of the transfer functions. On the upside, a high ratio will increase the peak force generated by the system. In the prototype, the actuator shafts are directly driving the tendons with no intervening capstans and r is close to 15:1.

3.2. Effect of k_e

To understand the effect of the sensor placement, the sensitivity function $S_{k_1}^G$ in the frequency domain with respect to k_1 was computed for various values of k_1, while the other parameters were set to values close to those of the actual prototype.

$$S_{k_1}^G = \frac{(r^2 I_M I_c^2 s^6 + r^2 I_c^2 B s^5 + 2r^2 I_c I_M k_z s^4 +}{r^2 I_M I_c^2 s^6 + r^2 I_c^2 B s^5 + (k_e I_c (k_e I_c + r^2 I_M (k_e + 2k_z))) s^4 + r^2 B k_e I_c (k_e +} \frac{+2r^2 I_c B k_z s^3 + r^2 I_M k_z^2 s^2 + + r^2 B k_z^2 s) \frac{k_2^2 k_1}{(k_1+k_2)^2}}{+2k_z) s^3 + k_z k_e (2 I_c k_e + r^2 I_M (k_z + k_e)) s^2 + r^2 B k_z k_e (k_e + k_z) s + k_z^2 k_e^2} \quad (4)$$

It can be seen in Figure 3 that a soft transmission with a stiff end portion ($k_1 >>$

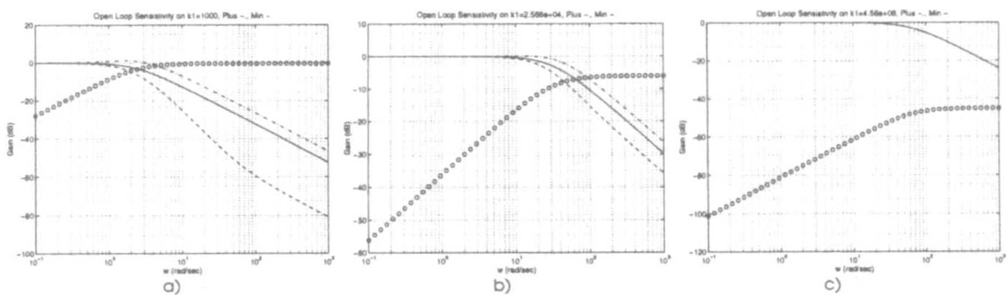

Figure 3. Plots for the transfer function, sensitivity function, and $G(s)(1 \pm S(s))$ with respect to k_1. Notation: Sensitivity function $S(s)$ ∘ ∘ ∘, Transfer Function $G(s)$ –, $G(s)(1+S(s))$ –. – . – ., $G(s)(1-S(s))$ – – –.
a) k_1=1 * 10^3, b) k_1=2.566 * 10^4, c) k_1=4.56 * 10^6

k_2) is very sensitive in the high frequencies to slight changes in k_z, while a stiff transmission with a soft end-portion, not only increases the response's bandwidth but also decreases the sensitivity to very small values. It can be concluded that k_e expressing the degree of collocation of the force sensor has a major influence on the ability for a transmission to be force-controlled.

This k_e-dependence has been noticed by many researchers while implementing force control on a manipulator [4]: a stiff force sensor clamped at the wrist and separated from the actuator by a soft transmission will make the control difficult, and if at all possible, the response will either be highly sensitive to the load variations (hence the hard contact bouncing so often discussed) or effective only in the very lowest range of the frequency domain. A force sensor located near the actuator and separated from the load by the structural elasticity of the manipulator has exactly the opposite property: the sensitivity to the load is low (so a single tuning will work for a wide range of loads but disturbance rejection is less good so it cannot be precise) and the response range is wide.

A parallel can be drawn between the effect of a gear ratio for position control and the effect of k_e in force control. A high gear ratio makes the position control insensitive to load variations and other disturbances (so it is easy to control), while a direct drive robot will be maximally sensitive to the same factors (so it can be accurate and the disturbance rejection can be good but is hard to control). From that viewpoint we may see that the co-location factor k_e plays for force control a role analogous to r for position control.

3.3. Effect of B

Damping B is important because this will inform the designer with the effect of changes of properties of the transmission.

$$S_B^G = \frac{-Br^2s(I_cs^2+(k_e+k_z))}{r^2I_MI_cs^4+r^2I_cBs^3+(k_eIc++r^2I_M(k_z+k_e))s^2+r^2B(k_e+k_z)s+k_zk_e} \tag{5}$$

The sensitivity curves for the nominal plant parameter values are extremely similar to the curves produced by k_e, so they are not reproduced here. The plant response is obviously mostly affected in the vicinity of the cut-off frequency (where half of the input signal is dissipated). The conclusion is evident: damping should be low and if it must be high, small changes will have big effects on the plant's response, possibly destabilizing the closed loop response.

3.4. Effect of k_z

We now consider the effect of load changes on the response.

$$S_{k_z}^G = \frac{k_z(k_e^2r^2I_Ms^2+k_e^2r^2Bs)}{r^2I_MI_c^2s^6+r^2I_c^2Bs^5+(k_eI_c(k_eI_c+r^2I_M(k_e+2k_z)))s^4+r^2Bk_eI_c(k_e+2k_z)s^3+k_zk_e(2I_ck_e+r^2I_M(k_z+k_e))s^2+r^2Bk_zk_e(k_e+k_z)s+k_z^2k_e^2} \tag{6}$$

It is seen in Figure 4 that the sensitivity to this parameter is very high for small

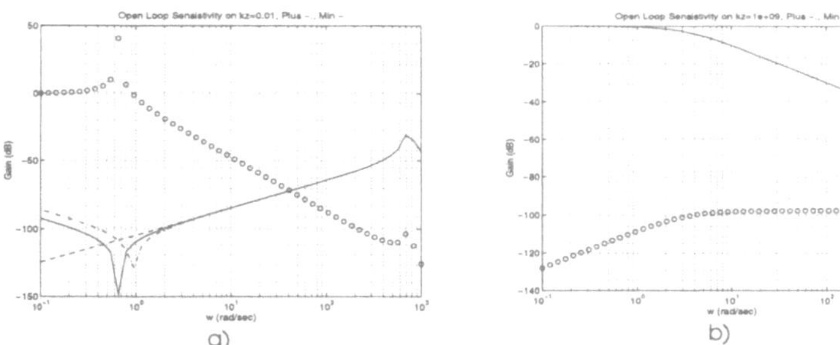

Figure 4. Plots for the transfer function, sensitivity function, and $G(s)(1 \pm S(s))$ with respect to k_z. Notation: Sensitivity function $S(s)$ o o o, Transfer Function $G(s)$ —, $G(s)(1+S(s))$ $-.-.-.$, $G(s)(1-S(s))$ $---$,
a) k_z=0.01, b) $k_z=1*10^9$

values, while it vanishes at high values. While this may seem obvious in retrospect, it is important to notice that the sensitivity has a resonant shape with the peak in the vicinity of the plant's first natural resonance. This can be seen in (2), the independent term, both in the numerator and the denominator, depends on k_z, so if k_z is small, the plant will have two zeros and one pole.

This means that the response is essentially unknown when the load is stiff. As a consequence it is imperative to consider feedback control to reduce sensitivity. The closed loop transfer function is as follows:

$$T(s) = \frac{k_eI_cs^2+k_zk_e}{r^2I_MI_cs^4+r^2I_cBs^3+(k_eIc(1+K)+r^2I_M(k_z+k_e))s^2+r^2B(k_e+k_z)s+k_zk_e(1+K)} \tag{7}$$

Recall that for a closed loop transfer function, where α is some parameter under study, $S_\alpha^T = S_G^TS_\alpha^G$ [5]. Since S_α^G was computed for B and k_z we only need to compute S_G^T. Setting $C(s) = K$ leads to $S_G^T = 1/(1+KG)$. To reduce the sensitivity with respect to parameter B or k_z we need to minimize S_G^T by setting K to some optimum value. Since the denominator of S_G^T is the same as $T(s)$, a value for K must be found that will also not only preserve stability but achieve a desired stability margin. Further analysis reveals that the choice of K only affects sensitivity with respect to B beyond the cut-off frequency of the closed loop system. For k_z, Figure 5 shows that for a small k_z, the single gain feedback controller neither improves the response, nor sensitivity. It is therefore concluded that more elaborate controllers must be considered. If we consider the next level in complexity which for example

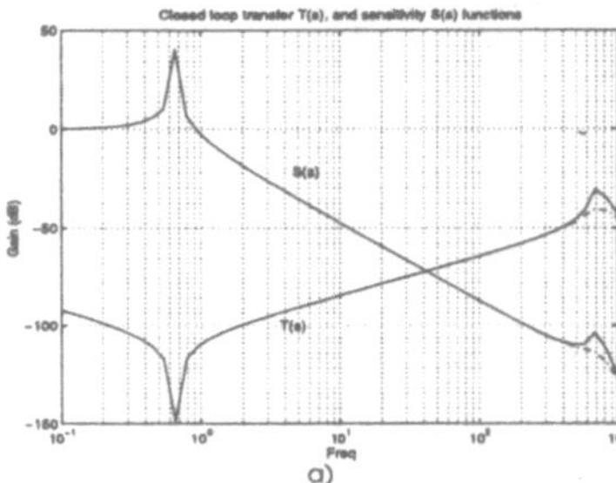

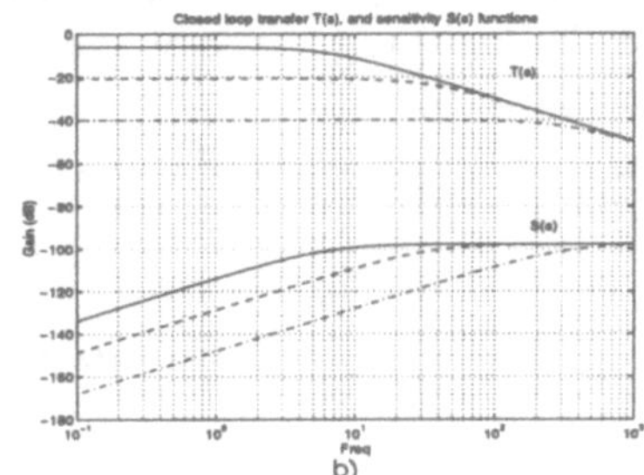

Figure 5. Plots for the transfer function T(s) and sensitivity function S(s) respect to k_z. a)$k_z = 0.01$, and b)$k_z = 10^9$ for a constant controller with values 1, 10, 100. Notation: K=1 —, K=10 - - -, K=100 -.-.-.

could be a lead compensator of the form $C(s) = (s+z)/(s+p)$, it can be found that the sensitivity function becomes:

$$S_G^T = \frac{(s+p)D(s)}{(s+p)D(s) + (s+z)N(s)} \tag{8}$$

Now, any change in p or z will affect sensitivity in the same order as a simple gain K with the disadvantage of complicating tuning.

3.5. Conclusion about design

In the absence of further information about the exact nature of the plant, the simple gain controller K should be preferred over a complex controller composition. This simple controller will improve the response, and decrease the sensitivity, while its tuning is particularly simple. The tuning will only involve raising the value of K under the worst conditions (smallest needed k_z) until the closed loop stability is compromised, while observing the response in the time domain, for example.

This is further indicative of the fact that the plant's non-linearities in fact play an important role in the system's response, and this vindicates the use of a single gain controller in the absence of additional information.

4. Controller Design

We now consider the design of a less conventional controller designed for the plant described in the previous section with a goal of improving the extent and the precision of the response of the system, and reducing the apparent friction of the transmission when it is back driven as well as the apparent inertia.

4.1. Discussion

The input-output behavior of the tendon transmission, although complicated because of the presence of non-linearities, might in fact be viewed as the combination of simpler subsystems, which once combined create an apparently complicated behavior. For illustration, Figure 6, shows the input force to output force relationship exhibiting a complex hysteretic behavior. A possible decomposition is suggested by

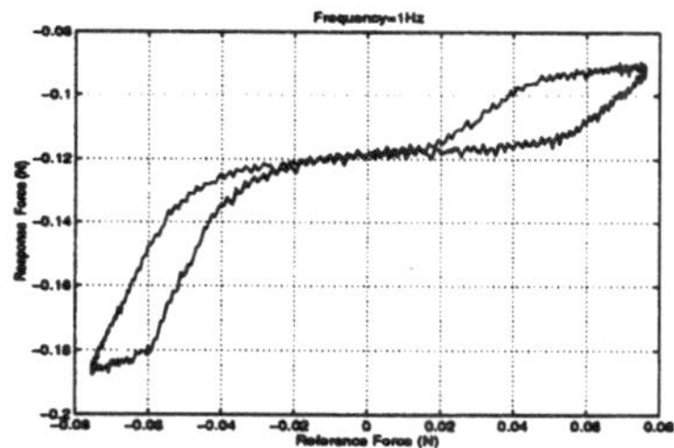

Figure 6. Hysteretic behavior of the plant.

the physical nature of the plant. The transmission includes a linear system representing elasticity, damping and inertia of its mechanical components. This was verified by measuring the transfer function of the system and then observing that the response is indeed well defined, including resonant characteristics that could be precisely identified for a given input amplitude. However, the dependency of the response with respect to the input presented the hallmarks of nonlinear characteristics such that, the linear part is camouflaged by the nonlinear distortion. Figure 7 shows in fact how the response may present a resonant peak shifting from 15 Hz to 30 Hz depending on the amplitude of the input. The response is nevertheless precise and was found not to change with time. This response was experimentally obtained with a very stiff load which is the worst case as shown in the previous section.

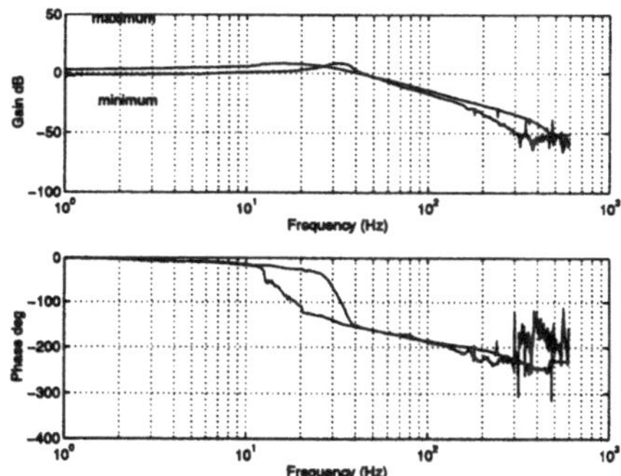

Figure 7. Open loop Bode plot.

It was further verified that nonlinear stiffening of the material used to make the tendons, a possible source of non-linearity, was not significant. The other likely cause for a non-linear response is obviously friction. Friction has been extensively studied and various models have been proposed. The reader is referred to the extensive survey by Armstrong-Hélouvry, Dupont, and Canudas De Witt [1], for a summary.

It was said that friction increases roughly linearly with the tendon tension. But more importantly, the observed friction does not exhibit noticeable stiction, often referred to as Stribeck friction. From this observation we can conclude that no significant potential energy is being stored by the occurrence of the friction phenomenon per se; only dissipation occurs. It can therefore be considered as memoryless and thus can be completely represented as a single valued input-output relationship.

In this paper, we adopt a simple representation of friction: the standard "breackaway" model. With this model, a transmission is an input-output device transmitting torque (or force), while motion is not considered explicitly. A force balance equation states that the transmission transmits the input torque to the output torque minus a torque lost in dissipation, with the exception that when the input torque is under a threshold (under the breakaway level), no torque is transmitted to the output (since no motion is observed) and the friction balances exactly the input torque. This results in an input-output force-force friction model represented as a dead-band as seen in Figure 8, which is a single valued relationship. The exact nature of this curve, whether it is even dependent or not from some other parameters is irrelevant to the rest of this discussion. All what matters is that its slope varies with the input. It is well known that an input-output non-linear relationship of

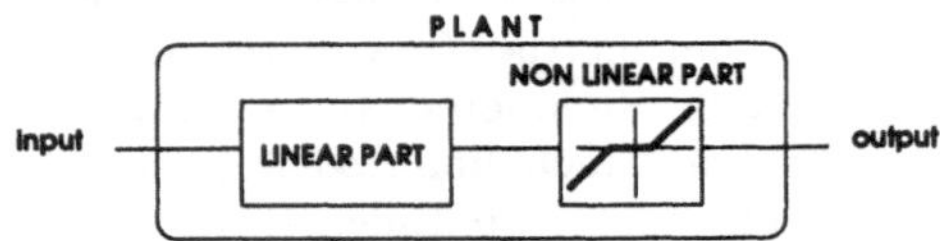

Figure 8. Plant representation: Wiener model [11]

the type just described can easily be "straightened" with the application of simple gain feedback and that no stability problem may occur since the closed loop system has no memory, no energy is stored. However, any controllers having dynamics, for example a PD, a PI, or any other filter for that matter, are liable to create complex behaviors including instabilities, limit cycles, or even chaotic patterns [9].

Returning to the physical structure of the transmission, recall that we may view it as a linear system resulting from a combination of springs, dampers and inertias forming a low pass filter, cascaded with a single valued deadband-like nonlinear relationship. Such a combination will certainly create a hysteretic-like behavior. This can easily be seen by considering a ramping input: while in the deadband, no signal is observed at the output; this has the effect of shifting the response on the right. When the input reverses, the system enters the deadband again, shifting the response to the left, and so on, forming a hysteresis-like loop.

4.2. Controller Synthesis

Recall that another objective is to extend the frequency response as far as possible, thus the transmission has to be stiffened by feedback. From the previous discussion, this also has the effect of correcting the hysteretic-like behavior of the plant.

4.3. Experimental Single Gain Controller

The experimental closed loop Bode plot is shown in Figure 9. As predicted by the previous analysis, it has a marked resonant characteristic. The tuning is trivial, a phase margin is chosen, and the gain follows from this choice. It must be said that despite the low pass nature of the transmission, an effective apparent friction

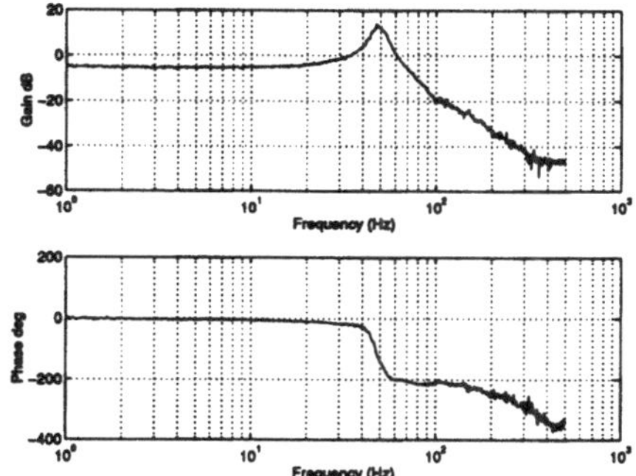

Figure 9. Closed loop Bode plot of the force response with a proportional controller.

reduction is achieved. The usable frequency range, which was 40 Hz open loop, is slightly improved. It is robust and noise free.

4.4. Approximate Plant Inversion

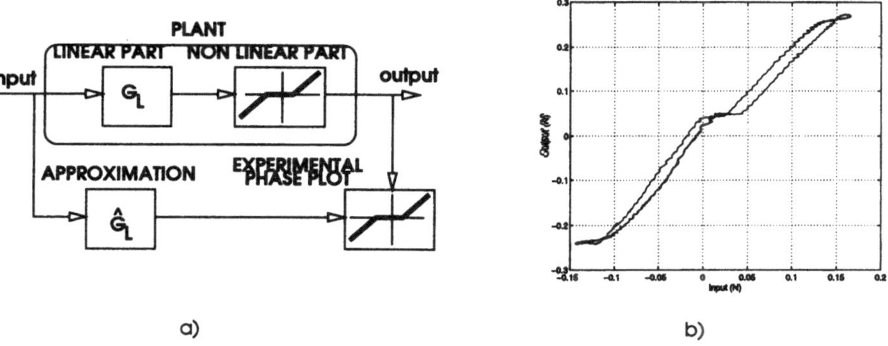

Figure 10. a) Theoretical representation, and b) Experimental phase plot at low velocity, $F = 0.2$ Hz and small amplitude: 0.31 N

Referring to Figure 10.a), the objective is to cancel the low pass dynamics of the plant in order to achieve stiffening, so that the feedback will only see a uni-valued input output relationship. This concept is represented in Figure 11. In order to verify that the transfer function actually decomposes in the needed fashion, a model $\hat{G}_L(s)$ of the plant is identified (using conventional identification methods) and the same input (of various kinds) is presented to the plant and to the model. The phase plot of the plant output is traced against the output of the model. The optimal model will minimize the area of the phase plot at all frequencies. See Figure 10.b) for the experimental result. Once the model is found, the ideal controller is simply $\hat{G}_L^{-1}(s)$; however, since the plant is lowpass, it would not have a proper transfer function and would not be realizable. We must therefore settle for an approximate inverse in the desired frequency range and poles are added to achieve this.

The resulting pole-zero cancellation control is effectively a non-robust design since it relies on a precise identification of the plant. In fact, because of its uncertain

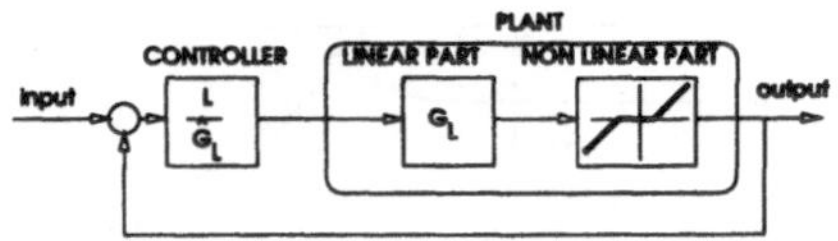

Figure 11. Loopshaping technique.

non-linearities, such identification is not possible.

Loopshaping technique was used to design a feedforward controller [3]. The idea, as is well known, is to choose a loop transfer function $L(s)$ so that we can achieve a robust performance, good robustness with $T(s)$ small at high frequencies, and disturbance rejection at low frequencies making S_G^T small, which is sometimes not possible to achieve just by modifying the system parameters. The condition to design a controller with robust performance are to have an internally stable plant and to enforce the following inequality:

$$|||W_1S| + |W_2T|||_\infty < 1 \tag{9}$$

W_1 is a weigthing function used to determine internal stability by enforcing nominal performance such that $||W_1S||_\infty < \epsilon$, where ϵ is the maximum amplitude of the error signal over the whole frequency range. W_2 is another weighting function to enforce robust stability, $||W_2T||_\infty < 1$. T is the closed loop transfer function and S the sensitivity function S_G^T. $L(s)$ can then be determined using a graphical method.

The controller $C(s)$ is obtained from $C(s) = \frac{L(s)}{P(s)}$, with $P(s) = \hat{G}(s)$. Again, the controller has to be proper and internal stability of the plant has to be ensured. This method is suitable for our purpose since the plant $G(s)$ is stable and minimum phase, as can be seen in Figure 7, and has all its poles and zeros in the right half plane. We chose $L(s)$ to behave as a second order system of the form:

$$L(s) = \frac{\omega_n^2}{s^2 + 2\zeta\omega_n s + \omega_n^2} \tag{10}$$

with a natural frequency $\omega_n = 40Hz$, and $\zeta = 0.5$. This can be considered as a good response in open loop and is what we can expect for this plant. Physically the plant may not achieve more than a few Hertz beyond its original natural frequency, for this reason we did not place the response of $L(s)$ further than 40Hz. Furthermore, in closed loop the bandwidth will increase. As we demonstrated experimentally, we can achieve almost the same bandwidth for an $L(s)$ with $\omega_n = 40Hz$ than for $\omega_n = 80Hz$. The difference is that the closed loop response has dithering behavior when we used a higher ω_n. This happens because we are amplifying the noise that appears after 40 Hz, and because we were breaking the condition needed to achieve a robust performance as specified by the loopshaping technique. The controller was designed using an approximation to the plant obtained with a large input amplitude, when nonlinear disturbance is minimized. It was also shown to be effective for all amplitudes. The experimental response of the system in closed loop using the controller described above is presented in Figure 12, where it can be observed that no matter which amplitude of the input we give to the closed loop system, the result is always almost the same, and also a very good noise rejection is achieved. The range of uncertainty (Figure 7) in amplitude was quite large, and now this range has been reduced for most of the low frequency, which for haptic interfaces is crucial.

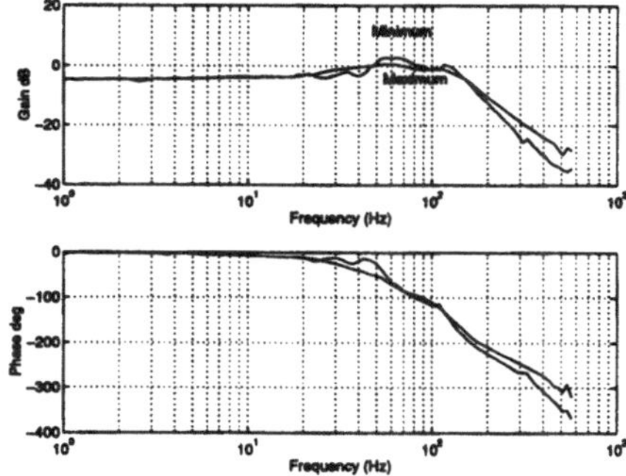

Figure 12. Closed loop Bode plot using $C(s) = \frac{L(s)}{P(s)}$.

We can look at the input-output relation of the closed loop system and compare how this behavior has been modified, see Figure 13.a. The deadband presented in

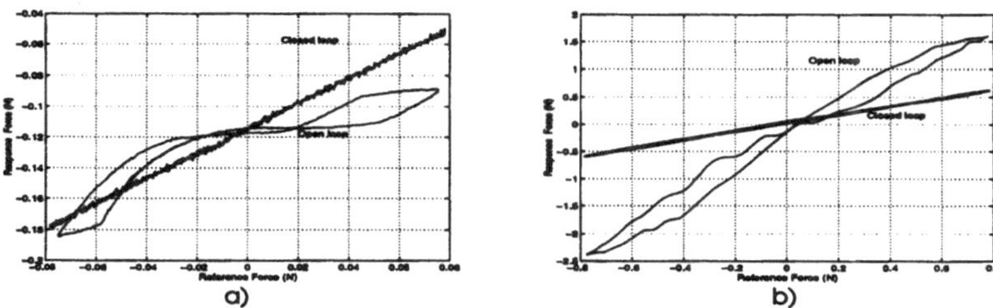

Figure 13. Comparison of open and closed loop. a) Small amplitude and F=0.2 Hz., and b) Large amplitude and F=1 Hz.

open loop as well as the hysteresis-like behavior is corrected to give a linear behavior. We present another curve of higher amplitude and different frequency as in Figure 13.b. The small area that appears in the closed loop signal is due to some phase shift between the input and the output and is not due to hysteresis. We have been able to improve the system response and compensate for nonlinearities, not just for some frequencies but for a wide range of them.

5. Conclusion

A model for tendon transmissions was presented. An extensive sensitivity analysis was carried out to understand how the parameters affect the behavior of the system. It was found that for the value of r, a tradeoff between friction rejection, inertia reduction and saturation in the actuator has to be achieved. k_e (the force dividing factor) was found to be an expression of the degree of collocation of the force sensor along the transmission. Sensitivity analysis enable us to show the effects of collocation on a transmission ablity to transmit forces. The effect of a variable load k_z can be reduced only by a feedback controller. The proportional controller is the only one which can reduce the sensitivity function S_G^T, without any complicated tuning.

The final design of the controller was done using the loopshaping technique, to ensure robustness and the criteria to choose an appropriate approximation of the plant were presented. This controller, because of its robustness and disturbance rejection, compensates for the nonlinearities that appeared in open loop and also reduces the uncertainty range of the response.

6. Acknowledgments

Initial funding for this project was provided by a research contract with the Canadian Space Agency (No. 9F009-1-1441/01-SR). Major funding is now provided by the project "Haptic Interfaces for Teleoperation and Virtual Environments" (AMD-5) funded by IRIS (second phase), the Institute for Robotics and Intelligent Systems part of Canada's National Centers of Excellence program (NCE). Additional funding was from an operating grant from NSERC, the National Science and Engineering Council of Canada.

The second author would like to acknowledge the generous support of Universidad Nacional Autónoma de México (UNAM) in the form of a postgraduate fellowship.

The authors wish to acknowledge contributions of Christopher Strong, Xianze Chen, and Jan Sinclair from MPB technologies Inc., Montreal, Canada.

References

[1] Armstrong-Helouvry, B. Dupont, P. Canudas De Wit, C. 1994. Survey of models, analysis tools and compensation methods for the control of machines with friction, *Automatica* 30(7), pp. 1083-1138.

[2] Bejczy, A. K., Salisbury, K. 1980. Kinesthetic coupling between operator and remote manipulator. Proc. *International Computer Technology Conference, ASME*, San Fransisco, pp. 197–211.

[3] Doyle, John C., Francis, B. A., Tannenbaum A. R. 1992. *Feedback Control Theory.* New York: Maxwell Macmillan International.

[4] Eppinger, S.D. and Seering, W.P. 1987 Understanding Bandwidth Limitations in Robot Force Control. *Proc. IEEE International Conference on Robotics and Automation,* Vol. 1, pp. 904–909.

[5] Frank, P. M. 1978. *Introduction to sensitivity analysis.* Academic Press.

[6] Goertz, R. C., Thompson, W. M. 1954. Electronically controlled manipulator. *Nucleonics*, 12(11), pp. 46–47.

[7] Hayward, V. 1995. Toward a seven axis haptic interface. *IROS'95, Int. Workshop on Intelligent Robots and Systems.* pp. 133–139.

[8] Jacobsen, S.C. and Ko, H. and Iversen, E.K. and Davis, C.C. 1989 Control Strategies for Tendon-Driven Manipulators *Proc. International Conference on Robotics and Automation,* Vol. 1, pp. 23–28,

[9] Towsend, W.T., Salisbury S.K. The effect of Coulomb friction and stiction on force control. *Proc. IEEE Conference on Robotics and Automation,* pp. 883–889.

[10] Vertut, J. 1976. Advance of the new MA 23 force reflecting manipulator system. Proc. *2nd International Symposium on the Theory and Practice of Robot and Manipulators,* CISM-IFToMM, pp. 307–322.

[11] Wiener, N. 1958. *Nonlinear problems in random theory.* New York: The Technology Press of The Massachusetts Institute of Technology and John Wiley and Sons, Inc.

Stiffness Isn't Everything

Gill A. Pratt, Matthew M. Williamson
Peter Dillworth, Jerry Pratt, Anne Wright
MIT Artificial Intelligence Laboratory and Laboratory for Computer Science
Cambridge, MA, 02139
gill@ai.mit.edu, matt@ai.mit.edu
chunks@ai.mit.edu, jpratt@ai.mit.edu, anarch@ai.mit.edu

Abstract

Most robot designers make the mechanical interface between an actuator and its load as stiff as possible[9][10]. This makes sense in traditional position-controlled systems, because high interface stiffness maximizes bandwidth and, for non-collocated control, reduces instability. However, lower interface stiffness has advantages as well, including greater shock tolerance, lower reflected inertia, more accurate and stable force control, less damage during inadvertent contact, and the potential for energy storage. The ability of series elasticity (usually in the form of a compliant coating on an end-effector) to stabilize force control during intermittent contact with hard surfaces is well known. This paper proposes that for natural tasks where small-motion bandwidth is not of paramount concern, actuator to load interfaces should be significantly less stiff than in most present designs. Furthermore, by purposefully placing the majority of interface elasticity inside of an actuator package, a new type of actuator is created with performance characteristics more suited to the natural world. Despite common intuition, such a series-elastic actuator is not difficult to control.

After an analytic treatment of the trade-offs and limitations of series elastic actuators, we present a simple hybrid feed-forward / feed-back control system for their use. We conclude with test results from a revolute series-elastic actuator being used in the arms of the MIT humanoid robot Cog[5] and also in the arm of a small planetary rover[1]. A similar concept, but with pulley driven series-elastic tendons, is presently being used in a 2-D walking biped named "Spring Turkey".

1. Introduction

Robot designers have traditionally maximized the interface stiffness between actuators and loads[19], and with good reason. Stiffness improves the precision, stability, and bandwidth of position-control. When either open-loop positioning or collocated feedback are used, increased interface stiffness decreases end-point position errors under load disturbances. In non-collocated feedback systems (where the position sensor is located at the load side of the interface), increased stiffness both lowers necessary actuator motion in response to load variations and raises the resonant frequency of the motor inertia and interface compliance. As a result, stiffer interfaces allow the bandwidth of a position control feedback loop to be raised without compromising stability[7][8].

But stiffness isn't everything. Most electric motors have poor torque density and thus can deliver high power only at high speed[15]. To provide high power to slowly moving

1. This work was supported by JPL contract # 959333, for which we are most grateful.

loads, gear reduction become necessary. Unfortunately, gears introduce friction and/or backlash, torque ripple, and noise. The use of N:1 gearing also causes an N^2 increase in reflected inertia so that shock loads cause very high stress on the teeth of the output gear, possibly resulting in failure. This increased reflected intertia and the typically high back-drive friction of high ratio gear trains can also cause damage to the robot or environment when unexpected contact occurs.

Reducing interface stiffness by inserting series elasticity can resolve many of these problems. The basic configuration of a series elastic actuator is shown below:

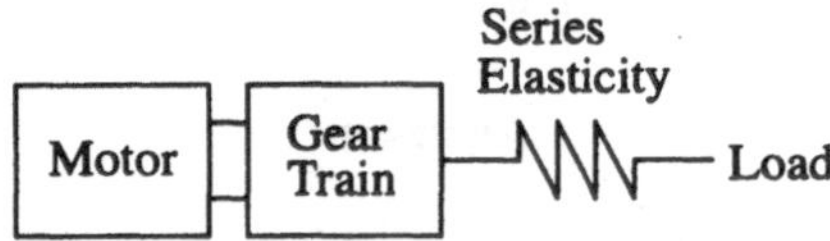

Fig. 1. Block Diagram of Series-Elastic Actuator

The first benefit of the series elasticity is to low-pass filter shock loads, thereby greatly reducing peak output gear forces. Although this also low-pass filters the actuator's output, we believe this is a place for an engineering trade-off, not the traditional "stiffer is better" minimization. The proper amount of interface elasticity can substantially increases shock tolerance while maintaining adequate small motion bandwidth for natural tasks like locomotion and manipulation.

Series elasticity also turns the force control problem into a position control problem, greatly improving force accuracy. In a series elastic actuator, output force is proportional to the position difference across the series elasticity multiplied by its spring constant. Because position is much more easy to control accurately through a gear train than force, the force errors usually caused by friction and torque ripple are reduced. Friction and backlash are usually a trade-off in gear train design. Series elasticity allows this trade-off to be driven much further towards high friction and low backlash, resulting in better position control at the gear train's output and thus better force control at the load. Importantly, high friction, low backlash gear trains can also be made inexpensively.

Increased series elasticity also makes stable force control more easy to achieve. Contrary to the case in position control, stable force control is easier to achieve when the frequency of interface resonances are **lowered**. This is because force feedback works well at low frequencies, creating a virtual zero-rate spring in series with the non-zero mechanical elasticity (i.e. a net spring rate of zero).

Finally, series elasticity provides for the possibility of energy storage. In legged locomotion, such energy storage can significantly increase efficiency[1]. By incorporating elasticity into the actuator package, efficiency benefits can be had despite the elasticity being hidden from the higher level control system. In other words, unlike methods that try to account for link elasticity at a systems level[19][20], the high level control system thinks it is controlling independent force actuators when in fact those actuators have internal springs that provide the aforementioned benefits.

Several authors have previously studied methods for controlling unavoidably flexible structures (such as those expected in space[4]), and the role of interface compliance in stabilizing force control during contact transitions[23]. But with the exception of systems where energy-storage is paramount (such as the legs of a hopping robot[18]), and some passive hand mechanisms[21][11], few have suggested that elasticity should be incorporated into general purpose robotic actuators. This seems strange, particularly for robots executing natural tasks, because elasticity is used for a wide variety of purposes in animals[1].

2. Performance Limits

Series elasticity creates the need for elastic deformation of the series element whenever force is modulated. This extra motion may add either constructively or destructively to the motion of the load. Thus, depending on the relative amplitude and phase of the load's force and motion waveforms, it is possible for the interface elasticity to either increase or decrease bandwidth.

Ignoring output inertia, a series-elastic actuator can be modeled as follows:

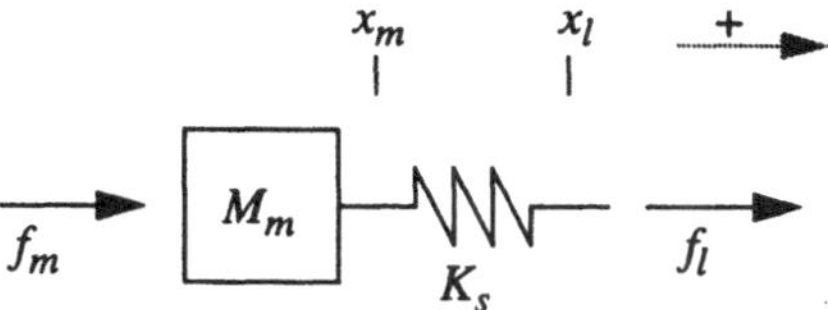

Fig. 2. Model of a series-elastic actuator

with the following frequency-domain system diagram:

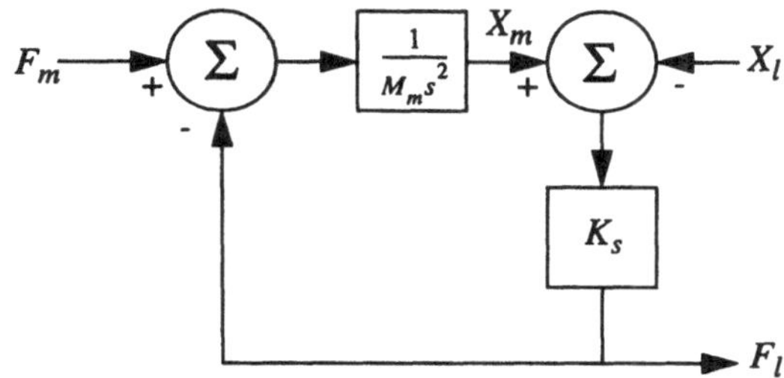

Fig. 3. Frequency Domain System Diagram

and the following variable definitions:

f_m, F_m	Magnetic Force Applied to Motor Rotor
f_l, F_l	Force Applied to Load
x_m, X_m	Position of Motor
x_l, X_l	Position of Load
M_m	Motor Mass
K_s	Elasticity Spring Rate

Fig. 4. System Variables

From the diagram above we can derive the following equations:

$$F_l = K_s(X_m - X_l) \tag{1}$$

$$X_m = \frac{F_m - F_l}{M_m s^2} \tag{2}$$

Setting $s = j\omega$ and solving for F_m, in terms of F_l and X_l we have:

$$F_m = F_l - \frac{M_m}{K_s}\omega^2 F_l - M_m \omega^2 X_l \tag{3}$$

As can be seen above, the motor force has three components. The first, F_l , is the force applied through the elasticity to the load. The second, $-\frac{M_m}{K_s}\omega^2 F_l$, is the force required to accelerate the motor's mass in order to change the deformation of the elasticity. The third, $-M_m\omega^2 X_l$, is the force required to accelerate the motor's mass so as to track motion of the load. Of these three terms, only the middle one is unique to the series elastic actuator.

Ignoring velocity saturation, we can compute performance by imposing a limit on the magnitude of F_m , i.e. $|F_m| < F_{max}$. For most motors, this translates into a bound on the maximum motor current. It is helpful to draw a vector diagram showing the magnitude and phase relationship of F_l and X_l, and the resulting F_m :

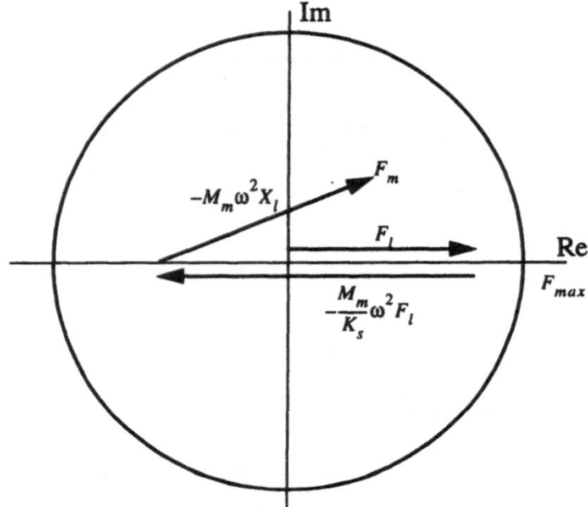

Fig. 5. Phase Diagram of Necessary Motor Force

Here we have arbitrarily aligned F_l with the real axis. To satisfy $|F_m| < F_{max}$, end point F_m must land inside the circle of radius F_{max}. Note that the series elasticity term $-\frac{M_m}{K_s}\omega^2 F_l$ opposes the F_l vector. Thus, for all frequencies below $\sqrt{2\frac{K_s}{M_m}}$, the series elasticity will bring the starting point of the $-M_m\omega^2 X_l$ vector closer to the circle's center and thus allow for a greater range of possible motion amplitudes and phases than would be possible with a stiff interface. If impedance control[12] is used, the $-M_m\omega^2 X_l$ term of the vector sum will point to the right when simulating positive rate springs, and thus the inclusion of series elasticity will improve actuator performance. At frequencies less than $\sqrt{2\frac{K_s}{M_m}}$, the maximum force amplitude of damping impedances, such as are used in damping control[22], is also increased.

It is also informative to consider actuator output force as a function of the output impedance $Z = \frac{F_l}{X_l}$ and motor force:

$$F_l = \frac{F_m Z}{\left(1 - \frac{M_m}{K_s}\omega^2\right)Z - M_m\omega^2} \quad (4)$$

This equation is plotted against actual test data later in the paper.

3. Control

Stable, accurate, force control can be obtained by using the architecture shown below, where F_d is the desired force:

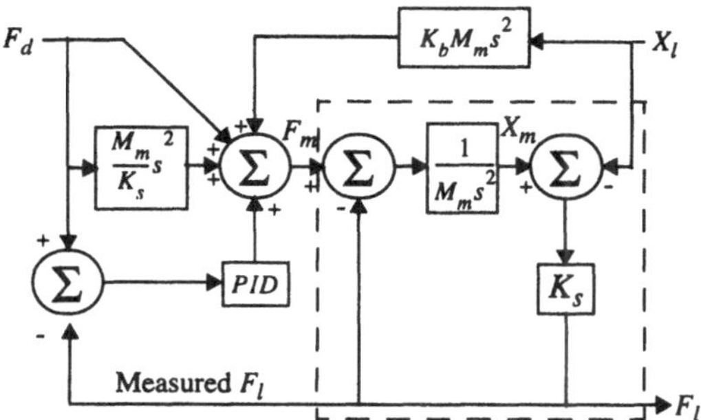

Fig. 6. Control Architecture

The feed-forward paths attempt to fully compensate for all three terms of equation 3, with the exception of the last (load motion) term, where a gain K_b is made less than 1 so as to prevent feedback inversion and instability.

Feedback to compensate for modeling errors and $K_b < 1$ is accomplished by an ordinary PID loop, operating on force error. This loop has a transfer function of:

$$PID(s) = K_p + \frac{K_d s}{1 + \tau_d s} + \frac{K_i}{1/\tau_i + s} \tag{5}$$

with parameters defined as follows:

K_p	Proportional Gain
K_i	Integral Gain
τ_i	Integral Roll-Off
K_d	Derivative Gain
τ_d	Derivative Roll-off

Fig. 7. Feedback Parameters

Stability can be analyzed by looking at the output impedance as a function of frequency $s = j\omega$ with a commanded force $F_d = 0$:

$$Z = \frac{F_l}{X_l} = \frac{K_s(1 - K_b) M_m \omega^2}{K_s(PID(j\omega) + 1) - M_m s^2} \tag{6}$$

If the imaginary part of this impedance is less than or equal to zero, than the actuator as whole will be passive and thus stable when interacting with any passive load[13][14]. The only imaginary component of the impedance comes from the PID term, which is in the denominator. Thus, for the impedance to have a negative imaginary part, the PID term must have a positive imaginary part, i.e.:

$$\text{imag}\left(K_p + \frac{K_d j\omega}{1 + \tau_d j\omega} + \frac{K_i}{1/\tau_i + jw}\right) \geq 0 \tag{7}$$

which is guaranteed for all ω when

$$\tau_i \leq \sqrt{\frac{K_d}{K_i}} \tag{8}$$

i.e., when the integral gain is rolled off below a sufficiently high frequency.

In a real system with motor saturation, the actuator will take on the natural impedance of the series elasticity at sufficiently high frequencies[10]. Thus, a light load mass may resonate with the series elasticity. To avoid this problem, placing a minimum mass on the load will lower the resonant frequency to where the control loop operates well. At this low frequency, the impedance of the series elasticity disappears from the overall impedance (which is very low), and resonance cannot occur.

4. Experimental Setup

To evaluate performance, a series elastic actuator, shown in the photograph below, was constructed:

Fig. 8. Experimental Series-Elastic Actuator

The motor used was a MicroMo 3557K (48V, 25W) with a 66:1 reduction planetary gearbox. The gearbox's output shaft was attached to a steel torsion spring, which formed the series elasticity. The actuator output was taken from the other end of the spring. The spring was of a cross-shaped cross-section, which was found to give the best stiffness v. strength characteristics. The inertia of the motor at the output of the gearbox was calculated to be 0.02 kgm^2 and the stiffness of the spring was 34 Nm/rad, making the natural frequency of the system 41 rad/s or about 7Hz. The twist in the spring was measured using strain gauges mounted on the flats of the spring.

The control loop used was similar to that shown in fig. 6, only the $\frac{M_m s^2}{K_s}$ term was not implemented. The control parameters were set as follows:

K_p	12.41
K_i	12.41
τ_i	0.08
K_d	0.124
τ_d	.0015

A controlled load impedance was implemented by connecting the series elastic actuator's output to a conventional position-controlled motor, as shown below:

Fig. 9. Dual actuator test rig

5. Results

Both force and position were commanded sinusoidally at the same frequency, while magnitude and relative phase were varied. The performance was measured by calculating the root mean square force error and normalizing with respect to the commanded force amplitude. By then limiting RMS force error to a specific value, plots of the maximum possible output force magnitude over a range of output impedences were made. These were compared to the theoretical maximums given by equation 4, modified to take into account motor efficiency. In the plots below, the left plots show measured performance, the right show theoretical predictions. In each plot, the horizontal plane is impedance and the vertical axis is maximum possible force magnitude. Tests were performed at 12, 25, 38 (resonance) and 44 rad/sec.:

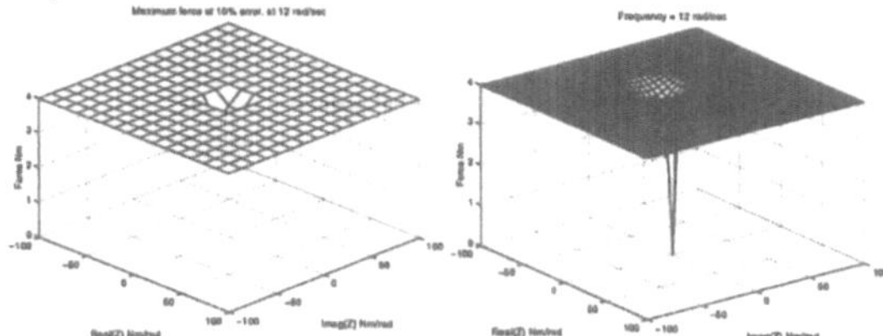

Fig. 10. Maximum output force vs. load impedance at 12 rad/sec

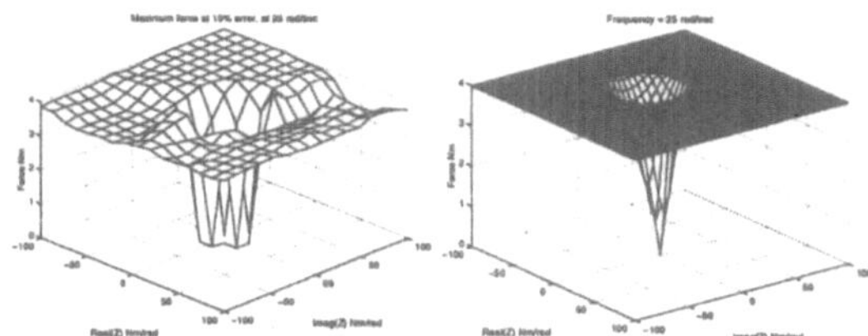

Fig. 11. Maximum output force vs. load impedance at 25 rad/sec

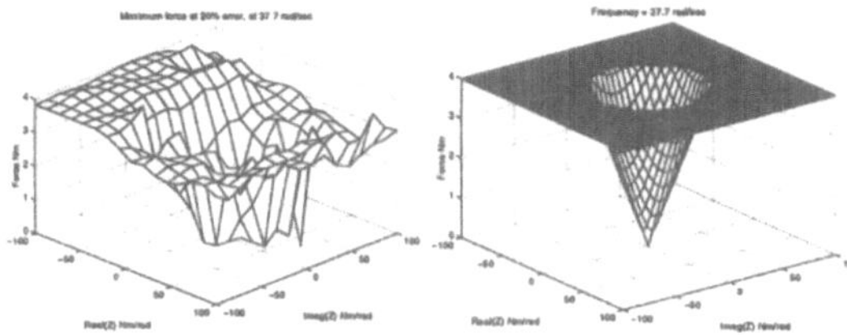

Fig. 12. Maximum output force vs. load impedance at 38 rad/sec

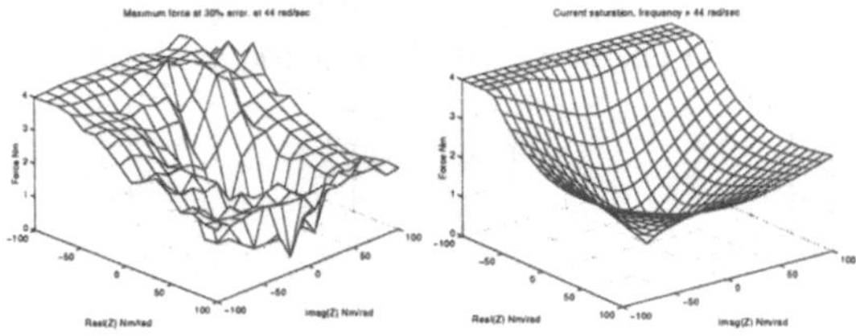

Fig. 13. Maximum output force vs. load impedance at 44 rad/sec

At low frequencies, performance is quite good The small downward spike corresponds to the lowest impedance that could be generated on the test rig without large-motion saturation. At resonance, performance at low impedances degrades, while at larger impedances performance is still good. Above resonance, it can clearly be seen that the actuator only performs well when its output impedance has a negative real part, which corresponds to positive spring-like behavior.

6. Conclusions

Series-elastic actuators are presently being utilized in two research robots, and a third is now under construction. The first robot is Cog[5], whose arms are powered by revolute series-elastic actuators very similar to those used in the above tests. Another robot - a planer biped walker named "Spring Turkey" - utilizes series-elastic tendons to drive its leg joints. The limbs of both of these robots are shown below:

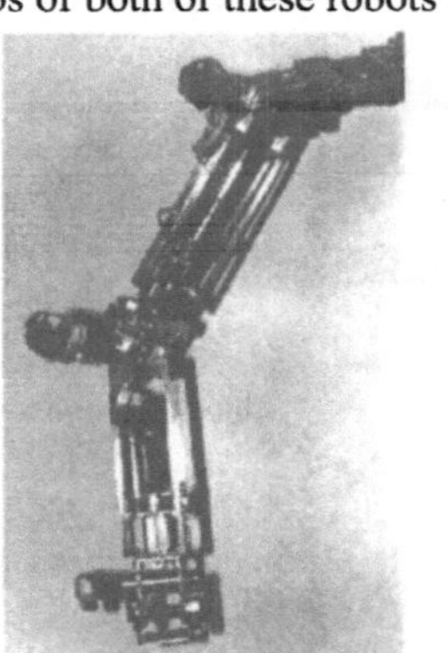

Fig. 14. One of Cog 's Arms (left) and Spring Turkey's Legs (right)

A series-elastic arm for a small planetary rover is presently under construction.

In early system tests, both Cog's arm and Spring Turkey's legs have demonstrated performance that verifies the advantages of series-elastic actuators. Both robots interact with the environment under force or impedance control without any instability during transient contact with hard surfaces. Both robots have (so far) been robust to shock (presently more often a result of control programming errors than the environment). Spring Turkey has recently taken a few steps, showing that walking with series-elastic actuators is feasible

We believe that for natural tasks (such as walking and manipulation), series elastic actuators provide many benefits when compared to traditional actuators. These benefits include shock tolerance, lower reflected inertia, more accurate and stable force control, less damage during inadvertent contact, and energy storage. Although zero motion force bandwidth is reduced, force bandwidth for many tasks that involve load motion is improved. This is particularly true for natural tasks that are spring- or damper-like in their impedance[22].

We have shown that a simple control system can generate a range of complex output impedances - not just that of the passive series elasticity, and have demonstrated experimentally that accurate, stable control is easy to achieve.

Several avenues are open for future work, including parallel connections that extend both dynamic range and bandwidth [17] and variable-rate springs whose modulation of bias point can effect changes in passive stiffness. This type of mechanism has been studied before[21] and a more sophisticated version is currently being investigated at MIT by Ken Salisbury's group and that of the authors.

7. References

[1] Alexander, R. McNeill, "Elastic Mechanisms in Animal Movement", Cambridge University Press, 1988.

[2] Angle, C.M. and Brooks, R.A., "Small Planetary Rovers", IEEE International Workshop on Intelligent Robots and Systems, Tsuchiura, Japan, July 1990 pp. 383-388.

[3] Asada, H. and Kanade, T. "Design of Direct-Drive Mechanical Arms", ASME J. of Vibration, Acoustics, Stress, and Reliability in Design 105(3), pp. 312-316.

[4] Balas, Mark J., "Active Control of Flexible Systems", Proc. of the 1977 Symposium on Dynamics and Control of Large Flexible Spacecraft, June 13-15, 1977, Blacksburg, VA pp 217-236.

[5] Brooks, R.A. and Stein, L.A., "Building Brains for Bodies", to appear in Autonomous Robots, (1:1), 1994.

[6] Brooks, R.A., "The L Manual", MIT AI Lab Internal, 1994.

[7] Cannon, Robert H. Jr., and Rosenthal, Dan E., "Experiments in the Control of Flexible Structures with Noncolocated Sensors and Actuators", J. Guidance, Vol. 7 No. 5 pp. 546-553, Sept.-Oct. 1984.

[8] Cannon, Robert H. Jr., and Schmitz, E., "Initial Experiments on the End-Point Control of a Flexible One-Link Robot", Int. J. of Robotics Research, Vol. 3 No. 3 (1984) pp 62-75.

[9] Eppinger, Steven D., and Seering, Warren P., "Three Dynamic Problems in Robot Force Control", IEEE Intl. Conf. on Robotics and Automation, 1989, pp 392 - 397.

[10] Eppinger, Steven D., and Seering, Warren P., "Understanding Bandwidth Limitation in Robot Force Control", IEEE Int. Conf. on Robotics and Automation, April 1987.

[11] Hashimoto, Minoru, and Imamura, Yuichi, "An Instrumented Compliant Wrist Using a Parallel Mechanism", Japan/USA Symp. on Flexible Automation, V. 1 ASME 1992, pp 741-744.

[12] Hogan, N., "Impedance Control: An Approach to Manipulation: Part I - Theory, Part II - Implementation, Part III - Applications", J. of Dyn. Syst., Measurement Contr., 107:1-24 (1985)

[13] Hogan, Neville, "On the Stability of Manipulators Performing Contact Tasks", IEEE J. of Robotics and Automation, V4 N6, Dec. 1988, pp 677-686.

[14] Hogan, Neville, and Colgate, Ed, "An Analysis of Contact Instability in term of Passive Physical Equivalents", IEEE Intl. Conf. on Robotics and Automation, 1989, pp 404-409.

[15] Hunter, Ian W., Hollerbach, John M., and Ballantyne, John, "A comparative analysis of actuator technologies for robotics", Robotics Review 2, MIT Press, 1991.

[16] Khatib, O., "Real Time Obstacle Avoidance for Manipulators and Mobile Robots", Int. J. of Robotics Res. V5 N1 (1986).

[17] Morrell, John B., and Salisbury, J. K., "Parallel Coupled Actuators for High Performance Force Control: A Micro-Macro Concept", Submitted to IROS 95.

[18] Raibert, M. H., "Legged Robots That Balance." Cambridge, Mass.: MIT Press (1986).

[19] Readman, Mark C., "Flexible Joint Robots", CRC Press, 1994.

[20] Salisbury, K., Eberman, B., Levin, M., and Townsend, W., "The Design and Control of an Experimental Whole-Arm Manipulator", Proc. 5th Int. Symp. on Robotics Research. (1989)

[21] Spong, M. W., "Modeling and Control of Elastic Joint Robots", Trans. of the ASME, Vol 109, Dec. 1987, pp 310-319.

[22] Sugano, S., Tsuto, S., and Kato, I., "Force Control of the Robot Finger Joint equipped with Mechanical Compliance Adjuster", Proc. of the 1992 IEEE/RSJ Int. Conf. on Intelligent Robots and Systems, 1992, pp 2005-2013.

[23] Whitney, Daniel E., "Force Feedback Control of Manipulator Fine Motions", J. Dyn. Syst. Measurement Contr. 98:91-97 (1977)

[24] Whitney, Daniel E., "Historical Perspective and State of the Art in Robot Force Control", Int. J. of Robotics Research, V6, N1, Spring 1987 pp 3-14.

In Pursuit of Dynamic Range: Using Parallel Coupled Actuators to Overcome Hardware Limitations

John B. Morrell & J. Kenneth Salisbury

Department of Mechanical Engineering
Artificial Intelligence Laboratory
Massachusetts Institute of Technology

Abstract

We report on a new actuator concept which combines two actuators to create a micro-macro actuator. The new actuator has improved force resolution and bandwidth. Unlike previous micro-macro robots which used actuators coupled in series to achieve positioning resolution, the actuators in this system are coupled in parallel using a compliant transmission to achieve *force* resolution. A parallel coupled micro-macro actuator (PaCMMA) which embodies these ideas is described. A control algorithm is presented. A series of tests were performed and the PaCMMA is compared to two single actuator systems. We have obtained small signal force bandwidth of 60 Hz and force resolution of 0.12% . The PaCMMA concept is shown to increase force range and reduce mechanical impedance while preserving small signal force bandwidth.

1. Introduction

Robot manipulation tasks have typically been specified as a collection of force and position trajectories, or impedances at the robot endpoint. Some manipulation tasks such as stroking and surface following require low impedance while other manipulation tasks such as pushing require high impedance. In general, it has been hard to utilize the same hardware in both types of tasks. Despite having relatively low bandwidth actuators, humans perform a wide range of tasks. One obvious difference between machine actuation and human actuation is the large range of forces that humans can controllably exert. For machine actuators, saturation, friction and inertia impose significant limits on the achievable force range and force bandwidth. This paper presents a dual actuator design which achieves higher force resolution than single actuator systems ($15\times$) while exhibiting high force bandwidth and low apparent inertia. This section presents some background information which will lead to our design.

Early work in force control identified two significant robot characteristics which make controlling force difficult. The first is robot impedance [5, 7], and the second is non-collocation of the force sensor and the actuator [2, 3]. Robot impedance is of obvious importance. When a robot with high impedance (large mass or stiffness) is in contact with an object, a small displacement of either the robot or the object will result in large contact forces. The large forces can result in damage to both the robot and the object. Thus, successful force control relies on the ability of the robot to control the contact impedance. This may be accomplished with active control, or through passive characteristics.

Active control has the benefit that the impedance may be changed as the task changes. Unfortunately, active control is limited by the bandwidth of the controller and by the intrinsic mechanical properties of the system, i.e. stiffness, mass, natural frequency etc. Passive characteristics can have the benefit that they require little control effort and exhibit real-time continuous response. For force control, good passive characteristics include low robot inertia, well-damped dynamics and frictionless, zero-backlash transmissions.

Non-collocation of the sensor and actuator is another dominant problem in force control [2, 3]. The achievable force bandwidth has been shown to be limited by the bandwidth of the transmission between the sensor and actuator. This fact has led many designers to conclude that "stiffer is better". Electric motors and many other actuators also require some kind of transmission to both maximize torque and to move actuator mass away from the endpoint. Friction and backlash in the transmission will lower the control bandwidth in these situations. As a result, robot design has moved in two directions: 1) direct drive actuators and 2) stiffer, low friction, low backlash transmissions.

Direct drive actuators solve the non-collocation problem by providing a stiff connection between the force source and the sensor [4]. The Whole-Arm-Manipulator at MIT is an excellent example of a robot with naturally low impedance [10]. In this case, the motors are connected to the links via efficient, low reduction ratio cable transmissions. The system is controlled open loop and disturbance rejection relies on low mass links and the low friction transmission. This works well, but the disturbance rejection is limited since the control is open loop.

The dual motor design presented in the following section uses these design paradigms to address three goals: 1) overcome the limits created by motor saturation and friction, 2) maximize force control bandwidth, and 3) minimize the actuator impedance (apparent inertia).

2. A Parallel Micro-Macro Actuator Concept

Presented with design trade-offs mentioned above, we have developed a solution using two actuators, coupled in *parallel.* Figure 1 shows a schematic of the actuator concept. A large actuator is coupled via a compliant transmission to the joint axis. A micro-actuator is directly coupled to the joint axis. We refer to this concept as a *Parallel Coupled Micro-Macro Actuator* (PaCMMA).

Consider the case where the stiffness of the transmission is zero, i.e., the micro-actuator is the only force acting on the output link. In this case, force control may be achieved at high bandwidth due to the proximity of the sensor and the actuator. The lower limit on controllable force is on the order of the brush friction of the micro-actuator, which causes a limit cycle or finite error as smaller and smaller forces are commanded. The upper limit on controllable force is limited by the saturation force of the micro-actuator.

Now allow the stiffness of the transmission to increase. If the transmission stiffness is considerably lower than the environment stiffness, then the micro-actuator

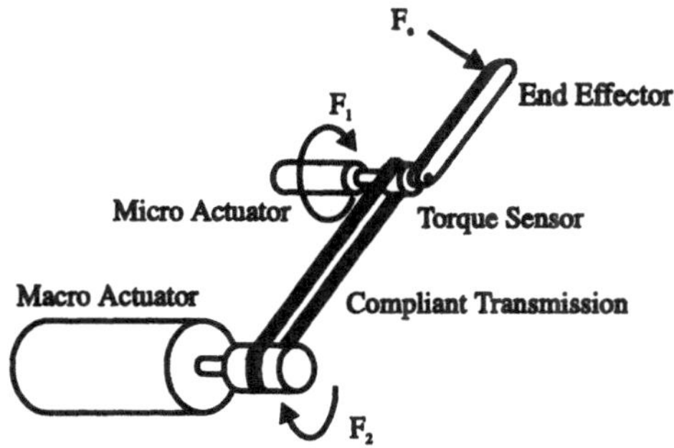

Figure 1. The Parallel Micro-Macro Actuator Concept

closed loop performance (and stability) will be dominated by the stiffness of the environment. Consequently, the transmission can exert a force on the endpoint which is summed with the force of the micro-actuator without affecting the dynamics of the micro-actuator. The macro-actuator can be used to impose a low frequency force bias on the endpoint, which will have little effect on the control performance (stability) of the micro-actuator. The result is that we can now exert forces near the maximum of the macro-actuator while controlling variations at the level of the micro-actuator. A hi-fi loudspeaker provides a metaphor for this concept – the two actuators are a woofer and a tweeter, coupled in parallel by a compliant transmission, air.

Several aspects of the design should be noted. First, the concept uses the "stiffer is better" paradigm on the direct drive part of the system (the micro-actuator), but diverges from this principle for the macro-actuator. In fact, the transmission between the two actuators must *not* be stiff for the concept to work. If the transmission were very stiff, the micro-actuator would be affected by the impedance presented by the macro-actuator, and there would be no improvement in performance. Second, the concept allows the use of a low performance actuator for the macro-actuator since its inertia and friction are "filtered" out by the transmission. Third, the concept allows a resolution bounded by the minimum controllable force of the micro-actuator and the maximum force of the macro-actuator. Micro-macro designs which are coupled in series can not achieve this kind of resolution – a series-coupled design is limited to the force range of the micro-actuator.

The PaCMMA concept may be implemented in a variety of ways. Figure 1 suggests that the micro-actuator should be placed proximal to the end effector. However, another instantiation of the design could place the actuators together, with the end effector at a remote location. As long as the end effector is coupled to the micro-actuator with the stiffest transmission possible, and the macro-actuator is coupled with a compliant transmission, the concept is the same. In fact, the two actuators could be inside one housing, provided the transmissions meet our design constraints.

In [8], we presented some initial findings on this concept. This paper reports our latest results with a prototype parallel coupled actuator. The remaining sections will discuss the control law and experimental findings.

3. Control Law

The control law for the PaCMMA must accomplish two tasks. First, it must provide stability for the actuator in both free space and in contact with stiff environments since manipulation typically requires operation in both of these states. Second, it should provide fast response which exploits the advantages of the two actuators. A lumped element model of the PaCMMA is shown in figure 2.

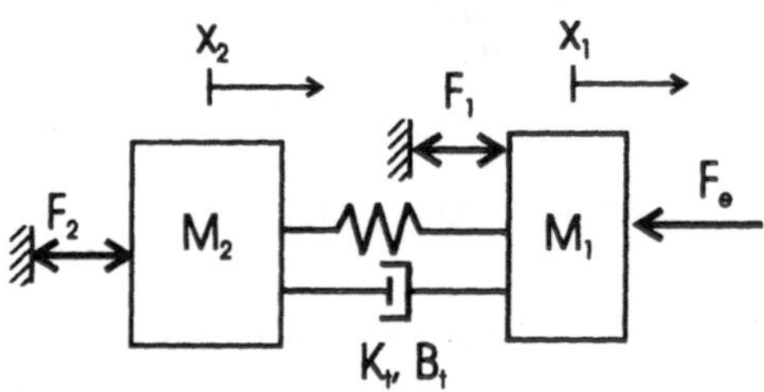

Figure 2. Lumped Parameter Model

For response to unknown transients, it is clear that the best initial condition is to have the micro-actuator at zero effort with the macro-actuator providing all of the necessary force. When a force transient (disturbance) is measured, the micro-actuator will have the best chance to reject this disturbance. Likewise for sinusoidal force disturbances, the load carried by the macro-actuator should be maximized so that the micro-actuator can respond more quickly to errors in either direction.

Considerable research on force control algorithms has been performed. Literature in force control has shown that operating above the bandwidth of the transmission between the torque sensor and the actuator is extremely difficult. For the PaCMMA, the micro actuator has a direct drive connection to the force sensor and very high bandwidth as a result (300 Hz). Integral force control has been experimentally shown to be stable and fast in contact with a wide variety of environments [11]. The micro-actuator is controlled using integral control with a low-pass filter to add gain margin:

$$F_1 = \frac{G_1}{s} \frac{\omega_{co1}^2}{s^2 + 2\zeta_1\omega_{co1} + \omega_{co1}^2}(F_{des} - F_e)$$

where G_1, ω_{co1} and ζ_1 are chosen to yield the desired phase and gain margin. We are able to obtain closed loop force control at bandwidths near 60 Hz with this controller.

Control of the macro-actuator is less obvious. The macro-actuator should provide the maximum possible component of force to the output, subject to stability and speed constraints. The control law for the macro-actuator is:

$$\begin{aligned} F_2 &= G_p F_1 + G_d(\dot{x}_1 - \dot{x}_2 + G_v \dot{F}_{des}) \\ &\quad + G_{ff}(\frac{\hat{M}_2}{\hat{K}_t}s^2 + \frac{\hat{B}_t}{\hat{K}_t}s + 1)F_{des} \end{aligned}$$

The control law has several components which are used to maximize performance. Feed-forward of the desired force ($G_{ff} < 1$) is used to account for plant dynamics with estimates of mass, stiffness and damping. However, backlash, friction and other unmodelled dynamics produce errors which require feedback terms. Gain G_p causes the macro-actuator to reduce the control effort of the micro-actuator (which represents the integrated force error). Gain G_d provides damping between M_1 and M_2 in addition to tracking the velocity of the endpoint $\dot{x}_1$ or changes in the desired force, $G_v \dot{F}_{des}$.

4. Experiments

A prototype actuator has been built and tested under a variety of operating conditions. In order to evaluate the effectiveness of the concept we also tested the performance of the macro-actuator and the micro-actuator separately. This section contains a description of our experimental apparatus, an explanation of our performance metrics and the results of the tests.

4.1. Performance Metrics

A wide variety of robot designs and tasks exist, and it is clear that some designs perform better at certain tasks. To quantify robot characteristics in force controlled tasks, we use several measures of performance.

Force Control Response: The frequency response (transfer function) of the system when the endpoint is held stationary, i.e.

$$H_s(\omega) = \left. \frac{F_e(\omega)}{F_{des}(\omega)} \right|_{X_e=0}$$

where
F_{des} = the desired force
F_e = the force exerted on the environment
X_e = the position of the end effector

This specification comes from the desire to quantify a robot's performance in quasi-static applications like slow manipulation. In this case, the ability to modulate forces applied to relatively motionless environment is of premium importance.

Impedance Response: The frequency response (transfer function) of the system to a position disturbance at the endpoint, i.e. the endpoint is connected to a position source while the desired force is commanded to be constant:

$$Z(\omega) = \left. \frac{F_{error}(\omega)}{X_{in}(\omega)} \right|_{F_{des}=\text{constant}}$$

where
$F_{error} = F_{des} - F_e$
X_{in} = the position disturbance

This specification is motivated by the realization that impedance of the robot endpoint is extremely important when the robot and the environment it contacts are in motion. In pure force control, an ideal actuator would present zero impedance across all frequencies, but in real systems, this number should be as small as possible. These two specifications use the frequency response function in lieu of time domain specifications for its density of information and because it can be obtained for many systems, even those exhibiting non-linear behavior. In two works [1, 9], experiments of this nature are performed. These measurements are also suggested in [6].

Finally, we use a third specification which emphasizes the range of forces that an actuator can exert:

Force Resolution: The ratio of the maximum force the robot can controllably exert to the minimum force that the robot can controllably exert (in percent):

$$\text{Force Resolution} = \left. \frac{F_{min}}{F_{max}} \right|_{\omega=0}$$

where
F_{min} = RMS steady state error
F_{max} = maximum continuous force

4.2. Apparatus

The apparatus consists of a one degree of freedom PaCMMA mounted on a rigid surface. The endpoint may be free floating, clamped, or connected to a position disturbance. The micro-actuator is high performance servo motor with maximum continuous torque of 80 mNm. The macro-actuator is the same motor fitted with a 36:1, 2-stage planetary gearhead. The endpoint is coupled directly to an in-line torque sensor (1.4 mNm resolution), which is connected directly to the micro-actuator. The micro and macro-actuators are fitted with pulleys which are used to connect the macro-actuator to the output shaft. A pair of conventional steel extension coil springs were used to vary the transmission stiffness.

4.3. Procedure

Force bandwidth data was obtained by clamping the endpoint and commanding a series of sinusoidal force trajectories. Impedance data was generated by using a third motor to apply a sequence of sinusoidal position disturbances to the actuator while commanding zero force. The force error was measured.

In order to facilitate the comparison of the PaCMMA concept to traditional designs, three configurations were tested. First, the macro-actuator was evaluated using a stiff cable transmission with the micro-actuator disconnected from the system, much like a traditional cable drive system. The micro-actuator was then evaluated as a direct drive transmission with the macro-actuator disconnected from the system. Finally, the PaCMMA concept was evaluated for two candidate transmission stiffnesses.

4.4. Results

The results of the tests are reported in a number of figures and tables. The frequency response obtained at various operating points depends on the mean value of the input, the degree of actuator saturation and other nonlinear factors (backlash). To emphasize this, data is shown for both large and small signal inputs. Space limitations prevent displaying data for all operating points. Table 1 contains the results of the tests. Force bandwidth is taken to be the 3db point from the force control response tests. Impedance is summarized by a best fit to the slope of the actuator impedance.

Figure 3 and figure 4 show the resulting force bandwidth for the two actuators alone and the PaCMMA system for two different force amplitudes.The gearhead on

	Macro Only	Micro Only	PaCMMA	PaCMMA
Large Signal Force Bandwidth	12 Hz	NA	5 Hz	5 Hz
Small Signal Force Bandwidth	10 Hz	56 Hz	56 Hz	56 Hz
Impedance (mNm/rad)	$12.6\omega^2$	$5.31 \times 10^{-5}\omega^3$	$1.32 \times 10^{-3}\omega^3$	$3.18 \times 10^{-3}\omega^3$
Force Resolution	1.9 %	1.7 %	0.12 %	0.12 %
K_t (mNm/rad)	41000	NA	1140	3000

Table 1. Summary of Performance Specifications

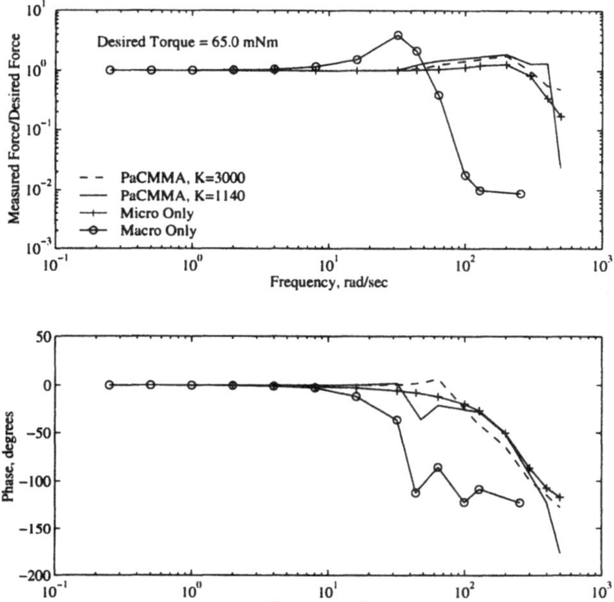

Figure 3. Small Signal Force Bandwidth, $H_s(\omega)$

the macro-actuator produced a resonance near 30 rad/sec. Since the tests were run with zero mean force, the backlash produced a pathological vibration which was not present when a bias force is applied. The data shown here are for the zero mean cases. Better performance can be achieved by using a backlash-free transmission or a direct drive macro-actuator.

Figure 5 and 6 show the resulting impedance response for two position disturbance amplitudes. Under perfect force control the impedance would be zero. Since there is always some small finite error (from torque sensor resolution and actuator torque errors) impedance is constant at some small value. The horizontal section of the impedance data represents the noise level of the system. As the frequency

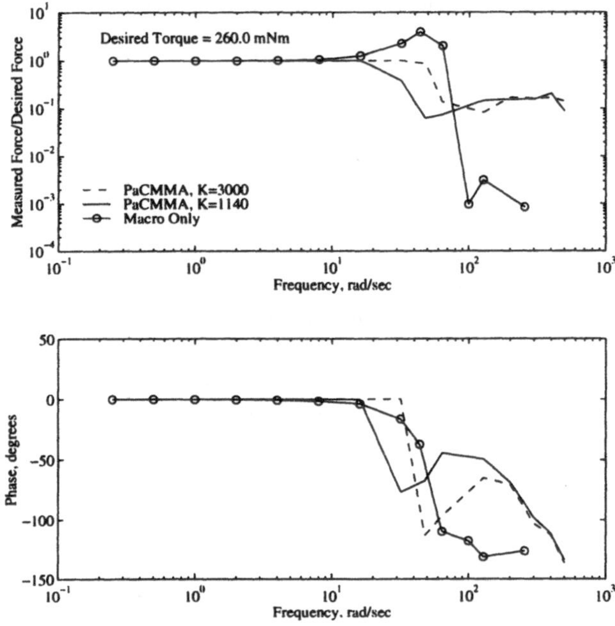

Figure 4. Large Signal Force Bandwidth, $H_s(\omega)$

of the position disturbance increases, the magnitude of the force required to cancel the mass of the end effector increases. At some point, the micro-actuator saturates and the mass of the end effector is no longer canceled. With no phase lag in control (perfect control), this occurs when $F_{1sat} = M_1 X_{in} \omega^2$. This break frequency is shown clearly on figures 5 and 6. The slope of the impedance line is determined by the control law and the transmission stiffness. The stiffer transmission creates a larger impedance than the softer transmission. Compared to the macro-actuator alone, the PaCMMA demonstrates a significant reduction in impedance, and makes accommodation of position errors possible while still maintaining torque capacity. Further, when the macro-actuator saturates in velocity or torque, the impedance is dominated by M_1 and K_t while the macro-actuator mass, M_2, has minimal effect on the contact dynamics.

The parameter space spanned in figure 7 contains all measured operating points in force and frequency. The region enclosed by each line represents the operating points with less than 10% error for the macro and micro actuators used individually. Figure 8 shows the performance two motors used together in the PaCMMA. The operating region available now includes both high frequency and high amplitude regions, with a transitional region. The shape of the transition region is determined by the transmission characteristics. Increasing the spring stiffness will increase the width of the vertical lobe, decreasing the stiffness will increase the height of the horizontal lobe. Further, by changing the ratio of the maximum accelerations (and/or torques) of the actuators, the designer may shape the operating region to meet a variety of performance specifications.

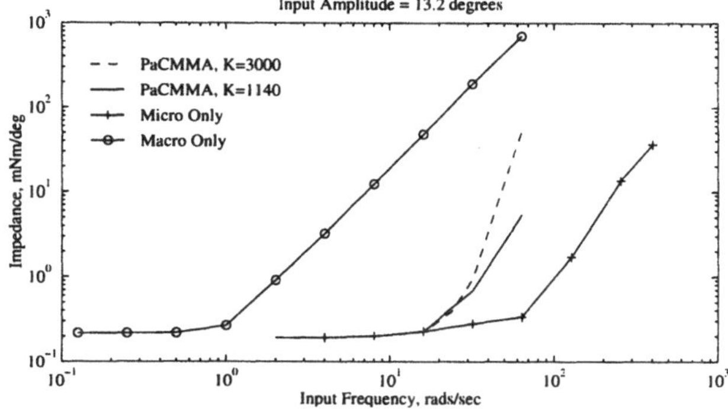

Figure 5. Small Amplitude Impedance Bandwidth, $Z_s(\omega)$

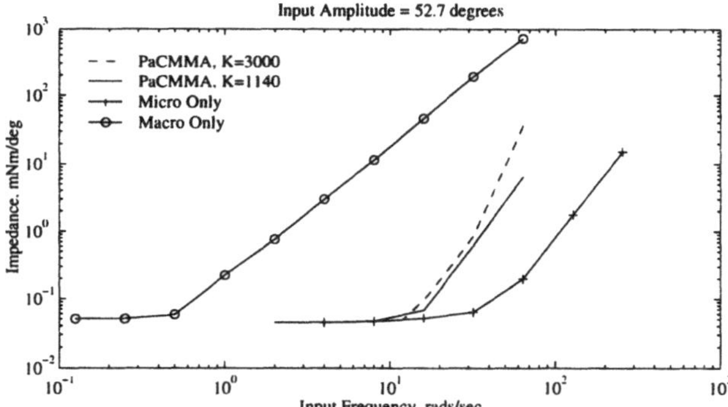

Figure 6. Large Amplitude Impedance Bandwidth, $Z_s(\omega)$

5. Conclusion

The use of a gearhead for the macro-actuator led to some harmonic distortion at 20-30 rad/sec. A better design would eliminate the backlash with either a bigger motor and/or a backlash free transmission. Loading the gearhead with a bias force helped to diminish the effects, but eliminating the backlash is a better solution.

Steel extension springs were used to modify the stiffness of the transmission. The maximum possible torque of the macro-actuator was capable of plastically deforming the springs. For this reason, we plan to use nonlinear stiffening springs in order to take advantage of the maximum torque of the macro-actuator, while still preserving low impedance at small forces.

A more formal relationship between maximum torque, mass and transmission stiffness should be generated to aid the designer in component selection. The upper limit on PaCMMA performance is governed by several actuator parameters, including mass, maximum torque, maximum allowable speed and nonlinear parameters such as friction and backlash.

The PaCMMA concept synthesizes several design paradigms (direct drive, low

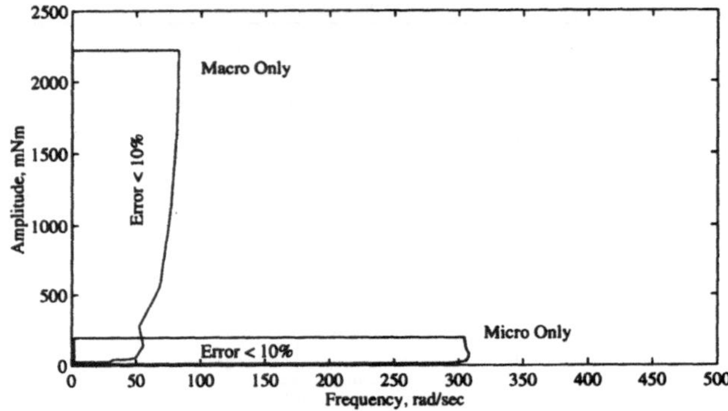

Figure 7. Operating Regions for Micro and Macro Actuators where Force Control Bandwidth attenuation is less than 10%

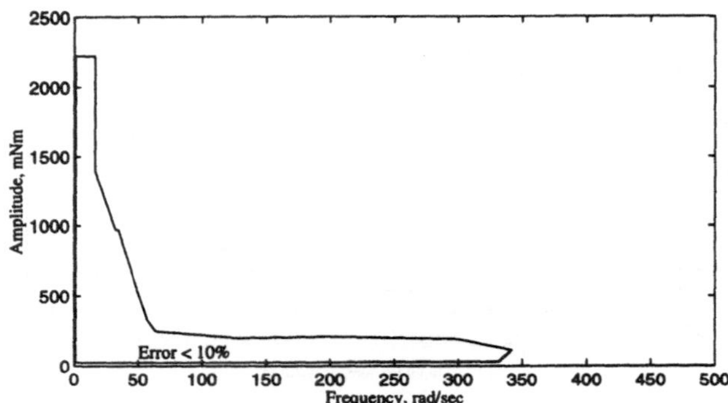

Figure 8. Operating Region for PaCMMA where Force Control Bandwidth attenuation is less than 10%

passive/natural impedance, low friction transmissions) to achieve a high level of performance. Parallel coupled micro-macro actuators show great promise in applications where low impedance, high force resolution and high force bandwidth are required.

6. Acknowledgements

This work was supported in part by the Office of Naval Research, University Initiative Program Grant N00014-92-J-1814.

References

[1] C. H. An. *Trajectory and Force Control of A Direct Drive Arm.* PhD thesis, MIT, Dept. of Elec. Eng. and Comp. Sci., 1986.

[2] R. H. Cannon and D. E. Rosenthal. Experiments in control of flexible structures with noncolocated sensors and actuators. *AIAA Journal of Guidance and Control*, 7(5), 1984.

[3] S. D. Eppinger and W. P. Seering. Understanding bandwidth limitations in robot force control. In *Proc. IEEE International Conference on Robotics and Automation*, 1987.

[4] K. Kanade H. Asada and I. Takeyama. Control of a direct-drive arm. *Transactions of the ASME*, 105:136–142, 1983.

[5] Neville Hogan. Impedance control: An approach to manipulation: Part i - theory. *ASME Journal of Dynamic Systems Measurement and Control*, 107(1), 1985.

[6] Stephen C. Jacobsen, Craig C. Smith, Klaus B. Biggers, and Edwin K. Iverson. Behavior based design of robot effectors. In Michael Brady, editor, *Robotics Science*. MIT Press, 1989.

[7] Oussama Khatib. Reduced effective inertia in macro-/mini-manipulator systems. In H. Miura and S. Arimoto, editors, *Robotics Research 5*, pages 279–284. MIT Press, 1990.

[8] John B. Morrell and J. Kenneth Salisbury. Parallel coupled actuators for high performance force control: A micro-macro concept. Accepted to *Intl. Conf. on Intelligent Robots and Systems*, 1995.

[9] Gill A. Pratt and Matthew M. Williamson. Series elastic actuators. Accepted to *Intl. Conf. on Intelligent Robots and Systems*, 1995.

[10] J. K. Salisbury, B. Eberman, M. Levin, and W. Townsend. The design and control of an experimental whole-arm manipulator. In *Robotics Research: The Fourth International Symposium*, 1989.

[11] Richard Volpe and Pradeep Khosla. An experimental evaluation and comparison of explicit force control strategies for robotic manipulators. In *Proc. IEEE International Conference on Robotics and Automation*, pages 1387–1393, 1992.

Total Least Squares in Robot Calibration

John Hollerbach and Ali Nahvi

Depts. Computer Science and Mechanical Engineering
University of Utah, Salt Lake City, UT 84112
jmh@cs.utah.edu

Abstract

The role of input noise is seldom considered in robot calibration. The methodology of total least squares may be applied to handle both input and output noise in robot calibration. Experimentally, we apply this method towards joint torque sensor calibration, and towards kinematic calibration of a redundant parallel-drive spherical joint in a variant called the implicit loop method.

1. Introduction

Previous work in robot calibration has almost exclusively employed Ordinary Least Squares (OLS), applied in nonlinear cases to the linearized equations iteratively. The premise for OLS is that the output or dependent variable in an explicit equation is the only source of measurement error. Yet errors in the input or independent variables are often significant, and the use of OLS in the presence of input error or noise is known to lead to bias errors in the estimates [10].

The issue of input noise is known in the statistics literature as the errors-in-variables problem [3]. Figure 1A illustrates the simple case of fitting a straight line $y = ax$. OLS assumes that the only error is in the ouput variable y; the input variable x is presumed known exactly. Hence the squared vertical distances $\Delta y^i = y^i - y^i_c$ from P measured data points (x^i, y^i) to the corresponding computed points (x^i_c, y^i_c) on the fitted line $y^i_c = ax^i_c$ should be minimized (Figure 1A(a)):

$$\min_{a,x^i_c,y^i_c} \sum_{i=1}^{P} (\Delta y^i)^2 \quad \text{subject to } y^i_c = ax^i_c \tag{1}$$

Besides the slope a, OLS also determines the computed "correct" point locations (x^i_c, y^i_c) on the fitted line for each point i, where $x^i_c = x^i$. The other coordinate y^i_c is eliminated from (1) by substituting the explicit constraint equation $y^i_c = ax^i_c$:

$$\min_{a} \sum_{i=1}^{P} (y^i - ax^i)^2 \tag{2}$$

This is the normal least squares formulation, whose origin from (1) is instructive for subsequent developments.

Suppose now that there is only input noise, i.e., in x^i. Then $y^i_c = y^i$ and we should minimize the horizontal fitting error $\Delta x^i = x^i - x^i_c$ (Figure 1A(b)):

$$\min_{a,x^i_c,y^i_c} \sum_{i=1}^{P} (\Delta x^i)^2 \quad \text{subject to } y^i_c = ax^i_c \tag{3}$$

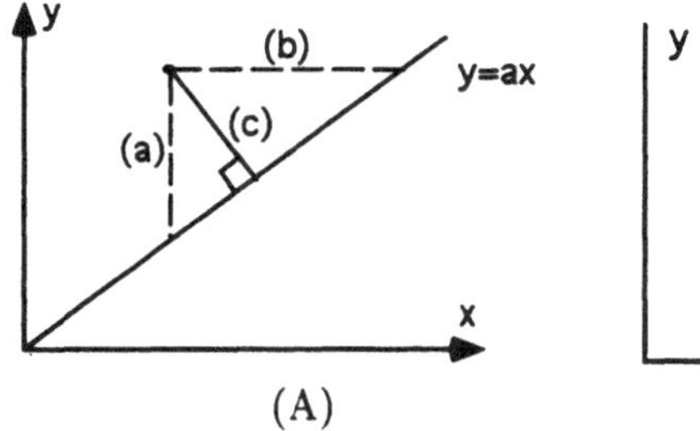

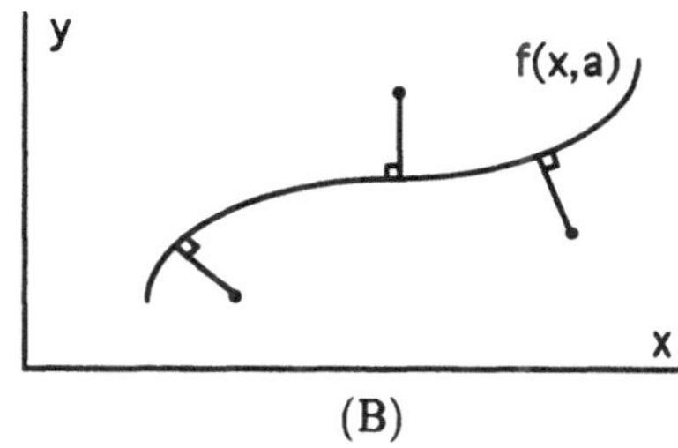

(A) (B)

Figure 1. (A) Fitting a straight line in case of (a) output noise only, (b) input noise only, and (c) equal input and output noise. (B) Nonlinear orthogonal distance regression.

The unknown line coordinate x_c^i is again eliminated by incorporating the constraint equation into the objective function:

$$\min_a \sum_{i=1}^{P} (x^i - y^i/a)^2 \tag{4}$$

This process has simply exchanged the role of dependent and independent variable.

Next suppose that there are equal errors in the x and y variables. Then we should fit the orthogonal distance from the data to the line (Figure 1A(c)):

$$\min_{a,x_c^i,y_c^i} \sum_{i=1}^{P} (\Delta x^i)^2 + (\Delta y^i)^2 \quad \text{subject to } y_c^i = a x_c^i \tag{5}$$

The orthogonal distance means that the line from (x^i, y^i) to (x_c^i, y_c^i) should be perpendicular to the fitted line $y_c^i = a x_c^i$. Incorporating orthogonality and the constraint equation yields the objective function:

$$\min_a \sum_{i=1}^{P} \frac{(y_i - a x_i)^2}{1 + a^2} \tag{6}$$

A final embellishment is to suppose that there are unequal uncertainties in the input/output variables, given by the standard deviations σ_x and σ_y. Then the weighted normal distances from points (x_i, y_i) to the line are minimized:

$$\min_{a,x_c^i,y_c^i} \sum_{i=1}^{P} \frac{(\Delta x^i)^2}{\sigma_x^2} + \frac{(\Delta y^i)^2}{\sigma_y^2} \quad \text{subject to } y_c^i = a x_c^i \tag{7}$$

After substitution of the constraint equation, one eventually gets

$$\min_a \chi^2 = \sum_{i=1}^{P} \frac{(y_i - a x_i)^2}{\sigma_y^2 + a^2 \sigma_x^2} \tag{8}$$

which is written as a χ^2 statistic [11], i.e., the least-squares sum has a χ^2 distribution with expected value $E(\chi^2) = P - 1$. The expected value says nothing more than that the experimentally measured standard deviations should be about the same as the *a priori* standard deviations. The number of measurements P is reduced by 1 to

$P - 1$ because the squared terms are not independent, as the equivalent of one has been used to determine a. The number $\nu = P - 1$ is called the *degrees of freedom* in the statistics literature (unfortunate conflict with robotics terminology), and reflects the excess of measurements over fitted parameters.

The solution to the orthogonal distance objective function (8) looks like a nonlinear optimization problem, and has been solved as such in [11]. This procedure is however unnecessary in the linear case, as analytic procedures exist. The solution to the errors-in-variables problem is known as *total least squares* in the numerical analysis literature [12] and as *orthogonal distance regression* in the statistics community [2]. The analytic solution in the total least squares approach uses singular value decomposition.

Orthogonal distance regression has been known since last century, and has been often rediscovered under various guises. In robotics and vision, one common application of orthogonal distance regression is to fit a plane to a set of measured 3D points; all 3 components of measured position have noise (e.g., [5]). The ties to the literature in the numerical analysis and statistics communities have unfortunately seldom been made. On the other hand, the major developments in these communities on handling input noise have been recent, especially for nonlinear optimization.

Recently, Wampler et al. [13] have introduced an approach towards nonlinear orthogonal distance regression for the problem of kinematic calibration, termed the *Implicit Loop Method* (ILM). Key features of this method that distinguish it from previous approaches are the use of an implicit constraint equation, based on a notion that kinematic calbration approaches can nearly all be viewed as closed-loop approaches with the sensor as a joint [7], and the use of *a priori* parameter estimates. In this paper, we apply this method to two different calibration problems: (1) the calibration of joint torque sensors, and (2) the calibration of a parallel-drive shoulder joint.

2. Implicit Loop Method

Consider fitting a scalar function $y = f(x, a)$ by adjusting parameters a. In case of equivalent error in x and y measurements, we should again minimize the Euclidean distances from the measurements to the fitted curve f (Figure 1B):

$$\min_{a, x_c^i, y_c^i} \sum_{i=1}^{P} (\Delta x^i)^2 + (\Delta y^i)^2 \quad \text{subject to } y_c^i = f(x_c^i, a) \tag{9}$$

This nonlinear constrained optimization problem is solved by linearization and iteration [2], but the solution is presented in a more general context next.

Implicit constraint equations can arise in kinematic calibration of mulitple-loop mechanisms. Even for serial chains, Hollerbach and Wampler [7] take the viewpoint that nearly all calibration methods can be cast as closed-loop methods. Consider an implicit loop-closure constraint equation

$$\mathbf{0} = \mathbf{f}(\mathbf{x}^i, \mathbf{p}) \tag{10}$$

All measurements, whether endpoint sensing or joint angles, are lumped into a measurement vector $\mathbf{x}^i$. The vector $\mathbf{p}$ contains all unknown parameters. Let $\mathbf{x}$ represent the stacked vector for all poses i and redefine $\mathbf{f}$ as well.

We separate the vector $\mathbf{x} = \mathbf{x}_0 + \mathbf{x}_e$, where $\mathbf{x}_0$ is the measurement and $\mathbf{x}_e$ is the measurement error. Also set $\mathbf{p} = \mathbf{p}_0 + \mathbf{p}_e$, where $\mathbf{p}_0$ are a priori estimates for the parameters. Incorporate $\mathbf{x}_0$ and $\mathbf{p}_0$ as constants into the loop closure equation (10), which is then a function $\mathbf{f}'(\mathbf{x}_e, \mathbf{p}_e) = \mathbf{0}$. Then minimize the optimization function

$$\min_{\mathbf{x}_e, \mathbf{p}_e} \mathbf{x}_e^T \mathbf{V}_x^{-1} \mathbf{x}_e + \mathbf{p}_e^T \mathbf{V}_p^{-1} \mathbf{p}_e \tag{11}$$

subject to $\mathbf{f}'(\mathbf{x}_e, \mathbf{p}_e) = \mathbf{0}$, where $E(\mathbf{x}_e \mathbf{x}_e^T) = \mathbf{V}_x$ and $E(\mathbf{p}_e \mathbf{p}_e^T) = \mathbf{V}_p$ are covariance estimates. Again, this function is a χ^2 statistic. Note that the objective function biases the solution to the a priori values.

Because of the implicit constraint equation, direct substitution into the objective function to eliminate some unknowns is not possible. Instead, the equations are linearized and variable elimination is achieved through QR decomposition [13]. Let the scaled variables $\mathbf{y}$ and $\mathbf{q}$ be defined via $\mathbf{x}_e = \mathbf{V}_x^{1/2}\mathbf{y}$ and $\mathbf{p}_e = \mathbf{V}_p^{1/2}\mathbf{q}$, where the superscript 1/2 means the symmetric square root. Iterate from an initial guess $\mathbf{y} = \mathbf{0}$ and $\mathbf{q} = \mathbf{0}$ to find corrections $\Delta\mathbf{y}$ and $\Delta\mathbf{q}$ to minimize

$$\chi^2 = (\mathbf{y} + \Delta\mathbf{y})^T(\mathbf{y} + \Delta\mathbf{y}) + (\mathbf{q} + \Delta\mathbf{q})^T(\mathbf{q} + \Delta\mathbf{q}) \tag{12}$$

subject to the linearized constraints

$$-\mathbf{f}'(\mathbf{y}, \mathbf{q}) = \frac{\partial \mathbf{f}'}{\partial \mathbf{y}}(\mathbf{y}, \mathbf{q})\Delta\mathbf{y} + \frac{\partial \mathbf{f}'}{\partial \mathbf{q}}(\mathbf{y}, \mathbf{q})\Delta\mathbf{q} \equiv \mathbf{J}_y\Delta\mathbf{y} + \mathbf{J}_q\Delta\mathbf{q} \tag{13}$$

Compute the QR-decomposition $\mathbf{Q}\tilde{\mathbf{R}} = \mathbf{J}_y^T$ and define

$$\mathbf{D} = \tilde{\mathbf{R}}^{-T}\mathbf{J}_y, \quad \mathbf{E} = \mathbf{Q}^T\mathbf{y} - \tilde{\mathbf{R}}^{-T}\mathbf{f}(\mathbf{y}, \mathbf{q}) \tag{14}$$

The variables $\Delta\mathbf{y}$ are eliminated and the step in $\mathbf{q}$ found from:

$$\begin{bmatrix} \mathbf{D} \\ \mathbf{I} \end{bmatrix} \Delta\mathbf{q} = \begin{bmatrix} \mathbf{E} \\ -\mathbf{q} \end{bmatrix} \tag{15}$$

and the updated error estimates are

$$\mathbf{y} + \Delta\mathbf{y} = \mathbf{Q}(\mathbf{E} - \mathbf{D}\mathbf{q}) \tag{16}$$

Later we will use the definition $\hat{\mathbf{D}} = [\mathbf{D}^T\mathbf{I}]^T$, to examine the effect of a priori estimates on singular values.

3. Torque Sensor Calibration

Ma et al. [8] developed an autonomous joint torque sensor calibration, which utilizes combinations of single-joint rotations and an arm's own gravity load. A key feature of the approach is that nothing is assumed known about the arm's inertial parameters or of the location of a reference load in the grasp. The method has been applied to the Sarcos Dextrous Arm Master and Slave (Figure 2). The method determines not only joint torque sensor gains and offsets, but also those of the joint angle sensors.

Individual joint rotation results in a sinusoidal torque versus angle relation:

$$\tau^i = A\sin(k\psi^i + \delta) + \gamma = f(\mathbf{p}, \psi^i) \tag{17}$$

with unknown parameters $\mathbf{p} = (A, k, \delta, \gamma)$ given P input/output measurements ψ^i, τ^i. From the fitted parameters one may extract torque sensor and joint angle sensor gains and offsets; the details may be found in [8]. The ILM was applied,

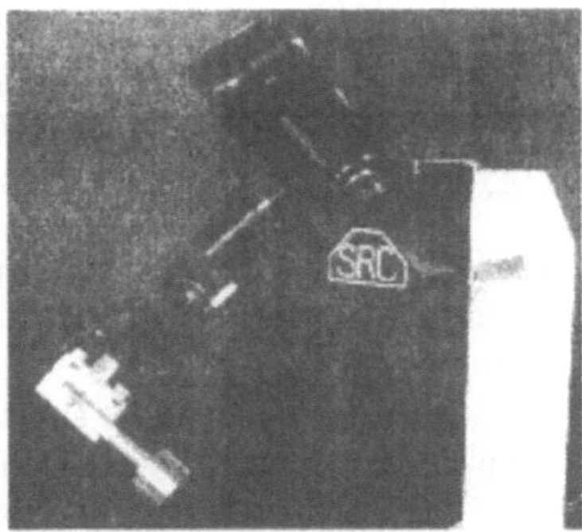

Figure 2. Automatic torque sensor calibration for the Sarcos Dextrous Arm Slave.

without however utilizing a priori parameter values. The weighted objective function is:

$$\min_{\mathbf{p},\psi_c^i,\tau_c^i} \sum_{i=1}^{P} \frac{(\Delta\psi^i)^2}{\sigma_\psi^2} + \frac{(\Delta\tau^i)^2}{\sigma_\tau^2} \quad \text{subject to } \tau_c^i = f(\mathbf{p}, \psi_c^i) \tag{18}$$

where $\Delta\psi^i = \psi^i - \psi_c^i$ and $\Delta\tau^i = \tau^i - \tau_c^i$ and the subscript c again stands for computed. The uncertains in the input/output measurements are σ_ψ and σ_τ. Again the constraint equation is substituted to eliminate the τ_c^i unknowns.

Results are presented in Table 1 for joint 1 of the Sarcos Master in the case of equal weighting for OLS versus ILM. The parameters are fairly close, because of low noise in the data. At present there is no independent method for ascertaining the accuracy of one parameter set over the other, except simulation studies.

Method	**A**	**k**	δ	γ
OLS	372.96	0.00076419	1.8730	2014.6
ILM	373.08	0.00076390	1.8735	2014.7

Table 1. Sinusoidal parameters for OLS versus ILM.

4. Kinematic Calibration of Redundant Parallel-Drive Robot

The mechanism is a 3-DOF platform type closed-chain mechanism with its output link constrained to undergo spherical motions (Figure 3A) [4]. Previously we performed ordinary least squares kinematic calibration [9]; the present section is a reanalysis of the data via the implicit loop method.

The kinematic model is shown in Figure 3B. $\mathbf{A}_i$, $i = 1, \ldots, 4$, represents a spherical joint at the stationary side of each actuator. d_i, $i = 1, \ldots, 4$, is the input of the mechanism and represents a pair of actuator and displacement sensor. $\mathbf{B}_1$ and $\mathbf{B}_2$ are universal joints and lie in the intersection of the centerlines of each two adjacent actuators. Plane $\mathbf{B}_1\mathbf{B}_2\mathbf{O}$ defines the end plate which should be placed into the desired orientation. k_1 and k_2 are imaginary links which are used in calibration loops.

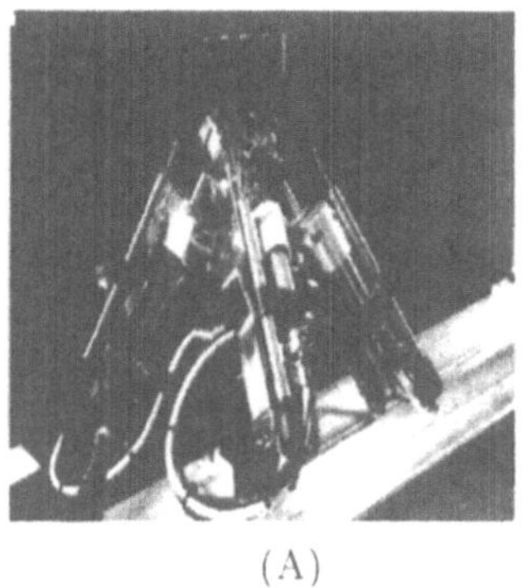

(A)

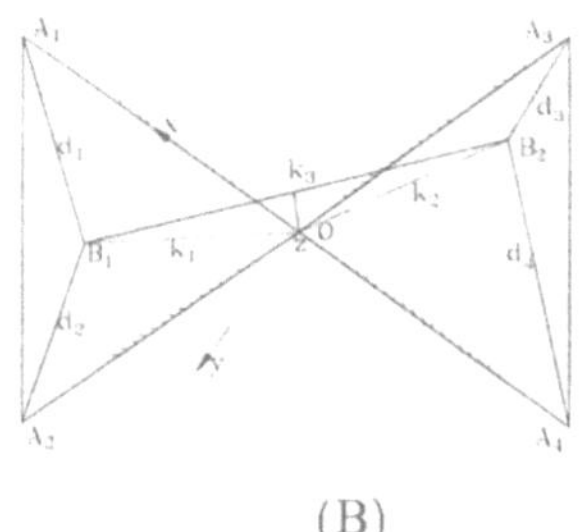

(B)

Figure 3. (A) 3-DOF redundant shoulder joint. (B) Kinematic model viewed from above.

4.1. Calibration Procedure

Since not all angles are sensed in this mechanism, the simplest way to formulate calibration equations is to use distance equations. We use measurement redundancy to establish our objective function which is to be minimized. Assume we move the mechanism into P different poses. Define the vector function $\mathbf{f}$ with components:

$$f(k) = (\mathbf{B}_1(k) - \mathbf{B}_2(k))^2 - k_3^2 + x(k) = 0 \tag{19}$$

where $k = 1, ..., P$ represents the pose number. $\mathbf{B}_1(k)$ and $\mathbf{B}_2(k)$ are the position vectors of the end plate universal joints in pose k. $x(k)$ is an error term and makes $f(k)$ equal to zero. We find $\mathbf{B}_1(k)$ by the intersection of the following three spheres:
Sphere 1: centered at $\mathbf{O}$ with radius k_1,
Sphere 2: centered at $\mathbf{A}_1$ with radius $d_1(k)$,
Sphere 3: centered at $\mathbf{A}_2$ with radius $d_2(k)$.

Then we find $\mathbf{B}_2(k)$ by the intersection of the following three spheres:
Sphere 1: centered at $\mathbf{O}$ with radius k_2,
Sphere 2: centered at $\mathbf{A}_3$ with radius $d_3(k)$,
Sphere 3: centered at $\mathbf{A}_4$ with radius $d_4(k)$.

Length $d_i(k)$ which is measured by LVDT i in pose k is obtained as follows:

$$d_i(k) = s_i * v_i(k) + d_{0i} \; (i = 1, ..., 4, k = 1, ..., P) \tag{20}$$

where s_i, $v_i(k)$, and d_{0i} represent the gain, output voltage, and offset of LVDT i respectively. It is worth mentioning that each time LVDTs are disassembled and then reassembled, d_{0i} may change considerably (a few millimeters) and the closed-loop calibration is a promising approach for finding new offsets.

The objective function and solution procedure were outlined in Section 2.

4.2. Simulation Results

The selection of the parameters to be calibrated is an important issue. While we would ideally identify all the kinematic parameters, we should first evaluate the limits of the approach by considering measurement errors due to resolution, nonlinearity, noise of sensors, and the numerical error of the method used. We may then decide which parameters should be assumed known and which should be identified.

Param.	True value	Initial guess	Result
d_{01}	90.0	80.0	89.9
d_{02}	90.0	95.0	90.2
d_{03}	90.0	85.0	90.0
d_{04}	85.0	90.0	85.0
s_1	3.659	3.660	3.659
s_2	3.656	3.660	3.659
s_3	3.658	3.660	3.661
s_4	3.660	3.660	3.662

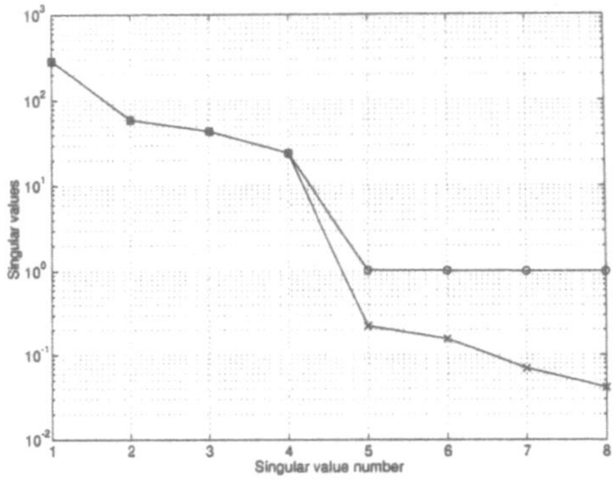

(A)

(B)

Figure 4. (A) Simulation results of the shoulder joint calibration. Units for d_{0i}: mm. Units for s_i: mm/v. (B) Singular values of the fitting matrices $\mathbf{D}$ (x) and $\hat{\mathbf{D}}$ (o).

In order to find the accuracy of the sensors, we moved each LVDT off-line using an accurate milling machine and found that the overall accuracy of each sensor is about 40mv (0.15 mm). We treat this error as noise, though it is due to a combination of sensor nonlinearity, noise, and resolution. On the other hand, the mechanical parts of this robot have a manufacturing tolerance better than 0.1mm. Considering the 0.15mm sensor noise, it is not useful to calibrate those kinematic parameters which already have a better accuracy in blue print data than the sensor noise. Thus we should only calibrate the sensor offsets and gains. The parameter vector $\mathbf{p}$ is defined as:

$$\mathbf{p} = [d_{01}\ d_{02}\ d_{03}\ d_{04}\ s_1\ s_2\ s_3\ s_4]^T \tag{21}$$

We added Gaussian noise of 40mv (0.15mm) to the sensor readings derived from simulation. Each pose set contained 50 poses. The standard deviation for sensor offsets was set to 7.0, for sensor gains to 0.005, and for the residual error function to 49.0. Thus the weighting matrices are:

$$\mathbf{V}_p = \mathrm{diag}[\frac{1}{7^2}, \frac{1}{7^2}, \frac{1}{7^2}, \frac{1}{7^2}, \frac{1}{.005^2}, \frac{1}{.005^2}, \frac{1}{.005^2}, \frac{1}{.005^2}] \tag{22}$$

$$\mathbf{V}_x = [\frac{1}{49^2}] \tag{23}$$

Figure 4A shows the results. It is seen that calibration results are close to true values, but we were suspicious about the accuracy of scale factors because the results were somewhat dependent on our initial guess. Figure 4B shows the singular values for the fitting matrices $\mathbf{D}$ and $\hat{\mathbf{D}}$. It is seen that the last four singular values of $\mathbf{D}$ are much smaller than 1, resulting in poor values for the gains. In other words, four singular values of $\hat{\mathbf{D}}$ are almost equal to 1, indicating no reduction in the uncertainty of our initial guess for gains. The χ^2 value is 49.5 which is near the expected value of 50.

Param.	Initial guess	Result
d_{01}	95.0	96.8
d_{02}	95.0	99.2
d_{03}	95.0	95.9
d_{04}	95.0	101.9
s_1	3.697	3.697
s_2	3.586	3.586
s_3	3.604	3.604
s_4	3.678	3.677

(A)

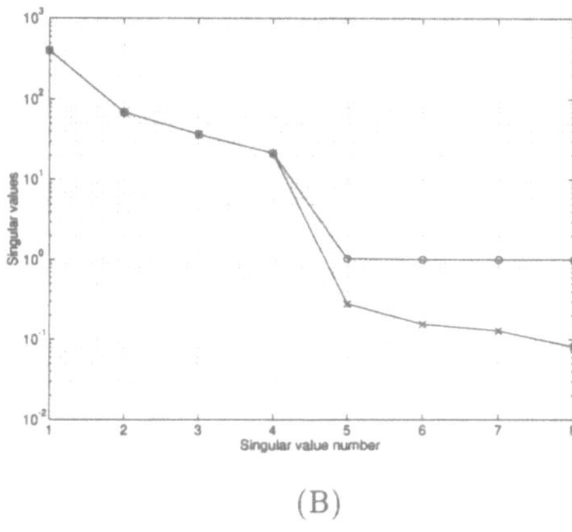

(B)

Figure 5. (A) Experimental results of the shoulder joint calibration. (B) Singular values of the fitting matrices $\mathbf{D}$ (x) and $\hat{\mathbf{D}}$ (o) resulting from experimental data of the shoulder joint.

4.3. Experimental Results

The end plate was moved into different orientations manually and data were acquired simultaneously from four LVDT's through 12-bit A/D converters with a sampling frequency of 20 Hz. 450 poses were recorded. We wrote an algorithm in MATLAB to select 50 poses out of 450 poses to have a good observability index [7]. The weighting matrices were defined as:

$$\mathbf{V}_p = \mathrm{diag}[\frac{1}{6^2}, \frac{1}{6^2}, \frac{1}{6^2}, \frac{1}{6^2}, \frac{1}{.005^2}, \frac{1}{.005^2}, \frac{1}{.005^2}, \frac{1}{.005^2}] \quad (24)$$

$$\mathbf{V}_x = [\frac{1}{36^2}] \quad (25)$$

Figure 5A shows the results. Again, we suspected about the accuracy of scale factors because the results were somewhat dependent on our initial guess. Figure 5B shows the singular values for the fitting matrices $\mathbf{D}$ and $\hat{\mathbf{D}}$. The results are very similar to simulation, indicating no improvement in the uncertainty of our initial guess for gains. We expect a χ^2 value of 50 with a standard deviation of 10. The χ^2 value obtained from the fit is 20.1 which falls within 3σ of its expected value.

5. Conclusions

Input noise is often significant in robot calibration problems, but the approach of total least squares may be applied to these problems. we have presented the application of one variant of this method, the implicit loop method [13], to two different calibration problems. One is torque sensor calibration, the other kinematic calibration of a rendundant parallel-drive spherical mechanism. Results indicate so far that significant although not large differences in parameter estimates are obtained over those based on OLS.

Acknowledgements. This work was supported by National Science Foundation grant MIP-9420352, and by Natural Sciences and Engineering Research Council of Canada Network of Centers of Excellence Institute for Robotics and Intelligent

Systems (IRIS) grant AMD-5.

References

[1] Bennett, D.J., and Hollerbach, J.M., "Autonomous calibration of single-loop closed kinematic chains formed by manipulators with passive endpoint constraints," *IEEE Trans. Robotics and Automation*, vol. 7, pp. 597-606, 1991.

[2] Boggs, P.T., Byrd, R.H., and Schnabel, R.B., "A stable and efficient algorithm for nonlinear orthogonal distance regression," *SIAM J. Sci. Stat. Comput.*, vol. 8, pp. 1052-1078, 1987.

[3] Fuller, W.A., *Measurement Error Models.* NY: John Wiley & Sons, 1987.

[4] Hayward, V., "Design of a hydraulic robot shoulder mechanism based on a combinatorial mechanism," *Preprints of the Third International Symposium on Experimental Robotics.* Kyoto, Japan:, 1993.

[5] Hollerbach, J.M., and Bennett, D.J., "Automatic kinematic calibration using a motion tracking system," *Robotics Research: the Fourth International Symposium*, edited by R. Bolles and B. Roth. Cambridge, MA: MIT Press, pp. 191-198, 1988.

[6] Hollerbach J.M., and Lokhorst, D.M., "Closed-loop kinematic calibration of the RSI 6-DOF hand controller," *IEEE Trans. Robotics and Automation*, vol. 11, pp. 352-359, 1995.

[7] Hollerbach, J.M., and Wampler, C.W., "The calibration index and taxonomy for robot kinematic calibration methods," *Intl. J. Robotics Research*, in press, 1996.

[8] Ma, D., Hollerbach, J.M., and Xu, Y., "Gravity based autonomous calibration for robot manipulators," *Proc. IEEE Intl. Conf. Robotics and Automation*, 1994.

[9] Nahvi, A., Hollerbach, J.M., and Hayward, V., "Calibration of a parallel robot using multiple kinematic closed loops," *Proc. IEEE Intl. Conf. Robotics and Automation*, pp. 407-412, 1994.

[10] Norton, J.P., *An Introduction to Identification.* London: Academic Press, 1986.

[11] Press, W.H., Teukolsky, S.A., Vetterling, W.T., and Flannery, B.P., *Numerical Recipes in C.* Cambridge, UK: Cambridge Univ. Press, 1992.

[12] Van Huffel, S., and Vandewalle, J., *The Total Least Squares Problem: Computational Aspects and Analysis.* Philadelphia: SIAM, 1991.

[13] Wampler, C.W., Hollerbach, J.M., and Arai, T., "An Implicit Loop Method for kinematic calibration and its application to closed-chain mechanisms," *IEEE Trans. Robotics and Automation*, vol. 11, pp. 710-724, 1995.

Symbolic Modelling and Experimental Determination of Physical Parameters for Complex Elastic Manipulators

Dipl.-Ing. H. Schütte, Dr.-Ing. W. Moritz

Mechatronics Laboratory Paderborn, MLaP
(Prof. Dr.-Ing. J. Lückel), University of Paderborn,
Pohlweg 55, D - 33098 Paderborn, Germany
email: schue@mlap.uni-paderborn.de

Abstract

After the introduction of a complete and compact multibody system (MBS) formalism for systems with discrete and distributed masses and elasticities, a specialization to tree-structured systems will be presented. A novel approach to the symbolic implementation of these formulas in MAPLE using non-commutative operators facilitates the introduction of symbolic simplifications.

The description of a model for a six-axis robot with elasticities in the gears, bearings and links is followed by the identification of this system. Static deflection measurements are used for the determination of the nonlinear (NL) gear stiffnesses. Frequency response measurements in several positions are employed for the calculation of the dynamic parameters of the system.

1. Introduction

The modelling of mechanical plants is one of the main elements of the description of mechatronic systems, which consist also of sensors, actuators, feedforward and feedback controls. The principle aim of this description of mechanical systems is to build up a solid basis for the controller design. From this aim two substantial demands can be concluded:

a) the model has to describe the static and dynamic properties of the system up to a certain bandwidth

b) a symbolic generation of the nonlinear (NL) equations of motion is desirable and allows:

- insight into the system structure
- derivation of controller structures
- efficient realtime realization of controllers

From point a) results that, apart from the use of an appropriate multibody system (MBS) formalism for the generation of the equations of motion, the model structure to be chosen plays a crucial part. Thus, there is a very close connection between the identification (of structure and parameters) and the generation of the dynamic equations. This fact is often ignored.

The modelling of the elastic robot system *tempo* [14] takes into account the following essential physical effects:

- NL dynamics
- NL gear elasticity
- elasticities in the bearings
- link elasticities

- Coulomb friction and viscous dampings

These values have been verified by several measurements at this modular robot system. It is interesting that the single effects are well known and often described but seldom used in models for the controller design. Works which consider complete elastic NL models for the identification in the frequency domain are rare.

This contribution is divided into two main chapters: The second chapter describes the basics of the MBS formalism employed and a special approach to the symbolic programming in view of the robotics application and the necessary simplifications. The following chapter deals with the identification of the dynamic parameters of this model.

2. Symbolic Modelling

The reasons for the symbolic treatment of MBS mentioned in the introduction yield a great number of program systems. They can be divided into two categories: on the one hand those which are based on general-purpose symbol-manipulation programs (e. g. Tab. 1) and on the other hand the stand-alone (e. g. Tab. 2) systems, also called special-purpose systems. The system presented in this paper belongs to the first one and is completely written in MAPLE [4]. The main reasons for the choice of MAPLE are

- efficient memory management
- extensive basic routines / libraries
- availability: UNIX, MS-DOS, Windows, ...
- tool coupling (e. g. MATLAB 4.2b)
- treatment of control problems

name/system	author(s)	algorith./remarks
ΣSYGMMAE (1993), MAPLE V	Capolsini	Kane, rigid robots
- (1992), REDUCE	Cetinkunt, Ittoop	Lagrange, assumed mode, non-recursive, elast. links
- (1989), MACSYMA	Ju, Mansour	various methods, simplifications, rigid robots
- (1989), REDUCE, MACSYMA	Cetinkunt, Book	non-recursive Lagr., assumed mode, elast. joints/links
Robotica™ (1989), Mathematica 2.1™	Ghorbel, Nethery, Spong	Euler Lagrange, rigid robots, animation
- (1986), MACSYMA	Nicosia, Tomei, Tornambè	generalized Lagr., assumed mode, elast. links/joints
DYMIR (1984), REDUCE	Cesareo, Nicolo, Nicosia	Euler Lagrange, elast. joints, homog. transf.

Table 1. MBS tools (with general-purpose systems)

name/language	author(s)	algorith./remarks
AUTOLEV2.0 (1991), FORTRAN	Levinson, Schaechter	Kane, elast. joints
MESA-VERDE (1986), PASCAL	Wolz, Wittenburg	Wittenburg
SD/EXACT (1985), (FORTRAN)	Rosenthal, Sherman	Kane, rigid robots, simplif.
ARM (1984), "C", LISP	Murray, Neumann	various methods, rigid robots, simplif.
NEWEUL (1977), FORTRAN (MAPLE)	Kreuzer, Schiehlen	Newton-Euler, elast. joints/links, closed kinem. chains, etc.

Table 2. MBS tools (special-purpose systems)

2.1 MBS Formalism

The aim of the modelling is the generation of (1). A frequent starting point for their derivation is the Lagrange equation (2). The kinetic and potential energies are calculated according to (3) and (4):

$$\mathbf{M}(\mathbf{q})\,\ddot{\mathbf{q}} + \mathbf{C}\,\mathbf{q} + \mathbf{h}(\mathbf{q},\dot{\mathbf{q}}) = \mathbf{f}(\mathbf{q}) \qquad (1)$$

$$\frac{d}{dt}\left(\frac{\partial T}{\partial \dot{\mathbf{q}}}\right) - \frac{\partial T}{\partial \mathbf{q}} + \frac{\partial U}{\partial \mathbf{q}} = \mathbf{f} \qquad (2)$$

$$T = \frac{1}{2}\sum_i (\mathbf{v}_i^T m_i \mathbf{v}_i + \omega_i^T \mathbf{I}_i \omega_i), \quad U = \frac{1}{2}\sum_i \Delta\mathbf{x}_i^T \mathbf{C}_i \Delta\mathbf{x}_i \qquad (3),(4)$$

Notations: (*i* denotes the body in question)

$\mathbf{v}_i, \omega_i$	absolute translational and rotational velocity (center of gravity); arbitrary coordinate systems (COS)
m_i	mass of body *i*
$\mathbf{I}_i$	$\in R^{3,3}$ inertia tensor (center of gravity)
$\mathbf{C}_i$	const. element stiffness matrix of spring *i*
$\Delta\mathbf{x}_i$	local deformation of spring *i*
$\mathbf{q}, \dot{\mathbf{q}}$	vector of generalized positions / velocities

These equations are often directly used for the evaluations, but this yields of cause an unnecessary extension of calculations. More efficient is the direct calculation of the Jacobian matrices of translation and rotation (5) and (6) and their concatenation (7):

$$\mathbf{v}_i = \frac{\partial \mathbf{r}_i(\mathbf{q})}{\partial \mathbf{q}^T}\dot{\mathbf{q}} = \frac{\partial \mathbf{v}_i(\mathbf{q})}{\partial \dot{\mathbf{q}}^T}\dot{\mathbf{q}} = \mathbf{J}_{Ti}\,\dot{\mathbf{q}} \qquad (5)$$

$$\omega_i = \frac{\partial \omega_i(\mathbf{q})}{\partial \dot{\mathbf{q}}^T}\dot{\mathbf{q}} = \mathbf{J}_{Ri}\,\dot{\mathbf{q}} \qquad (6)$$

$$\mathbf{J}_S(\mathbf{q}) = [\ldots, \mathbf{J}_{Ti}^T, \mathbf{J}_{Ri}^T, \ldots]^T \qquad (7)$$

By means of the introduction of the Jacobians J_{Ci} (8) and J_C (9) it is easy to get the kinetic and potential energies as a function of the generalized positions and velocities (see (10) und (11)).

$$\mathbf{J}_C(\mathbf{q}) = [\ldots, \mathbf{J}_{Ci}^T, \ldots]^T, \quad \mathbf{J}_{Ci} = \frac{\partial \Delta\mathbf{x}_i(\mathbf{q})}{\partial \mathbf{q}^T} \qquad (8),(9)$$

$$T(\mathbf{q}, \dot{\mathbf{q}}) = \frac{1}{2}\,\dot{\mathbf{q}}^T\,[\mathbf{J}_S^T\,\mathbf{M}^*\,\mathbf{J}_S]\,\dot{\mathbf{q}} = \frac{1}{2}\,\dot{\mathbf{q}}^T\,\mathbf{M}(\mathbf{q})\,\dot{\mathbf{q}} \qquad (10)$$

$$\mathbf{M}^* = \text{diag}\,[\ldots, \text{diag}(m_i), \mathbf{I}_i, \ldots]$$

$$U(\mathbf{q}) = \frac{1}{2}\mathbf{q}^T\,[\mathbf{J}_C^T\,\mathbf{C}_S^*\,\mathbf{J}_C]\,\mathbf{q} = \frac{1}{2}\,\mathbf{q}^T\,\mathbf{C}\,\mathbf{q} \qquad (11)$$

$$\mathbf{C}_S^* = \text{diag}[\ldots, c_i, \ldots]$$

Further evaluation of (2) yields:

$$\frac{d}{dt}\left(\frac{\partial T}{\partial \dot{\mathbf{q}}}\right) - \frac{\partial T}{\partial \mathbf{q}} = \mathbf{M}(\mathbf{q})\,\ddot{\mathbf{q}} + \frac{d}{dt}(\mathbf{M}(\mathbf{q}))\,\dot{\mathbf{q}} - \frac{\partial}{\partial \mathbf{q}}\left(\frac{1}{2}\,\dot{\mathbf{q}}^T\,\mathbf{M}(q)\,\dot{\mathbf{q}}\right)$$

$$= \mathbf{M}(\mathbf{q})\,\ddot{\mathbf{q}} + \mathbf{h}(\mathbf{q}, \dot{\mathbf{q}})$$

The centrifugal and Coriolis forces are calculated according to (12). This representation is advantageous if a pattern matrix for **A** is introduced. Moreover it is possible to neglect small terms which are not as important in the way that all small coordinates in **M** are neglected. The model of the system *tempo* contains only the centrifugal and Coriolis forces of the rigid manipulator.

$$\mathbf{h}(\mathbf{q},\dot{\mathbf{q}}) = \frac{d}{dt}(\mathbf{M}(\mathbf{q}))\,\dot{\mathbf{q}} - \frac{\partial}{\partial \mathbf{q}}\left(\frac{1}{2}\,\dot{\mathbf{q}}^T\,\mathbf{M}(\mathbf{q})\,\dot{\mathbf{q}}\right)$$

$$= \frac{\partial(\mathbf{M}(\mathbf{q})\,\dot{\mathbf{q}})}{\partial \mathbf{q}^T}\,\dot{\mathbf{q}} - \frac{1}{2}\left(\frac{\partial(\dot{\mathbf{q}}^T\,\mathbf{M}(\mathbf{q}))}{\partial \mathbf{q}}\,\dot{\mathbf{q}}\right)$$

$$\mathbf{h}(\mathbf{q},\dot{\mathbf{q}}) = \left(\mathbf{A} - \frac{1}{2}\mathbf{A}^T\right)\dot{\mathbf{q}} \quad \text{with} \quad \mathbf{A} = \frac{\partial(\mathbf{M}(\mathbf{q})\,\dot{\mathbf{q}})}{\partial \mathbf{q}^T} \tag{12}$$

The terms resulting from the potential energy (4) and the generalized forces (2) can be obtained via (13) and (14): ($\mathbf{f}_{ei}$, $\mathbf{m}_{ei}$: external forces / moments)

$$\frac{\partial U(\mathbf{q})}{\partial \mathbf{q}} = \frac{\partial}{\partial \mathbf{q}}\left(\frac{1}{2}\mathbf{q}^T\,\mathbf{C}\,\mathbf{q}\right) = \mathbf{C}\,\mathbf{q} \tag{13}$$

$$\mathbf{f} = \mathbf{J}_S^T\,\mathbf{f}^* \;,\quad \mathbf{f}^* = [\ldots,\mathbf{f}_{ei}^T,\mathbf{m}_{ei}^T,\ldots]^T \tag{14}$$

The translational and rotational velocities for every body i which are necessary for the Jacobians are calculated by means of (15) and (16) with respect to a body-fixed coordinate system. In this way it can be dealt with constant diagonal inertia tensors.

$${}^i\mathbf{v}_i = {}^i\dot{\mathbf{r}}_i + {}^i\tilde{\omega}_i\;{}^i\mathbf{r}_i \tag{15}$$

$${}^i\tilde{\omega}_i = {}^i\mathbf{S}_I^T\;{}^i\dot{\mathbf{S}}_I, \qquad \text{with } {}^i\mathbf{S}_j = \prod_{k=j}^{i}\mathbf{S}_k(q_k) \tag{16),(17}$$

Vectors are marked with two indices; the first one (upper left corner) describes the corresponding coordinate system and the second one (lower right corner) marks the body in question. The rotations (16) will be calculated according to equation (17); the character 'I' describes the inertial coordinate system. The rotation $\mathbf{S}_k(q_k)$ denotes a relative rotation between two bodies resp. two frames with the angle q_k.

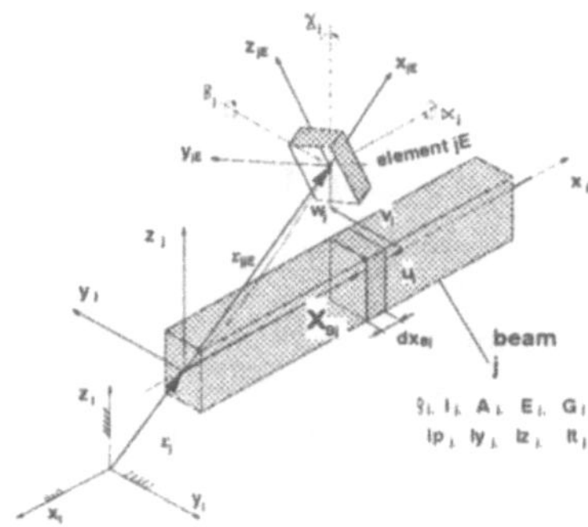

Figure 1 . Kinematic description of a beam

Systems containing beams are treated by the assumed mode method, which is e. g. used by Truckenbrodt [17] (see also Meirovitch [10]). At the basis of each beam a body-fixed tangential coordinate system is utilized. The distributed coordinates $(u_j, v_j, w_j, \alpha_j, \beta_j, \gamma_j)$ describe the small local deformations of the beam in dependency of the beam-coordinate x_{Bj}. For each deformation a product of the form: position-function $u(x_B)$ * generalized coordinate $q(t)$ is used.

$$ {}^j\mathbf{r}_{jE}(x_{Bj}) = {}^j\mathbf{r}_j + {}^j\mathbf{r}_{jjE}(x_{Bj}) \quad \Rightarrow \quad {}^j\mathbf{v}_{jE}(x_{Bj}) = {}^j\dot{\mathbf{r}}_{jE} + {}^j\tilde{\omega}_{jE}\, {}^j\mathbf{r}_{jE} \tag{18} $$

$$ {}^{jE}\mathbf{S}_I(x_{Bj}) = {}^j\mathbf{S}_I\, {}^{jE}\mathbf{S}_j \quad \Rightarrow \quad {}^{jE}\tilde{\omega}_{jE}(x_{Bj}) = {}^{jE}\mathbf{S}_I^T\, {}^{jE}\dot{\mathbf{S}}_I \tag{19} $$

$$ {}^{jE}\mathbf{S}_j = \begin{bmatrix} 1 & -\gamma_j & \beta_j \\ \gamma_j & 1 & -\alpha_j \\ -\beta_j & \alpha_j & 1 \end{bmatrix}, \quad \begin{matrix} \textit{with} \\ \beta_j = -w_j' - \beta_{Tj} \\ \gamma_j = v_j' - \gamma_{Tj} \end{matrix} \tag{20} $$

$$ \begin{bmatrix} {}^j\mathbf{r}_{jjE}(x_{Bj}, t) \\ \hline {}^j\varphi_{jE}(x_{Bj}, t) \end{bmatrix} = \begin{bmatrix} u_j \\ v_j \\ w_j \\ \hline \alpha_j \\ \beta_{Tj} \\ \gamma_{Tj} \end{bmatrix} = \begin{bmatrix} \sum_{k=1}^{k_{uj}} u_{jk}(x_{Bj}) q_k(t) + x_{Bj} \\ \sum_{k=k_{uj}+1}^{k_{vj}} v_{jk}(x_{Bj}) q_k(t) \\ \sum_{k=k_{vj}+1}^{k_{wj}} w_{jk}(x_{Bj}) q_k(t) \\ \hline \sum_{k=k_{wj}+1}^{k_{\alpha j}} \alpha_{jk}(x_{Bj}) q_k(t) \\ \sum_{k=k_{\alpha j}+1}^{k_{\beta Tj}} \beta_{Tjk}(x_{Bj}) q_k(t) \\ \sum_{k=k_{\beta Tj}+1}^{k_{\gamma Tj}} \gamma_{Tjk}(x_{Bj}) q_k(t) \end{bmatrix} \tag{21} $$

The kinematics of an Euler-Bernoulli beam j (with small deflections) is completely described by the equations (18) - (21). By introducing the Jacobians (22) the corresponding mass-matrix contribution of the beam can be obtained via integration along the beam-coordinate (24). The contribution to the stiffness-matrix is calculated in analogy (see (27)):

$$ \mathbf{J}_{TjE} = \frac{\partial\, {}^j\mathbf{v}_{jE}(x_{Bj})}{\partial \dot{\mathbf{q}}^T}, \quad \mathbf{J}_{RjE} = \frac{\partial\, {}^{jE}\omega_{jE}(x_{Bj})}{\partial \dot{\mathbf{q}}^T} \tag{22} $$

$$ \mathbf{M}^*_{Bj} = \mathrm{diag}[\rho_j A_j, \rho_j A_j, \rho_j A_j, \quad \rho_j Ip_j, \rho_j Iy_j, \rho_j Iz_j] \tag{23} $$

$$ \mathbf{M}_{Bj}(\mathbf{q}) = \int_{x_{Bj}=0}^{l_{Bj}} [\mathbf{J}^T_{TjE}, \mathbf{J}^T_{RjE}]\ \mathbf{M}^*_{Bj}\ [\mathbf{J}^T_{TjE}, \mathbf{J}^T_{RjE}]^T dx_{Bj} \tag{24} $$

$$ \mathbf{J}_{CTjE} = \frac{\partial [u_j', \beta_j', \gamma_j']^T}{\partial \mathbf{q}^T}, \quad \mathbf{J}_{CRjE} = \frac{\partial [\alpha_j', \beta_{Tj}, \gamma_{Tj}]^T}{\partial \mathbf{q}^T} \tag{25} $$

$$ \mathbf{C}^*_{Bj} = \mathrm{diag}[E_j A_j, E_j Iy_j, E_j Iz_j, \quad G_j It_j, G_j A_j \kappa_{zj}, G_j A_j \kappa_{yj}] \tag{26} $$

$$\mathbf{C}_{Bj} = \int_{x_{Bj}=0}^{l_{Bj}} [\mathbf{J}^T_{CTjE}, \mathbf{J}^T_{CRjE}]\ \mathbf{C}^*_{Bj}\ [\mathbf{J}^T_{CTjE}, \mathbf{J}^T_{CRjE}]^T\, dx_{Bj} \tag{27}$$

The centrifugal and Coriolis terms can easily be obtained (if necessary) through the equations (24) and (12).

2.2 Tree-Structured Systems

The formalism introduced supplies equation (1) via simple differentiations (Jacobians) of the vectors and matrices denoting the relative displacement and relative rotation between a body-fixed frame and the initial frame. For tree-structured systems with purely rotational joints (relative coordinates are used) simpler equations for the Jacobians can be obtained. By means of equations (17) and (28) the Jacobian for rotation can be obtained without any differentiation (29). The vectors $\mathbf{e}_i$ denote the unit-vectors in the direction of the rotation axis of the joint in question.

$$^i\omega_i = \sum_{k=1}^{i} {}^i\mathbf{S}_k^T \mathbf{e}_k \dot{q}_k \tag{28}$$

$$\mathbf{J}_{Ri} = [{}^i\mathbf{S}_1^T\mathbf{e}_1\,|\,{}^i\mathbf{S}_2^T\mathbf{e}_2\,|\,\dots\,|\,{}^i\mathbf{S}_i^T\mathbf{e}_i\,|\,0\,|\,\dots\,|\,0] \tag{29}$$

With (31) it is possible to reduce the calculation effort by using the well-known Jacobian $\mathbf{J}_{Ri}$ for the calculation of $\mathbf{J}_{Ti}$.

$$^i\mathbf{r}_i = {}^i\Delta\mathbf{r}_{iS} + \sum_{k=1}^{i-1} {}^i\mathbf{S}_{k+1}^T\, {}^k\Delta\mathbf{r}_{kE} \tag{30}$$

$$\mathbf{J}_{Ti} = \frac{\partial\, {}^i\mathbf{v}_i}{\partial \dot{\mathbf{q}}^T} = \frac{\partial}{\partial \dot{\mathbf{q}}^T}\left(\frac{\partial\, {}^i\mathbf{r}_i}{\partial \mathbf{q}^T}\dot{\mathbf{q}} + {}^i\tilde{\omega}_i\, {}^i\mathbf{r}_i\right) = \frac{\partial\, {}^i\mathbf{r}_i}{\partial \mathbf{q}^T} - {}^i\tilde{\mathbf{r}}_i \mathbf{J}_{Ri} \tag{31}$$

$^i\Delta\mathbf{r}_{iE}$ distance between the joints of the i-th body

$^i\Delta\mathbf{r}_{iS}$ distance of the first joint of the i-th body to the center of gravity (both body-fixed COS)

The contribution of each mass to the mass-matrix is easily obtained by (32) and (33).

$$\mathbf{M}_{Ri} = \mathbf{J}_{Ri}^T\, diag(Jx_i, Jy_i, Jz_i)\, \mathbf{J}_{Ri} \tag{32}$$

$$\mathbf{M}_{Ti} = m_i\, \mathbf{J}_{Ti}^T\, \mathbf{J}_{Ti} \tag{33}$$

The advantage of this representation is that the rotation matrices come "closely" together; this enables simplifications (see the next chapter) in an elegant way with reduced calculation effort.

2.3 Simplifications

The main simplifications during the evaluation of the equations (28) - (33) are given by the introduction of trigonometric equivalences (i) and (ii)

(i) $\sin^2(q_i)+\cos^2(q_i)=1$

(ii) $\sin(q_i)\cos(q_j) + \sin(q_j)\cos(q_i) = \sin(q_i+q_j)$

and the physically motivated simplification of small coordinates (e. g. deformations of bearing and beams). Admissibility of (iii) and (iv) for $q_i<<1$ is verified during the identification (see chapter 3).

(iii) $\sin(q_i)=q_i$, $\cos(q_i)=1$

(iv) $q_i^n=0$, $n\geq 2$, $q_i\, q_j=0$, $q_i\, \dot{q}_i=0$

The trigonometric identity (i) occurs e. g. if a rotation matrix is multiplied with their transpose (see (33)). Eq. (ii) can be applied if there are two or more parallel axes; this is even true if small deflections (bearings, beams) occur between these axes.

The symbolic consideration of (i) - (iv) brings about a very high calculation effort because these simplifications have to be introduced into very large terms (sq := sin(q), cq := cos(q)).

$$\mathbf{S}(x,q)=\begin{bmatrix}1 & 0 & 0\\ 0 & cq & -sq\\ 0 & sq & cq\end{bmatrix},\mathbf{S}(y,q)=\begin{bmatrix}cq & 0 & sq\\ 0 & 1 & 0\\ -sq & 0 & cq\end{bmatrix},\mathbf{S}(z,q)=\begin{bmatrix}cq & -sq & 0\\ sq & cq & 0\\ 0 & 0 & 1\end{bmatrix} \tag{34}$$

$$\frac{\partial \mathbf{S}(x,q)}{\partial q}=\mathbf{D}(x)\cdot\mathbf{S}(x,q),\qquad \mathbf{D}(x)=\begin{bmatrix}0 & 0 & 0\\ 0 & 0 & -1\\ 0 & 1 & 0\end{bmatrix} \tag{35}$$

$$\mathbf{e}^T(x)=[1\quad 0\quad 0],\quad \mathbf{e}^T(y)=[0\quad 1\quad 0],\quad \mathbf{e}^T(z)=[0\quad 0\quad 1] \tag{36}$$

In order to apply the simplifications on the level of rotation matrices and position vectors and not on the elements of these values, which would be the usual procedure, a MAPLE program has been developed which can directly deal with these values. The crucial point here is the definition of a non-commutative product operator '•'. This operator can deal directly with rotation matrices (34), differentiation matrices (35) and unit vectors (36). In order to evaluate the given formulas, additional operators for the transposition of a matrix $(.)^T$ and for the cross-product-matrix of a vector $(.)^{\sim}$ are implemented. Some calculation rules are exemplified in the following (**Id** = identity matrix).

$$\begin{aligned}
\mathbf{S}^T(x,q1)\bullet\mathbf{S}(x,q1)&=\mathbf{Id}\\
\mathbf{S}(x,q1)\bullet\mathbf{S}(x,q2)&=\mathbf{S}(x,q1+q2)\\
\mathbf{S}(dx,q1)&=\mathbf{Id}+\mathbf{D}(x)q1\\
(\mathbf{S}(x,q1)\bullet\mathbf{S}(y,q2))^T&=\mathbf{S}^T(y,q2)\bullet\mathbf{S}^T(x,q1)\\
\mathbf{D}(x)&=\tilde{\mathbf{e}}(x)\\
(\mathbf{S}(x,q1)\bullet\mathbf{e}(y))^{\sim}&=\mathbf{S}(x,q1)\bullet\tilde{\mathbf{e}}(y)\bullet\mathbf{S}^T(x,q1)
\end{aligned} \tag{37}$$

The three simplifications (i), (ii) and (iii) are implemented via the first three relations in (37). In this context 'dx' denotes a rotation with a small angle. Special emphasis is placed on the last equation in (37) because it is necessary for the efficient application of the other rules (see (33)). In addition to these relations and rules for the treatment of MBS problems it will be necessary to introduce hyper-matrices (38) and normal forms (39). The normalization of terms is important, e. g. for the recognition of equalities.

$$\mathbf{J}_i=[\mathbf{S}^T(x,q1)\bullet\mathbf{e}(y)\,|\,\mathbf{S}^T(y,q2)\bullet\mathbf{S}^T(x,q1)\bullet\mathbf{e}(z)\,|\,\ldots] \tag{38}$$

$$\begin{aligned}
\mathbf{e}(y)\bullet\mathbf{S}(z,q1)\bullet\mathbf{e}(y)\quad&\Rightarrow\quad\mathbf{e}(x)\bullet\mathbf{S}(z,q1)\bullet\mathbf{e}(x)\\
\mathbf{e}(z)\bullet\mathbf{S}(y,q1)\bullet\mathbf{e}(z)\quad&\Rightarrow\quad\mathbf{e}(x)\bullet\mathbf{S}(y,q1)\bullet\mathbf{e}(x)
\end{aligned} \tag{39}$$

Products of sums of rotation matrices cannot be expanded with the standard function of MAPLE ('expand(.)'), so it was necessary to implement a new 'expand'-procedure for these terms. Within this new procedure which is based on list-calculations small values (see (iv)) can be recognized early and eliminated. In this way the terms remain small from the beginning of the calculations.

In order to support further calculations (e. g. decoupling) and to reduce the evaluation effort all constant quantities (combination of physical parameters link mass, length, etc.) are substituted [14]; i. e. the mass-matrix depends only on these abbreviations and on the generalized coordinates. By means of a subsequent use of parenthesis the term can be shortened by about 30%. The ordering of the parenthesis is chosen according to the number of appearances.

2.4 Model of an Elastic Six-Axis Robot

For the system *tempo* shown in Fig. 2 a rigid body model with six degrees of freedom (dofs) with relative coordinates in the joints was derived in the first place. This model was successively modified and specialized. Due to the fact that a six-axis model with gear elasticities (analogous to the model given by Spong [16]) yields unsatisfactory correspondences with the frequency response measurements additional dofs for the description of the bearing in the links (e. g. L1 - L2) are used: occasionally two dofs for the first three axes and one dofs for axes 4 and 5. After the detailed measurement of the elasticities of the links and bearings (for the horizontal direction see Fig. 2) with the help of a high-precision and contact-free laser-sensor (1-2μm) the link segment marked in grey (not as usual from joint to joint) are modelled as an elastic beam. A simple parabolic or cubic shape function for the deformations $v_{2/3}$ and $w_{2/3}$ seems to be sufficient.

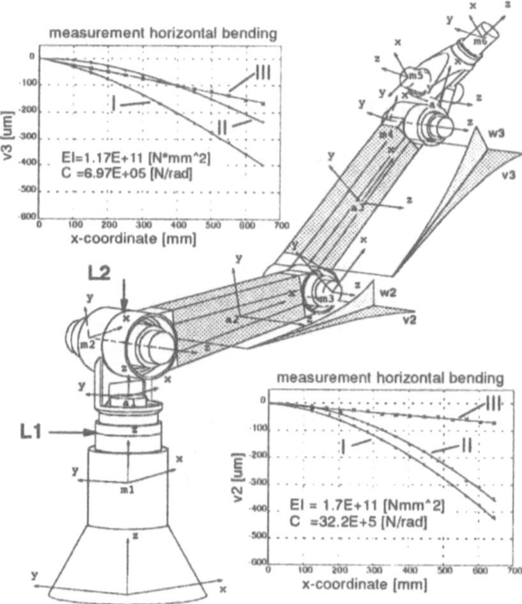

Figure 2 . Static deflections

In the end the entire system has 24 mechanical dofs. They are ordered in 12 coordinates for the description of motor and link-motion (q_1 - q_{12}) and 12 small coordinates for the description of the elasticities of bearings and links (q_{13} - q_{24}). The diagrams in Fig. 2 represent the static deformation of the system as a consequence of a load of 213 N applied to joint 4. Line III shows the portion of deflection of the bearings and line II the portion of the links respectively. The entire deformation is given by line I. The approximated stiffnesses of the bearings and the beams are specified.

3. Identification

The identification procedure has two principle aims: the first is the determination of the dynamical parameters (mass, inertia, stiffness, etc.) of a given model and the second is a successive adaption of the chosen model structure which is not ad hoc fixed for elastic systems.

On the identification of elastic NL manipulator-models in the frequency domain there are very few studies. One approach using experimental modal analysis in one position is shown by Behi and Tesar [1] for a planar elastic three-axis model. [5] considers a three-axis elastic model including linear gear and link elasticities for one position. In the majority of works concerning identification of robots the parameters of simplified rigid models and friction are determined in the time domain, e. g. [15].

3.1 Static Identification of Gears

In order to identify an elastic NL manipulator in the frequency domain it is necessary to determine the NL gear elasticities beforehand, as these elasticities will later be used for the linearization (under gravity load). Fig. 3 presents the measured torque-torsion characteristics and Fig. 4 the corresponding torsional stiffness of the first and second axis. It is important to point out that the entire load-range of the gears has to be measured, due to the fact that it is not possible extract this information from the data given by the Harmonic-Drive company (dotted line, Fig. 4).

The complete knowledge of the stiffnesses of all axes is important for the controller design because they are often used for the feedforward gains or in feedback decoupling schemes where they have a direct influence on the accuracy of the control.

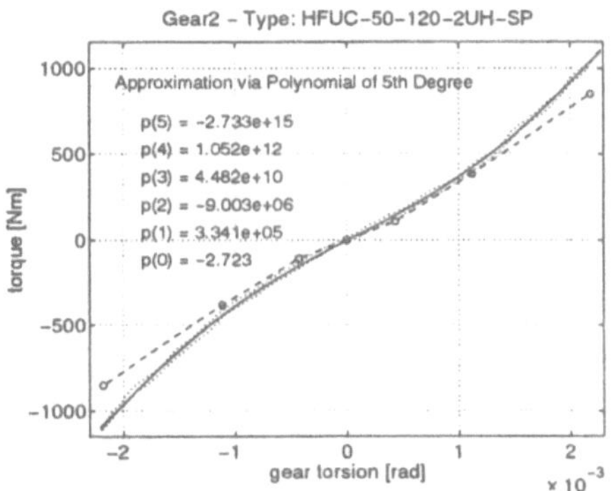

Figure 3 . Deformation of gear 1/2

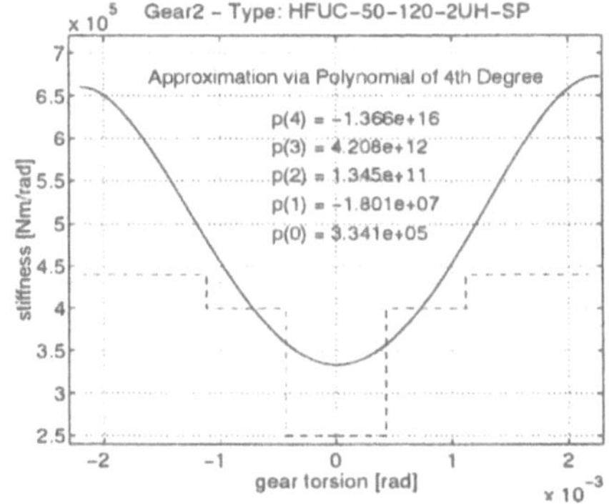

Figure 4 . NL stiffness of gear 1/2

3.2 Frequency Domain Identification

The identification of the other constant stiffnesses, masses and damping coefficients of the NL model is performed with the help of frequency response measurements, which are obtained in different positions. Measurement devices are available for the motor- and loadside positions of each joint and additionally an accelerometer is fixed to the endeffector (EE). A comparison of the calculated and measured frequency responses yields that the EE accelerations resp. the corresponding velocities are very well suited for the optimization.

For the structuring of the identification process the measurements are divided into three groups:

a) torsion (Fig. 5)

b) horizontal (e. g. Fig. 6 - 7)

c) vertical (e. g. Fig. 8 - 9)

This partitioning aims, among others, a reduction of the parameter space in order to optimize only the sensitive parameters of this direction in each group. By the simultaneous consideration of frequency responses in different positions it is possible to treat the varying NL characteristics of the system. In the group "torsion" two frequency responses with different transfer paths are adapted. The groups "horizontal" and "vertical" each collect five frequency responses in different positions and with different transfer paths which are simultaneously optimized. The adaptation of the frequency responses measured to those calculated by varying the physical parameters of the model are done up to a frequency of about 150 Hz depending on the quality of the measurements (criterion see [14]).

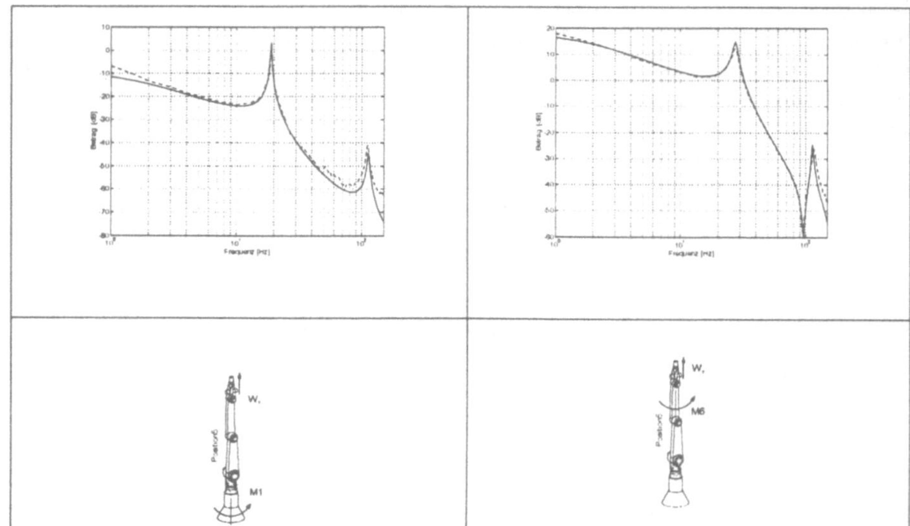

Figure 5 . Torsional vibrations

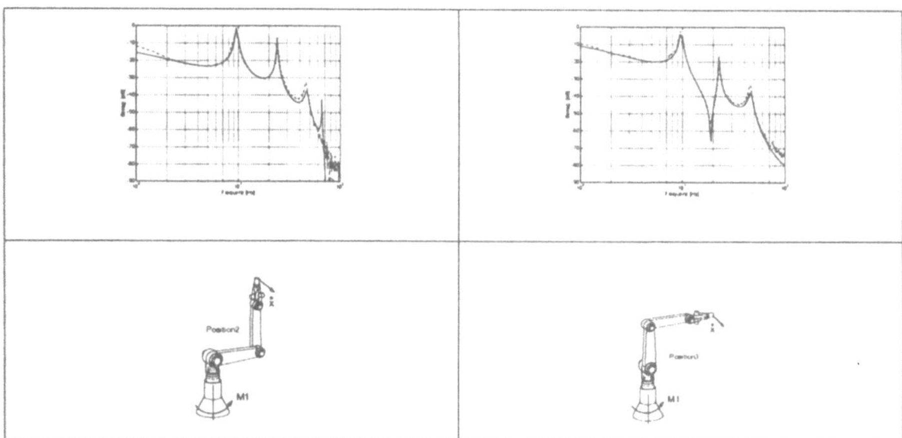

Figure 6 . Horizontal vibrations I

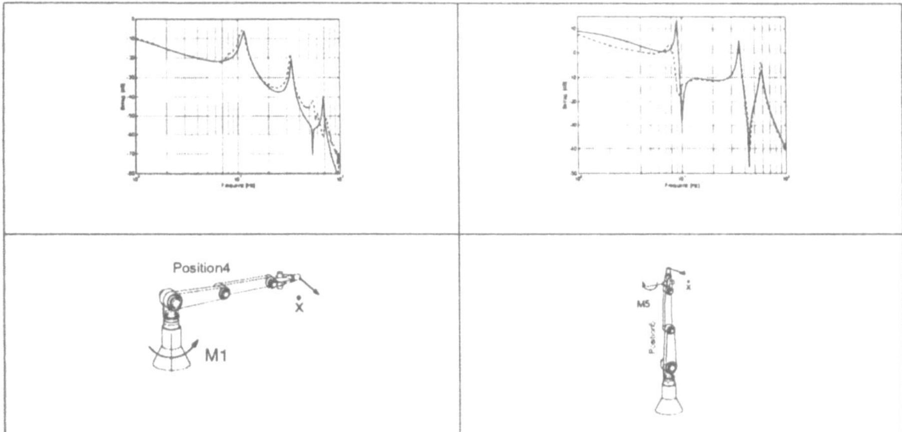

Figure 7 . Horizontal vibrations II

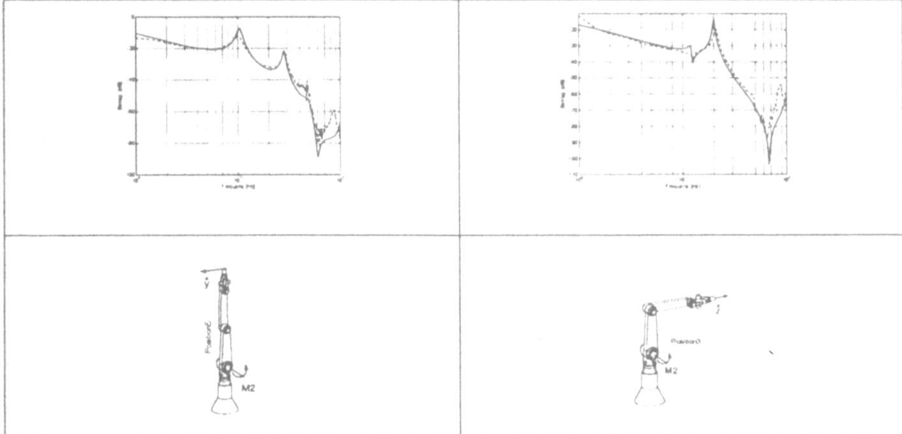

Figure 8 . Vertical vibrations I

The measurements are drawn with dotted lines and the calculations as solid lines. During the measurements each axis is excitated with band-limited white noise and for the other axes the brakes are set. For the model, this state can be achieved by setting very high motor-dampings for the fixed axes. In order to eliminate the influence of Coulomb friction (avoiding a change of the sign of velocity) during the measurements the joint in question has to be moved at constant velocity.

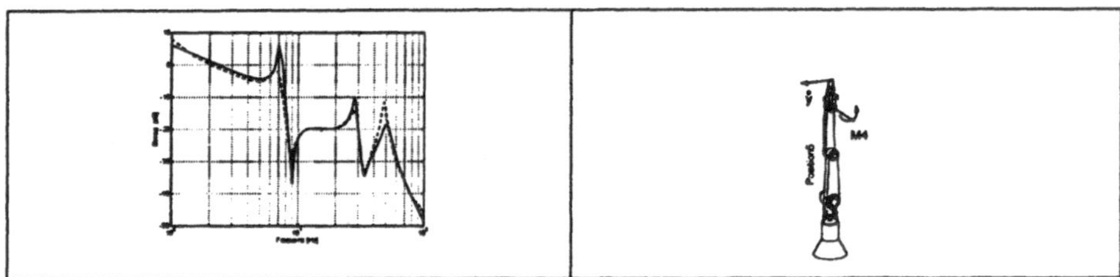

Figure 9 . Vertical vibrations II

Stabilization of this movement is achieved by a very small position-feedback. Furthermore, it is necessary to calculate the EE rotational and translational velocities (40). The corresponding jacobians for the velocities represented in the EE frame are symbolically calculated:

$$[{}^{EE}\mathbf{v}_{EE} \quad {}^{EE}\omega_{EE}]^T = \mathbf{J}_{EE}(\mathbf{q})\dot{\mathbf{q}} \tag{40}$$

Special emphasis is placed on the high correspondences up to 100 Hz. In all about 56 parameters has to be determined for this model (about 10 linear stiffnesses).

4. References

[1] **Behi, F., and Tesar, D.:** Parametric Identification for Industrial Manipulators using Experimental Modal Analysis, IEEE Transactions on Robotics and Automation, Vol. 7, No. 5, October 1991, pp. 642-652, 1991.

[2] **Capolsini, P.:** The Use of MAPLE for Multibody Systems Modeling and Simulation, Mathematical Computation with Maple V: Ideas and Applications, Tom Lee (ed.), Birkhäuser Boston, pp. 109-117, 1993.

[3] **Cetinkunt, S., and Book, W. J.:** Symbolic Modelling and Dynamic Simulation of Robotic Manipulators with Compliant Links and Joints, Robotics & Computer-Integrated Manufacturing, vol. 5, no. 4, pp. 301-310, 1989.

[4] **Char, B. W., Geddes, K. O. et al:** MAPLE, Reference Manual, Fifth Edition, University of Waterloo, Canada.

[5] **Henrichfreise, H.:** Aktive Schwingungsdämpfung an einem elastischen Knickarmroboter, Fortschritte der Robotik 1, Vieweg Verlag, Braunschweig/Wiesbaden 1989.

[6] **Hollerbach, J. M.:** A Recursive Lagrangian Formulation of Manipulator Dynamics and a Comparative Study of Dynamics Formulation Complexity, IEEE Transactions on Systems, Man, and Cybernetics, vol. SMC-10, no. 11, pp. 730-736, Nov. 1980.

[7] **Kalender, Th.:** Statistische Modellierung von Präzisionsgetrieben in elektromechanischen Antrieben, Fortschritt-Berichte, Reihe 1, Nr. 236, VDI Verlag, Düsseldorf, 1994.

[8] **Kircanski N., Goldenberg, A. A., and Jia, S.:** An Experimental Study of Nonlinear Stiffness, Hysteresis, and Friction Effects in Robot Joints with Harmonic Drives and Torque Sensors, Preprints of the ISER '93, Kyoto, Japan, Oct. 28-30, pp. 147-154, 1993.

[9] **Leu, M. C., and Hemati, N.:** Automated Symbolic Derivation of Dynamic Equations of Motion for Robotic Manipulators, Transactions of the ASME, Journal of Dynamic Systems, Measurement, and Control, vol. 108, pp. 172-179, Sept. 1986.

[10] **Meirovitch, L.:** Analytical Methods in Vibrations, Mcmillan Publishing Co., Inc., New York 1967.

[11] **Moritz, W., Neumann, R., and Schütte, H.:** Control of Elastic Robots Using Mechatronic Tools, Harmonic Drive International Symposium, Hotaka, Nagano, Japan, May 23-24, 1991.

[12] **Nethery, John F., and Spong, Mark W.:** Robotica: A Mathematica Package for Robot Analysis, IEEE Robotics & Automation Magazine, pp. 13-20, March 1994.

[13] **Schiehlen, W.:** Technische Dynamik, Teubner Verlag, Stuttgart 1986.

[14] **Schütte, H., Moritz, W., Neumann, R., Wittler, G.:** Practical Realization of Mechatronics in Robotics. Preprints of the ISER'93, Kyoto, Japan, Oct. 28-30, pp. 138-146, 1993.

[15] **Seeger, G., Leonhard, W.:** Estimation of rigid body models for a six-axis manipulator with geared electric drives, Proc. IEEE Int. Conf. Robotics Automation, Scottsdale, AZ, May 1989.

[16] **Spong, M. W., and Vidyasagar, M.:** Robot Dynamics and Control, John Wiley & Sons, New York, Chichester, Brisbane, Toronto, Singapore 1989.

[17] **Truckenbrodt, A.:** Bewegungsverhalten und Regelung hybrider Mehrkörpersysteme mit Anwendung auf Industrieroboter, Fortschritt-Berichte, Reihe 8, Nr. 33, VDI Verlag, Düsseldorf 1980.

[18] **Türk, S.:** Zur Modellierung der Dynamik von Robotern mit rotatorischen Gelenken, Fortschritt-Berichte, Reihe 8, Nr. 211, VDI Verlag, Düsseldorf, 1990.

[19] **Wolz, U., and Wittenburg, J.:** MESA-VERDE: A Program for the Symbolic Generation of Equations for Multibody Systems, Z. angew. Math. Mech., ZAMM 66 (5), T399-T400, 1986.

Chapter 7

Autonomy via Learning

One of the difficult problems in the utilization of robots has been the required description to the robot of *how* to perform a desired task. While there have been significant advances in developing formalisms to encode strategies, the complexity and diversity of real world tasks suggest that the development of systems which can *learn* the details of how to actually perform a task would be of great value. Several examples of this higher level approach to enabling robot autonomy are presented in this chapter.

Koeppe and Hirzinger employ neural networks to enable supervised learning of compliant tasks. The system observes a human performing the task once in a virtual environment and effects skill transfer to the network, thus enabling later autonomous performance of the task.

Watanabe, Yoshii, Masutani, and Miyazaki develop a vision-based motion controller for playing Japanese badminton, to which they add learning ability. They describe employing passive learning to update and smooth action commands based on a Gaussian kernel. Their experimental results show significant increase in performance with learning.

Li, Yang, and Asada describe an approach to learning control in which progressively higher frequency excitation to the system is used to enable gains to stably converge. The dynamics of the system are sufficiently complex that traditional methods for adaptively tuning gains via single frequency excitation fail. The authors' progressive approach is verified on a high-speed direct-drive robot and shown to enable high-precision chip placement.

Popović, Gorinevsky, and Goldenberg present a new fuzzy logic controller capable of performing precise positioning of mechanisms with stick-slip friction. Using this controller, they experimentally demonstrate accurate positioning of a robot to encoder resolution despite the presence of friction.

Learning Compliant Motions by Task-Demonstration in Virtual Environments

R. Koeppe G. Hirzinger

Deutsche Forschungsanstalt für Luft- und Raumfahrt (DLR)
Institute of Robotics and System Dynamics
Postfach 1116, 82230 Wessling, Germany
Ralf.Koeppe@dlr.de

Abstract We are proposing a supervised learning approach in robot force control which enables robot programming by demonstration through an operator. The learning element of the controller is based on a neural network prestructured to represent knowledge in terms of rules. Tasks are demonstrated in a virtual environment. Visual feedback using sensor ball devices and graphic display of forces as well as visual and proprioceptive feedback using haptic interfaces are being considered. Experimental results are shown for the "Put Block in a Corner of a Box Problem".

1. Introduction

Sensors in outer feedback loops are known to be the key to performing complex assembly tasks. Due to the fact that the teaching and programming of sensor controlled tasks is a time consuming tedious process, only a few advanced robots on present day industrial shop floors are equipped and operate with external sensory feedback. Therefore the need for easy and flexible programming techniques incorporating sensor information is obvious. A new assembly task should be programmed within 10 minutes [1].

To perform assembly operations with robots the planned motion of the end-effector has to be corrected according to the sensor data of the process. In the case of force/torque sensor data this problem is known as the Compliant Motion Problem. *Conventional approaches* for compliant motion control were proposed (e.g.[2]) and used to perform generic assembly tasks in space [3]. *Programming by demonstration* of the task by a human operator was used in addition to reduce the difficulties in explicitly writing programs for the force control strategy [4]. In [5] natural constraints, as well as desired motion and forces were calculated from recorded sensorimotion signals. A program representing a conventional controller was then generated automatically. *Supervised learning approaches* using neural networks or linear approximaters [6] [7] [8] were proposed. The training data has to consist of the desired correspondance between a measured force and the correction of motion. In the above cited learning approaches, the data is either generated analytically, or automatically by a system.

If we want to merge the learning with the programming by demonstration approach, the correspondance problem has to be solved first, since the correction of motion due to a force perception is delayed by the operators dead time. We are proposing an ***approach to learning compliant motions by task-demonstration***, where this correspondance problem is solved inherently by the controller. Since humans describe a task in a rule based way, using the concept of fuzziness to describe forces and commands, we have designed a neural network, prestructured to represent a fuzzy controller. We will show, that by training the neural network sensorimotion patterns recorded during the demonstration of the task in a virtual environment, the operators strategy acquistion problem - the first step in skill transfer - can be solved. We investigated two types of feedback on the operator: visual feedback using a force display and a sensor ball command device or both visual and proprioceptive feedback using a master/slave (haptic) interface.

2. Learning Compliant Motion

2.1. Perception-action model of a human operator

A human operator performing a task executes an action $\boldsymbol{u}(t)$, due to a received perception $\boldsymbol{S}(t-\tau)$. The dead time τ, which the operator needs to process the sensor signals, depends on the type of feedback. A human operator relies on exteroceptive (visual and auditory) and/or proprioceptive (force and touch) feedback. We describe the human task in terms of a perception-action model:

$$\boldsymbol{u}(t) = \boldsymbol{f}[\boldsymbol{S}(t-\tau_{extero}), \boldsymbol{S}(t-\tau_{proprio})]. \tag{1}$$

We assume that the operator's dead time τ is dominated during the demonstration of the task by one type of feedback, an assumption which will be validated in the examples shown by our experiments.

The change of the sensor pattern is caused by a change of the pose $\boldsymbol{x}$ of the human operator :

$$\begin{aligned} \Delta\boldsymbol{S}(t) &= \boldsymbol{S}(t) - \boldsymbol{S}(t-\tau) & (2)\\ &= \boldsymbol{h}[\boldsymbol{x}(t)] - \boldsymbol{h}[\boldsymbol{x}(t-\tau)] & (3)\\ &= \boldsymbol{h}^{\star}[\boldsymbol{x}(t), \boldsymbol{x}(t) - \boldsymbol{x}(t-\tau)]. & (4) \end{aligned}$$

Introducing the velocity by

$$\dot{\boldsymbol{x}}(t)\cdot\tau = \boldsymbol{x}(t) - \boldsymbol{x}(t-\tau) \tag{5}$$

yields

$$\boldsymbol{S}(t-\tau) \approx \boldsymbol{S}(t) - \boldsymbol{h}^{\star}[\boldsymbol{x}(t), \dot{\boldsymbol{x}}(t)\cdot\tau]. \tag{6}$$

Equation (5) implies that the velocity $\dot{\boldsymbol{x}}$ should be constant within the operators feedback dead time τ. With equation (6) and the inverse sensor function

$$\boldsymbol{x}(t) = \boldsymbol{h}^{-1}[\boldsymbol{S}(t)] \tag{7}$$

we rewrite the human perception action model

$$\boldsymbol{u}(t) = \boldsymbol{f}^{\star}[\boldsymbol{S}(t), \dot{\boldsymbol{x}}(t), \tau]. \tag{8}$$

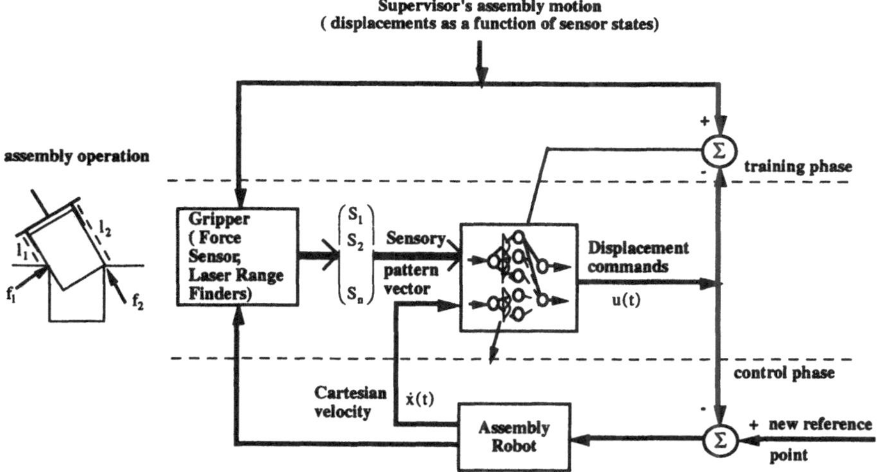

Figure 1. Learning controller with rule-based neural network

With τ a constant parameter (or a variable depending on $\boldsymbol{S}(t)$ and/or $\dot{\boldsymbol{x}}(t)$) the human perception-action model can be approximated by the function

$$\boldsymbol{u}(t) = \boldsymbol{g}[\boldsymbol{S}(t), \dot{\boldsymbol{x}}(t)], \tag{9}$$

which can be represented by a feedforward type neural network. We now want to discuss the assumptions made above:

Constant velocity within the operator's dead time:
Equation (5) requires that the time constant of the feedback loop is at least equal to the dead time of the operator. This is always true in case of force feedback, since the dead time $\tau_{proprioceptive} \approx 60ms$ is approximately the same as the time constant of the muscle dynamics [9]. In case of visual feedback ($\tau_{exteroceptive} \approx 250ms$) the sensor as well as the robot time constant contribute to meet the requirements of a slower overall time constant in the feedback path during demonstration. If this requirement can not be satisfied, the perception-action model has to be enhanced, which will lead to the use of feedback type neural networks, which are capable of representing a dynamic system.

Existance of the inverse sensor function:
Equation (7) implies that there exists an inverse sensor function $\boldsymbol{h}^{-1}$, which means, that no out-of-range values occur within the demonstration process.

Requirements on the dead time:
Equation (9) was derived by the assumption that the feedback dead time τ of the human operator is a function of the input variables $\boldsymbol{S}(t)$ and $\dot{\boldsymbol{x}}(t)$ of the approximation model. This restriction doesn't seem to be severe. In our experiments, the dead time of a human operator appeared to be constant. Research within this subject leads to the interdisciplinary study of the human perception system.

2.2. Learning control structure

The result obtained above, now allows the training of a neural network with input data samples $(\boldsymbol{S}(t), \dot{\boldsymbol{x}}(t))$ and output data samples $(\boldsymbol{u}(t))$ recorded by *demonstrating the task but once.* Note that, no reassembling of the data samples is necessary, taking into account which sensor pattern $\boldsymbol{S}$ perceived at time $(t - \tau)$ caused the operator to command an action $\boldsymbol{u}$ at a later time t. Nor is it necessary to determine the dead time τ of the human feedback. This is done by the network implicitly.

Therefore we propose the learning control structure shown in Figure 1. A sensor data vector $\boldsymbol{S}(t)$ and the actual cartesian velocity $\dot{\boldsymbol{x}}(t)$ are the input to the learning block which is a rule based network. The output of the learning block is the command to the robot, the cartesian increment vector $\boldsymbol{u}(t)$. *The approach allows sensor fusion*, since the sensor data can be composed of force/torque data, extracted features by a vision system and/or distance measures obtained from laser range finders.

2.3. Rule-based neural networks for function approximation

Learning the mapping of sensor and robot state data to motion commands leads to the formulation of an approximation problem. Various neural networks e.g. Multi-Layer Perceptrons (MLP) or Radial-Basis Function Networks are capable of learning nonlinear mappings and generalizing over a set of previously unseen examples very accurately. Rule-based neural networks implement a fuzzy logic system. After the training process the neural network can be interpreted and verified.

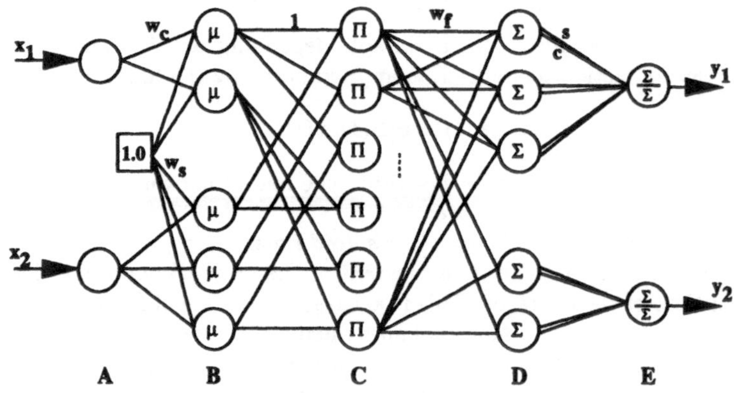

Figure 2. Rule-based neural network

Our rule based neural network (Figure 2) [10] uses a multi-layer network structure with transfer functions and links representing a fuzzy logic control system. Each input neuron of layer A represents a fuzzy variable, the neurons in layer B their fuzzy sets. The fuzzy set membership functions are implemented as Gaussian functions. The bias weights w_s determine the width of the Gaussian, the weights w_c to layer B its center. The neurons of layer C implement the rules and the neurons of layer D and E the membership functions of the output fuzzy sets and fuzzy variables. The weights w_f determine the truth value for each consequence and the weights s and c connecting layer D and E determine the shape and location of the Gaussian function of the output fuzzy set. The fuzzy sets are initialized according to the known data ranges of the fuzzy variables, therefore *including a priori knowledge* in the sense of what is a big, middle or small magnitude of the sensor, velocity or command value. Weights representing the truth values of the rules w_f are initialized by the value 0.5. The network is trained by a backpropagation process minimizing the squared error of all output neurons over all data sets. The width and centers of the Gaussian fuzzy sets are adjusted as well as the rule weights w_f. By analyzing the network weights it is possible to extract rules characterizing the human operator feedback.

3. Teaching Compliant Motion

A key technology in telerobotics [3] and sensor-offline-programming applications [8] is the use of high performance graphic computers for world and sensor modelling and task programming. The proposed approach can be seen as a new method in the field of skill transfer, by demonstrating a task in a virtual environment and transferring the acquired skill on a real robot system.

Human interface systems provide only subsets of the rich variety of possible feedbacks, e.g. a 2-dimensional visual display of a 3-dimensional world, or rely heavily on sensory substitution, e.g. the substitution of proprioceptive feedback by visual feedback (display of forces and torques). Proprioceptive feedback can be achieved by computing the force/torque vector of the contact scenario in the virtual environment and reflecting it to the operator using haptic interfaces or exoskeleton arms in combination with dextrous master hands. In both cases the characteristic of human task execution is changed drastically because the operator himself is part of the closed loop system.

Since the task demonstration process is greatly dependent on the man-machine interface in the command, as well as in the feedback path, we will describe the characteristics of both setups in further detail. The system sampling rate required for proper demonstration of the task has to be considered. Means of subsampling the recorded sensorimotion data have to be found to reduce the training time of the neural network, without loosing human skill information. One of the key measures for this is the bandwidth with which the human finger can apply force and motion commands comfortably. Its range is 5-10 Hz [11].

3.1. Teaching with visual feedback

Input commands to the virtual robot are generated by the sensor ball device, Space Mouse (Figure 3 (a)), reading forces and torques with a sampling rate of 16 Hz (current system setup). Using a stiffness matrix a pose change vector $\boldsymbol{u}$ is applied to the virtual robot. Forces and torques are calculated and displayed as bar diagrams within the graphic display of the scenario, which is updated with a rate of 30-50 frames/second. After a visual dead time of approximately 250 ms the user reacts to the force display.

Considering the sampling rate of the sensor ball as well as the bandwith of the human finger the rate of the control system is chosen to 20 Hz. Data cannot be subsampled since information would be lost.

3.2. Teaching with force feedback

Force feedback on the operator requires a haptic interface. We use the PHANToM, a desktop commanding and force feedback device, designed by Massie and Salisbury [12] (Figure 3(b)). Unlike the sensor ball, where increments are the command output, the PHANToM's position is read by the encoders and transformed into the virtual world. The virtual robot tracks the position of the tool center point of the haptic interface $\boldsymbol{x}_m$ by the following control law:

$$\boldsymbol{u} = \boldsymbol{k}_s \cdot (\boldsymbol{x}_m - \boldsymbol{x}_s). \tag{10}$$

Simulated forces $\boldsymbol{f}_s$ are transformed from the virtual world to the coordinate

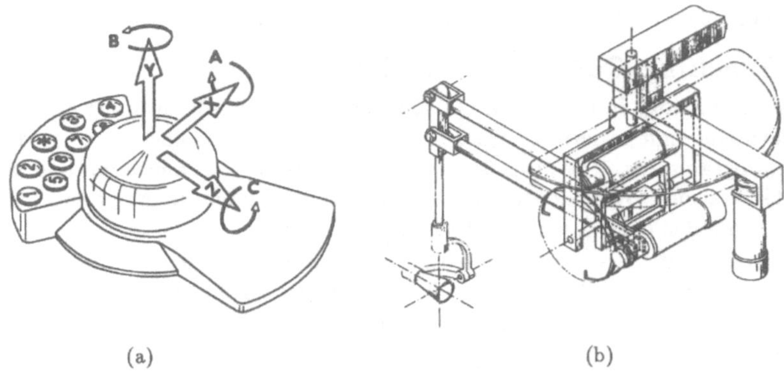

Figure 3. Human interface devices: (a) Sensor Ball (Space Mouse), (b) Haptic Interface (PHANToM).

system of the haptic interface and are tracked by the control law,

$$\boldsymbol{\tau} = \boldsymbol{J}^T(\boldsymbol{q})[\boldsymbol{k}_m(\boldsymbol{x}_s - \boldsymbol{x}_m) + \boldsymbol{f}_s] \tag{11}$$

where $\boldsymbol{x}_s$ is the position of the slave system and $\boldsymbol{\tau}$ the joint torque vector applied to the haptic interface. Motion commands and force feedback are therefore coupled by bilateral control [13]. The control loop runs at a sampling rate of 2 kHz and the force/torque contact simulation at approximately 1 kHz, enabling stable force reflection and a crisp feeling of contact. After a proprioceptive dead time of 60 ms the user reacts to the force.

Since the rate necessary for force feedback application is much higher than the rate carrying information about the human operator task strategy, the recorded data is filtered and subsampled to 20 Hz.

4. Experiments

The task in the experiment was to align a block in the corner of the box, by moving the tool center point of the robot first in $+y$, then in $-z$ and finally in $-x$ direction (Figure 4 (a)). The operator was asked to move into a new direction as soon as he/she perceived a medium force. A medium force was defined in a way, that the operator feels a firm contact by not having to push hard to keep the contact force. The bar magnitude in the visual display (Figure 4 (b)) was half of the interval. The robot and the surface in the virtual environment was stiff. Compliance was introduced by the simulated force sensor. Changes in orientations were not considered in this example.

Figure 5 shows the Task-Demonstration by the human operator and the Task-Execution by the learning controller for (a) the visual and (b) the force feedback case. Task-Demonstration in the visual feedback case took $16s$, in the force feedback case $2s$. The demonstration of compliant motions using only visual feedback of forces, required a skilled operator. In contrast to this, the task accomplishment with force feedback was intuitive. This proves clearly that the operator was dominated by proprioceptive, rather than by visual feedback, when both type of feedback were available.

The rule-based network was trained with the data of one demonstration run for

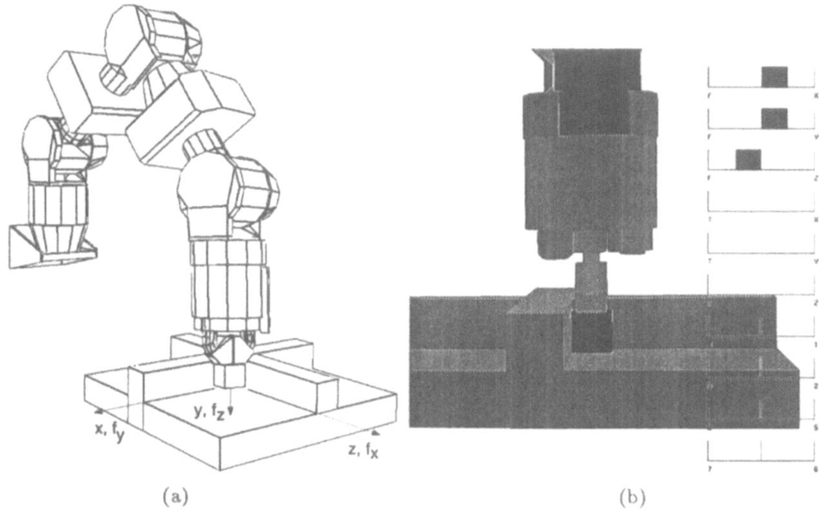

Figure 4. Virtual Robot with world and sensor coordinate system (a) with Visual display of forces by bar diagrams (in sensor coordinates) (b).

each feedback type. The network was structured as follows: 6 input neurons, the fuzzy variables $(F_x, F_y, F_z, \dot{x}, \dot{y}, \dot{z})$ with each data range divided into 3 fuzzy sets (forces: small, medium, big and velocities: negative, zero, positive); resulting in 729 rules, i.e. neurons in layer C; and 3 output neurons, fuzzy varaiables (u_x, u_y, u_z) with each data range divided into 3 fuzzy sets (displacement commands: negative, zero, positive). After 100 training steps the task was repeated with the trained network. The learned task was executed successfully by the robot in the simulation environment. The overshoots in force and command signals were reduced due to the generalisation capability of the network, counteracting already small increasing forces. This was not the case during the demonstration by the operator. The control law, determining the compliant motion, was acquired by the above described learning structure.

5. Conclusion

A control structure for learning compliant motion using sensory information has been presented. No explicit control gains or selection matrices had to be specified. The task was demonstrated only once. No data preprocessing besides compression was done.

The next step will be the transfer of the skill from a virtual environment to the real world. Transfer concepts to cope with sensor noise in the real world and with the different dynamic properties of the demonstration and execution system have to be found. Investigations in solving more complex tasks need to be made.

Acknowledgements: We would like to thank B.Brunner, J.Kiener and G.M.Prisco visiting doctoral student from Scuola Superiore S.Anna Pisa for their contributions and support.

References

[1] Brian Carlisle. New technology needs, Adept Technology, Inc. In *Sixth International*

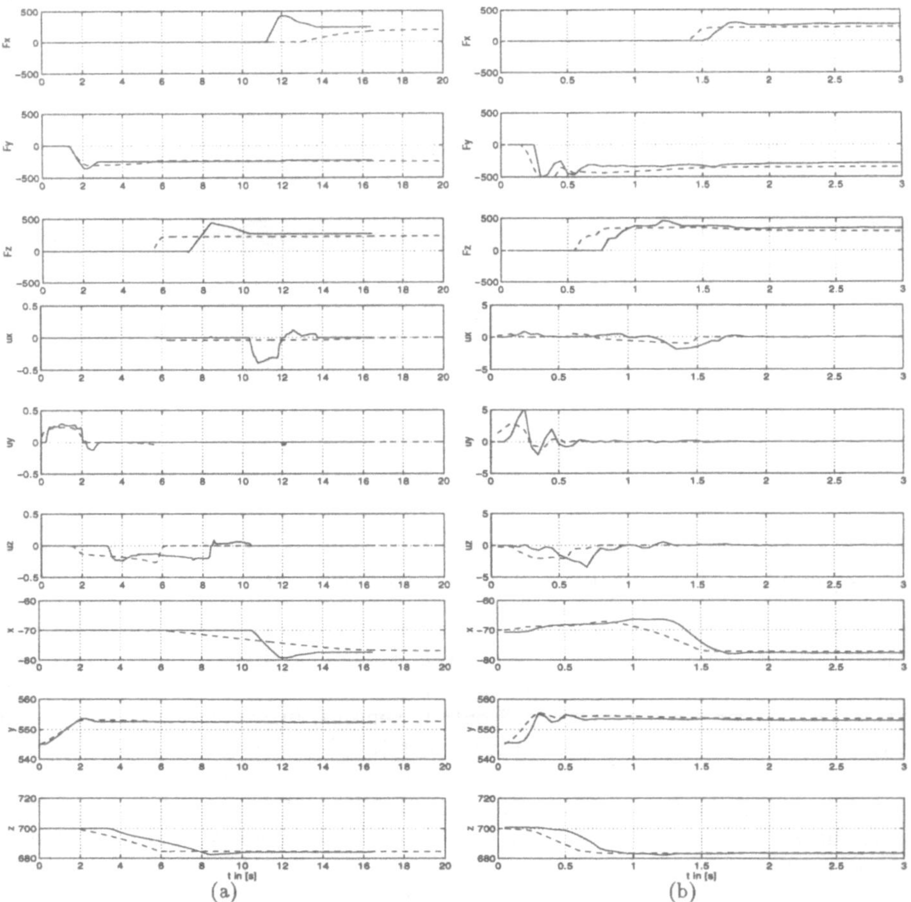

Figure 5. Experiment: Put Block in a Corner of a Box Problem. (a) Teaching with visual feedback, (b) Teaching with force feedback, Task-Demonstration by human operator are represented by solid lines, Task-Execution by learning controller by dashed lines. (Signals w.r.t. world coordinate system)

Symposium on Robotics Research, October 1993.

[2] Daniel E. Whitney. Historical perspective and state of the art in robot force control. *The International Journal of Robotics Research*, 6(1):3–14, 1987.

[3] Gerd Hirzinger, Bernhard Brunner, Johannes Dietrich, and Johann Heindl. Sensor-based space robotics - rotex and its telerobotic features. *IEEE Trans. on Robotics and Automation*, 9(5):649–663, October 1993.

[4] G. Hirzinger and J.Heindl. Sensor programming - a new way for teaching robot parts and forces/torques simultaneously. In *Proc. of 3rd Int. Conf. on Robot Vision and Sensory Controls*, pages 549–558, 1983.

[5] Haruhiko Asada and Haruo Izumi. Direct teaching and automatic program generation for the hybrid control of robot manipulators. In *Proc. of the IEEE Int. Conf. on Robotics and Automation*, pages 1401–1406, April 1987.

[6] Haruhiko Asada. Teaching and learning of compliance using neural nets: Representation and generation of nonlinear compliance. In *Proc. of the IEEE Int. Conf. on Robotics and Automation*, pages 1237–1244, May 1990.

[7] Guo-Quing Wei, Gerhard Hirzinger, and Bernhard Brunner. Sensorimotion coordination and sensor fusion by neural networks. In *Proc. IEEE Int. Conf. on Neural Networks, San Francisco*, 1993.

[8] Bernhard Brunner, Klaus Arbter, and Gerd Hirzinger. Task directed programming of sensor based robots. In *IEEE/RSJ Int. Conf. on Intelligent Robots and Systems, München, Germany*, September 1994.

[9] Gerhard Hirzinger. Robot-teaching via force-torque-sensors. In *Proc. EMCSR'82, 6th European Meeting on Cybernetics and Systems Res., Wien*, 1982.

[10] Johannes Kiener. Neuronale Netze und Fuzzy Systeme in der Regelung. Diplomarbeit, DLR Oberpfaffenhofen IB 515-94-15, July 1994.

[11] Grigore Burdea and Philippe Coiffet. *Virtual Reality Technology*. John Wiley Sons, Inc., 1994.

[12] Thomas H. Massie and J. Kenneth Salisbury. The phantom haptic interface: A device for probing virtual objects. In *Proc. of the ASME Winter Annual Meeting, Symposium on Haptic Interfaces for Virtual Environment and Teleoperator Systems, Chicago*, November 1994.

[13] Yasuyoshi Yokokohji and Tsuneo Yoshikawa. Bilateral control of master-slave manipulators for ideal kinesthetic coupling. In *Proc. of the IEEE Int. Conf. on Robotics and Automation*, 1992.

Motion Control for A Hitting Task: A Learning Approach to Inverse Mapping

Hiroshi Watanabe
Yoshinori Yoshii
Yasuhiro Masutani
Fumio Miyazaki

Faculty of Engineering Science
Osaka University
Toyonaka, Osaka 560, Japan
miyazaki@me.es.osaka-u.ac.jp

Abstract We describe our approach to the robot's Hanetsuki task (Japanese badminton), that is, to return the incoming ball to the human opponent with a racket. A learning algorithm that consists of updating action commands and smoothing them based on the Gaussian kernel is proposed to compensate for the insufficiency in a model-based approach. Experimental results obtained by using the developed Hanetsuki robot are also shown.

1. Introduction

Robot arms are similar to the average human arm in physical capabilities. However, everyone recognizes that humans are much superior to robots in their ability to perform a variety of tasks. Task variation results from the environment in which the task is performed. Humans skillfully adjust task performance to suit varying environments. Present robots, however, have only a limited capability in this area. Improving this capability of robots is one of the most exciting topics in the field of robotics today [1][2][3][4].

In this paper, we concentrate on the task of hitting a moving object. This is a "dynamic manipulation" task as defined by Mason [3], and the moving object can be considered as a kind of changing environment. To perform this task, a robot must choose an appropriate action according to the motion of the incoming object. Koditschek et al achieved spatial two-object juggling, the ability to bat two freely falling balls into stable, periodic, vertical trajectories with a single three degree of freedom robot arm using a real-time stereo camera system [5]. The falling balls are considered as a kind of changing environment, though the motion of each ball is primarily provided by the robot itself. The nonlinear feedback law introduced for the spatial two-object juggling generates appropriate actions in this changing environment. Andersson constructed a sophisticated robot system that can play ping-pong against humans [6]. Since the motion of a ping-pong ball hit by a human is rich in variation in comparison with juggling, more intelligent robot control methods are required. Andersson developed an expert controller for the ping-pong task. The controller was structured to support numeric processing as well as symbol manipulation. The expert controller performed task planning, updated the plan, and improved its quality as the sensor data changed. However, controller algorithms were developed based on situations a human expects to happen. A model of the changing environment plays an important role in these approaches. Specifically, the robot's action commands are generated based on this model. This may present a problem if an unmodelled change occurs in a real environment. Since unmodelled factors are inevitable in the real world and especially affect dynamic manipulation, a learning ability that copes with the problem due to model uncertainty is quite important.

Schaal et al [7] proposed a learning approach to the juggling task known as "devil sticking." Using locally weighted regression(LWR), they first built models of the world then exploited algorithms from optimal control to design controllers. However, it is difficult to apply this approach to tasks in which humans are involved in the environment, like playing ping-pong against humans, because the model employed in devil sticking does not assume human factors. Moreover, the learning applied in devil sticking is "active learning" in which the learner has complete choice in the information received. In other words, it premises that a robot can actively generate the change of environment.

The task we deal with in this paper is "Hanetsuki," Japanese badminton. In the Hanetsuki task, the immobile robot has to return the incoming ball to the human opponent with a racket. Ping-pong differs from Hanetsuki in that the ball bounces on a table. To achieve this task, we first construct a vision-based motion controller using a model based approach like Andersson's, and then add learning ability to it. The learning algorithm proposed in the paper consists of updating action commands and smoothing them based on the Gaussian kernel. The algorithm employs "passive learning" in which the learner plays no role in obtaining information as opposed to active learning mentioned previously.

Section 2 describes the Hanetsuki task and specifies the role of the robot. Section 3 outlines the flow of data processing from sensing to action. Section 4 presents a learning method and experimental results of its implementation.

2. Hanetsuki Task

In this paper, we assign the robot to play Hanetsuki (Japanese badminton) against a human (see Figure 1). The robot Hanetsuki task is to return the incoming ball to the human. Specifically, the robot with five degrees of freedom must return the ball to a point virtually set up in the restricted area while considering the ball flight conditions. Figure 2 shows the flowchart of the robot task.

The robot has to perform a series of subtasks in a short time (less than 800 msec in our system). These subtasks includes measuring the ball motion with a pair of cameras (every 1/30 sec), predicting the ball trajectory, deciding the hitting point and racket velocity, generating the robot's motion for the return shot, and monitoring the resultant return shot. Algorithms associated with these subtasks are given in the following sections.

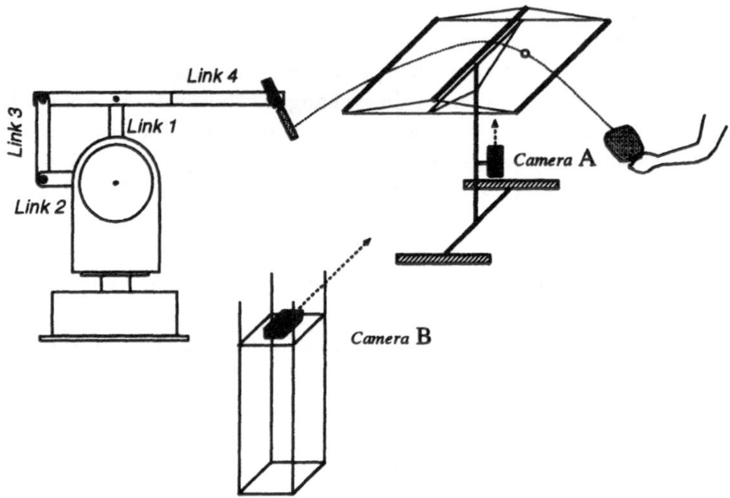

Figure 1. "Hanetsuki" task

3. Data Processing from Sensing to Action

3.1. Coordinate Transformation between Image and World Frames

Defining coordinate frames is the first step in generating the robot's actions. The coordinate origin of the world frame Σ_W is located at a point fixed on the robot's base. We also define the camera frames Σ_A and Σ_B such that each axis is parallel to the corresponding axis of the world frame, and image planes A and B corresponding to each camera as shown in Figure 3.

Let P be an arbitrary point whose position vector is given as $(x, y, z)^T$ with reference to Σ_W, and P_A and P_B be the projected points of P onto image planes A and B whose coordinates are $(x_A, y_B)^T$ and $(x_B, z_B)^T$. The coordinate transformation between camera and world frames is then represented as the following mapping:

$$\left\{ \begin{array}{l} (x_A, y_A)^T \\ (x_B, z_B)^T \end{array} \right. \Longrightarrow (x, y, z)^T.$$

Point P is located at the intersection of lines O_AP_A and O_BP_B. Strictly speaking, lines O_AP_A and O_BP_B are slightly twisted with respect to each other due to the quantization errors included in the detected coordinates of P_A and P_B. Considering the quantization errors, we regard the point P as the middle point of the line common normal to the lines O_AP_A and O_BP_B . The estimated world coordinates $\boldsymbol{p}^*$ of the point P are given by

$$\begin{aligned} \boldsymbol{p}^* &= \frac{1}{2}\left(\frac{-\|\boldsymbol{a}_B\|^2(\boldsymbol{b}_A - \boldsymbol{b}_B)\cdot\boldsymbol{a}_A + (\boldsymbol{a}_A\cdot\boldsymbol{a}_B)(\boldsymbol{b}_A - \boldsymbol{b}_B)\cdot\boldsymbol{a}_B}{\|\boldsymbol{a}_A\|^2\|\boldsymbol{a}_B\|^2 - (\boldsymbol{a}_A\cdot\boldsymbol{a}_B)^2}\boldsymbol{a}_A \right. \\ &\quad \left. + \frac{-(\boldsymbol{a}_A\cdot\boldsymbol{a}_B)(\boldsymbol{b}_A - \boldsymbol{b}_B)\cdot\boldsymbol{a}_A + \|\boldsymbol{a}_A\|^2(\boldsymbol{b}_A - \boldsymbol{b}_B)\cdot\boldsymbol{a}_B}{\|\boldsymbol{a}_A\|^2\|\boldsymbol{a}_B\|^2 - (\boldsymbol{a}_A\cdot\boldsymbol{a}_B)^2}\boldsymbol{a}_B + \boldsymbol{b}_A + \boldsymbol{b}_B \right) \end{aligned} \quad (1)$$

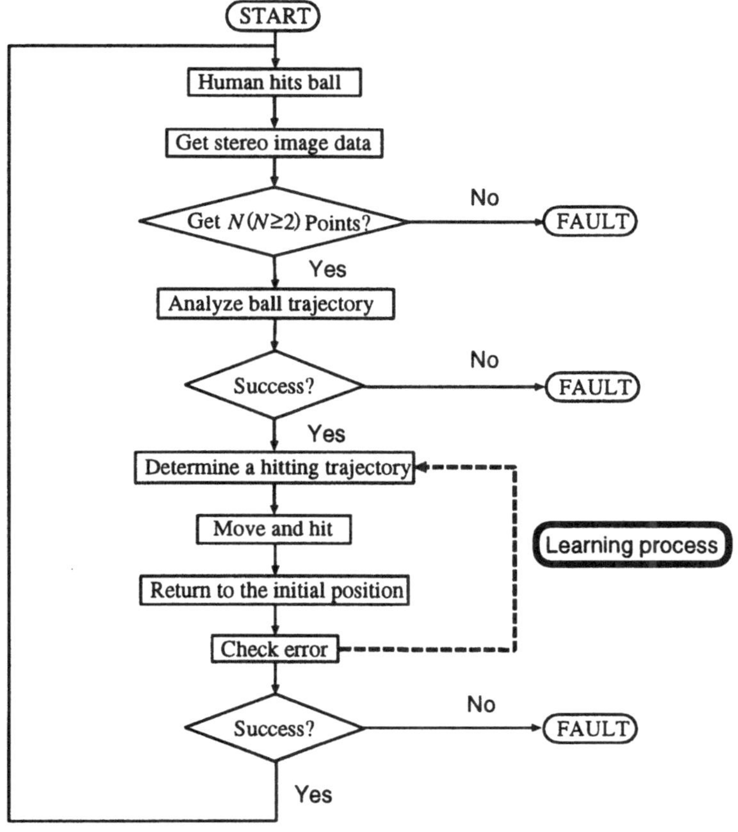

Figure 2. Flowchart of Robot's Task

where $\boldsymbol{a}_A = (x_A, y_A, f_A)^T$, $\boldsymbol{a}_B = (x_B, f_B, z_B)^T$, f_A and f_B are values of the focal length of each camera, $\boldsymbol{b}_A$ and $\boldsymbol{b}_B$ are position vectors of the origin O_A and O_B represented in terms of Σ_W. In the robot Hanetsuki task, this vector provides the position of the ball with reference to the world frame.

3.2. Motion Estimation of Ball

The next step in generating the robot's actions is to estimate the path of the moving ball. Let $\boldsymbol{p}_i$ be the position vector of the ball with reference to the world frame that is calculated by Eq.(1) with the i-th sampled image data. We estimate the path of the ball by using the sequence of calculated position vectors

$$\left\{ \begin{array}{l} \boldsymbol{p}_1, \boldsymbol{p}_2, \ldots, \boldsymbol{p}_N \\ t_1, t_2, \ldots, t_N \end{array} \right. \quad (N \geq 2) \tag{2}$$

and the equation of motion

$$\ddot{\boldsymbol{p}} = -\boldsymbol{g} - k\dot{\boldsymbol{p}} \tag{3}$$

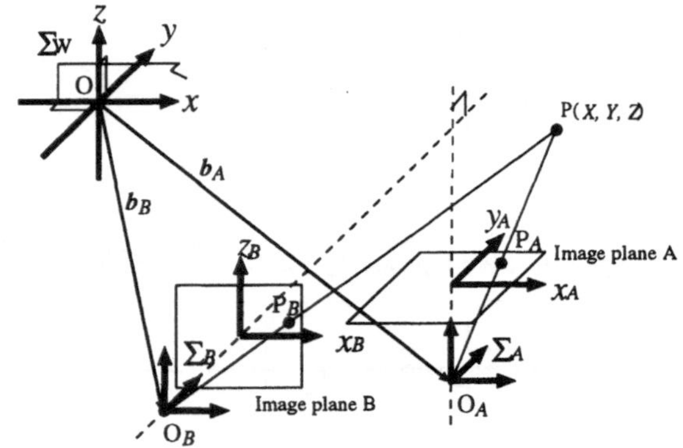

Figure 3. World and Image Frames

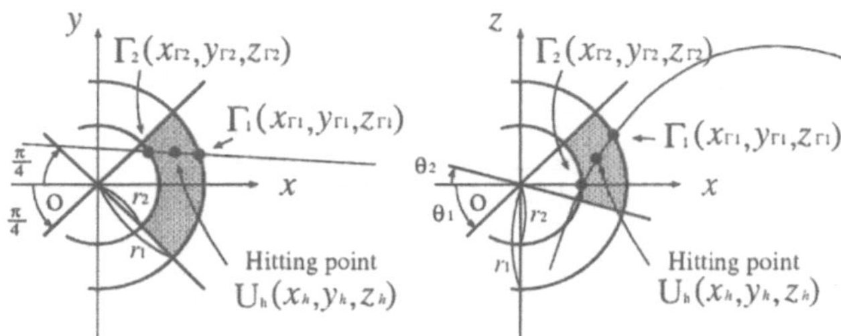

Figure 4. Robot's Permissible Workspace

where **g** is the acceleration vector of gravity and k is the coefficient of the air drag. If k is small, Eq.(3) can be approximated as

$$\boldsymbol{p} = -\frac{1}{2}(\boldsymbol{g} + k\dot{\boldsymbol{p}}_I)t^2 + \dot{\boldsymbol{p}}_I t + \boldsymbol{p}_I \tag{4}$$

where $\dot{\boldsymbol{p}}_I$ and $\boldsymbol{p}_I$ are the initial conditions of velocity and position vectors. The unknown vectors in Eq.(4), $\dot{\boldsymbol{p}}_I$ and $\boldsymbol{p}_I$, are estimated by the least square method with a sequence of position vectors given by Eq.(2).

3.3. Hitting Point and Time

Since the course of the ball hit by a human changes every trial, we first consider such problems as when and where the robot should hit the incoming ball in order to plan the whole hitting motion of the robot. To do so, we first define the robot's permissible workspace $\boldsymbol{V}$ as the shaded area in Figure 4. Of course, the permissible workspace $\boldsymbol{V}$ is included in the robot's physical workspace.

Let the intersection between the ballistic trajectory of the ball and the boundary of the permissible workspace $\boldsymbol{V}$ be $\Gamma_1(x_{\Gamma 1}, y_{\Gamma 1}, z_{\Gamma 1})^T$ $\Gamma_2(x_{\Gamma 2}, y_{\Gamma 2}, z_{\Gamma 2})^T$ (see Figure 4). We assume that the hitting point $U_h : \boldsymbol{u}_h = (x_h, y_h, z_h)^T$ is located in the middle between the point Γ_1 and Γ_2 with respect to their x components for the sake of safety. The time t_h when the robot hits the ball is then given by

$$t_h = \frac{[\dot{\boldsymbol{p}}_I]_x - \sqrt{[\dot{\boldsymbol{p}}_I]_x^2 - k[\dot{\boldsymbol{p}}_I]_x(x_{\Gamma 1} + x_{\Gamma 2} - 2[\boldsymbol{p}_I]_x)}}{k[\dot{\boldsymbol{p}}_I]_x} \tag{5}$$

where $[*]_x$ denotes the x component of a vector $*$. Using t_h yields the hitting point

$$\boldsymbol{u}_h = -\frac{1}{2}(\boldsymbol{g} + k\dot{\boldsymbol{p}}_I)t_h^2 + \dot{\boldsymbol{p}}_I t_h + \boldsymbol{p}_I. \tag{6}$$

3.4. Planning of Hitting Motion

Once the hitting point and time are determined, the remaining important problem is to decide the attitude and velocity of the racket at the hitting point such that the returned ball passes through a target point given a priori. The returned ball velocity $\boldsymbol{w}$ just after hitting can be easily obtained by using the motion equation of Eq.(3) and the condition on the ball trajectory.

The robot's hitting motion must be well planned to return the ball to specified points. Let $\boldsymbol{v}$ be the velocity vector of the ball just before hitting (this can be estimated from Eq.(6)), $\boldsymbol{n}$ be the unit normal to the racket surface, and $\boldsymbol{s}$ be the racket velocity at hitting (all are represented with respect to the world frame). From the definition of the coefficient of restitution e , we have

$$-\frac{(\boldsymbol{w} - \boldsymbol{s}) \cdot \boldsymbol{n}}{(\boldsymbol{v} - \boldsymbol{s}) \cdot \boldsymbol{n}} = e. \tag{7}$$

Assuming that the ball speed projected onto the horizontal plane of the racket surface is conserved just before and after hitting, we have

$$\boldsymbol{v} - \boldsymbol{s} - \{(\boldsymbol{v} - \boldsymbol{s}) \cdot \boldsymbol{n}\}\boldsymbol{n} = \boldsymbol{w} - \boldsymbol{s} - \{(\boldsymbol{w} - \boldsymbol{s}) \cdot \boldsymbol{n}\}\boldsymbol{n}. \tag{8}$$

Eq.(8)is rewritten as

$$\boldsymbol{w} - \boldsymbol{v} = \{(\boldsymbol{w} - \boldsymbol{v}) \cdot \boldsymbol{n}\}\boldsymbol{n} \tag{9}$$

which means $(\boldsymbol{w} - \boldsymbol{v})//\boldsymbol{n}$. Hence the attitude of racket $\boldsymbol{n}$ is of the form

$$\boldsymbol{n} = \frac{\boldsymbol{w} - \boldsymbol{v}}{||\boldsymbol{w} - \boldsymbol{v}||}. \tag{10}$$

From Eq.(7) , we can obtain

$$\boldsymbol{s} \cdot \boldsymbol{n} = \frac{1}{1+e}(\boldsymbol{w} + e\boldsymbol{v}) \cdot \boldsymbol{n}. \tag{11}$$

However, it is impossible to determine $\boldsymbol{s}$ uniquely from this equation. Because it is desirable for control to choose $||\boldsymbol{s}||$ as small as possible, we apply a condition to minimize $||\boldsymbol{s}||$, that is, $\boldsymbol{s}//\boldsymbol{n}$. We then obtain

$$\boldsymbol{s} = \frac{||(\boldsymbol{w} + e\boldsymbol{v})||}{1+e}\boldsymbol{n}. \tag{12}$$

Finally, we obtain the following results:

> The robot must return the ball at $t = t_h$ and at the point $\boldsymbol{u}_h$ with racket orientation $\boldsymbol{n}$ and velocity $\boldsymbol{s}$.

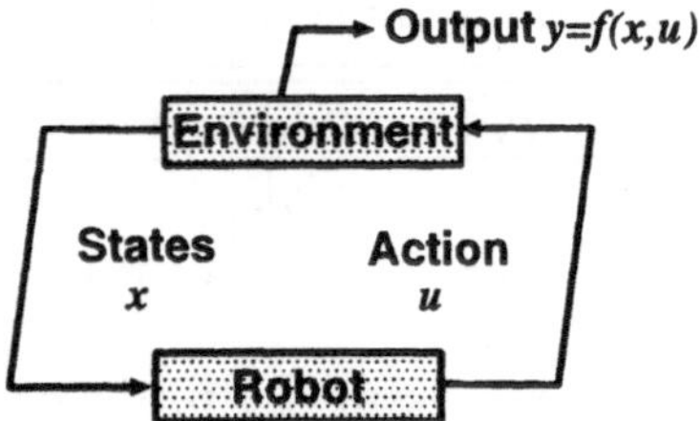

Figure 5. Relation between Robot and Environment

4. Learning

4.1. Learning Method

In the Hanetsuki task, many complex factors must be carefully characterized. Experiments we have done show that the model employed in the above-mentioned data processing is insufficient in some cases. Instead of refining the model, we attempted to improve execution of the Hanetsuki task by directly refining the commands to the robot. The basic idea of this type of learning is given below.

If the action determined by using a specific model is inappropriate, the robot must modify the action. We then have two problems to consider. The first is how to generate the modified action. The second is to give the robot the ability to use its experience when a similar situation occurs. Though using spline interpolation is a well-known technique for solving this problem, we originated a new method considering the memory requirement.

Let us assume the following situation(see Figure 5):

- The robot playing "Hanetsuki" against a human can detect the state of environment $\boldsymbol{x}$, that is, the location and velocity of a ball, only at a pre-defined moment.
- The robot affects the environment by an action $\boldsymbol{u}$ based on this state $\boldsymbol{x}$.
- As a result, the environment generates an output $\boldsymbol{y}$ with which task performance carried out by the robot can be evaluated.

Then we consider the following problem.

To decide an appropriate action $\boldsymbol{u}_d$ that results in a desired output $\boldsymbol{y}_d$ for every state $\boldsymbol{x}$.

If the map

$$\boldsymbol{y} = \boldsymbol{f}(\boldsymbol{x}, \boldsymbol{u}) \tag{13}$$

is known, $\boldsymbol{u}_d$ is decided by solving

$$\boldsymbol{y}_d = \boldsymbol{f}(\boldsymbol{x}, \boldsymbol{u}_d) \tag{14}$$

or its inverse map

$$\boldsymbol{u}_d = \boldsymbol{g}(\boldsymbol{x}, \boldsymbol{y}_d) \tag{15}$$

Now we propose a method to decide $\boldsymbol{u}_d$ when the map of Eq.(15) is unknown but its rough model

$$\overline{\boldsymbol{u}}_d = \overline{\boldsymbol{g}}(\boldsymbol{x}, \boldsymbol{y}_d) \tag{16}$$

which is explicitly described in the previous section is available. A key point of this method is the way to learn the residual inverse map

$$\Delta \boldsymbol{u}_d = \boldsymbol{g}(\boldsymbol{x}, \boldsymbol{y}_d) - \overline{\boldsymbol{g}}(\boldsymbol{x}, \boldsymbol{y}_d) \equiv \Delta \boldsymbol{g}(\boldsymbol{x}, \boldsymbol{y}_d). \tag{17}$$

Our method is summarized below.

Repeat the following process from $k = 0$.

- Read query $\boldsymbol{x}_q$.
- Calculate a nominal action $\overline{\boldsymbol{u}}_q$ by solving the inverse rough model
$$\overline{\boldsymbol{u}}_q = \overline{\boldsymbol{g}}(\boldsymbol{x}_q, \boldsymbol{y}_d). \tag{18}$$
- Look up $\Delta \boldsymbol{u}_q(k)$ in the k-th residual inverse map $\Delta \boldsymbol{u}(k)$ at $\boldsymbol{x}_q$.
- Modify $\overline{\boldsymbol{u}}_d$ in the form of
$$\boldsymbol{u}_q = \overline{\boldsymbol{u}}_q + \Delta \boldsymbol{u}_q(k). \tag{19}$$
- Perform the action $\boldsymbol{u}_q$ and measure the output of environment $\boldsymbol{y}(k)$.
- Update $\Delta \boldsymbol{u}_q(k)$ in the form of
$$\Delta \boldsymbol{u}'_q = \Delta \boldsymbol{u}_q(k) + \boldsymbol{G}^*(\boldsymbol{y}_d - \boldsymbol{y}(k)). \tag{20}$$
- Smooth the k-th residual inverse map using the Gaussian kernel of the form
$$\Delta \boldsymbol{u}(k+1) = \Delta \boldsymbol{u}'_q \exp[-(\boldsymbol{x}_q - \boldsymbol{x})^T \boldsymbol{M} (\boldsymbol{x}_q - \boldsymbol{x})] + \Delta \boldsymbol{u}(k) \tag{21}$$
where $\boldsymbol{M} = \mathrm{diag}(1/\nu_i)$.
- $k = k + 1$.

Figure 6 illustrates how to update and smooth the residual inverse map for a single state and single action variable. If we choose an appropriate gain matrix $\boldsymbol{G}^*$ in Eq.(20) and weighting parameters ν_i in Eq.(21) that determine local smoothing, the approximation of the residual inverse map is efficiently improved in some reasonable amount of time.

4.2. Application to Hanetsuki Task

We choose $\boldsymbol{v}$, the velocity of the ball just before hitting, and $\boldsymbol{w}$, the velocity of the ball just after hitting, as the state of the environment;

$$\boldsymbol{x} = (\boldsymbol{v}^T, \boldsymbol{w}^T)^T. \tag{22}$$

$\boldsymbol{w}$ is not an actual velocity vector but a velocity vector planned based on rough models and the state of environment. We adopt $\boldsymbol{w}$ instead of the position vector just before hitting as a part of the state because we desire to unify the unit of state variables in order to make the residual inverse map as smooth as possible. Of course,

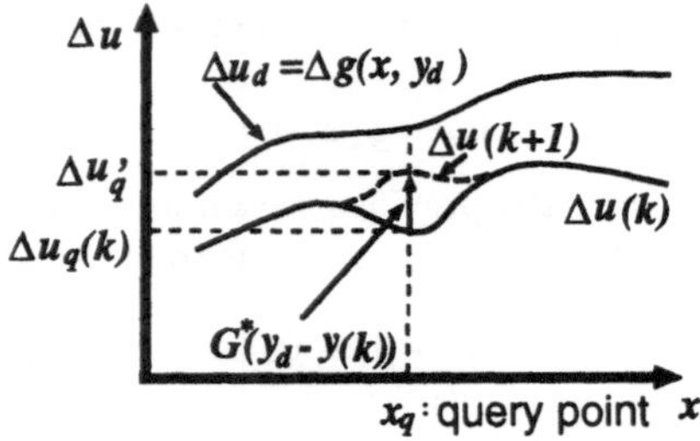

Figure 6. Updating Residual Inverse Map for SISO

the state defined as Eq.(22) uniquely represents the ball motion. As action variables, we choose the velocity command modification, $\Delta \boldsymbol{w}$:

$$\boldsymbol{u} = \Delta \boldsymbol{w}. \tag{23}$$

In practice, the attitude and velocity of the racket at hitting are calculated using $\boldsymbol{v}$ and $\boldsymbol{w} + \Delta \boldsymbol{w}$.

The output of environment $\boldsymbol{y}$ is defined as follows. The return path of a ball can be estimated in the same way as in the previous section. We consider a plane G parallel to the yz-plane of the world frame that includes the target point M_1 . Let M_1^* be the point located at the intersection between the return path and the plane G. Then we can define errors in the return path as the difference between M_1 and M_1^*, that is, (e_y, e_z) see Figure 7 . We choose these errors as the output of the environment;

$$\boldsymbol{y} = [e_y, e_z]^T \tag{24}$$

The action is modified so that the index of the form

$$J = \frac{1}{2}(e_y^2 + e_z^2) \tag{25}$$

decreases. This can be easily achieved by choosing the learning gain G^* in Eq.(20) based on the gradient method in combination with the fact that the ball trajectory is linearly dependent on $\boldsymbol{w}$ as shown in Eq.(4) where $\boldsymbol{w}$ corresponds to $\dot{\boldsymbol{p}}_I$.

4.3. Experimental Results

Figure 8 shows a change in the performance index J obtained during the robot's Hanetsuki task. Each point corresponds to the value of index averaged over five consecutive trials. From this figure, we can see that errors drastically decrease after the 10th trial owing to the learning capability.

5. Conclusion

In this paper we have described the Hanetsuki robot developed in our laboratory and the flow of data processing from sensing to action. Learning the inverse map is the key issue of the paper and we have proposed a learning algorithm consisting of updating action commands and smoothing them based on the Gaussian kernel. It has been also shown that this algorithm is effective especially to the case that the learning system has to deal with inputs and outputs with high dimensionality or sparse data like the Hanetsuki task against a human.

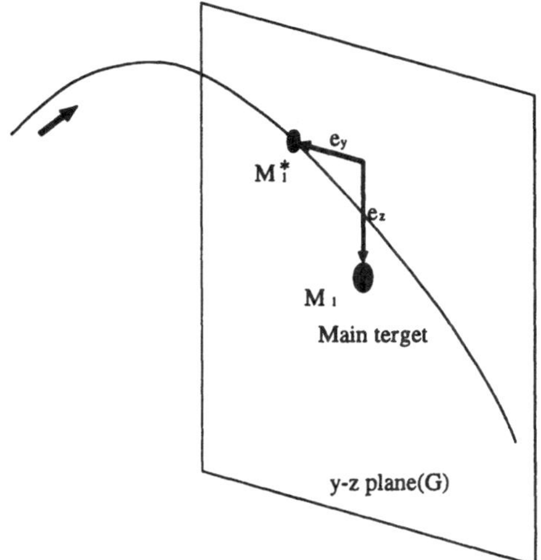

Figure 7. Definition of Errors e_y, e_z

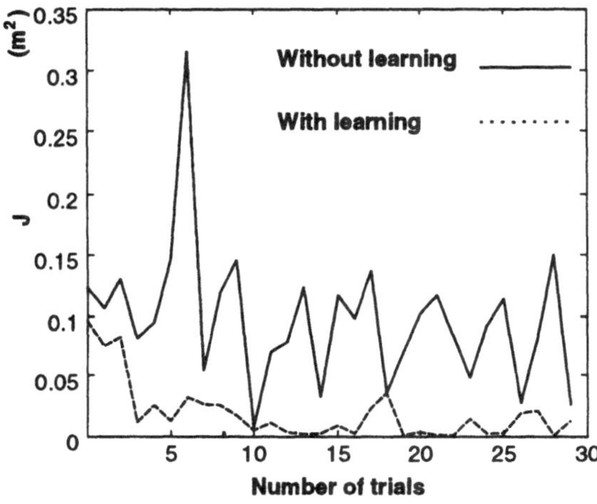

Figure 8. Change in Performance Index J

Our future work will focus on comparing our method with other learning algorithms such as those based on [7].

References

[1] Stefan Schaal and C. G. Atkeson: "Open Loop Stable Control Strategies for Robot Juggling", Proc.IEEE Int. Conf. on Robotics and Automation, pp.913–918, 1993.

[2] T. McGeer: "Passive Dynamic Walking", Int. J. Robotics Research, Vol.9, No.2, pp.62–82, 1990.

[3] M. T. Mason and K. M. Lynch: "Dynamic Manipulation", Proc. IEEE/RSJ Int. Cof.

on Intelligent Robots and Systems, pp.152–159, 1993.
[4] M. H. Raibert: "Legged Robots That Balance", Cambridge, MIT Press, 1986.
[5] A. A. Rizzi and D. E. Koditschek: "Further Progress in Robot Juggling: The Spatial Two-Juggle", Proc. IEEE Int. Conf. on Robotics and Automation pp.919–924, 1993.
[6] Russell L. Andersson: "A Robot Ping-Pong Player", The MIT Press 1988.
[7] Stefan Schaal and C. G. Atkeson: "Robot Juggling: Implementation of Memory-Based Learning", Control Systems (IEEE Control Systems Society), Vol.14, No.1, pp.57–71, 1994.

Experimental Verification of Progressive Learning Control For High-Speed Direct-Drive Robots with Structure Flexibility and Non-Collocated Sensors

Shih-Hung Li, Boo-Ho Yang, Haruhiko Asada

Massachusetts Institute of Technology
Cambridge, MA, U.S.A.
christ@mit.edu

Abstract

A novel approach to the learning control of non-collocated robotic systems is presented, and its feasibility and effectiveness are verified through experiments using an ultra-high speed direct-drive robot having structural flexibility. This approach, called "progressive learning," allows the system to learn parameters recursively and progressively, starting with the ones associated with low frequencies and moving up to the ones with a full spectrum. Even though the system has non-collocated sensors and actuators and the relative order is three or higher, the learning process is guaranteed to converge by exciting the system with a particular series of reference inputs having an appropriate frequency spectrum. A controller augmentation technique to accelerate the learning process is also presented.

1. Introduction

Non-collocated systems with a high relative order have been one of the systems to which adaptive and learning controls are difficult to apply. One of the most fundamental requirements for adaptive control to work is to "persistently" excite the system. However, because of the parameter uncertainty and large phase lag, the system might become unstable when persistently excited. Naturally, the key question is how to design the input frequency spectrum that will persistently excite the system without making it unstable. Our solution to this problem is to use what we called "Progressive Learning Method" [1].

Inspiration for the progressive learning theory naturally comes out of the human learning scheme. When a human learns new tasks, he starts slowly and meticulously at the beginning but speeds up the operation as he gains experience and skills. The Progressive Learning Theory suggests that a similar approach can be taken to satisfy the "full-excitation" requirements of stable adaptive control. Namely, progressive learning uses scheduled excitation inputs that allow the system to learn quasi-static, slow modes in the beginning, followed by the learning of faster modes. In [1], Yang has stated and proven the necessary and sufficient conditions for the input frequency spectrum that will guarantee the stability of adaptive control systems with a relative order greater than two.

In this paper, we will verify this learning method by applying to a high precision chip-placement machine that uses a high speed direct-drive robot with end-point sensor feedback. In this system, a PSD optical sensor with a laser beacon is mounted at the tip of the arm, which results in a non-collocated system. Our objective is to solve this non-collocated system problem by using the progressive learning theory. Also, based on the original progressive learning theory, we extend the theory to develop a learning schedule and to perform the trajectory design that links the assembly speed directly to the excitation frequency spectrum. The major point is that we can perform progressive learning to tune the system and, more importantly, perform actual assembly tasks at the same time while guaranteeing globally stable parameter convergence.

2. Theory of Progressive Learning

In this section, we review the progressive learning method in the context of model reference adaptive control. We first show stability conditions that relate frequency contents of the reference input to the stability property of the system. Based on this stability analysis, we provide the main theorem that guarantees the existence of a sequence of reference inputs that achieve the progressive convergence of the control parameters for the adaptive control system. The complete statement and proof of the progressive learning theory are found in [1].

A. Problem Formulation

Let us consider a model reference adaptive control(MRAC) scheme of the type treated in standard textbooks (e.g., [2]). The plant to be controlled is linear and time-invariant with input $u \in \mathbb{R}$ and output $y_p \in \mathbb{R}$ and its transfer function is $W_p(s) = k_p(Z_p(s)/R_p(s))$. The reference model to be followed is linear and time-invariant with input $r \in \mathbb{R}$ and output $y_m \in \mathbb{R}$ and its transfer function is $W_m(s) = k_m(Z_m(s)/R_m(s))$. $Z_p(s)$, $Z_m(s)$ and $R_m(s)$ are assumed to be Hurwitz. The objective of control is to find a differentiator-free control law $u(t)$ such that the output error $e_1 = y_p - y_m$ converges to zero asymptotically for arbitrary initial conditions and arbitrary piece-wise continuous, uniformly bounded reference signals $r(t)$.

The controller is described completely by the following differential equations and definitions:

$$\dot{w}_1 = \Lambda w_1 + lu, \quad \dot{w}_2 = \Lambda w_2 + l y_p \tag{1}$$

$$w \stackrel{\text{def}}{=} [r,\ w_1^T,\ y_p,\ w_2^T]^T \quad \theta \stackrel{\text{def}}{=} [k,\ \theta_1^T,\ \theta_0,\ \theta_2^T]^T \tag{2}$$

$$u = \theta^T w \tag{3}$$

where $\theta_1, \theta_2, w_1, w_2 \in \mathbb{R}^{n-1}$, $k, \theta_0 \in \mathbb{R}$, and $(\Lambda,\ l)$ is an asymptotically stable system in controllable canonical form with

$$\lambda(s) \stackrel{\text{def}}{=} \det(sI - \Lambda) = \lambda_0(s) Z_m(s) \tag{4}$$

for some monic Hurwitz polynomial λ_0 of degree $n - m - 1$.

Let us define $C(s)$ and $D(s)$ as

$$\frac{C(s)}{\lambda(s)} = \theta_1^T (sI - \Lambda)^{-1} l, \quad \frac{D(s)}{\lambda(s)} = \theta_0 + \theta_2^T (sI - \Lambda)^{-1} l. \tag{5}$$

Then, assuming that the control parameters are constant, the overall transfer function of the plant together with the controller can be expressed as

$$W_\theta(s) = \frac{kk_pZ_p(s)\lambda(s)}{(\lambda(s) - C(s))R_p(s) - k_pZ_p(s)D(s)}. \tag{6}$$

From this transfer function, the closed-loop characteristic polynomial can be given as

$$\Phi_\theta(s) = (\lambda(s) - C(s))R_p(s) - k_pZ_p(s)D(s). \tag{7}$$

It is well known that using the above control structure there exists a unique constant vector θ^* such that the closed-loop transfer function $W_{\theta^*}(s)$ matches $W_m(s)$ exactly. Namely, we can express the reference model as the plant $W_p(s)$ with the same controller at $\theta = \theta^*$. Let $\Phi_m(s)$ be the model characteristic polynomial, that is, the closed-loop characteristic polynomial when $\theta = \theta^*$, and it can be derived that

$$\Phi_m(s) = Z_p(s)\lambda_0(s)R_m(s). \tag{8}$$

In progressive learning, we use the gradient descent rule, often referred to as the MIT rule for adaptation. The idea of the MIT rule is to reduce e_1^2 by adjusting θ along the direction of steepest descent. Namely, the MIT rule can be expressed as

$$\dot{\theta}(t) = -\frac{1}{2}\alpha\frac{\partial e_1^2}{\partial \theta}^T = -\alpha e_1\frac{\partial e_1}{\partial \theta}^T. \tag{9}$$

B. Main Results

Based on the above formulation, the theory of progressive learning proved the following theorem stating that the overall stability of the system depends on the frequency contents of the reference input [1]:

Theorem 1 (Stability Condition) *Suppose that the reference input r is a summation of sinusoidal signals with N distinct frequencies such as*

$$r = \sum_{i=1}^{N} R_i \sin(\omega_i t), \quad R_i > 0 \ \text{for all } i. \tag{10}$$

Then, the overall adaptive system is exponentially stable, if

$$|\arg\{\Phi_m(j\omega)\} - \arg\{\Phi_\theta(j\omega)\}| < \frac{\pi}{2} \qquad \text{for all } \omega_i, \tag{11}$$

This theorem states that instability can be avoided if the system is excited in a frequency range in which the output of the closed-loop system is in phase with that of the reference model.

The idea of progressive learning is that the system is excited in low frequencies in the beginning to avoid the instability and the frequency range for stability is expanded gradually according to the progress of learning. Defining Ω_θ be the frequency range of stability as

$$\Omega_\theta = \left\{\omega \mid |\arg\{\Phi_m(j\omega)\} - \arg\{\Phi_\theta(j\omega)\}| < \frac{\pi}{2}\right\} \tag{12}$$

the main theoretical result that supports the above argument is given as:

Theorem 2 (Progressive Excitation Theorem) *Suppose, for a given* $\theta^{(h-1)}$, *there exists* $\omega^{(h)}$ *such that*

$$\Omega_{\theta^{(h-1)}} = \{\omega | 0 \leq \omega \leq \omega^{(h)}\}. \tag{13}$$

Also suppose that the parameter vector converges from $\theta^{(h-1)}$ *to* $\theta^{(h)}$ *by a stable adaptation law with the above reference input. Then, there always exists* $\varepsilon > 0$ *such that*

$$\omega^{(h+1)} = \omega^{(h)} + \varepsilon, \quad \textit{and} \tag{14}$$

$$|\arg\{\Phi_m(j\omega)\} - \arg\{\Phi_{\theta^{(h)}}(j\omega)\}| < \frac{\pi}{2} \quad \textit{for all} \quad \omega \in \{\omega|\ 0 \leq \omega \leq \omega_{\theta^{(h+1)}}\}. \tag{15}$$

See [1] for the proof. The above theorem states that the frequency range for stability can be always expanded if the system is excited fully within the current frequency range for stability. Namely, for given plant, reference model, adaptation rule, and initial parameter values, there always exist a sequence of excitation frequencies such that the system is maintained stable, and the control parameters and the output error converge to the true parameters and zero respectively.

3. Application to High-Speed Chip-Placement

3.1. Task Description

In this paper, the progressive learning method is applied to tune the controller for a high-speed chip-placement machine. A pair of two-link direct-drive robots with end-point sensors are used for chip placement [3]. The second link is driven by a motor in the base through a steal belt for each robot. To meet a target positioning accuracy of 100 μm, a PSD optical sensor with a laser beacon is mounted at the tip of the arm. At low frequencies, the robot system can be modeled as a rigid-body system, while at high frequencies the structural flexibility becomes prominent, and thereby the endpoint sensor feedback results in a complex non-collocated control system. The tuning of such a system is a difficult task, but the progressive learning method allows for stable, effective tuning.

The task that the chip-placement machine has to complete can be viewed as a trajectory tracking task. The trajectory is specified by a finite set of points in space. It is required to go through all the specified points, and come back to the original point, but the intermediate points are not specified. Therefore, there are certain degrees of freedom in generating an actual path. Also unspecified is the total cycle time required to complete one cycle of trajectory tracking. As learning proceeds, the cycle time should be shortened as much as possible, but during the learning phase, the system is allowed to track the trajectory at slow speeds, as long as all the specified points are visited correctly.

In progressive learning, we begin with a slow tracking speed, where the frequency spectrum of the trajectory is low enough to satisfy the initial stability conditions. In this low frequency range, the phase lag between the motor rotor and the joint axis due to the belt compliance is negligible. Therefore, the whole movable part can be modeled as a single rigid body. Both the joint encoder and the endpoint sensor provide basically the same positional information, as long as the robot tracks a trajectory at a low speed and the frequency spectrum of the trajectory is within a low frequency range. Tuning of control parameters is rather straight forward for this collocated system. At this low frequency range, instability of the end-point feedback does not occur under normal conditions. Instability, however, may occur in adaptation since the strictly positive realness condition, the standard stability condition for adaptive systems, is not met. Progressive learning is an efficient way for stabilizing the overall adaptive control system.

3.2. Controller Augmentation

After the learning process converges fo the initial slow speed and the tracking accuracy has been significantly improved, the tracking speed is increased and the learning process is repeated for the higher speed. Although the geometric trajectory and its spatial frequency spectrum remain the same, the temporal frequency spectrum shifts towards higher frequencies as the tracking speed increases. As a result, high frequency vibration modes associated with the steel belt compliance may be excited at this time. The robot dynamic model is then extended to the one including high-frequency modes. At the same time, the controller is augmented to include an observer to estimate the state variables in accordance with the augmented dynamic model. Progressive learning is carried out for this augmented controller comprising a larger number of parameters. The key point here is that the stability limit has been expanded after the completion of the initial learning with the slow tracking speed. Therefore, the learning of the augmented controller can be stabilized for the higher speed.

The traditional learning approach is to use a full-order control structure based on the full-order dynamic model of the robot from the beginning of learning. However, the simultaneous tuning of all the control parameters is a time-consuming process and requires a large number of sample data, which may result in an slow convergence. It is expected that the learning process will be accelerated by starting a reduced-order controller and gradually augmenting the control structure in accordance with the excitation level. This argument is verified by experiments later in this paper.

3.3. Trajectory Synthesis

In progressive learning, the stability condition is provided in terms of frequency contents of the reference input as shown in Section 3.2. Therefore, unless we can relate the task to the frequency domain, the system cannot perform any meaningful task during the learning process. To solve this problem, we developed a method that translates the desired trajectory in the time domain to the frequency domain that will satisfy the reference input requirement. Suppose that the task specification is provided in terms of targets points to follow. The basic idea is to find a spatial trajectory that not only includes those target points but also contains a pre-determined frequencies which satisfy the stability condition.

Suppose that the trajectory is specified to follow M target points, $y_1, y_2, \cdots, y_M$, where M is an even number. We also specify at least $M/2$ distinct frequencies $\omega_1, \omega_2, \cdots, \omega_{M/2}$ that satisfy the stability condition. Then, the trajectory $f(t)$ is generated by

$$f(t) = \sum_{m=1}^{M/2} [A_m \sin(\omega_m t) + B_m \cos(\omega_m t)] \tag{16}$$

where A_m and B_m, $m = 1, \cdots, M/2$, are calculated from

$$\sum_{m=1}^{M/2} [A_m \sin(\omega_m t_j) + B_m \cos(\omega_m t_j)] = y_j, \quad j = 1, 2, \cdots, M$$

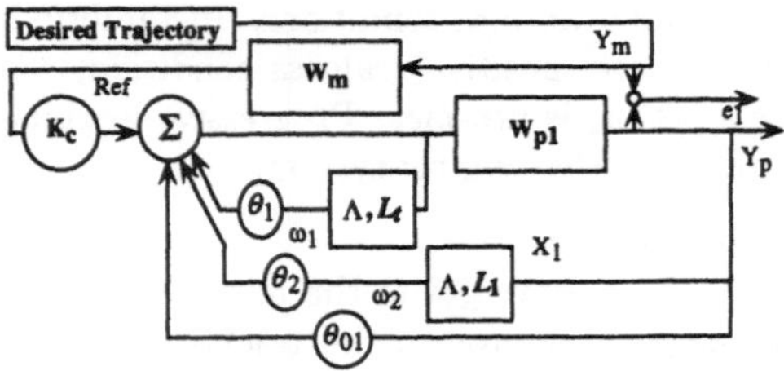

Figure 1. The overall adaptive control block diagram

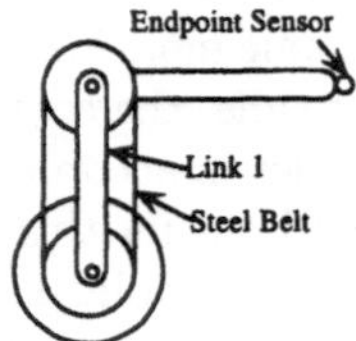

Figure 2. Experimental Setup

Finally, the reference input to the learning system is calculated from the inverse of the reference model. The overall control structure is shown in Figure 1.

4. Experiment

4.1. Experimental Setup

we have built a two-link direct-drive robot for high-precision chip-placement, and implemented the above progressive learning method to tune the control parameters. Figure 2 shows the experimental setup for the system. As shown in the figure, the second link is driven through a steel belt. An encoder is mounted on each joint and a PSD laser pointer is mounted at the tip of the second link for the use of a 2D PSD sensor that is located on a target trajectory. For accurate trajectory control, we feedback the end-point sensor signals to control the direct-drive robot. Therefore, our system results in a non-collocated system.

For simplicity, we assume the following conditions to the experimental setup:

1. All parameters involved are time-invariant,
2. The maximum operation speed is less than the excitation level of the flexible mode of the link but high enough to excite the transmission belt dynamics,
3. The first link is maintained immobile,
4. The sensor noises are small, and
5. The assembly process and the system are all deterministic.

Based on these assumptions and simplifications, Figure 3 shows possible lumped parameter models of the second joint control system. As explained in the previous section, at a low excitation level, the whole movable part can be modeled as a single rigid body, as shown in Figure 3A. However, as the frequency spectrum expands, the system model must be updated to the one involving the belt compliance, resulting

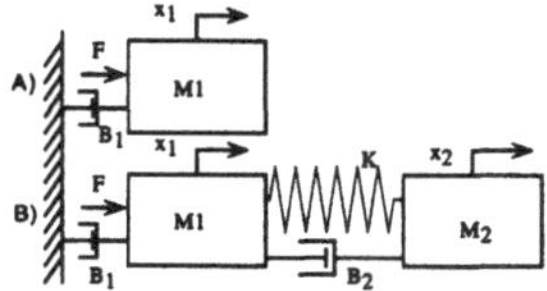

Figure 3. Lumped parameter models with different system orders

in a higher order system. Therefore, as shown in Figure 3B, the single rigid body is now splitted into the motor rotor side inertia, m_1, and the arm link side inertia, m_2. Both masses are connected by the belt stiffness k, whereas b_1 represents the viscous damping at the base of the second motor and b_2 represents the viscous damping of the belt structure.

Based on this simplified model, one can easily derive the following transfer functions:

$$\frac{x_2}{F} = W_{p_1} = \frac{\left(\frac{b_2}{m_1 m_2}\right)\left(s + \frac{k}{b_2}\right)}{s^4 + \left[\frac{b_1+b_2}{m_1} + \frac{b_2}{m_2}\right] s^3 + \left[\frac{k}{m_1} + \frac{k}{m_2} + \frac{b_1 b_2}{m_1 m_2}\right] s^2 + \frac{b_1 k}{m_1 m_2} s} \tag{17}$$

4.2. Plant and Control Structure

At low frequency, the transmission dynamics can be ignored so the system is simplified to be a second order system. At high frequency, however, the transmission dynamics can no longer be ignored. Thus, the system model has to be changed into a fourth-order system model. Using a system identification method based on the actual data provided from various data sheets and experiments, we obtained the plant model using the second-order approximation and the fourth-order approximations follows:

$$W_{p_1}(s) = \frac{26418}{s^2 + 301.13s}, \quad W_{p_2}(s) = \frac{2.46 \times 10^7 (s + 0.0076)}{(s^4 + 391.320s^3 + 57289s^2 + 980350s)} \tag{18}$$

A reference model for the adaptive control has to be chosen so that it represents a desired dynamics while the relative order of the model is equal to that of the plant. To meet these requirements, we chose the following second-order and fourth-order reference models, which are used for the second order and fourth order approximations of the plant respectively:

$$W_{m_1}(s) = \frac{101710}{s^2 + 314s + 2467}, \quad W_{m_2}(s) = \frac{9.94 \times 10^7 (s + 10)}{(s^2 + 314s + 2467)(s^2 + 120s + 11700)}. \tag{19}$$

The pre-determined feedback component $\lambda(s)$ is chosen for each of the plant approximations as:

$$\lambda_1(s) = (s + 1), \ \lambda_2(s) = (s + 1)(s + 1)(s + 10).$$

Based on the above transfer functions, the reference model characteristic equations are obtained as:

$$\begin{aligned} \Phi_{1_m}(s) &= s^3 + 314s^2 + 24676s + 123 \\ \Phi_{2_m}(s) &= s^7 + 435.17s^6 + (0.0745s^5 + 6.712s^4 + 295s^3 + 310s^2 + 868s + 0.659) \times 10^6. \end{aligned}$$

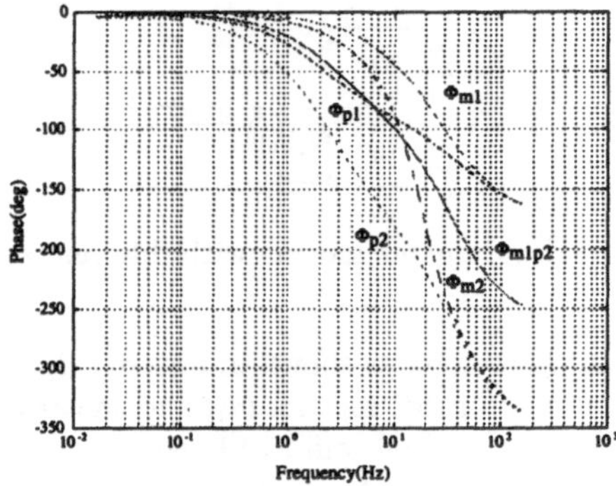

Figure 4. Phase plot of various transfer functions

Let $R_{p_1}(s)$ and $R_{p_2}(s)$ be denominators of $W_{p_1}(s)$ and $W_{p_1}(s)$ respectively. The closed-loop plant characteristic equation are obtained for the second-order and fourth-order MRAC in terms of control parameters as follows:

$$\begin{aligned}\Phi_{1_p}(s) &= (\lambda_1(s) - \theta_{11}R_{p_1}(s) - k_p(\theta_2 + \lambda_1(s)\theta_0)\\ \Phi_{2_p}(s) &= \left[\lambda_2(s) - (\theta_{11} + \theta_{12}s + \theta_{13}s^2)\right]R_{p_2}(s) - k_{p_2}\left[(\theta_{21} + \theta_{22}s + \theta_{23}s^2) + \lambda_2(s)\theta_0\right]\end{aligned}$$

where the control parameter vector for the second-order approximated plant is defined as:

$$\theta_{2nd} \stackrel{\text{def}}{=} [\theta_1,\ \theta_2,\ \theta_0,\ k_{c_1}]^T \tag{20}$$

and the control parameter vector for the fourth-order approximation plant is defined as:

$$\theta_{4th} \stackrel{\text{def}}{=} [\theta_{11},\ \theta_{12},\ \theta_{13},\ \theta_{21},\ \theta_{22},\ \theta_{23},\ \theta_0,\ k_{c_2}]^T \tag{21}$$

The parameters are tuned based on the progressive learning method. The true parameters based on the estimated second-order system parameters are: $k_{c_1}^* = -3.86$, $\theta_1^* = -13.0328$, $\theta_2^* = -0.1461$, $\theta_0^* = -0.7752$. The true parameters based on the estimated 4th-order system parameters are: $k_{c_1}^* = -4.07$, $\theta_1^* = [9.1827, -86.5461, -32.8461]^T$, $\theta_2^* = [-10.4583, -16.7737, -8.9612]^T$, $\theta_0^* = -0.1301$. The feedforward gain k_{c_1} and k_{c_2} were all initialized to 1 whereas the other control parameters were initialized to zeros. Thus, the closed-loop characteristic polynomials for both second-order and fourth-order systems are defined respectively as follows:

$$\Phi_{1_0}(s) \stackrel{\text{def}}{=} R_{p_1}\lambda_1(s) = s^3 + 302.1264s^2 + 301.1264s \tag{22}$$

$$\begin{aligned}\Phi_{2_0}(s) \stackrel{\text{def}}{=} &R_{p_2}\lambda_2(s) = s^7 + 402.32s^6 + (0.00616s^5 + 0.1616s^4 +\\ &1.154s^3 + 1.446s^2 + 2.9411s) \times 10^7\end{aligned} \tag{23}$$

Figure 4 shows the phase shift plots of the reference model characteristic functions and the initial closed-loop characteristic functions.

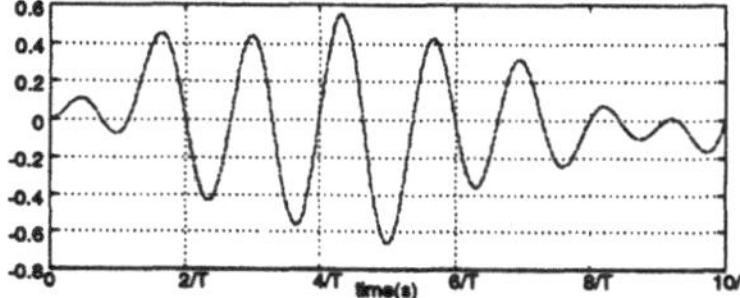

Figure 5. A typical reference trajectory used for the given reference target points

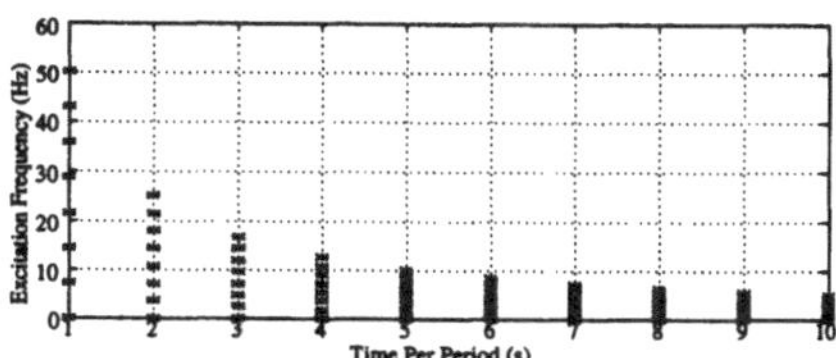

Figure 6. Excited frequency bandwidth for various speed

4.3. Learning Procedure

The desired trajectory as shown in Figure 5 is generated based on the following set of target points.

$$\begin{aligned} \text{targ pts} \;=\; & [0,\ 0.106,\ -0.073,\ 0.439,\ -0.437,\ 0.443, 0.576, 0.537,\ -0.652,\ 0.416, \\ & -0.365,\ 0.319,\ -0.253,\ 0.071,\ -0.114,\ 0.012,\ -0.167,\ 0.012] \end{aligned} \tag{24}$$

In the experiments, a friction compensation was also employed independently from the MRAC control loop in order to alleviate the non-linearity of the frictional effect as much as possible.

In order to demonstrate the effectiveness of the progressive learning, we ran the experiments in three series: 1) uses only the second-order reference model, 2) uses the fourth-order reference model, 3) starts with the second-order reference model and switches to the fourth-order reference model after the excitation level reaches the bandwidth of the fourth-order dynamics. For each series, the desired trajectory generated from the target points as presented earlier increases its speed. This, in turn, increases the excited frequency bandwidth. The notation we used to denote the speed is specified by its numerator; i.e., $10s/P$ denotes that each period is completed in 10 sec. Figure 6 shows the excitation frequency bandwidth with various speed and Table 1 shows how each series is run.

Table 1. History of the Experiments

Series No.	10s/P	8,6,5 s/P	4 s/P	3,2 s/P
#1:1:2nd-od	4x	2x	2x	n/a
#2:4th-od	4x	2x	2x	2x
#3:2,4th	4x	2x	4x	2x

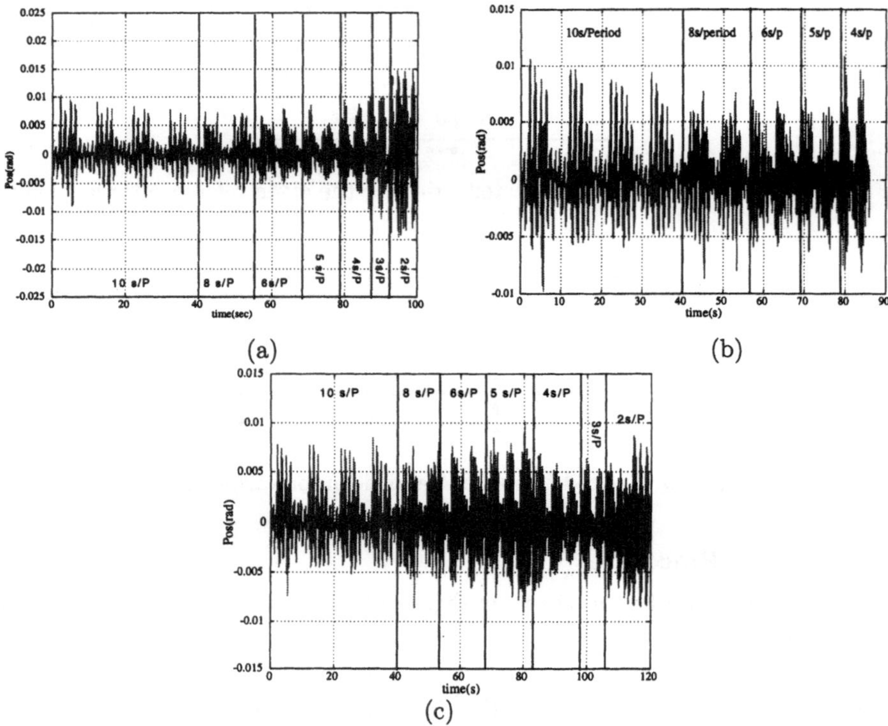

Figure 7. Error plots

4.4. Results and Discussions

Figures 7 and 8 show the results for the three sets of experiments that we described earlier. Based on these results, we can clarify three major points: bound on tracking error, stability of the progressive learning, and parameter convergent speed with the use of gradual model augmentation method.

Figure 7-(a) shows the error plots for the first set of experiments which were conducted under the assumption that the plant is a second-order system, and Figure 7-(b) shows the error plots of the second set of experiments which assumes that the plant is a fourth-order system, while Figure 7-(c) exhibits the errors of the third set of experiments which started with the second-order plant model and later augmented to the fourth-orderplant model. Our primary concern for this research is the tracking accuracy. Although the desired trajectory was synthesized based on the inverse reference model transfer function independently of the plant, the tracking errors quickly converged to within the tolerance. Small errors, however, remain due to other non-linearity effect of the system and sensor noise that we ignored in modeling.

Figure 8-(a) shows the learning curves of the control parameters for the first set of experiments which assume the plant is a second-order system. As shown in this figure, the stability and parameter convergence were acquired by the progressive

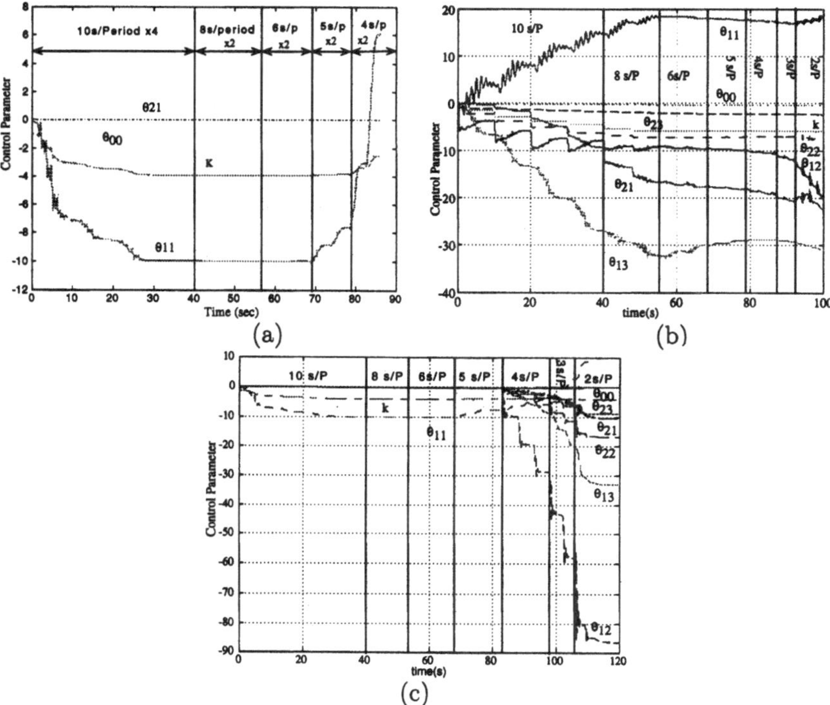

Figure 8. Control parameter histories

learning. By gradually exciting the system, we have achieved the stable adaptive control even though the excitation signals are band-limited with respect to the overall system bandwidth and the relative order of the plant is two or higher.

Figure 8-(b) shows the learning history of the eight control parameters for the second set of experiments where the controller were constructed based on the fourth-order plant assumption. Due to the fact that the phase discrepancy between the desired and actual system is large as shown in Figure 4, the control parameter convergence is quite slow. This result is consistent with the stability analysis we have shown earlier. Figure 8-(c) shows the history of the control parameters for the third set of experiments. In this particular set, we constructed a low-order controller based on the assumption that the plant is a second-order system, and, therefore, only four control parameters were adapted. As we increased the excitation level through the increased speed of the trajectory tracking command, the belt dynamics was excited. Once the controller sensed this change, the original second-order reference model was augmented to a fourth-order reference model. Thus, the number of the control parameters to be adappted increased from 4 to 8. By comparing the results shown in Figures 8-(b) and Figure 8-(c), it is found that the convergence was accerelated by this method of augmenting the controller as well as the reference model. One explaination for this observation is that after learning a trajectory at a slow speed, the stable range of excitation frequency has been expanded. More specifically, in the beginning of experiments, since the number of the control parameters is much less with a low-order assumption, it is easier for the parameter to converge. Graphically, from Figure 4, we see the low-order reference model allows the system adapts to an intermediate state, $\Phi_{m_1p_2}$, where only the low-order poles are being modified. Once the switching is made, the controller has already some information about the lower-order system dynamics; thus, adjustment for the controller to expand from four control parameters to eight is quite easy. Graphically, as seen in Figure 4, the phase discrepancy between the desired and actual plant model is below 90 degrees, the maximum allowed phase shift to maintain parameter adaptation stability and convergence. Thus, we can tune the system by fully exciting the system up to the system bandwidth while parameter convergent stability is guaranteed and tracking performance is maintained. Figure 8-(c) shows clearly that eight control parameters converged smoothly and quickly to their true values as compared with the results in Figure 8-(b). These results clearly demonstrated the effectiveness of the model augmentation method. Consequently, we achieved both trajectory control and control parameter convergence by using progressive learning and trajectory synthesis as demonstrated in these three sets of experiments.

5. Conclusion

In this paper, we showed through experiments that progressive learning is an efficient method for tuning a high order, non-collocated robotic system. The main idea of progressive learning is to excite the system gradually in accordance with the progress of the adaptation. By incorporating a trajectory synthesis, we developed a method of generating a series of tracking trajectoryies that satisfy the stability conditions of progressive learning. A controller augmentation method associated with progressive learning was also presneted and its usefulness was validated through experiments.

References

[1] B.-H. Yang and H. Asada, "Progressive Learning - Part I: Stability Analysis," *Proc. of the 1995 ASME International Mechanical Engineering Congress and Exposition*, San Francisco, November, 1995,

[2] K. S. Narendra and A. M. Annaswamy, Stable Adaptive Systems, Prince Hall, 1989

[3] Shih-Hung Li, Noriyuki Fujiwara, and Haruhiko Asada, "An Ultrahigh Speed Assembly Robot System: Part I. Design," *Proc. IEEE 1994 Japan-USA Symposium on Flexible Automation*, Japan, 1994

Accurate Positioning of Devices with Nonlinear Friction Using Fuzzy Logic Pulse Controller*

M.R. Popović D.M. Gorinevsky A.A. Goldenberg

Robotics and Automation Laboratory
University of Toronto
Toronto, Canada
e-mail: popovic@me.utoronto.ca

Abstract

In this paper a new fuzzy logic controller capable of performing precise positioning of mechanisms with nonlinear (stick-slip) friction is proposed. The controller is a deadbeat sample-data controller that uses short rectangular torque pulses to further reduce or eliminate steady-state position error generated by conventional position controllers. It uses a fuzzy logic model of the system response to rectangular torque pulses to calculate a pulse which is sent to mechanism's actuator in order to move the mechanism to a desired position. It was experimentally demonstrated that the developed controller is capable of accurately positioning a direct-drive mechanism up to the limit of the position encoder resolution.

1. Introduction

Conventional controllers which are used to perform precise positioning of mechanisms are incapable of performing this task satisfactory under the influence of stick-slip effect [1, 3, 4, 5, 9, 11]. Stick-slip motion is caused by nonlinear friction, in particular, negative damping effect that makes mechanical systems inherently unstable at low velocities. This instability is manifested as limit cycles or steady state position errors when conventional controllers, such as PID controllers, are used to control precise positioning of the mechanisms.

As precise positioning is needed for such devices as micro manipulators, assembly robots and surgical tools, a number of different control strategies have been proposed to overcome problems caused by stick-slip effect. The most relevant strategies presented in the literature are: friction compensation using dither signal [1]; friction compensation through position or force control using high gain feedback [3]; adaptive feedback friction compensation [4]; robust nonlinear friction compensation [11]; model-based feedforward friction compensation [1, 9]; and precise positioning using pulse control [6, 12]. Even though these strategies are incapable of performing precise positioning under the influence of negative damping friction, the last two groups of methods have shown some promising results.

*Supported by a grant from MRCO, Ontario, Canada

The model-based feedforward friction compensation method proposed by Armstrong [1] uses a detailed nonlinear friction model in the form of a look-up table to calculate force pulses as a function of the desired displacement which are used to move the mechanism to a desired position. The main problem with this method, besides that it requires an excessive amount of computer memory for the look-up table storage, is that it generates pulses which cannot perform displacements smaller than 10 encoder increments [1]. This significantly reduces the positioning accuracy of the mechanical system and limits the effectiveness of the proposed controller. Yang and Tomizuka [12], and Higuchi and Hojjat [6] have designed controllers that use short force pulses to perform precise positioning of mechanisms. Both methods use linear models of friction to calculate the force pulses as a function of the desired displacement, which are used to move the mechanism to a desired position. As these two methods do not take into consideration the negative damping phenomenon they can not be used to control very small displacements where nonlinear friction properties are strongly manifested.

In this paper, a novel design of a torque-pulse controller that is able to cope with nonlinear friction phenomena and, unlike Armstrong's method, does not require massive data storage, is proposed. The proposed controller uses a fuzzy logic approach to generate rectangular torque pulses used for precise positioning of the mechanism. The fuzzy logic approach provides a convenient means for fast computation of the torque pulse which depends on the desired displacement of the mechanism. In experiments with a direct-drive mechanism, the designed controller proved to be very effective. The accuracy of positioning the direct-drive mechanism was improved up to the encoder resolution limit. The proposed approach is rather general and requires minimum of device-dependent data.

The organization of the paper is as follows. Section 2 describes the experimental setup. The model of the system response to short torque pulses is given in Section 3 and the experiments which validate this model are presented in Section 4. Design of the fuzzy logic controller is outlined in Section 5. Section 6 presents a practical example of the proposed controller design and shows the experimental results obtained with this controller. Conclusions are given in Section 7.

2. Experimental Setup

Precise positioning, in particular performing very small displacements, is especially difficult with direct-drive mechanisms. Direct-drive mechanisms are devices without transmission between the drive and the link. The absence of a transmission mechanism combined with the existence of nonlinear (stick-slip) friction usually creates problems with precise positioning. The control strategy presented in this paper was tested experimentally on a two-link planar direct-drive (DD) manipulator designed at the Robotics and Automation Laboratory (RAL), University of Toronto [8, 9]. This is a typical direct-drive mechanism, which is actuated by two Yokogawa brushless permanent magnet DC motors DMA 1050 and DMB 1035 [13]. In experiments, only the first joint (DMA 1050) was used.

The main goal of this work is to develop a controller that applies short torque pulses to a mechanical system in order to achieve very accurate positioning of the system despite nonlinear friction. Short torque pulse were selected in order to perform positioning in a fast and robust way. Experiments conducted for several different pulse shapes have shown that there is little difference in positioning accuracy for different pulse shapes, while computational requirements increase as more complicated pulse shapes are used. Therefore, simple rectangular pulse shape was found to be the best choice for the purpose of controlling precise positioning of the experimental mechanism.

It is important to mention that conventional controllers (such as PID controllers) that were used in RAL in other experiments for positioning the experimental DD-arm are capable of performing this task with an accuracy of up to ± 30 encoder increments [8, 9]. The fuzzy logic controller presented in this paper was intended to further reduce the error after the conventional controller brings the arm to the vicinity of the desired position. Thus, the design specification considered in the controller development was that the fuzzy logic-based controller should work within displacements of ± 100 encoder increments.

3. Modelling the Response to Short Torque Pulses

In this section, a statistical model of the system response to short rectangular torque pulses is developed. The model proposed in this section as well as the theoretical analysis based on this model are general, and they are applicable to a broad class of mechanical systems. This model was confirmed in experiments with the DD-arm, which are discussed later on. By introducing this model, the theoretical background for the subsequent presentation is prepared.

Let us first introduce $h(t)$ as the notation for a step function (Heaviside function): $h(t) = 0$ for $t < 0$, and $h(t) = 1$ for $t \geq 0$. Let us further denote by A the amplitude of the pulse, τ the pulse duration, Q the torque applied by the drive, and t the time since the controller has been engaged. Then, a simple rectangular torque pulse shown in Figure 1.a can be presented in the following form:

$$Q(t) = \varphi(t; A, \tau) \equiv A \cdot [h(t) - h(t - \tau)], \quad t \in [0, T] \tag{1}$$

In (1) and further on in this section, time t is considered to be limited to the time interval $[0, T]$. The period T is selected in such a way that τ is always less than T, and all transient effects have settled down at $t = T$, as shown in Figure 1.b. As we only need to know the steady-state position of the mechanism after the torque pulse was applied to it, and since we know that the mechanism is in steady-state by the time $t = T$, it is sufficient to observe the behaviour of the system for the period of time $t \leq T$. Subsequently, when pulse control is discussed, the period T is the time between two consecutive engagements of the controller.

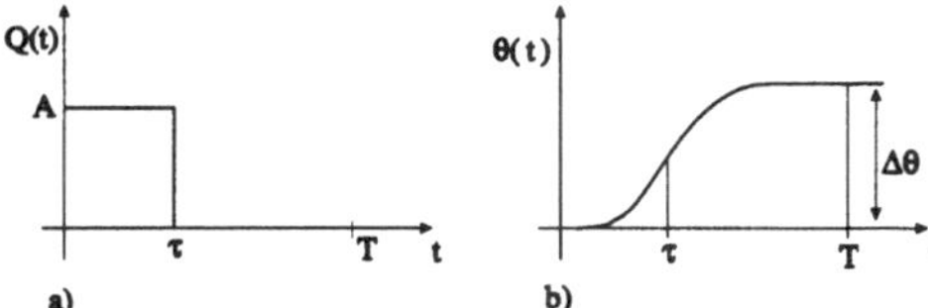

Figure 1. a) A rectangular pulse shape applied to the experimental mechanism; b) Transient response of the mechanism to the rectangular pulse shape

Let us denote by $\Delta\theta$ the difference between initial and final steady-state position of the mechanism after the torque pulse (1) was applied to it (see Figure 1.b). If we now denote by $F[Q(\cdot)]$ a nonlinear operator which defines the dependence of the steady-state displacement $\Delta\theta$ on the pulse shape $Q(\cdot)$ (1), then the following expression can be written:

$$\Delta\theta = F[Q(\cdot)], \tag{2}$$

where the operator $F[Q(\cdot)]$ depends on the entire pulse shape $Q(\cdot)$ and not only on its instantaneous value $Q(t)$, $t \in [0,T]$. Note that equation (2) assumes that the mechanism displacement does not depend on the initial position. This is true in most cases [6, 12], in particular for our experimental setup, as discussed later in the text.

It is well known from literature [1, 9] that the low velocity friction contains a significant random component which is superimposed on the static friction and the negative damping friction. Physically, the random friction is caused by establishing and breaking of asperity contacts between sliding surfaces during static friction and negative damping friction regimes (boundary lubrication regime) that occur at low velocities [1, 9]. If we take into consideration the random component of the friction and some measurement error, then model (2) holds only statistically. In fact, (2) should be given in the following form:

$$\Delta\theta = F[Q(\cdot)] + \epsilon_1, \tag{3}$$

where ϵ_1 is a bounded random variable which models a combined influence of the random friction and the position measurement error, and function $F[Q(\cdot)]$ describes the deterministic part of the measured displacement $\Delta\theta$. Equation (3) is a statistical model that describes a nonlinear dependence of the mechanism displacement on the rectangular torque pulse (1).

Before some of the properties of the random variable ϵ_1 are highlighted, let us define $\Delta\theta_{max}$ and $\Delta\theta_{min}$ as the upper and lower extremes of the variable $\Delta\theta$ respectively, where $\Delta\theta_{min} \le \Delta\theta \le \Delta\theta_{max}$, and R as the range of the variable $\Delta\theta$ such that $R = \Delta\theta_{max} - \Delta\theta_{min}$ [7]. Let us also define $\Delta\theta_{ctr}$ as the centre of distribution (centre of displacement) of the variable $\Delta\theta$ such that $\Delta\theta_{ctr} = \frac{1}{2}(\Delta\theta_{max} + \Delta\theta_{min})$ [7]. If we now select the centre of distribution $\Delta\theta_{ctr}$ to represent a deterministic part of measured displacement $\Delta\theta$, then the function $F[Q(\cdot)]$ in (3) could be used to model $\Delta\theta_{ctr}$ as a function of pulse shape (1). In that case the random displacement variation ϵ_1 models the difference between the variable $\Delta\theta$ and centre of distribution $\Delta\theta_{ctr}$. The centre of distribution $\Delta\theta_{ctr}$ was chosen to represent the deterministic part of the displacement $\Delta\theta$ because of convergence considerations discussed later in the text.

The probability distribution function for the random displacement variability ϵ_1 is not know, yet it can be shown that, in general, the following bound can be used to describe ϵ_1:

$$|\epsilon_1| \le \xi |\Delta\theta|, \quad (0 < \xi < 1), \tag{4}$$

where parameter ξ can be determined experimentally [10]. The inequality (4) establishes that the relative variation of the mechanism displacement under pulse control is bounded. This inequality is of major importance for the control design and closed-loop stability analysis discussed in the following sections of the paper.

4. Open-loop Experiments

Experiments that are reported in this paper were conducted with the first joint of the DD-arm described in Section 2. The second joint of the arm was immobilized. In the experiments, a torque pulse of rectangular shape (1) was sent to the motor, and the time history of the motor displacement $\Delta\theta(t)$ was recorded. Extensive experiments were performed with pulse amplitudes A varying from -20 Nm to 20 Nm in steps of 1 Nm, and pulse durations τ varying from 0 to 0.01 sec in steps of 0.001 sec. In other words, 451 pulse shapes which correspond to different pairs (A, τ) were experimentally studied. For each pulse shape, 10 experiments were performed (4510 experiments in all). As a result of these experiments dependence (2) of the centre of distribution displacement $\Delta\theta_{ctr}$ on the pulse parameters A and τ was obtained, as shown in Figure 2.

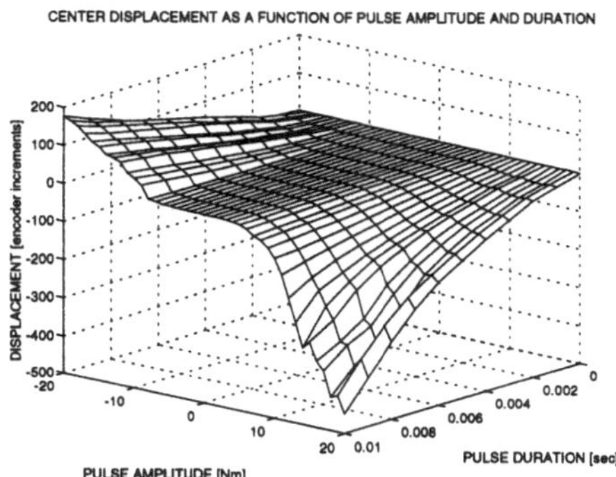

Figure 2. Experimentally obtained centre of distribution displacement $\Delta\theta_{ctr}$ as a function of pulse amplitude A and pulse duration τ

The experimental results obtained from the described experiments show that it is possible to move the mechanism between 1 to 140 encoder increments in each direction (100 encoder increments are required by design specifications) using a rectangular pulse shape (1) with pulse amplitude A less than 40% of the maximum motor torque (50 Nm), and duration τ less than 10 msec. The experimental results also show that the mechanism displacement does not depend on the initial position, and that the relative deviation $\sigma/|\Delta\theta_{ctr}|$ of the measured displacement (where $\sigma = \frac{1}{2}R$) increases as $|\Delta\theta_{ctr}|$ decreases. In particular, for $|\Delta\theta_{ctr}| \geq 4$ encoder increments, the relative deviation $\sigma/|\Delta\theta_{ctr}|$ is very small and often has a value close to 0. On the other hand, for $|\Delta\theta_{ctr}| \leq 3$ encoder increments, the relative deviation significantly increases and in some cases it reaches the maximum value 1. This relatively large variation for small displacements is caused by the relatively large position measurement error at the limit of the encoder resolution and by the random friction. Therefore, the experimentally designed controller described in Section 6 is expected to have, conservatively, a deadzone of ± 4 encoder increments.

The experiments also confirmed the fact that displacements of direct-drive mechanisms caused by torque pulses are strongly influenced by two friction related phenomena: **Asymmetry**[1] and **Dwell time**[2] [1, 9]. In this paper problems caused by asymmetry were overcome by introducing distinct friction models for different directions of motion, and a "warming-up" procedure was used to eliminate the influence of dwell time effect [1, 10].

---

[1] A phenomenon that results in dissimilar friction characteristics for different motion directions.

[2] Dwell time is a period of time during which friction characteristics of the mechanism significantly change because the mechanism is motionless.

5. Controller Design

5.1. Problem Statement

The proposed controller is a deadbeat sampled-data controller that should be used after a conventional controller brings the system to the vicinity of a desired position. The controller would work as follows. First, the difference between desired and actual positions of the system is determined. Second, a rectangular pulse shape (1) is calculated as a function of the difference, and the computed torque pulse is sent to the motor in order to move the mechanism to the desired position. After sending the pulse to the motor, the controller waits until a new steady-state of the mechanism is reached. If the difference between the newly measured and the desired position is within the bounds of the required precision, the control action is terminated. If not, a new pulse is calculated and a new control cycle is initiated.

The proposed deadbeat controller is designed to move the mechanism to the desired position after a single engagement, in absence of disturbances. As various disturbances, such as random friction and measurement error discussed in Section 5, are affecting the system, the controller is often unable to position the mechanism right at the desired position after the first engagement. Instead, it needs a few control cycles to place the mechanism within the required precision bounds of the desired position. The controller presented in this subsection ensures that a small finite number of control cycles are needed to make the system converge to the desired position.

It is also important to emphasize that the controller developed herein is a sampled-data controller. In other words, once the torque pulse is calculated, the controller does not have any influence on the system behaviour until a new sampling instant. Between the sampling instances, the system works in an open-loop configuration. The main concerns are whether or not the system will always perform bounded motions during the pulse control, and whether or not it will move unpredictably between any two consecutive controller engagements. Friction, which acts as a breaking device and limits the motion of the mechanical systems, ensures that the motion between the sampling instances is always predictable and bounded. This is an interesting concept, because the controller is overcoming problems induced by the low velocity friction, and at the same time it uses friction to ensure that the motion of the mechanism is bounded.

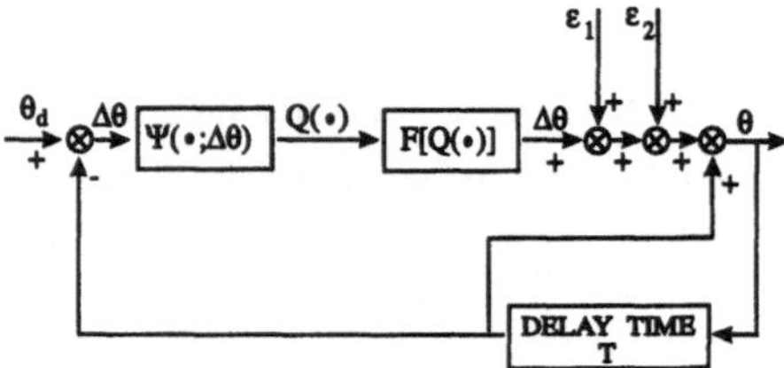

Figure 3. Schematic diagram of the closed-loop system

As already mentioned above, the proposed controller measures the difference between desired and the actual positions of the mechanism $\Delta\theta$, and based on this difference it calculates a rectangular torque pulse (1) needed to move the mechanism to the desired position. This procedure can be described using a nonlinear operator

$\Psi(t\ ;\Delta\theta)$ (see Figure 3) which gives the time dependence of the calculated torque pulse $Q(t)$ needed to perform a desired displacement. The operator $\Psi(t\ ;\Delta\theta)$ is defined for all displacements $\Delta\theta$ whose absolute value is not larger than the maximum displacement $\Theta_\star$ the controller is designed to perform. For our experimental mechanism $\Theta_\star = 100$ encoder increments (see Section 2). Therefore, for all $|\Delta\theta| \leq \Theta_\star$ the rectangular pulse computed by the controller is:

$$Q(t) = \Psi(t\ ;\Delta\theta), \tag{5}$$

Let us now define θ_d as the desired final position of the mechanism, θ_n as the actual position of the mechanism at the control cycle n, and $\Delta\theta$ as the desired displacement at the control cycle n such that $\Delta\theta = \theta_d - \theta_n$. Let us also assume that there are no disturbances affecting the system. If the displacement $\Delta\theta$ is supplied to the controller $\Psi(\cdot\ ;\Delta\theta)$ (5), then the output of the controller is the rectangular torque pulse $Q(\cdot)$ such that $F[Q(\cdot)] = \Delta\theta$ (2). If we now take into consideration the random variation of the mechanism's displacement ϵ_{1_n} at the control cycle n (3) and the computational error ϵ_{2_n} caused by the controller at the cycle n, then the closed-loop system (Figure 3) can be described by the following equation:

$$\theta_{n+1} = F[\Psi(\cdot\ ;(\theta_d - \theta_n))] + \theta_n + \epsilon_{1_n} = \theta_d + \epsilon_{1_n} + \epsilon_{2_n}, \tag{6}$$

In (6) the computational error ϵ_{2_n} is generated by the controller $\Psi(\cdot\ ;\Delta\theta)$ which calculates the torque pulse $Q(\cdot)$ depending on the desired displacement $\Delta\theta$. As the controller $\Psi(\cdot\ ;\Delta\theta)$ is designed by interpolating experimentally obtained data, the inaccuracy of the interpolation causes errors in calculating the torque pulse $Q(\cdot)$ which after being applied to the mechanism $F[Q(\cdot)]$ causes a displacement slightly different from the desired one. In (6) this difference in displacement is modelled using the computational error ϵ_{2_n} which is in essence the approximation error.

In order for the controller (5) to be useful, (6) has to converge to the desired position θ_d. In other words, the difference $|\theta_n - \theta_d|$ has to converge to zero as the number of control cycles n increases. It can be shown that the closed-loop system (6) monotonously converges toward a desired position as long as the following conditions hold [10]:

$$\begin{aligned} &|\epsilon_{1_n}| < \beta_1|\theta_d - \theta_n|, \quad |\epsilon_{2_n}| < \beta_2|\theta_d - \theta_n|, \\ &\beta_1 + \beta_2 \leq \alpha, \qquad\qquad \alpha < 1 \end{aligned} \tag{7}$$

In the following subsection a design of a controller (5) is presented that satisfies the conditions (7).

5.2. Fuzzy Logic Controller for Pulse Shape Computation

The controller $\Psi(\cdot\ ;\Delta\theta)$ is designed using the knowledge about the mechanism response to rectangular torque pulses, defined by the relationship (3). During the development of the control function $\Psi(\cdot\ ;\Delta\theta)$ the following considerations should be taken into account: simplicity of the design; minimal time and knowledge required to design the controller; closed-loop convergence requirements (7); and the computational complexity of the control function. A framework that conveniently satisfies all of the above requirements is given by the fuzzy logic approach.

The controller (5) is designed in the following way. As a first step, open-loop exploratory experiments identical with those presented in Section 4 are conducted for a number of selected rectangular torque pulses (1). These pulses are selected in such way that, for every displacement in the range $[-\Theta_\star, \Theta_\star]$, there exists at list one corresponding rectangular pulse which causes that particular displacement ($\Theta_\star$ is the maximum displacement the controller is designed to perform). Note that the given range has both negative and positive displacements, which implies that displacements made in one direction of rotation are given as positive values and the displacements made in the other direction as negative values. This enables us to take into consideration the asymmetry effect.

In the second phase of controller design, using the above experimental results, interpolation nodes which are needed to design the controller, are calculated. The interpolation nodes consist of two parameters $(\Delta\theta_j, Q_j(\cdot))$, $j = 1, \ldots, J$, where $\Delta\theta_j$ represents a desired displacement that should be generated by the rectangular pulse $Q_j(\cdot)$. There are two criteria used to select interpolation nodes. The first criterion is that the displacements $\Delta\theta_j$ of the selected nodes cover the range $[-\Theta_\star, \Theta_\star]$ in such a way that when they are arranged in increasing order ($\Delta\theta_1 < \ldots < \Delta\theta_j < \Delta\theta_{j+1} < \ldots < \Delta\theta_J$), the largest displacement in the negative rotational direction $|\Delta\theta_1|$ is greater than $\Theta_\star$ and the largest displacement in the positive rotational direction $\Delta\theta_J$ is greater than $\Theta_\star$ ($\Delta\theta_1 < -\Theta_\star$ and $\Theta_\star < \Delta\theta_J$). Note that $\Delta\theta_j$ denotes the centre of distribution of the displacement obtained for the pulse $Q_j(\cdot)$. The second criterion is that the interpolation nodes satisfy convergence conditions (7). For the experimental system, the procedure describing how the interpolation nodes are calculated is given in Subsection 6.1.

After the interpolation nodes $(\Delta\theta_j, Q_j(\cdot))$ have been obtained we proceed with designing the fuzzy logic algorithm. As the selected nodes $(\Delta\theta_j, Q_j(\cdot))$ do not include all arbitrary displacements in the range $[\Delta\theta_1, \Delta\theta_J]$ and as the corresponding control function $\Psi(\cdot\,;\Delta\theta)$ has to be defined for all such displacements, the control function must be obtained by interpolating nodes $(\Delta\theta_j, Q_j(\cdot))$. This interpolation is done using the fuzzy logic approach.

In accordance with the fuzzy logic approach to controller design, linguistic variables associated with interpolation nodes $(\Delta\theta_j, Q_j(\cdot))$ are created. One linguistic variable of the form "[DESIRED DISPLACEMENT IS $\Delta\theta_j$]" is associated with displacements $\Delta\theta_j$ and the other linguistic variable of the form "[APPLY TORQUE PULSE $Q_j(\cdot)$]" is associated with pulses $Q_j(\cdot)$. The derivation rule which maps one linguistic variable into another is defined as follows:

$$\textbf{if}\ [\text{DESIRED DISPLACEMENT IS } \Delta\theta_j],\quad \textbf{then}\ [\text{APPLY TORQUE PULSE } Q_j(\cdot)] \tag{8}$$

In order to make the linguistic variables fuzzy, membership functions $\mu_j(\Delta\theta)$ associated with the linguistic variables are introduced. Each of the membership functions is associated with one derivation rule (8). The membership functions $\mu_j(\Delta\theta)$ are selected in such way that they have a triangular form as shown in Figure 4, such that $\mu_j(\Delta\theta_j) = 1$, $\mu_j(\Delta\theta_{j-1}) = 0$ and $\mu_j(\Delta\theta_{j+1}) = 0$ for $j = 1, \ldots, J$. In addition, the membership functions $\mu_1(\Delta\theta)$ and $\mu_J(\Delta\theta)$ associated with the derivation rules that describe displacements at the limits of the interpolation range $\Delta\theta_1$ and $\Delta\theta_J$, respectively, have constant values equal to 1 outside the bounds of the interpolation range, i.e., $\mu_1(\Delta\theta) = 1$ for $\Delta\theta < \Delta\theta_1$ and $\mu_J(\Delta\theta) = 1$ for $\Delta\theta > \Delta\theta_J$ (see Figure 4).

The fuzzy logic controller designed with the above method works in the following way. First, a desired displacement $\Delta\theta$ at the control step n is determined as $\theta_d - \theta_n$. Next, all membership weights $\mu_j = \mu_j(\Delta\theta)$ associated with the derivation

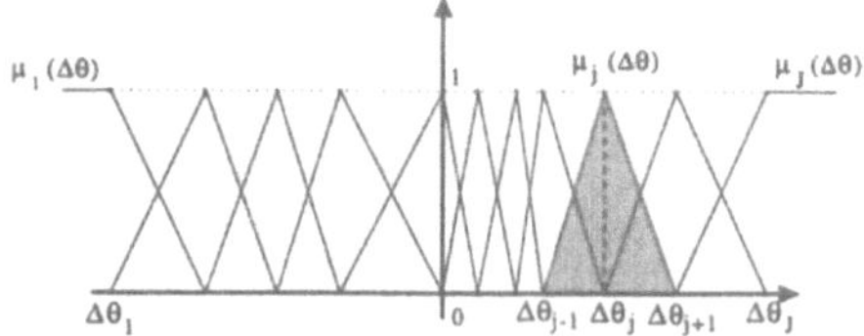

Figure 4. Schematic diagram of the fuzzification procedure

rules (8) are calculated for the desired displacement $\Delta\theta$. This procedure is called fuzzification. Then, the calculated weights μ_j in accordance with the derivation rule (8) and the corresponding node pulses $Q_j(\cdot)$ are used, based on the (Centre-of-Area) defuzzification method, to produce a "crisp" rectangular pulse shape which is the output of the controller $\Psi(\cdot\,;\Delta\theta)$, as shown with the following expression:

$$\Psi(\cdot\,;\Delta\theta) = \frac{\sum_{j=1}^{J}\mu_j(\Delta\theta)Q_j(\cdot)}{\sum_{j=1}^{J}\mu_j(\Delta\theta)}, \tag{9}$$

6. Experiments with Pulse Control

6.1. Controller Design

In this subsection the controller design for our experimental setup is briefly described. The controller design is based on selecting the interpolation nodes ($\Delta\theta_j$, $Q_j(\cdot)$) which are used to generate the controller output as a function of the desired displacement (see Section 5). The interpolation nodes are calculated from the experimental results which describe the response of the mechanism to rectangular torque pulses, using convergence conditions (7).

For the purpose of calculating the interpolation nodes the experimental results presented in Section 4 are used. As a first step, a number of curves which represent centre displacement $\Delta\theta_{ctr}$ as a function of pulse amplitude A for constant pulse duration τ are selected from Figure 2. A minimal number of these curves should be chosen such that for any arbitrarily selected displacement $\Delta\theta$ in the range $[-\Theta_*, \Theta_*]$ there exists at least one curve that describes the selected displacement as a function of pulse amplitude A and pulse duration τ. One such set of curves is shown in Figure 5, where solid lines represent the centre displacement of the mechanism as a function of the pulse amplitude for the specified pulse durations ($\tau = 1$ msec, $\tau = 3$ msec, $\tau = 5$ msec, and $\tau = 9$ msec). The dashed lines in Figure 5 represent upper and lower extremes of the displacement.

From these experimental results the interpolation nodes are further calculated as follows. Let us denote the expected displacement by $\Delta\theta_{exp}$, that is, the centre displacement which should be obtained if the pulse $Q(\cdot) = \Psi(\cdot\,;\Delta\theta_{ctr})$, computed by the designed controller, is sent to the system. In other words:

$$\Delta\theta_{exp} = F[\Psi(\cdot\,;\Delta\theta_{ctr})] \tag{10}$$

It is expected that the displacement $\Delta\theta_{exp}$ will not be exactly equal to the desired value $\Delta\theta_{ctr}$ due to the interpolation error of the fuzzy logic controller. Therefore, the interpolation error ϵ_2 in (6) can be defined as $\epsilon_2 = \Delta\theta_{ctr} - \Delta\theta_{exp}$. Also, by definition $\epsilon_1 = \Delta\theta - \Delta\theta_{ctr}$. In accordance with the discussion in Section 3, the following relationship holds:

$$|\epsilon_1| = |\Delta\theta - \Delta\theta_{ctr}| \le \frac{R}{2} \equiv \Delta\theta_{max} - \Delta\theta_{ctr} \tag{11}$$

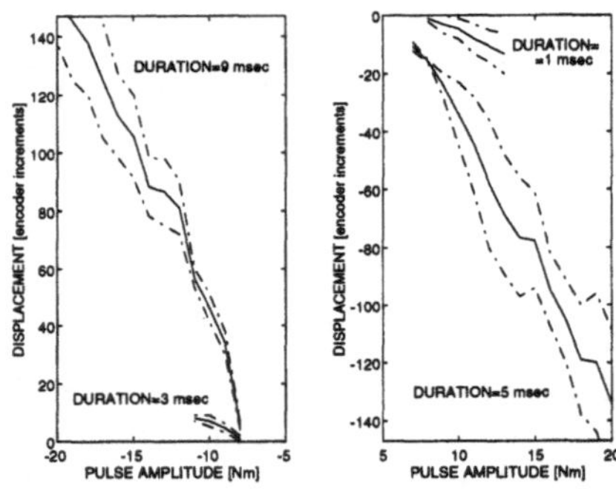

Figure 5. Experimental results used to design the controller

Node	1	2	3	4	5	6	7	8	9	10	11	12	13
$\Delta\theta$	-135	-21	-14	-13	-3	-1	0	1	5	8	9	34	151
A	20	8	7.3	13	9	8	0	-8	-9	-11	-8.1	-9	-20
τ	5	5	5	1	1	1	0	3	3	3	9	9	9

Table 1. Interpolation nodes for the designed controller: $\Delta\theta$ is given in encoder increments, A in Nm, and τ in msec

where R denotes the range of the variable $\Delta\theta$. Then, the convergence conditions (7) can be given in the following form:

$$\frac{|\Delta\theta_{max} - \Delta\theta_{ctr}| + |\Delta\theta_{ctr} - \Delta\theta_{exp}|}{|\Delta\theta_{exp}|} < \alpha, \tag{12}$$

where $\alpha < 1$. Inequality (12) gives a convenient way of expressing convergence conditions given by (7).

Now the interpolation nodes are calculated using the inequality (12) and the results from Figure 5. This is done as follows. First, a number of rectangular pulses $Q_j(\cdot)$ are selected from Figure 5 such that their central displacements cover the range $[-\Theta_\star, \Theta_\star]$ (see Subsection 5.2). Second, for those pulses $\Delta\theta_{exp}$ values are chosen satisfying inequality (12). Selected $\Delta\theta_{exp}$ values and corresponding torque pulses define the interpolation nodes $(\Delta\theta_j, Q_j(\cdot))$. At this point it has to be checked whether or not the selected nodes are properly chosen. This is done by linearly interpolating $\Delta\theta_j$ values which define the interpolation nodes, and by checking whether or not all the values $\Delta\theta_{exp}$ obtained through the interpolation satisfy inequality (12). If they do, the interpolation nodes are properly selected, and if not, a new set of pulses is chosen and the same procedure is repeated once again. The interpolation nodes calculated for the experimental mechanism are presented in Table 1.

The interpolation nodes displayed in Table 1 completely define the designed fuzzy controller, as explained in Subsection 5.2. Using inequality (12), it is possible to further determine the achievable accuracy of the closed-loop system (6) for the selected nodes. In our case, the convergence conditions (7) hold for $|\Delta\theta| \geq 4$, meaning that the accuracy of the system is determined to be ± 4 encoder increments, conservatively. The controller design is concluded.

6.2. Closed-Loop Pulse Control Experiments

The designed controller was tested experimentally in closed-loop, as follows. The experimental mechanism was placed in a randomly selected position and a desired final

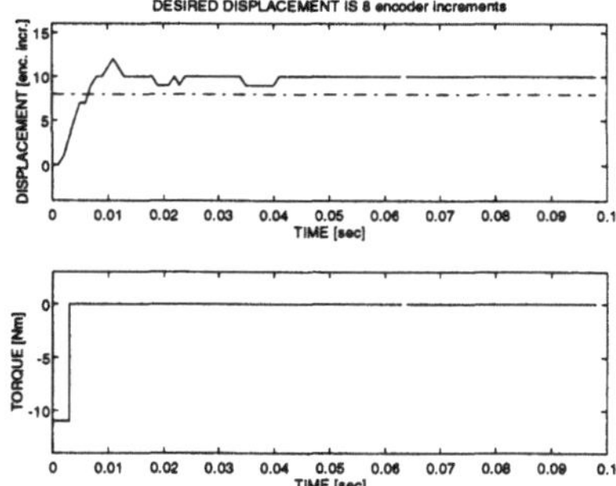

Figure 6. Control action and displacement of the arm as a function of time. Experimental results

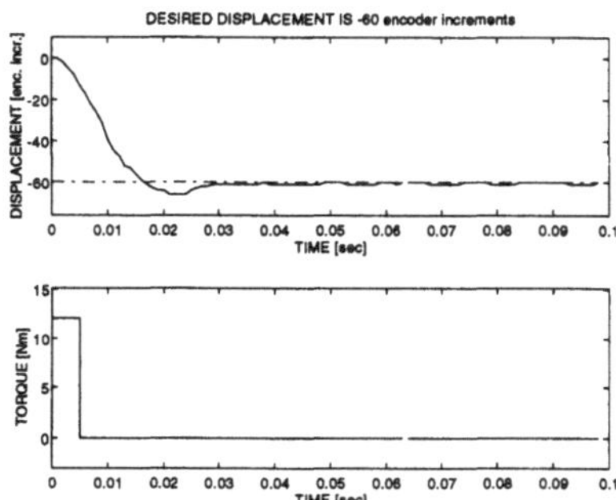

Figure 7. Control action and displacement of the arm as a function of time. Experimental results

position was selected within the range of ±100 encoder increments from the initial position. After that, the controller was activated and a sequence of torque pulses was sent to the motor until the desired position was reached with the predefined accuracy (±4 encoder increments).

The extensive testing has demonstrated that the developed controller is very robust and that it generally needs no more than three pulses to move the DD-arm to any desired position within the range of 100 encoder increments. With the 100 msec sampling period between the pulse applications it takes less than 0.3 sec to position the experimental mechanism. In many cases the controller is capable of positioning the mechanism at the desired position with a single torque pulse. The controller is capable of performing all necessary calculations in less than 0.5 msec, which eliminates the need for an additional delay between two consecutive pulses.

Figures 6, and 7 show experimental closed-loop results obtained with the designed fuzzy logic controller. In Figure 6, the desired displacement is 8 encoder increments, and in Figure 7, -60 encoder increments. In both cases, the error becomes less or equal to four encoder increments after a single controller engagement.

7. Conclusions

In the paper a novel approach to very accurate positioning of mechanical systems with nonlinear (stick-slip) friction was proposed. The designed controller applies narrow rectangular torque pulses to achieve a desired displacement of a mechanism. The torque pulses are computed through fuzzy logic approximation of the dependence between the desired displacement and the torque pulse. The convergence conditions of the proposed controller were derived taking into account an influence of the random variation of friction values. A detailed experimental study of the behaviour of response to the rectangular torque pulses, and the detailed controller design, were presented for a direct-drive manipulator setup. It was experimentally demonstrated that the developed controller achieves precise positioning within the limits of position encoder resolution.

References

[1] Armstrong B., Control of Machines with Friction, Kluwer Academic Publishers, Boston, USA, 1991.

[2] Driankov D., Hellendoorn H., and Reinfrank M., An Introduction to Fuzzy Control, Springer-Verlag, N.Y., USA, 1993.

[3] Dupont P.E., "Avoiding Stick-Slip in Position and Force Control Through Feedback," *Proc. IEEE Int. Conf. on Robotics and Automation*, 1991, pp.1470-1475.

[4] Canudas de Wit C., Adaptive Control for Partially Known Systems, Elsevier, Boston, USA, 1988.

[5] Hess D.P. and Soom A., "Friction at a Lubricated Line Contact Operating at Oscillating Sliding Velocities," *J. of Tribology*, Vol.122(1), 1990, pp.147-152.

[6] Higuchi T. and Hojjat Y., "Application of Electromagnetic Impulsive Force to Precise Positioning," *IFAC 10th Triennial World Congress*, 1987, pp.283-288.

[7] Gumbel E.J., Statistics of Extremes, Columbia University Press, New York, USA, 1966.

[8] Liu G.J. and Goldenberg A.A., "Experiments on Robust Control of Robot Manipulators," *Proc. IEEE Int. Conf. on Robotics and Automation*, 1992, pp.1935-1940.

[9] Popović M.R., Shimoga K.B., Hui R.C., and Goldenberg A.A., "Model-Based Compensation of Friction in Direct-Drive Robotic Arms," *J. of Studies in Informatics and Control*, Vol.3, No.1, March/1994, pp.75-88.

[10] Popović M.R., Gorinevsky D.M., and Goldenberg A.A., "Fuzzy Logic Controller for Accurate Positioning of Direct-Drive Mechanism Using Force Pulses," *Proc. IEEE Int. Conf. on Robotics and Automation*, Vol.1, 1995, pp.1166-1171.

[11] Southward S.C., Radcliffe C.J., and MacCluer C.R., "Robust Nonlinear Stick-Slip Friction Compensation," *ASME Winter Annual Meeting*, 1990, ASME paper No. 90-WM/DAC-8, 7 pages.

[12] Yang S. and Tomizuka M., "Adaptive Pulse Width Control for Precise Positioning Under the Influence of Stiction and Coulomb Friction", *Trans, ASME, J. Dynam. Syst. Meas. and Control*, September 1988, Vol. 110, No3, pp.221-227.

[13] Yokogawa, Product Literature on DMA and DMB Series Direct-Drive Motors, Yokogawa Co. of America, Lake Geneva, MI, 1989.

Chapter 8

Vehicle Navigation

Navigating a vehicle through a geometrically complex, changing environment requires the ability to estimate the vehicle's motion relative to the ground and relative to objects in the environment. Further, operations in the real-world involve moving a vehicle to a specified goal or maintaining or reaching a specified relationship with other vehicles. This requires a high level of integration of perception, validation and action capabilities to enable safe, reliable execution.

Daviet and Parent describe developments that enable several vehicles to autonomously follow a lead vehicle via visual information. They employ a linear camera on each following vehicle which senses in real-time the position of the proceeding vehicle (on which is mounted an illuminated infrared target). Their experiments demonstrate success of the approach on an electric car operating in city driving conditions at speeds up to 50 km/hr.

Stevens and Durrant-Whyte describe a system which uses rapidly stearable sonar detectors to track environmental features to achieve continuous localization of a moving vehicle. Using known targets, a multi-level architecture allows targets to be locked-upon, validated, and tracked during motion. Environment maps and odometric information are fused with the sonar target tracking data via an extended Kalman filer algorithm. The paper describes experimental trials in the laboratory and in an active industrial environment.

Ben Amar and Bidaud analyze the motion of off-road robot vehicles, taking into account slip, soil shear deformation, soil compaction and elastic wheel deformation. Combining these effects with vehicle dynamic, they show in simulation the effectiveness of their analysis in predicting actual vehicle motion, and suggest how this capability can be used to plan vehicle motions in difficult terrain.

Durrant-Whyte describes an autonomous guided vehicle (AGV) system designed to transport cargo containers in seaport environments. It focuses on the use of vehicle-mounted millimeter-wave radar sensors that detect beacons placed at known locations and on the algorithms used to reliably distinguish beacons from environmental clutter. The paper reports on experiments in which the system was used guide large vehicles operating in a cluttered environment and docking with gantry cranes.

Platooning for Small Public Urban Vehicles

Pascal Daviet
Michel Parent
INRIA,BP 105, 78153 Le Chesnay Cedex, FRANCE
E-Mail : Pascal.Daviet@inria.fr,Michel.Parent@inria.fr

Abstract

This paper presents a vision technique for platoon driving of a fleet of homogeneous electric cars. These cars, under supervision from a central computer, will form a novel public transportation system now under development in France. In the first version of the system, the public cars are driven by their users but their operation will be automated in some instances when no-one is on board. For later versions, fully automated driving (with passengers) is considered on dedicated tracks and on low-traffic road networks at slower speed. The paper will focus on the vision and control techniques now developped for the first version. We wish to be able to drive a "train" of empty public cars with only one driver (in the front car). This fonction is needed to pick up cars which have been abandonned in various locations and to redistribute them to appropriate locations. Trains can be up to six cars long and the distance between cars will vary between 0 and 5 meters depending on the speed.

This paper will present the vision sensor based on a linear camera and the longitudinal and lateral controlers developped to implement this function. The sensor is now operationnal and has been integrated in an electric car under full computer control. The technique works very efficiently under city driving conditions with speeds up to 50 km/h .

1. Introduction

1.1. Background

Now under test is the technology to implement cooperative driving of a platoon of vehicles, only the first car being driven by an operator. This function is essential to move easily the empty vehicles from one location to another. The displacement of a vehicle can be decomposed into three phases :

- a rendez-vous phase between a single car and a passing platoon,
- cooperative driving in a platoon at speeds up to 50 kmph and distances up to 5 m,
- exit from a platoon to reach a new parking space.

This paper will present the problem and our solution for longitudinal and lateral control for cooperative driving.

A similar problem has already been studied for cooperative driving and in particular for ICC (Intelligent Cruise Control) for regular cars on highways. However, our context is quite different in the following aspects :

- no one will be in the cars which are in automatic mode,
- the driver in the first car will assume responsibility for the platoon,
- the speed will not exceed 50 kmph,
- the distance between cars should be very small,
- all the cars can be equipped with identical markers.

With these constraints in mind, we have investigated the technical possibilities to implement the longitudinal and lateral control of a platoon of several cars and in particular, what are the sensors needed for the function.

1.2. Particular constraints

The first constraint is linked to the performances of the vehicles in speed and acceleration. It is clear that the lead vehicle should impose on itself the least performances of the fleet. On the other hand, we want to be able to maintain performances compatible with normal city driving. Therefore we have decided to limit the speed to $50kmph$, the acceleration to $2m/s/s$ and the deceleration (in emergency situations) to $5m/s/s$.

the second is linked to the regulation of the platoon. The platoon should be stable asymptotically, that is, without any amplification from one car to the next which would rapidly invalidate the previous constraint.

2. Formulations

The problem can be formulated as the control in acceleration and turning radius of a vehicle with respect to a previous one which it tries to follow as close as possible. We have no a priori knowledge of the behavior of the preceding vehicle which can therefore chose its own path and its own speed profile. In normal driving conditions, the state of each vehicle can be characterized by its location and orientation in the plane (three degrees of freedom constrained) and its longitudinal and angular speeds (linked through its front wheel angle).

The control we want to apply to the automatic vehicle concerns the acceleration (or deceleration), that is the motor torque (positive or negative) and eventually the brake pressure for emergency situations, and the changes in wheel angle.

This control will receive as input, estimates of the state of the preceding vehicle obtained through sensors in order to follow as closely as possible this preceding vehicle in a safe way. The only assumptions we will make about the behavior of the previous vehicle is that its performances in acceleration and turning radius will never exceed those of the following vehicle.

In order to simplify the problem, we have decided to separate it into two distinct problems : a longitudinal control and a lateral control. We think that these two problems are sufficiently independent in normal driving conditions to validate this assumption.

2.1. Longitudinal Controller

The main constraint in platooning is that you can't allow any amplification from one car to the next and it is well know that without constant and rapid communication of the speed of the first vehicle to all the vehicles in the platoon (which we do not want to assume), it is not possible to maintain a constant distance between the vehicles.

So we have chosen to set a linear relation between the distance and the speed of the vehicles :

$$X_l - X_f = d\min + h \cdot V_f$$

where X_l represent the position of leading vehicle and X_f the position of the following vehicle, or after differentiation :

$$V_l\_V_f = h \cdot A_f$$

or in Laplace form :

$$V_f = \frac{V_l}{1 + h.s}$$

So we can see that this choice is equivalent to filter the speed with a low pass constant of h seconds and it is asymptotically stable. We are already experimenting with h=0.35s and dmin=1m and our goal is to reduce these values to h=.2s and dmin=0.5m.

2.1.1. Constant Gain Linear Corrector

Because no one will be in the cars which are in automatic mode, we choose not to take a Jerk saturation into account. This means that we take the acceleration as the command variable and this leads to the following controller:

$$A_{fc} = C_v \Delta V + C_p (\Delta X - hVf - d\min)$$

with $\Delta V = V_l - V_f$ and $\Delta X = X_l - X_f$.Indeed, we want to maintain $\Delta X = hVf + d\min$ as best as we can. The closed loop transmittance with no acceleration saturation is :

$$\frac{X_f(s)}{Xl(s)} = \frac{C_v.s + C_p}{s^2 + (C_v + C_p.h)\,s + C_p}$$

since we want :

$$\frac{X_f(s)}{Xl(s)} = \frac{1}{1 + h.s}$$

Which leads to the following choices : $C_v = \frac{1}{h}$ and $Cp \leq \frac{1}{h^2}$ .The larger Cp is,the faster the system is. Hence we are tempted to choose $Cp = \frac{1}{h^2}$. This would be optimal if saturations in acceleration are not taken into account. However they are important :$|A_f| < A_{\max} = 2m/s^2$with the_electric motor in traction or brake. So we must study the influence of saturations.

2.1.2. Linear corrector with variable coefficients

Let Vo be the initial speed of the vehicle. We can compute the minimal stopping distance D which satisfies the limitations :

$$D(V_0) = \frac{V_0^2}{2.A_{\max}}$$

We can define a safety zone where : $\Delta X \geq D(Vf) - D(Vl)$. An optimal control can be obtained by sliding along the security curve : $\Delta X + D(Vl) - D(Vf) = 0$. After linearization, we obtain : $\Delta X + \frac{\partial D(Vf)}{\partial Vf}\Delta V = 0$.

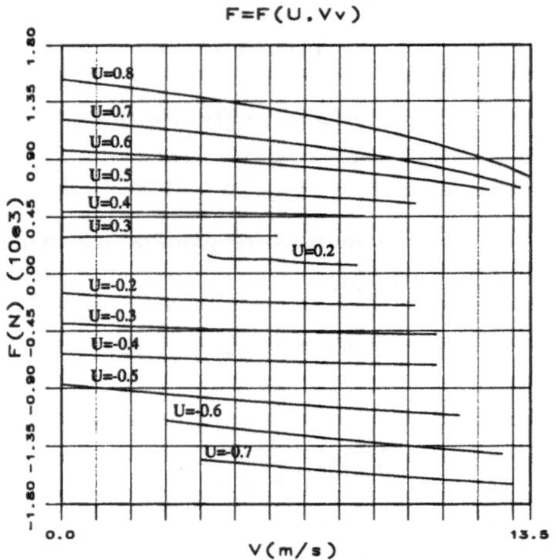

Figure 1. Electrical motor force

One finds a corrector of the form :

$$\Delta v + Kp.\Delta x = 0 \text{ with } , Kp = \frac{A\max}{Vf}$$

and this gives us a variable gain corrector with :

$$A_{fc} = \frac{1}{h}\left(\Delta V + K_p\left(\Delta X - hVf - d\min\right)\right)$$

with $K_p = \min(\frac{1}{h}, \frac{A\max}{Vf})$.This controller can bear great initial errors since the position gain Kp decreases when the speed increases.

2.1.3. Acceleration control

Now we have to control the vehicle acceleration A_f at the desired value A_{fc}.We have two actuators :

-An electrical motor controlled by a voltage U which provides a force $F = F(U, V_f)$. (You can see the experimental measurements fig1) : its response time is very short.

-Brakes controlled in pressure by a PID with a piston whose response time is long ($\simeq 0.3s$). So we used them only in emergency cases when the required deceleration $A_{fc} \leq -2m/s^2$. In that case we add hydraulic braking to motor deceleration.

-The longitudinal dynamics are described by the following equations:

$$M.A_f = F_t - k\left(V_f + V_{wind}\right)^2 + Mg\sin\left(\theta\right)$$

$$J\omega_w^\circ = r_w\left(F_m - F_b - Ft\right)$$

which gives with a good approximation at our low speed (less than 50kmph)

$$M.A_f \simeq F_m - F_b + Mg\sin\left(\theta\right)$$

the main problem in our urban situation is the estimation of the road gradient θ.

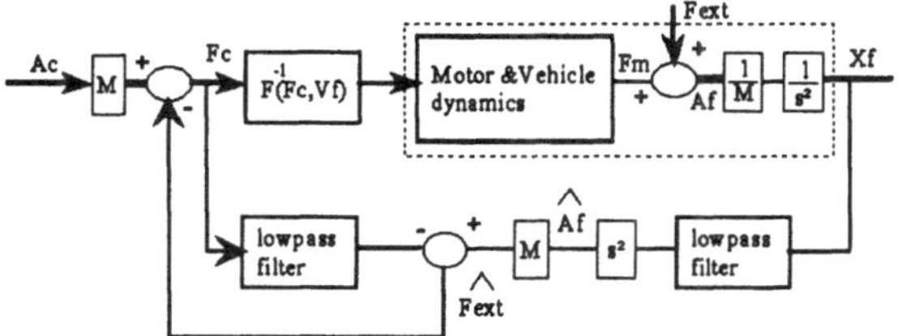

Figure 2. Acceleration controller

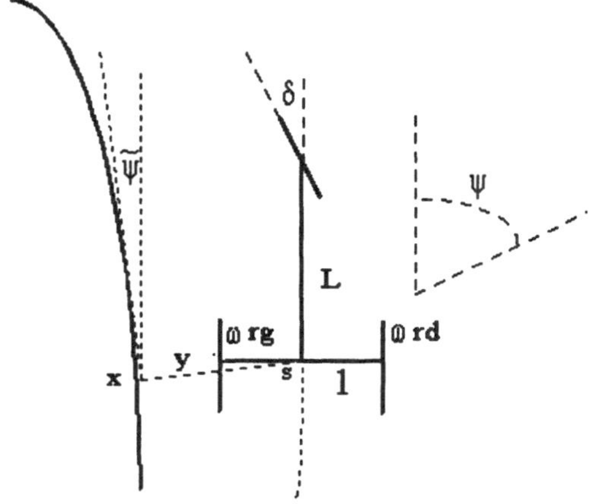

Figure 3. lateral model and errors

So we designed the acceleration controller with the inverted characteristics of the electrical motor : $U = F^{-1}(F_m, V_f)$ and a gradient estimator, it gives the scheme presented in (fig 2):

2.2. Lateral Controller

At the moment , we are using a simple cinematic model for the lateral part which is in the cartesian space :

$$\begin{pmatrix} \mathring{X}_0 = V\cos(\Psi_0) \\ \mathring{Y}_0 = V\sin(\Psi_0) \\ \mathring{\Psi}_0 = V\frac{\tan(\delta)}{L} \end{pmatrix}$$

We have very precise encoders on the rear wheels which allow us to estimate the speed Vf , the derivative of course $\mathring{\Psi}_0$ and the path of the car in the cartesian space.

$$\begin{pmatrix} \mathring{\Psi}_0 = r_w\frac{\omega_{rr}-\omega_{rl}}{l} \\ V = r_w\frac{\omega_{rr}+\omega_{rl}}{2} \end{pmatrix}$$

The simplest control we can think of is to select the wheel angle as equal to the direction of the leading vehicle.

$$\tan(\delta) = \frac{\Delta Y}{\Delta X - L}$$

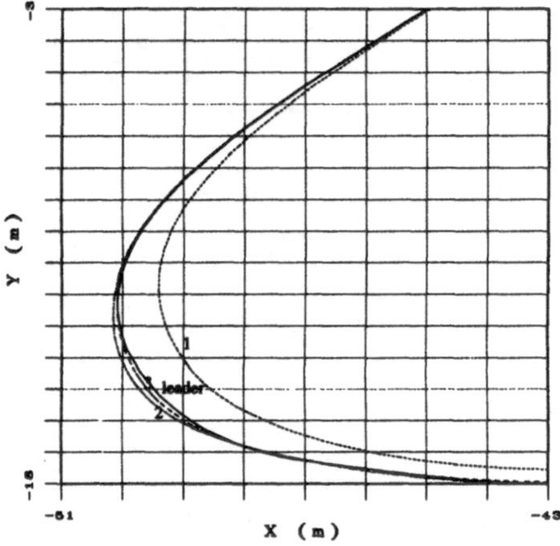

Figure 4. Instantaneous following

This is the so-called tractor model and it is very stable although it leads to a following vehicle which cuts corner and this might be a problem with a train of several cars in city situations.

Some more complex lateral control are at the moment in progress, our goal is to be as close as possible to the leader path. We propose two others instantaneous controls with polynomials curves and a method which memorizes the trajectory of the previous vehicle and servos the automated vehicle on this trajectory.

With a third degree polynomial which verify the positions and courses of the two cars we obtain :

$$\tan(\delta) = 2L \cdot \frac{3\Delta Y - \Delta X \cdot \tan(\Delta\Psi)}{(\Delta X)^2}$$

with a fourth degree polynomial which verify the positions and courses of the two cars and the curvature Cf continuity of the follower path we obtain :

$$\overset{\circ}{\delta} = \frac{6V_f \cdot L \cdot b}{1 + \tan^2(\delta)} \text{ with}$$

$$b = \frac{4\Delta Y - \Delta X \cdot \tan(\Delta\Psi) - C_f \cdot (\Delta X)^2}{(\Delta X)^3} \text{ and}$$

$$C_f = \frac{\tan(\delta)}{L}$$

with simulation we can compare this three different methods : the like-truck one (1) , the cubic one (2) and the quadric one(3) (fig 4).

The other method is based on the following of a path: the memorized path of the leader car. We want to minimize the distance between the car and the path: y (fig 3). We work in a spatial state and we make the derivative on the curvilign absissa : s. , it gives the new equations relative to the path :

$$\begin{pmatrix} x' = \frac{dx}{ds} = \frac{\cos(\widetilde{\Psi})}{1-C(x)*y} \\ y' = \sin\left(\widetilde{\Psi}\right) \\ \widetilde{\Psi}' = C_f - x'.C(x) \end{pmatrix}$$

We choose k_1, k_2, k_3 in order to place stable eigen values in the spatial state.

$$y^{(3)} + k_3 \cdot y^{(2)} + (k_1 + k_2) \cdot y' + k_1 \cdot k_3 \cdot y = 0$$

So the control is :

$$\overset{\circ}{\delta} = \frac{V_f \cdot L \cdot b}{\left(1 + \tan^2(\delta)\right) \cos\left(\widetilde{\Psi}\right)}$$

with

$$b = y^{(3)} + C_f^2 \sin \widetilde{\Psi} - 3x' C(x) \cdot \widetilde{\Psi}' \sin \widetilde{\Psi} + \left(x'\right)^3 \frac{dC(x)}{dx}$$

This method is under experiment and we think is the best one to avoid cutting the curves.

3. Experiment

3.1. Sensors

From what we have seen previously, it can be gathered that we need for each vehicle a sensor capable of measuring the distance to the previous vehicle, their relative speed and the angle at which this previous vehicle is located with respect to the automated one. Furthermore, these measurements are needed with a noise and at a rate compatible with the constraints of the servo loop. This rate has been estimated through simulations at about 50 Hz

The sensor which has been developed at INRIA capable of these performances is based on a vision approach with targets located at the rear of each vehicle. The camera we selected is a linear camera with 2048 pixels capable of operating at 1000 Hz equipped with a spherical lens and a cylindrical lens in order to adjust to the changes of the relative angle between the two cars in the vertical plane. We have added an infrared filter and a polarized filter to minimize the influence of ambient light and sun reflections.

The target is made of three sets of LED organized in vertical lines and non co-linear(fig 5). This arrangement allows us to compute the three degrees of freedom of the previous car in the horizontal plane. Given the design parameters f, e and h, simple geometrical considerations (but involving many trigonometric computations), give us the three degrees of freedom that we are looking for : the distances Dx,Dy and the angle Dψ. The precision obtained is very high: 5mm at 10 m and through adequate filtering, we can obtain the relative speed that we also need.

We also have velocity sensors based on encoders for the motor and the rear wheels to know the longitudinal and angular velocity of the vehicle.

Other sensors will be developed soon like ultrasound ones for parking.

3.2. Experimental Results

In normal case, the hydraulic brakes are not needed ; the motor brake gives a deceleration of $2m/s/s$. As we can see (fig6) the regulated error stays under 30 cm and the distance between the two cars is equal to 4,5m at a 10 m/s speed.

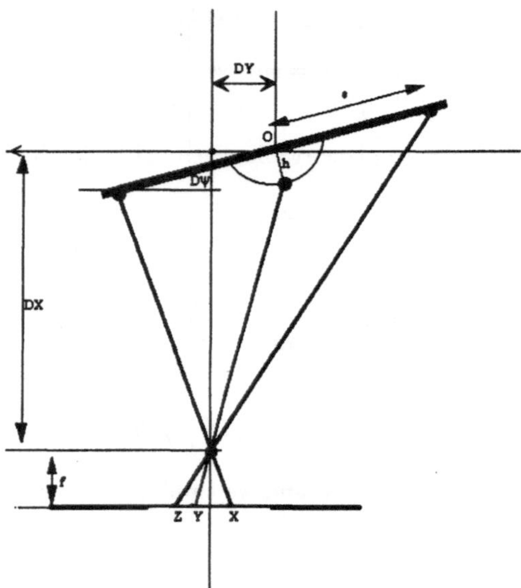

Figure 5. Target and camera scheme

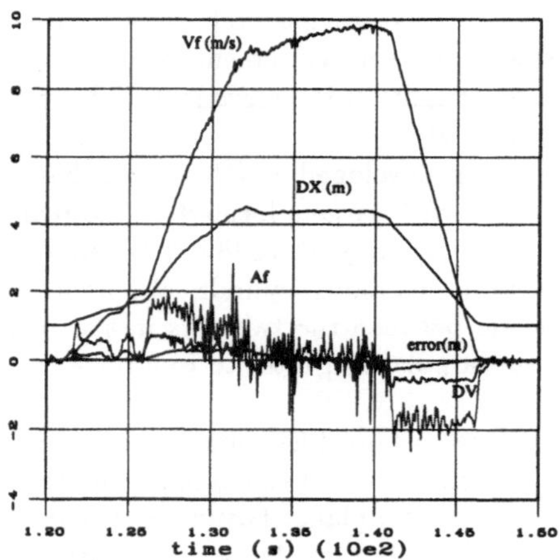

Figure 6. Experiment : normal case

We can notice the regulated error tends towards zero when speed is stabilized which shows that there is no static error.

In emergency case, the hydraulic brake (added to the motor one) gives a deceleration of $5m/s/s$ and the regulated error stays under 50 cm.(fig7) which is correct and guarantees train security.

As we memorize the leader path, we can measure the lateral error y between the two car paths. We verify (fig8) that like-truck following cuts the curves (up to 60 cm).

We are experimenting cubic following : first results are hopeful, indeed the error

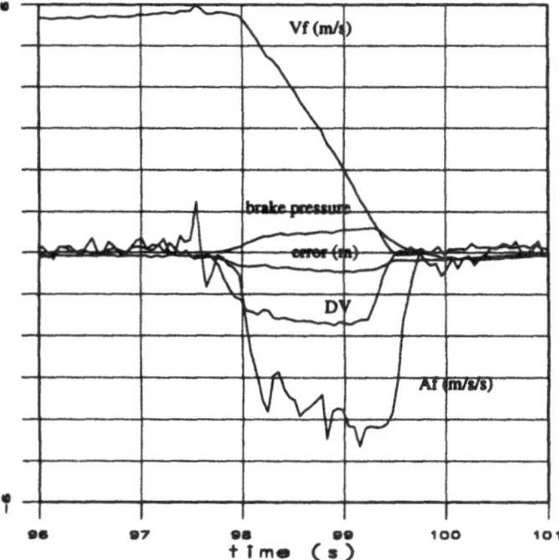

Figure 7. Experiment : emergency case

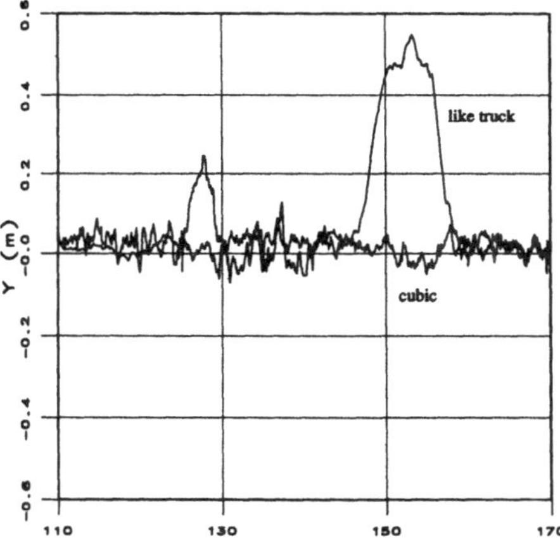

Figure 8. lateral errors

do not exceed 8 cm but the stability must be reinforced.

4. Conclusion

Using this new vision tool, we have demonstrated that it is possible to implement very close platooning of vehicles. This technique is essential for the concept of self-service vehicles because we need to move empty vehicles between distant locations.

However, this technique could also be used for the automatic highway where it is needed to build platoons of cars very closely spaced in order to increase the throughput. In fact, we have demonstrated that we can use the same sensor to locate the rear reflectors of any regular car if we send a stroboscopic infra-red light from the camera.

In a soon future, we are going to reduce the time between the vehicles to 0.2s

Figure 9. Train of vehicles

and validate the exact following of the leader path.

References

[1] Parent,M.,Daviet,P.(INRIA)."Automatic Driving for Small Public Urban Vehicles." Intelligent Vehicle Symposium Tokyo, July 14-16, 1993.

[2] Parent,M.,Daviet,P.(INRIA)." Automatic Driving in Stop and Go Traffic". Intelligent Vehicle Symposium Paris, october 24-26, 1994.

[3] Parent,M.,Daviet,P.,Abdou,S.(INRIA)."Vision technique for Platoon Driving",Applications of Advanced Technologies in Transportation Conference, Capri,Italy, June 1995.

[4] Kappelos,K.,Abdou,S.,Jourdan,M.,Espiau,B. "Formal Verification and Implementation of Tasks and Missions for an autonomous Vehicle" ISER95 Stanford,USA,June 1995.

[5] Schladover S.E. "Longitudinal Control of Automated Guideway Transit Vehicles Within Platoons". Trans. of the ASME. Vol.100. Dec. 1978.

[6] Tan,H. S. And Tomizuka, M., "A Discrete Time Robust Vehicle Traction Controller Design", ACC proceedings 1989.

[7] McMahon, D.H.,Hedrick, J.K. and Schladover, S.E." Vehicle Modelling and Control for Automated Highway Systems", ACC proceedings 1989

[8] Chiu H.Y., Stupp G.B. and Brown S.J. "Vehicle-Follower Control with Variable-Gains for Short Headway Automated Guideway Transit Systems". Journal of Dynamic Systems, Measurement and Control. Sept. 1977.

[9] Samson Claude. "Commande de véhicules non-holonomes pour le suivi de trajectoire..."Automatique pour les Véhicules Terrestres 22-sept-1993 Amiens-France.

[10] Slotine J.J. "Sliding controller design for non-linear systems." International Jour. on Control, 1984, vol. 40, p421-434.

ROBUST VEHICLE NAVIGATION

Michael Stevens, Andrew Stevens and Hugh Durrant-Whyte
Department of Engineering Science
University of Oxford, Oxford OX1 3PJ United Kingdom
{michael,as,hugh}@robots.ox.ac.uk

Abstract

This paper presents experimental results obtained in 'OXNAV', a 3-year research project investigating autonomous vehicle navigation systems. The project aimed to develop navigation systems for low-speed, continuously moving, indoor vehicles suitable for industrial use. Research addressed on two main issues: validation of distributed control algorithms for wheeled vehicles, and development of distributed filtering and sensor management algorithms capable of ensuring robust reliable operation in the presence of noise and clutter. The work presented in this paper focuses on the results obtained in lengthy experimental trials at active industrial sites and in the laboratory. Details of the system architecture and theoretical advances in distributed Kalman filtering that contributed to the results presented here have been previously published in [1,3,4,5,8,9,14].

1. OXNAV Localisation System

The OXNAV system's primary sensor is differential time of flight sonar [4,6,7,10]. Continuous localisation of a moving vehicle is achieved with this relatively slow sensing technology by using a directed sensing strategy. Instead of scanning the whole environment, the system uses rapidly stearable sonar detectors to lock onto and track a few particularly well defined features. These **target** sonar features (or, sonar **targets**) are selected from a prior map of the environment provided by the user. Thus, although no beacons are required, the OXNAV system navigates by observing the positions of fixed, known, targets. Location estimates are derived by fusing target observations, prior map data, and odometric information using an extended Kalman filter (EKF) [2] algorithm. The localisation filter is supported by sophisticated sensor management and data-association algorithms that control sensor data gathering. These ensure that only well defined targets likely to produce good positional information are selected (reducing noise sensitivity), and that sensor data likely to be associated with features other than the selected targets is rejected (reducing sensitivity to clutter).

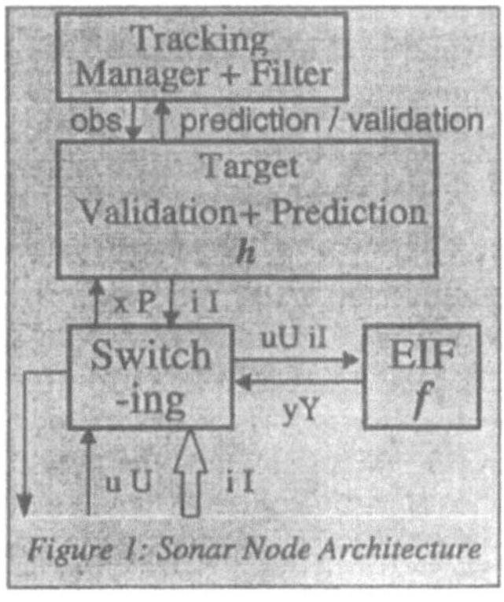

Figure 1: Sonar Node Architecture

1.1 Architecture

The OXNAV localisation filter is implemented as a distributed extended information filter (EIF) [2,8]. This allows it to 'partitioned' to support an efficient distributed implementation across seven processor nodes - one node for each sonar sensor or wheel controller. The processor nodes are physically integral components of the sensor / actuator modules. The sensor sub-system of each sonar module comprises four major components: the tracking filter, target predictor, target validator, and tracking manager. The target predictor's job is sensor management. It selects sonar features as targets for tracking that are **(a)** likely to be detected and **(b)** placed so that observations will minimise positional uncertainty. It accomplishes this task using a novel information-theoretic algorithm based on estimation of the mutual information between estimated vehicle location and target observations. The OXNAV target validator is responsible for data-association. It detects and rejects sensor data likely to be inconsistent with the target selected for observation. It employs a combination of geometric checks and a standard information-theoretic validation gate [2] using the sensor observation model of the localisation filter. The Tracking Manager (TM) triggers the selection of a new target when tracking is lost, data association is unreliable, or the gimbal limits of the sensor are reached. In performing this task the TM employs a focus of attention strategy: once established, a target is tracked until it is repeatedly rejected by the target validator and an alternative found by the target predictor.

Figure 1 shows the internal architecture of a sonar processing node indicating. The computations is partitioned between so that computations involving the sensor model **h** (filter correction, target prediction and validation) are all performed together by a single server process. This allows the (expensive) computation of $\nabla_x \mathbf{h}$ to be shared between the filter calculation and sensor management and permits the prediction and correction phases of the filter calculation to be performed partly in parallel for maximum throughput.

Target selection, and validation algorithms are presented in detail along with data motivating the TM strategy.

2. Experimental Results and Methods

System testing through extensive trials in active industrial sites formed a major part of the OXNAV project. These trials utilised a modular AGV developed specifically for the OXNAV project [1] (see Figure 2) and involved over 100 hours of vehicle running time. Through these trials we have obtained a great deal of data on the performance of the OXNAV system, both as a whole and its various hardware and software components.

2.1.1 Trials Programme

The trials sites covered a postal sorting office **S**, a residential nursing home **R**, a heavy vehicle factory **F**, and a working office **O**. An office corridor **C** was also used. The trials sites were deliberately selected to be as varied as possible, to avoid the possibility of particular environmental characteristics hiding flaws in the navigation system. The residential environment, for example, provided a mix of targets ranging from easily navigable corridors to sonar 'no go' areas surrounded by soft furnishings with no visible permanent targets. The industrial site, by contrast, provided a pitted concrete floor with a very high level of ambient noise. In between the two, the office was characterised by many good targets but also significant clutter. The sorting office had many impermanent sonar features.

Figure 2: Trials Vehicle

A typical trial comprised the mapping of several hundred square metres of trials space, followed by two or more days testing a wide variety of runs through this space. Example trial site maps are shown in Figure 3. All sites were 'live' and in constant use during the trials period, and thus subject to normal levels of change and human clutter. Operating speeds for the OXNAV system are currently restricted to 0.25 m/s, with 0.20 m/s usual for trials. The RCDDs are, however, capable of tracking targets while moving at up to 0.4 m/s, sufficient for commercial indoor applications. Higher speed would, however, require a more complex plant model then is currently implemented.

2.1.2 Filter Performance Assessment

Conventionally the most important objective performance measure for a Kalman filter is the whiteness (or otherwise) of the innovation sequence associated with the filter estimates. In the case of the OxNAV navigation system this cannot easily be applied. The set of features being observed changes continually as the vehicle moves. There are no lengthy innovation sequences based on comparable sets of observations that can be analysed for whiteness. Thus, a variety of alternative performance measures had to be developed to assess the performance of the navigation system.

This subjective analysis revealed a variety of useful observations regarding the performance of the OxNAV navigation system. Firstly, despite the relatively crude sonar's used in the RCDDs, inaccuracy in localisation exceeded 5 cm only in exceptional circumstances (e.g. when one or no targets were trackable for an extended period). Localisation accuracy was, furthermore, remarkably consistent between environments, with successful navigation accurate to 2-3 cm achieved at all trial sites. The only notable variation between trials was a modest increase in target tracking failures in the noisy industrial environment. The percentage of tracked targets validated for the various sites are shown in Table 1.

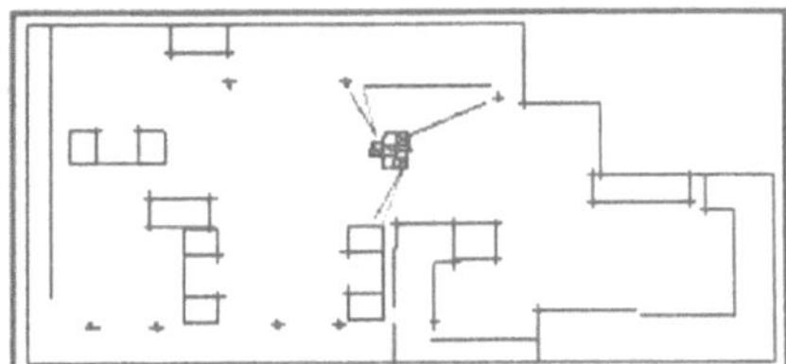

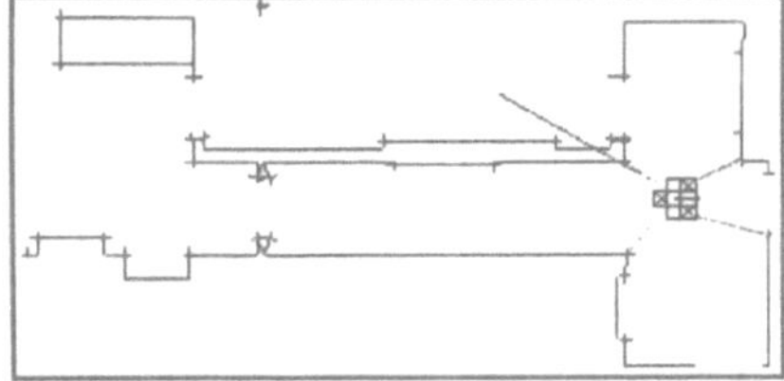

Figure 3: Target maps and tracking display for sites **O** *and* **R**

The most important source of information for overall systems performance was subjective comparison of estimated state covariance and positions from a graphical front-end (Figure 3) against externally observed positional error. The occurrence of significant inconsistencies was used to assess overall sensing precision and detect failure scenarios.

Performance proved to be highly robust. Trials in busy areas often resulted in obstacles being placed between the vehicle and its targets. The system coped easily with this type of clutter, and remained stable even when much of the mapped environment was obscured. Failure only occurred if multiple RCDDs were obscured and could be readily detected by a sharp fall in correct data-associations.

2.1.3 Performance Measures

Site	O	R	F	S
Mapped Targets	32	125	76	136
Validated Targets	39%	41%	64%	49%
$\lambda_{1\sigma}$ (cm)	1.78	1.47	2.57	2.55

Table 1: Trial Site Statistics

A variety of objective performance measures were obtained from the logs of state-covariance and data association rates. The most useful of these proved to be λ, the square-root of the largest eigenvalue of the estimated state covariance **P**. Expressed as a distance this represents a single standard deviation (1 sigma limit - λ_{1_σ}) in the largest component of positional uncertainty. Analysis of λ-logs (see Table 1 and Figure 4) revealed robust system performance up to location covariance of 5 cm. Beyond this level of positional uncertainty, system failure would occur frequently due to mis-assocation of targets, and the estimated position would diverge from the vehicle's actual position. The total uncertainty given by the determinant of **P** was also used, but proved a less useful indicator.

Figure 4 shows a sample run with the filter set-up to behave poorly where, after around 40 seconds, no targets are matched and λ grows. In this case the vehicle is 'lucky' and recovers after correctly associating new targets. Data association performance was also judged subjectively by comparing front-end logs against the actual locations of the vehicle and sonar targets.

The systems sensitivity to its prior map (measured by hand) was evaluated using deliberately corrupted maps. This revealed that although accuracies of a few centimetres were sufficient, it was very important to achieve consistency in the mapping of targets visible from the same location. Inconsistent maps significantly increased the probability of navigation failure. This phenomenon is clearly illustrated in Figure 4. After 90 seconds an incorrectly mapped target is tracked causing others to be rejected and the vehicle to become lost with λ rising steadily.

The performance of the sensor management system in selecting targets was assessed using a variety of statistical measures: percentage time on target, percentage successful target predictions, and prediction request rates. Figure 5 shows the identity of the validated targets for a single RCDD (sensor number 4) during a short trial run. It can be seen that several distinct targets are tracked during the run and target tracking fails every few seconds but is often re-established.

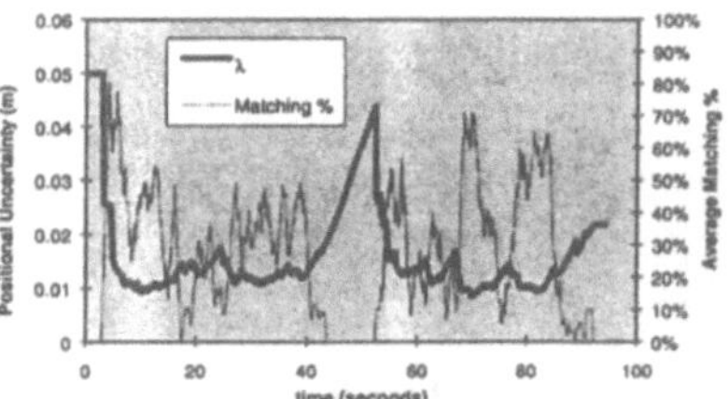

Figure 4: Data-association performance

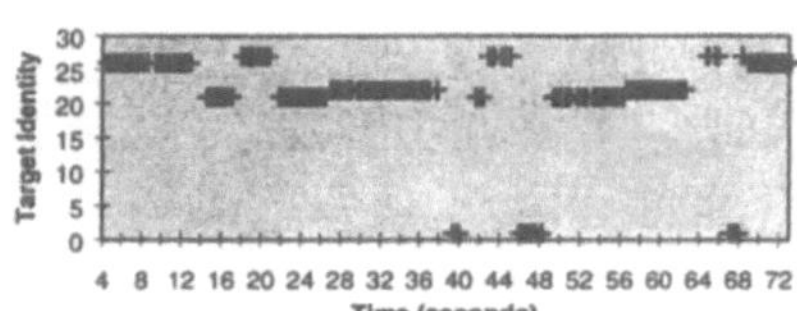

Figure 5: Change of Target Identity over time

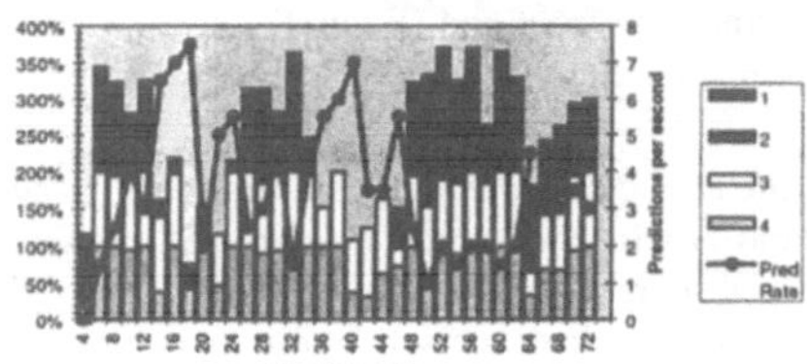

Figure 6: Time on Target and Prediction Rate

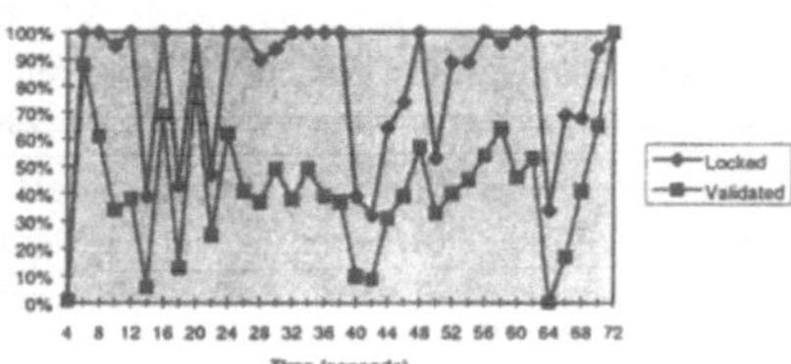

Figure 7: Locking and Validation Performance

The percentage of time an RCDD spends locked onto a target provides a useful measure of target visibility for a particular sensor and how successful tracking is. Figure 6 shows the percentage time each RCDD spends locked on targets throughout the same short run displayed in Figure 5. RCDDs 3 and 4 spend most of their time locked onto targets. RCDD 2 is not very successful with only the period between 48 and 60 seconds being spent stably locked. The prediction rate, the number of target predictions being made per second, is also shown. It can clearly be seen that the prediction rate increased during the periods when one or more RCDD are spending little time locked.

The track manager determines when lock is broken and a new target prediction is requested based on differential tracking and the validation of observations. Although an RCDD may be locked, only some proportion of observations will be validated. Figure 7 details the percentage time RCDD 4 is locked and also the percentage time validated readings are produced. It can be seen that the RCDD spends most of it time locked with occasional target change over, see also Figure 5. Half its time is spent producing valid observations, more than adequate for position update, indicating the TM is successfully dwelling on the locked targets.

2.1.4 Summary Statistics

Table 2 gives the statistics summarising system performance for whole runs at various sites. The trial runs at the office site **O** use newer tracking and management algorithms, significantly improving overall performance. For comparison two set of data are given for this site including the run shown in Figure 4 were the filter becomes lost. The major statistics, lock and valid percentage times and predictions per second, summarise Figure 6 and Figure 7 as discussed previously. The very low valid observation time in $\mathbf{O_{fail}}$ clearly indicates repeated target validation failure which causes the system eventually to become lost. The ultimate cause of failure proved to be the inconsistent mapping of a prominent target.

Site	O_{good}	O_{fail}	C	S	R
lock (% time)	60%	22%	36%	29%	29%
valid (% time)	30%	14%	26%	18%	19%
predictions / s	0.87	2.04	1.45	1.32	1.80
predictions locked	32%	31%	62%	52%	48%
re-prediction	19%	24%	47%	42%	38%
re-lock	38%	65%	72%	72%	69%
servo time (s)	0.43	0.23	0.32	0.38	0.29
lock time (s)	0.03	0.03	0.02	0.03	0.02

Table 2: Summary Statistics

Surprisingly the percentage of predictions locked is lower for **O** then for the other sites. It is assumed that this is due to the relatively large amount of clutter in this site obscuring some of the best targets.

The re-prediction and re-lock percentages are included to evaluate the TM's performance. Ideally the TM should only break lock from a target when it is no longer observable, and therefore unlikely not be re-predicted. In **O** only 19% are re-predicted of which 38% re-lock indicting the lock was incorrectly broken.

Servo time is the time on average it takes for the RCDD to servo given a new prediction. This is dead time as no observations can be made. This is artificially low for the older algorithm due to the large percentage of re-locks where relatively little servoing is required. Lock time is the time required before observations of a predicted target are validated after the servoing period. Generally if a target locks it does so almost immediately giving the very low figure.

We believe the performance measures devised in the OXNAV trials provide a useful basis for the systematic testing of the robustness of robotic system.

2.2 RCDD wide-beam tracking sonar

The main sensor employed by the OXNAV navigation system is the RCDD 'Region of Constant Depth Detector' differential time of flight tracking sonar. The final RCDD design (illustrated in Figure 8) mounts a pair of Polaroid electrostatic transducers on a simple panning frame driven by a stepper motor. Two transducers are used so that by using both as receivers for a single sonar pulse differential time of flight data can be obtained. This provides a means of obtaining the real-time target bearing information needed for target tracking. A single sonar transducer cannot provide real-time bearing information of this type. Sonar features produce returns on a range of bearings corresponding to the sonar beam width. Thus, since beam width is typically about 20 degrees, scanning even the smallest sonar targets produces an arc of returns ('region of constant depth') rather than a well defined point [4,5,7,10].

Trials results (see section 2.1.1) demonstrated that this (low-cost) sensor was completely adequate for indoor navigation. However, some significant short-comings could be noted. Firstly, of the standard Polaroid driver electronics imposed an inconveniently long blanking period after transducer firing. This blanking period - needed to allow transducer ringing to subside - equates to a minimum sensor range of around 50 cm. In narrow corridors this proved a significant handicap as otherwise excellent sonar targets on nearby walls had to be rejected in favour of much more distant targets. An improved sensor design with modified driver electronics could employ a separate transmit-only transducer with two receive-only transducers to allow target ranges down to around 15 cm to be measured. This would greatly improve performance in long corridors with smooth walls lacking occasional corner reflectors.

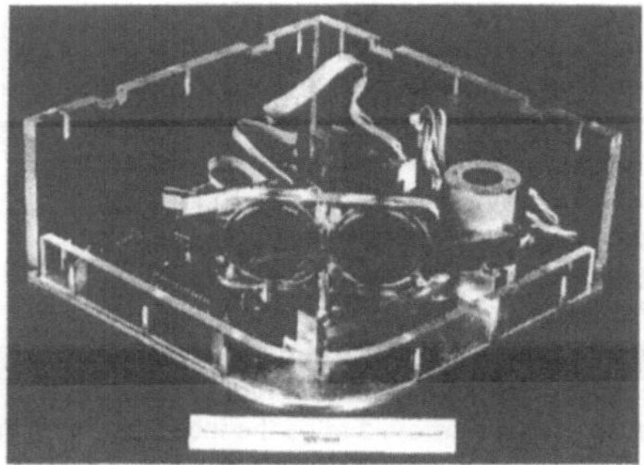

Figure 8: RCDD Sonar Sensor

Another significant limitation of the RCDD is the presence of significant bias in the bearing information obtained from differential time of flight readings. The Polaroid receiver trigger circuit thresholds an integral of total received sonar energy. Weaker returns will thus typically trigger the receiver a little later than strong returns. This is most undesirable as unpredictable acoustic effects such as diffraction or target irregularities often result in significantly different signal return strengths at an RCDD's two transducers. The net result is that the bearing information obtained from an RCDD often exhibits a significant bias that varies unpredictably between different targets and target aspects. This unpredictable bias in the bearing information makes it unsuitable for data-fusion. The vehicle position estimates are, in consequence, based only on the range data obtained from the RCDD's. The bearing data is used only for target tracking and data-association. Improved results might be obtained using multi-hypothesis filtering techniques to obtain estimates of bearing bias so as to permit both bearing and range information to be used in obtaining position estimates. Alternatively, using more advanced receiver circuitry with echo detection based on matched filters would greatly reduce this problem [10].

2.3 Navigation Filter

The numerical stability of the distributed extended information filter calculation [8] proved to be an important issue in the implementation of the system. This formulation of the extended Kalman filter presupposes that all the distributed filter nodes maintain the exactly the same estimate of vehicle location. In the estimation phase of the filter computation the estimate at each node is updated with a correction that is the *sum* of the partial corrections computed at each node. Thus, if estimates differ between nodes only the summed error is corrected for from sensor observations. This has important practical implications: if numerical rounding errors differ between nodes, but largely cancel each other out the error at individual nodes may accumulate over time. Such increasing errors in a node's local estimate will eventually produce errors in error covariance estimates etc. that lead to filter divergence. During our trials runs this pathological behaviour appears to have cause occasional failures of the navigation system after very long periods of continuous operation.

Ensuring that all nodes use the same values for the state estimate and error covariance solves this problem. Crude but effective measures include broadcasting the values of a single distinguished single node or broadcasting the values for all nodes and using their mean. Alternatively, if all nodes use processors with identical arithmetic units, it is suffices to ensure the computations at each node are performed in the precisely the same order. In practice this either entails buffering or the use of a fair low-latency prioritisation scheme [12].

2.4 Environment Maps

A sensor based navigation system such as OxNAV requires a representation of the environment in which it is to operate. In OxNAV, the primary assumption is that the stable features in an indoor environment can be held in a geometric map. Maps of actual environments are included for our trails sites discussed later. These maps are used to validate and fuse all sensor observations, and are never altered by the navigation system. The maps represents all the *prior* information required concerning the environment.

In practice walls, door jams, and any permanent furniture are included in the map. Any unmapped objects are termed *clutter* and the validation system in OxNAV will reject any sensor observations of these objects. OxNAV uses a simple geometric representation of joined line and arc (of variable radius) segments. Where lines follow from each other corners are assumed. Solid objects are represented by closed paths, with the object only being visible from the outside. This representation was found to be fully adequate for the types of objects described above.

Internally OxNAV automatically transforms the given map prior to navigation into a segmented target list. For sonar, a specular target model is used. Ultrasound of the frequency used in the directed time of flight sensor (40kHz) reflects primarily specularly 4,5,7]. Each target generated contains the mapped position and a list of visibility criteria. These indicate from which positions the target would be visible if not occluded by another. To allow rapid prediction of which targets can be seen from any given position the map can be segmented into a grid structure using the given position and visibility criteria. The OxNAV maps can also be transformed into a diffuse target model for use by other sensors, primarily Infra Red rangers [9]. Although the maps are inclusive in character, for different target models it is found that some objects are more relevant to a particular model type. For example a piece of conduit on a wall forms an excellent sonar target in its own right and should be included in the map. However for Infra Red, it can usually be removed from the target list as it makes little difference to the wall depth.

The system's sensitivity to its prior map (measured by hand) was evaluated using deliberately corrupted maps. This revealed that although accuracies of a few centimetres were sufficient, it was very important to achieve consistency in the mapping of targets visible from the same location. This is because inconsistent target map locations can result in inconsistent corrections to the location estimate that partly cancel each other out. Long periods with under-corrected position estimates inevitably result in an increased probability of the navigation system failing due to the filter diverging. Inconsistencies between widely separated regions of the map have no such effect as, relative to the 'frame' defined by a local set of map features filter estimate corrections are consistent. Isolated, badly erroneous, map features are also benign in that they are simply ignored as any returns will fail to validate against the map. The need to properly allow for 3D effects in the (2D) target map also provided some difficulties. Predicting the appearance of sloped or stepped vertical surfaces to the sonar sensor proved very difficult. Both map consistency and 2D/3D problems are, of course, eliminated or greatly reduced if maps derived using autonomous sonar map-building techniques are used [4,11].

3. Conclusions and Analysis

The systematic assessment of system / component performance in realistic trials proved crucial to the success of the research programme. Without the feedback from the statistical measures we developed many significant flaws would have been missed and many useful insights lost. With hindsight it also seems clear that the provision of some means of accurately determining absolute position during trials would have been invaluable. Without accurate absolute position data it has proved difficult to properly asses the adequacy of the sonar sensor model in the field. Accurate position information would also have made (timely) detection of filter divergence easier and would thus have greatly simplified debugging during system development.

Overall, the OXNAV system successfully achieved near-industrial levels of robustness and reliability. Despite the relatively crude sonar's employed, a localisation accuracy of 2-3 cm was reliably achieved at all trials sites. Performance was also highly robust with runs of 1/2 an hour or more readily achieved. The only notable variation between trials was a modest increase in target tracking failures in the noisy industrial environment. The system coped easily with clutter from moveable obstacles. Failure only occurred if multiple sonars were obscured for extended periods. Failure could be readily detected by a sharp fall in correct data-associations. The system's robustness is largely attributable to selective tracking of the best available targets combined with the EIF's ability to fuse data over time via its plant model.

The system's main weakness was the relatively low quality of the vehicle bearing estimates. The information on position and heading obtained from an observation of range is cross-correlated. Thus, given observations of only 1 or 2 targets (as is typical) the filtering calculations cannot completely decorrelate estimates of position and heading. In more recent laboratory trials considerably improved performance was achieved through the addition of a low-cost solid-state gyro compass (Murata Gyrostar [13]) to prove uncorrelated heading information. The addition of this device greatly increased the system's resilience to disturbances from bumps and minor collisions when tracking only a few targets.

4. Bibliography

[1] T. Burke and H.F. Durrant-Whyte. Modular mobile robot design. *1st IFAC International Workshop on Intelligent Autonomous Vehicles*, Southampton, UK, 1993

[2] Y. Bar-Shalom and T.E. Fortmann. *Tracking and Data Association.* Academic Press, 1988.

[3] S. Grime, H.F. Durrant-Whyte, and P. Ho. Communication in decentralized sensing. *Proc. American Control Conference*, 1992

[4] J.J. Leonard and H.F. Durrant-Whyte. *Directed Sonar Navigation*. Kluwer Academic Press, 1992

[5] J.M. Manyika and H.F. Durrant-Whyte. *Data Fusion and Sensor Management*. Ellis Horwood, 1994

[6] J. Crowley. World modelling and position estimation for a mobile robot using ultra-sonic ranging. *Proc. IEEE Int. Conf. Robotics and Automation*, 1985

[7] B. Barshan and R. Kuc. Differentiating sonar reflections from corners and planes by employing and intelligent sensor. *IEEE Trans. Pattern Analysis and Machine Intelligence*, 12(6):560-569, 1993

[8] A.G.O. Mutambara and H.F. Durrant-Whyte. Non-linear information space. *The Intl. Soc. for Optical Eng. (SPIE) '94*, 1994

[9] Stephen Borthwick, Michael Stevens and Hugh Durrant-Whyte. Position Estimation and Tracking using Optical Range Data. *Proc. IEEE Int. Conf. Intelligent Robots and Systems,* 1993

[10] L. Kleeman and R. Kuc. An optimal sonar array for target localisation and classification. *IEEE International Conference on Robotics and Automation.* 1994. San Diego USA.

[11] I.J. Cox and J.J. Leonard. Probabilistic Data Association for Dynamic World Modelling: A Multi-Hypothesis Approach. *Proc. International Conference on Advanced Robotics (ICAR), Pisa, Italy.* 1991

[12] A.Stevens, M.Stevens. Low-latency ordering for distributed data-fusion. *Correspondence. Dept. of Engineering Science, University of Oxford.*

[13] Murata Manufacturing Co Ltd. Gyrostar ENV-05 Piezoelectric Vibrating Gyroscope.

[14] A. Stevens, M. Stevens and H.F. Durrant-Whyte. 'OxNav': Reliable Autonomous Navigation. *Proc. IEEE Int. Conf. Robotics and Automation*, 1995

Dynamic Analysis of Off-Road Vehicles

Faïz BEN AMAR and Philippe BIDAUD
Laboratoire de Robotique de Paris - Centre Universitaire de Technologie
10-12 avenue de l'Europe - 78140 Vélizy - FRANCE.
e-mail : amar@robot.uvsq.fr

Abstract

This paper deals with the simulation of off-road robots while taking into account the mechanical behavior of the locomotion system and its interaction with its environment. This interaction is studied and discussed for different behaviors of wheel-soil contact. The analysis considers phenomena of slips, of soil shear deformations, of soil compaction and wheel elastic deformation. Wheel-soil contact models are expressed by analytical relationships which link each contact force components (radial, longitudinal and lateral forces) to relative displacements (radial displacement, longitudinal slip ratio and side slip angle). These laws are then coupled to the dynamic equations of the mechanical system in order to characterize the behavior of the whole system. These models are implemented on a graphic simulation system which provides a basic tool for the design of the mechanical system, for the path planning and for the definition of control schemes.

1. Introduction

Today, the automation of off-road vehicles has many applications such as in agriculture, in construction, in planetary exploration or for military applications. Their common function is to realize displacements of loads or of work tools. The performance of the vehicle depends on the intrinsic dynamics of the mechanical system and on the behavior of the vehicle-soil contact. This interaction is very complex to analyze because it introduces characteristics both of friction and deformation. Moreover, the topology of the vehicle's mechanical system is relatively complex and can present a large number of mobilities.

Advanced engineering of these systems for optimizing the mechanical design, for the definition of the control schemes, or for the mission planning must be based on a powerful simulation system.

We present in this paper, a simulation system for off-road wheeled vehicles considering both the mechanical behavior of wheel-soil contacts and that of the locomotion system considered as an articulated structure.

The navigation problem on uneven terrain has been addressed in a limited number of works. Some of them concern path planning by considering geometric or simple dynamic constraints [7] [6]. In other respects, some mechanical concepts for off-road robots are proposed in [4], [3], but these studies remain in qualitative form. In this paper, we choose a classical mechanical approach for vehicle behavior modeling as in [2] contrary to [13] and [14] who use a physical modeling approach.

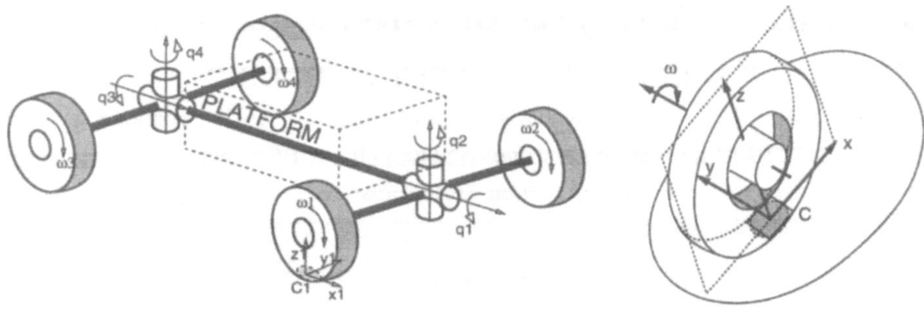

Figure 1. Schematic view of 4 wheeled vehicle Figure 2. Contact frame

Vehicles considered in this paper consist of n axles ($n \geq 2$). each axle joins two conventional wheels (cf. fig.(1)(7)). Two contiguous axles are articulated along the yaw direction to permit vehicle steering, while articulations along roll and pitch directions allow their suspension on three-dimensional soils. The joints, including those of wheel rotation, can be passive (free joint, spring-damper system) or active.

We choose B-spline functions to represent the soil-surface (fig.(7)), because of their differential properties and their easy use. The B-spline control points are specified using a graphic interface.

In this paper, we will first study the wheel-soil contact problem. We will show in this section how we establish a closed form contact model. Next, these local models are introduced in the dynamics of the vehicle. Finally, we will give some simulation examples to illustrate the dynamic behavior of the vehicle.

2. Wheel-soil contact

The aim of this section is to establish closed-form model of wheel-soil contact which could later be integrated into the dynamic equations of the locomotion system. A rolling wheel always slips under external forces and torques. A contact behavior model is represented by a set of relationships which express contact force components as a function of the relative motion between the wheel and the soil.

At each wheel-soil contact C, we define a local frame $\mathcal{R}_C = (C, x, y, z)$ (cf. fig.2), where z is the external normal to the contact plane, x is the longitudinal tangential vector and y is the lateral tangential vector.

In the general case of rolling with sliding, the wheel displacement velocity is defined by two sliding parameters:

- the longitudinal slip ratio s which is defined as:

$$s = \begin{cases} (V_x - V_r)/V_x & \text{while braking} \\ (V_x - V_r)/V_r & \text{while traction} \end{cases} \tag{1}$$

  where V_x is the longitudinal displacement velocity of the wheel center and $V_r = R\omega$ (ω expresses the wheel rotation velocity and R its radius). We differentiate braking and traction cases in order to consider the two critical cases: (1) when the wheel is locked without rotating $V_r = 0$, (2) when the wheel spins without traveling $V_x = 0$,

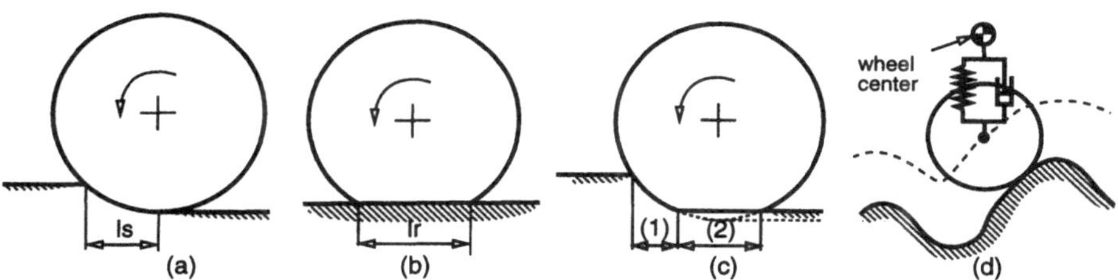

Figure 3. (a),(b),(c) Different contact surfaces considered for computing tangential forces, (d) normal behavior of wheel-soil contact

- and the side slip angle α defined by:

$$\alpha = \arctan\left(V_y/V_x\right) \tag{2}$$

V_y is the lateral component of wheel displacement velocity.

Contact forces are obtained by summing normal and tangential stresses along the contact patch. By considering wheels with a conventional width, momentum along z axis can be neglected. The resulting force can be decomposed into 3 components along the axis of $\mathcal{R}_C$: the normal force F_z, the longitudinal force F_x and the lateral force F_y. The longitudinal force F_x is controlled directly by the wheel torque and can then be positive or negative. There is another longitudinal component, called rolling resistance. This force, denoted by $\tilde{F}_x$, is always opposite to the travel direction and thus has a negative value. This force is due to soil compaction or to the energy dissipation in the tire carcass.

2.1. Tangential behavior

The interaction between a wheel and the soil is mainly governed by deformations of bodies in contact. A realistic model must take into account the material behavior laws of this bodies. It seems judicious to investigate the three following cases (fig.3):

1. a rigid wheel on soft soil,
2. a flexible wheel on rigid soil,
3. and a flexible wheel on soft soil.

We thus practically cover all possibilities for a macroscopic observation of contact behavior. The case of a rigid wheel on a rigid soil is not developed because, except for railway vehicles, it is not interesting in practice.

Rigid wheel on soft soil : This case, not very useful, is interesting to study in order to deduce the general case. We note that a pneumatic wheel with a hight inflation pressure on a relatively soft soil can be assumed to behave as a rigid wheel. For agricultural or military vehicles we use in general empirical models for the prediction of the soil mobility and the vehicle performances. These models are based on the characterization of the soil physics by the cone index value [12]. The latter is measured by a cone penetrometer. This method is well known to be insufficient to characterize the terrain mobility. The other method to characterize the terrain physical properties is that developed by Bekker [5]. This method uses the bevameter technique which is composed of a sinkage test and a shear test. The first test provides the sinkage parameters k_c, k_Φ and n_s defined in the following relationship [5]:

$$\sigma = (k_c/w + k_\Phi)z^{n_s} \tag{3}$$

where σ is the contact normal pressure, z the soil sinkage, w the contact width. The second test measures the shear parameters c, Φ and K which are defined in the Mohr-Coulomb law:

$$\tau = (c + \sigma \tan \Phi)(1 - \exp -j/K) \tag{4}$$

where τ is the shear stress and j the soil tangential displacement. These stresses are summed along the contact patch in order to express the tangential forces [1]. Their analytical expression are given in appendix A.

Elastic wheel on rigid soil : The wheel is considered to have an elastic behavior with different stiffness values along the 3 directions. The damping effect of a tire are in general neglected for computing the tangential forces. There are many works which attempt to model this case because of the general use of the pneumatic tyres. We choose the analytical model given by Gim et al. [8][9] who use the parameterization of contact motions and forces defined in paragraph 2. In this model, the portion of the wheel in contact with the soil is supposed to be composed of a set of bristles which are flexible in the two directions of the tangent plane. The relationships are given in appendix A.

Elastic wheel on soft soil : This case does not have an analytical solution. We propose a solution by a simple combination of the 2 previous cases. We assume that the contact is established into 2 regions (cf. fig.3(c)) : (1) a first region where the wheel has a rigid behavior on a soft soil, (2) a second one where the wheel has an elastic behavior on a rigid soil. Tangential components are then given by the sum of components exerted at each region (cf. appendix A).

The different models, developed here, are homogeneous since they use an identical parameterization. To sum up, these forces are expressed as a function of the longitudinal slip s, the side slip angle α, the normal force F_z and then they can be written as follows:

$$F_x = f(s, \alpha, F_z, ...) \tag{5}$$
$$F_y = g(s, \alpha, F_z, ...) \tag{6}$$

2.2. Normal behavior

The normal force depends on normal wheel and soil deformations. The computation of the contact surface between an elastic wheel and a soft soil locally irregular needs a refinement of the contact by considering a finite-element mesh [10]. In order to provide a realistic and useful model, we choose to characterize the normal behavior as a rigid tread band, connecting to the wheel center by a parallel joint of a spring and a damper, and rolling on a rigid geometric profile of the soil. This is illustrated in figure 3(d). If δ is the relative displacement between the wheel center and that of the rigid tread band, the normal force (including the case of contact breaking off) can be expressed by:

$$F_z = \max(0, k_z\delta + a_z\dot{\delta}) \tag{7}$$

where k_z and a_z express the radial (or normal) stiffness and damping of the wheel.

3. Dynamic equations

The development techniques of dynamic equations for wheeled or walking robots has been widely discussed. However, these equations are based on no-sliding condition. This assumption implies a mechanical topology with parallel chains, then the dynamic model is defined by a differential-algebraic system.
The topology considered in this paper is fundamentally different since there are not kinematic constraints at wheel-soil contact (3 relative rotations and 3 relative translations since we consider the longitudinal and the lateral slips and the normal displacement). The mechanical topology can be then considered as to be composed of a set of open chains (the wheel-soil contacts is not a kinematic link but a dynamic interaction). Therefore, dynamic model is then represented by the differential equations which are deduced from the application of the Newton-Euler formulation. The assumptions are:

- the bodies system composed by (1 axle, 2 wheels) can be considered as a gyrostat since its inertia properties (mass center, inertia momentum) are constant in this system of bodies [11]

- the links between axles and plate-form (fig.1) or between modules (fig.7) are considered massless.

By substitution of forces exerted at internal links and by including contact forces (function of the relatives velocities and displacements), the motion equations system can be written as follows:

$$\mathbf{A}\ddot{U} = B(\dot{U}, U, \Gamma, \ldots) \tag{8}$$

where $\dot{U}$ is the vector of kinematic parameters, $\mathbf{A}$ is mass matrix and Γ is the torque vector at internal links. For the example of the 4 wheeled vehicle given in figure 1, $\dot{U} = (u, v, w, p, q, r, \omega_1, \omega_2, \omega_3, \omega_4, \dot{q}_1, \dot{q}_2, \dot{q}_3, \dot{q}_4)$ where (u, v, w, p, q, r) are the kinematic parameters of the reference body (plate-form).

The direct dynamic problem concerns the prediction of the behavior of the mechanical system under the effect of a given actuator torque. Some examples, illustrating this problem, will be discussed in the next section. The inverse dynamic problem is not studied because vehicle motion with slipping and skidding is instantaneously uncontrollable and then there is often no solution of actuator torques which produce a given accelerations on the plate-form.

4. Simulation

For a given set of initial conditions (kinematic and geometric parameters of the system) and a given set boundary conditions (geometric and physical properties of the terrain), we resolve at each instant the equation (8) which provides the reference body and the joint accelerations. By means of two successive numerical integrations, we deduce kinematic and geometric parameters. This simulator is interfaced with a graphical modeler ACT[1] running on a Silicon Graphics Workstation.

As an example, figure (5) shows the difference between the kinematic trajectory based on no-sliding conditions and the trajectory considering side slip angles which are given in figure (4). This is obtained for the 4 wheeled vehicle of figure (1) and for a given steer angle profile of the front axle (q_2) which is shown in figure (4). The side slip angles of the front wheels are bigger than that of rear wheels. This produces a difference between lateral forces (called steering forces and are quasi-proportional to the side slip angle α for small values) and consequently an angular acceleration of the vehicle.

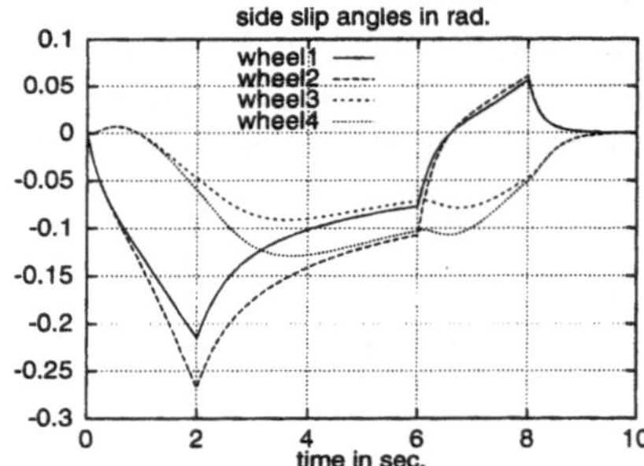

Figure 4. Side slip angles for a given steering angle profile for the 4 wheeled vehicle of figure (1), speed=4m/s

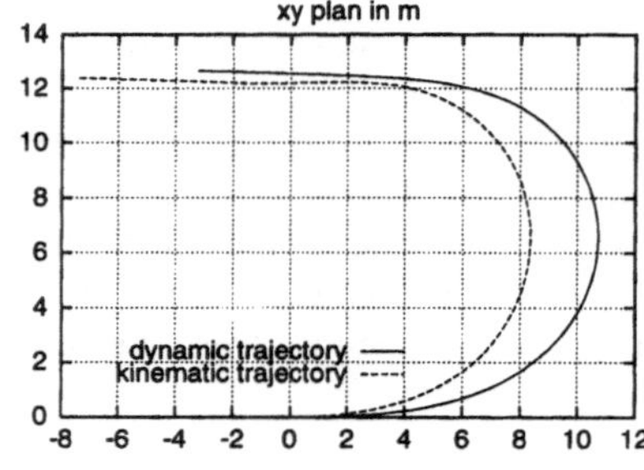

Figure 5. Dynamic and kinematic trajectories for the given steering angle profile

Figure (6) shows the vehicle behavior when it crosses over a lateral bump at 4m/s speed. The vehicle has a spring-damper suspension on front and rear axles which are articulated in the roll direction. We notice that the contact deformations are not represented graphically. The normal behavior model used here appears to be insufficient for simulating the crossing over hight obstacles. This model must be completed by other spring-damper elements arranged along the wheel radial directions.

An answer to the control problem is tested for a quasi-static motion and for a small slipping assumption. In this case, actuator torques are computed in order to have an optimal distribution of contact loads. In the other side, the command velocities are computed from a trajectory tracking procedure. For a given parametric soil and by means of an iterative analysis, we can then deduce the trajectories which satisfy the main operating conditions such as stability, adherence and manoeuvrability (fig.7).

5. Conclusion

A method for modeling and simulating the complex mechanical behavior of off-road wheeled vehicle has been presented in this paper. Models take into account at once the dynamic of the mechanical system and particularly the mechanical interaction between the wheels and the soil which could have different behaviors. This simulator will be exploited to define a set of strategies and rules for crossing difficult regions or for the negotiation of typical obstacles in order to integrate them into the local

[1]ACT: by Aleph-Technologies

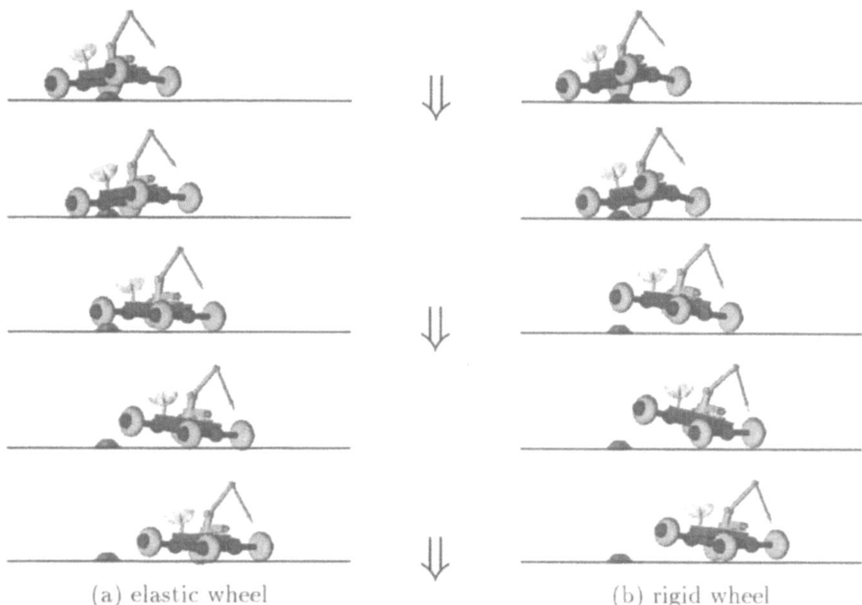

Figure 6. Dynamic negotiation of a lateral bump at 4m/s speed

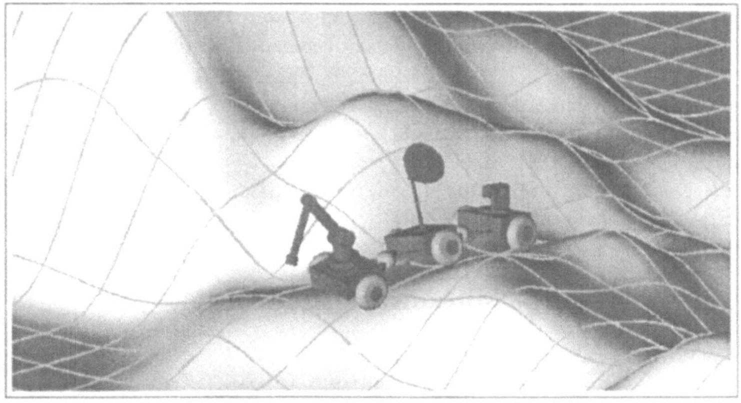

Figure 7. Simulation of quasi-static motion for a six wheeled articulated vehicle

motion planner of the robot. The main problem remains the definition of the control schemes of the locomotion sub-systems (traction/braking, steering and suspension). The necessity of low computation time and the uncertainty or the change of some parameters (as that of the soil) require the introduction of adaptative and reactive modes into control schemes.

References

[1] F. BenAmar *Modèles de Comportement des Véhicules Tout Terrain pour la Plani-*

fication Physico-Géomét-rique de Trajectoires, Thèse de Doctorat de l'Université de Paris VI, Juillet 94.

[2] F.X. Potel, F. Richard and P. Tournassoud. *Design and execution control of locomotion plans*, Proc. of the Int. Conference on Advanced Robotics, Pisa, 1991.

[3] F. Littman, E. Villedieu. *VAP Concepts de l'Organe Locomoteur*, Reports of VAP program, CEA, 1991.

[4] K.J. Waldron, V. Kumar and A. Burkat. *An Actively Coordinated Mobility System for a Planetary Rover*, Proc. Int. Conf. on-Advanced Robotics, 1987.

[5] M.G. Bekker. *Introduction to Terrain-Vehicle Systems*, The University of Michigan Press, 1969.

[6] Z. Shiller and J.C. Chen. *Optimal Motion Planning of Autonomous Vehicles in Three Dimensional Terrains*, Proc. of IEEE Int. Conferences on Robotics an Automation, Cincinnata, 1990.

[7] T. Siméon, B. Dacre-Wright. *A Practical Motion Planer for All-Terrain Mobile Robots*, IEEE/RSJ Int. Conferences on Intelligent Robots and Systems, Yokahama, Japan, 1993.

[8] G. Gim and P.E. Nikravesh. *An Analytical Model of Pneumatic Tires for Vehicle Dynamic Simulations. Part 1: Pure Slips*, In. J. of Vehicle Design, Vol. 11, No. 6, pp. 589-618, 1990.

[9] G. Gim and P.E. Nikravesh. *An Analytical Model of Pneumatic Tires for Vehicle Dynamic Simulations. Part 2: Comprehensive Slips*, In. J. of Vehicle Design, Vol. 12, pp. 19-39, No. 1, 1991.

[10] K. M. Captain, A. B. Boghani, D. N. Wormley. *Analytical Tyre Models for Dynamic Vehicle Simulation*, Vehicle System Dynamic, No. 8, 1979.

[11] J. Wittenburg. *Dynamic of Systems of Rigid Bodies*, B. G. Teubner Stuttgart, 1977.

[12] J.Y. Wong. *Terramechanics and Off-Road Vehicles*, Elsevier, 1989.

[13] M. Cherif, Ch. Laugier, Ch. Milesi-Bellier, B. Faverjon *Combining Physical and Geometric Models to Plan Safe and Executable Motions for a Rover Moving on a Terrain*, Third Int. Symposium on Experimental Robotics, Japan, 1993.

[14] S. Jiminez, A. Luciani, Ch. Laugier. *Predicting theDynamic Behavior of a Planetary Vehicle using Physical Modeling*, IEEE/RSJ Int. Conferences on Intelligent Robots and Systems, Yokahama, Japan, 1993.

Appendix A: Wheel-soil contact models

We define two slip ratios :

$$S_s = \mid s \mid$$

$$S_\alpha = \begin{cases} \mid \tan\alpha \mid & \text{while braking} \\ (1 - S_s) \mid \tan\alpha \mid & \text{while traction} \end{cases}$$

Case 1 : rigid wheel on soft soil

The soil sinkage is expressed by Bekker [5] :

$$\delta_s = \left[\frac{3F_z}{w(3 - n_s)(k_c/w + k_\Phi)\sqrt{2R}} \right]^{2/(2n_s+1)}$$

where w is the contact width and R the wheel radius. Then contact length can be approximated by $l_s \simeq \sqrt{2R\delta_s}$

if $\sqrt{S_s^2+S_\alpha^2} \leq K/l_s$ (case of small slip values)

$$F_x = -sgn(s)\frac{l_s}{2K}\left(Ac + F_z \tan\Phi\right) S_s$$
$$F_y = -sgn(\alpha)\frac{l_s}{2K}\left(Ac + F_z \tan\Phi\right) S_\alpha$$

else

$$F_x = -sgn(s)\left(Ac + F_z \tan\Phi\right)\left(1 - \frac{l_p}{l_s} + S_s\frac{l_p^2}{2Kl_s}\right)$$
$$F_y = -sgn(\alpha)\left(Ac + F_z \tan\Phi\right)\left(1 - \frac{l_p}{l_s} + S_\alpha\frac{l_p^2}{2Kl_s}\right)$$

where l_s is the contact length, $A = wl_s$ the contact area and $l_p = K/\sqrt{S_s^2+S_\alpha^2}$
The rolling resistance due to soil compaction is given by :

$$\tilde{F}_x = -w\left(\frac{\delta_s^{n_s+1}}{n_s+1}\right)\left(k_c/w + k_\Phi\right)$$

Case 2 : elastic wheel on rigid soil

We denote by k_x,k_y longitudinal and lateral stiffness of the wheel and by w, l_r width and length contact. Gim et al. [8] show the existence of critical slips, defined by :

$$S_{sc} = 3\mu_c F_z/C_s$$
$$S_{\alpha c} = \frac{C_s}{C_\alpha}\left(S_{sc}^2 - S_s^2\right)^{1/2}$$

where μ_c is the wheel-soil friction coefficient, and $C_s = k_x w l_r^2/2$, $C_\alpha = k_y w l_r^2/2$.
if $S_s < S_{sc}$ and $S_\alpha < S_{\alpha c}$

$$F_x = -sgn(s)\left(C_s S_s l_n^2 + \mu_c S_s\left(S_s^2+S_\alpha^2\right)^{-1/2} F_z(1 - 3l_n^2 + 2l_n^3)\right)$$
$$F_y = -sgn(\alpha)\left(C_\alpha S_\alpha l_n^2 + \mu_c S_\alpha\left(S_s^2+S_\alpha^2\right)^{-1/2} F_z(1 - 3l_n^2 + 2l_n^3)\right)$$

else

$$F_x = -sgn(s)\mu_c S_s\left(S_s^2+S_\alpha^2\right)^{-1/2} F_z$$
$$F_y = -sgn(\alpha)\mu_c S_\alpha\left(S_s^2+S_\alpha^2\right)^{-1/2} F_z$$

where $l_n = l_a/l_r$ and

$$l_a = l_r\left(1 - \frac{1}{3\mu_c F_z}\left[(C_s S_s)^2 + (C_\alpha S_\alpha)^2\right]^{1/2}\right)$$

Case 3 : elastic wheel on soft soil

$$F_x = F_x^{(1)} + F_x^{(2)}$$
$$F_y = F_y^{(1)} + F_y^{(2)}$$

where $F_x^{(i)}$ and $F_y^{(i)}$ denote longitudinal and lateral forces exerted in the ith region.

An Autonomous Guided Vehicle for Cargo Handling Applications

Hugh F. Durrant-Whyte

Department of Mechanical and Mechatronic Engineering
University of Sydney, Sydney NSW2006, Australia
hugh@tiny.me.su.oz.au

1. Introduction

This paper describes an Autonomous Guided Vehicle (AGV) system designed to transport ISO standard cargo containers in port environments. The AGV consists of a large 17.5 tonne chassis, which is driven and steered through a diesel-hydraulic power set. The vehicle drives on pneumatic wheels over unprepared road surfaces at speeds up to 6m/s. The navigation system is based on the use of millimeter-wave radar sensors detecting beacons placed at known locations in the environment. The navigation system reliably achieves accuracies of better than three centimeters. The planning and control systems allow the AGV to operate in the quay and yard area, to dock with gantry cranes and can achieve a duty cycle comparable to that of a conventional manned vehicle. This paper concentrates on describing the design of the overall vehicle and on-board systems architecture, and the experimental programme leading to validation of the vehicle system.

2. The Vehicle System

2.1. Vehicle Design

A photograph of the complete vehicle with load is shown in Figure 1. The essential mechanical layout of the vehicle is shown in Figure 2. The vehicle is designed to carry either one ISO 40 foot container (with additional space for the new planned 50 foot containers), or two ISO 20 foot containers. This constrains the principle dimensions of the vehicle to resemble a conventional trailer. The main chassis is 15.5m long, 2.9m wide and 1.6m high. The distance between front and rear axles is 9m, and each of the front and rear bumpers extend approximately 1.3m from the main chassis. The main chassis has two structural I-bars running front to back. The loading platform, motor and all other components are mounted from these bars. The vehicle dry weight is 17.5 tonnes. In addition, the vehicle carries 1 tonne of diesel, 0.5 tonnes of hydraulic oil, and a maximum container load of 60 tonnes.

Figure 1. Photograph of completed AGV with 40 foot container load. One of the two radars can be seen at the front of the vehicle below the main deck.

The vehicle is driven through two axles, each having four conventional trailer tyres. Each axle is independently steered. This means that when the wheels are steered in opposite directions the vehicle can turn in a short baseline, and when the wheels are steered in the same direction the vehicle can move sideways without a change in orientation (crabbing). The wheels on each axle are mechanically linked and driven by a single set of hydraulic rams with a maximum lock of 30°. The front right wheel and the rear left wheel are powered. The front left wheel and rear right wheel

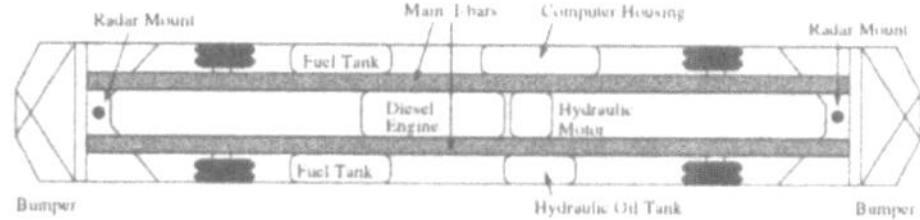

Figure 2. Mechanical layout of main AGV chassis. The chassis is based around two structural I-bars running front to back. All main vehicle components are fixed to these bars. The front and rear axles are independently steered. The front right wheel and rear left wheel are driven, the front left wheel and rear right wheel are instrumented with incremental encoders. The vehicle is thus diagonally symmetric and mechanically identical when traveling both forward and backward.

are instrumented with incremental encoders. This means that the vehicle is diagonally symmetric and mechanically identical when travelling both forward and backward.

The vehicle drive and steer systems are diesel hydraulic. The choice of hydraulic over electrical was made purely on cost grounds, the hydraulic option being approximately half the cost of the equivalent diesel-electric generator set. A 168Kw constant speed diesel engine powers a hydraulic pump. The pump swash plate is used to control the volume flow rate and pressure of the hydraulic fluid [2]. The hydraulic drive motors are mounted in the wheel hubs. This avoids the need for a drive shaft (a considerable cost saving). The drive motors are controlled using swash plates identical to those in the pump. Each steer axle is driven by a pair of conventional hydraulic rams. The rams are controlled by electro-mechanical valves. All low-level control is accomplished using proprietary PID control boards. The maximum full-load speed of the vehicle is restricted by the power of the diesel hydraulic unit to 6m/s.

2.2. System Architecture

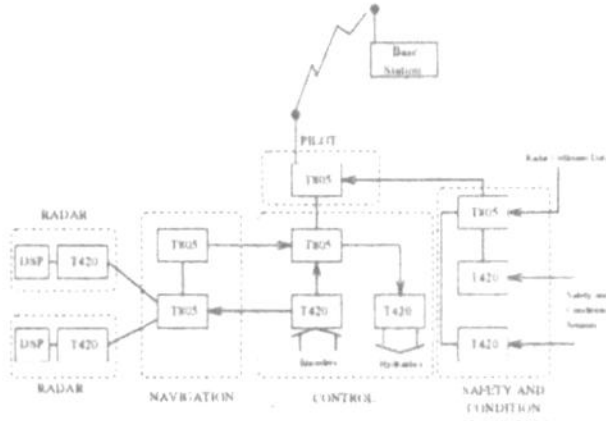

Figure 3. Overall systems architecture showing base station, navigation, control and safety and condition monitoring system.

Broadly, the AGV system architecture consists of five functional components: an off-board base-station linked by a radio-data system to the vehicle, the navigation system, the pilot, the control system, and a safety and condition monitoring system. A functional diagram of the system architecture is shown in Figure 3. The system architecture is based around a network of eleven Transputers.

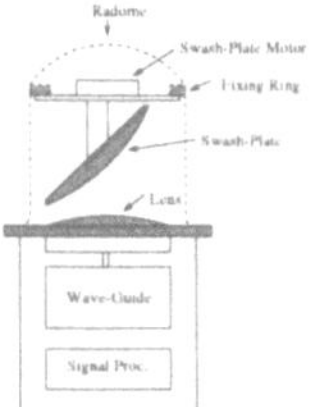

Figure 4. Schematic Showing Arrangement of Radar Components. The waveguide generates a vertical beam which is scanned in azimuth by a rotating 45° swash plate.

3. Navigation System

3.1. Navigation Sensing

The vehicle employs two Millimeter Wave Radar units as the primary navigation sensors The choice of navigation sensor was dictated by a need for high-speed, long range, accurate all weather operation. These units are mounted at the front and rear of the vehicle below the main cargo deck, at a height of approximately 1.2m from the ground as shown in Figures 1 and 2. Each unit consists of a wave-guide and electronic assembly transmitting a beam vertically upwards through a lens antenna (Figure 4). The vertical beam is reflected off a 45° swash-plate, mounted above the antenna, into a horizontal beam. As the swash-plate rotates, so the beam is scanned in azimuth. The swash-plate scan rate is approximately 6Hz. The swash-plate is instrumented with an encoder to provide measurements of beam bearing. The radar units are 77GHz (European automotive band) Frequency Modulated Continuous Wave (FMCW or chirp) devices. The radars transmit left-circular polarized radiation but detect both left and right circular polarizations. This allows the use of polarization information to identified objects and reject ground clutter. The radars have a swept bandwidth of 600MHz, and sweep time (pulse duration) of 500μs. This fixes the range resolution of the radar at 25cm which may be increased through a super-resolution algorithm to approximately 10cm. The radars each have a quasi-optic lens antenna providing 42dB of gain, resulting in a beam-width of less than 1°. The maximum practical range of the radars is restricted to approximately 200m.

The radar units provide range information from the chirp signal. Figure 5(a) shows a typical IF signal consisting of 1024 discrete samples. To obtain frequency information from this signal, a discrete (fast) Fourier transform (FFT) is performed yielding the result in Figure 5(b). The figure shows a complete range plot out to 150m obtained from one pulse. Bearing information is obtained from encoder measurements of the swash-plate. The navigation system works by detecting the range and bearing to a set of beacons placed at known, mapped locations about the environment. These beacons are radar trihedrals (effectively internal corner reflectors). The radar is also used as a collision detection sensor. In this case, the cross-polarized pulse return is recorded out to 35m and thresholded to provide an indication of first reflection from a target.

3.2. The Navigation Algorithm

The navigation algorithm is based on the extended Kalman filter. The Kalman filter algorithm is a widely used algorithm both in the robotics community and in many other areas. The **process model** used on the vehicle is

$$\begin{aligned} x(k+1) &= x(k) + \Delta T R(k)\omega(k)\cos[\phi(k)+\gamma_f(k)] \\ y(k+1) &= y(k) + \Delta T R(k)\omega(k)\sin[\phi(k)+\gamma_f(k)] \\ \phi(k+1) &= \phi(k) + \Delta T \frac{R(k)\omega(k)}{B}[\sin\gamma_f(k) - \sin\gamma_r(k)] \\ R(k+1) &= R(k), \end{aligned} \tag{1}$$

where $x(k), y(k), \phi(k)$ describe the location and orientation of the vehicle, $\omega(k)$ the mean angular wheel rate, $\gamma_f(k), \gamma_r(k)$ the front and rear wheel steer angles, and $R(k)$ the mean wheel radius.

Errors are injected into the AGV system by three primary sources; forward drive signals, steer angle signals and changes in wheel radius. The forward drive error is modelled as a combination of additive disturbance error $\delta\omega(k)$ and multiplicative slip error $\delta q(k)$

$$\omega(k) = \overline{\omega}(k)[1+\delta q(k)] + \delta\omega(k),$$

where $\overline{\omega}(k)$is taken to be the mean measured wheel rotation rate as recorded by the vehicle encoders, and $\omega(k)$ is defined to be the true mean wheel rotation rate defined through Equation 1. The steer drive error is similarly modelled as a combination of an additive disturbance error $\delta\gamma(k)$ and a multiplicative skid error $\delta s(k)$, assumed to be from an identical source for both front and rear axles

$$\gamma_f(k) = \overline{\gamma}_f(k)[1+\delta s(k)] + \delta\gamma(k),$$

$$\gamma_r(k) = \overline{\gamma}_r(k)[1+\delta s(k)] + \delta\gamma(k).$$

where $\overline{\gamma}_f(k)$and $\overline{\gamma}_r(k)$are taken to be the mean measured axle steer angles as recorded by the steer encoders, with $\gamma_f(k)$ and $\gamma_r(k)$ defined to be the true mean axle steer angles defined through

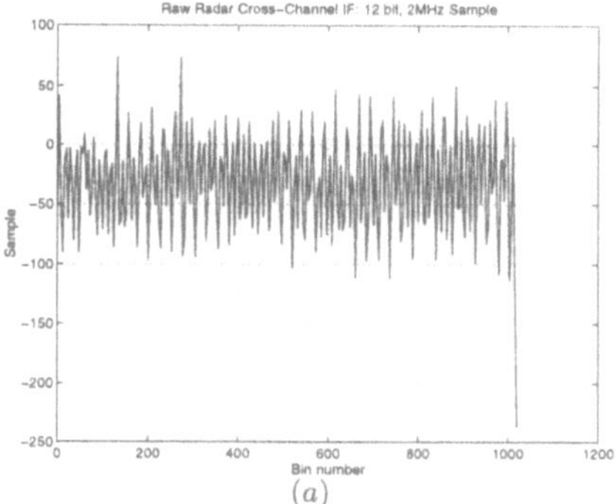

(a)

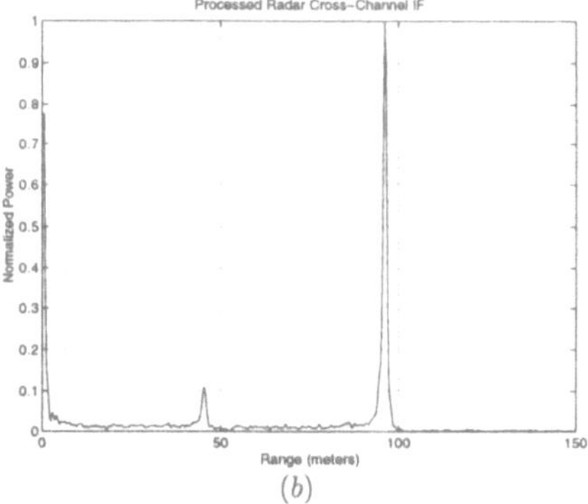

(b)

Figure 5. (a) Raw (IF) Radar Signal Return. (b) Frequency Content of return signal. The small return is a person, the larger return is a 100m^2 cross section radar reflector

Equation 1. The error in wheel radius is modelled as a discrete additive disturbance rate error (a random walk) so that

$$R(k) = \hat{R}^{+}(k) + \Delta T \delta R(k).$$

The source errors $\delta q(k)$, $\delta\omega(k)$, $\delta s(k)$, $\delta\gamma(k)$, and $\delta R(k)$ are modeled as constant, zero mean, uncorrelated white sequences, with variances σ_q^2, σ_ω^2, σ_s^2, σ_γ^2 and σ_R^2 respectively.

The error models for forward drive and steer signals are designed to reflect two important features. First, the multiplicative component of the error models reflect the increased uncertainty in vehicle motion as speed and steer angles increase (slipping and skidding). Second, the additive component of the error models is designed to reflect both stationary uncertainty and motion model errors such as axle offsets. The additive error is also important to stabilize the estimator algorithm. The random walk model for wheel radius is intended to allow adaptation of the estimator to wheel radius changes caused by uneven terrain and by changes in vehicle load (the chassis and wheels deform substantially at loads of 30-60 tonnes).

The essential observation information used by the vehicle consists of measurements of range and bearing made by the radar units to a number of beacons placed at fixed and known locations in the environment. Processing of observations occurs in four stages.

1. The measurement is converted into a Cartesian observation referenced to the vehicle coordinate system.

2. The vehicle-centered observation is transformed into base-coordinates using knowledge of the predicted vehicle location at the time the observation was obtained.

3. The observation is then matched to a map of beacons maintained by the AGV in base-coordinates.

4. The matched beacon is transformed back into a vehicle centered coordinate system where it is used to update vehicle location according to the standard extended Kalman filter equations.

4. Implementation and Validation

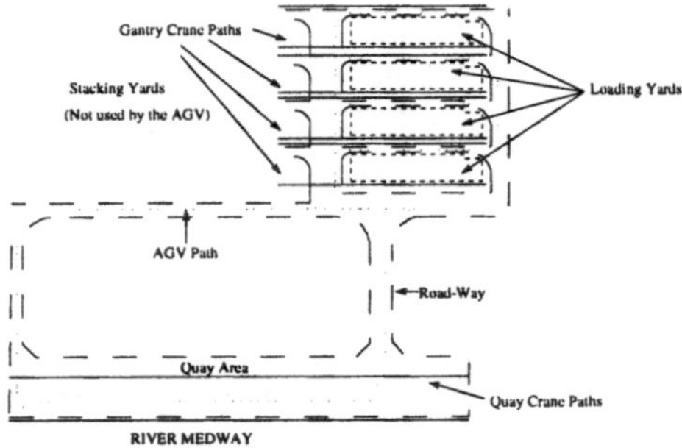

Figure 6. Map of a container port showing the quay, yard and stacking area, and main AGV paths.

Figure 6 shows a map of a container terminal on the Medway estuary in southern England. Shown on this map is the quay area and the stacking yards. A set of paths is also shown. These paths are simple straight line specifications joined by turns of known radius. The map is maintained as both a graph of nodes corresponding to each segment and curve, and as a geometric description of length, curvature start and end location. Paths may be added deleted or changed from the base station.

4.1. Validation Process

The validation of the navigation system, although theoretically straightforward, turned out to be quite difficult in practise. This was for two main reasons: First, except when the vehicle is stationary, it is impossible to know what the "true" location of the AGV is at any given time. Second, standard innovation-type measures (the sequence of errors between predicted observations and true measurements) of performance are difficult to use as the observation prediction process depends heavily on which beacon is being observed.

To overcome these problems, the validation of the navigation systems and software was accomplished in 3 phases.

1. Independently calibrated trials of the radar units were performed. The radars were placed on a buggy on a 200m long railway line, calibrated for absolute distance to 5mm. A number of different beacon layouts were tested at all ranges.

2. The radars were placed on the vehicle. The vehicle was equipped with a paint drip and required to move up and down in a straight line of about 200m. The paint tracks were then surveyed in to the beacon map. This provided a coarse absolute measure of straight-line absolute accuracy. Repeat tests with container clutter were performed.

3. Repeatability tests were performed in which the vehicle was required to execute a path, coming to a halt at specific places where location and orientation were measured.

The validation process was undertaken over a six-month period. It highlighted some major problems in the generation of absolute performance measures and validation procedures for this type of AGV system.

4.2. System Analysis

To explain the operation and performance of the vehicle system, we concentrate on describing a specific example trajectory in detail. The trajectory considered and the layout of beacons in this instance is shown in Figure 7.

Figure 8(a) shows the trajectory estimated by the navigation filter. It should be emphasized again that only the start and end locations of the trajectory can be independently validated (by

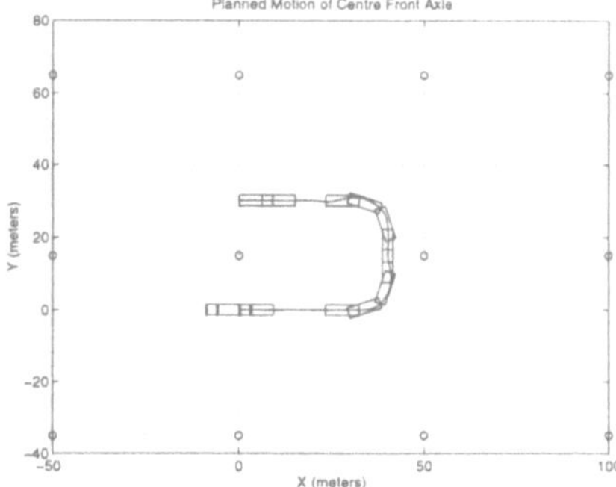

Figure 7. The example trajectory and beacon layout. The figure shows the path of the vehicle front axle and an outline of the vehicle at each of the hit-boxes generated along the trajectory. The beacons are shown by the symbol 'o'.

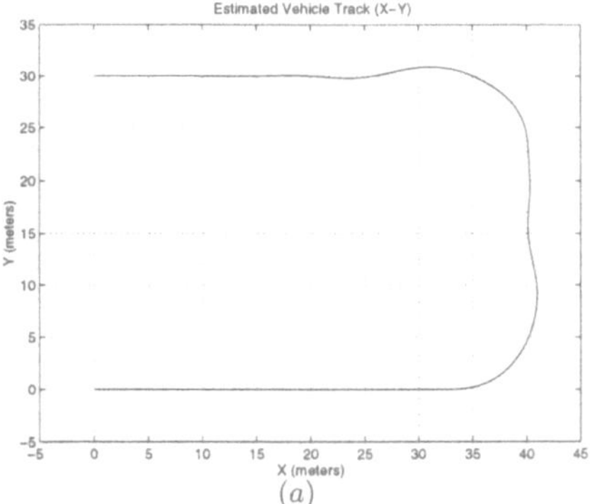

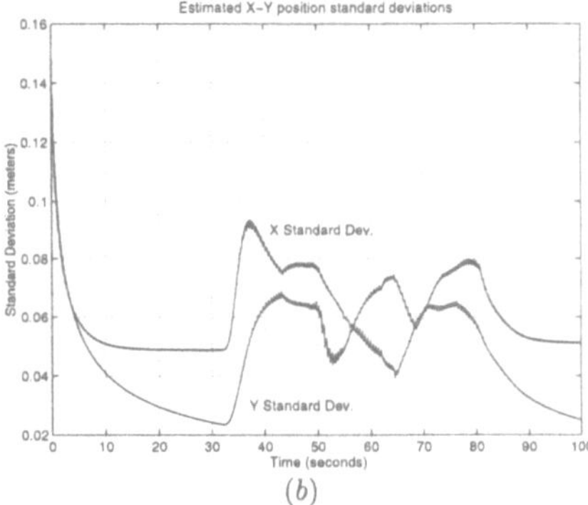

Figure 8. (a) The estimated trajectory of the center front axle and (b) the estimated standard deviations in x and y coordinates of the estimated front axle location as computed by the navigation system.

theodolite), and that the trajectory shown is an estimated path and not an actual path. The standard deviations in x and y location estimates are shown in Figure 8(b). From 0–32 seconds the standard deviation estimates converge from initial to steady state values. The steady-state values depend on the layout of the observed beacons in the neighbourhood of the vehicle. The y standard deviation is lower than the x value at this start location because the vehicle is oriented along the x axis: The along axis process model noise source in the direction of vehicle travel is larger than the cross-axis process noise source. At 32 seconds when the vehicle starts moving, both x and y standard deviations rise rapidly. This is for two reasons; first the additional injection of multiplicative process noise in to (primarily) along-axis motion estimates, and second due to the effect of large wheel radius uncertainty affecting motion estimates. After an initial transient, the in-motion standard deviation estimates reach a steady state at approximately 45 seconds. The increase in standard deviation at this time as compared to stationary values is almost all due to the injection of multiplicative noise. Note that this increase in standard deviation increases linearly with vehicle speed. As the vehicle enters the first curve both x and y standard deviations *reduce*.Once the steer signal reduces close to zero (at approximately 65 seconds), the x and y standard deviations return to a straight-line motion value in which x and y values are reversed from their values prior to the turn.

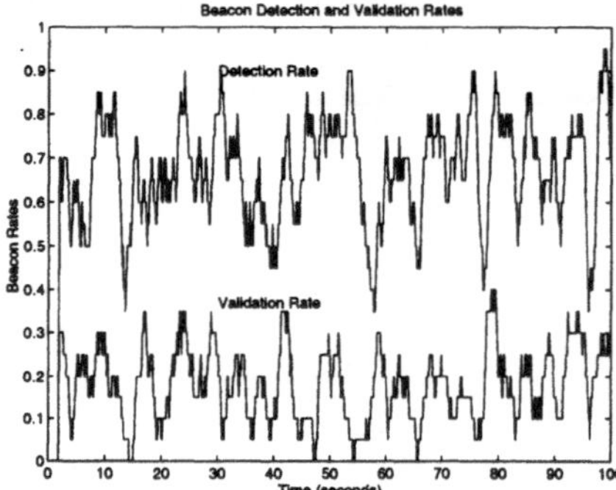

Figure 9. 10 second moving average of beacon detection and beacon validation rates.

One of the most serious problems in operating the radar sensor in environments such as container terminals is the elimination of false-alarm or clutter signals due to large objects such as containers, cranes and buildings. The problem is minimized by providing the beacons with a 'unique' polarization pattern unlikely to be found in other environment features. However, this does not completely eliminate the problem as some error must always be tolerated in matching polarization patterns and this consequently always leads to false targets being matched. False targets are eliminated using a conventional normalized innovation validation gate. The size of the gate is a compromise between having a small enough gate to reject false alarms and having a large enough gate to ensure a high matching rate to correct beacon detections. Figure 9 shows a 10 second moving average of the rate of beacon detection and the rate of beacon validation with this gate-size. The beacon detection rate corresponds to a detection probability and indicates the fraction of correct beacon observations correctly validated. The beacon validation rate corresponds to one minus the false alarm probability and shows the overall fraction of observations validated. The beacon detection rate indicates that approximately 80% of correct observations are used by the navigation filter. The beacon validation rate shows that approximately 80% of all measurements are considered to be false alarms.

4.3. Application Trials

The vehicle system described in this paper was installed at an operating container port, Thamesport, on the Medway estuary in southern England. The vehicle was logistically integrated with the manned vehicle fleet and interfaced with central logistics and with both gantry and quay crane systems. Installation took place over a 6 month period and was completed in October 1993. We briefly describe here some of the main points of the installation and the main problems experienced during operations so far.

The installation area consisted of a main stacking area, with 6 rail-mounted gantry cranes covering 3 stacking yards, and a quay area with 3 post-panamax (90m) quay cranes. The total installation area is approximately 2 square kilometers, with a quay length of about 600m (an outline map of the port is shown in Figure 6). The AGV was operated in one allocated stack (with other manned vehicles), over a bridge connecting the yards with quay, and along the eastern end of the quay itself. Typical cycles from yard to ship are about 1-2Km. A total of 150 beacons were set up to cover the operation area. In the yard area, these were placed along roadways between container stacks at intervals of approximately 40m. Elsewhere, the beacons were placed so as to tessellate the ground area with a triangle grid of base dimension 50m (the triangular grid maximizes location accuracy). Typically, 3 beacons are available for viewing at any location.

The most immediate problem faced was the very high level of clutter experienced by the radar in a container environment. The AGV had to move accurately through roadways between stacks of metal containers 18m tall. A great deal of time was spent adjusting radar sensitivity and improving gating procedures to overcome this problem. A second important problem was the detection of navigation faults due to mis-matching and drifting of system components and estimates. The difference in estimates pre and post update were used to identify abrupt faults due to mismatching and sudden load changes. However, low frequency drifts in component values (radar calibration, steer slippage, geometry change with temperature, etc), proved very difficult to isolate. A final issue is that of safety. The vehicle system has proved to be reliable and has operated safely during fault conditions. However, a continual worry has been the need to provide some guarantee of vehicle safety, something that ultimately is very difficult to measure.

5. Conclusions

The development of the AGV has highlighted a number of important issues that will need to be addressed by the robotics community as techniques developed in the laboratory mature to become commercial systems. In particular, the need to develop and build reliability in the context of autonomous systems is seen as crucial. This requires a careful re-think of the way in which techniques and systems are implemented, and the development of quantifiable measures of performance and reliability at both the sub-system level and as part of an overall vehicle.

References

[1] P.S. Maybeck. *Stochastic Models, Estimaton and Control, Vol. I.* Academic Press, 1979.

[2] T.J. Viersma. *Analysis, Synthesis and Design of Hydraulic Servosystems and Pipelines.* Delft University Press, 1990.

Chapter 9

Autonomy via Teaching

For some environments (particularly unstructured), it is perhaps unrealistic to expect robots to be fully autonomous in the foreseeable future. Where humans can intercede to guide and refine robot behavior, robots can be augmented to perform tasks requiring human judgment, planning and perceptual abilities. Advancements in this area will allow us to manage robots with a strategic rather than a task level focus.

Konolige identifies a set of useful robot/human interaction modes, attending, taking advice and tasking, each involving complex underlying activities in sensing and planning. Using gesture and speech recognition as input modalities, the paper reports on a demonstration in which SRI's Flakey robot is guided by an untrained human supervisor to enter an office environment and perform delivery tasks.

Gat discusses the lack of formal, statistically significant, and well controlled experiments in mobile robotics research. He then presents an approach to statistical analysis of experiments with real robots and reports on results from a series of experiments with JPL's Rocky 3.2 robot.

Coste-Manière, Perrier, and Peuch discuss programming issues that relate to the control of autonomous vehicles to which human intervention and high level command may be applied. The paper presents an illustrative mission performed by the VORTEX vehicle.

Kapellos, Jourdan, Espiau, and Abdou show how formal specification and verification methods can be used for checking the behavioral and temporal correctness in the high level control of complex robotic systems. This is illustrated by an autonomous vehicle navigation experiment.

Robots that Take Advice

Kurt Konolige

Artificial Intelligence Center
SRI International
333 Ravenswood Avenue
Menlo Park, CA 94025
konolige@ai.sri.com

Abstract

Mobile robots, if they are to perform useful tasks and become accepted in open environments, must interact with humans. Robot response to humans at the most basic level means not injuring them directly or indirectly, e.g., treating humans as obstacles to be avoided in the performance of a task. Robots will be more useful if they can interact in more sophisticated ways. Here we identify a set of useful robot/human interaction modes: attending, taking advice, and tasking. All three involve complex sensing and planning operations on the part of the robot, including the use of visual tracking of humans, gesture recognition, and speech recognition and understanding. We show how these capabilities are integrated in the Saphira architecture on Flakey, a mobile robot testbed. We demonstrate a scenario in which an untrained supervisor is able to introduce Flakey to an office environment, and command it to perform delivery tasks.

1. Introduction

In March 1994 we were approached by the producers of the science show "Scientific American Frontiers," who were interested in showcasing the future of robotics. After some discussion, we decided on a scenario in which our robot, Flakey, would be introduced to the office environment as a new employee, and then asked to perform a delivery task. To perform the scenario we needed at least the following capabilities:

Attending and Following. A supervisor would introduce Flakey to the office by leading it around and pointing out who inhabited each office. Flakey would have to locate and follow a human being. It would also have to know if a human was present by speech, e.g., going to an office door and inquiring if anyone was present.

Taking advice. Advice from the teacher would include map-making information such as office assignments and information about potential hazards ("There's a possible water leak in this corridor."). It would also include information about how to find people ("John usually knows where Karen is").

Tasking. Flakey would have to perform delivery tasks using its learned knowledge. The task was chosen to illustrate the different types of knowledge Flakey had: maps, information about office assignments, general knowledge of how to locate people.

The scenario was made more difficult by three factors: there were only 6 weeks to prepare; the supervisor would have no knowledge of robotics (it was Alan Alda, the program host); and the scenario had to be completed in one day. We were fortunate in that most of the perceptual and control architecture we needed already existed, as described below.

2. The Flakey Testbed

Flakey is a mature (some might say old) robot designed and built in the mid-1980's. In its current incarnation, sensors include a ring of 12 sonar sensors on the bottom, and a stereo camera pair mounted on a pan/tilt head. Flakey also has a speaker-independent continuous speech recognition system called CORONA, developed at SRI, and a standard text-to-speech program for speech output.

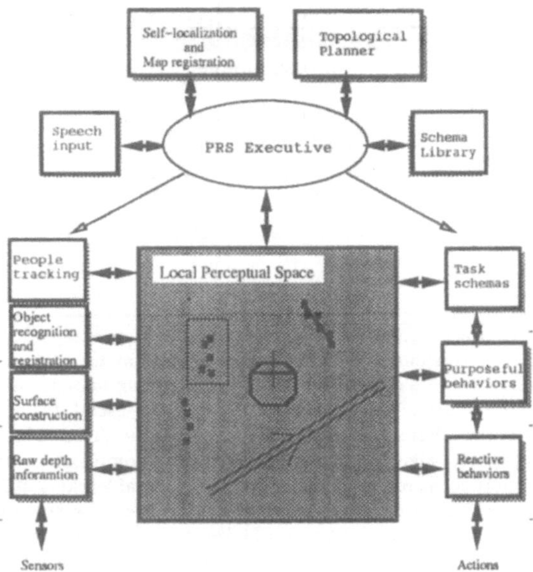

Figure 1. Saphira system architecture. Control is coordinated by the Procedural Reasoning System, which instantiates routines for task sequencing and monitoring, and perceptual coordination.

All systems run on-board, using a 2-processor Sparcstation configuration. One processor is dedicated to speech and robot control, the other runs the visual interpretation programs. These include stereo algorithms [14], which give full-frame dense stereo maps at a rate of about 2.5Hz. We use the results in two ways.

1. to identify surfaces that could be possible obstacles by matching their height against the ground plane, and
2. to find and track person-like objects.

Tracking objects is done at an 8Hz rate, by using only a partial image located around chest height.

Flakey was used in early experiments with Stan Rosenschein's Situated Automata theory [11], which eschews representational models of the environment. In subsequent years we have develop a more classical environmental model we call the Local Perceptual Space (LPS). The LPS is an egocentric, geometric representation of space. It combines several technologies:

- a grid-based representation similar to Moravec and Elfes' occupancy grids [9];
- more analytic representations of surface features such as linear surfaces; and
- fully-interpreted structures such as corridors or doorways (*artifacts*).

We describe this architecture, and related control issues, in what follows.

2.1. Perceptual and Control Architecture

The Saphira control architecture has been written up in several places [12, 13, 3]; here a basic description will suffice. At the center is the Local Perceptual Space (LPS), as just described (see Figure 1). The LPS gives the robot an awareness of its immediate environment, and is critical in the tasks of fusing sensor information, planning local movement, and integrating map information. The perceptual and control architecture make constant reference to the local perceptual space.

In Brooks' terms [1], the organization is partly vertical and partly horizontal. The vertical organization occurs in both perceptual (left side) and action (right side). Various perceptual routines are responsible for both adding sensor information to the LPS and processing it to produce surface information that can be used by object recognition and navigation routines. On the action side, the lowest level behaviors look mostly at occupancy information to do obstacle avoidance. The basic building blocks of behaviors are fuzzy rules, which give the robot the ability to react gracefully to the environment by grading the strength of the reaction (e.g., turn left) according to the strength of the stimulus (e.g., distance of an obstacle on the right).

More complex behaviors that move to desired locations are used to guide the reactive behaviors, and utilize surface information and artifacts; they may also add artifacts to the LPS as control points for motion. At this level, fuzzy rules blend possibly conflicting aims into one smooth action sequence. Finally, at the task level complex behaviors are sequenced and their progress is monitored through events in the LPS. The horizontal organization comes about because behaviors can choose appropriate information from the LPS. Behaviors that are time-critical, such as obstacle avoidance, rely more on very simple processing of the sensors because it is available quickly. However, these routines may also make use of other information when it is available, e.g., prior information about expected obstacles that comes from the map.

One important feature of the LPS is that as Flakey moves, information from the wheel encoders is used to update all information in the LPS with respect to the robot. New sensor readings are automatically registered with respect to old ones, so that a composite picture, based on many sensor readings, emerges. At the level of grid-based representations, which are useful for obstacle avoidance, the new measurements are merged using the techniques of occupancy grids [9]. Interpreted features such as surfaces and objects require more complex handling, which we consider in the section on registration.

2.2. Controlling Executive: PRS-lite

Behaviors are coordinated by a decision system called PRS-lite, a realtime version of SRI's Procedural Reasoning System [4]. The basic component of PRS-lite is an *intention schema*, a finite-state machine whose arcs are labeled with conditions to check or goals to achieve. Each schema embodies a *strategy*: a sequence of perceptual checks and behavior instantiations that accomplishes a goal. Some schemas are fairly simple, for example, to detect closed doors, a monitoring schema is fired up every time Flakey attempts to go through a doorway, and if no progress is made after a fixed amount of time, or if the sensors detect that the doorway is closed, the schema halts the current door-crossing behavior, updates a global map with the new information (more on this below), and signals the plan executive that the current plan has failed.

PRS-lite is interesting because it incorporates a small amount of deliberation in its schemas, but is still able to react to contingencies in realtime. It does this by throwing away some of the more costly (and often useful) features of full PRS, especially the database of facts and goals that is used to trigger new schemas. Instead, every schema must be triggered explicitly by another schema, which gives a much less flexible system. However, the advantage is that PRS-lite is very fast: written in C, it has a cycle time of 100ms, during which every executing schema is updated. This is very fast, considering that at any given moment there may be 10 or 15 schemas operating, monitoring various conditions and coordinating behaviors.

Interestingly, we have not found a need for a larger planning capability to control Flakey. PRS-lite is the main controlling agent, and issues calls to higher-level functions such as a navigation planner or registration mechanism when it requires their services.

2.3. Maps, Navigation and Registration

To navigate through extended regions, Flakey uses a global map that contains imprecise spatial knowledge of objects in the domain, especially walls, doorways, and junctions of corridors. Using a map depends on reliable extraction of object information from perceptual clues, and we (as well as others) have spent many frustrating years trying to produce object interpretations from highly uncertain sonar and stereo signatures (see, for example, [6, 2, 9]). The best method we have found is to use extended aperture sonars readings, perhaps augmented with depth information from the stereo system. As Flakey moves along, readings from the side sonars are accumulated as a series of points representing possible surfaces on the side of the robot. This gives some of the resolution of a sensor with a large aperture along the direction of motion. By running a robust linear feature algorithm over the data, we can find walls segments and doorways with some degree of confidence. False-positive rejection, the bane of sonar sensors, is excellent for wall surfaces, and reasonable for doorways, although we do get some of these when the robot rotates quickly.

Extracting wall and doorway features makes it easy to build a global map automatically, by having Flakey explore an area. The map is imprecise because there is error in the dead-reckoning system, and because the models for spatial objects are linear, e.g. corridors are represented as two parallel, straight lines. As features are constructed and categorized, they are placed into the LPS as *artifacts*, that is, representations of objects that are assumed to exist in the world.

We call the problem of identifying previously-found artifacts on the basis of current perceptual information, the *registration* problem. In Flakey, the structure of decision-making for the registration problem takes the following form.

feature	(identified)	$\Rightarrow$	hypothesis
hypothesis	(matched)	$\Rightarrow$	update position
	(unmatched)	$\Rightarrow$	create new artifact

As features are identified, they are converted to hypotheses about the presence of objects such as doors or corridors. These hypotheses are matched against artifacts existing in the LPS, put there either from a priori knowledge of the map, or by finding new artifacts. If they match against an artifact, the match produces information for updating Flakey's position with respect to the artifact. If there is no match, then a new artifact is created and added to the map. Conversely, if an artifact has no perceptual support when it should (i.e., when it is in the range of the cameras or sonars), then it will eventually disappear from the map.

The registration process is sufficiently robust that the original map the robot makes does not have to be very precise (and it can't be, if dead-reckoning is poor). The robot doesn't have to survey the area, just make a map that has approximately the right dimensions. The registration process will keep the robot's position updated with respect to the objects in the environment. This assumes, of course, that the robot can find and match objects of the right sort. For instance, going down a long corridor with no features, the robot will stay correctly registered in the center of the hallway, but the error in its longitudinal position will grow. So it's important to find doorways or breaks in the corridor at reasonable intervals.

Our current ideas for keeping Flakey localized are turning towards correlation-based algorithms using the stereo vision system. By keeping track of a number of small surface patches that are sufficiently distinctive, we hope to be able to use the same registration algorithms to reliable localize the robot in both indoor and outdoor environments.

2.4. Tracking people

Flakey incorporates a simple people-tracker based on stereo information. The tracking algorithm does not perform correlations on successive temporal frames to keep track of the object. Rather, it looks for person-like objects in each successive frame without reference to previous ones, formulates a hypothesis, and then passes it to the LPS registration routines. Again, the registration algorithm makes a decision about whether the hypothesized person is the same as one represented by an artifact, and either updates the artifact position or creates a new one. We keep a simple velocity model of person artifacts, which helps the tracker to position the camera, and also helps the registration process to make better matching decisions. If an artifact does not receive support for a period of time, it is removed from the LPS.

Interestingly, the registration process makes it possible to follow people around corners and through doorways, where they may be temporarily lost from sight. The artifact represents their most likely position, and the vision routines keep searching this area until the person is re-acquired, or cannot be found.

2.5. Gestures

We were ambitious enough to want gesture recognition as part of the attention process. The idea was to use a mixture of speech and gesture for reference, e.g., "This office (***point to the right or left***) is Karen's." We did manage to extract enough information from the stereo system to recognize left or right-pointing arms, but did not have enough time to integrate it correctly with the speech input, even for simple reference. This would be a good project for future work, with a more elaborate natural-language understanding system.

2.6. Speech input

We were fortunate to have an excellent speech-recognition system, developed at SRI, called CORONA. CORONA has speaker-independent recognition models that need no training, although it will have better accuracy with individually-tuned models. CORONA accepts a BNF grammar of phrases, and produces strings of the words spoken. On a Sparc 10-51, it operates at about twice realtime on the simple grammar we used, i.e., a 5-second sentence would take about 10 seconds to recognize.

One of the hardest problems in voice communication is letting the supervisor know when Flakey has heard an utterance, and what the state of its understanding is. We employed several techniques:

Keying. There can be a lot of extraneous chatter during interactions. The grammar was created with a keyword initiation, so that only phrases beginning with the word "Flakey" were recognized. This worked extremely well; it was natural to address commands and statements to the robot in this fashion. A nice additional feature would be to use directional interpretation of sound sources for getting Flakey's attention; we currently do not have this capability.

Process status. CORONA employs an "endpointer," a period of low speech energy, to signal the end of a speaker phrase. From this point there is a delay until the speech is processed; the speaker can be confused about whether his input was received, and whether it was recognized. We implemented a simple end-of-speech flag: Flakey says "um" when the endpoint is reached. Thus the speaker can tell that his input was received and is being processed.

Results. It is important to give feedback to the speaker about the interpretation of his input. For example, if the speaker says "turn left," and it is interpreted as "go forward" (a not altogether unknown occurrence), the speaker will be mystified by the robot's behavior. So Flakey would acknowledge each spoken command, either by paraphrasing it, or nodding its cameras in recognition. If the phrase was not recognized, Flakey said "duh?" or "what?" Crude, but effective.

3. Communicating with Flakey

We implemented a simple but surprisingly sufficient set of movement schemas to perform the scenario. They fall into several categories: direct motion commands ("turn around"), sensor-based movement ("follow me", "follow the corridor"), tasks ("get the file from Karen"), and information ("this is John's office").

3.1. Direct Motion

These are direct commands to move forward or turn, or to look (move the cameras) in certain directions, e.g., "look behind you." Direct motion commands are implemented as simple sequential behaviors with well-defined ending conditions and short timeouts. For example, if Flakey is told to move forward while facing a wall, it will turn away from the wall and move forward about 1 meter, since collision avoidance is active at all times. It would be smarter, of course, to recognize that it is not possible to move forward, and state this fact. But in our current implementation Flakey does not check for possibility of carrying out commands; it only tries to do the best it can.

3.2. Attending and Following

These commands involve coordination of sensing and acting, mediated by artifacts in the LPS. Even very simple movement schemas, when coordinated with sensing, can give the appearance of a well-trained robot with human-like capabilities.

The Attending schema finds the closest person-like object and brings Flakey to face the person at a distance of about 1 meter. The command "look at me" or "find me" triggers this schema. Flakey performs a scan from the current camera position to the extreme left and right position of the pan/tilt head, giving a full 360 degree field of view. The first person-like object detected is placed in the LPS, and a movement schema position the robot to face the object and keep the cameras centered on it. Success of the Attending schema is indicated by saying "here I am;" failure, after a full scan, by "I can't find you!".

The Following schema uses the same kind of movement/sensor control to track a person. If there is no person currently in view, the Attending schema is invoked. Once found, the person is kept centered in the cameras, as described in the previous section. The movement schema keeps Flakey about 1 - 1.5 meters from the person as much as possible (top speed of 300mm/sec), while avoiding obstacles. If the person is lost, Flakey looks in the most likely area for awhile, but does not attempt to re-attend, since if there are other people around, it may acquire the wrong one. Flakey's vision system cannot distinguish individuals.

Both these behaviors worked well in practice. Mr. Alda was able to lead Flakey through several corridors and doorways at SRI, with only minor problems in re-acquisition.

3.3. Information

This is one of the most interesting areas for robot/human interaction. The hard problem is relating robot-centered representations of the world with human-centered ones. For example, when the supervisor says "..this office...", the robot must understand that there is an object in the current focus of dialogue that is being referred to. The point of contact, for the robot, is in the artifacts of the LPS. Flakey has enough background knowledge to infer that doors can lead to offices, even though it cannot recognize an office *per se*. So the phrase "this office" is linked to the nearest doorway in the robot's perceptual space.

In general the problem of reference resolution can be phrased in terms of abductive inference: find an object that, if assumed as the referent, would make the speaker's phrase carry reasonable information [5]. For the scenario demonstration, we did not use a general inference method to determine reference, since we decided beforehand that we would only refer to offices, people, and certain other objects. But in general the problem of reference for robots will require some of the same tools that have been used in computational linguistics. However, robots do enjoy one advantage: since they also perceive the world, they can draw on a repertoire of perceived objects as referents (in philosophical terms, they are *situated* in the word). Instead, the problem becomes one of matching the artifacts of the robot's representation with the linguistic expressions of the speaker. Karen Myers and I have formulated a theoretical foundation for this interchange [10]; but it was not applied in the scenario. Other more complex theories of knowledge and action have been developed in a logical framework [8, 7]; we expect to see more application of these theories as robots become more sophisticated in their interactions.

Another type of information used by the robot is knowledge of how to locate people, since delivery tasks often involve finding someone. In general this is a planning problem, with reasoning about where someone is, or who might have information about where that person is. For the scenario, we implemented a simple routine that would first check the person's office (using speech output and waiting for a reply), and then start looking for people who might know where the person is. More sophisticated routines might check a personal calendar, or compile a list of likely places, or perform inference about knowledge prerequisites for action, and so on. Here again the problem of reference emerges. Phrases such as "Karen is around the corner" would be hard to interpret, so we settled on a set of place-names that could be used: offices, corridors, and open areas such as the library. These were all places that Flakey could learn from the supervisor, since there were artifacts (doors, corridors, junctions) that could be identified with the area.

One lesson we learned from attempting to give Flakey advice of this sort is that the closer a match between the robot's perceptual categories and the human's, the easier and more foolproof the exchange of information. For example, if Flakey had no concept of a corridor, it would be almost impossible to tell it the name of the corridor (e.g., "J-wing") and have it internalized in an effective way. Suppose Flakey were to store its current position in association with the corridor name. Then the command "Go to J-wing" would cause Flakey to go back to the same location, even if another location in the corridor were much closer. Commands such as "follow the corridor" wouldn't make any sense at all.

3.4. Tasking

A robot is supposed to perform useful tasks. For the scenario, Flakey was a delivery robot, whose main task was to deliver messages and manuscripts. While performing these tasks, Flakey maintained the heterostatic goals of obstacle avoidance and localization. These required no overt planning; PRS invoked the requisite behaviors automatically. In more complicated situations, for example where the robot must explore to find a new route, there might be explicit planning for localization.

Navigation plans were computed by graph search on the learned topological map, and then executed by PRS. More complicated procedures were constructed as conditional navigation plans: find person X, ask him/her where person Y is, and then go to that location. These plans were simple enough that we just created PRS schemata for them. A major failure in the plan (e.g., person X not found) caused PRS to instantiate an alternative schema. The most complicated plan Flakey executed was going to Karen's office and finding out she wasn't there; then going to John's office, asking him where Karen was, finding and getting a report from her, and returning to deliver it.

4. Conclusion

Integrating all the functions necessary for autonomous, interactive behavior is a challenge. In general, simple techniques that work reliably and give rise to predictable behavior are desirable. Automatic achievement of heterostatic goals — invariants of the robot and its environment — make the planning task much easier. We look forward to integrating more complicated natural language understanding and planning theories into Saphira.

References

[1] R. A. Brooks, A layered intelligent control system for a mobile robot, in: *Proceedings of the IEEE Conference on Robotics and Automation* (1986) 14–23.

[2] M. Drumheller, Mobile robot localization using sonar, A. I. Memo 826, Massachusetts Institute of Technology (1985).

[3] C. C. et. al., CARMEL versus FLAKEY: A comparision of two winners, *AI Magazine* **14** (1) (1993) 49–57.

[4] M. P. Georgeff and F. F. Ingrand, Decision-making in an embedded reasoning system, in: *Proceedings of the Conference of the American Association of Artificial Intelligence*, Detroit, MI (1989) 972–978.

[5] J. R. Hobbs, M. Stickel, D. Appelt, and P. Martin, Interpretation as abduction", *Artificial Intelligence* **63** (1–2) (1993) 69–142.

[6] J. Leonard, H. Durrant-Whyte, and I. J. Cox, Dynamic map building for an autonomous mobile robot, in: *IROS* (1990) 89–95.

[7] Y. Lespérance and H. J. Levesque, Indexical knowledge and robot action — a logical account, *Artificial Intelligence* **73** (1–2) (February 1995).

[8] R. C. Moore, Reasoning about knowledge and action, PhD thesis, Massachusetts Institute of Technology, Cambridge, MA (1980).

[9] H. P. Moravec and A. E. Elfes, High resolution maps from wide angle sonar, in: *Proceedings of the 1985 IEEE International Conference on Robotics and Automation*, Washington, D. C. (1985) 116–121.

[10] K. Myers and K. Konolige, Reasoning with analogical representations, in: B. Nebel, C. Rich, and W. Swartout, eds., *Principles of Knowledge Representation and Reasoning: Proceedings of the Third International Conference (KR92)*, San Mateo, CA (Morgan Kaufmann, 1992).

[11] S. J. Rosenschein, The synthesis of digital machines with provable epistemic properties, Technical Note 412, SRI Artificial Intelligence Center, Menlo Park, California (1987).

[12] A. Saffiotti, E. H. Ruspini, and K. Konolige, Integrating reactivity and goal-directedness in a fuzzy controller, in: *Procs. of the 2nd Fuzzy-IEEE Conference*, San Francisco, CA (1993).

[13] A. Saffiotti, E. H. Ruspini, and K. Konolige, A multivalued logic approach to integrating planning and control, SRI Tech Report 533, SRI International (1993).

[14] R. Zabih and J. Woodfill, Non-parametric local transforms for computing visual correspondence, in: *3rd European Conf. Computer Vision*, Stockholm (1994).

Towards Principled Experimental Study of Autonomous Mobile Robots

Erann Gat
Jet Propulsion Laboratory
California Institute of Technology
4800 Oak Grove Drive
Pasadena, CA 91109
gat@robotics.jpl.nasa.gov

ABSTRACT

We review the current state of research in autonomous mobile robots and conclude that there is an inadequate basis for predicting the reliability and behavior of robots operating in unengineered environments. We present a new approach to the study of autonomous mobile robot performance based on formal statistical analysis of independently reproducible experiments conducted on real robots. Simulators serve as models rather than experimental surrogates. We demonstrate three new results: 1) Two commonly used performance metrics (time and distance) are not as well correlated as is often tacitly assumed. 2) The probability distributions of these performance metrics is exponential rather than normal, and 3) a modular, object-oriented simulation accurately predicts the behavior of the real robot in a statistically significant manner.

1. Introduction

There is an apparent and unfortunate dichotomy in autonomous mobile robotics research between theory and practice. Published reports in this particular area of robotics seem to fall largely into one of two categories: theoretical work with little or no experimental verification (except, on occasion, in simulation), and anecdotal experimental results from implemented systems with little or no formal theoretical foundation. It is rare to find a formal theoretical prediction verified (or refuted) by independently reproducible experiments performed on a real robot. It is even rarer to find such results supported by an analysis of their statistical significance. Control experiments are nearly unheard of.

For example, in a cursory survey of 44 papers in the mobile robotics track of the 1994 International Conference on Robotics and Engineering (track 5, excluding four papers on legged robots) we found seventeen papers that describe work done on an actual mobile robot. Of these, only three reported quantitative results from more than one experimental trial. (A few papers claim to have produced such results but do not actually report them.) Of these three, only one deals directly with autonomous control (and that one is the work of the present author). A complete list of citations can be found in [Gat95].

Standard theoretical approaches tacitly assume that most of these independent variables can be ignored. For example, there is a vast theoretical literature on the path-planning problem, which is almost invariably posed as a purely geometrical problem where the quality of a solution is measured in terms of path length, e.g. [Latombe91]. Such formulations routinely ignore such factors as computational costs, sensor noise, occlusions and resolution limits, and mechanical interactions between a robot and a supporting surface, including friction and surface deformation.

Likewise, in order to make experimentation tractable, issues such as controlling for extraneous effects and statistical significance of results are routinely ignored. It is rare to find a description of an experimental setup that is sufficiently detailed to allow the experiment to be independently reproduced. The choice of the number of experiments to conduct is usually made on a purely *ad hoc* basis (reporting the result of a single experimental trial is common), and control experiments and statistical analysis are all but nonexistent. As a result there is a lot of passionate debate, but no objective basis for evaluating the relative merits of different approaches to the problem of autonomous control.

We can no longer afford to sweep these issues under the rug. In 1996 NASA will launch an autonomous mobile robot to explore the surface of Mars. This robot represents a substantial expenditure of taxpayer money, and so it is important to accurately assess the reliability of our control methodology before launch. Furthermore, the harsh realities of the Martian environment do not permit us the luxury of making arbitrary simplifying assumptions in our theories, even if those assumptions appear intuitively plausible. Reality, not theorems, is our ultimate arbiter of truth.

2. Approach

Our approach is to treat the experimental study of mobile robots in the manner of a natural science or an empirical engineering discipline. The natural sciences (e.g. biology) regularly study the interactions of systems that are as complex or more than the environments mobile robots interact with. It is usually impossible to model such complex systems starting from first principles. Instead, probability theory is used to model system components as random processes. Experiments are designed to measure sampling distributions of the resulting random variables, and statistical methods are used to analyze the results.

One important result of our work is that two commonly used performance metrics turn out to have probability distributions that appear to be exponential rather than normal. Most of the standard techniques used in the natural sciences assume normal distributions. We will therefore be forced to rely on some non-traditional analysis methods, known as non-parametric methods, which do not rely on the probability distribution having any particular shape.

Although we will emphasize experimental results (we hope to describe our apparatus, methods, and models in sufficient detail to allow independent replication), we will also construct models of our robot systems. These models will take the form of simulations. The measure of our simulations, however, will be how well they predict real experimental results, not whether they reflect a particular mathematical formalism. This is an important distinction. In the mobile robot literature simulators are commonly used as experimental testbeds to "demonstrate" the validity of mathematical models derived from first principles. In our case, the simulator is the model, and the validation comes from experiments on a real robot.

We also wish to highlight the *process* leading to our results, and to contrast that process with current practice. In our view, the current practice places emphasis on *getting things to work*, resulting in the widespread use of iterative design interleaved with *ad hoc* evaluation. This approach often results in working systems, but it does not yield an understanding of the limitations of these systems. In particular, it provides little assurance that a system will continue to operate when environmental parameters are changed. This is of particular concern to NASA because we cannot test our robot in the actual conditions under which it will operate (since we don't know what they will be).

Our approach therefore emphasizes *reliable prediction of system performance*. We intend our approach to complement the current practice rather than supplant it. Ultimately our goal is to build systems that work. Reliable measurement of system performance is a necessary component of the process, but cannot replace the current iterative design methodologies, which we continue to advocate.

3. Apparatus

The robot under study is Rocky 3.2 [Matthies95], a rebuilt version of Rocky 3 [Gat94]. This robot has essentially the same chassis design, size, computer and sensor suite as those designed for the actual flight rover, which is known as the MFEX (Mars Flight EXperiment) rover. Both rovers are six-wheel rocker-bogie type vehicles. Each has an 8085 processor with 1/4 megabyte of bank-switched RAM, most of which is used to hold image data. The primary sensor on the robot is a structured-light range sensor. The rover has wheel encoders and a rate gyroscope, which it uses to keep track of its position through dead reckoning. The rate gyroscope tends to drift, and the wheels slip in loose soil, resulting in dead reckoning errors. The rover also has articulation sensors and inclinometers, but these were not used in the work described here.

All experiments were performed indoors in a 4 m by 12 m enclosure filled with loose sand. The enclosure (which we refer to as the sandbox) was instrumented with a tracking system comprising four overhead CCD cameras, whose field of view covered 10 m of the sandbox's length. The rover was equipped with a visual target that could be easily identified and located in the images. Calibration test results showed that the rover's location could be accurately determined to within about 1 cm and 1 degree. The overhead tracking system was used only for gathering experimental data, and was not accessible to the rover navigation software.

The sandbox was made as large as the available space would allow. Unfortunately, this turned out not be large enough. Initial experiments revealed that the rover would often approach the edges of the sandbox when avoiding obstacles. The rover would then detect the sandbox side rails as obstacles, which would affect the experimental results in unrealistic ways (since long, straight obstacles like the side rails are unlikely to exist in Mars).

To address this problem we implemented a "virtual sandbox" in the experiment management software. During a run if the rover came within 90 cm of an edge (as measured by the overhead tracking system) the experiment manager program would command the rover to stop and turn in place so that its heading was reflected about an axis parallel to the edge of the sandbox. The goal location was also reflected about this same axis. The net effect was to create a "virtual sandbox" adjacent to and a mirror image of the original sandbox. This reflection could be repeated to produce a sandbox of effectively infinite size, but in practice only one reflection on either side of the physical sandbox has been used to date.

This technique allows runs of unlimited length in a sandbox of finite size. The supervisory program keeps track of the rover's "virtual position", and issues the proper commands whenever the rover nears an edge. The main limitation of the method is that the terrain in the "infinite" sandbox is just repeated mirror images of the terrain in the original sandbox, and so the obstacle distributions in the virtual sandbox are not quite random. Nevertheless, this is a useful technique for gathering data in an experimental area which is not quite big enough by itself for realistic tests.

4. Method

In our experiments we measured the values of two performance metrics: traverse distance and elapsed time to reach a goal. These are commonly used metrics, but in fact the choice of these metrics is based more on convenience than on sound theoretical considerations. Time and distance are easy to measure. There are many other performance metrics we could have chosen, many of which affect the outcome of a mission much more directly than time or distance (energy consumption, for example). It is often tacitly assumed that all performance metrics are correlated, and that if one optimizes, say, path length then traverse time, energy consumption, etc. will also be optimized. We will show that this is not the case. The present experiments could be improved by measuring energy consumption, but we currently do not have the means to do so.

To conduct an experiment we first generate a test course by placing rocks of various sizes in the sandbox. The size and placement of these rocks is chosen according to a published model of the rock distributions on Mars, to be described shortly. We then instruct the rover to travel from one end of the sandbox to the other, between two predetermined locations that are 7.6 meters apart. (This is the longest traverse that is possible within the constraints imposed by the field of view of the overhead cameras.) The route that the rover takes is recorded by tracking the rover using the overhead cameras. The rover's position and orientation are recorded at regular intervals, along with a time stamp.

A test course is constructed by first choosing an obstacle density. This density can be chosen arbitrarily (for example, to study the effects of gradually increasing obstacle densities on the performance of a particular navigation algorithm), or it can be chosen according to the standard model of Martian terrain, developed by Moore [Moore89]. According to Moore's model, the number of rocks per square meter with a diameter less than or equal to a given size D is:

$$N=kD^{-2.66} \tag{1}$$

where k is a parameter that varies according to the particular location on the surface. This model is accurate for D > 14 cm, which is fortuitous since this is about the smallest size rock that the laser ranging sensor will detect as an obstacle. Thermal inertia data indicate that the modal value of k across the entire surface of the planet is approximately 0.00415; this case is referred to as "Mars-nominal" terrain.

Locations for obstacles are generated using a random number generator. Rocks are placed at the prescribed location using a tape measure. Setting up a test course is time-consuming work, so each course was used for four runs, two in each direction. This also allows some interesting statistical analysis to be done to determine, for example, how repeatable (and thus how predictable) the rover's performance is in a given terrain.

The rover moves in discrete steps, where each step is either a turn in place or a forward movement of approximately one wheel radius (7 cm). After each step a supervisory program running on an off-board workstation recorded the rover's position and heading as computed by dead reckoning, the rover's absolute position, the state of the rover's obstacle detectors, and a time stamp. Datasets were indexed to records of the terrain layout in which they were run.

5. Results

The rover is quite slow, moving at an average speed of less than 1 cm/s. Most of the time is spent processing the data from the laser range finder. We were able to complete a total of about 100 runs over the course of a summer, of which 40 were performed in Mars-nominal terrain. The remainder were performed under a variety of other obstacle densities, including zero obstacles as a control case. (The zero-obstacle case was also used to evaluate the rover's dead-reckoning performance [Matthies95].)

The raw data consisted of a complete record of the rover's path for each run. We reduced this data by computing two performance metrics for each run: total path length and total traverse time (corrected for delays introduced by the overhead tracking system).

The reduced data for the Mars-nominal case are depicted as a scatter plot in figure 1. Each point on the plot corresponds to one run. The two axes represent the two different performance metrics. This figure illustrates our first result: path length and traverse time are poorly correlated; the correlation coefficient is 0.69. In this case the poor correlation is easily explained by the fact that the rover occasionally turns in place as part of its navigation strategy. Nevertheless, in the path planning literature it is often tacitly assumed that all performance metrics of interest will be well correlated with path length, an assumption we see here to be false.

The cumulative distribution function for the reduced distance data is shown as the bold line in figure 2. (This figure shows the simulation results superimposed on the results from the real robot — see section 6. The distribution function for time looks virtually identical, but with a different scale on the y-axis.) This figure illustrates our second result: the distribution functions are not normal. In section 7 we will show that they are in fact exponential (or, at least, that an exponential distribution is a good fit) but until we do this we cannot assume any particular shape. This will complicate our analysis somewhat because most standard statistical methods assume a normal distribution.

There is to date insufficient data from the real robot in obstacle densities other than Mars nominal to allow strong quantitative conclusions to be drawn. As one would expect, performance does appear to drop off as obstacle densities increase. There is significant degradation at densities above about one obstacle per square meter, which is cause for some concern, since that is approximately the obstacle density expected at the MFEX landing site. See [Matthies95] for further discussion of these results.

6. Simulation

We constructed a simulation of the rover to serve as a reference model for making predictions about the rover's performance. Because the design of the rover is evolving we designed our simulation to be extremely flexible. Our simulator is written in Common Lisp using the Common Lisp Object System (CLOS) [Steele90]. The code is object-oriented to an extreme. Everything in the simulator is a software object, including environments, objects in the environment, robots, sensors and actuators. Simulations are constructed by "installing" robot objects and passive objects (e.g. obstacles) into a world object. Robot objects are constructed by installing sensor and actuator objects into a "robot chassis" object. The software is designed to make it easy for users to add new sensor and actuator models, as well as new models of object interactions. The simulator naturally supports multiple-robot simulations; all that is required is to install more than one robot into the world.

The only assumption imposed by the simulator itself is that all objects with physical extent (e.g. robots, but not drive mechanisms) have that extent described by a two-dimensional polygon. (A three-dimensional model could be used if desired.) The root functionality provided by the simulator is simply efficient computation of polygon intersections. This may seem like a severe restriction, but in turns out to provide a tremendous amount of power. Furthermore, it does not preclude certain types of three-dimensional modeling. Because the interactions of objects can be defined by the user, it is possible, for example, to make a "rough terrain" object, whose physical extent is the entire world, and whose interactions with a robot are governed by a model of rough-terrain traversal provided by the user. However, we have found this to be unnecessary; a much less sophisticated model is adequate for describing the behavior of our robot in the relatively sparse obstacle fields in which we have conducted our experiments to date. A description of the particular model used in our experiments can be found in [Gat95].

Despite the relatively simple nature of this simulation it produces remarkably high quality results. Qualitatively, the behavior of the simulated rover is virtually indistinguishable from that of the real one. However, this sort of gestalt assessment is precisely the sort of informal, anecdotal result that we have criticized so severely. We therefore now proceed to demonstrate formally that we have captured some of the relevant aspects of the real rover's behavior in our simulator model.

We duplicated the sandbox experiment on the simulator in two separate sets of experimental trials. In the first set of trials we did 100 runs in each of nine different terrain densities. (A random field of obstacles was generated for each run, so there were a total of 900 different obstacle fields used.) The results from these trials (excluding failures — see section 7.4) are shown in figure 2 as cumulative probability distributions. The results from the real robot are superimposed as a bold line. The Moore-model parameter ranges from 0.00015 to 0.00815 in even increments of 0.001.

In the second set of simulator trials we ran 804 runs using a Moore model parameter of 0.006. (See section 7.4 for an explanation of the apparent discrepancy between this and the value of 0.00415 used for the real experiments.) The results of these runs (again excluding failures) are shown in figure 3, superimposed with the real data and a best-fit exponential curve. By visual inspection, the fit of all three curves appears to be quite good. Selected results when failures are not excluded are shown in figure 4. As usual, the real data are superimposed as a bold line. The results do not match the real data nearly as well, indicating that the simulator's emergent failure model may be flawed.

In the next section we will formally analyze these informal observations.

7. Analysis

To draw conclusions about our data we employ statistical tests. There are a number of subtle issues in the use of statistical tests and in the interpretation of their results which are not common knowledge among mobile robotics researchers. It is therefore worthwhile to digress for a moment to discuss statistical tests in general before returning to the analysis of our data. A reader familiar with statistical methods should feel free to skip to section 7.2.

7.1 Digression: On the nature of statistical tests

Statistical analysis is related to probability theory in that both deal with phenomena that, by assumption, contain elements that are unmodelable *a priori*. However, unlike pure probability theory, which attempts to derive probabilities from first principles, statistics deals with the problem of drawing conclusions about probability distributions by examining sets of data points drawn from those distributions.

In general, a statistical analysis proceeds as follows: given a set D of samples drawn from a probability distribution P, a computation is performed on the elements of D (of which there may be any number) which yields a single result S. This result is called a statistic[1]. The familiar mean, median, mode and variance are all examples of statistics.

Statistics are themselves random variables; a statistic computed on two separate sets of samples drawn from one distribution will generally not yield the same value. However, certain statistics can be shown to have probability distributions that, under certain conditions — the so-called *null-hypothesis* conditions — are independent of the underlying distribution P which generated the datasets on which the statistic is computed. If S turns out to have a value which is very unlikely under these conditions we can confidently *reject the null hypothesis*, i.e. conclude that the conditions under which the distribution of S is known do not hold.

It is important to note that the converse reasoning is faulty. If the value of S is a likely one it does not follow that the null-hypothesis conditions are true. It is possible that the null hypothesis is false, but in a way that does not alter the distribution of S. Lack of evidence that a hypothesis H is false is not the same as evidence that H is true.

---

[1]Common usage implies that the word statistic means an element of D, e.g. "The car crash victim became just another statistic." This is incorrect; a statistic is the result of a reducing computation performed on the elements of D.

7.2 Analysis 1: exponential distribution

We wish to formally test the hypothesis that the data generated by the real robot and the simulator are drawn from exponential distributions. To do this we employ the Smirnov statistic [Lehmann75] (sometimes called the Kolmogorov-Smirnov statistic), defined as:

$$KS = Max_x(|F(x) - D(x)|) \qquad (2)$$

where F and D are cumulative probability distributions. When F and D are the same (the null hypothesis condition) the distribution of KS is independent of that of F and D. The Smirnov statistic can be used either to compare two sampling distributions (in which case the null distribution of KS depends on the number of data points in each of the two samples) or it can be used to compare a sampling distribution with an *a priori* closed-form expression. Here we will use the second method.

When we test the hypothesis that the cumulative distribution functions for our data are of the form:

$$P(x) = 1-e^{-k(x-x_0)} \qquad (3)$$

where x is the random variable (distance or time in this case), x_0 is the smallest value of x, and k is a parameter chosen for best fit. For distance, the best fit value of k is 0.4, and for time the best fit value of k is 0.0014, yielding values of KS of 0.11 and 0.12 respectively. (The best-fit exponential curve for distance is shown in figure 3 along with the real distance data and the results of the second set of simulator runs.)

For n=40 (the number of data points) the null-distribution probabilities of the above values of KS are 0.68 and 0.60, i.e. the value of KS is expected to be at least as large as the observed values 68% and 60% of the time for sets of 40 data points drawn from the hypothesized distributions. Thus, there is no basis for rejecting the null hypothesis. (Even better fits are possible if we choose x_0 to be a value slightly less than the smallest observed values.) Note that this does not mean that the distributions *are* exponential, just that we can't distinguish any differences that there might be on the basis of the data we have.

By way of contrast, if we test the hypothesis that the distributions are normal with mean and variance equal to the sampling means and variances of the two datasets, we obtain values of KS of 0.23 for distance and 0.18 for time. The corresponding null-distribution probabilities for 40 data points are 0.025 and 0.13. Thus we can conclude with better than 95% confidence that the distribution for distance is not normal, and better than 85% confidence[2] that the distribution for time is not normal.

7.3 Analysis 2: comparison of simulated and real results

To test the second hypothesis we have three options. First, if we assume that the distributions are exponential we could employ a parametric analysis and estimation theory to derive a numerical solution (with error bounds) for the distribution functions and compare them. However, the evidence that they are in fact exponential is pretty thin, and such an assumption could lead us seriously astray. The second alternative is to use the discrete form of the Kolmogorov-Smirnov test to compare them. The third alternative is to employ a different test altogether. It turns out that for comparing two sampling distributions there are better methods available. We will use a statistic advocated by Lehmann, the Wilcoxon-Mann-Whitney (WMW) statistic.

The WMW statistic is computed as follows. Let D1 and D2 be sets of m and n data points drawn respectively from probability distributions P1 and P2. The data in D1 and D2 are combined and sorted. Each datum is then ranked according to its position in the sorted list; the

[2]Usually, a confidence level of at least 90% is required for a result to be considered statistically significant.

first number in the list is assigned the rank 1, the second number the rank 2, etc. The ranks are then separated according to which distribution (D1 or D2) its corresponding datum was drawn from. The separated lists of ranks are then summed to produce two numbers, S1 and S2.

It can be shown that if P1 and P2 are the same, then S1 and S2 have normal probability distributions whose parameters are independent of P1 and P2. Instead, the parameters depend on the number of data points, m and n:

$$E(S1) = n(m+n+1)/2 \tag{4}$$

$$E(S2) = m(m+n+1)/2 \tag{5}$$

$$Var(S1) = Var(S2) = mn(m+n+1)/12 \tag{6}$$

where E(X) denotes the expected value (mean) of a random variable X, and Var(x) denotes its variance. Because the variances are the same, the quantity:

$$S = S1-n(m+n+1)/2 = S2-m(m+n+1)/2 \tag{7}$$

is sometimes used instead of S1 and S2. The variance of S is the same as that of S1 and S2, and the mean of S is, of course, zero.

Since we know that if P1 and P2 are the same then S is drawn from a normal distribution with known variance, any large deviation (relative to the variance) of S away from zero can be taken as evidence that P1 and P2 are different. If we compute:

$$p = |\Phi(S) - \Phi(-S)| \tag{8}$$

$$= 2\Phi(|S|)-1$$

where Φ is the error function (i.e. the integral of the normal distribution), the result is the probability that the magnitude of a sample from the null distribution of S (i.e. when P1=P2) is less than or equal to the observed value S. So, for example, if p=0.95 then there is a 95% probability that P1 and P2 are different, and only a 5% probability that P1 and P2 are the same, and that the observed value of S is due to chance.

We compared the data generated on the real rover with that generated on the second set of simulator experiments. When failures are ignored, the resulting value of p is 0.056 and 0.088 for distance and time respectively, indicating an excellent fit. (The value of p would need to be 0.90 to reject the null hypothesis.) The Smirnov probabilities for the same datasets are both greater than 99%, illustrating that the WMW test is more sensitive to differences than the Smirnov test.

7.4 Failures

Throughout the analysis we have been ignoring failures produced by the simulator. If failures are not ignored, the data produced by the simulator does not match the real robot data, indicating that the simulator's failure model is faulty. We do not have enough failure data from the real robot to properly calibrate the simulator's failure model, so we have made no attempt to correct the situation. Instead, we use this mismatch to our advantage to illustrate how our statistical procedures can detect faulty models.

Figure 4 show the distance data from the sets of three simulator runs with the best fits to the real data when failures are ignored. In this figure failures are treated as infinite distances (presumably, taking an infinitely long path to the goal is tantamount to failure), and it is clear that the match is not nearly as good as before.

Using the WMW test, the value of p for the lower curve (804 simulator runs, 175 of which were failures) is 0.98, indicating a highly statistically significant difference. The value of p for the other two curves are 0.62 and 0.64, which is not quite high enough to confidently reject the fit. (The fact that these curves have a better fit is due to the fact that the failure model was adjusted for these runs.) We would need about four times as much data from the real robot to distinguish the second two simulator distributions from the real distribution with 95% confidence, assuming that the results are reproducible. (This result was obtained by computing p for datasets consisting of multiple copies of the real data.) We are currently beginning experiments to gather this data.

Finally, although we have not yet culminated our research by verifying a statistical prediction made by our model, we have made one very interesting postdiction[3]. The best fit of the simulator data and the real data is obtained when the Moore model parameter in the simulator is set to approximately 0.006. However, the value of the parameter in the real experiments was 0.00415, the Mars Nominal value. This mismatch caused some consternation until it was discovered that the minimum rock diameter, D_0, in the real experiments was set to 10 cm rather than 14 as in the simulator. Moore's model is very sensitive to the value of D_0 in the 10-15 cm range, and this difference resulted in enough additional obstacles to make the overall obstacle density approximately the same as the simulator.

8. Conclusions and Future Work

We are working towards rigorous experimental study of autonomous mobile robots. Our approach is to employ the methods of the natural sciences in our investigations. We use simulations, but we treat them as models rather than as the system under study. The value of a simulation is measured by how well it predicts the behavior of a physical system. We use statistical methods to evaluate our experimental data.

To date we have carried out only a portion of our research program. We have constructed a simulation and verified that it postdicts the behavior of a real robot in a statistically meaningful way. We have also used the simulator to generate predictions about the behavior of the rover under conditions in which it has never been tested. The final step of our research, to be completed this summer, is to test these predictions by performing a second series of experiments.

Our work offers two central contributions. We offer the first solid experimental evidence that certain performance metrics, often tacitly assumed to be well correlated, can in fact be highly uncorrelated in practice. In retrospect this is fairly obvious; nevertheless, it is a fact routinely ignored in the literature. Hopefully our results will help change this unfortunate state of affairs.

Our second contribution is the introduction of statistical rigor to the evaluation of experimental results. This is complicated by the fact that the probability distributions involved are not normally distributed. We adopt non-parametric methods for comparing probability distribution functions. This allows us to draw quantitative conclusions about the probabilities of certain events without knowing a priori the shape of the probability distribution.

A complimentary line of research is to develop a mathematical theory to explain the observed shapes of the probability distribution functions for the performance metrics we have chosen. They appear to be exponential, indicating that rover navigation in rough terrain is a Poisson process. However, it can be argued on theoretical grounds that rover navigation cannot be a Poisson process because of the non-independence of obstacle encounters and the form of the termination condition. A diffusion process might be a better model. Arguments can also be

---

[3]This is a postdiction rather than a prediction because the result was obtained after the experiment was completed. Experimental data collected before the formulation of a hypothesis is weaker evidence for that hypothesis than data collected afterwards. Thus, a postdiction is not as interesting as a verified prediction. However, in this case the postdiction showed that what we thought was a negative result was, in fact, a positive one by predicting a mistaken assumption in our analysis.

made that the expected probability distributions should be two-tailed distributions (such as a chi-square), especially in denser terrains. (We have collected some preliminary data that indicate that this is not the case; even in very dense terrain the exponential distribution appears to persist.)

Finally, we also need to combine the present analysis with a similar probabilistic analysis of the performance of the robot's lookahead sensor. Measuring the performance of the lookahead sensor is currently in progress.

Acknowledgments

Reid Harrison originated and developed the support software for the virtual sandbox and performed the experimental trials on Rocky 3.2. Todd Litwin, Larry Matthies, Brian Wilcox and Wayne Zimmerman constructed the sandbox and overhead tracking system. This work was performed at the Jet Propulsion Laboratory, California Institute of Technology, under a contract with the National Aeronautics and Space Administration.

References

[Gat94] Erann Gat, et al., "Behavior Control for Robotic Exploration of Planetary Surfaces," *IEEE Transactions on Robotics and Automation*, August 1994.

[Gat95] Erann Gat, "Towards Principled Experimental Study of Autonomous Mobile Robots I," submitted to *Autonomous Robots*. (This is an expanded version of the present paper.)

[Horswill94] Ian Horswill and Masaki Yamamoto. "A $1000 Active Stereo Vision System," in Proceedings of the IAPR/IEEE Workshop on Visual Behaviors, W. Martin, ed. Seattle, WA. IEEE Press, 1994.

[Latombe91] Jean-Claude Latombe, *Robot Motion Planning*, Kluwer Academic Publishers, 1991.

[Lehmann75] E. L. Lehmann, *Nonparametrics: Statistical Methods Based on Ranks*, Holden-Day, 1975.

[Matthies95] Larry Matthies, et al., "Mars Microrover Navigation: Performance Evaluation and Enhancement," *Autonomous Robots*, to appear.

[Moore89] H.J. Moore and B.M. Jakosky, "Viking Landing Sites, Remote Sensing Observations, and Physical Properties of Mars Surface Materials," *Icarus*, 81:164-184, 1989.

[Steele90] Guy L. Steele Jr., *Common Lisp: The Language*, Second Edition, Digital Press, 1990.

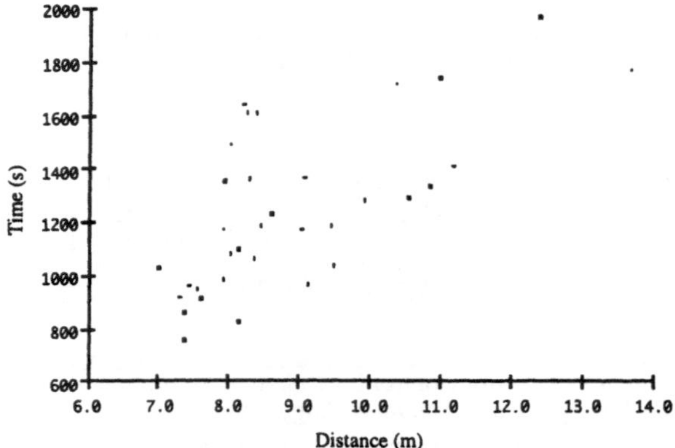

Figure 1: Scatter plot of elapsed time and distance travelled for forty runs of the real robot in Mars-nominal terrain.

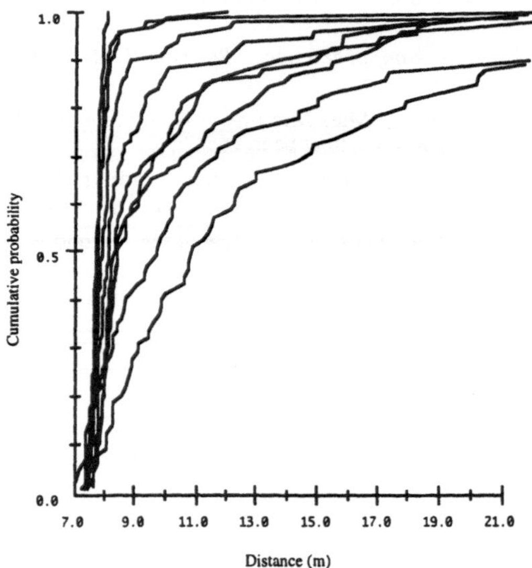

Figure 2: Cumulative probability distributions of the distance metric for a goal 7.6 meters from the starting location at various obstacle densities, superimposed on data from the real robot (bold line). The curve furthest to the left is for a Moore model parameter of 0.00015, and each successive curve increments this value by 0.001.

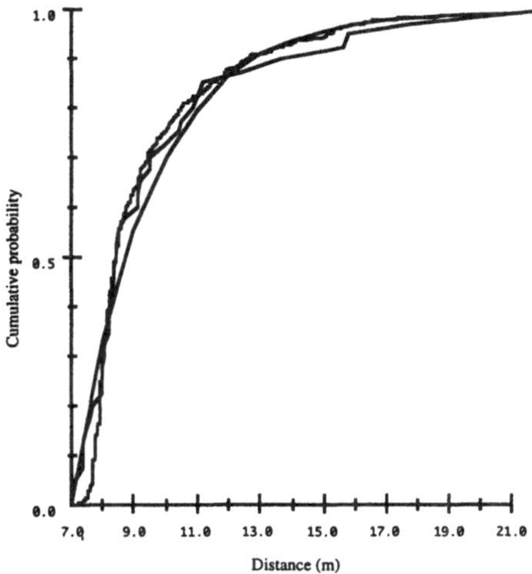

Figure 3: Comparisons of three cumulative probability functions for distance: 40 data points from the real rover, 629 data points from the simulator, and an exponential curve.

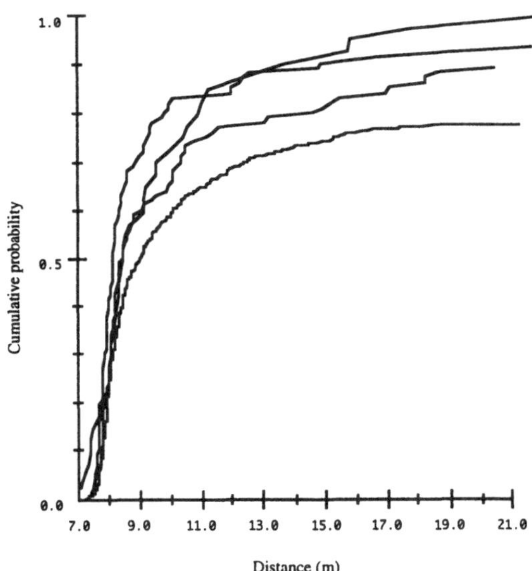

Figure 4: Simulator results when failures are not ignored superimposed on data from the real robot.

Mission Programming: Application to Underwater Robots

Eve Coste-Manière
INRIA Sophia Antipolis
BP 93
06902 Sophia Antipolis cdx – FRANCE
Eve.Coste_Maniere@sophia.inria.fr

Michel Perrier, Alexis Peuch
IFREMER Toulon
BP 330
83507 La Seyne Sur Mer cdx – FRANCE
peuch,mperrier@ifremer.fr

Abstract

In this paper, mission programming issues for autonomous vehicles with a gateway to teleoperation for online user intervention in dynamic environments are addressed. We define the various requirements of a generic and yet efficient mission programming method. The proposed solution can be integrated into various existing control architectures. To focus the presentation, special emphasis is place on the programming of a complex and generic mission executed by the VORTEX *underwater vehicle. Experimental results illustrate the* "Find Target and Operate with Obstacle Avoidance" *mission.*

1. Introduction

Robotics systems achieve their fundamental scope and usefulness thanks to their continuous interaction with their external world. They attain a high degree of autonomy and robustness while performing missions if they can plan and execute actions, build and update maps, sense and adapt their behaviour in response to the environment, and detect and react to defined events. Obviously, this cannot be performed with a single, unique control law, however complex, that is dedicated to a particular control situation. In order to carry out complex applications, a well-formalized structure or "control architecture" that handles automatic control aspects, real-time issues, reactivity, discrete-event management and control-law switching within a formalized framework is therefore required.

In this context, this paper addresses some critical issues that, when resolved, will promote the genericity of programming methodologies at the mission level to be used on more than the individual systems for which they were first developed. This ability to use a specific programming methodology on multiple hardware architectures should be possible, with no loss of efficiency, provided that theoretically well-founded concepts such as those introduced in ORCCAD [11] and identified services between abstraction levels are available on the host architecture.

The paper is organized as follows: in section 2, we describe the various requirements of a generic, mission-programming methodology and their proposed solutions. After the description of our experimental setup (the VORTEX vehicle), a generic underwater mission, which has been adopted by various underwater institutions [13] to evaluate mission programming techniques, is defined in section 4. The underwater mission requires the sequencing of up to ten different control laws implemented on the PIRAT [8] real-time control software. The mission also exercises the physical capabilities of our testbed with the different fundamental types of actions that may be required during a "real" mission at sea. Then, section 5 provides a tutorial of our mission programming techniques as applied to the generic mission. Actual results from the experimental execution of the mission by VORTEX are provided afterwards. Finally, we will conclude with some guidelines for further research and experiments.

2. Mission Programming Technic

2.1. Requirements and proposed solutions

In order to deal with the global realization of a mission requiring planning, control and execution, successful architectures will integrate the planning aspects of top-down approaches with the real-time and physical aspects of the bottom-up approaches. In this research, we investigate some key issues of prime importance at the mission level where the merging should be done.

The input of a mission controller can be created in two different ways. Either, it can be generated directly by a planner (developed by the artificial intelligence community), or a user can create manually the controller by using a programming method/language that expresses the mission as compositions of "elementary" or "atomic" robotic actions. In this paper, we are interested in the following points that address issues within the last context of mission programming languages:

- The actions required to define a complex mission and handled by the language primitives must be well characterized. The robot-task concept, first introduced in the ORCCAD [11] system, is adopted here to describe an elementary action. Its simplified scheme is composed of two parts: a parameterized, control law executed by the system and a logical behaviour which handles the start, stop or typed exceptions during the life of the control law.
  Specifically, we encapsulate control laws implemented in PIRAT [8] (the real-time control software of the VORTEX vehicle) with a logical behaviour imposed by the set of events that govern its execution. Then, a logical and temporal arrangement of these elementary actions, when performed incrementally, carries out a complex mission. Missions can be programmed directly or within the robot-procedure context [6]. The set of input/output events constitutes the logical behaviour of the global mission program as defined for reactive systems [4]. Observation functions are used to trigger events and to interface the program with its environment (see below).
- Various formalisms to describe the action's behaviour can be envisioned: for example, declarative, data flow, or synchronous, depending on the formal representation adopted. Because a robotic system can be seen as a reactive system [4], the synchronous formalism [1] has been adopted [3] through the use of the

ESTEREL [2] language which provides an imperative style and a rich development environment. A mission program written in ESTEREL is then translated into a form understandable by the mission controller. Because of the choice of the synchronous assumption, the translation format corresponds to a finite automaton (see section 6) which ensures the global control of the application and the robot-tasks.

- In order to allow the user to program robotic actions (seen as atomic actions), the mission programming language must include the following classical functionalities: sequencing, parallelism, conditional branching, and iteration. Also some control structures that specify reaction to defined events and others dedicated to synchronization of flow are required. The management of cooperating robots may eventually require the extension of the proposed structures.
  The program created by the mission programming language is translated into a format that the mission controller uses to coordinate robot actions to complete a mission. With a proper formalism, an automated mission planner can generate output in the same format for input into the mission controller.
- We believe that whatever format that has been developed for use by the mission controller, the actions encoded by this format must be verifiable. Since we have adopted the synchronous formalism, we directly benefit from tools that check for and verifies logical correctness [10]. Hybrid and temporal extensions should be investigated as proposed in [5] in the ORCCAD context.
- When verification methods are not available or sufficient, simulation tools should be available to perform the corresponding trials.

2.2. Interfacing mission control and execution

At the mission execution level, robotic actions are executed and observation functions monitor the state of the robot and its environment. The mission controller directs the execution of the robot actions and the observation functions. Since we have chosen ESTEREL to implement the mission controller, it is now well established [3] that its connection with the execution level requires the following parts:

1. A reactive kernel that manages the execution of the ESTEREL's deterministic automaton produced at compile time which in turn drives the computation of the control laws and observation functions.
2. Most of the algorithms (e.g. control laws) that reside at this level have foundations in classical control theory and must be computed. But classical control theory does not cover how control actions are started or stopped and how the system can be made to switch from one control mode to another. It is the role of the interface between the mission control level and the execution level to ensure such services.
3. The interface also transforms asynchronous events from various origins to logical events to be fed to the automaton and vice versa;

The interface hides the implementation details of handling the robot actions and observation functions from the mission controller. We consider the interface to these objects to be a set of input/output events with or without parametric values that represents all the ways that an object can be manipulated. Based on the information returned from execution level through the interface layer, the mission controller will proceed to execute the high-level program. The interface layer is event driven. It does no work unless events trigger it. These events may be global and periodic such as a clock event or dependent on what is executing on the robot at the current instant. The following is an example set of functions that can be called by the interface layer upon request of the mission program. These functions implement services that handle the robot actions and the observation functions:

- initialize, start, stop, re-parameter the robot actions with the corresponding parameters.
- register, start, stop, re-parameter, an observation function with its given parameters.

Symmetrically, the execution layer sends events (with or without parameters) to the interface layer. These events may be generated by an observer function started by the higher level for the control of the mission (e.g. "depth reached"), by a robot action which was itself started by the higher level (e.g. "post-conditions reached" or "exception encountered"), or by the execution layer itself (e.g. "hardware failure"). All of these events call a mechanism to pass the event up to the interface layer with possibly some associated parameters.

3. The Experimental Set up

VORTEX is an experimental ROV-like vehicle designed by IFREMER to test in a 15×6 m pool the autonomous functions that may be used by future AUVs [8]. Mechanically, VORTEX is made of a tubular structure on which up to eight propellers can be set up at any location (figure 4.a). The on-board electronics package and the proprioceptive sensors reside at the center of this structure. Modular foam blocks providing buoyancy are fixed on the top of the structure. The set of sensors includes attitude sensors (two inclinometers and three piezoelectric angular-rate sensors coupled with a fluxgate compass), a depthmeter, a tilt-orientable video camera, a panoramic sonar and an eight-element lateral sonar belt. The camera is used for target tracking tasks, and the sonar belt is used for tasks such as wall following and obstacle detection and avoidance.

Each sensor and actuator has an on-board microcontroller used for data collection and transmission. The microcontrollers are arranged in a modular, network architecture so that it is very easy to add or remove components on the vehicle. At any time, the vehicle control can be switched to a manual, joystick control mode for obvious safety reasons. The new extension of the PIRRAT [8] real-time, control software, developed at IFREMER, provides the different services required by the interface layer between the mission control and execution levels as described in the previous section. Object-oriented techniques are used and PIRAT runs on a surface VME system. The vehicle and the surface computer are linked by a tether used for energy and data transfer. Video and sonar images are transferred using a separate fiber-optic link within the tether. Additional data are sent on a FDDI fiber-optic loop. The VME controller, the development workstation, and the operator workstation running the LIVE interface are connected on an Ethernet link.

4. The Chosen Mission

Two main issues are under consideration when defining the experimental mission:

a. The VORTEX experimental vehicle

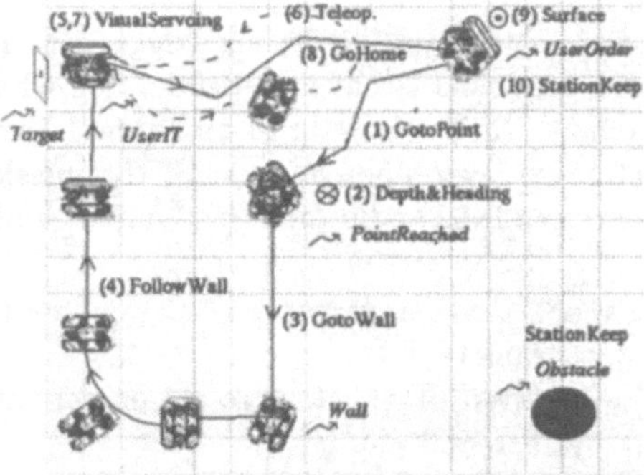

b. Top view of the nominal mission

- We study realistic missions profiles related to subsea intervention that could be effectively performed at sea with complex and heterogeneous systems. In the field of subsea manipulation with a "free" swimming vehicle (ROV and AUV), missions involve structured or unstructured tasks that require the integration of aspects of teleoperation together with full autonomous control within the same framework. Reactivity handling and reliability are, of course, of prime importance (e.g. obstacle avoidance, user interruptions). For our mission, we have only considered structured tasks such as: Survey, inspection, maintenance and repair, or assembly tasks [12, 9].
- Since our long term goal is to compare different architectures controlling similar robots to perform the same task or mission, the proposed mission must be feasible by various vehicles [13]. Thus, the mission scenario will be limited by the physical capabilities of the three targeted vehicles: the VORTEX, the OTTER (MBARI/STANFORD ARL) and the PHOENIX (NPS).

The following mission scenario (figure 4) involving autonomous and teleoperated structured tasks has been chosen: The vehicle transits to its starting position upon user command. Then it proceeds towards a wall. There, it starts following the pool wall in any direction. A target (such as a pipe to be inspected) is located on the wall. When the target appears within the vehicle's camera's field of view, the user (who in the future could be replaced by an intelligent observation function within the vehicle controller itself) commands the vehicle to visually servo upon the target (in motion or motionless) simulating, for example, an inspection task. Then, upon the user's request, the vehicle is teleoperated to follow a path which replaces an operational action such as "go and fetch the required tool". When the target reappears within the workspace, the vehicle is again servoed visually upon the target image, simulating a manipulation task. After a while, the user triggers the vehicle to go back home.

The whole set of actions is performed under supervision of the user who can interrupt the mission causing the vehicle to start a homing procedure. During the execution of the mission, the presence of an obstacle will cause the vehicle to station keep until the obstacle is removed from its nominal path. Then, the vehicle proceeds and continues the mission. More drastic exceptions (such as alarms for water leakage) can occur at any time during the mission execution.

5. Mission Programming: A Tutorial

Three steps are required to create a program to effectively execute this mission. First, automatic control considerations are taken into account to design the various control laws needed at the execution level. Next, logical behavior based on discrete events are encoded in the robot-task framework together with the design of the program in charge of their global coordination. Finally, the integration of the automatic control laws with the logical behavior is completed.

5.1. Description of the required actions

Eight robot-tasks have been derived from seven control laws of various complexity including sonar and vision-based control. Robust control is used for several of the control laws in order to reject some perturbations due to external disturbances (e.g. when the vehicle is close to the water surface) or due to the effects of the coupling between several controlled states.

Because of the current thruster arrangement, the roll angle of the VORTEX is not controllable. However, for the remaining five degrees-of-freedom, the vehicle is considered to be holonomic. During complex tasks, such as *'FollowWall'* or *'VisualServoing'*, four degrees-of-freedom of the vehicle are controlled. The remaining uncontrolled state is the pitch angle which is passively stable. A nonlinear, PID control approach [7] has been successfully adopted for robust control considerations.

The *Depth&Heading* and *Surface* robot-tasks are built upon two different control laws which only involve information about the proprioceptive state of the vehicle such as its depth or its heading angle. With the first control law, the vehicle is controlled to reach a desired depth and heading at the same time. In the second case, only the depth is controlled and reaching the desired depth meets the postconditions that indicate the end of the task. *Depth&Heading* is in a critical situation if the depth of the vehicle is ever close to the depth limit of the pool. If this extreme condition is encountered, a typed exception is raised to be handled by the mission program to stop the mission.

The *GotoWall* and *FollowWall* robot-tasks In the *'FollowWall'* task, the purpose is to control the vehicle to move facing a pool wall at a specified depth and distance, with no particular orientation. The relative position and orientation of the vehicle are controlled using telemetric distances provided by the two frontal acoustic sounders. The vehicle depth is controlled using the depthmeter, and the lateral motion of the vehicle is controlled using the sounder corresponding to the specified direction (left or right). A particular strategy is used when the vehicle has to go around a pool corner. *GotoWall* is stopped when the vehicle has reached the desired distance to the wall. *FollowWall* is executed for a specified duration. For both *GotoWall* and *FollowWall*, a typed exception is raised when the distance to the wall indicated by the front sounders is smaller than a given threshold for safety reasons, and the mission is stopped.

The *VisualServoing* robot-task controls the vehicle to stabilize in front of a given target (motionless or not) at a specified distance, with no particular orientation. The control law involves telemetric distances provided by acoustic sounders to control the relative range and orientation. The two remaining degrees-of-freedom, which correspond to the motion of the vehicle in a plane parallel to the target, are controlled using information provided by the image processing unit. This unit feeds the control law with the position error information of the target with respect to the center of the video image. A typed exception exists if the target is lost.

The *GotoPoint* and *Homing* robot-tasks Here, the same control law is used with different parameters. These tasks are used to drive the vehicle to different pool positions and are stopped when the specified position is reached. The control law uses information not directly measured by sensors, but instead, estimated from other measured data. This control law is fed by the Cartesian position of the vehicle expressed in the pool frame and estimated by fusioning telemetric distances provided by the acoustic sounders and the vehicle heading angle. The depthmeter is used to control the depth.

The *StationKeep* robot-task Here the vehicle is controlled for a given duration to keep its current position and orientation when the robot-task is started. A typed exception is triggered if the vehicle shifts from this position beyond a maximum tolerance range.

The *Teleoperation* robot-task We have chosen to incorporate the teleoperation action within the robot-task skeleton to insure the overall control of the mission by the operator. The precondition is triggered by the user while the postconditions are satisfied when the vehicle is within working distance of the target.

5.2. The Mission Program

The set of primitive goals associated with the previous mission and the sequence of their execution result from an application of goal-driven-like reasoning by the mission developer. The mission originates with a single high-level goal: *"Find Target and Operate with obstacle avoidance"*. This top-level goal is then reduced by introducing simpler but more specific subgoals.

Four subgoals have been identified: *"GotoPoint&Depth"*, *"FindTarget"*, *"Teleoperation/Manipulation"* in coherence with structured missions that can be found in the offshore industry, and finally *"Homing"*. Taken together, these subgoals will satisfy the top goal. They widely influenced our design of the global mission program which handles a judicious composition of the instances of the ten robot-tasks plus a teleoperation mode together with the various exceptions raised by the environment or by the actions themselves.

The user starts the nominal evolution of the mission which is sequential. The four subgoals are completed one after the other during the nominal execution of the program. Each subgoal consists of the sequencing of various robot-tasks guarded by some safety conditions. The mission is built incrementally with each increment being a standalone program.

We will only detail the minimal phase required to tackle the particular situation of obstacle handling. We need to envelope each robot-task with a logical handling of the obstacle occurrence. That is to say: robot-tasks are iteratively started while checking in parallel for an obstacle. If an obstacle is detected, the currently executing robot-task is stopped and the station-keeping action is started. When the obstacle disappears, the original robot-task is restarted. Each phase of the mission is guarded by this behaviour.

The mission controller was directly programmed "by hand" in ESTEREL since our goal is to validate the adequacy of the interface layer between execution and mission-control levels with this programming methodology. The ESTEREL programming and proofs environment was extensively used during the design phase of the program. For example, as shown figure 1, we checked, using MAUTO and AUTO-

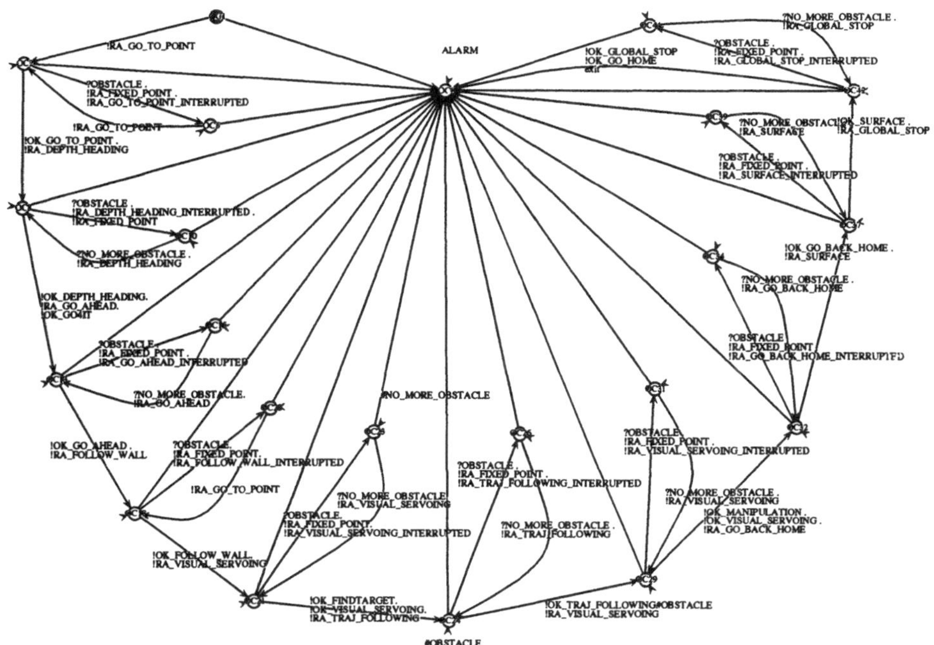

Figure 1. The control automaton: observation of obstacles & management of actions

GRAPH [10], that the logical behaviour induced by the occurrence of an obstacle was the expected one, first for all robot-tasks and then in all mission phases.

5.3. Interface with the execution level

The real-time control software PIRAT was extended using object-oriented techniques to offer an interface between the execution level and the mission program that follows the structure described in section 2. The services identified to start and stop robot actions and observation functions were successfully implemented and used to convert the significant information into logical events that can be understood by the ESTEREL program driving the application.

6. Experimental Results

The mission was effectively and iteratively performed by the VORTEX vehicle in its testpool under nominal and sub-nominal conditions (e.g. with obstacles and user intervention).

During the experiments, once the mission is launched, the LIVE [8] interface provides the end-user with the vehicle global state while recording all the corresponding data. Thus, the user can check that the expected functionalities are being performed and can intervene to abort the mission during critical situations. The teleoperation task is performed by manually driving the vehicle with a joystick or by clicking in the interface on the appropriate entry point. Also the occurrences of the events that direct the execution of the mission are indicated by popup windows in the interface. This added functionality allows the user to get a rough knowledge of what is going on in the mission from a discrete-event point-of-view.

After the experiment is over, two methods are used to qualitatively analyze the

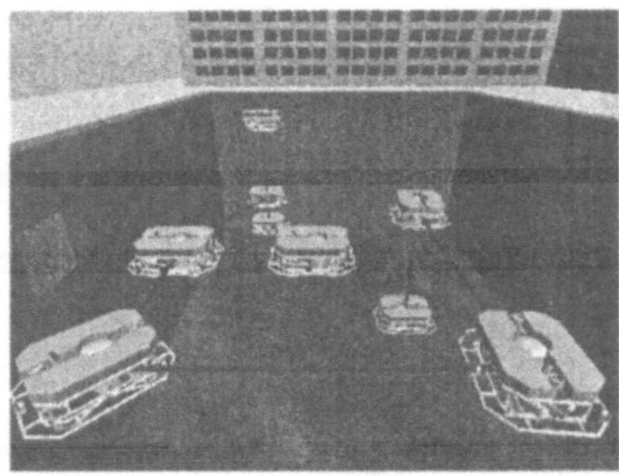

a. The nominal trajectory in VESUVE

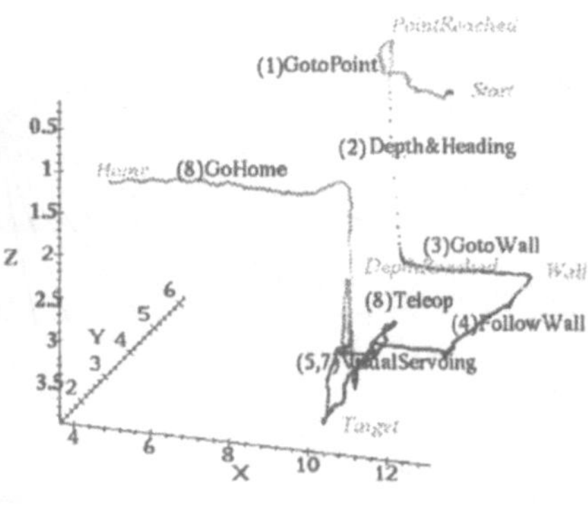

b. The experimental trajectory in 3d.

collected results. First, the vehicle's trajectory can be re-played using the VESUVE 3D Graphical tool shown in figure 6 with a symbolized nominal trajectory. Second, to allow a deeper study of how events occur in conjunction with the execution of the control laws, the results can be plotted in 2D and versus time. The performance of the control laws can be checked for further enhancements (such as filtering to over come noisy measurements induced by the sonars). The sequence of events during the nominal mission execution is studied by jointly analyzing figure 6 and the circle of mission states in figure 1. Furthermore, the data can be loaded into the vehicle simulator (developed within the Matlab/Simulink environment) and analyzed in order to enhance the control laws or the data filtering or to update the models of the vehicle, sensors, and actuators.

7. Summary and Trends for the Future

The problem domain discussed in this paper deals with research in control architectures. The focus is on the mission programming level for robotic systems operating in a dynamic environment. The objective is to provide–at the task level–generic and efficient programming methodologies for rigorous mission specification with a gateway to teleoperation for online user intervention.

Some features of this methodology were presented and successfully used to perform a complex and generic mission for underwater vehicles. The mission is adapted from another mission script defined jointly by two French (INRIA, IFREMER) and three American (MBARI, Stanford ARL, NPS) institutions to compare, contrast and enhance the results of separate research programs in robotic control architectures [13]. In this framework, the next step will be to perform the same mission scenario with the mission programming methodology developed by this research on a completely different control system and vehicle.

The programming methodology will allow to jointly specify and check an intermediary language for reliable and efficient connections between the execution level and ultimately, planning systems. When available, it will effectively force the user to develop his desired mission within a well-founded framework. Unresolved issues include the extensions required to handle synchronization and communication between cooperating robots or within robots run by multiple controllers and/or multiple control laws.

Real, operational ROV missions will make the basic scenario more complicated. Clever obstacle observation functions and avoidance algorithms should be integrated. When the VORTEX vehicle is fitted with a 6 degree-of-freedom arm, it will be possible to effectively carry out an assembly task (as mentioned in the European UNION basic research project [9]).

References

[1] A. Benveniste and G. Berry. The synchronous approach to reactive and real-time systems. *Proc. of the IEEE, Special Issue*, 79(9):1270–1282, Sept. 1991.

[2] F. Boussinot and R. de Simone. The ESTEREL language. *Proceedings of the IEEE, Special Issue*, 79(9), September 1991.

[3] E. Coste-Manière. A synchronous/asynchronous approach to robot programming. In *Proc 5th Euromicro Workshop on Real-Time Systems*, pages 268–273, Oulu, Finland, 22-24 June 1993.

[4] D. Harel and A. Pnueli. On the development of reactive systems, logic and models of concurrent systems. *Springer-Verlag*, pages 477–498, 1985.

[5] M. Jourdan. Integrating formal verification methods of quantitative real-time properties into a development environment for robot controllers. Technical Report 2540, INRIA, April 1995.

[6] K. Kapellos. *Environnement de programmation des applications robotiques réactives.* PhD thesis, Ecole des Mines de Paris, November 1994.

[7] M. Perrier, V. Rigaud, C. C. de Witt, and R. Bachmayer. Performance oriented robust nonlinear control for subsea robots: Experimental validation. In *IEEE International Conference on Robotics and Automation*, page to appear, San Diego, May 1994.

[8] A. Peuch, E. Coste-Manière, D. Baticle, M. Perrier, V. Rigaud, and D. Simon. An advanced control architecture for underwater vehicles: the mission execution system of VORTEX. In *Proceedings of IEEE OCEANS'94*, Brest, France, Septembre 1994.

[9] V. Rigaud and e.a. Underwater intelligent operation and navigation: Union main objectives of the project. In *Proc of the joint US/Portugal workshop : Intl Prgm Dev in Undersea Robotics and Intelligent Control*, Lisbon, Portugal, March 1995.

[10] V. Roy and R. de Simone. Auto and autograph. In R. Kurshan, editor, *Workshop on Computer Aided Verification*, New Brunswick, June 1990.

[11] D. Simon, B. Espiau, E. Castillo, and K. Kapellos. Computer-aided design of a generic robot controller handling reactivity and real-time control issues. *IEEE Transaction on Control Systems Technology*, 1(4):213–229, December 1993.

[12] C. Thorpe. Report of the us/french workshop on underwater robotics. pages 17–19, Toulon, France, May 1992.

[13] H. H. Wang, E. Coste-Manière, A. Peuch, M. Lee, S. Rock, and T. Healey. A proposal for a joint mission for evaluation of mission-level programming. Proposal stanford arl/mbari, inria, ifremer, naval postgraduate school, November 1994.

Specification, Formal Verification and Implementation of Tasks and Missions for an Autonomous Vehicle

Konstantinos Kapellos, Muriel Jourdan, Bernard Espiau
INRIA Rhône-Alpes
41, rue Félix-Viallet
38031 Grenoble Cedex - France
{Firstname.Name}@inria.fr

Sofiane Abdou
INRIA Rocquencourt
Domaine de Voluceau
78150 Le Chesnay Cedex - France
Sofiane.Abdou@inria.fr

Abstract

This paper describes the use of modern approaches of formal verification of behavioral and temporal properties in an experimentation of automatic vehicle following. We firstly recall the key concepts of the proposed specification method: the robot-task, which allows to describe in a structured way an automatic control law and the associated discrete-event handling; the robot-procedure, which is the specification of complex missions from the logical point of view. We then present the used environment (ORCCAD) and verification tools, based on the use of synchronous languages for the reactive part of the system. This is followed by a description of the conducted experiments, which involves two electric cars, within the PRAXITÈLE project. Results and foreseen future works are given in the conclusion.

1. Introduction

Robotic systems are hybrid systems operating in real time and handling events as well as "continuous computations". Reliable and easy programming of these systems at the "task level" requires a systematic method for the specification of the missions, formal verification of their execution from a continuous and discrete-time point of view and efficient implementation over the target architecture. Models proposed in the literature cannot handle in an *unified* way discrete events parts and highly complex continuous time expressions, like the control equations used in advanced robots. Nevertheless, for every component of the system, wellknown theories could be applied: automatic control theory for the design and analysis of the control law, reactive systems theory [2] for the discrete events aspects. In the robotic area the first theory is widely used, while this is not the case for the second one. As a consequence, a discrete events controller cannot in general be clearly isolated in an application. It is therefore not possible to apply formal verification methods to check its correctness, although this be required, for truely autonomous robots, in order to be as sure as possible before launching that the system will behave correctly.

The work presented here constitutes an extension of the ORCCAD ([7]) framework, which allows the specification, the simulation and the implementation of robotic elementary actions (called *Robot-tasks*), by harmoniously integrating discrete and continuous aspects. In this paper we propose a method for the specification, the validation by using *formal* verification methods (not simulation), and implementation of the events handling parts of robotic missions. This is achieved by composing Robot-tasks using a set of operators, the final result being called a *Robot-procedure.* The systematic translation of a *Robot-procedure* into suitable synchronous languages (see below) provides the robot "program" with a nice semantics. This allows us to methodicaly verify a large set of behavioral and quantitative temporal properties, including crucial properties of liveliness and safety, as well as the conformity with missions requirements.

In our approach, the controller behaviour is considered as a reactive system: *a system which maintains a continous interaction with its environment by sending outputs in reaction of inputs.* The family of synchronous languages, which aim at describing the complex ordering and causality relations between the inputs and the corresponding outputs of a reactive system, has been a very important contribution to the domain. These languages may be compiled into a wide class of models, usually labeled transition systems. The models are then used to perform static verification of *properties* (order relations between inputs and outputs) of the system.

We illustrate the concept of *Robot-procedure*, its use as a specification of complex robotic missions and the formal verification methods we use, on an example taken from the area of automatic vehicle driving. The long-term objective is to specify, validate and implement a virtual "train" of electric vehicles: each car is expected to closely follow the previous one automatically using dedicated sensors — a vision system is used to locate the previous vehicle in distance and angle — and a computerized system. The first vehicle would be the only one with a human driver. This work is a part of the PRAXITELE Project ([8]).

The paper will be organized as follows. In a first part, we will briefly present the ORCCAD system and its central entity the *Robot-task* (RT). In a second part, we will introduce the concept of the *Robot-procedure* (RP) which allows to combine RTs in a structured way in order to express complex controllers. The formal verification methods will be exposed in a third part. All these concepts and methods will be illustrated through an automatic vehicle driving example. Finally, we will describe the current status and the results of experiments.

2. The Orccad system

ORCCAD is a development environment for specification, validation by formal methods and by simulation, and implementation of robotic applications.

The formal definition of a robotic action is a key concept in the ORCCAD framework. It is based on two principles: a) most physical actions to be performed by robots can be stated as automatic control problems which can be efficiently solved in real-time, b) the characterization of the physical action is not sufficient for fully defining a robotic action: starting and stopping times must be considered, as well as reactions to significant events observed during the task execution. In order to capture coherently the diversity of elements involved in these two issues, two entities are defined: the RT, representing an elementary robotic action, where automatic control aspects are predominant although coherently merged with behavioral ones, and the RP, the RTs are the basic element of which, and where only behavioral aspects are considered.

The RT entity has now been well studied and validated through multiple experiments (see [6, 9, 7]). In the following, we therefore only present its behavioral aspects, since they are used to compose RPs, which are more extensively presented in section 2.2.

2.1. The Robot-task

A RT is formally defined as the *parametrized specification of an elementary control law, i.e. the activation of a control scheme structurally invariant along the task duration, and a logical behavior associated with a set of signals(events) which may occur just before, during and just after the task execution.*

From the control point of view, RTs specification requires the explicitation of the set of functions, models and parameters appearing in the analytical expression, in continuous time, of the control outputs to be applied to the actuators in order to perform the desired physical action. Besides, the specification of the logical behavior is obtained by setting the events to be considered and their treatment. In order to make that specification easier to people from the robotics community who are not familiar with this approach, the events and the associated processings are typed. We distinguish:

• the pre-conditions: their occurence is required for starting the servoing task. A temporal watchdog can be associated with each pre-condition.

• the exceptions: they are generated during the execution of the servoing task and indicates a failure detection. Their processing is as follows:

- type 1: the reaction to the received exception is limited to the modification of the value of at least one parameter within the control scheme,
- type 2: the exception requires the activation of new RTs. The reaction consists in killing the current one and reporting the causes of the disfunction to the adequate level. The recovering process to activate is known and specified, as discussed in section 2.2,
- type 3: the exception is considered as fatal. The overall application is stopped and the robotic system must be driven to a safe position.

• the post-conditions: often related to the environment, they are handled as conditions for a normal termination of the RT. Watchdogs can be associated to their waiting. Normal or forced termination of the RT are accompanied by the diffusion of specific synchronization signals.

Finally, a RT is completely specified by the explicitation of implementation aspects including temporal properties, like the sampling rates. This is done by implementing each RT in terms of communicating real-time computing tasks, called Module-tasks, which each implement a part of the control law or of the behavior.

The ORCCAD system provides also a graphical environment that offers a series of functionalities, ranging from specification of RTs by instanciation of its object-oriented model and automatic generation of behavior-related programs, to its validation using SIMPARC ([7]) and realtime code generation running under VxWORKS.

2.2. The Robot-procedure

The aim in designing the RP entity is to be able to define a representation of a robotic action that could fit any abstraction level needed by the mission specification system. In its simplest expression, it coincides with a RT, while the most complex one might represent an overall mission. Briefly speaking, it specifies in a structured way a logical and temporal arrangement of RTs in order to achieve an objective in a context-dependent and reliable way, providing with predefined corrective actions in the case of unsuccessful execution of RTs. More formally a RP is the full specification of

- a main program, which characterizes the nominal execution of the action and is composed of RTs, RPs and conditions,
- a set of triplets (exception event, processing, assertion), which specifies the processing to be applied for handling the exception, and the information to transmit to the planning level (if any), and
- a local behavior defining the logical coordination of the previously considered items.

The composition of RTs and RPs in the main program is obtained through operators which express the sequence, the parallelism, the conditional, the iteration, the rendez-vous and various levels of preemption.

2.2.1. The Application

Let us informally describe the considered application. We have two electric vehicles, the leader being driven and the second having to follow it like in a virtual train. Initially, the undriven car tries to catch the right signal in order to locate the first one. When done, the expected nominal execution is that the undriven car follows the driven one until all be stopped by an operator intervention. During the execution the video signal may be lost or irrelevant. In this case a parking manoeuvre is started. The driven car is supposed to come back and the train to be reformed. In addition, other situations must be monitored like physical damaging of crucial components.

This informal specification could be translated into a RP named FOLLOWME (see fig.1). Its nominal execution is described in the RP main program specified as an infinite loop the body of which begins with the test of the external condition *TargetFound* which indicates that the driven car is detected by the second one. Whenever it is satisfied, within a specified elapsed time, a RP named GUARDEDFOLLOW, detailed below, is activated to control the second car. Let us emphasize that the programming is structured, in the sense that we can use a RP inside the definition of another one. Before starting the FOLLOWME nominal execution, a set of three preconditions must be satisfied before the indicated delays; the initialization phase (motors, sensors, ...) must have been achieved without detecting errors, the automatic mode must be activated and the "human" operator has to give the start order. The nominal execution of this main RP is stopped in two cases: either the supervisor gives a stop order or the manual mode is activated. In the first case, the RP ends normally; in the second one it is interrupted by a global T3 exception.

In this example, there are six involved RTs: two for the driving and steering motors of the car using exteroceptive sensor information, which virtually links the driven leader car to the non driven following one (named SENSLOC and SENSDIR

```
Name : FOLLOWME
Pre-cond:OkInit[30ms],AutoMode[30ms],Start[5mn]
Main program:
      Loop
         wait TargetFound [5mn]
         start (GUARDEDFOLLOW)
      EndLoop
T3 exceptions: Auto2man, MecFaill, ...
Post-cond : Stop [60mn]
```

Figure 1. FOLLOWME RP specification

```
Name: GUARDEDFOLLOW
Nominal execution :
   Parallel
      start (SENSLOC)
      start (SENSDIR)
      Loop
       if  MoreBrake then start (BRAKE)
      EndLoop
   EndParallel
T2 Exception: (TargetLost, start (PARKING))
```

Figure 2. GUARDEDFOLLOW RP specification

respectively); two allowing the undriven car to track a reference trajectory on the basis of odometry information only (respectively nammed CARTLOC and CARTDIR); one (named BRAKE) using the foot-brake of the car in order to impose a desired deceleration. The RT SENSLOC is choosen to be presented in section 4.

The RP GUARDEDFOLLOW (see fig.2) is built from the parallel composition of the three RTs SENSLOC, SENSDIR and BRAKE . The RT BRAKE is activated every time the leading car imposes strong decelerations, indicated by *MoreBrake* event. Let us note that the first two RTs may be forced to stop if the exception of type 2 concerning the loss of the video signal between the two cars is detected. The RP GUARDEDFOLLOW handles this situation by starting a recovery program made of the RP PARKING, the specification of which is analog to GUARDEDFOLLOW RP. However, RTs based on odometry information only are used instead of the sensor based ones.

The RP formalism allows an user to program at the "task-level", without worring about the coding of tricky things, like signal exchange between elementary tasks. A systematic translation to the ESTEREL language ([1]) is provided, minimizing in that way the risk of errors. Even so, the complexity of programmed applications and their critical character require the use of formal methods for ensuring the correctness at the *specification* level.

3. Formal verification in Orccad

In the ORCCAD framework formal verification is used to prove the largest possible set of assertions about the execution of the RTs and RPs and, ideally, directly from their definition. A complementary objective is to provide the end-user, who is not supposed to be an expert in verification, with a systematic way to express the property to verify, to interpret the results and therefore to analyse the behavior of complex robotic actions.

Verification methods we used are model-based. They suppose that 1) the system is modelled by a labelled transition system; 2) the property is expressed either by another labelled transition system — in which case the property and the model are compared modulo a well-chosen equivalence relation — or by a formula from a temporal logic — in which case the formula are evaluated by computation on the model.

For the first point, the use of synchronous languages is very helpful, since their semantics are expressed in terms of labelled transition systems. This explains why we translate the RT and RP formalisms into a suitable synchronous langage. It is important to choose the right one since the type of properties we would like to check depends on the form of the model used to express the semantics of the language. Moreover their environment does not offer exactly the same type of verification tools.

We have chosen to translate our formalisms into: 1)ESTEREL, because it is friendly interfaced with AUTO, a tool aimed to prove behavioral properties by reducing/comparing boolean automata; 2) the timed extension of ARGOS [5], since its semantics is expressed in terms of temporized automata (automata extended with real variables) which allows to prove time dependent properties ([4]). This is achieved by using KRONOS [3] which is a model-checking tool for the real-time temporal logics TCTL.

We now present how behavioral properties on RTs and RPs, and time-dependent properties on RPs can be verified. This last is justified by the fact that a lot of explicit delays appear in a RP specification.

3.1. Robot-task

Concerning the RT, we proved ([6]) that its specified behavior is non-blocking and satisfies the liveliness property (a successful termination of the RT can be reached from any state of its evolution), and the safety property (any fatal exception is appropriately handled by the emission of a specific signal driving the system to a safe situation). Taking advantage of the synchronous assumption on which Esterel is based, the RT behavior is encoded in a generic way, i.e. independently of any particular specification, a dedicated instance being obtained by an appropriate substitution of input/output signals. These properties are verified on the *generic* specification by observing the resulting behavior. Therefore the basic objects to be composed for designing complex robotic actions are ensured to have "good" properties.

3.2. Robot-procedure

Contrary to the RT behavior, which is proved to be always correct, the user is in charge of most of the RP verification. Nevertheless, the **crucial properties** of safety and liveness, and a time-dependent property which states that the maximal execution time of the RP is bounded, can be checked automatically.

The very important class of properties representing the **conformity** of the RP specification to the mission constraints must be handled by the user. In general, such constraints can be expressed in terms of relations between events and events, events and actions or actions and actions. Let us illustrate that by a few examples from our case study. Mission specification requires that the target is searched each time after its loss. This could be tested by specifying a behavioral property which is of the event-event type and involves *TargetLost* and *TargetFound* events. The

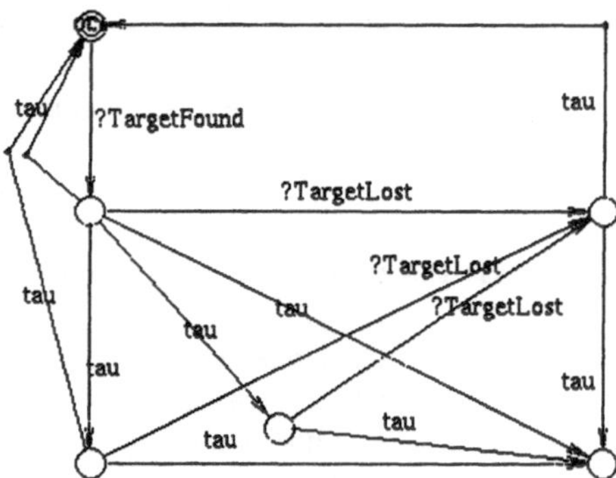

Figure 3. Automaton Resulting from the Verification Process

property expression and the verification procedure can be automatically derived by simply indicating the relevant events. It is left to the user to interpret, visually in this case, the results of the method (fig.3). Another example, which is of the type action-action and is time-dependent, consists of the proof that the time-lag between the starting points of the wheel execution control law and of the motor execution law is bounded. This property could be expressed by instantiating a well-chosen generic TCTL formula.

Finally, we consider verification methods also as a way of obtaining **abstract views** during the stage of RPs specification, in order to help the user in its mission design. Actually, every mismatch in the behavior specification is reflected by the resulting automaton. So, by projecting a complex behavioral activity in a simpler one at a relevant level of abstraction, useful (i.e easy to understand) views of the overall behavior can be given.

4. Experimental Issues

4.1. Description of the Facilities

We use two LIGIER electrical cars. A set of infrared emitters is mounted on the back of the first one, while the second is equipped with a vision system. This last consists of a linear CCD camera with acquisition frequency 1 Khz and a PC with an image processing board. It is connected to a VME rack, which includes a Motorola MVEM 162 board with four IP-modules (IP-DAC, IP-ADC, IP-QUADRATURE, IP-DIGITAL) and is dedicated to the car control [1].

The ORCCAD environment is running on a Sun Sparc workstation; the generated realtime code is loaded on the on-board resources through ETHERNET.

4.2. SensLoc Robot-task design

In order to understand how the elementary parts of the application are designed and how automatic driving is achieved, we now present the SENSLOC RT. The aim of this RT is to move the non-driven car longitudinally using information on the speed and on the distance between the two cars obtained through the vision system. The required action is: to start, after having checked that the connexion with the driving motors is OK and that the automatic mode is on. During its execution it is also required to detect a possible change of mode (switch to manual mode), a

[1]this work is made in cooperation with Aleph Technologies, Grenoble, France

possible defect in the system and to verify that the leader car is always in sight of the camera. Let us now detail some aspects of its specification:

Continuous-time specification The control input a_c is the driving motors acceleration, given by ([8]): $a_c = (\frac{1}{h})\Delta v + \frac{1}{h^2}(\Delta x - hv - d_{min})$ where v is the velocity of the following car, d_{min} is the minimal distance and h, Δx, Δv denote differences between the two cars, in time, position and speed respectively. For its computation we use the following functional modules:

- PC: interprets the results of the image processing in order to compute Δx and Δv,
- EST: estimates the slopes,
- GT: computes the desired acceleration a_c,
- CO: converts the a_c in current and send it to the DAC,
- VOIT: reads the car state (wheels position, speed).

The event-based behavior of the task is expressed through two pre-conditions and two exceptions (PREC, EXC, ATR modules). Let us notice that the *TargetLost* exception is handled by a type-2 processing, which consists in stopping the RT and activating the recovery program specified during the design of the GUARDEDFOLLOW RP.

Time constrained specification For the implementation, a sampling period of 10ms is assigned to all periodic tasks, except for CO which runs at 5ms. In addition, non-blocking message passing mechanisms are selected for inter-tasks communication.

The full specification is validated by simulations using SIMPARC and, after that, effectively executed.

4.3. Experimental results

Software Issues The RTs used in the application are gathered in a library. The RP FOLLOWME was translated in a methodic way to an ESTEREL program. This last produces after compiling the automaton which controls the RTs. Its size is of 300 states and corresponds to 78751Kb of "C" code produced by the ESTEREL V4.40 compiler. The realtime software used for the execution is running under VXWORKS. In addition, a prototype environment integrating tools for the specification, analysis and verification of RPs is currently designed.

A Run Let us finally describe informally the evolution during a 10mn experimentation in the INRIA domain. Some of the results are given fig. 4.

Initially, the inter-car distance was of 9m, the leading car being motionless. After a while the visual target was detected, so TARGETFOUND condition was satisfied, and therefore the GUARDEDFOLLOW RP was activated. Using the visual information, the undriven car after having reached the minimum inter-car distance (1.5m) followed the driven one with a velocity of 30 km/h and a range of environs 4m. At the third mn, the MOREBRAKE event was broadcasted because of a sudden deceleration of the leader vehicle; the RT BRAKE was activated increasing the pressure of the brakes of the following vehicle and therefore decreasing its speed (fig.4). The same situation reproduced later inducing the reactivation of the RT BRAKE. The loss of visual information, signaled by the TARGETLOST event, was immediatly handled by the activation of the recovery program PARKING, which simply drived smoothly the following vehicle at a motionless position. The driven car came back, the visual target was again detected, and the train reformed. The experiment finished when the user emitted the STOP signal.

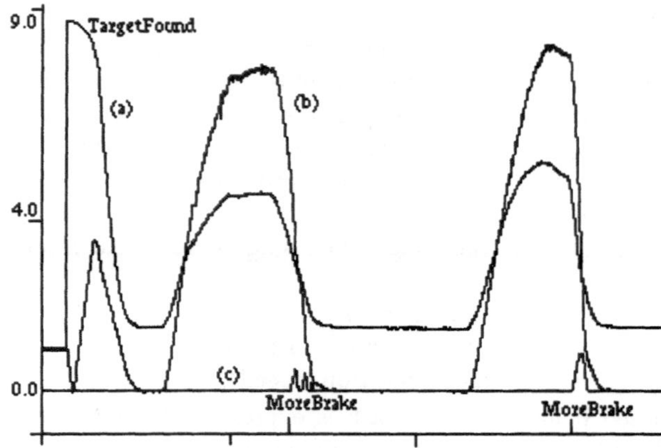

Figure 4. (a) Inter-car distance, (b) Speed of the following car, (c) Brakes pressure

5. Concluding Remarks

We have shown in this paper how formal specification and verification methods can be used for checking the behavioral and temporal correctness in a complex robotics application. This was illustrated by a nontrivial experiment of automatic vehicle driving in an outdoor environment. From that experimental study, we may draw some conclusions which open future research directions. A first thing is that, even with a sophisticated environment like ORCCAD, it is not obvious for a non-expert user to handle verification tools. It is therefore necessary to greatly improve their interfaces, in order for example to avoid the need for writing TCTL formulas. A second point is that the performances of the compilers and verification tools have to be also improved if we want to consider highly complex applications like underwater or space ones.

Provided that these issues are addressed, we guess that the relevance of the proposed approach is not questionable. It has been proved that specification and programming of rather difficult applications is done with a better robustness, is more efficient and allows easier modifications and evolutions with the proposed framework than using a classical method.

References

[1] F. Boussinot, R. de Simone: *The* ESTEREL *Language*, Proceedings of the IEEE, 79(9), pp 1293-1304, September 1991.

[2] D. Harel and A. Pnueli: *On the Development of Reactive systems*, in Logic and Models of Concurrent Systems, NATO Advanced Study Institute on Logics and Models for Verification and Specification of Concurrent Systems, Springer Verlag, 1985.

[3] T. Henzinger and X. Nicollin and J. Sifakis and S. Yovine :*Symbolic Model-Checking for Real-Time Systems*, LICS'92, IEEE Computer Society Press, 1992.

[4] M. Jourdan :*Integrating Formal Verification Methods of Quantitative Real-time Properties into a Development Environment for Robot Controllers*, INRIA Research Report no 2540, April 1995.

[5] M. Jourdan, F. Maraninchi, A. Olivero :*Verifying Quantitative Real-time Properties of Synchronous Programs*, 5th International Conference on Computer-aided Verfication, LNCS 697, Springer Verlag, 1993.

[6] K. Kapellos, *Environnement de Programmation des Applications Robotiques Réactives*, PhD thesis, Ecole des Mines de Paris, November 1994.

[7] D. Simon, B. Espiau, E. Castillo, K. Kapellos :*Computer-aided Design of a Generic Robot Controller Handling Reactivity and Real-time Control Issues*, IEEE Trans. on Control Systems and Technology, vol 1 no 4, December 1993.

[8] Parent, M. and Daviet, P., *Automatic Driving for Small Public Vehicles*, Intelligent Vehicle Symposium ,Tokyo, July 14-16, 1993.

[9] P. Rives, R. Pissard-Gibollet, and K. Kapellos, *Development of a Reactive Mobile Robot Using Real Time Vision*, in ISER'93, Kyoto, Japan, October 1993.

[10] V. Roy, R. de Simone :*Auto and autograph*, In R. Kurshan, editor, *Proc. of Workshop on Computer Aided Verification*, New Brunswick, June 1990.

[11] V. Roy, R. de Simone :*Auto and autograph*, In R. Kurshan, editor, *Proc. of Workshop on Computer Aided Verification*, New Brunswick, June 1990.

[12] L. Lynch, H.B. Weinberg: *Proving Correctness of a Vehicle Maneuver: Deceleration*, Second European Workshop on Real-Time and Hybrid Systems, Grenoble, France 1995.

Chapter 10

Dynamics and Control

Control, dynamics, and design are close allies in the field of robotics. Proper design of a robot relies on understanding the effect of mechanism dynamic properties on performance. Conversely, good quality control requires thorough understanding of the robot's dynamics. The papers in this chapter seek to bridge these disciplines in service of better robot performance.

Yoshikawa and Matsudera deal with the problem of modeling and controlling flexible robot arms. They approximate the arm by introducing virtual rigid links and passive joints and then derive an LQ controller. Their experimental results show that the number of virtual joints and links introduced does not need to be high to enable satisfactory control of an arm.

Yoshida, Mavroidis, and Dubowsky address the control of rigid manipulators supported on a compliant base, which are subjected to impact disturbances. Using the system's dynamic model they demonstrate a control method which minimizes the impact effect and support structure vibration.

Russakow, Rock, and O. Khatib deal with the problem of control of a multi-arm free-flying vehicle. The paper presents an extended operational space formulation for dealing with the coordination and control of this vehicle/multi-arm system. The resulting control capability is demonstrated on a planar robot.

Chung, Nakamura, and Sordalen show how nonlinear control theory can be applied to the design and control of non-holonomic systems. They present a novel non-holonomic "gear" design which is incorporated into a prototype robot arm. The reported results illustrate and validate their control approach.

Fraisse, Dauchez, Pierrot, and Cellier consider the problem of manipulating fragile objects with a mobile robot. To solve this problem, they employ sliding and impedance control. The paper presents experimental results in which the vehicle is subjected to disturbance forces.

Experimental Study on Modeling and Control of Flexible Manipulators Using Virtual Joint Model

Tsuneo Yoshikawa* and Katsuki Matsudera
Department of Mechanical Engineering
Kyoto University, Kyoto 606, Japan
* yoshi@mech.kyoto-u.ac.jp

Abstract

Experimental study of a dynamic model of flexible manipulators is presented in this paper. The basic idea of this modeling is to model each flexible link of the manipulator by several virtual rigid links and passive joints. First, the modeling scheme is described. The model parameters are identified by measured data of the real arm and the dynamic behavior of the models are compared with that of the real arm. The obtained results shows the validity of this modeling scheme. It also indicates that the number of the virtual joints does not need to be large in the case of our experimental flexible arm. Then, an LQ controller based on this model is proposed. The effectiveness of the proposed controller is verified by experimental results.

1. Introduction

The requirement for light and fast industrial robots has reached to a point where the oscillation due to elasticity of links and joints cannot be ignored. For space applications of manipulators, long arms are needed that are light in comparison to their length and load. Their elasticity demands compensation for arm deformation and vibration.

In the field of applied mechanics, there have been many studies on predicting and simulating dynamic behavior of flexible beams, link mechanisms, and structures, with applications to large mechanical structures, aircrafts, and spacecrafts in mind. Their approaches are mostly based on modal analysis (for example, Kane et al.[1], Banerjee and Lemak[2]) or finite segmentation (for example, Sadler and Sandor[3], Amirouche and Huston[4]) of flexible bodies. The former is based on distributed parameter models described by partial differential equations. The latter is based on finite segment models given by a finite number of masses fixed to their representative nodes and by massless springs and dampers connecting these nodes. Most of these studies have not been for the purpose of active control of vibrations.

In the robotics field, similar approaches have also been applied for modeling flexible manipulators. Distributed parameter models are used in, for example, Book et al.[5] , Cannon and Schmitz[6], and Pfeiffer[7]. Finite element method is used in Sunada and Dubowsky[8]. The distributed parameter models can describe the arm dynamics precisely, but can only describe simple beams. It is difficult to extend to multi-link arms, because of its complicated boundary conditions. The finite element method approach can model a real arm precisely by using many elements, but it needs a lot of calculation. In the robotics field, suppression or active control of vibration of flexible manipulators is usually the main concern. To this end, simple dynamic models that allow real-time computation have been proposed recently, for example, in Yoshikawa et al.[9] and Yoshikawa and Hosoda[10].

The approach proposed by Yoshikawa and Hosoda[10] is to model a flexible link by several virtual rigid links and virtual passive joints with springs and dampers. The parameters of virtual links and joints are identified from measured data of the real flexible link. In this approach, non-uniform links can be easily modeled, and the whole flexible arm can be modeled by just connecting models of individual links. It does not need complex boundary conditions.

The dynamic structure of this model is similar to that of finite element method and the method of Huang and Lee[11], but the way to determine the parameters of the model is different. This modeling scheme identifies the model parameters from measured data on the static and dynamic characteristics of the real links, while the finite element method and Huang and Lee[11] use local estimated characteristics such as material constants or shape of cross-section.

In this paper, experimental study on modeling a planar flexible manipulator using the virtual joint model (Yoshikawa and Hosoda[10]) is presented. First, the outline of the modeling scheme is given. Then a flexible link is modeled by using three different models, that is, one-, two-, and three-virtual-joint models. The model parameters are identified by measured data of the real arm and the step response of the models are compared with that of the real arm. The obtained experimental result shows the validity of this modeling scheme. Then, based on this model, an LQ controller is proposed. The effectiveness of the proposed controller is verified by experimental results.

2. Modeling Scheme Using Virtual Joint Model

2.1. Outline of Modeling Procedure

The proposed model is shown in **Fig.1**. Physical parameters of this model are length, mass, center of mass, and inertia matrix of the virtual links, and spring constants and damping coefficients of the virtual joints.

Various approaches will be possible for determination of these parameters of the model. An identification method of these parameters proposed in Yoshikawa and Hosoda[10] is to measure typical static characteristics (e.g. deformation of end-

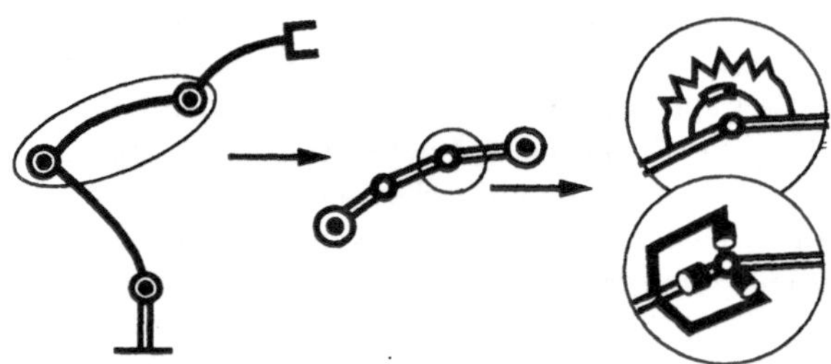

Figure 1. Modeling of flexible link using virtual rigid links and passive joints

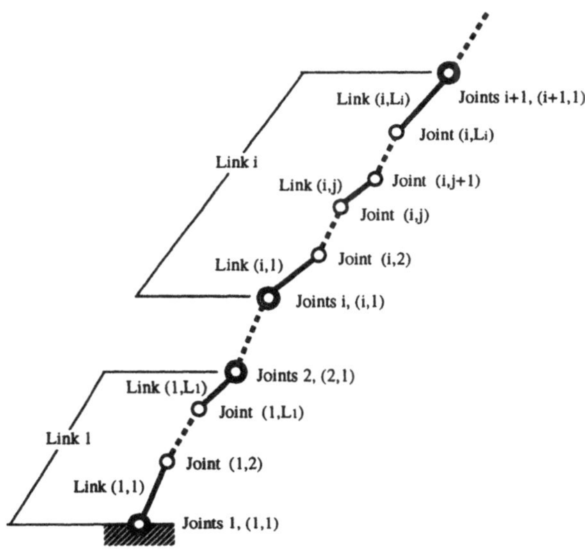

Figure 2. Virtual links and passive joints

tip under certain load) and dynamic characteristics (e.g. natural frequency) of the real flexible links and to find the parameter values of the model that make the corresponding characteristics of the model as close as possible to these measured characteristics for the real arm. To make the calculation simpler, it was assumed that, for any lengths of the virtual links, we can calculate the mass, the center of mass, and the inertia matrix from the information at hand on the shape and mass distribution of the flexible links.

2.2. Parameters of Model

We consider an N-joint flexible manipulator. The joints and links are numbered $1, 2, \cdots, N$, starting from the base side as in **Fig.2**. Flexible link i is divided into L_i virtual rigid links. These virtual links and passive joints are numbered $(i,1), (i,2), \cdots, (i,L_i)$ from the base side. Note that virtual passive joint $(i,1)$ and active joint i which is driven by a joint motor are both placed at the base side of link i.

We define several coordinate frames as shown in **Fig.3**. The base coordinate frame Σ_0 is fixed to the ground. Joint coordinate frame Σ_i has its origin at the joint i, z_i axis along the rotational direction of the motor, x_i axis along the tangent of the link i, and y_i axis which completes a right hand coordinate system. The displacement of the motor is denoted by θ_i. Virtual link frame Σ_{ij} has its origin at the passive joint (i,j). The directions of axes of Σ_{ij} are the same as those of Σ_i, when the flexible link i has no deformation.

The variables which represent the angular deformation of passive joints along the axes x_{ij}, y_{ij}, z_{ij}, are denoted by $(\phi_{ij})_x$, $(\phi_{ij})_y$, $(\phi_{ij})_z$, respectively. The corresponding spring constants are denoted by $(k_{ij})_x$, $(k_{ij})_y$, $(k_{ij})_z$, and damping coefficients are $(d_{ij})_x$, $(d_{ij})_y$, $(d_{ij})_z$. Let the deformation vector be $\boldsymbol{\phi}_{ij} = [(\phi_{ij})_x, (\phi_{ij})_y, (\phi_{ij})_z]^T$, the

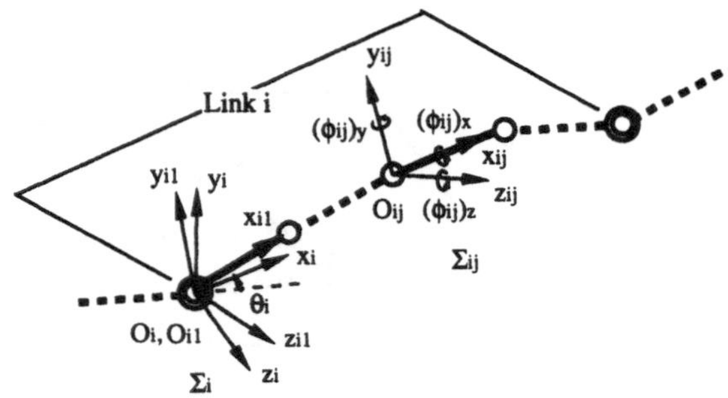

Figure 3. Coordinate frames of virtual links

spring constant vector be $\boldsymbol{k}_{ij} = [(k_{ij})_x, (k_{ij})_y, (k_{ij})_z]^T$, and the damping coefficient vector be $\boldsymbol{d}_{ij} = [(d_{ij})_x, (d_{ij})_y, (d_{ij})_z]^T$, where T represents the transpose of matrix or vector. Then the physical parameters of the virtual passive joints are given by $\boldsymbol{k}_{ij}$ and $\boldsymbol{d}_{ij}$.

Let m_{ij} represent the mass of virtual link (i, j), $\boldsymbol{p}_{ij}$ the vector from O_{ij} to O_{ij+1} expressed in Σ_{ij}, $\boldsymbol{s}_{ij}$ the vector from O_{ij} to the gravity center of virtual link (i, j) expressed in Σ_{ij}, and $\boldsymbol{I}_{ij}$ the inertia tensor of virtual link (i, j) expressed in Σ_{ij}, respectively. Note that the superscript on the left denotes the reference frame. Then the physical parameters of the virtual rigid links are given by m_{ij}, $\boldsymbol{p}_{ij}$, $\boldsymbol{s}_{ij}$, and $\boldsymbol{I}_{ij}$,

2.3. Determination of Model Parameters

In this subsection, the determination of model parameters is outlined for a simple case(Yoshikawa and Matsudera[12]). The model parameters are identified to make the model characteristics as close as possible to the measured characteristics of real manipulators.

The axes of Σ_{ij} are assumed to coincide with the principal axes of moment of inertia of the virtual link, and y_i and z_i elements of $\boldsymbol{p}_{ij}$ and $\boldsymbol{s}_{ij}$ are assumed to be zero. From now on, since only the motion of link i in $x_i - y_i$ plane is considered, we let $m_j = m_{ij}$, $p_j = (\boldsymbol{p}_{ij})_x$, $s_j = (\boldsymbol{s}_{ij})_x$, $I_j = (\boldsymbol{I}_{ij})_{zz}$, $\phi_j = (\phi_{ij})_z$, $k_j = (\boldsymbol{k}_{ij})_z$, and $d_j = (\boldsymbol{d}_{ij})_z$, for the convenience of notation.

First, the dynamic characteristics of link i are calculated. As for small vibration in $x_i - y_i$ plane for the case where one end is fixed and the other is free (shown in **Fig.4**), since non-linear terms and second or higher order terms of vibration are negligible. Damping coefficient is assumed to be small enough. The dynamics equation of the model is

$$\boldsymbol{M}\ddot{\boldsymbol{\phi}} + \boldsymbol{K}\boldsymbol{\phi} = 0 \tag{1}$$

where $\boldsymbol{\phi} = [\phi_2, \phi_3, \cdots, \phi_{L_i}]^T$ is the angular deformation vector, $\boldsymbol{M} = [M_{ij}]$ is the inertia matrix in $x_i - y_i$ plane, and $\boldsymbol{K}$ =diag$\{k_2, k_3, \cdots, k_{L_i}\}$ is the spring constant matrix. The entries M_{ij} of $\boldsymbol{M}$ are functions of parameters of the virtual rigid links.

From (1), the natural frequencies ω_{zi}, $(i = 1, 2, \cdots, L_i - 1)$ of the model satisfy the equation

$$\det(-\boldsymbol{M}w_{zi}^2 + \boldsymbol{K}) = 0 \tag{2}$$

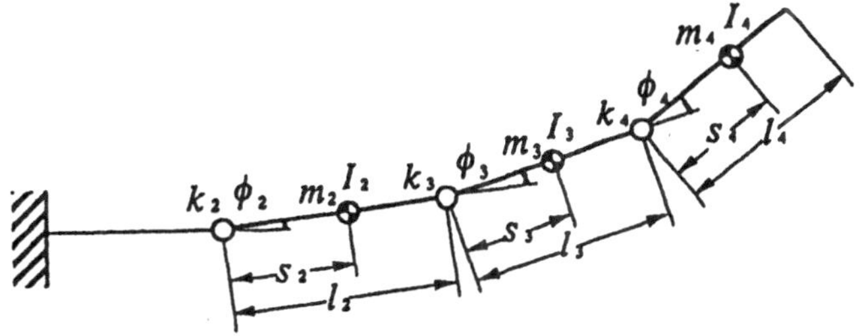

Figure 4. Flexible link model with three virtual joints

Second, for static characteristics, static deformation in $x_i - y_i$ plane, when some force or moment is acting on the free end of the link, is considered. When force P is acting on the free end, let linear deformation of the free end be u_P and angular deformation be ϕ_P. When moment M is acting on the free end, let linear deformation be u_M and angular deformation be ϕ_M. Then the static characteristics of the model are obtained by

$$\boldsymbol{\alpha}_{ms} = \begin{bmatrix} u_P/P \\ \phi_P/P \\ u_M/M \\ \phi_M/M \end{bmatrix} = \boldsymbol{H} \begin{bmatrix} 1/k_2 \\ 1/k_3 \\ \vdots \\ 1/k_{L_i} \end{bmatrix} \tag{3}$$

where

$$\boldsymbol{H} = \begin{bmatrix} \{\sum_{j=2}^{L_i} p_j\}^2 & \{\sum_{j=3}^{L_i} p_j\}^2 & \cdots & p_{L_i}^2 \\ \sum_{j=2}^{L_i} p_j & \sum_{j=3}^{L_i} p_j & \cdots & p_{L_i} \\ \sum_{j=2}^{L_i} p_j & \sum_{j=3}^{L_i} p_j & \cdots & p_{L_i} \\ 1 & 1 & \cdots & 1 \end{bmatrix} \tag{4}$$

Even when the assumptions made at the beginning of this subsection do not hold, the same type of equations as (1) and (3) can be obtained, although they are more complicated.

A procedure for determining the model parameters of link i is given as follows:

1. Initial values of virtual link lengths $p_2, p_3, \cdots, p_{L_i}$ are given.

2. Other physical parameters of the virtual rigid links, m_j, s_j, and I_j, are calculated from $p_2, p_3, \cdots, p_{L_i}$ by cutting the real link virtually.

3. Spring constants $k_2, k_3, \cdots k_{L_i}$ are calculated from (2), using the measured natural frequencies, ω_{zi}.

4. The value of the following criterion J' based only on the static characteristics is calculated.

$$J' = \sum_{k=1}^{L_i} w'_k \frac{(\alpha_{rsk} - \alpha_{msk})^2}{\alpha_{rsk}^2}, \tag{5}$$

where α_{rsk} and α_{rmk}, $k = 1, 2, 3, 4$, are elements of $\boldsymbol{\alpha}_{rs}$ and $\boldsymbol{\alpha}_{ms}$, respectively, and w'_k are weighting coefficients. $\boldsymbol{\alpha}_{rs}$ and $\boldsymbol{\alpha}_{ms}$ are the static characteristics of the real arm and the model, respectively.

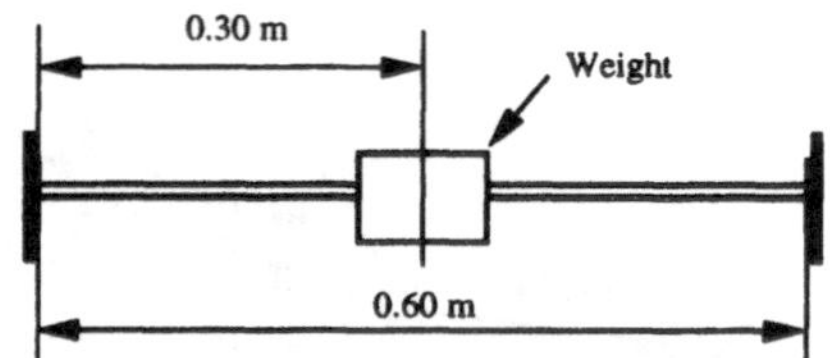

Figure 5. Flexible link used for experiment

5. By changing the values $p_2, p_3, \cdots, p_{L_i}$ by small amounts and by performing the steps 2 through 4 repeatedly, the optimal values of lengths $p_2, p_3, \cdots p_{L_i}$ that minimize the criterion J' and the corresponding spring constants $k_2, k_3, \cdots, k_{L_i}$ are determined.

6. The damping coefficients d_j are determined by trial and error so that the measured step response of the real link and that of the model coincide as much as possible.

3. Experiment of Identification

3.1. Experimental Setup and Measured Data

A flexible link shown in **Fig.5** is modeled to study the effect of the number of virtual joints on the quality of the model in terms of the step response. The link is a simple beam whose cross section is circle, made of steel, having length 0.60m, diameter 0.013m and mass 0.629kg. A weight of 1.2kg is attached to the middle of the link. Note that the weight makes the link non-uniform and thus difficult to model by the mode shape approach.

The characteristics of the link are measured. For dynamic characteristics, the first, second, and third natural frequencies are measured. For static characteristics, linear deformation and angular deformation of the link tip are measured when certain force or moment is applied on the link tip. The result is: $\omega_{z1} = 14.25$[Hz], $\omega_{z2} = 71.0$[Hz], $\omega_{z3} = 310.0$[Hz], $u_P/P = 0.000294$[m/N], $\phi_p/P = 0.000689$[rad/N], $u_M/M = 0.000689[\mathrm{N}^{-1}]$, and $\phi_M/M = 0.00262[\mathrm{rad/N\cdot m}]$.

A step response of this link is measured by cutting a string supporting a weight of 0.4kg at the tip. The response taken in terms of the acceleration at the link tip is shown in **Fig.6(a)**.

3.2. Identification Results

We have obtained the model of the above described link for the cases of one, two, and three virtual joints (that is, $L_i = 2, 3, 4$) to see the effect of the number of virtual joints on the quality of the model. The procedure of determining the parameters of the model is described in the previous section. Note that we do not use the datum ω_{z3} in the two virtual joint case (ω_{z2} and ω_{z3} in the one virtual joint case).

The obtained model parameters for $w'_k = 1, k = 1, 2, 3, 4$, are shown in **Table 1** for $L_1 = 3$ (cases $L_1 = 2$ and 4 are not shown due to lack of space). The static characteristics of the three models are compared with the real arm characteristics in **Table 2**.

Then, to show the validity of this model, the step responses of the model and the real arm are compared.

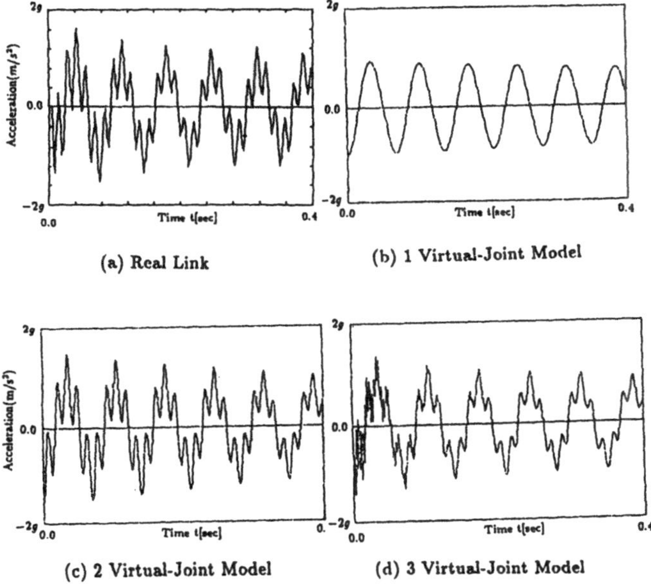

(a) Real Link (b) 1 Virtual-Joint Model

(c) 2 Virtual-Joint Model (d) 3 Virtual-Joint Model

Figure 6. Link-tip acceleration in step response

Table 1. Parameters of the 2-Joint Model

a)Virtual rigid links

	Link-1	Link-2	Link-3
Length[m]	0.1419	0.1981	0.2600
Gravity center[m]	0.0743	0.1494	0.1785
Mass[kg]	0.14871	1.40772	0.41857
Inertia[kg·m^2]	2.89×10^{-4}	3.96×10^{-3}	3.38×10^{-3}

b) Virtual passive joints

	Joint-2	Joint-3
Spring constant[N·m/rad]	845.5	1220.9
Damping coefficient[N·m·s/rad]	0.15	0.020

Figs.6(b), (c), and **(d)** show the link-tip acceleration in the step response for the one-, two-, and three-joint models. Comparing these figures with **Fig.6(a)**, we can see that the one-joint model does not reflect the effect of the second mode of oscillation and the two-joint and three-joint models simulate the step response rather well. The main difference between the latter two is the higher frequency oscillation at the beginning of the step response that lasts only less than 0.1 sec. Hence the two-joint model seems to be good enough for control purposes of most of flexible

Table 2. Static Parameters of Real Link and Models

	Real link	1-joint model	2-joint model	3-joint model
μ/P[m/N]	0.000294	0.000314	0.000304	0.000288
ϕ/P[rad/N]	0.000689	0.000766	0.000755	0.000688
μ/M[N^{-1}]	0.000689	0.000766	0.000755	0.000688
ϕ/M[rad/N·m]	0.00261	0.00187	0.00200	0.00262

manipulators.

4. Controller Design using Virtual Joint Model

4.1. Formulation of State Equation

In this section, we derive an LQ controller using the virtual joint model. First, we formulate a state equation. Using Lagrange's method and neglecting the centrifugal and Coriolis forces, the equation of motion of the virtual joint model is derived as follows:

$$\boldsymbol{u} = \boldsymbol{M}\ddot{\boldsymbol{q}} + \boldsymbol{F}\dot{\boldsymbol{q}} + \boldsymbol{K}\boldsymbol{q} \tag{6}$$

where $\boldsymbol{q}$ is the joint displacement vector of real and virtual joints, and $\boldsymbol{u}$ is the driving force vector of real joints, given by

$$\begin{aligned} \boldsymbol{q} &= [\theta_1\ \phi_{11}\cdots\phi_{1L_1}\ \cdots\ \theta_N\ \phi_{N1}\cdots\phi_{NL_N}]^T \\ \boldsymbol{u} &= [\tau_1\ 0\cdots0\ \cdots\ \tau_N\ \ 0\cdots0]^T \end{aligned} \tag{7}$$

$\boldsymbol{M}$ is the inertia matrix. $\boldsymbol{K}$ and $\boldsymbol{F}$ are respectively the stiffness matrix and the damping matrix of virtual joints given by

$$\begin{aligned} \boldsymbol{K} &= \mathrm{diag}[0\ k_{11}\cdots k_{1L_1}\ \cdots\ 0\ k_{N1}\cdots k_{NL_N}]^T \\ \boldsymbol{F} &= \mathrm{diag}[0\ d_{11}\cdots d_{1L_1}\ \cdots\ 0\ d_{N1}\cdots d_{NL_N}]^T \end{aligned} \tag{8}$$

A state equation derived from (6) is

$$\dot{\boldsymbol{x}} = \boldsymbol{A}\boldsymbol{x} + \boldsymbol{B}\boldsymbol{u} \tag{9}$$

where $\boldsymbol{x} = [\boldsymbol{q}^T\ \dot{\boldsymbol{q}}^T]^T$ and

$$\boldsymbol{A} = \begin{bmatrix} \boldsymbol{0} & \boldsymbol{I} \\ -\boldsymbol{M}^{-1}\boldsymbol{K} & -\boldsymbol{M}^{-1}\boldsymbol{F} \end{bmatrix}\ \boldsymbol{B} = \begin{bmatrix} \boldsymbol{0} \\ \boldsymbol{M}^{-1} \end{bmatrix} \tag{10}$$

In the case of the virtual joint model, the displacements of virtual joints cannot be directly measured. To estimate the displacements of virtual joints, we attach accelerometers at the tip of link $i(i = 1\cdots N)$. The relation between the output of the accelerometers and joint acceleration is described as

$$\boldsymbol{a} = \tilde{\boldsymbol{J}}\ddot{\boldsymbol{q}} \tag{11}$$

Here, $\boldsymbol{a}$ shows the acceleration at the tip of link i with respect to Σ_i. $\tilde{\boldsymbol{J}}$ is a jacobian matrix of the tip of link i with respect to $\boldsymbol{q}$. Substituting (6) into (11), we have

$$\boldsymbol{a} = \begin{bmatrix} -\tilde{\boldsymbol{J}}\boldsymbol{M}^{-1}\boldsymbol{K} & -\tilde{\boldsymbol{J}}\boldsymbol{M}^{-1}\boldsymbol{F} \end{bmatrix}\boldsymbol{x} + \tilde{\boldsymbol{J}}\boldsymbol{M}^{-1}\boldsymbol{u} \tag{12}$$

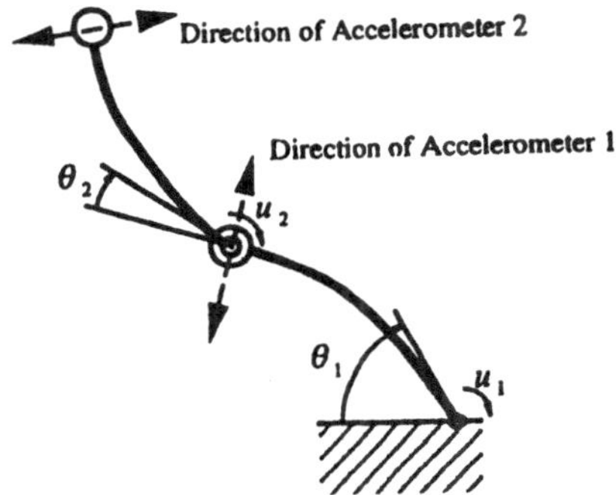

Figure 7. Direction of Measured acceleration for 2 d.o.f. flexible manipulators

Hence the equation of observation is given by

$$\boldsymbol{y} = \boldsymbol{C}\boldsymbol{x} + \boldsymbol{D}\boldsymbol{u} \tag{13}$$

where $\boldsymbol{y} = [\boldsymbol{a}^T \ \theta_1 \cdots \theta_N \ \dot{\theta}_1 \cdots \dot{\theta}_N]^T$.

We can easily show that, in the case of 2 d.o.f. flexible manipulators, the pair (9) and (13) is observable, if the accelerations in the directions shown in **Fig. 7** are measured. We will derive a state observer using (9) and (13) in the next subsection.

4.2. Control Algorithm

In this subsection, an LQ control algorithm is derived. We assume that the performance index is given by

$$J = \int_0^\infty \{(\boldsymbol{x} - \boldsymbol{x}_d)^T \boldsymbol{Q}(\boldsymbol{x} - \boldsymbol{x}_d) + \boldsymbol{u}^T \boldsymbol{R}\boldsymbol{u}\} dt \tag{14}$$

where $\boldsymbol{Q}$ and $\boldsymbol{R}$ are weighting matrices. $\boldsymbol{x}_d$ is a desired value of $\boldsymbol{x}$. The state feedback control law that minimizes this performance index is given by

$$\boldsymbol{u} = \boldsymbol{P}(\boldsymbol{x} - \boldsymbol{x}_d) \tag{15}$$

where $\boldsymbol{P}$ is the optimal feedback gain matrix. Since the state $\boldsymbol{x}$ is not available, we use the estimate $\boldsymbol{z}$ of $\boldsymbol{x}$ from the state observer given by

$$\dot{\boldsymbol{z}} = (\boldsymbol{A} - \boldsymbol{G}\boldsymbol{C})\boldsymbol{z} + (\boldsymbol{B} - \boldsymbol{G}\boldsymbol{D})\boldsymbol{u} + \boldsymbol{G}\boldsymbol{y} \tag{16}$$

where $\boldsymbol{G}$ is the observer gain matrix.

4.3. Experiment

To verify the effectiveness of the proposed LQ controller, we performed experiments. A 2 d.o.f. flexible manipulator shown in **Fig. 8** was used for experiment. In this manipulator, link 1 is the same as that we used in the experiment of identification. At the middle of each link, a weight is attached. Each joint has a DC servo motor with harmonic drive gear. We used PC with i386+387 as the host computer.

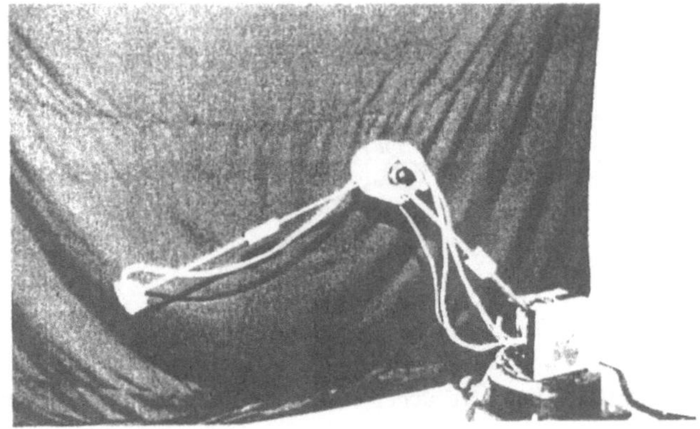

Figure 8. A photograph of 2 d.o.f flexible manipulator used for experiment

To show the validity of the controller, PTP control is performed. The initial and desired joint angles are $[\pi/6 \quad -\pi/6]$ and $[\pi/4 \quad -\pi/4]$, respectively. The sampling period is 3.75[msec].

For the purpose of comparison, a PD feedback control of joint displacements was also performed. We choose the PD feedback gain as close as possible to those of the corresponding gains of optimal regulator.

From the experimental results shown in **Fig. 9**, we can see that in the case of the proposed optimal regulator, vibration is suppressed well. showing the effectiveness of the proposed controller.

5. Conclusion

Validity of the virtual joint model of flexible manipulators has been shown by some preliminary experiments in this paper.

The authors would like to express their thanks to Dr. Koh Hosoda, now at Osaka University, for his contribution in the early stage of this research and to Mr. Kensuke Harada for his support.

References

[1] Kane, T.R., R.R. Ryan and A.K. Bnerjee, Dynamics of a Cantilever Beam Attached to a Moving Base. *J. Guid. Control Dyn.*, 10(2):139–151, 1987.

[2] Banerjee, A.K. and M.E. Lemak, Multi-Flexible Body Dynamics Capturing Motion-induced Stiffness. *Trans. of ASME, J. of Appl. Mechanics*, 58:766–775, 1991.

[3] Sadler, J.P. and G.N. Sandor, A Lumped Parameter Approach to Vibration and Stress Analysis of Elastic Linkages. *Trans. of ASME, J. of Eng. Industry*, 95:549–557, 1973.

[4] Amirouche, F.M.L. and R.L. Huston, Dynamics of Large Constrained Flexible Structures. *Trans. of ASME, J. of DSMC*, 110:78–83, 1988.

[5] Book, W.J., Maizza-Neto, O., and Whitney, D.E., Feedback Control of Two Beams, Two Joint System with Distributed Flexibility. *Trans. of ASME, J. of DSMC*. 97(4):424-431, 1975.

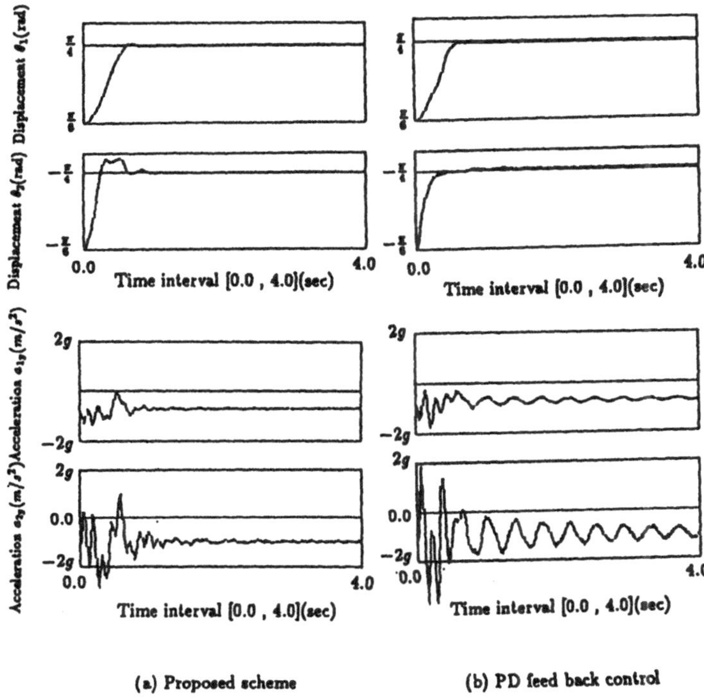

Figure 9. Experimental results

[6] Cannon Jr., R.H. and Schmitz, E., Initial Experiments on the End-Point Control of a Flexible One-Link Robot. *Int. J. of Robotics Research*, 3(3):62-75, 1984.

[7] Pfeiffer, F., A Feedforward Decoupling Concept for the Control of Elastic Robots. *J. of Robotics Systems*, 6(4):407-416, 1989.

[8] Sunada, W.H. and Dubowsky, S., On the Dynamic Analysis and Behavior of Industrial Robotic Manipulators with Elastic Members. *Trans. of ASME J. of DSMC*, 105(1):42-51, 1983.

[9] Yoshikawa, T., Murakami, H., and Hosoda, K. Modeling and Control of a Three Degree of Freedom Manipulator with Two Flexible Links. *Proc. of IEEE Int. Conf. on Decision and Control*, pp. 2532–2537, 1990.

[10] Yoshikawa, T. and Hosoda, K., Modeling of Flexible Manipulators Using Virtual Rigid Links and Passive Joints. *Proc. of IEEE/RSJ Int. Workshop on Intelligent Robots and Systems(IROS'91)*, pp. 967-972, 1991

[11] Huang, Y. and Lee, C.S.G., Generalization of Newton-Euler Formulation of Dynamic Equations to Nonrigid Manipulators. *Trans. of ASME J. of DSMC*, 110:308-315, 1988.

[12] Yoshikawa, T. and Matsudera, K., Experimental Study in Modeling of Flexible Manipulators using Virtual Joint Model. *Proc. of SYROCO'94*, 1994

Experimental Research on Impact Dynamics of Spaceborne Manipulator Systems

Kazuya Yoshida
Dept. of Aeronautics and Space Engineering
Tohoku University
Aoba-campus, Sendai 980-77, JAPAN
yoshida@space.mech.tohoku.ac.jp

Constantinos Mavroidis
Steven Dubowsky
Dept. of Mechanical Engineering
Massachusetts Institute of Technology
Cambridge, MA 02139, U.S.A.
dmavro@mit.edu, dubowsky@mit.edu

Abstract

The problem of impact dynamics of space robotic systems that consist of a rigid manipulator supported by a flexible deployable structure is addressed. Due to joint back-drivability and the dynamic coupling between the manipulator and its supporting structure, unknown motion of the system occurs after it makes impulsive contact with the environment. A method that uses the system's dynamic model is proposed to estimate the motion of the system after impact and used to minimize the impact effect and vibrations of the supporting structure. This method is verified experimentally using the MIT Vehicle Emulation System (VES II). The experimental results show that the impact force and the system motion after impact can be reduced if the manipulator configuration prior to impact and the controller gains are properly selected.

1. Introduction

Robotic systems supported by flexible long-reach deployable structures have been proposed for future space projects. The Special Purpose Dexterous Manipulator (SPDM) mounted on the Space Station Remote Manipulator System (SSRMS) (Figure 1) [1] and the Japanese Experiment Module Remote Manipulator System (JEMRMS) proposed by the Japan's NASDA [2], are examples of long-reach space manipulator systems now being developed.

While promising, the development of long reach space manipulator systems requires the solution of fundamental technical problems. A key problem is dynamic coupling between the manipulator and its flexible supporting structure. This causes uncontrolled motion of the manipulator supporting structure when the manipulator performs a task. This undesired base motion can degrade the system's performance, including its dexterity.

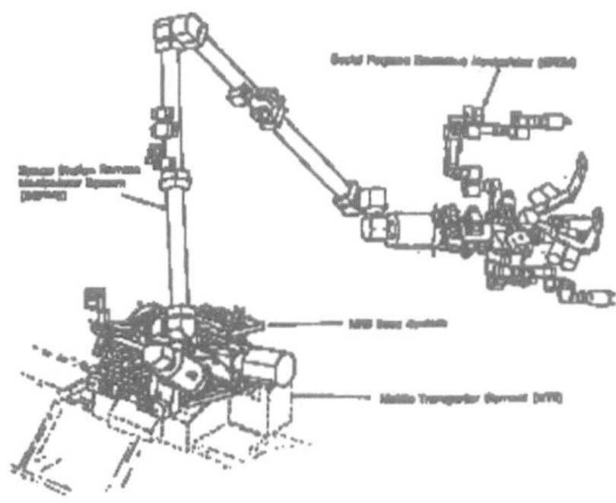

Figure 1 An example of a spaceborne long-reach manipulator system: The Canadian SPDM and SSRMS [1]

The problem of vibrations of long reach manipulator systems when these are excited by the system's internal inertial forces has been studied [3], [4]. However, little has been done when the vibrations of the supporting structure are excited by external disturbance such as impact forces that act on the system when it makes contact with the environment. This problem can be critical when a long reach manipulator catches a passive free-floating object, such as a satellite. In this operation, substantial impact forces to the manipulator system can excite vibration of the flexible supporting structure.

The mechanical behavior of manipulators under impact is not well understood when the joints show *some* back-drivability even if they are ground-fixed. In ideal conditions, when the joints are either free-to-move or break-locked, the dynamics are well defined. However in real situations, due to friction, stiffness, damping, and inertia of the actuator and gear train, unknown manipulator joint motion is produced after impact. In addition, when the manipulator supporting structure is flexible, dynamic interaction between the manipulator and its base will produce unknown vibratory motion of the system after impact.

A method called Extended-Inversed Inertia Tensor (Ex-IIT) has been proposed to obtain estimations of the impact forces and a space system's motion after impact [5]. These estimations can be extended in the system design phase to check its structural safety under impact, and in the planning phase to select optimal manipulator configurations to catch a free floating object so that impact forces are reduced. The Ex-IIT method can be applied in the case of manipulator systems with fixed, free-floating or flexible supporting structures and with joints that can be free, break-locked or back-drivable. The Ex-IIT method has been validated with simulations and verified experimentally using a planar two degree of freedom model of a free-floating manipulator system [5].

In this paper the problem of impact dynamics of long reach manipulator systems is studied experimentally in conditions that are close to reality. The Ex-IIT method is experimentally verified, using the MIT Vehicle Emulation System mod II (VES-II) testbed [6]. This system can emulate in real time, the spatial six degree of freedom (three translations and three rotations) of any free floating or flexibly supported manipulator system, in ground or microgravity conditions [7]. It consists of a 6 DOF hydraulically driven Stewart platform, a six degree of freedom force/torque sensor and any manipulator that can be mounted on the platform. For the experiments performed during this work a PUMA 560 was mounted on the platform. Impact conditions were produced at the tip of the manipulator end effector using a specially designed impact device. Experiments were performed using various controller gains and manipulator configurations. In a first phase the manipulator supporting structure was kept fixed. Then, in a second phase the VES emulated the motion of a flexible supporting structure for the manipulator. The motion of the manipulator system and that of its base after impact was recorded and compared to the estimation given by the Ex-IIT method. The experimental results verified Ex-IIT and showed that the manipulator configuration and the joint controller gains affect the magnitude of the impact force and the motion of the flexible supporting structure.

2. Modeling of Impact Dynamics

Collision of rigid bodies is a classical problem in mechanics ([8]). This phenomenon becomes very complex when the collided systems are subject to multibody or free-floating dynamics, or to friction and compliance effects. For ground based manipulator systems the problem of reducing the impact force has been addressed [9]-[13]. Recently a method has been proposed to model impact dynamics of manipulator systems with friction and compliance characteristics at their joints supported by free-floating or ground-fixed bases, using an Extended-Inversed Inertia Tensor (Ex-IIT) [5].

Let $\mathcal{F}$ be an impact force/torque at the end-effector of a fixed base manipulator system, $\boldsymbol{\tau}_f$ be a passive joint torque due to friction, and $\boldsymbol{\tau}_s$ be an active joint torque due to servo control. The manipulator system dynamic equation takes the form:

$$\mathbf{H}\ddot{\boldsymbol{\phi}} + \mathbf{c}(\boldsymbol{\phi}, \dot{\boldsymbol{\phi}}) + \boldsymbol{\tau}_f + \boldsymbol{\tau}_s = \mathbf{J}^T \mathcal{F}, \tag{1}$$

where $\mathbf{H}$ is the system's inertia matrix, $\mathbf{c}$ is a non-linear velocity dependent term, $\boldsymbol{\phi}$ is the manipulator joint angle vector and $\mathbf{J}$ is the Jacobian matrix.

Consider the integral of equation (1) for the impact period from t to $t + \delta t$, in the idealized case where $\delta t \to 0$:

$$\lim_{\delta t \to 0} \int_t^{t+\delta t} (\mathbf{H}\ddot{\boldsymbol{\phi}} + \mathbf{c}(\boldsymbol{\phi}, \dot{\boldsymbol{\phi}}) + \boldsymbol{\tau}_f + \boldsymbol{\tau}_s)\mathrm{dt} = \lim_{\delta t \to 0} \int_t^{t+\delta t} \mathbf{J}^T \mathcal{F}\mathrm{dt} \tag{2}$$

As collision is an instantaneous, high-frequency phenomenon, terms such as $\int_t^{t+\delta t}(\mathbf{c}(\boldsymbol{\phi}, \dot{\boldsymbol{\phi}}) + \boldsymbol{\tau}_f + \boldsymbol{\tau}_s)\mathrm{dt}$ that can obtain a finite value during the infinitesimal impact period $\delta t \to 0$, can be neglected. During this period, the expressions $\int_t^{t+\delta t} \mathbf{H}\ddot{\boldsymbol{\phi}}\mathrm{dt}$ and $\int_t^{t+\delta t} \mathbf{J}^T \mathcal{F}\mathrm{dt}$ can take an infinite value and dominate the impact dynamics. Using the notation $(\ddot{\circ}) \to (\Delta\dot{\circ})$ to replace an instantaneous acceleration by the corresponding velocity change before and after the impact, equation (2) takes the form:

$$\mathbf{H}\Delta\dot{\boldsymbol{\phi}} = \mathbf{J}^T \overline{\mathcal{F}} \tag{3}$$

where $\overline{\mathcal{F}}$ represents impulse.

In reality, the impact duration is not zero but has a very small finite value ϵ. In this case, equation (2) takes the form:

$$\lim_{\delta t \to \epsilon} \int_t^{t+\delta t} (\mathbf{H}\ddot{\boldsymbol{\phi}} + \mathbf{c}(\boldsymbol{\phi}, \dot{\boldsymbol{\phi}}) + \boldsymbol{\tau}_f + \boldsymbol{\tau}_s)\mathrm{dt} = \lim_{\delta t \to \epsilon} \int_t^{t+\delta t} \mathbf{J}^T \mathcal{F}\mathrm{dt} \tag{4}$$

It is assumed that the joint torque effect now can not be neglected during this period, then (4) is written as:

$$\mathbf{H}\Delta\dot{\boldsymbol{\phi}} + \mathbf{T} = \mathbf{J}^T\overline{\mathcal{F}} \tag{5}$$

where $\mathbf{T}$ represents the joint effect due to the term $\int_t^{t+\delta t}(\boldsymbol{\tau}_f + \boldsymbol{\tau}_s)\mathrm{dt}$. $\mathbf{T}$ models higher frequency phenomenon than ordinary, non-impulsive damping and stiffness effects. It is not an inertial effect and is not directly related to acceleration. $\mathbf{T}$ can be related to the impulsive velocity change using the following relationship:

$$\mathbf{T} = \boldsymbol{\lambda}\Delta\dot{\boldsymbol{\phi}}, \tag{6}$$

Then, equation (5) is written as:

$$(\mathbf{H} + \boldsymbol{\lambda})\Delta\dot{\boldsymbol{\phi}} = \mathbf{J}^T\overline{\mathcal{F}} \tag{7}$$

The matrix coefficient $\boldsymbol{\lambda}$ is called Virtual Rotor Inertia and is considered to be an additional joint inertia, or an increased gear-ratio connected to the rotor of a joint actuator. The impact dynamics equation with respect to the manipulator end-effector is written in the following form:

$$\mathbf{G}^{*-1}\Delta\dot{\mathbf{x}}_h = \overline{\mathcal{F}} \tag{8}$$

where $\Delta\dot{\mathbf{x}}_h$ is the difference of the end-effector velocities before and after the impact, and $\mathbf{G}^*$ is called the Extended-Inversed Inertia Tensor. It is defined as:

$$\mathbf{G}^* \equiv \mathbf{J}(\mathbf{H} + \boldsymbol{\lambda})^{-1}\mathbf{J}^T \tag{9}$$

for ground-fixed manipulators.

It can be shown that for free-floating or flexible based manipulator systems $\mathbf{G}^*$ takes the form:

$$\mathbf{G}^* \equiv \mathbf{J}^*(\mathbf{H}^* + \boldsymbol{\lambda})^{-1}\mathbf{J}^{*T} + \mathbf{R}_{gh}\mathbf{M}^{-1}\mathbf{R}_{gh}^T \tag{10}$$

where $\mathbf{J}^*$ and $\mathbf{H}^*$ are the augmented forms of the system's Jacobian matrix and inertia matrix respectively [14]. The matrices $\mathbf{M}$ and $\mathbf{R}_{gh}$ are equal to: $\mathbf{M} = \begin{bmatrix} w\mathbf{E} & \mathbf{0} \\ \mathbf{0} & \mathbf{H}_g \end{bmatrix}$ and $\mathbf{R}_{gh} = \begin{bmatrix} \mathbf{E} & -\tilde{\mathbf{r}}_{gh} \\ \mathbf{0} & \mathbf{E} \end{bmatrix}$ where w and $\mathbf{H}_g$ are the total mass and inertia of the moving system with respect to its centroid respectively, $\mathbf{r}_{gh}$ is a moment arm from the system centroid to the end-effector, and $\mathbf{E}$ is the 3×3 identity matrix. For detailed derivation of $\mathbf{G}^*$, see [5].

Consider the case where the manipulator end-effector collides with a moving object of mass m_b and velocity $\mathbf{v}_b$. The manipulator end-effector velocity is $\mathbf{v}_a$. At the moment of collision an impact force $\mathbf{f}$ is exerted at the manipulator end-effector at the contact point in a certain direction $\mathbf{n}$. It is assumed that no moment is developed at the manipulator end-effector during the impact. Using the principle of linear momentum conservation, equation (8) is written in the form:

$$\mathbf{G}_{33}^{*-1}\Delta\mathbf{v}_a = \overline{\mathbf{f}} = -m_b\Delta\mathbf{v}_b \tag{11}$$

where $\mathbf{G}_{33}^*$ is the top-left 3×3 quadrant of the Ex-IIT associated with the linear motion.

If equation (11) is multiplied (inner product) with a unit vector indicating the direction of the impact force, $\mathbf{n}$, the following scalar formula is obtained:

$$m_a^*(v_a' - v_a) = m_b(v_b - v_b') \tag{12}$$

where v_a and v_b are the projections of $\mathbf{v}_a$ and $\mathbf{v}_b$ on $\mathbf{n}$, $\{'\}$ indicates the velocity after collision, and

$$m_a^* \equiv \frac{|\mathbf{f}|}{|\dot{\mathbf{v}}_a|} = \frac{1}{\mathbf{n}^T \mathbf{G}_{33}^* \mathbf{n}} \tag{13}$$

is the manipulator effective mass that expresses the inertial characteristics of the manipulator end-effector in the $\mathbf{n}$ direction [15][16].

The elastic restitution coefficient e of the end-effector is defined by:

$$v_a^{'} - v_b^{'} = e(v_b - v_a). \tag{14}$$

From equations (12) and (14), the post-impact manipulator end-effector velocity and the magnitude of impulse can be obtained using pre-impact information:

$$v_a^{'} = \frac{(1+e)m_b v_b + (m_a^* - e m_b)v_a}{m_a^* + m_b} \tag{15}$$

$$\overline{\mathbf{f}} \cdot \mathbf{n} = \frac{(1+e)m_a^* m_b (v_b - v_a)}{m_a^* + m_b} \tag{16}$$

3. Experiment Methodology

The Extended-Inversed Inertia Tensor (Ex-IIT) method is experimentally verified using the MIT Vehicle Emulation System mod II (VES-II) [6]. This system consists of a 6 DOF hydraulically driven Stewart platform, a force/ torque sensor and a PUMA 560 mounted on the top of the platform. When the manipulator moves or when it makes a contact with the environment, the resulting interaction forces and moments between the manipulator and its supporting structure are measured with the force/torque sensor. Using a computer dynamic model of a free-floating or flexible manipulator supporting structure is calculated under the measured load. Then the platform is commanded to reproduce this spatial six degree of freedom motion in real time. The VES can also be used to emulate the motion of space manipulators in micro-gravity conditions [7].

An impact device has been built to study impulsive contacts between the system and its environment (see Figure 2). The device is a pendulum with a steel hammer head. It is equipped with a piezo-electric sensor to measure the impact force and an encoder to be able to calculate the velocity of the impact head before and after impact.

In our experiments, the impact device hits the tip of the manipulator end-effector which is wrapped in a soft material. Results are presented for experiments with the manipulator base fixed and the VES emulating the motion of a flexible long-reach system. In each case, three manipulator configurations and three sets of joint control gains are tested. The three configurations are chosen to be representative of the manipulator workspace (see Figure 3.) For example in configuration 2 the arm is relatively stretched while in configuration 3 it is relatively folded. The three sets of gains for the joint PID controller are characterized as High, Low and Zero and they represent situations where the arm is stiff, compliant or free to move. The specific parameters that describe the configurations and the sets of control gains are shown in Table 1.

For each configuration and set of gains, impacts with various impact velocities in the range of 1.0-2.5 [m/s] are applied. The values of the impact force exerted at the manipulator end-effector vary between 30-70 [N]. Only the first three joints of the manipulator are active. The impact force direction is kept parallel to the -x direction of the inertial reference system. In the case of Zero and Low gains, a constant torque is applied to

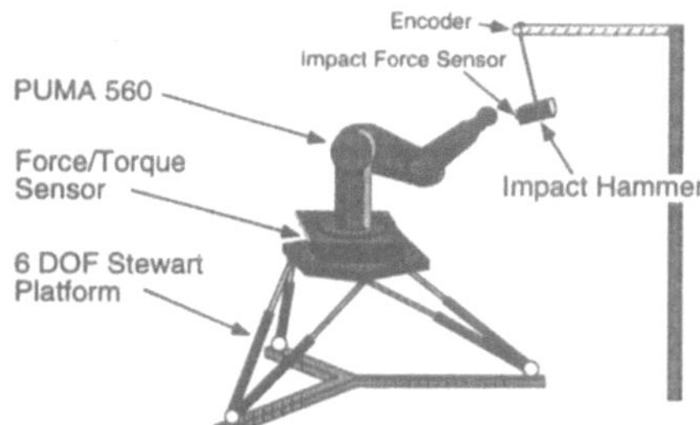

Figure 2 The MIT VES-II and the impact device

Table 1 Parameters of the experimental conditions

Configuration	joint 1	2	3
1	0 [deg]	150	90
2	0	172	40
3	0	135	135

Servo-gain	P	D	I
Zero	0	0	0
Low	5000	40	0
High	500000	4000	1500

(counts in a custom PID controller)

the manipulator joints to compensate for gravity to assist the arm in holding its nominal position.

During the experiments, the manipulator system moves due to the impact. The joint angles of manipulator arm, the angle of the impact device, the base/manipulator interaction forces and moments, the impact force, the emulated moving base passive motion are all measured variables. The velocities of the end-effector and of the impact device before and after impact v_a, v_b, v_a', v_b' are calculated after the experiment by numerical differentiation. From equations (12) and (14) the manipulator effective mass and the restitution coefficient are calculated.

The purpose of these experiments is:

a) To calculate experimentally the manipulator effective mass and restitution coefficient and observe the way they depend on the manipulator configuration and controller gains.

b) To confirm the idea derived by the Ex-IIT method that the joint friction and controller gains increase the manipulator effective mass.

c) To show that the manipulator joint motion and the motion of its flexible supporting structure when there is an impact with the environment depend on the manipulator configuration prior to impact and the controller gains.

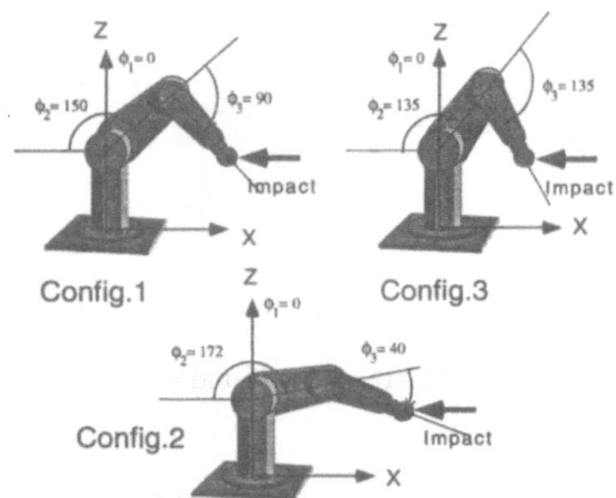

Figure 3 Arm configurations and impact direction

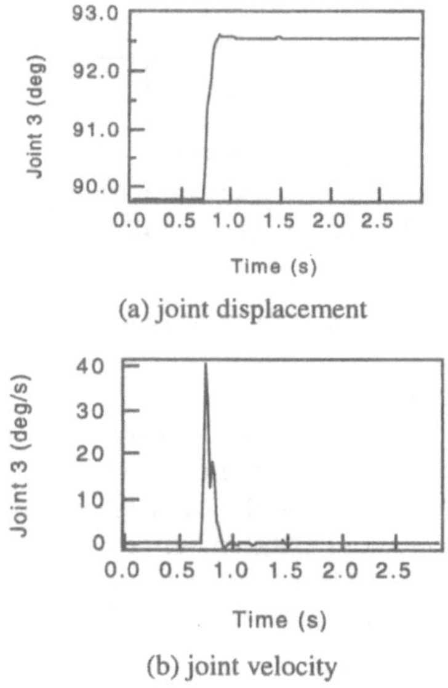

Figure 4 Joint 3 motion after impact in Configuration 1 with Zero gains

4. Experimental Results

4.1. Fixed-Base Experiments

The first experiments had the manipulator base fixed. The manipulator mass and restitution coefficient are calculated in each experiment and the manipulator joint motion after impact is observed. For example, in Figure 4 the motion of joint 3 after impact is shown. The manipulator prior to impact is at configuration 1, with zero gains. Due to impact the joint almost instantly moves about 2.5 degrees with a high velocity that is very quickly damped

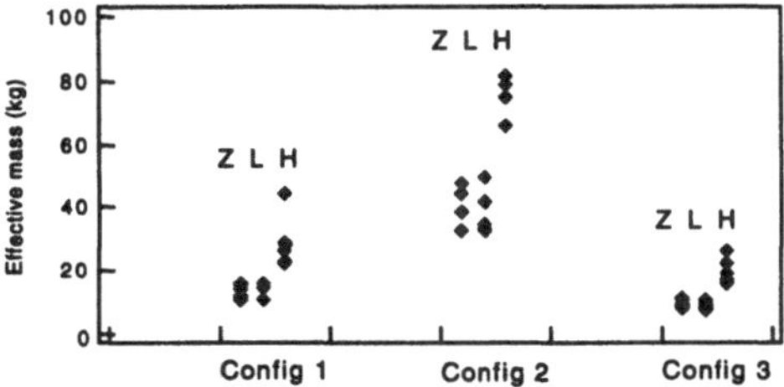

Figure 5 Observed effective mass

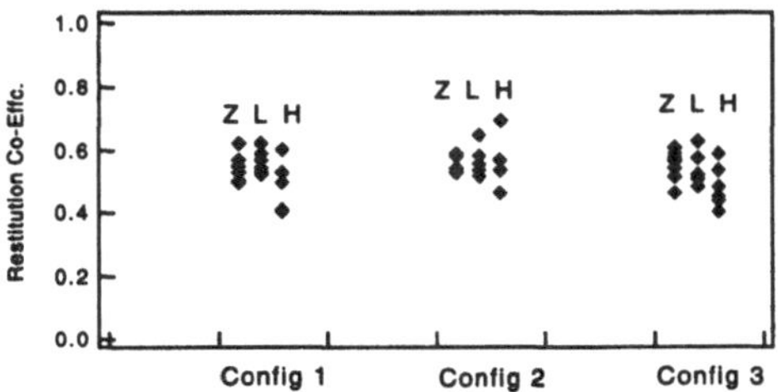

Figure 6 Observed restitution coefficient

Table 2 Identified effective mass versus joint condition (for fixed-base experiments)

Config.	Effective mass in [kg]			Ratio	
	Simulation	**Experiments**			
	Link Dynamics (m_1) (free joint)	Zero and Low gains (m_2) (+ joint friction)	High gain (m_3) (+ servo torque)	m_2/m_1	m_3/m_1
1	6.99	12.98	28.45	1.86	4.07
2	19.66	38.14	76.45	1.94	3.89
3	4.96	8.91	19.11	1.80	3.86

out. The sampling interval of this joint angle measurement is 20 [ms]. Using equations (12) and (14), the effective mass and restitution coefficient values are calculated and their values are shown in Figures 5 and 6.

The results show that the effective mass depends on both, the configuration and controller gains. In Configuration 3 (folded configuration) the effective mass is lower. In Configuration 1 which is characterized as intermediate, the effective mass increases. The effective mass obtains its higher value in Configuration 2 (stretched configuration). In the cases of Zero and Low gains the effective mass takes similar values which are much lower than in the case of High gains. This result supports the concept of Virtual Rotor Inertia that joint torques effect to increase the system's virtual mass. The result also shows that the impact effect can be reduced by properly selecting the manipulator configuration and by using lower controller gains.

In Figure 6 the result show that the restitution coefficient is not a function of the manipulator configuration or the controller gains. It depends on the elasticity of the material of the contact point. For our experimental system it has an average value of 0.55. The variation of the values of the restitution coefficient as it is shown in Figure 6, during the experiments, depends to a small variation of the impact point and of the surface condition.

In Table 2 a comparison among effective masses of the idealized model and the experimental measurements, is presented. In the first column, the values of the effective mass are displayed as these are calculated using the Ex-IIT method with $\lambda = 0$ that represents an idealized situation for the system with no joint friction and no controller. The next two columns show experimental results of the system, with friction with or without control feedback. In the last two columns the ratio of experimental calculated values to the model based estimated values are shown. A general statement can be made that for the system used in these experiment, the existence of friction in the manipulator joints doubles the impact effect. The same statement can be made in the case of high gain joint PID control. These two results taken together, show that the impact effect becomes four times higher if friction and high gain control are present on the same time.

4.2. Flexible Base Experiments

Experiments were performed with the VES-II system emulating the motion in microgravity conditions of a vertical flexible cylindrical beam with ($EI = 2.5 \times 10^8 [\text{Pa} \cdot \text{kgm}^2]$) that is assumed to be the supporting structure of the PUMA 560 (see Figure 7.)

Figure 8 shows the displacement from the initial position. Results are shown for configurations 1 and 2, for High gains and Low gains. Note that the motion is not a simple sinusoidal wave because the contact point is off-centered from the plane of symmetry of the three dimensional beam.

The experimental results show that the motion of the system supporting structure after impact is smaller if configuration 1 and/or Low control gains are used. On the other, the motion is much bigger if configuration 2 and/or High gains are used. The experimental data show a reduction of 20-30% of the maximum amplitude of base vibrations after impact when Low gains are used compared to High gain motion. These results indicate that there are configurations and gains that can reduce the system motion after impact.

Table 3 shows a comparison between the manipulator effective mass calculated in simulations using equation (10) and λ estimated by the fixed-base experiments, and the effective mass calculated experimentally using equation (12). These values show a good agreement and confirm the Ex-IIT method when it is applied in long reach manipulator systems.

Finally, an important observation is that the motion of the flexible based manipulator is smaller when the effective mass is small. This means that the effective mass can be used as an index to reduce vibrations of the supporting flexible beam.

5. Conclusions

Impact experiments with manipulators supported by fixed or flexible supporting structures have been performed using the MIT Vehicle Emulation System (VES-II). The experimental data revealed the following conclusions:

a) Fixed-base experiments:

- The restitution coefficient is not a function of the manipulator configuration or its joint control gains.
- The effective mass at the manipulator end-effector depends on the manipulator configuration and joint control gains. Joint friction and servo torque increase the effective

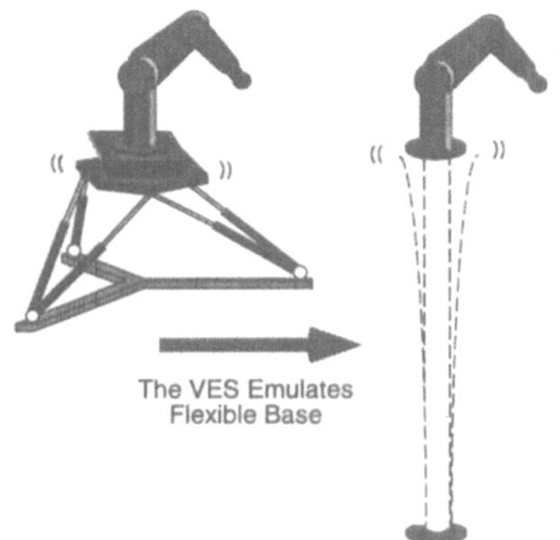

Figure 7 Flexible-base experiments

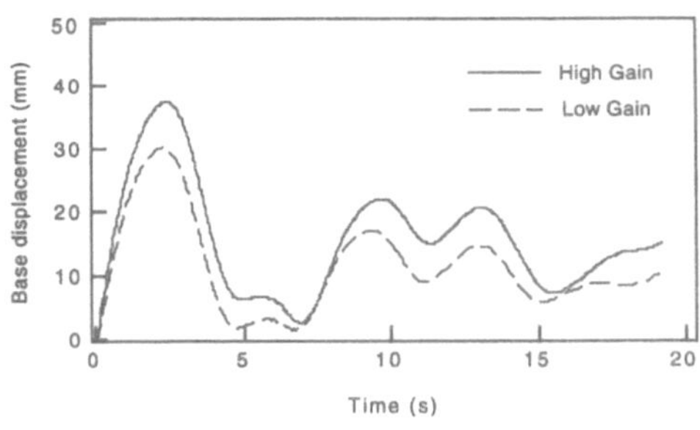

(a) Configuration 1

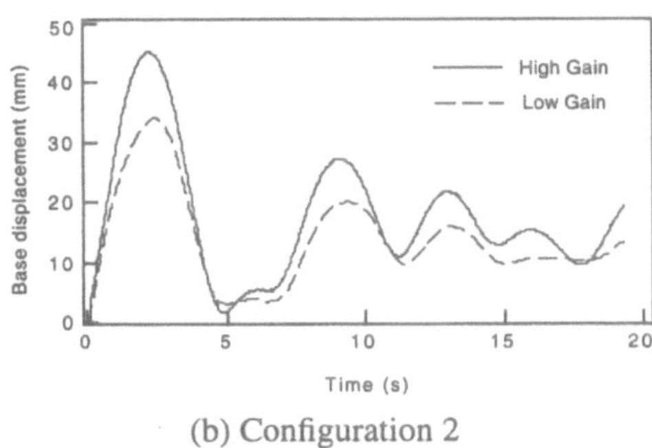

(b) Configuration 2

Figure 8 Manipulator base motion after impact

Table 3 Effective mass for flexible-base manipulator

Config	Gain	Effective mass in [kg]	
		Simulation using estimated λ	Experiment
1	Low	13.90	14.21
	High	21.89	20.98
2	Low	28.96	27.48
	High	38.99	39.33

mass and therefore the impact effect. The experimental data support the Ex-IIT method.

b) Flexible base experiments:

- The amplitude of the supporting structure vibrations when these are excited by impulsive contacts of the system to the environment depends on the manipulator configuration and its control gains.
- The manipulator effective mass can be used as an index to reduce the motion of the supporting structure that results from impact.

Further on, the fact that a set of lower values of joint PD gains are effective to reduce, or absorb, the vibration of the supporting structure suggests relevance to the Psuedo-Passive Energy Dissapation concept proposed by Torres [17]. Also an optimal direction of the impulsive manipuator motion in terms of minimum excitation of the supporting structure could be discussed with the Coupling-Map method [17][18].

Acknowledgments

The support of this work by NASA Robotics Branch, Grant NAG 1-801 is acknowledged.

References

[1] Brimley, W., Brown, D. and Cox, B., "Overview of International Robot Design for Space Station Freedom," *Teleoperation and Robotics in Space, ed. by Skaar and Ruoff*, pp.411-441, Progress in Astronautics and Aeronautics vol.161, AIAA, 1994.

[2] Kuraoka, K. et al., "Design and Development Status of the JEMRMS," *Proc. of Int. Symp. on AI, Robotics and Automation in Space (i-SAIRAS'90)*, pp.23-26, Kobe, Japan, November, 1990.

[3] Dubowsky, S., "Dealing With Vibrations in the Deployment Structures of Space Robotic Systems," *5th International Conference on Adaptive Structures*, Sendai, Japan, December, 1994.

[4] Mavroidis, C., Rowe, P. and Dubowsky, S., "Inferred Endpoint Control of Long Reach Manipulator Systems," *Proc. of IEEE/RSJ Int. Symp. on Intelligent Robot System (IROS'95)*, Pittsburg, PA, August, 1995.

[5] Yoshida, K., "Impact Dynamics Representation and Control with Extended-Inversed Inertia Tensor for Space Manipulators," *Robotics Research 6, ed. by Kanade and Paul*, pp.453-463, IFRR, Cambridge, 1994.

[6] Dubowsky, S. et al., "A Labratory Test Bed for Space Robotics: The VES II," *Proc. of IEEE/RSJ Int. Symp. on Intelligent Robot System (IROS'94)*, Munich, Germany, pp.1562-1569, 1994.

[7] Corrigan, T. and Dubowsky, S., "Emulating Microgravity in Laboratory Studies of Space Robotics," *Proc. of the 23rd ASME Mechanisms Conference*, MN, USA, 1994.

[8] Wittenburg, J., *Dynamics of Systems of Rigid Bodies*, Stuttgart: B.G.Teubner, 1977.

[9] Zheng, Y.F. and Hemami, H., "Mathematical Modeling of a Robot Collision With its Environment," *Journal of Robotic Systems*, Vol.2, No.3, pp.289-307, 1985.

[10] Youcef-Toumi, K. and Gutz, D.A., "Impact and Force Control," *Proc. 1989 IEEE Conf. on Robotics and Automation*, Scottsdale, AZ, pp.410-416, May 1989.

[11] Gertz, M.W., Kim, J-O. and Khosla, P.K. "Exploiting redundancy to reduce impact force," *Proc. 1991 IEEE/RSJ Int. Workshop on Intelligent Robots and Systems (IROS'91)*, Osaka, Japan, pp.179-184, Nov. 1991.

[12] Walker, I.D., "Impact Configurations and Measures for Kinematically Redundant and Multiple Robot Systems," *IEEE Trans. on Robotics and Automation*, vol.10, no.5, pp.670-683, 1994.

[13] Lin, Z., Patel, R. and Balafoutis, C., "Impact reduction for Redundant Manipulators Using Augmented Impedance Control," *Journal of Robotic Systems* Vol. 12, No. 5, pp.301-313, 1995.

[14] *Space Robotics: Dynamics and Cotrol*, ed. by Xu and Kanade, Kluwer Academic Publishers, 1993.

[15] Khatib, O. and Burdick, J., "Motion and Force Control of Robot Manipulators," *Proc. 1986 IEEE Int. Conf. on Robotics and Automation*, San Francisco, CA, pp.1381-1386, April 1986.

[16] Asada, H. and Ogawa, K., "On the Dynamic Analysis of a Manipulator and Its End Effector Interacting with the Environment," *Proc. 1987 IEEE Int. Conf. on Robotics and Automation*, Raleigh, NC, pp.751-756, March-April 1987.

[17] Torres, M., *Modeling, Path-Planning and Control of Space Manipulators: The Coupling Map Concept*, PhD Thesis, Department of Mechanical Engineering, MIT, February 1993.

[18] Torres, M., Dubowsky, S. and Pisoni, A.C., "Path-Planning for Elastically-Mounted Space Manipulators: Experimental Evaluation of the Coupling Map," *Proc. 1994 IEEE Int. Conf. on Robotics and Automation*, San Diego, CA, pp.2227-2233, May 1994.

An Operational Space Formulation for a Free-Flying, Multi-Arm Space Robot

Jeffrey Russakow Stephen M. Rock Oussama Khatib

Aerospace Robotics and Computer Science Robotics Laboratories
Stanford University
Stanford, California 94305

Abstract

An experimental validation of the Extended Operational Space Formulation for a two-arm, free-flying space robot is presented. Following a brief summary of the formulation and a quantitative description of the robot's macro-mini characteristics, the concepts of *macro-mini coordination* for a multi-arm system, *Dynamically Consistent Force/Torque Decomposition* for redundant manipulator control, and *Internal Motion Control* are demonstrated experimentally. The impact of actuator discretization at the robot base on the accuracy and dynamic decoupling of the robot is analyzed and discussed.

1. Background

Predictions of the number of hours of on-orbit assembly and maintenance that would be required to construct and service space-based platforms have highlighted the important role that space robotics can play. Tasks that robots will perform on a future space platform will include assembly, repair, maintenance, and service. Many tasks will require robotic systems that possess characteristics such as mobility, multiple arms, autonomy and dexterity.

Several researchers have experimentally investigated issues in multi-arm, space robots. Ullman developed a cartesian space, computed torque controller for a two-arm, free-flying space robot [1]. Koningstein [2], Papadopoulos and Dubowsky [3], and Yoshida [4] have studied the dynamics of similar one and two-arm space robots.

Most recently, Russakow, Khatib, and Rock [5] have proposed an Extended Operational Space Formulation for multi-limb, redundant robots. Under this formulation, the control of a redundant, serial-to-parallel (branching) multi-end-effector system is decomposed into those sets of torques that dynamically affect the end-effectors and those that do not. The advantages of this dynamic decomposition are: 1) it permits exploitation of the macro-mini nature of a branching system to enable high performance control at the end-effectors despite slower dynamics in the common lower links; and 2) it enables control of the internal (redundant) degrees-of-freedom of the system that will not dynamically couple into the performance of the end-effectors.

These two benefits of dynamic decomposition are particularly important to free-flying, multi-arm robotic systems in space. Such systems are highly coupled and dynamic, and are typically characterized by redundant macro-mini structures in

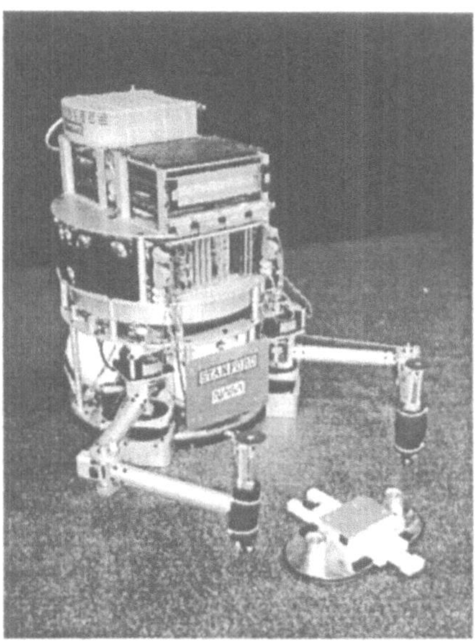

Figure 1. **ARL Free-Flying Space Robot**

which the macro (base) exhibits sluggish, inaccurate dynamic performance. Yet for space systems, high-bandwidth, accurate performance of the end-effector(s) in inertial space relative to a work site, not relative to the moving manipulator base, is critical. Moreover it is essential that the redundant degrees of freedom of the robot be able to reposition without introducing undesired accelerations to the dynamic performance at the end-effectors during manipulative tasks.

The Extended Operational Space Formulation has been experimentally validated on a two-arm, free-flying space robot at the Aerospace Robotics Laboratory (ARL) at Stanford University. The laboratory has three free-flying space robot prototypes, such as that shown in Figure 1. The robots float using a frictionless air bearing over a granite surface plate. Each robot features thrusters and a momentum wheel for locomotion, two DC motor-driven manipulators with pnuematic grippers, force sensing, real-time vision, on-board computation and power, and wireless communications.

Section Two of this paper summarizes the basics of the formulation for serial-to-parallel (branching) manipulators. **Section Three** documents the hardware characteristics for the ARL free-flying space robot and highlights its macro-mini nature. Lastly, **Section Four** experimentally demonstrates dynamic decomposition and internal motion control for the space robot while grasping a free-floating object.

2. Extended Operational Space

The Extended Operational Space Formulation is developed fully in [5] for both 1) open-chain serial-to-parallel systems, in which each end-effector is independent; and 2) open/closed-chain serial-to-parallel systems, in which some end-effectors grasp common objects.

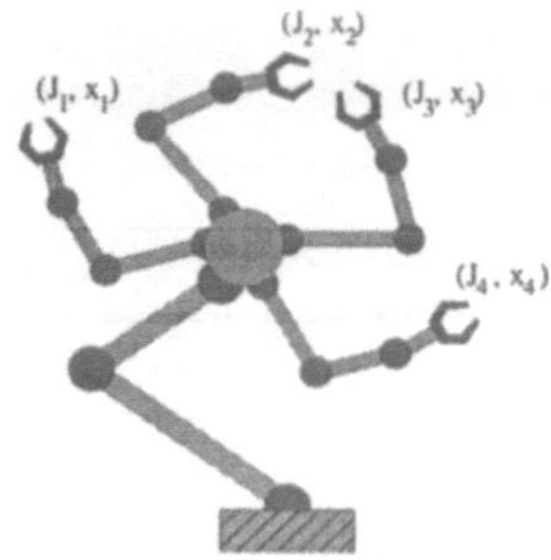

Figure 2. **A Serial-to-Parallel Manipulator**

2.1. Open-Chain Systems

For open-chain systems the operational space is the superset of the domains of the operational space of each manipulator. The position and orientation of the system end-effectors are defined by $\mathbf{x} = [\mathbf{x}_1^T \mathbf{x}_2^T \ldots]^T$, where $\mathbf{x}_i$ is the position and orientation of the ith end-effector. $\mathbf{q}$ is the vector of generalized joint coordinates for the entire branching manipulator system.

The operational space equations of motion of the system are:

$$\Lambda(\mathbf{q})\ddot{\mathbf{x}} + \mu(\mathbf{q}, \dot{\mathbf{q}}) + \mathbf{p}(\mathbf{q}) = \mathbf{F}_{op} \tag{1}$$

$\Lambda(\mathbf{q})$ is the kinetic energy matrix of the system with respect to the operational point, $\mathbf{x}$. $\mu(\mathbf{q}, \dot{\mathbf{q}})$ represents the Coriolis and centrifugal forces acting at the same operational point, and $p(\mathbf{q})$ depicts the gravitational forces also expressed at that point. $\mathbf{F}_{op}$ is the generalized force vector expressed in the operational space.

$\Lambda(\mathbf{q})$, $\mu(\mathbf{q}, \dot{\mathbf{q}})$, and $p(\mathbf{q})$ are defined by:

$$\Lambda(\mathbf{q}) = (J(\mathbf{q})A^{-1}(\mathbf{q})J^T(\mathbf{q}))^{-1}, \tag{2}$$

$$\mu(\mathbf{q}, \dot{\mathbf{q}}) = \overline{J}^T(\mathbf{q})\mathbf{b}(\mathbf{q}) - \Lambda(\mathbf{q})\dot{J}(\mathbf{q})\dot{\mathbf{q}}, \tag{3}$$

$$\mathbf{p}(\mathbf{q}) = \overline{J}^T(\mathbf{q})\mathbf{g}(\mathbf{q}), \tag{4}$$

$$\text{and } \overline{J}(\mathbf{q}) = A^{-1}(\mathbf{q})J^T(\mathbf{q})\Lambda(\mathbf{q}), \tag{5}$$

where $A(\mathbf{q})$ is the joint-space inertia matrix and $\mathbf{b}(\mathbf{q})$ and $\mathbf{g}(\mathbf{q})$ are the Coriolis/centrifugal and gravity force vectors in joint space. $J(\mathbf{q}) = [J_1^T(\mathbf{q})J_2^T(\mathbf{q})\ldots]^T$ is the Jacobian relating joint space velocities to the operational space velocity, where $J_i(\mathbf{q})$ is the basic Jacobian corresponding to the ith end-effector. And $\overline{J}(\mathbf{q})$ is the *dynamically consistent* generalized inverse of $J(\mathbf{q})$.

The dynamically consistent relationship between joint torques and operational forces for redundant manipulators is:

$$\mathbf{\Gamma} = J^T(\mathbf{q})\mathbf{F}_{op} + [I - J^T(\mathbf{q})\overline{J}^T(\mathbf{q})]\mathbf{\Gamma}_o \tag{6}$$

This relationship, which is the basis for the control of the redundant manipulator, provides a decomposition of joint torques into two dynamically decoupled control vectors: joint torques corresponding to forces acting at the end-effector(s), $J^T(\mathbf{q})\mathbf{F}_{op}$, and joint torques that only affect internal motions, $[I - J^T(\mathbf{q})\overline{J}^T(\mathbf{q})]\mathbf{\Gamma}_o$. Under this decomposition, internal motions of the system are guaranteed not to introduce accelerations at *any* of the end-effectors.

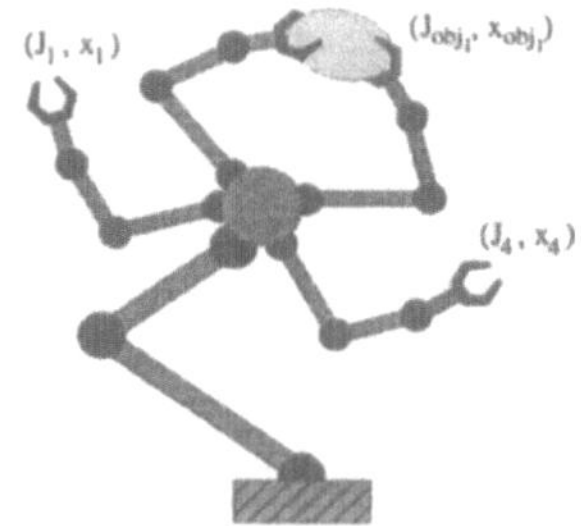

Figure 3. **A Serial-to-Parallel Manipulator Grasping an Object**

2.2. Open/Closed-Chain Systems

When some end-effectors grasp common objects, kinematic chains are closed and the dynamics of the system become constrained. The dimension of the operational space decreases, and it becomes necessary to consider a reduced, or constrained, set of dynamics:

$$A_c(\mathbf{q})\ddot{\mathbf{q}}_c + \mathbf{b}_c(\mathbf{q},\dot{\mathbf{q}}) + \mathbf{g}_c(\mathbf{q}) = \Gamma_c \tag{7}$$

where

$$A_c(\mathbf{q}) = C^T(\mathbf{q})A(\mathbf{q})C(\mathbf{q}) \tag{8}$$

$$\mathbf{b}_c(\mathbf{q},\dot{\mathbf{q}}) = C^T(\mathbf{q})\mathbf{b}(\mathbf{q},\dot{\mathbf{q}}) + C^T(\mathbf{q})A(\mathbf{q})\dot{C}(\mathbf{q})\dot{\mathbf{q}}_c \tag{9}$$

$$\mathbf{g}_c(\mathbf{q}) = C^T(\mathbf{q})\mathbf{g}(\mathbf{q}) \tag{10}$$

$\mathbf{q}_c$ is a minimal set of generalized coordinates that describes the configuration of the full system. $C(\mathbf{q})$, the constraint matrix, is defined by $\dot{\mathbf{q}} = C(\mathbf{q})\dot{\mathbf{q}}_c$. $\mathbf{\Gamma}_c$ is the reduced set of generalized forces that corresponds to the choice of $\mathbf{q}_c$.

Again, the operational space equations of motion for the system are the same as in Equation (1). Only now, the structure of $\mathbf{x}$ and $\mathbf{F}_{op}$ are determined by which end-effectors are grasping common objects and which are not. For the example shown in Figure 3, where the second and third end-effectors are grasping a common object, $\mathbf{x}$ and $\mathbf{F}_{op}$ will assume the form: $\mathbf{x} = [\mathbf{x}_1^T \mathbf{x}_{obj_1}^T \mathbf{x}_4^T]^T$ and $\mathbf{F}_{op} = [\mathbf{F}_{op_1}^T \mathbf{F}_{obj_1}^T \mathbf{F}_{op_4}^T]^T$.

$\mathbf{x}_i$ denotes the operational space coordinates of independent end-effectors, as before, and $\mathbf{x}_{obj_k}$ denotes operational space coordinates of objects grasped by two or more end-effectors. The notation convention for $\mathbf{F}_{op}$ is identical.

The Jacobian for the constrained system, $J_c(\mathbf{q})$, will be the Jacobian relating the choice of $\dot{\mathbf{q}}_c$ to $\mathbf{x}$: $J_c(\mathbf{q}) = [J_1(\mathbf{q})^T J_{obj_1}(\mathbf{q})^T J_4(\mathbf{q})]^T$.

$\Lambda(\mathbf{q})$, $\mu(\mathbf{q},\dot{\mathbf{q}})$, $p(\mathbf{q})$, and $\overline{J}_c(\mathbf{q})$ are now defined by:

$$\Lambda(\mathbf{q}) = (J_c(\mathbf{q})A_c^{-1}(\mathbf{q})J_c^T(\mathbf{q}))^{-1}, \tag{11}$$

$$\mu(\mathbf{q},\dot{\mathbf{q}}) = \overline{J}_c^T(\mathbf{q})\mathbf{b}_c(\mathbf{q}) - \Lambda(\mathbf{q})\dot{J}_c(\mathbf{q})\dot{\mathbf{q}}_c, \tag{12}$$

$$\mathbf{p}(\mathbf{q}) = \overline{J}_c^T(\mathbf{q})\mathbf{g}_c(\mathbf{q}), \tag{13}$$

$$\text{and } \overline{J}_c(\mathbf{q}) = A_c^{-1}(\mathbf{q})J_c^T(\mathbf{q})\Lambda(\mathbf{q}). \tag{14}$$

The structure of these equations is identical to Equations (2)-(5). The operational space dynamics are once again given by Equation (1).

The control of the robot system in this case will be similar to that of Equation (6) developed for the open-chain case. In this case, however, the manipulator system will have some closed chains which will have redundant actuation. The redundancy in the actuation may be used to control the internal forces of the grasped object or to minimize a performance criterion, e.g. minimizing actuator effort [6]. Williams and Khatib [7] have proposed a physically-based model, the Virtual Linkage Model, for the characterization and control of object internal forces. Using this Virtual Linkage Model, the control equation for the system is:

$$\mathbf{\Gamma} = J^T(\mathbf{q})G^{-1}\begin{bmatrix}\mathbf{F}_{op}\\ \mathbf{F}_{int}\end{bmatrix} + [I - J^T(\mathbf{q})\overline{J}^T(\mathbf{q})]\mathbf{\Gamma}_o, \tag{15}$$

$J(\mathbf{q})$ is the Jacobian for the full system described in Equation (11). $\mathbf{F}_{int}$ is a vector associated with the control of internal forces on the grasped objects. G is the grasp matrix which relates end-effector forces to resultant forces and internal forces, $\mathbf{F}_{op}$ and $\mathbf{F}_{int}$, exerted on the grasped objects.

Like Equation (6), the dynamic decomposition in Equation (15) can be used to decouple the dynamics of the end-effectors/grasped objects from the dynamics of the redundant, internal motions of the system.

Control of the motion redundancy of the system may then be achieved via the second term in Eqiation (15) without introducing undesired accelerations into the operational space.

3. Experimental Hardware

The control relations (6) and (15) have been implemented on the ARL free-flying space robots to evaluate the Extended Operational Space Formulation. The effectiveness of the implementation of dynamic decomposition and internal motion control have been experimentally demonstrated. These results will be presented in the following section, **Experimental Results**. In this section, we will present the characteristics of the space robot prototypes used for the experiments and discuss their capabilities and bounds of performance.

3.1. System Characteristics

The inertial and geometric parameters of the ARL space robots and two free-floating objects are shown in Table 1. The actuation characteristics of the robots are similarly depicted in Table 2.

component	mass	inertia	length
base	68kg	2kgm^2	0.5m (dia.)
upper arms	1.9kg	0.032kgm^2	0.3048m
lower arms	0.8kg	0.029kgm^2	0.2967m
object 1	0.98kg	0.011kgm^2	0.214m (dia.)
object 2	1.68kg	0.0217kgm^2	0.222m (dia.)

Table 1. **Inertial and Geometric Characteristics**

A common feature of free-flying space robots is that their bases are actuated by discrete thrusters. That is, the base may not exert arbitrary thrust values, but rather must select one of several discrete thruster combinations. The use of discrete actuation therefore introduces both actuation error and deadband into the system.

As the bandwidth of these thrusters is quite high, however, the base actuation can approximate continuous value performance within some error.

In addition, the ARL space robots do not have force/torque feedback on each joint. The momentum wheel and arm actuators are therefore subject to errors of a few percent.

joint	actuation method	actuation range
1,2	discrete thrusters	-2,-1,0,1,2 (N)
3	momentum wheel	-2 to 2 (Nm)
4,7	DC Motor	-0.8 to 0.8 (Nm)
5,8	DC Motor	-0.475 to 0.475 (Nm)
6,9	None	0 (Nm)

Table 2. **Actuator Characteristics**

The magnitude of these actuation errors on end-effector performance depends directly upon the inertial and geometric properties of the robot/object and the manipulator configuration, and can be computed using the dynamically consistent generalized inverse of the Jacobian. For the closed-chain case:

$$\begin{bmatrix} \mathbf{F}_{op} \\ \mathbf{F}_{int} \end{bmatrix}_{\text{max\_error}} = G\bar{J}^T(\mathbf{q})\mathbf{\Gamma}_{\text{max\_error}}, \tag{16}$$

$$\epsilon_x = K_p^{-1}\Lambda^{-1}(\mathbf{q})\mathbf{F}_{op\text{max\_error}} \tag{17}$$

$\mathbf{\Gamma}_{\text{max\_error}}$ is the vector of generalized joint torques which represents the worst possible actuation error. $\mathbf{F}_{op\text{max\_error}}$ represents the worst possible force error that can occur at a commonly grasped object in free space or in force control at a configuration, $\mathbf{q}$, due to actuation error. K_p^{-1} is the diagonal matrix of position gains used in the object position control law. And ϵ_x represents the maximum position error that can occur at the object grasped due to actuator error.

For a system with the inertial characteristics of Table 1, maximum actuation errors of 0.5N on the first two joints and 10 percent on joint three have minimal effect on object performance: 0.012N force error and 0.09mm position error at the center of the robot workspace. This is in large part due to the large inertia of the base. In contrast errors of 2 percent on the torques executed by the arm actuators theoretically account for as much as 0.05N error in force and 1mm in position at the workspace center. For a space robot of this nature, these errors are very small, and the performance of the robot is not significantly degraded.

4. Experimental Results

4.1. Macro-Mini Characteristics and Bandwidth

Figures 4 and 5 highlight the macro-mini nature of the free-flying space robot. Figure 4a plots the maximum effective bandwidth of the robot base (macro) in position control. Maximum effective bandwidth is defined here as the approximate frequency at which the phase lag of the system reaches 10 percent of the period of the commanded sinusoid input.

Figure 4b plots the frequency at which the arms of the robot (mini) can move Object 1 in position while the base of the robot is firmly fixed. Lastly, Figure 4c plots the maximum effective bandwidth at Object 1 for the full macro-mini system.

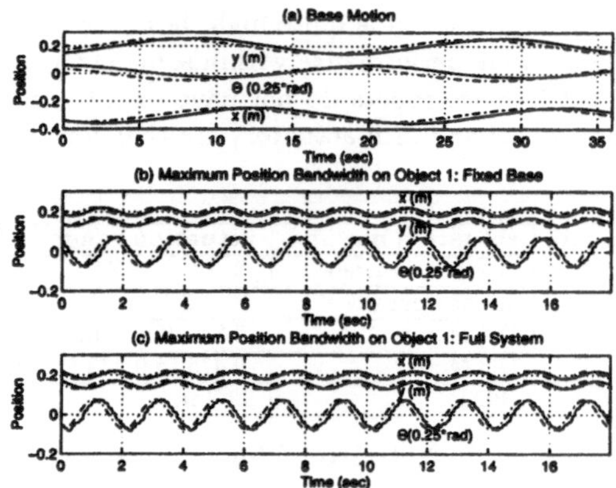

Figure 4. **Position Tracking Bandwidth of Arms, Base, and Full System**

It is identical to the bandwidth of the arms alone within the measurement accuracy of the system.

This result confirms the reduced effective inertia property [8] – namely, that the inertial properties at the object are dominated by those of the arms (mini). It also suggests that a system controlled under the Extended Operational Space Formulation, which treats the free-flying space robot as a redundant system, can properly achieve accurate control at frequencies near the bandwidth of the mini despite the slower dynamics of the macro.

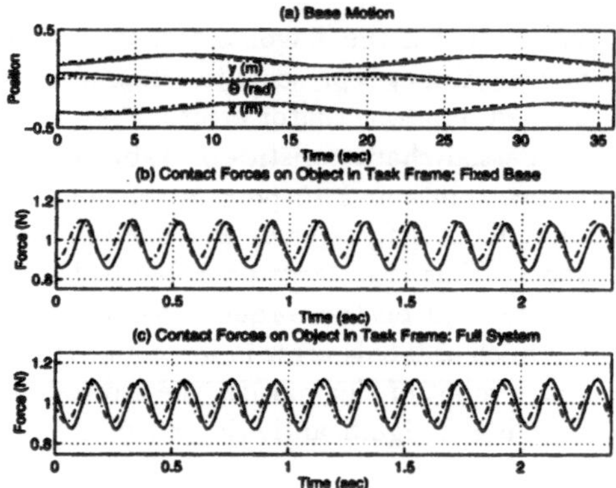

Figure 5. **Force/Position Tracking Bandwidth of Arms, Base, and Full System**

Figure 5 demonstrates a result similar to Figure 4, only for force control on Object 2 in contact with a wall. Notice that the bandwidth of the system in contact is an additional order of magnitude higher than the bandwidth when in free motion control.

The approximate bandwidth limitations of the space robot base, arms, and complete system found in the above experiments are summarized in Table 3.

Object Control	Bandwidth		
	Base	*Arms*	*Full System*
Position Control	0.05Hz	0.5Hz	0.5Hz
Force Control	–	5Hz	5Hz

Table 3. **ARL Space Robot Effective Bandwidth**

4.2. Dynamic Consistency

The nature of the control decomposition of Equation (15), which provides decoupling between the control of the end-effectors and the control of the internal motions of the system, is made clear in the position step response plotted in Figure 6.

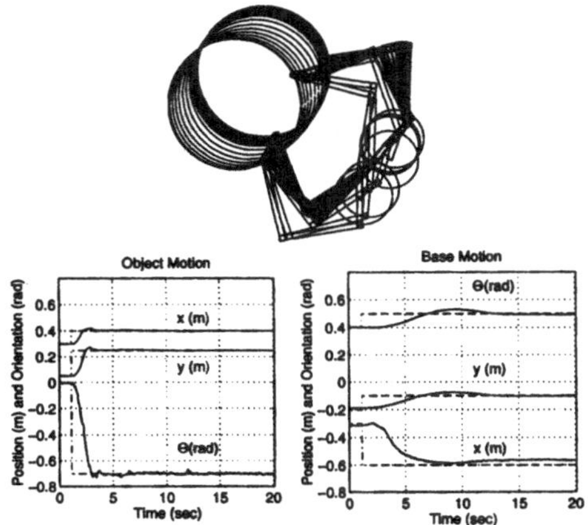

Figure 6. **Position Step Response Under Extended Augmented Object Control**

The entire system acts in concert to move the object quickly to a new desired location and regulate it there upon arrival. At the same time the system moves its redundant, internal degrees of freedom to a new desired location using only torques that will not cause accelerations at the object. This internal motion is dominated by the slower dynamics of the base of the robot. Notice that the system is able to regulate the object position well while the base is still in the midst of a large transient response.

Figure 7 further demonstrates the ability of the formulation to dynamically decouple the control of the system. Here, the free-flying robot executes a force step response of Object 2 against a stationary wall while the internal degrees of freedom of the system settle to a new desired configuration. The non-zero force signal prior to the step is due to inertial forces in free-space while the object was slewed toward the wall. Control switches from position to force when a contact greater than 0.75N is sensed.

4.3. Internal Motion Control

A final interesting property of the dynamic decoupling of a system into end-effector and internal motions is that sophisticated internal motion control schemes become possible.

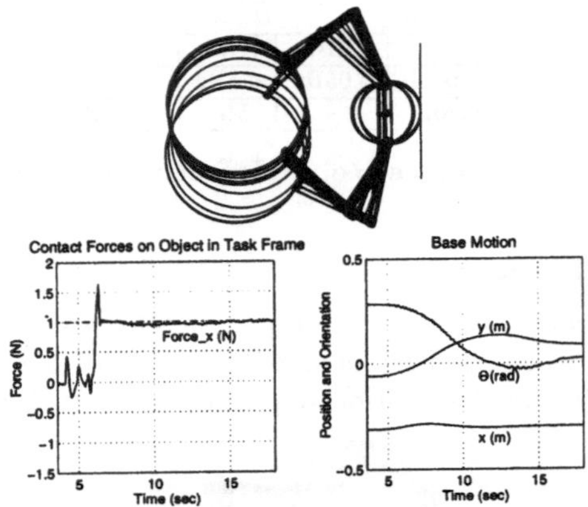

Figure 7. **Force Step Response Under Extended Augmented Object Control**

Figure 8 demonstrates an example of a pure internal motion. The object grasped by the free-flying robot is ordered to remain fixed in space while joints 1, 2, and 3 (the base degrees of freedom) are commanded to execute a large step response within the dynamically consistent nullspace of the system. The amount of motion by the object is a measure of how well the system has decoupled internal motion from end-effector. These results match expectations well, given the sensing and actuator limitations of the system.

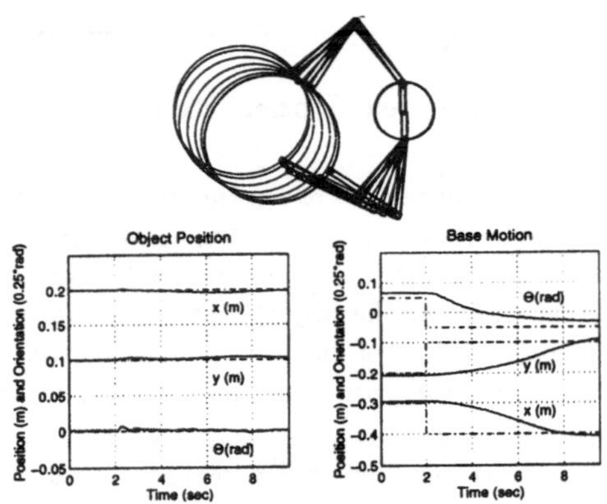

Figure 8. **Internal Motion of Joints 123 under Extended Augmented Object Control**

Figure 9 shows an internal motion in which joints 1, 2, and 4 are commanded to perform a step response. The only change required to perform this motion is that elements 1, 2, and 4 of Γ_o take the form, $-k_{p_i}(q_{i_{des}} - q_i) - k_{d_i}\dot{q}_i$, rather than joints 1, 2, and 3. All other elements of Γ_o take the form $-k_{d_i}\dot{q}_i$ to add damping in the dynamically consistent null space.

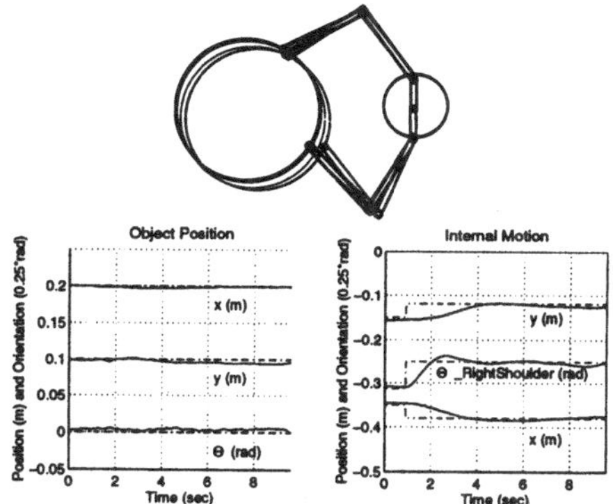

Figure 9. **Internal Motion of Joints 124 under Extended Augmented Object Control**

5. Conclusions

This paper has investigated the performance of a free-flying space robot controlled under the Extended Operational Space Framework. After characterizing the physical properties and limitations of the space robot, and illustrating the macro-mini nature of such a system, the concepts of *Dynamically Consistent Force/Torque Decomposition* and *Internal Motion Control* of a multi-arm, branching redundant manipulator were demonstrated experimentally. Both position and force control, as well as two internal motion control schemes, were presented under this framework.

References

[1] M. A. Ullman. *Experiments in Autonomous Navigation and Control of Multi-Manipulator, Free-Flying Space Robots.* PhD thesis, Stanford University, Stanford, CA 94305, March 1993.

[2] R. Koningstein, M. Ullman, and R. Cannon. Computed torque control of a free-flying cooperating-arm robot. In *Proc. of the NASA Conf on Space Telerobotics*, pages 235–243, Pasadena CA, January 1989.

[3] E. Papadopoulos and S. Dubowsky. Coordinated manipulator/spacecraft motion control for space robotic systems. In *Proc. of the IEEE Intl Conf on Rob and Autom*, pages 1696–1701, Sacramento CA, April 1991.

[4] K. Yoshida R. Kurazume and Y. Umetani. Dual arm coordination in space free-flying robot. In *Proc. of the IEEE Intl Conf on Rob and Autom*, pages 2516–2521, Sacramento CA, April 1991.

[5] J. Russakow O. Khatib and S. Rock. The extended operational space formulation for serial-to-parallel (branching) manipulators. In *Proc. of the IEEE Intl Conf on Rob and Autom*, Nagoya JP, May 1995.

[6] O. Khatib. Object manipulation in a multi-effector robot system. In B. Bolles and B. Roth, editors, *The Intl Symp of Rob Res*, pages 137–144, Santa Cruz CA, August 1987. MIT Press.

[7] D. Williams and O. Khatib. The virtual linkage: A model for internal forces in multi-grasp manipulation. In *Proc. of the IEEE Intl Conf on Rob and Autom*, pages 1025–1031, Atlanta GA, 1993.

[8] O. Khatib. Reduced effective inertia in macro-/mini-manipulator systems. In H. Miura and S. Arimoto, editors, *Robotics Research 5*, pages 279–284, Cambridge, 1990. MIT Press.

Experimental Research of A Nonholonomic Manipulator

Woojin Chung, Yoshihiko Nakamura
Department of Mechano-informatics, University of Tokyo
7-3-1 Hongo, Bunkyo-ku, Tokyo 113, Japan
chung@mech.t.u-tokyo.ac.jp

Ole Jakob Sørdalen
ABB Corporate Research, N-1361 Billingstad, Norway
ojs@nocrc.abb.no

Abstract

Nonholonomic systems are mechanical systems with nonintegrable constraints. It is shown how nonlinear control theory can be applied to design nonholonomic mechanical systems. Exploiting the unique features of nonholonomic systems, we designed a nonholonomic manipulator which is a controllable 2-input, n-joint manipulator. To create nonholonomic constraints, a special type of velocity transmission, called a nonholonomic gear, was used. In this paper, we show the theoretical design of a nonholonomic manipulator and experimental results with the fabricated prototype.

1. Introduction

One of the special features of nonholonomic systems is that the dimension of controllable configuration space is higher than that of the input space. It is due to the fact that the dimension of the system under the nonintegrable constraints cannot be necessarily reduced by the constraints. One typical example is a car with n trailer system. Although it has only two inputs, the system is controllable in the $n+3$ dimensional configuration space [1]. Exploiting this property, we designed a nonholonomic manipulator [2, 3, 4]. We can control n joint manipulator with only two actuators by the procedure presented later. A special type of velocity transmission, called a nonholonomic gear, is used as a key component. In this paper, after the introduction of the nonholonomic manipulator in section 2, conversion of the kinematic model to the chained form is presented in section 3. In section 4, fabricated prototype of a nonholonomic manipulator is presented. Control schemes and experimental results with the prototype is shown in section 5.

2. Nonholonomic Manipulator

2.1. Nonholonomic Gear

In order to transmit velocities we design a gear as illustrated in *Figure 1*. The basic components of this gear are a ball and wheels. Suppose that IW is an input wheel. If IW is forced to rotate, the ball in contact with wheels starts to rotate. The

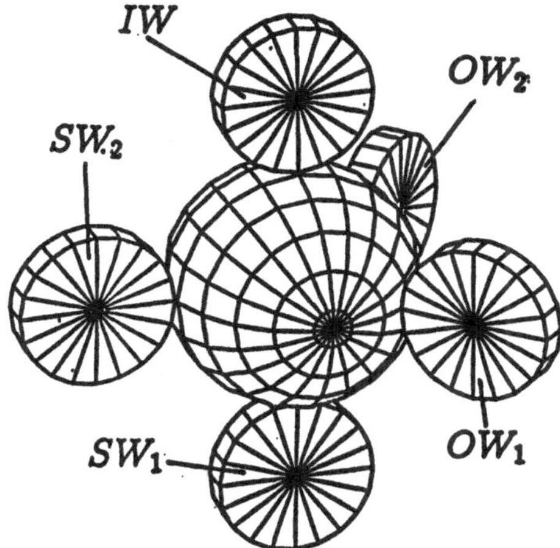

Figure 1. Nonholonomic gear at joint i with supporting wheels. The rotation axes of IW,SW_1 are fixed to link $i-1$. The rotation axes of OW_1,OW_2,SW_2 are fixed to link i.

velocity constraints of the ball are only due to point contacts with the wheels. The rotation of a ball makes the other wheels (OW_1, OW_2, SW_1, SW_2) rotate due to the rolling contact between the solid bodies. Hence, velocity is transmitted from IW to the other wheels. OW and SW indicate Output Wheel and Supporting Wheel respectively. For details of the principle of the velocity transmission, see [2, 3]. A nonholonomic gear is a single-input, multiple-output CVT (Continuously Variable Transmission).

2.2. Kinematic Model

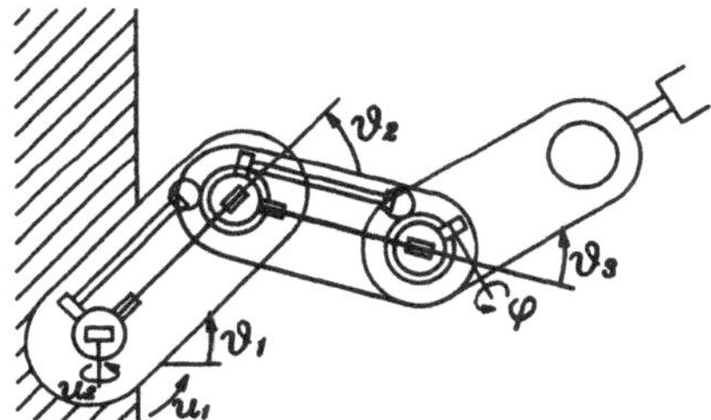

Figure 2. Configuration of a three-joint nonholonomic manipulator.

In *Figure 2*, an example of a planar nonholonomic manipulator, consists of three joints, is presented. Two actuators are both located at the base. One input u_1 drives the angular velocity of the first joint, and the other input u_2 drives the Input Wheel of the nonholonomic gear at the first joint. With the nonholonomic gear at each joint, angular velocities are transmitted. A nonholonomic gear is used here as a single-input, two-output velocity transmission. For details, see [2, 3].

The model has three joints and four state variables. φ is a rotation angle of the output wheel of the joint 3. By choosing φ as a state variable, kinematic model of a nonholonomic manipulator can be converted to the chained form, which is discussed later. Since φ is not a joint angle, it can be treated as a internal parameter for control.

A kinematic model of a n joint nonholonomic manipulator is as follows:

$$\dot{\theta}_1 = u_1 \quad (1)$$

$$\dot{\theta}_i = k_i \sin\theta_{i-1} \prod_{j=1}^{i-2} \cos\theta_j u_2, \qquad i \in \{2,\dots,n\} \quad (2)$$

$$\dot{\varphi} = k_\varphi \prod_{j=1}^{n-1} \cos\theta_j u_2 \quad (3)$$

where k_i are gear ratios given by the velocity transmission.

3. Conversion to Chained Form

To control the nonholonomic manipulator, the kinematic model (1)–(3) is converted to the chained form implying that existing control laws for chained form can be applied. The chained form considered here is given as follows [5]:

$$\dot{z}_1 = v_1 \quad (4)$$

$$\dot{z}_2 = v_2 \quad (5)$$

$$\dot{z}_i = z_{i-1}v_1, \qquad i \in \{3,\dots,m\} \quad (6)$$

where m is the dimension of the system.

The kinematic model (1)–(3) can be rewritten locally, for $\theta_j \in (-\frac{\pi}{2}, \frac{\pi}{2})$, where $j \in \{1,\cdots,m\}$, as follows:

$$\dot{\theta}_1 = u_1 \quad (7)$$

$$\dot{\varphi} = \mu_2 \quad (8)$$

$$\dot{\theta}_i = k_i \tan\theta_{i-1} \frac{1}{k_n \prod_{j=i}^{n-1} \cos\theta_j} \mu_2, \qquad i \in \{2,\dots,n\} \quad (9)$$

$$\mu_2 \triangleq k_n \prod_{j=1}^{n-1} \cos\theta_j u_2 \quad (10)$$

Modified kinematic model (7)–(9) is a triangular structure, presented in [2]. Systems with a triangular structure can be converted to the chained form by nonlinear coordinate transformation $z = h(x)$ and input feedback transformation $v = g(x)u$ as follows:

$$z_m = h_m(x_m) \quad (11)$$

$$z_i = h_i(\underline{x}_i) \triangleq \frac{\partial h_{i+1}(\underline{x}_{i+1})}{\partial \underline{x}_{i+1}} \underline{f}_{i+1}(\underline{x}_i) \quad (12)$$

$$z_1 = h_1(x_1) \triangleq x_1 \quad i \in \{2,\dots,m-1\} \quad (13)$$

$$v_1 = u_2 \quad (14)$$

$$v_2 = \frac{\partial h_2(\underline{x}_2)}{\partial \underline{x}_3} \underline{f}_3(\underline{x}_2) u_2 + \frac{\partial h_2(\underline{x}_2)}{\partial x_2} u_1 \quad (15)$$

In [6, 2], proof by induction is presented.

Notice that since the conversion to the chained form was carried out locally, we cannot control the converted system in the entire configuration space. However, it was shown that a nonholonomic manipulator is globally controllable in the m dimensional configuration space [2]. Controllability of a nonholonomic manipulator can be proved by calculating the dimension of the accessibility distribution.

Figure 3. Prototype of A Nonholonomic Manipulator

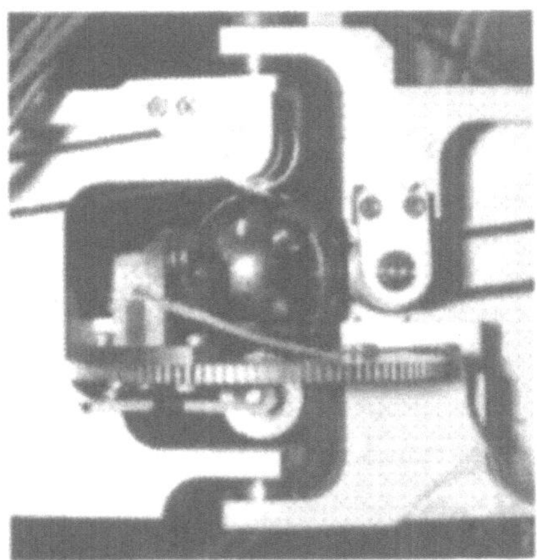

Figure 4. Prototype of A Nonholonomic Gear

4. Prototype of A Nonholonomic Manipulator

Since the nonholonomic manipulator was designed from the viewpoint of kinematic constraints, mechanical implementation is important in practice. The principle of velocity transmission at the nonholonomic gear is due to the frictional force at the contact zone. Power transmission using the frictional force between the solid bodies is called "Friction Drive" and it is one of important current issues in tribology. Various issues of the mechanical design and analysis were discussed in [3].

We designed and constructed a four-joint prototype nonholonomic manipulator,

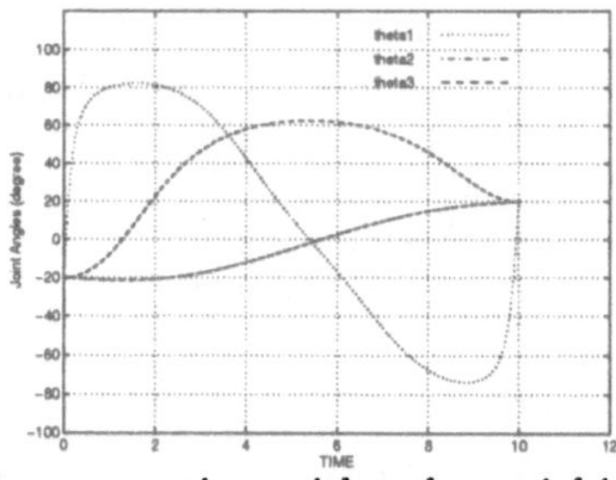

Figure 5. Joint angles versus time with polynomial inputs: Simulation 1.

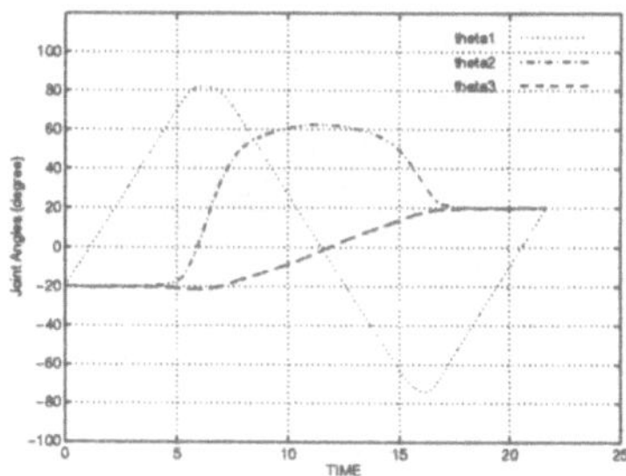

Figure 6. Joint angles versus time with a time scaling : Simulation 2.

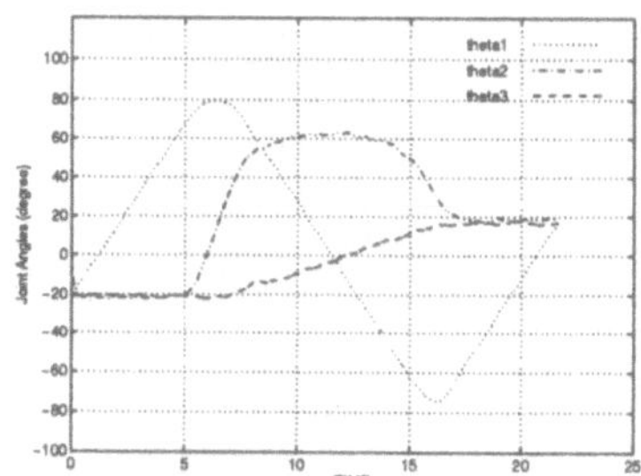

Figure 7. Joint angles versus time with time scaled polynomial inputs.: Experiment A.

as shown in *Figure 3*. *Figure 4* shows a prototype nonholonomic gear.

5. Experiments

5.1. Open Loop Control

We computed the joint and motor trajectories by applying the open loop control law using the time polynomial inputs to the chained form [7, 4]. In *Figure 5*, joint angles versus time are plotted (*Simulation* 1). The total time was set to 10 seconds. The gear ratios were selected as $[k_1, k_2, k_3] = [1, 1/10, 1/10]$. Joint angles of initial configuration $\theta(0)$ and desired configuration θ_d are $\theta(0) = [-20, -20, -20]^T(deg)$, $\theta_d = [20, 20, 20]^T(deg)$. From *Figure 5*, We see that the joint angles reach their desired values.

However, in practice, the motion of a nonholonomic manipulator has physical

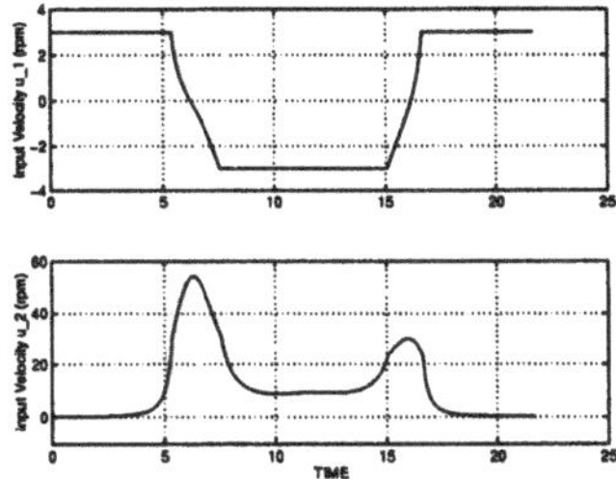

Figure 8. Input angular velocities versus time with a time scaling : Simulation 2.

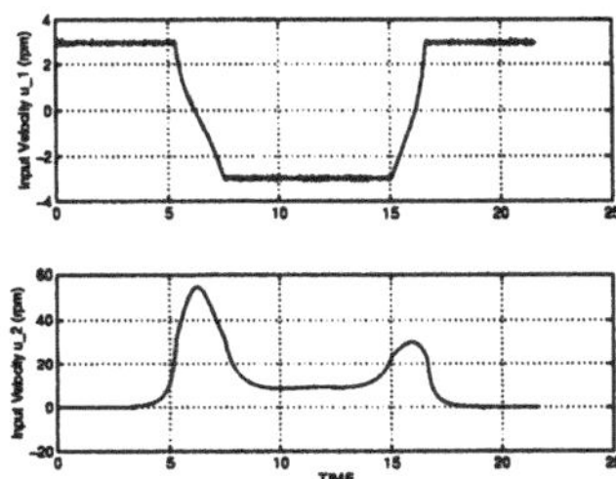

Figure 9. Input angular velocities versus time with time scaled polynomial inputs.: Experiment A.

and mechanical limitations. Therefore, we need to plan realizable trajectories. To this end, we proposed to make time scaling [4]. We exploit the time scaling to bound one of the input angular velocities. We set the desired velocity limit of u_1, and reshaped the trajectories computed by time polynomial inputs. *Figure 6* shows the joint angle trajectories after the time scaling (*Simulation* 2). The absolute value of the desired velocity limit of u_1 was selected as $0.3141(rad/sec)$. The total time need to reach the goal was 22 seconds, it became 12 seconds longer than the case of the original time polynomial inputs. The same motion was experimentally tested by applying velocity control to two motors for their predetermined trajectories (*Experiment A*). *Figure 7* illustrates the results of experiment. In *Figure 8*, input angular velocities of two actuators are presented. u_1 is bounted by $0.314(rad/sec)$ $= 3(rpm)$. *Figure 9* shows the inputs of the experiment. We can see that velocities of the actuators are precisely controlled. We define the index of error of the joint angles as follows:

$$E = \sum_{i=1}^{3} |\theta_i - \theta_{di}| \tag{16}$$

where θ_i is a joint angle after completing control, θ_{di} is a desired joint angle. At the end of experiment A, E was 5° . The backlash at the planar gear and the low stiffness of the long shaft would have caused error.

5.2. Feedback Control With Exponential Convergence

Feedback control would be important to live with various uncertainties. Fortunately, the equation of motion of the nonholonomic manipulator has the corresponding chained form [5]. We can apply any known feedback control laws developed for systems with the chained form [1, 6]. Such control schemes include piecewise analytic

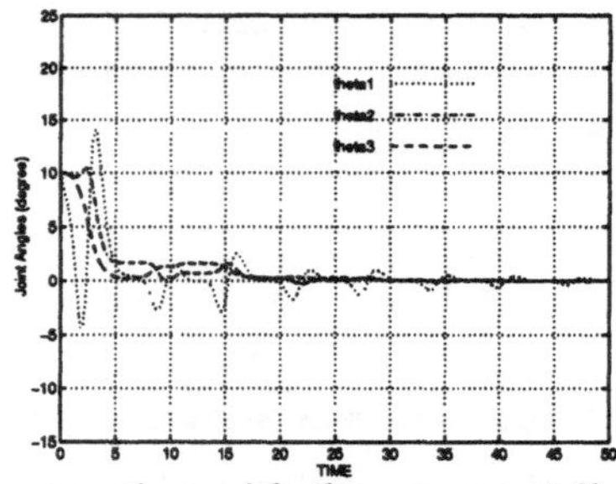

Figure 10. Joint angles versus time with the exponentially convergent feedback law: Simulation 3.

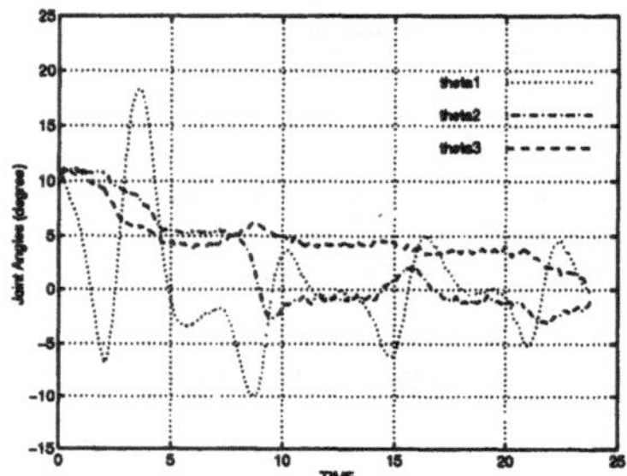

Figure 11. Joint angles versus time with the exponentially convergent feedback law: Experiment B.

control law and time dependent feedback control law.

In *Figure 10*, joint angles versus time are plotted (*Simulation* 3). Joint angles of initial configuration $\theta(0)$ and desired configuration θ_d are $\theta(0) = [10, 10, 10]^T(deg)$ $\theta_d = [0, 0, 0]^T(deg)$. From *Figure 10*, We see that the joint angles are stabilized to the desired values with exponentially convergent speed.

Experiment was carried out with the same condition (*Experiment B*). *Figure 11* shows the result. We set the condition of terminating the control as $E < 1°$ (Eq. (16)). Profiles of the joint angle trajectories are similar to those of the first 24 seconds of *Figure 10.*

One problem of this control law is that the equation of the feedback law becomes highly complicated as the number of the state variables increases and sometimes caused computational divergence with the four-joint model.

5.3. Feedback Control of a Pseudo-Linearized System

If we set a constant a_0 to the input v_1 of the chained form of Eq. (4) -(6), then the system is described as follows:

$$\dot{z}_1(t) = a_0 \tag{17}$$

$$\begin{bmatrix} \dot{z}_2(t) \\ \dot{z}_3(t) \\ \vdots \\ \dot{z}_n(t) \end{bmatrix} = A \begin{bmatrix} z_2(t) \\ z_3(t) \\ \vdots \\ z_n(t) \end{bmatrix} + \begin{bmatrix} 1 \\ 0 \\ \vdots \\ 0 \end{bmatrix} v_2 \tag{18}$$

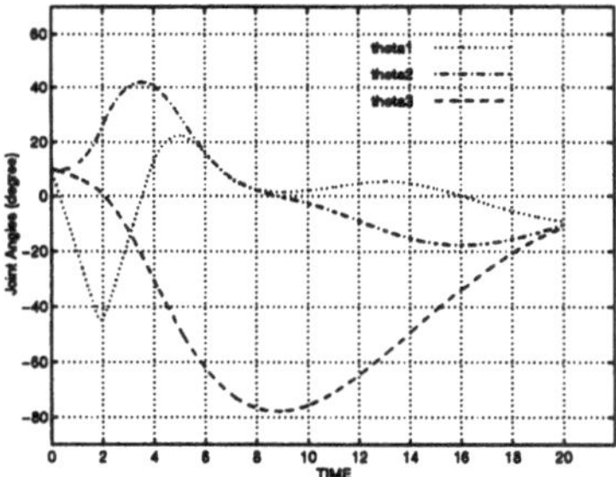

Figure 12. Joint angles versus time with the feedback control of a pseudo-linearized system: Simulation 4.

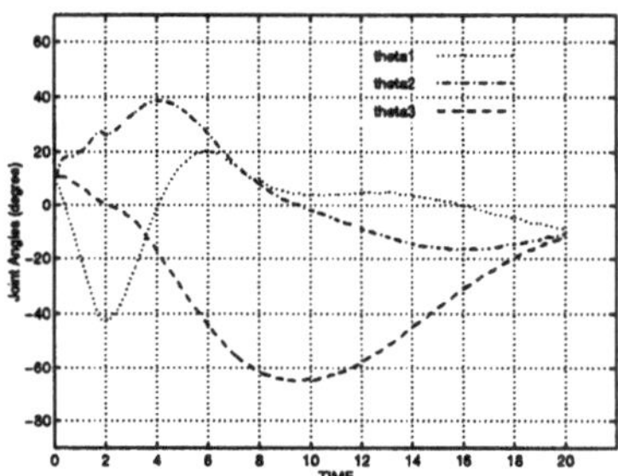

Figure 13. Joint angles versus time with the feedback control of a pseudo-linearized system: Experiment C.

$$A_{i,j} \triangleq \begin{Bmatrix} 0 & , & i-j \neq 1 \\ a_0 & , & i-j=1 \end{Bmatrix}$$

Equation (18) is a state equation of a linear system. We can apply the control scheme based on the linear control theory. z_1 in Eq. (17) is controlled with open loop. Since z_1 corresponds to φ in *Figure 2*, we can set the arbitrary initial value of z_1 as we wish. Therefore, by setting the proper a_0, z_1 goes to 0 with linear convergence. This determines a finite time to control.

When we stabilize the joint angles to the arbitrary values, the goal should be an equilibrium point. Equation (18) has its equilibrium point at $z_i = 0$, where $i \in \{2, 3, \cdots, n\}$, and the arbitrary change of its location is not easy with the standard linear control theory. However, we can use the nonlinear coordinate transformation in [6] to stabilize the chained system to the arbitrary point. Let the desired configuration $z^p = [z_1^p, z_2^p, \cdots, z_m^p]^T$, where $\dot{z}^p$=0. Then the coordinate transformation is given as follows:

$$z_i^r \triangleq z_i^p + \sum_{j=2}^{i-1} z_j^p \frac{1}{(i-j)!}(z_1 - z_1^p)^{i-j} \tag{19}$$

$$\bar{z}_i = z_i - z_i^r \tag{20}$$

where $i \in \{1, \cdots, m\}$. This coordinate transformation is valid for any control strategy to control a system with the chained form towards the origin. The linearized system of Eq. (18) can be easily stabilized to the origin by a state feedback scheme.

In *Figure 12*, joint angles versus time resulted from the feedback control scheme of a pseudo-linearized system, are plotted (*Simulation* 4). Joint angles of initial configuration $\theta(0)$ and desired configuration θ_d are same as those of *Simulation* 3. v_1 was $-0.327(rad/sec)$ and finite time of convergence was 20 seconds. Feedback gains were $[g_1, g_2, g_3]^T = [3.713, -5.787, 3.162]^T$. We see that the joint angles reach the desired values after the finite time. An experiment was carried out with the same condition (*Experiment C*). *Figure 13* shows the result. The index of errors at the 20 seconds was $E = 1^\circ$. Profiles of the joint angle trajectories are almost same as those of *Figure 12.* Since the control scheme introduced here requires only simple computation, it can be used for a nonholonomic manipulator with greater number of joints.

6. Conclusion

By exploiting nonholonomic constraints, it has been shown that a manipulator with n rotational joints can be controlled using only two actuators. The validity of the theoritical design and control schemes were verified with the fabricated prototype.

Application of nonlinear control theory leads to developing a new innovative mechanism that never existed. By paying attention to design the systems with triangular structure, we can make a controllable nonlinear system.

Acknowledgement

This research was supported in part by NTT Co., Ltd., Japan Society of the Promotion of Science, the Center of Maritime Control Systems at NTH/SINTEF and the Grant in Aid of Scientific Research from the Ministry of Culture and Education (General Research (B)04452153 and 07455110).

References

[1] O. J. Sørdalen, *Feedback Control of Nonholonomic Mobile Robots*, PhD thesis, The Norwegian Institute of Technology, February 1993, ITK-rapport 1993:5-W.

[2] O. J. Sørdalen, Y. Nakamura, and W. J. Chung, "Design of a nonholonomic manipulator", *in Proc. 1994 IEEE International Conference on Robotics and Automation*, pp. 8–13, San Diego, California, May 1994.

[3] W. J. Chung, Y. Nakamura, and O. J. Sørdalen, "Prototyping a nonholonomic manipulator", *in Proc. 1995 IEEE International Conference on Robotics and Automation*, pp. 2029–2036, Nagoya, Japan, May 1995.

[4] O. J. Sørdalen, Y. Nakamura, and W. J. Chung, "Control of a nonholonomic manipulator", *in Preprints of the Fourth IFAC Symposium on Robot Control'94*, pp. 323–328, Capri, Italy, September 1994.

[5] R. M. Murray and S. S. Sastry, "Nonholonomic motion planning: Steering using sinusoids", *IEEE Transactions on Automatic Control*, vol. 38, pp. 700–716, 1993.

[6] O. J. Sørdalen and K. Y. Wichlund, "Exponential stabilization of a car with n trailers", *in Proc. 32th Conference on Decision and Control*, pp. 978–983, San Antonio, December 1993.

[7] D. Tilbury, R. Murray, and S. Sastry, "Trajectory generation for the n-trailer problem using goursat normal form", *in Proc. 32th Conference on Decision and Control*, San Antonio, December 1993.

Mobile Manipulation of a Fragile Object

Philippe Fraisse, Pierre Dauchez, François Pierrot, Lise Cellier
LIRMM, UMR 9928 Université Montpellier II / CNRS
161 rue Ada
34392 Montpellier Cedex 5 - FRANCE
fraisse@lirmm.fr

Abstract

We present in this paper some experimental results obtained with a mobile manipulator carrying a fragile object. Different control structures are compared in order to determine the most efficient. One solution is based on "high gain" control (sliding mode). The other solutions are based on impedance control. From acceleration measurements on the object, we present and analyze the effects of bumps induced by the mobile base.

1. Introduction

Autonomous, semi-autonomous or teleoperated mobile manipulators have many potential applications in unstructured or dangerous environments, such as space, subsea [1], mines, nuclear plants [2]. We also believe they could be of great help on earth for manipulating dangerous objects, such as explosives, either in a military or in a civil context: the object shouldn't be shaken, despite the unknown disturbances induced by the motion of the vehicle on a rough terrain. This paper presents some preliminary experiments for which only the manipulator is controlled and tries to reject these disturbances. More precisely, we experimentally compare three control schemes for performing such a task: a robust hybrid position/force control scheme and two impedance control schemes. The experiment presented in this paper consists in carrying a box supposed to contain dangerous matter. When the vehicle is moving, we want the robot to absorb the disturbances and to act as a kind of intelligent shock absorber. In this case, we may not necessarily need a robust controller (i.e. with small position and force errors) for which the time response could be too long and the jerks too large, which is precisely what we would like to avoid. The results presented in this paper confirm this guess. We also compare them with the results given by "classical" impedance controllers. Two accelerometers mounted on the object inform us about the evolution of the forces acting on the object.

2. Control schemes

We want to analyze the ability of the control schemes to absorb the disturbances induced by the mobile base. Several solutions are presented. Firstly, we propose a hybrid robust controller that we have developed, based on sliding mode control [3]. Secondly an impedance control with an external force loop based on Hogan's formulation [4]. Thirdly, an impedance control scheme with an external acceleration loop. The vectors or matrices that we use in this section are expressed in a mobile platform frame.

2.1. Robust controller

The Variable Structure Control law is basically a robust control scheme which allows us to avoid the knowledge of the robot's dynamic parameters. The system is forced to dynamically behave according to an equation defined *a priori*: this equation is represented in the state space by a surface called *sliding surface*. For a unique actuator controlled in position, the usual VSC technique can be summarized by:

$$u = +V_m \; if \; S(x) > 0, \quad u = -V_m \; if \; S(x) < 0 \tag{1}$$

Where u is the control parameter, V_m is the motor maximum input voltage, x is the position of the actuator, S is the sliding surface. The sliding surface is defined by the following equation $S = \dot{e} + \lambda e$, e is a tracking error. We have shown [3] that it was more efficient to choose another equation to compute S: that is what we call VSC-HF (High Frequency). The results we obtained in position control showed a very good robustness with respect to the variations of the robot dynamics. We have extended these results to several degree-of-freedom manipulators and we have also defined a hybrid robust control scheme. The sliding surface that we use in the cartesian space is then defined as follows:

$$\mathbf{S} = (\mathbf{S_{el}}\mathbf{S_p} + (\mathbf{1} - \mathbf{S_{el}})\mathbf{S_f}) \tag{2}$$

where $\mathbf{S_{el}}$ is a 6×6 diagonal selection matrix, $\mathbf{S_p}$ is a 6×1 position sliding vector with $S_{p_i} = \dot{e}_{p_i} + \lambda_{p_i} e_{p_i}$ where e_{p_i} is the ith component of the tracking position error vector $\mathbf{e_p}$. $\mathbf{S_f}$ is a 6×1 force sliding vector with $S_{f_i} = \dot{e}_{f_i} + \lambda_{f_i} e_{fi}$ where $\mathbf{e_f}$ is the tracking force error. The control vector $\mathbf{u}$ is defined as:

$$u_i = sign(S_{q_i} + H_{f_i}(u_i)) \;\; i = 1..6 \tag{3}$$

where $\mathbf{S_q} = \mathbf{J}^+(\mathbf{q})\mathbf{S_p}$, with $\mathbf{J}^+$ the pseudo-inverse of the Jacobian matrix. The nonlinearity (*sign* function) induces oscillations of the state variables (position and velocity of the manipulator). $\mathbf{H_f}$ is a high frequency term which increases the frequency of these oscillations far over the bandwidth of the controlled system. The well-known chattering phenomenon of the VSC thus disappears. We finally obtain a classical hybrid control scheme with a nonlinear gain in joint space which ensures the robustness (i.e. small tracking errors) with respect to external disturbances.

2.2. Impedance controller

The impedance controller we have tested is a classical one where cartesian position and velocity gains are regarded as stiffness and damping matrices respectively [4]–[5]. The behavior of the robot is impedance-like if we implement a dynamic decoupling. The dynamic model of the manipulator is defined by:

$$\mathbf{F_c} - \mathbf{F_{ext}} = \mathbf{\Lambda}(\mathbf{q})\ddot{\mathbf{X}} + \mu(\mathbf{q}, \dot{\mathbf{q}}) + \mathbf{p}(\mathbf{q}) + \mathbf{\Gamma_f}(\mathbf{q}, \dot{\mathbf{q}}) \tag{4}$$

where $\mathbf{F_c}$ is the 6×1 force control vector, $\mathbf{F_{ext}}$ is the external force vector, $\mathbf{\Lambda}(\mathbf{q})$ is the inertia matrix, μ represents the Coriolis and centrifugal effects, $\mathbf{p}$ is the gravitational force vector and $\mathbf{\Gamma_f}$ the friction effects. The control vector $\mathbf{F_c^*}$ which allows dynamic decoupling is such that:

$$\mathbf{F_c^*} = \tilde{\mathbf{\Lambda}}\mathbf{M_d^{-1}}(\mathbf{B_d}\dot{\epsilon}_{\mathbf{x}} + \mathbf{K_d}\epsilon_{\mathbf{x}}) + \tilde{\mu}(\mathbf{q}, \dot{\mathbf{q}}) + \tilde{\mathbf{p}}(\mathbf{q}) + \tilde{\mathbf{\Gamma}}_{\mathbf{f}}(\mathbf{q}, \dot{\mathbf{q}}) \tag{5}$$

where $\epsilon_{\mathbf{x}}$ is the tracking error vector defined as: $\epsilon_{\mathbf{x}}=\mathbf{X_d}-\mathbf{X}$, $\mathbf{X_d}$ is the 6×1 desired position/orientation vector, $\mathbf{X}$ is the actual position/orientation vector, $\mathbf{M_d}$ is the desired mass matrix, $\mathbf{B_d}$ and $\mathbf{K_d}$ are the gain matrices which define the impedance (damping and stiffness matrix respectively), $\tilde{\mathbf{p}}$, $\tilde{\mathbf{\Gamma}}_{\mathbf{f}}$ and $\tilde{\mu}$ are the estimated matrices of the dynamic model. These assumptions are verified if and only if $\dot{\epsilon}_{\mathbf{x}}$ and $\epsilon_{\mathbf{x}}$ tend towards zero. The behavior of the robot is then given by the following equation:

$$-\mathbf{F_{ext}}=\mathbf{M_d}\ddot{\mathbf{X}}-\mathbf{B_d}\dot{\epsilon}_{\mathbf{x}}-\mathbf{K_d}\epsilon_{\mathbf{x}} \quad (6)$$

For our experiments,we have modified this basic control scheme by adding an external force control loop to obtain a position/force control scheme [5]. The force loop control law is an integral gain which modifies the desired trajectory (see equation 7).

$$\mathbf{X_d^*}=\mathbf{X_d}+\mathbf{\Upsilon K_f S_f}\int_0^t(\mathbf{F_d}-\mathbf{F})\mathbf{dt} \quad (7)$$

The force vector $\mathbf{F}$, the selection matrix $\mathbf{S_f}$ and the desired force vector $\mathbf{F_d}$ are defined in the task frame. The matrix $\mathbf{\Upsilon}$ allows the transformation between the task and the reference frame.

2.3. Modified impedance controller

In order to compensate for the accelerations induced by the vehicule on the object, we propose to modify the control equation defined in (5) so as to introduce an external acceleration loop. Thus, an acceleration term is added to the desired position vector X_d to obtain the new desired position vector:

$$\mathbf{X_d^*}=\mathbf{X_d}+\mathbf{S_a}\,\mathbf{K_a}(\ddot{\mathbf{X}}_{\mathbf{d}}-\ddot{\mathbf{X}}) \quad (8)$$

where $\ddot{\mathbf{X}}_{\mathbf{d}}$ = [$\ddot{X}_{d_1}$ $\ddot{X}_{d_2}$ $\ddot{X}_{d_3}$ 0 0 0] is the 6×1 desired acceleration vector, $\ddot{\mathbf{X}}$ = [$\ddot{X}_1$ $\ddot{X}_2$ $\ddot{X}_3$ 0 0 0] is the acceleration vector computed in the robot frame, $\mathbf{S_a}$ is a diagonal selection matrix composed of 1's or 0's depending upon the acceleration components we want to compensate for, $\mathbf{K_a}$ is a 6×6 diagonal gain matrix such that $\mathbf{K_a}=$ *diag*[K_{a_1} K_{a_2} K_{a_3} 0 0 0]. We only modify the desired position components. The desired orientations components are unchanged. The equation of the control vector $\mathbf{F_c^*}$ becomes:

$$\mathbf{F_c^*}=\tilde{\mathbf{\Lambda}}\mathbf{M_d^{-1}}(\mathbf{B_d}\dot{\epsilon}_{\mathbf{x}}^*+\mathbf{K_d}\epsilon_{\mathbf{x}}^*) \quad (9)$$

where $\epsilon_{\mathbf{x}}^*=\mathbf{X_d^*}-\mathbf{X}$ is a new 6×1 tracking error vector. The external loop which allows us to compensate for the acceleration of the object does not actually control the acceleration. This loop generates a motion when a transient acceleration appears. In this case an acceleration produces a motion and an acceleration. However if the object is uniformely accelerating then the action of the loop is cancelled (see equation 8).

3. Experiments

In this section we present our experimental setup. We define afterwards the experimental conditions. Finally we present and analyze the experimental results obtained with the different controllers.

3.1. Experimental setup

Our real setup (Figure 1) is composed of a PUMA 560 mounted on a mobile platform and controlled via the UNIMATE industrial amplifiers by a dSPACE system based

Figure 1. Mobile manipulator

on a TMS320C40 Digital Signal Processor (DSP). Since we use the UNIMATE controller, we have to provide the system with AC voltage via an electric cable. The mobile vehicle is a 6-wheel-drive electric golf car. Each wheel is driven by a DC motor and the 6 motors are run simultaneously by a speed controller. The control system is composed of a DSP C40 board including a dual port memory which allows to communicate with our PC compatible computer (which is used for developing programs and supervising their execution). More precisely, we have developed a program which is the interface between the user and the DSP control system. Via the dual port memory we can give orders such as motion execution, gain modification or reading the system state in real time. An A/D converter, a 16-bit D/A converter and a digital counter constitute the interface between the control system, the robot manipulator and the vehicle. The sampling period is 5ms when we use the impedance controller and 3ms for the VSC-HF (the sampling rate of the impedance control law is greater than the VSC-HF because the cartesian dynamic model is calculated at each period). The object is rigidly attached to the end-effector and the accelerometers

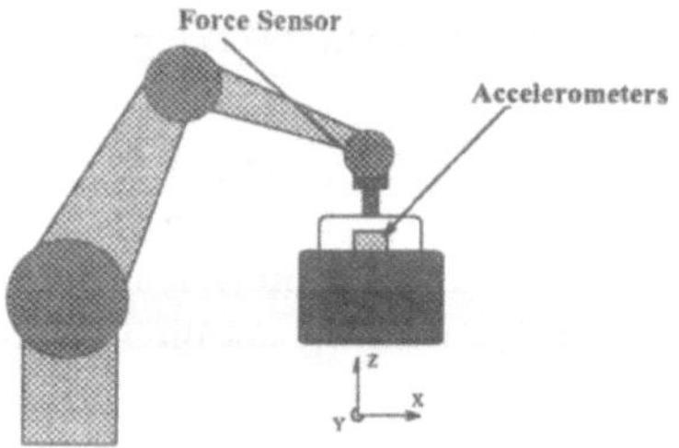

Figure 2. Mobile manipulator carrying an object

are fixed on the object. There are only two accelerometers, one along x axis and the other one along z axis (Figure 2). Actually, only measurements along z are significant and are reported in this paper.

3.2. Experimental conditions

In order to compare the control laws, we have defined a specific experimentation. This experimentation consists of doing a straight motion with the vehicle (Figure 3). The total path is 24 meters long. There is a 2-centimeter thick wooden plank on the

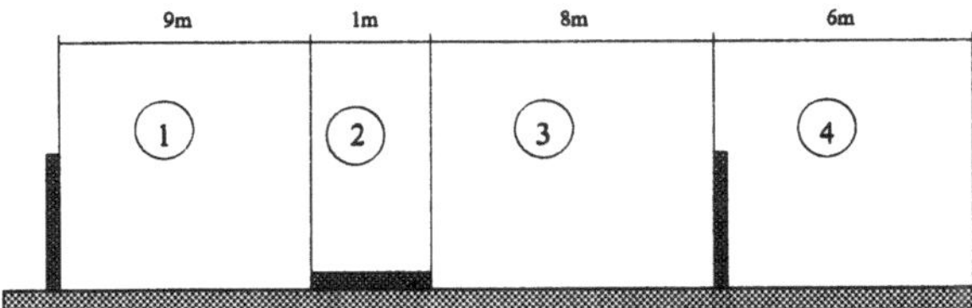

Figure 3. Trajectory of the mobile manipulator

path which produces shocks on the vehicle. An operator drives the vehicle during the experimentation. When the operator starts, the force component along the x motion axis (measured by a wrist HITACHI force sensor) activates the recording of the acceleration, force, position and control vector on the dual port memory. The velocity of the vehicle during phases 2 and 3 is almost 20km/h. The beginning of phase 1 is the acceleration period, phase 4 is the breaking period. The total motion lasts 8 seconds. Every two sampling periods, the force, acceleration and position components are recorded.

3.3. Experimental results

We have divided the presentation of the experimental results in three parts. First of all, we present the VSC-HF curves, afterwards the impedance control curves with several impedance values and finally the impedance control curves with the acceleration loop. Obviously, the performance depends on an evaluation criterion. We have chosen the criterion defined as:

$$I_{a_c} = \frac{1}{T_f} \sum_{n=0}^{T_f} \ddot{X}_c^2 T \qquad (10)$$

where T is the sampling period, T_f the final time of the trajectory. I_{a_c} is the performance criterion calculated on the component c. We have defined a reference test which consists of recording the acceleration along the z axis (Figure 2) without servoing the manipulator (its brakes are activated for this task). The performance

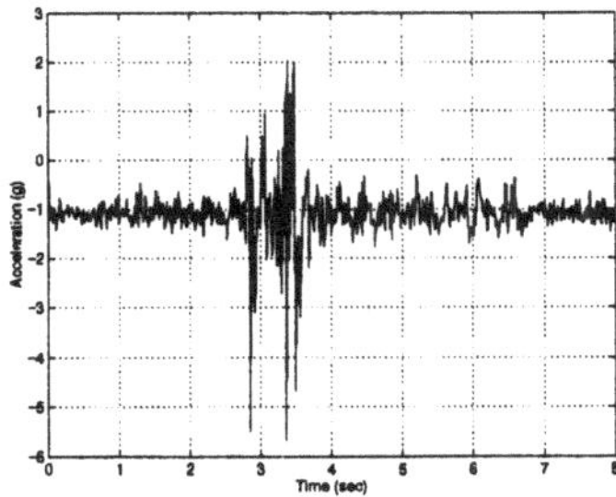

Figure 4. Reference test

criterion is then $I_{a_z} = 1.36 m^2/s^4$. The acceleration peaks observed in figure 4 correspond to the times where the vehicle axles run on the plank.

3.3.1. *Robust control*

The test with the VSC-HF is presented in figure 5. The performance criterion is $I_{a_z} = 1.95m^2/s^4$. We obtain in this case a bad performance criterion in comparison

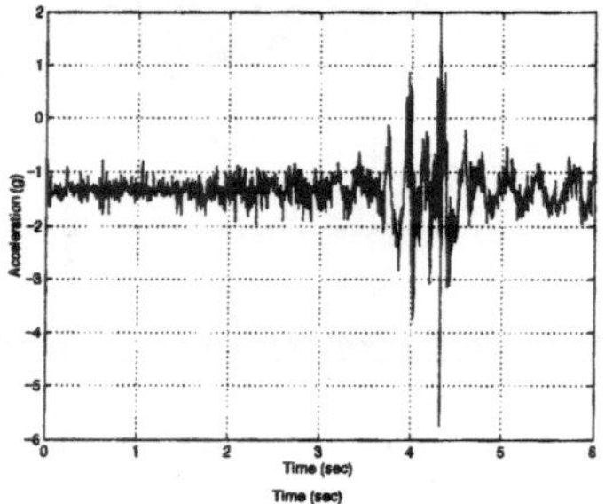

Figure 5. VSC-HF test

with the reference test. This result was foreseeable because the manipulator is extremely rigid and the disturbances are not absorbed. The position of the effector along the z axis in the vehicle frame is shown in figure 6. It can be seen that the

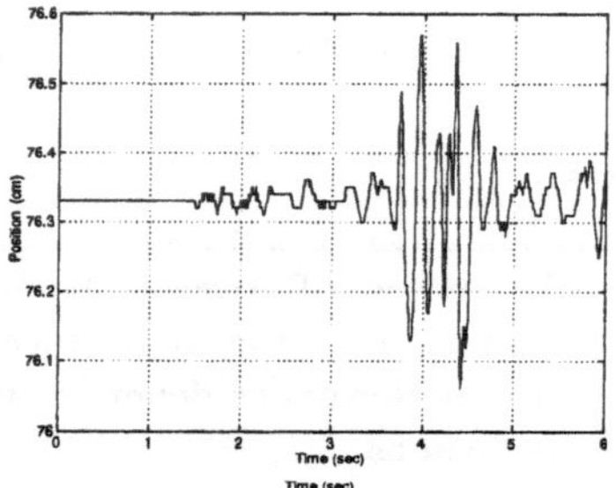

Figure 6. Position of the effector along z axis

position error remains small thanks to the robustness of the VSC-HF controller. Obviously, this produces high shocks on the object, which is not desirable.

3.3.2. *Impedance control*

The impedance control test is presented in figure 7 in the case of $K_{d_z} = 961rd/s^2$ and $B_{d_z} = 9rd/s$. The performance criterion is $I_{a_z} = 1.35m^2/s^4$. The acceleration peaks are smaller than for the reference test. The motion along z axis (Figure 8) is of course greater than for the VSC-HF curves. We present in figure 9 different performance criterions obtained with different K_{d_z} and B_{d_z}. Curves (a) and (b) are obtained with $B_{d_z} = 9rd/s$ and $B_{d_z} = 4.5rd/s$ respectively, and $16 < K_{d_z} < 961rd/s^2$. We experimentally observe an *obvious* phenomenon: globally speaking, the performance criterion increases when the stiffness increases (for a given damping coefficient) and with the damping coefficient (for a given stiffness). As an example, Figure 10 shows that a very small acceleration is acting on the fragile object for a good tuning of the gains, namely $B_{d_z} = 4.5rd/s$ and $K_{d_z} = 100rd/s^2$.

3.3.3. *External acceleration loop*

The results of the experimentation with the external acceleration loop are presented in figures 11 and 12. The performance criterion is larger than for the impedance control without acceleration loop (Figure 12). The external acceleration loop does not compensate for the vibrations induced by the vehicle because the frequency of these vibrations is higher than the bandwidth of the system. If the frequency of the vibrations were in the bandwidth of the manipulator robot, the impedance control should be able to compensate for these disturbances. When we analyze the frequency spectrum in the case of figure 7, we observe some components in the low-frequency domain (1Hz-20Hz) in figure 13. The impedance control allows to eliminate these components (Figure 10). However, the components in the high-frequency domain (>20Hz) do not disappear (Figure 14) because this frequency domain is far beyond the bandwidth of the manipulator.

4. Conclusion

We have presented some experimental results with a mobile manipulator carrying a fragile object. We have observed that the most efficient control law was the impedance control. Actually, several problems exist. What happens when the payload of the object is modified? We should be able to adapt the impedance to obtain a robust behavior. A force sensor could be used to estimate the mass of the payload. This is one goal of our future work. There is also the problem of the compensation of high frequency vibrations. We will look for a method for reducing the amplitude of the high frequency vibrations such that the absolute value of the acceleration along z axis be smaller than 0.3g.

References

[1] McMillan S., Orin D.E., McGhee R.B., "Efficient Dynamic Simulation of an Unmanned Underwater Vehicle with a Manipulator," Proc. of the 1994 IEEE Int. Conf. on Robotics and Automation, San Diego, USA, May 1994, pp. 1133-1140.

[2] Desbats P., Sgarbi F., Pottier P., Cammoun R., Gelin R. , " BRABAM: An open Manipulation System for the Development of Advanced Nuclear and Industrial Applications using a Manipulator Robot Embedded on a Vehicle," Proc. of the 24th ISIR, Tokyo, Japan, November 1993, pp. 827-833.

[3] Fraisse P., Delebarre X., Dauchez P., Pierrot F., "Towards Robust Hybrid Control for Two-Arm Robots," Proc. IEEE IROS'91, Osaka, Japan, November 1991, pp. 331-336.

[4] Hogan N., "Impedance Control: an Approach to Manipulation", Transactions of ASME, Journal of Dynamic Systems, Measurement and Control, vol. 107, March 1985, pp. 1-24.

[5] Morel G., "Programmation et Adaptation de l'Impdance de Manipulateurs au Contact", Thse de Doctorat, Universit Pierre et Marie Curie, Paris, France, June 1994 (in French).

[6] Fraisse P., Pierrot F., Dauchez P., " Virtual Environment for Robot Force Control," Proc. of the 1993 IEEE Int. Conf. on Robotics and Automation, Atlanta, USA, May 1993, pp. 219-224.

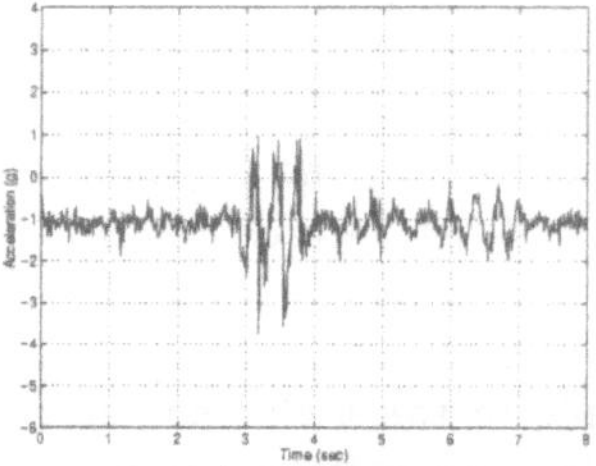

Figure 7. Impedance control

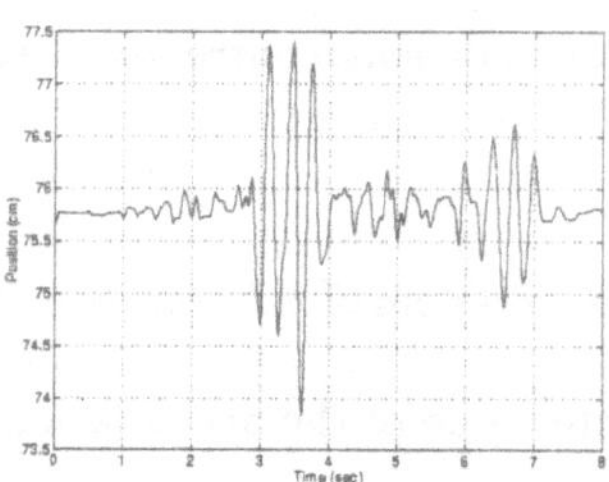

Figure 8. Position of the effector along z

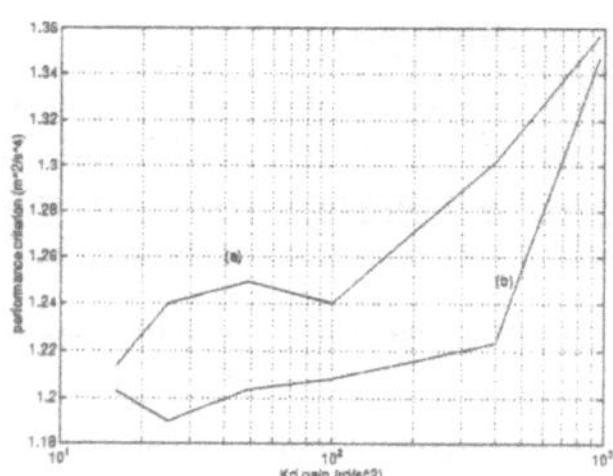

Figure 9. Performance criterion

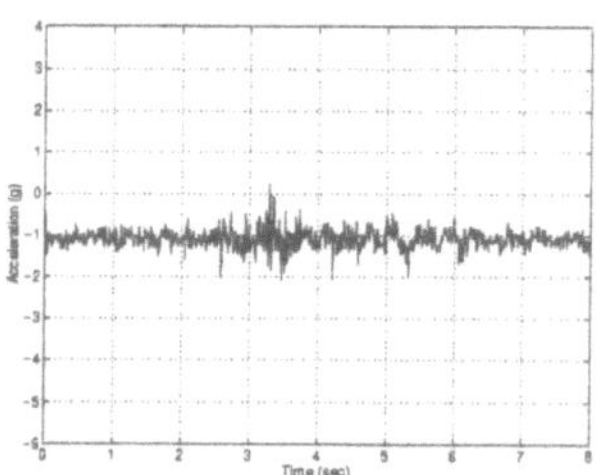

Figure 10. Impedance control

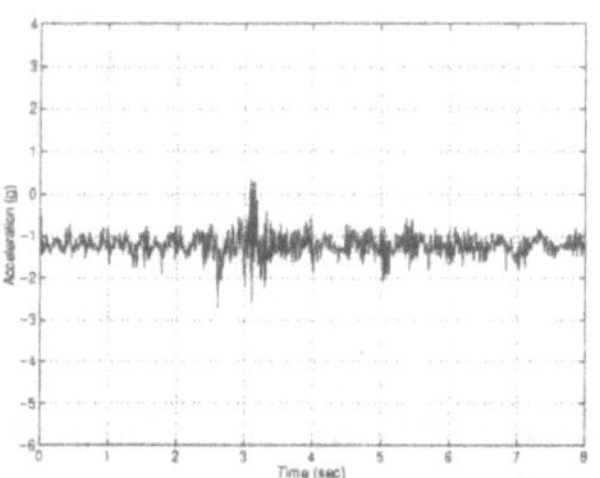

Figure 11. Modified impedance control

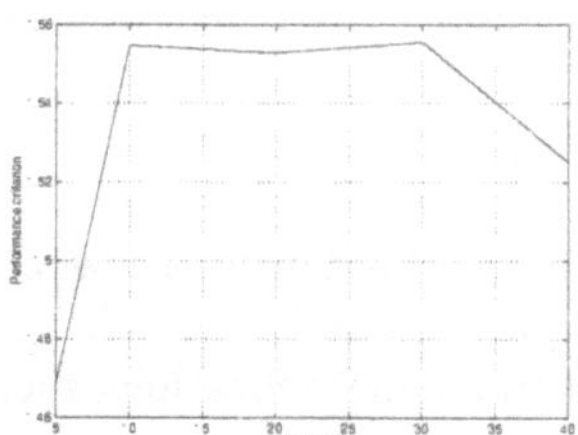

Figure 12. Performance criterion

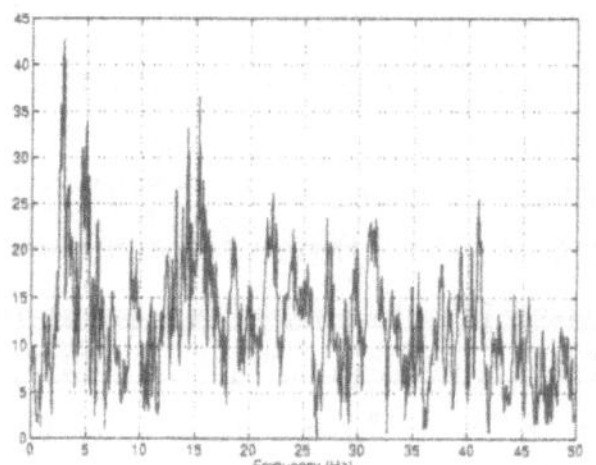

Figure 13. Fourier Transform

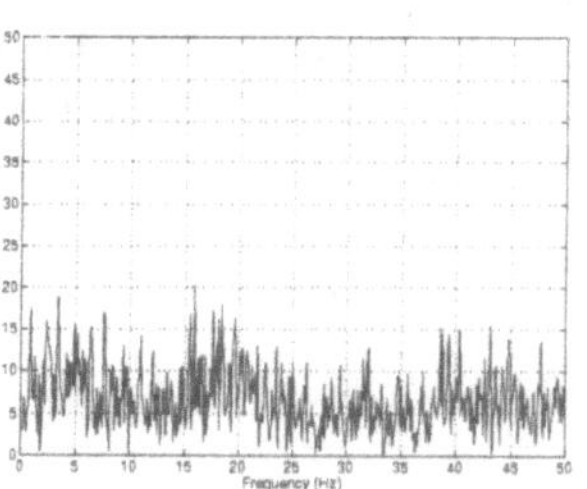

Figure 14. Fourier Transform

Chapter 11

Fine-Motion Planning and Control

Fine motion occurs when a robot approaches or engages in contact with the environment, as in assembly tasks. The impacts and partial constraints that occur at and after contact require careful attention to motion planning and accurate management of contact states. Papers in this chapter address mechanisms and algorithms to enable planning and control of fine motion in a variety of contexts.

Brost and Christiansen apply the planning approach developed by Lozano-Pérez, Mason, and Taylor (LMT) to a number of real world tasks. They demonstrate cases where it works and a case where it fails.

Goldberg, Craig, Carlisle, and Zanutta describe a simulator which permits testing a part feeding system that uses a robot to orient randomly presented objects. An efficient feeder throughput estimator is devised and shown to be applicable to predicting robot cycle times.

Reboulet, Plihon, and Briere propose a "dual hybrid teleoperation scheme" which is shown to improve the effectiveness of teleoperation even in the presence of time delay. They partition the control into force and position subspaces, according to task constraints.

De Schutter, Torfs, Dutré, and Bruyninckx employ a compliant end-effector that is actively damped to solve the problem of achieving stable motion in the presence of partial constraint. The approach is implemented (and shown to work) on a KUKA robot equipped with a compliant wrist force sensor.

Williams and O. Khatib describe a new method that allows joint torque feedback to be implemented in the absence of joint-torque sensors. Using a wrist force sensor they have implemented the method and validated the approach on a PUMA 560 robots. Experimental results confirm the method's effectiveness.

Empirical Verification of Fine-Motion Planning Theories

Randy C. Brost
Intelligent Systems and Robotics Center
Sandia National Laboratories
Albuquerque, NM 87185
rcbrost@isrc.sandia.gov

Alan D. Christiansen
Department of Computer Science
Tulane University
New Orleans, LA 70118
adc@cs.tulane.edu

Abstract

Successful robot systems must employ actions that are robust in the face of task uncertainty. Toward this end, Lozano-Pérez, Mason, and Taylor developed a model of manipulation tasks that explicitly considers task uncertainty. In this paper we study the utility of this model applied to real-world tasks. We report the results of two experiments that highlight the strengths and weaknesses of the LMT approach. The first experiment showed that the LMT formalism can successfully plan solutions for a complex real-world task. The second experiment showed a task that the formalism is fundamentally incapable of solving.

1. Introduction

The identification of robust manipulation strategies in the presence of uncertainty is a crucial problem facing robotics research. Because no robot sensing or control system is perfect, the presence of sensing and control errors is inevitable in all robot systems. These errors can lead to task failures, requiring the identification of special strategies that are robust in the face of these errors. This was first recognized by [8], and further clarified by [9]. Since the inception of robotics, robot designers and programmers have used *ad hoc* methods to identify these strategies when implementing robot systems. If we are ever to achieve the longstanding goal of task-level robot planning, we must replace these *ad hoc* methods with well-founded automated methods of synthesizing robust task solutions in the presence of uncertainty.

Lozano-Pérez, Mason, and Taylor [11] proposed a mathematically sound theory of manipulation in the presence of uncertainty, which explicitly represents the set of possible trajectories that can result from executing a given action. Depending on the nature of the task mechanics and the uncertainty, there may be a *set* of initial states from which a commanded action will succeed, even in the presence of imperfect sensing, control errors, and non-deterministic task mechanics. This set is the *strong backprojection* of the task goal under the commanded action [7]. If the strong backprojection persists after shrinking for uncertainty in position control, then a robust task solution is available: Command a position-controlled motion to a point inside the shrunk strong backprojection, and then execute the action.

There is a substantial body of work extending the original LMT formalism and studying its mathematical properties; a representative sampling would include [7,

6, 5, 1, 10]. Each of these papers has examined the problem using xy-generalized damper dynamics, which is a simple abstract model of action especially well-suited to theoretical study. In this paper, we examine the result of applying the LMT formalism to real tasks that include rotational degrees of freedom, and non-linear task mechanics that cannot be described using closed-form equations. Can the LMT formalism be successfully applied to tasks of this complexity?

This paper describes two primary results:

- First, we demonstrate the efficacy of the LMT approach by showing successful analysis of a pushing task. Even though the mechanics of this task are quite complex, a carefully-implemented planner based on the LMT formalism was able to produce tight discrimination between reliable and unreliable actions.

- Second, we show a task that the LMT approach is fundamentally incapable of solving. Because of the non-determinism due to the task's inherent mechanics, no guaranteed task solution exists, regardless of the precision available in sensing and control. Tasks of this nature require a more powerful theory for planning, such as the formalism proposed in [4].

Taken together, these experiments serve to clarify both the strengths and weaknesses of the LMT approach, and also to emphasize the importance of assuring that a planner's model of the task includes all of the salient physical phenomena that appear during task execution.

The experiments reported in this invited paper were reported previously in [2, 3].

2. Successful LMT-Based Planning

First, we demonstrate a successful use of this theoretical foundation to solve a complex pushing task. In this task, one object is used to push another object until they are in a desired relative position (Figure 1). The objects both have complex polygonal shapes, and there is uncertainty in the objects' mass properties, the coefficient of friction, the pushing direction, and the initial position of the pushed object. Further, the support pressure distribution of the pushed object is assumed to be unknown. This task is characterized by uncertainty in all of the task parameters except object shape, involves rotations during task execution, and has non-linear task dynamics that cannot be described using closed-form equations.

To apply the LMT formalism to this problem, we constructed a planner that analyzes the mechanics of pushing operations to construct an approximation to the strong backprojection of the task goal. The planner constructs a conservative approximation to the strong backprojection, in that the constructed set is always strictly inside the true strong backprojection. Figure 2 shows the constructed strong backprojection for our example pushing task; the planner is described in [2].

We now ask two questions:

- Are the returned results correct?

- Are the returned results overly conservative?

These questions are addressed by experiments in the following two sections.

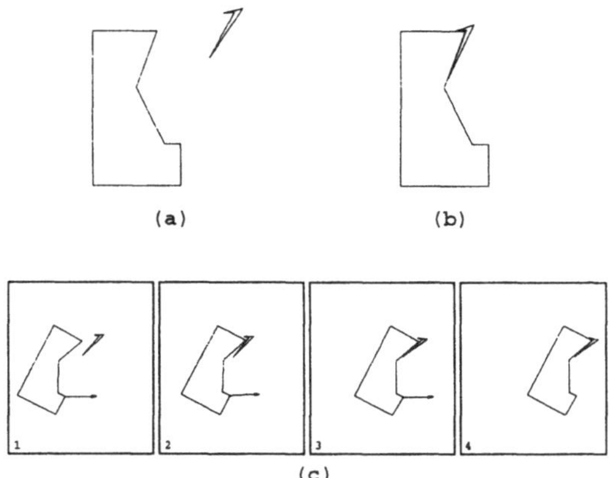

Figure 1. A pushing task. (a) The task objects. (b) The goal configuration. (c) A successful pushing action.

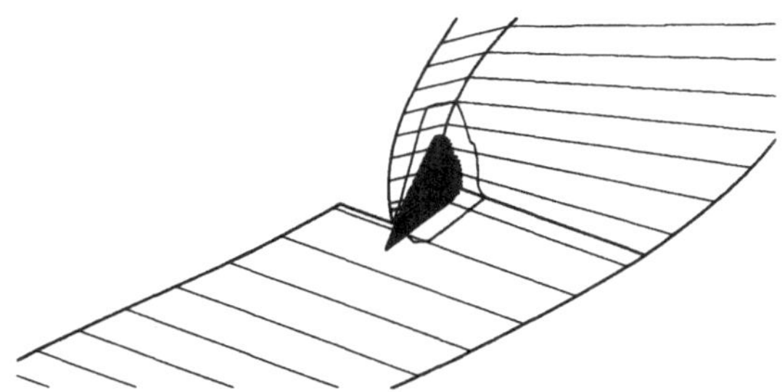

Figure 2. The planner's approximation of the strong backprojection for the pushing task in Figure 1.

2.1. Exploring the Worst-Case Conditions

The strong backprojection shown in Figure 2 represents a set of commanded starting positions that will produce the desired final configuration, even under the worst-case combination of uncertainty conditions. The goal of this experiment is to test that assertion.

An ideal test would be to consider every point in the strong backprojection, and then for each such point, execute a pushing operation under all possible combinations of uncertainty conditions consistent with the input uncertainty parameters. If the planner's analysis is correct, then the goal configuration should be achieved in every trial.

Of course, this ideal experiment is impossible, because it requires an infinite number of trials. A more practical experiment would be to choose a finite number of points in the volume, and test each point under a finite number of uncertainty conditions. These points and uncertainty conditions could be randomly chosen, but a better strategy is to choose points that lie on the boundary of the volume, and apply the maximum amount of uncertainty, making each trial as difficult as possible.

Using this strategy, we selected 25 points distributed over the boundary of the strong backprojection, and used an industrial manipulator to execute 48 pushing operations for each point. In each operation, we added a controlled amount of uncertainty, within the limits specified as input to the planner. The 48 pushing operations executed for each point corresponded to all combinations of extremal uncertainty conditions in five variables: support friction distribution, object position, object orientation, robot orientation, and robot pushing direction. The magnitude of some of these errors was reduced slightly from the input value to compensate for the presence of natural uncertainty in the experimental apparatus. Errors in the object center of mass location, coefficient of friction, and robot position were not controlled in the experiment, because the input uncertainty parameters closely matched the true uncertainty of the physical apparatus.

These choices produced a set of 1200 experimental trials; if the planner's output is correct, all 1200 trials should successfully achieve the goal configuration. All 1200 trials did succeed, confirming the validity of the planner for this example problem.

2.2. Is the Analysis Too Conservative?

In fact, the executed actions were so reliable that observing the actions was very boring. This immediately suggested the question: Are there actions outside the estimated strong backprojection that are also reliable? If there are such actions, then the planner is being overly conservative.

The strong backprojection returned by the planner delineates a set of points that *will* succeed; meanwhile, points outside the strong backprojection *might* fail, for some choice of uncertainty conditions. Thus, if we can show that points just outside the strong backprojection have a possibility of failing, then we can conclude that the planner is not overly conservative.

Figure 3 shows three such points. Each action trace corresponds to a point chosen a small distance outside the strong backprojection, with a set of uncertainty conditions that bears witness to the point's possibility of failure. Note that each point gives rise to a different failure mode; these are the task behaviors observed in physical trials.

3. A Task that Foils LMT

The previous experiments show that for at least some tasks, the LMT method is an effective method of planning robust manipulation actions. In this section, we demonstrate a fundamental limitation of the LMT approach by showing a task for which the LMT method cannot identify a robust solution.

As the task uncertainty increases, the strong backprojection necessarily shrinks. If the task uncertainty is too high, then the strong backprojection vanishes, and the LMT formalism fails to find a reliable action. In [4], we extended the LMT formalism by developing the notion of a *probabilistic backprojection,* which describes the action success probability as a function of task initial state. This paper also showed that this new theoretical formalism subsumes the original LMT formalism, because all of the LMT set constructions can be produced as special cases of our probabilistic formalism. This new formalism has the desirable property that the reliability of the returned actions degrades gracefully as the task uncertainty increases, avoiding the discontinuity that occurs in the LMT formalism when the strong backprojection vanishes.

This gives rise to an immediate question: Is this additional formal machinery really necessary? Perhaps it would be possible to successfully apply the LMT approach by improving the precision of the sensing and control system, or by synthe-

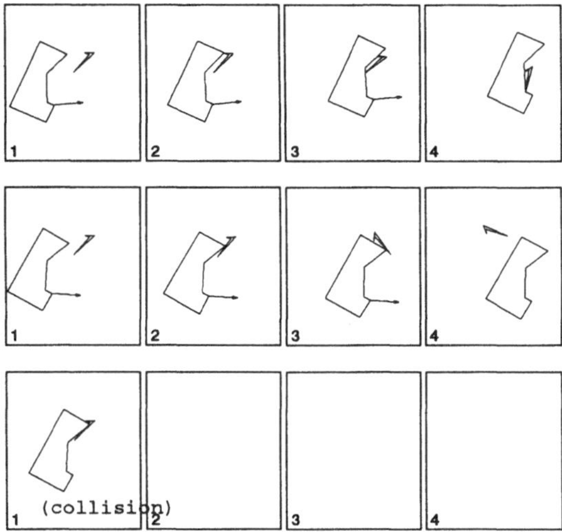

Figure 3. Failure modes observed for points outside the constructed volume. In the third example, the objects collided as the large object was lowered onto the plane along the z axis.

Figure 4. An example task chosen for empirical study.

sizing multi-step actions. In this section we will show a task that does not submit to these approaches, because the strong backprojection vanishes due to inherent non-determinism in the task mechanics. Consequently, the LMT approach will always fail for this task.

Our example task is a grasping operation that appears as one step in a larger assembly procedure; this task is shown in Figure 4. The task goal is to grasp the gear in the configuration at the bottom of Figure 5, using a SCARA robot with a parallel-jaw gripper. To perform a grasping operation, the robot opens the gripper fingers, moves the gripper to an initial position near the gear, and then closes the fingers. Figure 5 shows the motion of the gear that occurs during a typical successful operation. Figure 6 shows examples of the types of failure that can occur.

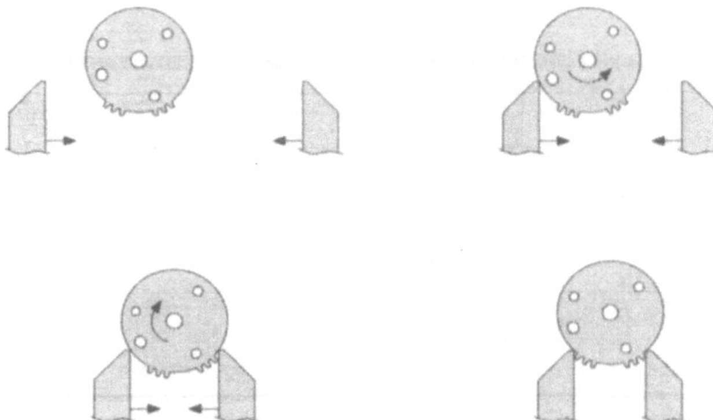

Figure 5. A typical grasping action, showing four phases of contact: No contact, single-finger pushing, two-finger squeezing, final grasp.

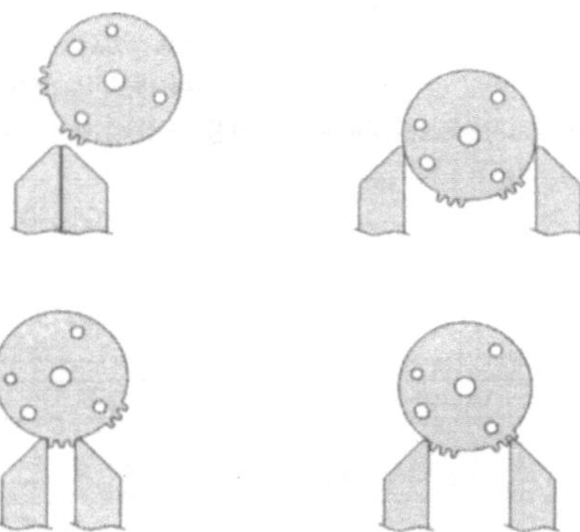

Figure 6. Example grasp failures.

Whether the action is guaranteed to succeed or not depends on whether the initial configuration at the beginning of the grasp is within the task strong backprojection. Alternatively, the success probability for a given action depends on the corresponding probability value in the probabilistic backprojection.

We can characterize the strong and probabilistic backprojections for this task by repetitively executing grasping operations at different initial positions, checking success after each operation. In this experiment, we developed an automatic robot control procedure that allowed the execution of a large number of trials. For each desired grasping operation, the procedure initially locates the gear using computer vision, calculates the desired gripper position in world coordinates, moves to the appropriate initial position, executes the grasping operation, and then checks the success of the operation using computer vision. The outcome of the operation is then recorded, and the next desired grasping operation is executed.

A number of precautions were taken to improve the accuracy of the collected data. Lighting sources were designed to provide even illumination with minimal robot shadows, and the support surface was chosen to provide a crisp high-contrast image of the gear. Edge-based feature localization algorithms were used to measure

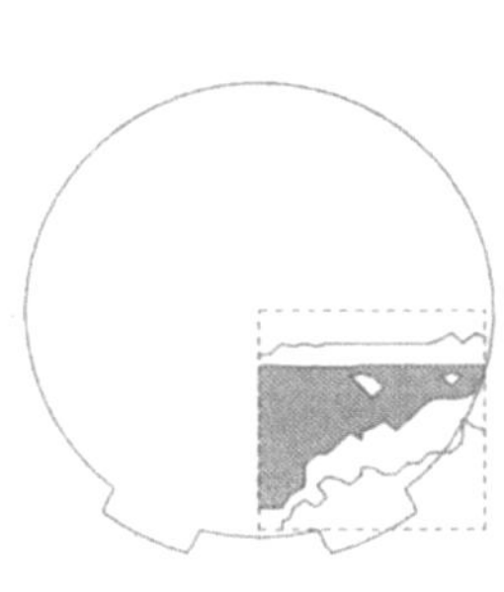

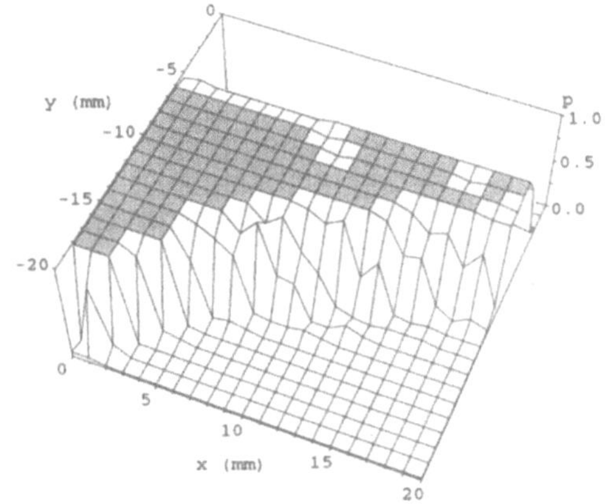

Figure 7. The result of exploring the space of grasping actions. In this experiment, the support surface was clean, and the fingers closed slowly.

the gear position; these were found to be much more robust and precise than typical blob analysis techniques. Measured image locations were adjusted to compensate for lens distortion using a radially-symmetric distortion model; the coefficients of this model were determined from calibration data obtained from a 21×21 grid of dots. We estimate the accuracy of the gear position measurements to be within ± 1mm and $\pm 3°$. The gear diameter is 41.48mm.

This apparatus was used to study the reliability of grasping operations throughout a region of initial positions. The region of interest was a 20mm square with the upper left corner at the center of the gear; the operations corresponded to a 1mm grid covering this region. Each grid point corresponds to an initial placement of the center point between the fingertips relative to the gear center. The initial orientation of the gear relative to the gripper was the same in all of the trials; only the initial (x, y) position was varied. Thus each experimental run consisted of $21 \cdot 21 = 441$ grasping operations, each with a different initial position.

We performed thirty experimental runs, for a total of 13,230 grasping operations. The results were then tabulated to form a histogram of successful grasping operations associated with each initial position. These data are shown in Figure 7, in two forms. At the top of the figure, the gear outline is drawn actual size, with the region of sampled points shown dashed. The shaded region outlines the set of points that succeeded thirty times out of thirty; this is the *observed strong backprojection* of this task. Surrounding the shaded region is another outline, which delineates the set of points that were observed to succeed at least once; this is the *observed weak backprojection* of this task. Points outside this outline failed thirty times out of thirty. At the bottom of the figure, the histogram of observed successes is drawn as a three-dimensional plot, showing the variation from complete failure to reliable success. The surface of this plot approximates the probabilistic backprojection. The shaded boxes of this plot identify squares where all four corners succeeded thirty times out of thirty; the ensemble of these regions approximates the probabilistic backprojection. The plot axes correspond to an (x, y) coordinate system whose origin is at the center of the gear.

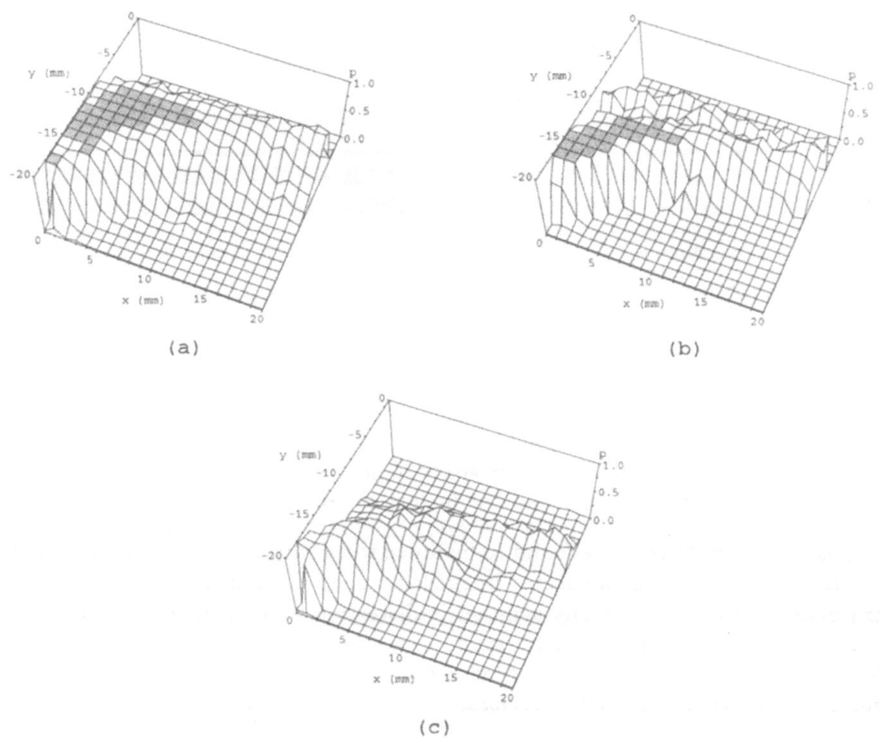

Figure 8. The results of the grasping experiment repeated under varying conditions. (a) With sand contaminating the support surface, and slow finger closing speed. (b) With a clean support surface and fast finger closing speed. (c) With sand contaminating the support surface, and fast finger closing speed.

This entire experiment was repeated three times, under varying conditions. The resulting data are shown in Figure 8. In part (a), a small amount of sand was added to the table surface, simulating the effect of grit or a less structured environment. In part (b), the finger speed was doubled, with no sand present. In part (c) the finger speed was doubled and sand was present. All four of the conditions studied represent reasonable simulations of practical situations that may be encountered in industry. As can be seen from the figure, these varying conditions had a dramatic effect on the observed strong and probabilistic backprojections for the task. For a more detailed discussion of this experiment and its results, see [4].

These experiments demonstrate that the LMT formalism is sometimes appropriate for analyzing manipulation tasks, and sometimes not. For the well-behaved control case with slow finger speed and a clean support surface, the strong backprojection is large and well-defined. On the other hand, for the case with a fast closing speed and a contaminated surface, the strong backprojection essentially vanishes due to inherent non-determinism in the task mechanics. For this case, the LMT formalism can never correctly identify a reliable action, because there is none. However, a probabilistic analysis could identify a large region of actions that have better than 80% probability of success.

4. Conclusion

The experiments described in this paper show that the LMT formalism may be successfully applied to real manipulation tasks. At the same time, the experiments show that the LMT method is fundamentally incapable of solving some tasks, and that probabilistic analysis could yield useful task-planning information in these cases. These results highlight the importance of finding an appropriate match between a planner's task model and the physical properties of the real task.

Acknowledgements

The authors would like to thank John Craig for inviting this paper to ISER, and Matt Mason, Bruce Donald, Mike Erdmann, and Ken Goldberg for many useful discussions during the development of these ideas.

References

[1] A. J. Briggs. An efficient algorithm for one-step planar compliant motion planning with uncertainty. Technical Report TR 89-980, Cornell University Department of Computer Science, March 1989.

[2] R. C. Brost. Dynamic analysis of planar manipulation tasks. In *Proceedings, IEEE International Conference on Robotics and Automation*, pages 2247–2254, May 1992.

[3] R. C. Brost. Natural sets in manipulation tasks. In K. Goldberg, D. Halperin, J.C. Latombe, and R. Wilson, eds, *The Algorithmic Foundations of Robotics*, pages 127–136. A. K. Peters, Boston, MA, 1995.

[4] R. C. Brost and A. D. Christiansen. Probabilistic analysis of manipulation tasks: A computational framework. *Intl. J. of Robotics Research*, 15(1), 1996.

[5] J. F. Canny and J. Reif. New lower bound techniques for robot motion planning. In *Proceedings, 28th Annual Symposium on Foundations of Computer Science*, pages 49–60. IEEE, October 1987.

[6] B. R. Donald. A geometric approach to error detection and recovery for robot motion planning with uncertainty. *Artificial Intelligence*, 37:223–271, 1988.

[7] M. A. Erdmann. Using backprojections for fine motion planning with uncertainty. *Intl. J. of Robotics Research*, 5(1):19–45, Spring 1986.

[8] H. A. Ernst. *MH-1, A Computer-Operated Mechanical Hand.* PhD thesis, MIT Department of Electrical Engineering, December 1961.

[9] H. Inoue. Force feedback in precise assembly tasks. Memo 308, MIT Artificial Intelligence Laboratory, August 1974.

[10] J. C. Latombe. *Robot Motion Planning.* Kluwer Academic Publishers, Boston, 1991.

[11] T. Lozano-Pérez, M. T. Mason, and R. H. Taylor. Automatic synthesis of fine-motion strategies for robots. *Intl. J. of Robotics Research*, 3(1):3–24, Spring 1984.

Estimating Throughput for a Flexible Part Feeder

Ken Goldberg*

University of Southern California
Los Angeles, USA
goldberg@usc.edu

John Craig†
Silma, Inc.
Cupertino, USA
jjc@silma.com

Brian Carlisle
Rob Zanutta
Adept Technology, Inc.
San Jose, USA

Abstract

To rapidly feed industrial parts on an assembly line, Carlisle *et. al.* [4] proposed a flexible part feeding system that drops parts on a flat conveyor belt, determines their pose (position and orientation) with a vision system and uses a high-speed scara robot to move them to a pallet in a desired pose. Such a feeder can be rapidly configured and reconfigured to handle a variety of parts. To facilitate rapid set-up of assembly lines, a simulator can provide visualization and realistic timing estimates. This paper focuses on estimating feeder throughput, which determines the "heartbeat" of an assembly line.

The simulator takes as input a CAD model of the part and its center of mass, as well as current position of the robot, conveyor belts, cameras, etc. An estimate feeder throughput depends on overall part arrival rate, the distribution of stable states in which parts arrive, and the time required to carry parts from the feeder conveyor belt. This paper formulates throughput estimation and focuses on an $O(n^2)$ algorithm for estimating the statistical distribution of stable poses. We also describe how robot cycle times can be estimated based on statistical sampling.

*goldberg@usc.edu. This work was supported in part by Adept Technology, Inc.
†jjc@silma.com

1. Introduction

To automate the assembly of mechanical components, parts must be precisely oriented prior to packing or insertion. A *parts feeder* is a machine that orients parts. Currently, the design of parts feeders is a black art that is responsible for up to 30% of the cost and 50% of workcell failures [13, 3, 7, 17, 18].

> *In [robot] assembly, two-thirds of the system costs are typically for peripherals such as supporting fixtures, special grippers, and part presentation devices. These peripherals are generally neither versatile nor programmable.* [6]

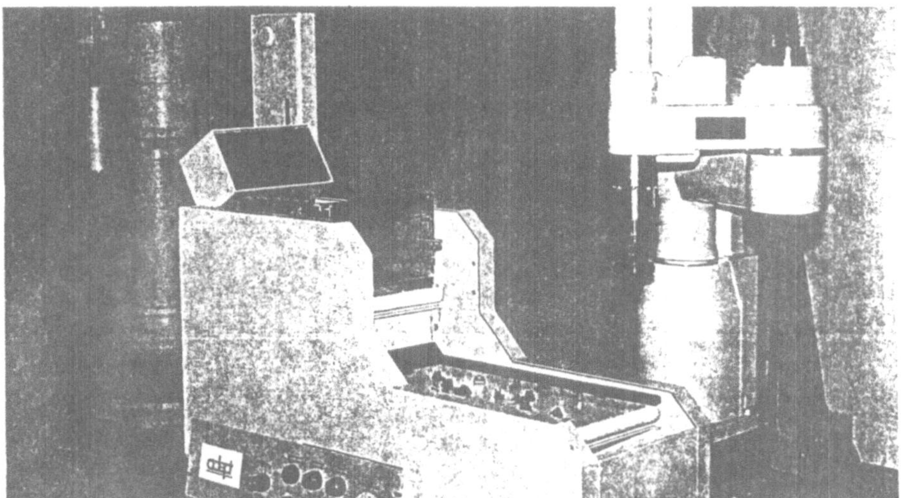

Figure 1. A flexible parts feeder using machine vision, a high-speed robot arm, and pivoting gripper.

Carlisle *et. al.* [4] proposed a flexible part feeding system that combines machine vision with a high-speed robot arm. In contrast to custom-designed mechanisms, this system can be programmed based on the type of part to be fed. During operation, a collection of like parts are randomly scattered on a flat worktable where they are subject to the force of gravity. An overhead vision system determines the pose (position and orientation) of each part. The robot arm then picks up each part and moves it into a desired final pose as illustrated in Figure 1. Rao, Kriegman, and Goldberg [16] considered the problem of planning optimal grasps for such a pivoting gripper and gave an $O(m^2 n \log n)$ algorithm to generate pivot grasps for a part with n faces and m stable configurations.

In this paper we focus on estimating throughput of this feeder. Feeder throughput is critical to load balancing and the layout of assembly lines. For example, it may be necessary to add an additional feeder to match the expected throughput of some assembly stage, or two feeders can be replaced with one if throughput is sufficient. In effect, part feeders provide the "heartbeat" of an assembly line.

Given part geometry and center of mass, a simulator that can predict feeder throughput can be useful during part design, since a minor change in part geometry or mass distribution may have pronounced effects on the stability of part poses and thereby effect throughput. Designers might experiment offline with part geometry to improve feeder throughput. Since fabricating and testing a series of physical parts is impractical, the simulator can facilitate "design for assembly".

Let p_i be the probability that a part arrives on the conveyor belt in pose i, and t_i be the time (in seconds) it takes the arm to move the part from pose i on the belt to the pallet and return to a neutral position. (Certain poses are difficult or impossible to reorient due to kinematic limitations; by setting $t_i = \infty$, the system will not attempt to reorient parts in pose i, allowing them instead to be recycled through the feeder.)

In industrial parlance, to "convert" part is to move it from some initial pose into the desired pose. If N parts arrive on the conveyor belt per second, we expect Np_i parts in pose i. Let α_i be the "throughput": number of parts in pose i that are converted per second.

Our goal is to estimate overall throughput, $\sum \alpha_i$. Overall throuput can be optimized subject to the availability of parts in pose i:

$$\alpha_i \leq Np_i,$$

and the available time:

$$\sum t_i \alpha_i \leq 1.$$

This optimization problem has some resemblance to the notorious $0-1$ knapsack problem. It can be efficiently solved because the "value" of each pose depends directly on the time to convert it. The optimal solution is found with a greedy algorithm that ranks poses in order of increasing t_i. During online operation, always convert the part with lowest t_i, breaking ties arbitrarily.

Thus the critical values for estimating throughput are pose probability, p_i, cycle time, t_i, and N, overall part arrival rate. In the body of the paper we discuss methods for estimating these values.

In the next section we discuss related work. We then review an $O(n^2)$ algorithm that yields a quasi-static estimate of p_i and compare estimated probabilities against experiments with a prototype part feeder. The last two sections discuss approaches to estimating N and t_i.

2. Related Work

It is notoriously difficult to model the dynamics of collisions with friction [10]. Anyone with an exact model should immediately travel to the craps tables at Las Vegas. Microscopic models are described in the graphics and simulation literature. These yield believable simulations [1, 12] but generally run too slowly to be sampled to predict pose probabilities.

In this paper we consider a "quasi-static" model of part motion that neglects inertial forces. The basic idea is to project the part's center of mass onto a sphere and consider the relative projected area of each face of the part's convex hull. If we assume the part makes contact with the conveyor belt in an orientation distributed uniformly on $SO(3)$, then the relative area yields the probability that the part will initially land on this face [2].

When some faces of the part's convex hull are unstable, we must recursively propagate the initial landing probability onto appropriate stable faces. [19] improved on Boothroyd's method by adding this and characterizing the computational complexity of the algorithm. Ngoi, Lim, and Lee [14] reported experimental results using a variant of the method from [19].

3. Estimating Pose Probability: p_i

Given

P_0 : vertices of an n-sided polyhedral part.

m : the part's center of mass.

Find

p_i : the probability that the part is in pose i for $i = 1$ to n.

Assumptions

- The part is a polyhedron with a known center of mass,
- The worksurface is flat and much larger than the part,
- The part does not collide with other parts,
- The orientation of the part upon contact is a uniform random variable on $SO(3)$,
- part motion is quasi-static (we ignore dynamics).

The algorithm

1. Compute the convex hull of the polyhedral part, call it P.

   Note that it is sufficient to analyze the convex hull of P_0 since any stable pose will correspond to a face of P resting against the worksurface. [15] gives a variety of methods for computing convex hulls; a divide-and-conquer strategy can be used to compute the convex hull in time $O(n \log n)$.

2. For each face of the convex hull, compute the initial probability. Let p_i be the probability of landing on face i

   Since all faces of P are convex, it is sufficient to project each face out from the center of mass onto a unit sphere centered at m.. The relative areas of projection give the relative probabilities. For a triangular facet of the convex hull, let $\beta_0, \beta_1, \beta_2$ be the interior angles of the projected facet. Its relative area is

$$A = \frac{\beta_0 + \beta_1 + \beta_2 - \pi}{4\pi} \tag{1}$$

   To compute β_i, let c be the center of mass and d_{ci} be the distance from c to vertex i. Using standard notation for triangles, let δ_2 be the arc that results from projecting the line from vertex 0 to vertex 1 onto the sphere. To compute δ_i, use the cosine law:

$$d_{02}^2 = d_{c0}^2 + d_{c2}^2 - 2 d_{c0} d_{c2} \cos \delta_1 .$$

   Find β_1 using the spherical cosine law:

$$\cos \delta_1 = \cos \delta_0 \cos \delta_2 + \sin \delta_0 \sin \delta_2 \cos \beta_1 .$$

3. To handle the unstable faces of P, we build a directed acyclic graph G. Let G have n vertices corresponding to the faces of P. Attach a directed arc if the polygon sitting on face i will topple (quasi-statically) onto face j. To decide if a face will topple, we project m onto the plane of face i. Let the projected point be m'. If m' lies outside the face, then gravity causes it to generate a moment on an edge of the contacting face. This will cause the part to rotate about this edge onto a new face. Non-generic scenarios can occur where m' lies directly above an edge or vertex of the contacting face, in which case the pose is quasi-stable, or where m' lies outside the face but is equidistant from two edges that share a common vertex.

4. To propagate probabilities to the appropriate stable faces, we recursively traverse the graph down to stable faces.

Henceforth we only consider the stable faces, those with $p_i > 0$.

The number of vertices and edges of P are both $O(n)$. We compute the projected areas in $O(n^2)$ time, since for each face we need to compute $O(n)$ interior angles. The DAG can be built in $O(n^2)$ since for each face the center of mass must be checked against $O(n)$ edges. Finally to traverse the DAG we are required to visit each node (face) only once so this requires $O(n)$ time. The total computation time is thus $O(n^2)$.

4. Implementation

The pose estimator was implemented in RAPID, a product under joint development by SILMA Inc. and Adept Technology Inc. We tested three industrial parts, the first is an insulator cap purchased at a local hardware store. The second is part of an assembly housing, and the third is a pushbutton designed for a commercial car stereo system.

When a CAD model of a part is available, RAPID can read it directly from a variety of native formats, or using the IGES neutral-file standard. For our particular test parts, since CAD models were unavailable, we entered part coordinates by hand and used a crude balance to estimate each part's center of mass.

The tables below compare the estimated probabilities with those measured with a prototype flexible feeder at Adept. We note that in the experiments, the top conveyor belt was moving so that parts fell from a stable pose in bulk to a new stable pose on a second fast-moving belt used to singulate. This resulted in inter-part collisions

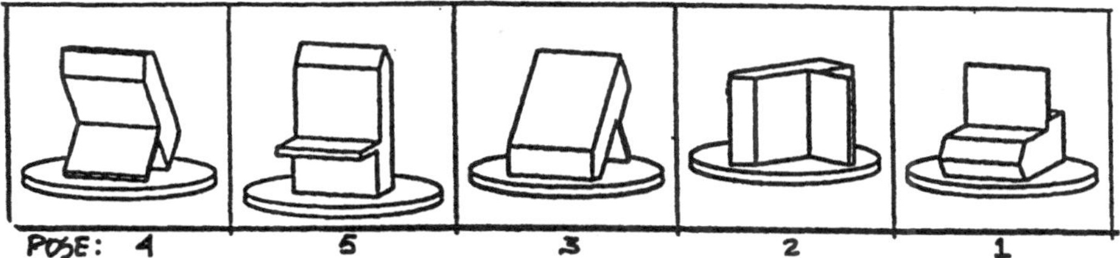

Figure 2. The "orange" insulator cap in each stable pose. Quasi-Static estimates of pose probabilities are given in Table 1.

and vibration which were not modelled in the quasi-static model.

Pose	Estimated	Measured
1	31 %	45.95 %
2	37 %	27.12 %
3	20 %	19.69 %
4	8 %	5.02 %
5	4 %	2.22 %

Table 1. Orange insulator cap (1036 trials).

The distribution compares reasonably well with the exception of pose 5, which occurs 14% more frequently than estimated. This is the most stable pose in terms of potential energy, so it is understandable that kinetic energy might shift the distribution toward this pose.

Pose	Estimated	Measured
1	40 %	52.5 %
2	32 %	39.4 %
3	14 %	4.5 %
4	7 %	1.9 %
5	7 %	1.6 %

Table 2. Assembly housing (not pictured) (627 trials).

Poses 4 and 5 were susceptible to toppling due to inter-part collisions and belt motion.

In the RAPID simulator, the above stable pose predictions are used in simple algorithms called *Drop*, *Bounce*, and *Jiggle*. For a given part in the simulated world, *Drop* determines which surface it will first strike as it falls under gravity. *Drop* also may modify the stable pose probabilities as a function of drop height. For example, a part "dropped" from a small height (say, 1 mm) above a surface has a high probability of staying in its current stable state. Above a critical height, probabilities are taken directly from the quasistatic model. *Bounce* uses a simple

Pose	Est-1	Measured
1	35 %	55.9 %
2	16 %	24.5 %
3	17 %	13.6 %
4	8 %	4.4 %
5	11 %	1.4 %
6	7.5 %	0.3 %
6	4.3 %	0.0 %

Table 3. Stereo Pushbutton (not pictured) (1099 trials).

radial scatter density function in which maximum scatter distance is a function of drop height. Finally, *Jiggle* is used to "knock down" low probability states in the case that the part is supported by a conveyor or other moving support surface.

5. Estimating N, parts per second

To estimate the average number of parts per second that will appear in the vision window, we must take into account factors such as the relative area of parts and belt velocity. Let:
A_v, viewing area
A_p, average area per part (including margin)
t_b, belt time (time to move A_v into view)
t_s, settling time for parts on belt
t_v, vision processing time per part (depends on processor, vision algorithm, part complexity)

The average number of parts per second visible to the camera is then

$$N = A_v/A_p(t_b + t_s + t_v).$$

As an example, we consider the following values for the Adept feeder with a typical industrial part:
$A_v = 38,400$ mm x mm
$A_p = 1400$ mm x mm
$t_b = .32$ sec
$t_s = 1$ sec
$t_v = 35$ msec,

which yields $N = 12$ parts/second.

Note that if all of these parts were accessible for transfer, no existing industrial robot would be able to maintain the necessary cycle time. The direct drive Adept 550 robot can transfer at most 1.6 parts per second with a single part gripper. Thus if more than 10% of the parts are in the desired pose, cycle time will be limited by arm velocity.

6. Robot Motion: Estimating Cycle Time: t_i

In this section we describe a method for accurately estimating the time required for gross manipulator motions. The approach is to use actual data collected from experiments to numerically fit a set of parameters in a parametric timing model of robot motion. Within this general scheme, many choices can be made in terms of model complexity, amount of data collected, and numerical technique used to do the fitting of parameters. We have made choices that we feel are consistent with providing the user with an accurate (within 10%) and fast (interactive on a workstation or fast PC) motion estimate for arbitrary motions with arbitrary payloads.

There are several reports on computing accurate timing estimates for commercial robotic mechanisms for use in an off-line planning tool, e.g. [5, 8]. These generally do not include dynamic effects such as load-dependent settling times.

Assembly and material handling robots push the limits of current available robotic simulation packages because they are typically run at high speeds at which the effects of inertia variation in payload and motor saturation cannot be neglected. Unlike many uses of industrial robots, in material handling and small parts assembly, it is typical for the robot programmer to tune robot motion parameters for maximum speed, and by so doing, frequently the robot is run at or near motor saturation during significant portions of the programmed motions. Hence, emulation of these motions must take into account the inertial properties of the payload since these influence the onset of motor saturation.

The inability of current commercial systems to accurately predict timing due to these effects has limited their useful application to this area. Experienced robot programmers seem able to increase part throughput in robotized material handling systems by 10% to 30% by careful adjustment of motion parameters (and feeder parameters). This final performance improvement is often crucial to the economic viability of a proposed robotic solution. Hence, unless the simulation system can accurately predict cycle times which are dependent on the details of the feeding system and robot motion, it will not constitute a viable planning tool.

Fully capable dynamic simulators certainly exist which are capable of predicting details of robot motions including the effects of payload mass variation and motor saturation. However, the real problem is obtaining sufficiently accurate data to build the required dynamic model. This is difficult at best, and in practice has been an impediment to the use of such simulation in industrial settings. A second problem is the processing required for such dynamic simulators. Generally, they are quite computationally intensive and do not offer the kind of user interaction required for acceptance by the potential user community.

Our approach is to create a system in which the collection and processing of actual robot motion data is semi-automatic. This allows the procedure to be repeated as needed to track changes made to the robot control system, or to be easily applied to a wide variety of robot designs. A suite of motions is executed by the robot using varying motion parameter settings and various payload masses. Motion data is recorded by the robot controller and output in an ASCII file format. This file is read and processed by parameter identification software which is a part of the RAPID system. Much of this collection and processing is automatic and easily accomplished.

A second element to our approach is to avoid a complete (and complex) dynamic model in favor of a simplified parametric form to which we optimally fit parameters with the identification method mentioned above. In particular, we split the emulation of motion into three levels:

- Level 1: Emulation of the desired trajectory as presented to the servo system.

For example, many motion systems use motions which exhibit a trapezoidal velocity profile. In this case, parameters which describe the relationship between user-supplied motion settings (sppeds, acceleration, blend zone sizes, etc) and these motion profiles are identified.

- Level 2: Emulation of settling at the ends of motions. This is a function of end-effector and payload mass as well as arm configuration. The simulator tags all end-effectors and parts with inertial properties so this information is available. A simplified parametric model describing the envelope of settling oscilations and settling time is used to fit parameters based on the collected data.

- Level 3: A simplified dynamic model of the robot (and end-effector and payload) is used to predict motor saturation. Note that this model is used only in the "forward dynamics" formulation to predict torques from motion. No numerical integration of any kind is needed to do the inverse. Hence, we predict the onset of motor saturation in order to warn the user, or to perform automatic adjustment of motion parameters, but we do not emulate motion once in saturation.

Level 1 emulation is accomplished by choosing a parametric form for the trajectory generation algorithm in the robot controller,and then fitting parameters based on actual collected data in order to minimize prediciton error. At the highest level of abstraction, the trajetory generator can be parameterized as

$$\Theta(t) = F_i(\Theta_s, \Theta_f, U, \Phi, t),$$

where $\Theta(t)$ is the time function of the vector of joint angles for a motion, Θ_s and Θ_f are start and final joint positions, U is a vector of user-supplied constraints (e.g. speed and acceleration settings, but possibly also intermediate via points), and Φ is a vector of unknown constant parameters used in the trajectory generation algorithm F_i. The trajectory generation algorithm F_i is itself a member of a set of various parametric algorithms. The identification problem is to choose the best F_i from the set and then optimally fit a vector Φ of parameters.

Level 2 emulation is accomplished by choosing a parametric form for the settling transient at the end of a motion. For this purpose we propose a form which corresponds to a second order system which is subject to a jerk in acceleration at the end of a tracking motion. We assume the second system is tuned to be critically damped for an effective mass of $\hat{m}$ but is operating with an actual effective mass of m. We assume that the servo stiffness is set at k_p. With these assumptions, the transient error at a given joint at the end of a motion is the impulse response of the differential equation

$$\ddot{e} + 2\frac{\hat{m}}{m}\sqrt{k_p}\dot{e} + \frac{\hat{m}}{m}e = 0.$$

with an impulse of size $(1 - \frac{\hat{m}}{m})||\ddot{\theta}_d||$, where e is the deviation of a single joint angle θ from the desired value θ_d.

Level 2 emulation uses knowledge of the manipulator configuration and mass properties of the end-effector and grasped part to compute the reflected inertia at the joint, m. Over a large suite of motions using various ending decelerations and various payloads, parameters appearing in the above model (such as k_p) are optimally fit. After this "tuning" from actual data, the model of settling transient above may be used to predict settling times as a function of arm configuration, deceleration at the end point, and payload of the end-effector and part.

Level 3 emulation will use a simplified dynamic model of the robot and a model of actuator capability in order to check for motor saturation. If saturation is predicted the user will be warned.

While the above discussion focuses on predicting gross motion times, our simulator also estimates gripper operation times and other effects which contribute to the overall cycle time. As of the writing of this paper, only the first of the three "levels" of emulation has been implemented, with the latter two constituting future work.

In particular, this timing estimation approach can be used to estimate the time required to move a part in pose i from a conveyor belt to a pallet. In most cases a part can be reoriented as it is being moved, so the time is dependently primarily on the time required to move from a neutral position, accelerate, move, and decelerate the arm into position above part, grasp the part, then move to the pallet, and return to a neutral position. Some poses may be impossible to reorient due to kinematic limitations of the arm. In such cases we set $t_i = \infty$.

6.1. Experimental Results

In one experiment using an Adept 550 manipulator a suite of 56 motions were designed which exercise the robot throughout its workspace and for various desired speed and acceleration settings. Times for the individual motions ranged from 0.125 seconds to 8.6 seconds. Over this suite of motions, the average timing error was 4.66%. This was achieved using only "Level 1" emulation, so these results will only be valid with small payloads. Experimental results with varying payloads await the final implementation of Level 2 and Level 3 emulation.

RAPID includes a feature called *Autoplace* which automatically determines the time optimal location of the manipulator relative to the workcell. A search scheme tries potential manipulator locations in the cell (at some discrete tesselation) and displays a color-coded locus of possible locations of the robot. The accuracy of *Autoplace* is directly set by the accuracy of the timing estimates for robot motion. This kind of optimization algorithm generally provides a superior design of the cell layout compared to manual trial and error methods. While others have reported this concept, e.g. [Hoang and Fenton], they have not reported an accurate cycle time emulation on which to base the automatic placement algorithm.

7. Summary

We present an $O(n^2)$ algorithm to estimate the distribution of stable resting poses for a given n-sided polyhedral part based on quasi-static mechanics. We compare estimates with measured frequencies and find that the quasi-static estimate is within 10-20%.

This could be improved with a dynamic model, for example that treats poses that require little kinetic energy to dislodge them as unstable. This is an important area for future research [11].

This simulator can be useful during the design cycle, to permit designers to visualize stable poses and design for "feedability" [9]. A more immediate application is to permit visualization of the flexible feeder with user-supplied parts. Accurate estimates of cycle times can be used to determine how many robots and assembly lines are required to meet specified production rates. This can greatly reduce the time required to set-up or changeover automated factories and hence allow new products to be more rapidly brought to market.

Acknowledgments

We acknowledge Anil Rao and Yan Zhuang for contributions to the quasi-static algorithm reviewed in section 2. Versions of the [19] algorithm were implemented by Goldberg's students at USC: Wiegley in 1992, Zheng Yeh in 1993, and Yan Zhuang in 1994. The convex hull routine used in our current implementation is due to Ioannis Emiris and John Canny. We also thank Canny and Brian Mirtich for sharing preliminary results from their dynamic simulator.

References

[1] David Baraff. Issues in computing contact forces for non-penetrating rigid bodies. Algorithmica, 10:292–352, August 1993. Special Issue on Computational Robotics.

[2] G. Boothroyd, A. H. Redford, C. Poli, and L. E. Murch. Statistical distributions of natural resting aspects of parts for automatic handling. SME Manufacturing Engineering Transactions, 1:72, 1972.

[3] Geoffrey Boothroyd, Corrado Poli, and Laurence E. Murch. Automatic Assembly. Marcel Dekker, Inc., 1982.

[4] Brian Carlisle, Ken Goldberg, Anil Rao, and Jeff Wiegley. A pivoting gripper for feeding industrial parts. In International Conference on Robotics and Automation. IEEE, May 1994. Also available as USC Techreport IRIS-93-316.

[5] John J. Craig. Issues in the design of an off-line programming system. In Proceedings of the 4th International Symposium of Robotics Research, August 1987.

[6] Peter Dewhurst. Design for robotic assembly. Manufacturing Engineering, June 1991.

[7] Gregory T. Farnum and Bill Davis. Delivering the part. Manufacturing Engineering, March 1986.

[8] K. Hoang and R.G. Fenton. Determination of robot's location in manufacturing cell. In Proceedings of the 23rd International Symposium on Industrial Robots, October 1992.

[9] Gerald Kim, George Bekey, and Ken Goldberg. A shape metric for design-for-assembly. In International Conference on Robotics and Automation. IEEE, May 1992.

[10] Matthew T. Mason, Ken Goldberg, and Yu Wang. Progress in robotic manipulation. In 15th Grantees Conference on Production Research and Technology. National Science Foundation, January 1989.

[11] B. Mirtich, Y. Zhuang, K. Goldberg, J. Craig, R. Zanutta, B. Carlisle, and J. Canny. Estimating pose statistics for robotic part feeders. In International Conference on Robotics and Automation, 1996 (submitted).

[12] Brian Mirtich and John Canny. Impulse-based dynamic simulation. In The First Workshop on the Algorithmic Foundations of Robotics. A. K. Peters, Boston, MA, 1995.

[13] James L. Nevins and Daniel E. Whitney. Computer-controlled assembly. Scientific American, 1978.

[14] B. K. A. Ngoi, L. E. N. Lim, and S. S. G. Lee. Analyzing the probabilities of natural resting for a component with a virtual resting face. ASME Journal of Engineering for Industry, (to appear), 1995.

[15] F. P. Preparata and M. I. Shamos. Computational Geometry: An Introduction. Springer-Verlag, 1985.

[16] A. Rao, D. Kriegman, and K. Goldberg. Complete algorithms for reorienting polyhedral parts using a pivoting gripper. In International Conference on Robotics and Automation. IEEE, May 1995.

[17] Berhard J. Schroer. Electronic parts presentation using vibratory bowl feeders. Robotics, 3, 1987.

[18] Neil C. Singer. Utilizing dynamic and static stability to orient parts. Master's thesis, MIT, 1985.

[19] Jeff Wiegley, Anil Rao, and Ken Goldberg. Computing a statistical distribution of stable poses for a polyhedron. In The 30th Annual Allerton Conference on Communication, Control, and Computing. University of Illinois at Urbana-Champaign, October 1992.

Interest of the dual hybrid control scheme for teleoperation with time delays for Proceeding of ISER'95

Claude Reboulet, Yann Plihon, Yves Briere
CERT/DERA
2, Av. Edouard BELIN
31 055 TOULOUSE Cedex
reboulet@cert.fr
plihon@cert.fr

Abstract

It is well known that classical teleoperation schemes are unstable and even useless when time delays become important. In this paper we propose a new concept called "dual hybrid teleoperation scheme" which really improves the ergonomy of a teleoperation system, even with time delays. Thanks to this concept, the operator can exert forces or perform displacements from the "master" to the "slave". These forces are the ones that the human operator wants to apply for the current task. This is possible thanks to a special law coupling the two robots, each of them having an hybrid position-force control.

This scheme also offers kinesthetic feedback, without compromising the stability, even with time delays. So the operator can feel some useful sensations, as "moving feeling". Besides, this scheme allows to split a complex task into subtasks, during which the operator only manages few degrees of freedom. The attention required to perform a given task is really decreased and the operator can focus on a limited set of degrees of freedom.

The experiments done with an system based on this new concept show that it can be apply even if time delay becomes superior to few seconds.

1. Introduction

A robot with autonomous manipulation capabilities does not always allow to solve complex tasks as space assembly or space maintenance tasks, and specially for the fine manipulations involving contact with the environment. It is not always possible in this case to foresee all the necessary actions for solving any task. Owing to the imperfect knowledge of the real world and the complex interaction with the remote environment, an ability of recovery, which can only be brought by a human being, must be introduced. The teleoperation remains the most interesting way for performing work remotely in some applications.

A perfect teleoperation system would be the one where the operator feels the remote object as if he manipulated it directly (as through a virtual mechanical link, with weak inertia and infinite stiffness). Then the interface between the operator and the remote manipulation would be totally transparent. As the operator cannot be in the remote site, a solution consists of a most exact kinesthetic feedback. It is well known that the kinesthetic feedback is as important as the visual one, in particular when the tasks imply a contact with the environment. The kinesthetic sense is the sense transmitted by bodily movement and pressure, therefore kinesthetic feedback is provided by motion and forces exerted by the master robot in the operator's hand. This force feeling really improves the telepresence. It is well known that the operator, being in the loop of the teleoperation system, realizes a task in terms of a feeling coming from the environment. A small delay in such a feedback loop can create unstabilities.

But the tasks with only visual feedback must be distinguished from the tasks with force feedback. The visual delay creates a disturbance the operator can ignore and avoid instability with an "move and wait" strategy. But the kinesthetic feedback with a delay creates a disturbance in the operator hand that he cannot ignore and which is responsible for unstabilities. Time delays are prejudical to the quality of kinesthetic feelings, and direct teleoperation with force feedback is impossible with a delay of one second and more [1].

Performances of such a system are due to the quality of the mechanical conception of master-slave devices, and more particularly of the master, but the control schemes play an important role in the performances of a teleoperation system. Many schemes have been proposed to solve this difficult problem.

The classical solutions to deal with contact forces in teleoperation are described in part 2. We present then in part 3 our new teleoperation scheme based on hybrid control for both the master and the slave robot. Part 4 described the experimental set-up designed to implement this scheme before the conclusion.

2. Force Control in Teleoperation

2.1. Bilateral Feedback Concepts

Two schemes could be considered without time delay:

-Bilateral position concept

In this case, both robots are position controlled and the inputs of each control loops are the positions measured at the other site. With such a scheme, the force felt by the operator is due to the position errors in the remote manipulator's control loop. In an ideal case, the operator should only sense the forces due to external forces applied on the slave robot. However, in the reality, the unavoidable position errors due to position control loop are responsible for viscous friction.

-Force feedback concept

In this case, this scheme is based on the measurement, with a force sensor, of the force exerted by the slave robot on its environment. This scheme generally increases telepresence because the closed loop takes into account the mechanical imperfections such as friction and flexibilities.

But these methods can only be use when the delay remains low. When the time delay is not too important (inferior to 0.1 second) the system can still be stabilized [2], [3], but the bandwidth of the closed loop system decreases drasticly.

Many schemes have been proposed when the delays become important. One of them, [4], proposes to return the predictive force by using an open-loop model-based prediction. The method used in [5] (called teleprogramming) is based on the generation of commands to the telerobot by moving the teleoperator master while getting both force and visual feedback from a computer-based model slave. The semi-autonomous command proposed in [6] allows the operator to interfere in the autonomous fonctions in a progressive way. Others technics are based on shared control [7] and use a remote and local force control.

2.2. Teleoperation with Local Autonomy Concepts

Two types of methods are considered to control the slave robot: the ones are based on active compliance and the others are based on hybrid control. The first type of methods induces a decrease of the gains so as to the slave is not stiffly position-servoed to the master. [8] proposes a compliant system in which the slave robot is compared to a spring with a programmable stiffness. This method allows to maintain the stability still for delays not exceeding one second.

Hybrid control described in [9] is based on a different idea. The choice of the cartesian space partition depends on the local interactions between the slave robot and its environment. The slave is position controlled in the directions unconstrained by the environment while it is force controlled in the constrained directions. This idea initially proposed for others applications can find an interest in many teleoperation schemes. In particular a stiff device can be used as in [10] to transmit commands to the remote manipulator.

3. The Dual Hybrid Position-Force Concept

3.1. General Approach

Let us consider the ideal case of a perfectly force controlled master robot. By setting a zero force input command along its six degrees of freedom, the master device is theoritically weightless. If the operator exerts a force he will encounter almost no resistance, so he can move the master hand freely in all the master robot workspace. It can be used as a single joystick device to transmit position commands to the position controlled remote manipulator. Let us now consider the ideal case of a stiff master device. If this system is equipped with a six degrees of freedom force sensor, it can be used to measure forces exerted by the operator. These measures can be used as input commands for the force controlled remote manipulator.

The concept of our teleoperation scheme combines these two possibilities [11], [12]. To apply this concept, the master robot must be equipped with a hybrid position-force command, dual to the slave hybrid command. It is the reason why we called the system "dual hybrid teleoperation system". Every operation conducted by this teleoperation system needs to separate the cartesian space into two orthogonal subspaces, according to concepts presented in [13]. In the first subspace, called χ_1, the master is zero force controlled in order to appear as transparent as possible. In this subspace, the operator transmits position commands to the slave robot (position controlled in this subspace). In the second subspace, called χ_2, the master robot is position controlled, and the force exerted by the operator are used as input commands for the slave robot (force controlled in this subspace).

This master system can be seen as a dual hybrid sensor. In the subspace where the position is kept constant, it is used as a force sensor. In the complementary subspace, it is used as a position sensor.

The idea of coupling two hybrid robots is already used for the cooperation between two robots handling the same object [14], but not yet proposed for teleoperation application. The figure 1 allows to understand the basic idea on which relies the dual hybrid concept.

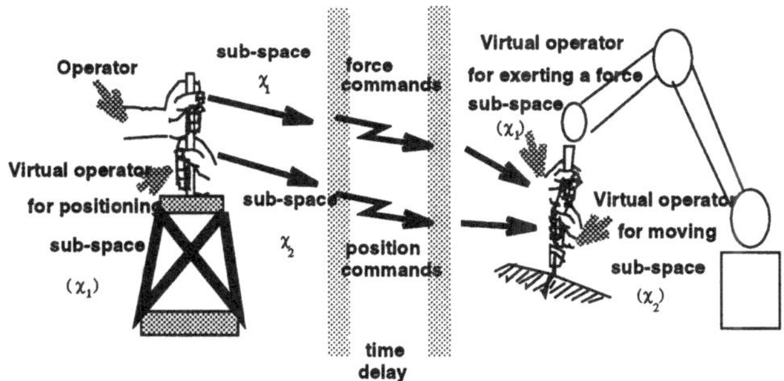

Figure 1. dual coupling (basic idea)

3.2. Definitions

The formalism using this scheme is quite simple. An hybrid task can be described with two frames: a compliant frame and a task frame. Those frames are defined relatively to two other frames, the fixed frame and the mobile frame:

- $\mathcal{F}_f$, the fixed frame, attached to the fixed base of the robot.
- $\mathcal{F}_m$, the mobil frame, attached to the end effector of the robot.
- $\mathcal{F}_t$, the task frame.
- $\mathcal{F}_c$, the compliant frame.

We now define X and F the position and the force measured. The same notations are adopted for both master and slave robots, measured and desired values. The superscript m or s is used to distinguish either the master or the slave robot. The subscript m or d is used to distinguish either the measured or the desired values. X is a six dimensional vector defining the position and the orientation of $\mathcal{F}_c$ relatively to $\mathcal{F}_t$. F is the force wrench in O_c (origine of $\mathcal{F}_c$), written in $\mathcal{F}_t$.

The space partition is described by S a six by six diagonal matrix, the same as defined in [9].

The behavior of a hybrid robot is completely defined by:

-The different frames defined above: $\mathcal{F}_t$ and $\mathcal{F}_c$.

-The configuration matrix: S.

-The position and forces desired (In fact, the teleoperation controller takes only into account the positions in the position controlled subspace, and the forces in the force controlled subspace, thanks to the matrix S).

Besides, a set of maximun speed V_{max}, maximun force F_{max} and maximun position X_{max} are defined for each degrees of freedom.

3.3. Dual Coupling of Two Hybrid Controlled Robots

The configuration definition for one hybrid robot can be extended for a couple of hybrid robots:

- Master: $\mathcal{F}_c^m$, $\mathcal{F}_t^m$, S^m, X_d^m, F_d^m, V_{max}^m, X_{max}^m, F_{max}^m
- Slave: $\mathcal{F}_c^s$, $\mathcal{F}_t^s$, S^s, X_d^s, F_d^s, V_{max}^s, X_{max}^s, F_{max}^s

A connection between the master and the slave along any of the 6 degrees of freedom links the measurements (force or position) of one robot to the commands of the other. In theory, all the connections are possible, hence we have a lot of possibilities. In opposition to the classical teleoperation system, coupling two hybrid robots leads to a lot of configurations, since any axis can be force or position controlled.

In reality, a lot of connections are out of interest. The dual coupling is obtained by simply setting:

- S^s=I-S^m where I is the identity matrix.
- $F_d^s = F_m^m$
- $X_d^s = X_m^m$

The space partitions of the master robot and the slave robot are dual. For any value of the matrix S^m, there is a dual value for the matrix S^s, defining a configuration characteristic of a special behaviour of the master-slave couple.

The configurations corresponding to a dual coupling are:

-The master robot is position controlled at a constant value in the subspace χ_1. It is then used as a force sensor.

-The master robot is force controlled with zero input in the subspace χ_2. It is then used as a position sensor.

Many other configurations can be proposed. For example, the input of some control loop can be provided dy autonomous task (this configuration is similar to the "shared-control" concept). The connections described above transmit the data from the master to the slave (forward connections), but the data can also be transmitted from the slave to the master (backward connections). The various possible connections are despicted on figure 2.

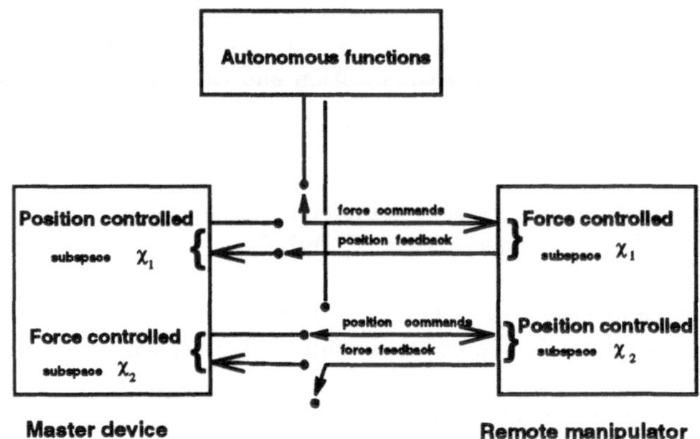

Figure 2. backward and forward connections

We will see in a next part that some of these connections are possible even with time delays.

The following example shows the interest of this new scheme. Only the connections from the master to the slave are taken into account.

In order to illustrate this remark, let us consider the task which consists in sliding a parallelepipedic object on a plane (figure 3) while maintaining a plane on plane contact.

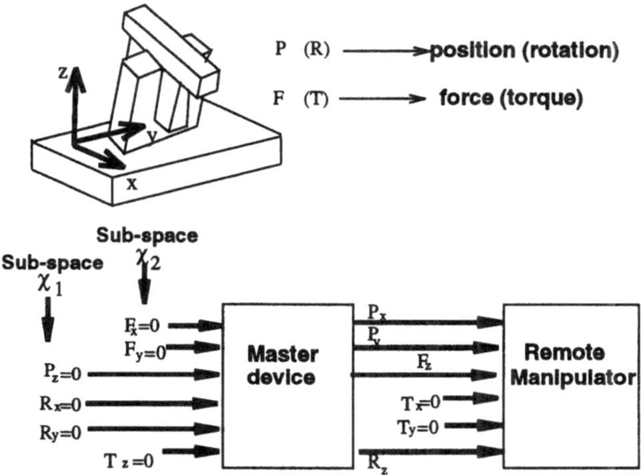

Figure 3. Example: plane on plane contact

Three master degrees of freedom are position controlled in the subspace χ_1 (orientation around x and y axes, and translation along z axis). Only the force measure along z axis is sent to the slave robot. The others are not used. In this example the torques, constantly set to zero, constraint the object to remain in a plane on plane contact situation with the environment. But they could also come from the master system.

The three complementary degrees of freedom are zero force controlled in the subspace χ_2 (translation along x and y axes, and orientation around z axis). The master robot is used as a position sensor for the slave robot.

The operator must only take into account four degrees of freedom, and two of them are critical: they correspond to the displacement in the plane.

4. Interest of the Dual Hybrid Position-Force Scheme for Complex Task Realization

The first important point is the possibility of backward connections. Indeed, time delays create instabilities in a force feedback teleoperation loop in a classical scheme. The position commands are sent from the master to the slave and the force commands are sent from the slave to the master.

In our teleoperation scheme, the master robot is position controlled in the subspace χ_1 and allows to transmit force commands from the master to the slave. The loop can be then closed and slave position measurements can be returned to the master robot, without disturbing the stability.

Indeed, the coupling between the master and the slave is not very strong in this case. In this subspace χ_1, the force exerted by the operator on the master hand is not disturbed by a modification of the position command coming from the slave.

However, this feeling of "moving", even weak, gives important information to the operator on the task success.

One can imagine the interest of this possibility when an insertion is added to the last example. The displacement of the slave robot during the insertion phase is then returned in the operator hand. The progression of the insertion could be felt with time delay of course, if there are communication delays.

A second important point deals with the decrease of the operator attention. Indeed, when working in teleoperation, the operator has difficulties to manage six degrees of freedom. Our scheme allows to split a complex task into subtasks, for which the operator manages only the critical degrees of freedom. Many of the complex assembly tasks can be split into subtasks, which require no more than two or three degrees of freedom at the same time.

5. Experimental Set-Up

We developed an experimental teleoperation set-up composed of two robots, each of them having six degrees of freedom. The slave robot is RCE1, a prototype developed at the CERT-DERA between 1984-1986. It is controlled by an hybrid position-force control scheme. RCE1 is a macro/mini device: the macro robot is a SCARA type robot and the mini is a fully parallel six degrees of freedom wrist. The advantage of this prototype comes from the hybrid position-force control scheme based on the macro/mini architecture, as shown in [15] and [16]. This architecture combines the advantage of both a serial robot (extented workspace) and a parallel robot (lightness and high quality force control). The mini robot is controlled by six pneumatic linear actuators driven by servo-valves. Four processors (68000) host the hybrid control of RCE1 with the real time system CESAR, also developed at CERT. The experimental set-up is depicted in Fig. 4.

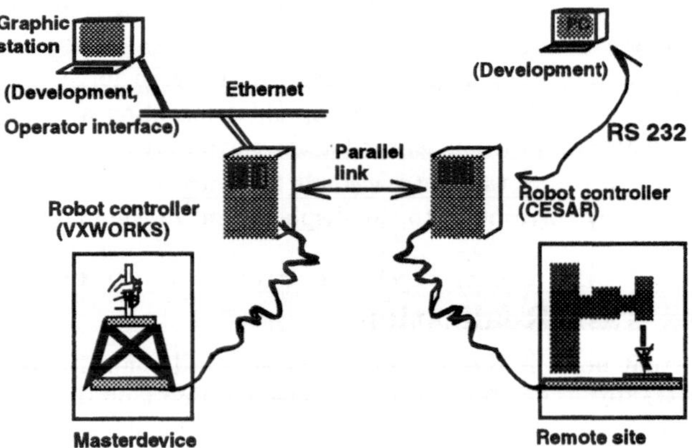

Figure 4. experimental set-up

The master robot is a fully parallel device having six degrees of freedom.The real time system, VXworks, is supported by two processors running in parallel (one 68040 MVME167 and one 68030 MVME147) and mounted on a VME rack. The sampling period of the two robots is about 3ms. A third processor is used for master-slave communication. Operator interface and development are done on a INDY graphical workstation.

6. Conclusion

In the classical teleoperation schemes, the forces are sent from the slave to the master. In this case, the force displayed on a screen or returned in the operator hand gives a true feeling about the interactions between the manipulated objects by the slave and its environment. As these schemes are instable and useless when the time delays are important, a new scheme, called "dual hybrid teleoperation scheme", is proposed. In this scheme, the operator can send the forces from the master to the slave. In fact, at the inverse of classical schemes, the forces which are sent, are the forces that the operator wants to impose during the current task.

This is possible thanks to a coupling between the two robots. But this scheme requires two robots equipped with hybrid position-force control.

This scheme also offers some possibilities of feedback connections, without creating instabilities in spite of time delays. A position command can be returned in the operator hand, which creates useful kinesthetic feelings. These kinesthetic feelings are less important that the ones obtained without time delays in the classical schemes, but they allow to better appreciate the success of each subtask and improve significantly the operator's feelings.

Besides, this scheme allows to split a complex task into subtasks, during which the operator only manages few degrees of freedom. The intellectual work is really decreased and the operator can focus on the critical degrees of freedom.

The experiments conducted in our laboratory show that these new concepts allow to achieve difficult tasks, such as insertions with very small clearance.

References

[1] W.R. Ferrel, "Delayed force feedback", IEEE Trans. Human Factors in Electronics, 449-455, October 1966.

[2] R.J. Anderson and M.W. Spong, "Asymtotic stability for force reflecting teleoperators with time delay", Proceeding of the International Conference on Robotics and Automation, Scottsdale, Arizona, pp 1618-1625, 1989.

[3] C. Andriot, R. Fournier, J. Vuillemey, "on the bilateral control of teleoperators with flexible joints and time delays by the passive approach", Proceeding of the International conference on Advanced Robotics, Pisa, Italy, vol. 1, pp 231-236, june 1991.

[4] F. Buzan and T.B. Sheridan, "A model-based predictive operator-aid for telemanipulators with time delay", Proceeding of the International Conf. Syst. Man and Cybern., Cambridge, MA., pp 14-17, nov 1989.

[5] J. Funda and R.P. Paul, "Efficient control of a robotic system for time-delayed environments", Proceeding of the International Conference on Advanced Robotics, Pisa, Italy, vol. 1, pp 219-224, june 1991.

[6] Y. Yokokohji, A. Ogawa, H. Hasunuma and T. Yoshikawa, "Operation modes for cooperating with autonomous functions in intelligent teleoperation systems", Proceeding of the International Conference on Robotics and Automation, Atlanta, Georgia, pp 510-515, 1993.

[7] S. Hayati, T. Lee, K. Tso and P. Backes, "A testbed for a unified teleoperated-autonomous dual-arm robotic system", Proceeding of the International Conference on Robotics and Automation, Cincinnati, Ohio, pp 1090-1095, 1990.

[8] A.K. Bejczy and W.S. Kim, "Predictive displays and shared compliance control for time-delayed manipulation", Proc. IEEE Int Workshop Intelligent Robots Syst., IROS, 1990.

[9] M.H. Raibert and J.J. Craig, "Hybrid position/force control of manipulator", ASME, Journal of Dynamic Systems, Measurement and Control, June 1981, pp 1-10.

[10] G. Hirzinger, J. Heindl and K. Landzettel, "Control Structures in sensor-based telerobotic systems", Proceeding of the International Conference on Advanced Robotics, Pisa, Italy, vol. 1, pp 267-276, june 1991.

[11] Y. Briere, "Teleoperation en presence de retard: le concept de teleoperation hybride duale", these, Ecole Nationale Superieure de l'Aeronautique et de l'Espace, 1994.

[12] Y. Briere, Yann Plihon, Claude Reboulet, "The dual hybrid position-force concept for teleoperation", Euriscon 94, Malaga, Spain, 21-26 August, 1994.

[13] M.T. Mason, "Compliance and force control for computer controlled manipulators", IEEE Trans. Systems, Man and Cybernetics SMC-11, 6 (1981), pp 418-432.

[14] P. Dauchez, "Description de taches en vue de la commande hybride symetrique d'un robot manipulateur a deux bras", These d'etat, Universite des Sciences et Techniques de Languedoc, 1990.

[15] C. Reboulet, A.Robert, "Hybrid control of a manipulator equipped with an active compliant wrist", Proc. of the third International Symposium on Robotics Research, Gouvieux, France, pp 237-241, October 1985.

[16] C. Reboulet and R. Pigeyre "Hybrid control of a 6 dof in parallel actuated micromanipulator mounted on a Scara robot", International Journal of Robotics and Automation, Vol 7, Isue 1, 1992.

Robot Force Control Experiments with an Actively Damped Compliant End Effector

Joris De Schutter, Dirk Torfs*, Stefan Dutré and Herman Bruyninckx

Katholieke Universiteit Leuven, Dept. of Mechanical Engineering
Heverlee, Belgium
joris.deschutter@mech.kuleuven.ac.be

Abstract

A new method based on modal decoupling is proposed for the hybrid force/position control of a robot with compliant end effector during constrained motion. This method incorporates results from control of flexible robots in order to damp out undesired vibrations of the end effector through active control. The contact dynamics during constrained motion are described by vibrational modes. These are either unconstrained real vibrational modes, or constrained, hence degenerated modes.

1. Introduction

Many researchers concentrate on finding stable and robust contact force control schemes for a rigid robot in contact with a rigid environment, i.e. the so called case of hard contact. This turns out to be a very hard problem. It involves the formulation of the highly nonlinear dynamics of the constrained robot, it results in very high bandwidths, and hence requires very high sample rates in the case of digital control. Also it is very difficult to find solutions which are robust with respect to modeling errors, e.g. friction in the drive systems, or poor knowledge of the actual robot dynamics, contact stiffness, contact geometry, etc. The problem gets much more easy—at least in the case of contact—if there is some passive compliance in the system. However most researchers avoid this solution because the introduction of passive compliance has some drawbacks for motion in free space: it decreases position accuracy and results in lowly damped, low frequency vibrations of the end effector. The latter phenomena are the subject of research on flexible robots, a quite distinct research area so far. On the other hand, a much more pragmatic approach has always been to indeed provide a substantial amount of passive compliance in the system, e.g. in the robot end effector. This passive compliance tends to decouple the robot dynamics from the force interaction with the environment. Hence, control of the interaction forces reduces to a motion control problem for the robot: the force controller transforms a force error into a motion command in order to compress or relax a multidimensional spring, [1]. Or, force control loops are closed around the motion control system. This cascaded approach results in *simple* and *robust* control systems. Simple, because a simple internal motion controller, e.g. without

*Currently at TRASYS Space, Zaventem, Belgium.

dynamic robot model, may be sufficient to achieve good force control behavior; robust, because disturbances in the robot actuation system, e.g. nonlinear friction, are first suppressed by the motion controller before they affect the contact force. With this approach industrial applications have been shown feasible [2] and also the first robot in space used this approach [3]. The low frequency vibrations are circumvented by using a passive compliance which is mechanically blocked during motions in free space, and released again just before establishing contact. Note that this blocking is non-selective: the degrees of freedom of the compliance, usually six, are either all blocked, or all free.

This paper starts from the second, pragmatic approach, but combines it with recent results obtained in control of flexible robots, more specifically, control of a robot with a compliant end effector, [4,5]. The aim is to achieve an *active* and *selective* damping of the passive compliance: during partially constrained motion the passive compliance is damped in the motion freedom space only.

The invariant approach adopted in this paper comes closest to [6,7]: [6] describes how to decompose a general stiffness matrix in physically meaningful *eigen-screws*, and to use it to design a hybrid control law for static contact forces. [7] looks into the problem of eigenscrews for singular compliance and stiffness matrices, i.e., compliant elements that are constrained by contacts with the environment. [6] and [7] consider only the stiffness or compliance component of the end effector. Since they do not introduce the inertia of the manipulated object, they are unable to describe its vibrational modes, as we do in this paper.

This paper is organized as follows. Section 2 describes the contact dynamics during contrained motion as a MIMO system. Using modal decoupling, this MIMO system is decomposed into its vibrational modes, i.e. a set of decoupled SISO systems. The modes are either unconstrained real vibrational modes, or constrained, hence degenerated modes. Both are easy to control, as explained in section 3, in order to yield the desired contact forces together with vibration free motion of the manipulated object. Section 4 presents experimental results for a partially constrained motion. Finally, section 5 draws the conclusions.

2. System Model

This section describes the dynamic model of a velocity controlled robot equipped with a wrist force/torque sensor and a compliant end effector during partially constrained motion. More details are given in [8,9]. Consider a robot with compliant end effector holding a manipulated object (MO).

The MO is constrained in its motion by contacts with other objects in the environment. Following assumptions are made: (1) the robot, the MO, and the objects in the environment are considered infinitely rigid; (2) the end effector is the main source of compliance in the system, and its compliance is concentrated; hence part of the end effector mass is incorporated into the robot mass; the other part is incorporated into the MO; (3) the robot is equipped with a very stiff velocity controller; hence the robot dynamics may be neglected, because they are much faster than the dynamics of the contact interaction, or: the actual robot velocity is assumed equal to the desired robot velocity.

Following notation is used. Translational and rotational velocities $\boldsymbol{v}$ and $\boldsymbol{\omega}$ are grouped into a twist vector $\boldsymbol{t} = [\boldsymbol{v}^T\ \boldsymbol{\omega}^T]^T$. Similarly, forces and torques $\boldsymbol{f}$ and $\boldsymbol{n}$ are grouped into a wrench vector $\boldsymbol{w} = [\boldsymbol{f}^T\ \boldsymbol{n}^T]^T$. Small displacements are represented by an infinitesimal displacement twist $\boldsymbol{t}_\Delta$ which contains infinitesimal translational and rotational components. The end effector compliance is assumed to be linear, and is characterized by its stiffness matrix $\boldsymbol{K}_{6\times6}$, and damping matrix $\boldsymbol{C}_{6\times6}$. This paper deals with the case of little or no damping, i.e. $\boldsymbol{C} = \boldsymbol{O}$, since this causes the undesired oscillatory behavior of the end effector. The MO is characterized by its mass m and inertia tensor $\boldsymbol{I}_{MO}$, combined in an inertia matrix $\boldsymbol{M}$:

$$\boldsymbol{M}_{6\times6} = \begin{bmatrix} m\boldsymbol{I}_{3\times3} & \boldsymbol{O}_{3\times3} \\ \boldsymbol{O}_{3\times3} & \boldsymbol{I}_{MO} \end{bmatrix},$$

as expressed in a centroidal reference frame. The robot twist $\boldsymbol{t}_R$ is given by

$$\boldsymbol{t}_R = \boldsymbol{J}_R\, \boldsymbol{u}, \tag{1}$$

where $\boldsymbol{J}_R$ is the robot Jacobian, and $\boldsymbol{u}$ is the vector of joint velocities (desired or actual), i.e. the input to the system. During constrained motion the twist of the MO can be written as

$$\boldsymbol{t}_{MO} = \boldsymbol{J}\boldsymbol{\mathcal{T}}, \tag{2}$$

where $\boldsymbol{J}_{6\times n_u}$ is the so-called contact twist Jacobian, containing a set of basis twists of the unconstrained twist space, i.e. the space of twists allowed by the contact; $\boldsymbol{\mathcal{T}}$ is the vector of dimensionless twist space coordinates, and n_u is the dimension of the unconstrained twist space, $n_u \leq 6$. [8,9] show that the transfer matrix between the input joint velocities $\boldsymbol{u}$ and the wrench $\boldsymbol{w}_m$ measured in the force/torque sensor is given by

$$\boldsymbol{w}_m = \boldsymbol{K}\left(\boldsymbol{J}\boldsymbol{J}_M^\dagger \boldsymbol{M}^{-1}\boldsymbol{K} + s^2\boldsymbol{I}\right)^{-1} s\boldsymbol{J}_R\, \boldsymbol{u}. \tag{3}$$

where the pseudo-inverse $\boldsymbol{J}_M^\dagger$ is defined by

$$\boldsymbol{J}_M^\dagger = \left(\boldsymbol{J}^T\boldsymbol{M}\boldsymbol{J}\right)^{-1}\boldsymbol{J}^T\boldsymbol{M}. \tag{4}$$

$\boldsymbol{J}\boldsymbol{J}_M^\dagger\boldsymbol{M}^{-1}\boldsymbol{K}$ is a 6×6 matrix whose eigenvalues are represented by a diagonal matrix

$$\boldsymbol{\Omega}^2 = diag(\omega_i^2), \quad i = 1, \dots, 6,$$

and whose eigenvectors are represented by a 6×6 matrix:

$$\boldsymbol{T}_\Delta = \left[\boldsymbol{t}_{\Delta,i}\right], \quad i = 1, \dots, 6. \tag{5}$$

$\boldsymbol{t}_{\Delta,i}$ and $\boldsymbol{\omega}_i^2$ are solutions of the eigenvalue problem

$$\boldsymbol{J}\boldsymbol{J}_M^\dagger\boldsymbol{M}^{-1}\boldsymbol{K}\boldsymbol{t}_{\Delta,i} = \omega_i^2\boldsymbol{t}_{\Delta,i}, \quad \text{or} \quad \boldsymbol{J}\boldsymbol{J}_M^\dagger\boldsymbol{M}^{-1}\boldsymbol{K}\boldsymbol{T}_\Delta = \boldsymbol{T}_\Delta\boldsymbol{\Omega}^2. \tag{6}$$

For motion in free space, or unconstrained motion, all $\omega_i^2 > 0$. For constrained motion, $n_c = 6 - n_u$ of the ω_i^2's are equal to zero, since the span of projection operator $\boldsymbol{J}\boldsymbol{J}_M^\dagger$ has dimension n_u. ω_i are the mechanical resonance frequencies of the constrained system; the corresponding eigenvectors, $\boldsymbol{t}_{\Delta,i}$, are called *modal vectors.* They form a set of basis vectors for the (infinitesimal) displacement twist space, i.e. $\boldsymbol{T}_\Delta$ has full rank. Every modal vector in the displacement space $\boldsymbol{t}_{\Delta,i}$, has a corresponding modal vector, $\boldsymbol{w}_i$, which can serve as a basis vector for the wrench space:

$$\boldsymbol{w}_i = \boldsymbol{K}\boldsymbol{t}_{\Delta,i}, \quad \text{or} \quad \boldsymbol{W} = \boldsymbol{K}\boldsymbol{T}_{\Delta}, \tag{7}$$

where

$$\boldsymbol{W} = [\boldsymbol{w}_i], \quad i = 1, \ldots, 6.$$

Now, express $\boldsymbol{w}_m$ using basis $\boldsymbol{W}$, and express $\boldsymbol{t}_R = \boldsymbol{J}_R\boldsymbol{u}$ using basis $\boldsymbol{T}_{\Delta}$:

$$\boldsymbol{w}_m = \boldsymbol{W}\boldsymbol{\gamma}, \quad \boldsymbol{t}_R = \boldsymbol{T}_{\Delta}\,\boldsymbol{\phi}. \tag{8}$$

$\boldsymbol{\gamma}$ is a dimensionless 6-vector of so-called wrench coordinates; $\boldsymbol{\phi}$ is a 6-vector of twist coordinates with units $\frac{1}{s}$. Introducing (6), (7), and (8) in (3) leads to

$$\boldsymbol{\gamma} = s(s^2\boldsymbol{I} + \boldsymbol{\Omega}^2)^{-1}\boldsymbol{\phi}. \tag{9}$$

This is equivalent to six decoupled SISO systems

$$\gamma_i = \frac{s}{s^2 + \omega_i^2}\phi_i, \quad i = 1, \ldots, 6. \tag{10}$$

In case $\omega_i = 0$, i.e. in the degenerated or constrained modal directions, this reduces to

$$\gamma_i = \frac{1}{s}\phi_i. \tag{11}$$

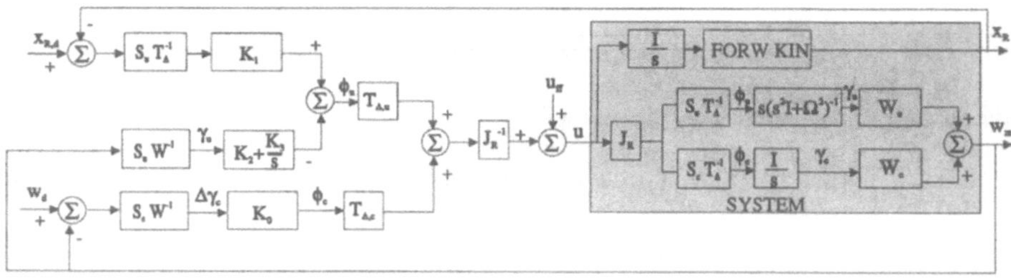

Figure 1. Hybrid control scheme for partially constrained robot.

3. Control

A set of decoupled controllers is designed for each of the modal directions. Figure 1 shows the resulting control scheme. In this figure, $\boldsymbol{S}_{\boldsymbol{u}(n_u \times 6)}$ and $\boldsymbol{S}_{\boldsymbol{c}(n_c \times 6)}$ are physically dimensionless selection matrices which are used to partition the twist and wrench coordinates $\boldsymbol{\phi}$ and $\boldsymbol{\gamma}$ into coordinates corresponding to unconstrained modes ($\boldsymbol{\phi}_u$ and $\boldsymbol{\gamma}_u$) and coordinates corresponding to constrained modes ($\boldsymbol{\phi}_c$ and $\boldsymbol{\gamma}_c$) respectively. For example, if the first n_u eigenvectors in (5) correspond to the unconstrained modes, then

$$\boldsymbol{S}_u = \left[\boldsymbol{I}_{n_u \times n_u}\ \boldsymbol{O}_{n_u \times (6-n_u)}\right], \quad \boldsymbol{S}_c = \left[\boldsymbol{O}_{n_c \times (6-n_c)}\ \boldsymbol{I}_{n_c \times n_c}\right].$$

Using $\boldsymbol{S}_u$ and $\boldsymbol{S}_c$ the system can be split up into its unconstrained and constrained modes, as shown in the shaded area on the right of Figure 1. The control of both types

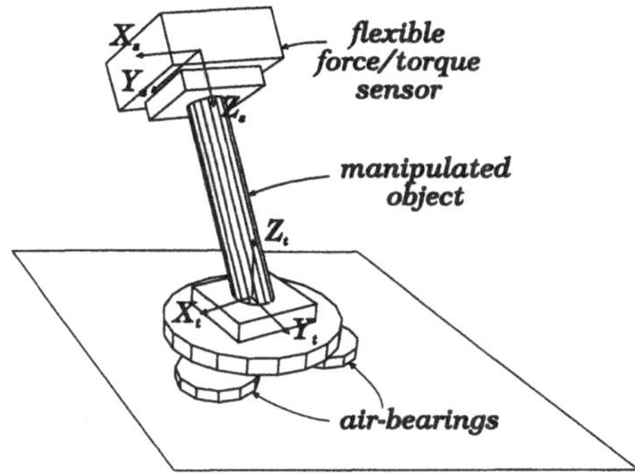

Figure 2. The manipulated object in contact with the environment.

of modes proceeds as follows. In the constrained modal directions, which basically correspond to pure integrators, one set of proportional feedback constants suffices (see lowest feedback loop in Figure 1). It transforms an error in the constrained model wrench coordinates $\Delta\boldsymbol{\gamma}_c$ to modal twist coordinates $\boldsymbol{\phi}_c$ in the constrained directions. Here $\boldsymbol{w}_d$ represents the desired contact wrench. Hence $\boldsymbol{K}_0$ is a $n_c \times n_c$ diagonal feedback gain matrix with units $\frac{1}{s}$. In the unconstrained modal directions the approach follows [5]. The controller output $\boldsymbol{\phi}_u$ consists of two terms: feedback of the *rigid body* mode and feedback of the *flexible* mode. As for the rigid body mode, the robot position error is first transformed into its modal coordinates. Here $\boldsymbol{x}_R$ and $\boldsymbol{x}_{R,d}$ represent actual and desired robot position. Then the coordinates in the unconstrained space are transformed to desired twist coordinates $\boldsymbol{\phi}_u$ by proportional feedback gains $\boldsymbol{K}_1$, as shown by the upper feedback loop in Figure 1.

As for the flexible mode, the unconstrained modal wrench coordinates $\boldsymbol{\gamma}_u$ are fed back using proportional and integral gains $\boldsymbol{K}_2$ and $\boldsymbol{K}_3$ as shown by the feedback loop in the middle. $\boldsymbol{K}_1$, $\boldsymbol{K}_2$ and $\boldsymbol{K}_3$ are diagonal $n_u \times n_u$ matrices with units $\frac{1}{s}$, $\frac{1}{s}$ and $\frac{1}{s^2}$ respectively. As shown in [5], positive $\boldsymbol{K}_2$-gains increase the damping of the modes, whereas positive $\boldsymbol{K}_3$-gains increase their frequencies. It is easy to prove that, using this controller, the unconstrained and constrained modes remain uncoupled, at least under the assumptions accepted in this paper.

4. Experiments

The test setup consists of an industrial KUKA IR 361/8 robot equipped with a six component force/torque sensor. This sensor consists of four linear springs. The springs are mounted between two aluminum plates; one plate is attached to the robot, the other is connected to the MO. The sensor compliance is instrumented with an optical 6 component deformation measurement. Hence, the forces and torques are obtained by multiplying the measurement with the stiffness matrix. The manipulated object (MO) is attached to the force/torque sensor. The robot has a very stiff velocity controller (bandwidth of 30 to 40 Hz), so that its dynamics may be neglected. On the other hand the robot has unmodeled structural dynamics; the lowest resonance frequency is about 20 Hz. The control frequency is 100 Hz.

Figure 2 gives the configuration of the end-effector, the manipulated object and the environment used in the experiments. Air-bearings are used in order to eliminate damping caused by friction. Axes X_t, Y_t, Z_t configure a task frame which is rigidly

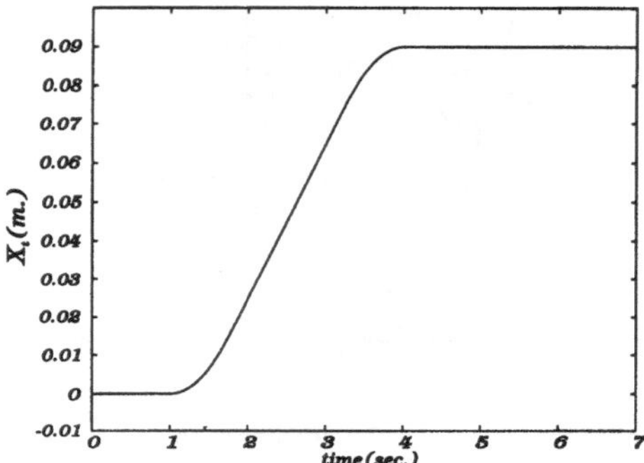

Figure 3. Desired position trajectory.

connected to the MO. The task is to move the MO over a flat surface, while maintaining desired contact wrenches in the constrained directions (forces f_{zt} and torques n_{xt} and n_{yt}). The stiffness and inertia matrix expressed in the task frame, $\boldsymbol{K}_t$ and $\boldsymbol{M}_t$ are given in the Appendix. The twist Jacobian, used in Eq. (2), expressed in the task frame, is

$$\boldsymbol{J} = \begin{bmatrix} 1\frac{m}{s} & 0 & 0 \\ 0 & 1\frac{m}{s} & 0 \\ 0 & 0 & 0 \\ 0 & 0 & 0 \\ 0 & 0 & 0 \\ 0 & 0 & 1\frac{rad}{s} \end{bmatrix}.$$

There are three nonzero resonance frequencies, which are computed from the eigenvalues of $\boldsymbol{J}\boldsymbol{J}_M^{\dagger}\boldsymbol{M}_t^{-1}\boldsymbol{K}_t$: 6.9, 14.6 and 20.5 Hz. The corresponding eigenvectors, $\boldsymbol{T}_{\Delta,u}$, are the basis vectors for the unconstrained twist subspace and are shown in the Appendix. In order to simplify the interpretation of the contact wrenches, the basis vectors for the constrained wrench subspace are chosen as unit vectors along f_{zt}, n_{xt}, n_{yt}:

$$\boldsymbol{W}_c = \begin{bmatrix} 0 & 0 & 0 \\ 0 & 0 & 0 \\ 1N & 0 & 0 \\ 0 & 1Nm & 0 \\ 0 & 0 & 1Nm \\ 0 & 0 & 0 \end{bmatrix}. \tag{12}$$

Finally, bases $\boldsymbol{T}_\Delta$ and $\boldsymbol{W}$ are completed as

$$\boldsymbol{T}_\Delta = \begin{bmatrix} \boldsymbol{T}_{\Delta,u} & \boldsymbol{K}_t^{-1}\boldsymbol{W}_c \end{bmatrix}, \qquad \boldsymbol{W} = \begin{bmatrix} \boldsymbol{K}_t\boldsymbol{T}_{\Delta,u} & \boldsymbol{W}_c \end{bmatrix}.$$

The experiments concentrate on active damping of the first unconstrained mode at 6.9 Hz only, for two reasons. First, this mode is quite dominant, and it is much more excited than the higher modes at 14.6 and 20.5 Hz. Also, these higher modes are influenced by the unmodeled structural dynamics at nearby frequencies. As a result active damping of these modes is much more difficult to attain. The first mode

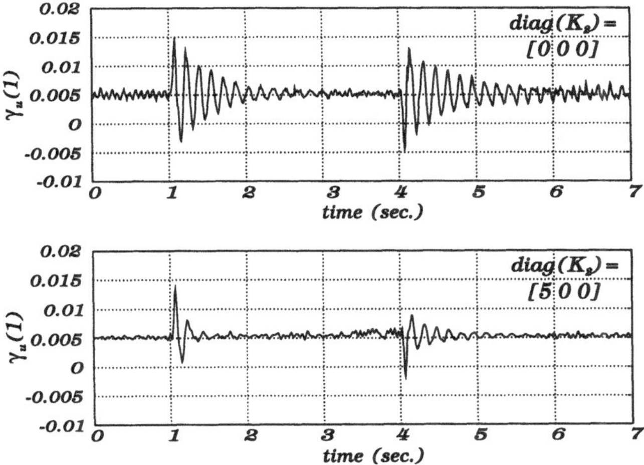

Figure 4. Unconstrained modal wrench coordinate corresponding to the first resonance mode, $\gamma_u(1)$. Top: without damping; bottom: with damping.

is excited by moving the MO in the X_t direction. Figure 3 shows the desired position trajectory. The trajectory has a trapezoidal velocity profile with a maximum velocity of 0.04 m/sec. The desired wrench trajectory with respect to the task frame consists of a constant contact force of 80 N in the Z_t direction and constant zero torques about the X_t and Y_t axes.

Figure 4 shows the resulting unconstrained modal wrench coordinate corresponding to the first mode at 6.9 Hz, $\gamma_u(1)$. The upper part of the figure shows $\gamma_u(1)$ in case the flexible part of the controller is turned off, i.e. $\boldsymbol{K}_2 = \boldsymbol{K}_3 = \boldsymbol{O}$. In the lower part of the figure $diag(\boldsymbol{K}_2)$ is chosen [5 0 0] in order to increase the damping of the first mode; $\boldsymbol{K}_3$ stays $\boldsymbol{O}$. In producing these plots a low pass filter with a 18 Hz cut off frequency has been used in order to remove the influence of the unmodeled structural dynamics.

Figure 5 shows the constrained modal wrench coordinates, γ_c, i.e the decomposition of the measured wrenches with respect to basis $\boldsymbol{W}_c$, Eq. (12) for the damped case. Again, these results are filtered using the same low pass filter. This figure confirms that the unconstrained and constrained subspaces are almost completely decoupled. It shows only a minor influence of the unconstrained subspace on the constrained subspace, e.g. at t=1 sec and t=4 sec.

5. Conclusion

A new method has been presented to perform independent control of both motion and contact forces during constrained robot motion using a compliant end effector. This method incorporates results from control of flexible robots, in order to damp out undesired low frequency vibrations through active control. By using the technique of modal decoupling the multidimensional system is reduced to a set of independent one-dimensional systems. These one-dimensional systems are either constrained or unconstrained. Experimental results have been presented for a motion with a planar constraint. They confirm the theoretical results, i.e. the introduction of end effector damping through active control, and the decoupling between motion and force con-

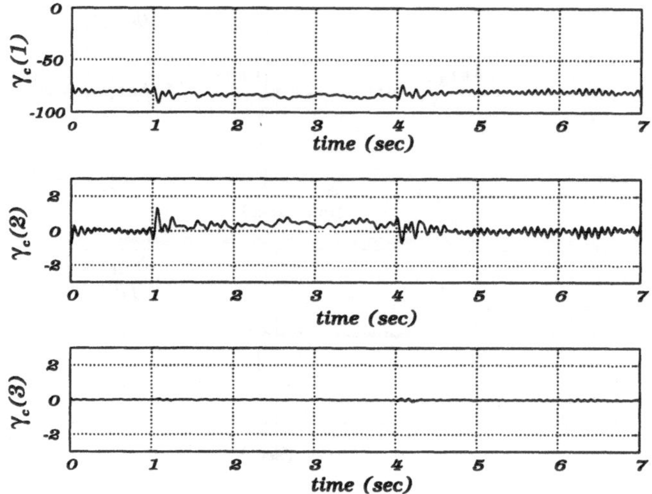

Figure 5. Coordinates of the constrained wrench basis vectors, γ_c.

trolled subspaces. In particular the first unconstrained mode is damped out very well; the two other unconstrained modes are more difficult to control because they are influenced by unmodeled structural dynamics at nearby frequencies.

Acknowledgment

This text presents research results of the Belgian programme on Interuniversity Poles of Attraction by the Belgian State, Prime Minister's Office, Science Policy Programming. This work was also sponsored by the *SECOND* project of the European Community (ESPRIT Basic Research Action 6769). H. Bruyninckx is Post-Doctoral Researcher of the NFWO (the Belgian National Fund for Scientific Research). The scientific responsibility is assumed by its authors. The authors would also like to thank F. Al Bender, L. Vanneste and L.-T. Van Waes for their help in performing the experiments.

Appendix

$$
\boldsymbol{M}_t = \begin{bmatrix}
5.3kg & 0 & 0 & 0 & 0.6kgm & 0.3kgm \\
0 & 5.3kg & 0 & -0.6kgm & 0 & 0 \\
0 & 0 & 5.3kg & -0.3kgm & 0 & 0 \\
0 & -0.6kgm & -0.3kgm & 0.2kgm^2 & 0 & 0 \\
0.6kgm & 0 & 0 & 0 & 0.16kgm^2 & 0.07kgm^2 \\
0.3kgm & 0 & 0 & 0 & 0.07kgm^2 & 0.07kgm^2
\end{bmatrix},
$$

$$
\boldsymbol{K}_t = \begin{bmatrix}
40991\frac{N}{m} & 0 & 0 & 0 & 11793\frac{N}{rad} & 5554\frac{N}{rad} \\
0 & 45079\frac{N}{m} & -8742\frac{N}{m} & -11784\frac{N}{rad} & 0 & 0 \\
0 & -8742\frac{N}{m} & 55173\frac{N}{m} & -4961\frac{N}{rad} & 0 & 0 \\
0 & -11784N & -4961N & 4240\frac{Nm}{rad} & 0 & 0 \\
11793N & 0 & 0 & 0 & 3571\frac{Nm}{rad} & 1583\frac{Nm}{rad} \\
5554N & 0 & 0 & 0 & 1583\frac{Nm}{rad} & 949\frac{Nm}{rad}
\end{bmatrix},
$$

$$T_{\Delta,u} = \begin{bmatrix} -0.1621m & -0.0305m & 0 \\ 0 & 0 & 1m \\ 0 & 0 & 0 \\ 0 & 0 & 0 \\ 0 & 0 & 0 \\ 0.9868rad & -0.9995rad & 0 \end{bmatrix}.$$

References

[1] De Schutter, J., A Study of Active Compliant Motion Control Methods for Rigid Manipulators Based on a Generic Scheme, *Proc. IEEE Conf. Rob. Automation*, Raleigh, pp. 1060–1065, 1987.

[2] De Schutter, J., Katupitya, J., Vanherck, P. and Van Brussel, H., Active Force Feedback in Industrial Robotic Assembly: A Case study, *Int. J. Adv. Manuf. Techn.*, 2(4):27–40, 1987.

[3] Hirzinger, G., Brunner, B., Dietrich, J. and Heindl, J., Sensor Based Space Robotics—ROTEX and Its Telerobotic Features, *IEEE Trans. Rob. Automation*, 9(5):649–663, 1993.

[4] Torfs, D. and De Schutter, J., Comparison of Control Results of a Flexible One-Link Robot Equipped with a Velocity or Torque Controlled Actuator, *Proc. IEEE Int. Conf. Rob. Automation*, pp. 230–235, 1993.

[5] Torfs, D. and De Schutter, J., Tuning of an Add-on Flexible Mode Controller for a Robot driven by a Velocity Controlled Actuator, *Proc. IFAC Symp. Robot Control*, pp. 735–740, 1994.

[6] Griffis, M. and Duffy, J., Kinestatic Control: A Novel Theory for Simultaneously Regulating Force and Displacement, *Trans. ASME, J. Mech. Des.*, 113:508–515, 1991.

[7] Patterson, T. and Lipkin, H., Duality of Constrained Elastic Manipulation, *Proc. IEEE Int. Conf. Rob. Automation*, pp. 2820–2825, 1991.

[8] De Schutter, J., Torfs, D. Dutré S. and Bruyninckx, H., Invariant Hybrid Position/Force Control of a Velocity Controlled Robot with Compliant End Effector using Modal Decoupling. Accepted for publication in the *Int. J. Rob. Research.*

[9] De Schutter, J., Torfs, D., and Bruyninckx, H., Combination of Robot Force Control with Active Damping of a Flexible End Effector, *Proc. Int. Conf. Recent Adv. Mechatronics*, Istanbul, pp. 1006–1011, 1995.

Improved Force Control for Conventional Arms Using Wrist-Based Torque Feedback

David Williams and Oussama Khatib

Robotics Laboratory
Department of Computer Science
Stanford University, Stanford CA 94305
david,ok@flamingo.stanford.edu

Abstract

Torque control is an effective technique for improving robot performance. However, most conventional robots are not equipped with torque sensors, and they are difficult to add without modifying the robot's mechanical structure. To address this problem, we have developed a new method that allows torque feedback to be implemented in the absence of joint-torque sensors. This technique estimates the torque components acting in the force-control subspace, using a conventional wrist force sensor. We have implemented and experimentally validated this approach using a PUMA 560 manipulator. Our results confirm the effectiveness of this method in dealing with the limitations of conventional actuation/transmission systems.

1. Introduction

Robot performance, especially in tasks involving force control, depends primarily on the accurate application of joint torques. However, industrial robots have many nonlinearities in their actuator/transmission systems which significantly affect the torques at their joints. The result is a situation similar to that shown Figure 1. Here, disturbances such as friction, backlash, and motor cogging all work to modify the applied torques. Unmodeled transmission dynamics may also be excited at certain frequencies, further degrading torque accuracy and perhaps even causing instability. When we include the effect of sensor noise and environmental disturbances, it isn't surprising that many robotic systems have limited performance in fine motion, force control, and cooperative manipulation tasks.

To improve performance in these tasks, a model of the actuator/transmission system should be used to design and implement joint torque control, as illustrated in Figure 2. This direct control of joint torques allows rejection of disturbances and improves the overall performance of the system.

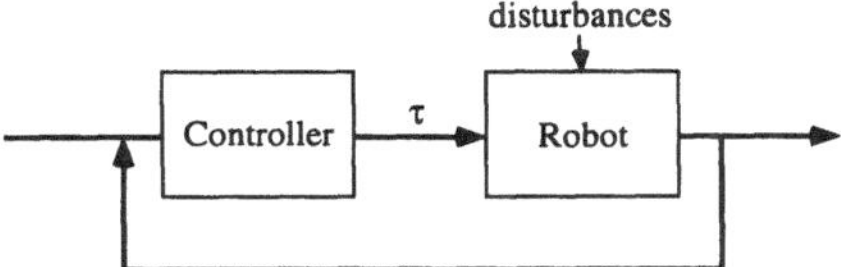

Figure 1. A Control System With No Torque Feedback

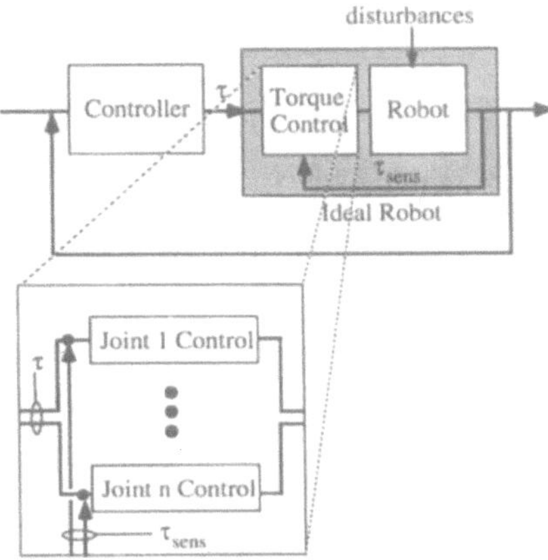

Figure 2. Using Torque Feedback to Improve Performance

2. End-Effector Sensing

The difficulty with implementing torque feedback is that most robots are not equipped with joint-torque sensors. To address this problem, we have developed a new method that allows implementation of torque feedback in the absence of joint-torque sensors. The method works by estimating joint-torque components acting in the force-control subspace, using a conventional wrist force sensor.

When the robot is rigidly connected to its environment and is completely constrained, a good estimate of its joint torques can be obtained from the sensed forces and moments at the end-effector, using the transpose of the Jacobian matrix

$$\mathbf{\Gamma}_{sens} = \mathbf{J}^T(\mathbf{q})\mathbf{F}_{sens}. \tag{1}$$

The assumption here is that the dominant robot flexibilities are located in the actuator/transmission system.

For tasks involving combined motion and force control, however, the manipulator is only partially constrained. In this case, some portion of the sensed force is due to motion and another part is due to contact. Therefore, in the directions of motion, the estimates given above no longer correspond to the torques applied at the links. Instead, end-effector forces in these directions, and the corresponding torques, are due to acceleration.

Our approach to torque feedback is to estimate and control only those torques that correspond to the constrained subspace. This estimate is obtained by a projec-

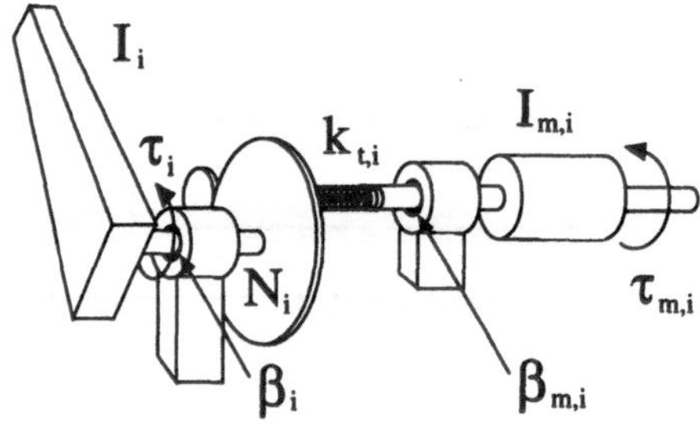

Figure 3. The Transmission Model

tion of the force sensor output into the force-control durections using the *generalized selection matrix* [3]

$$\overline{\mathbf{F}}_{sens} = \overline{\mathbf{\Omega}}\mathbf{F}_{sens}. \tag{2}$$

Equation 2 gives only the sensed forces due to contact. Since $\overline{\mathbf{\Omega}}$ is a part of the task specification, these forces are easy to calculate. The corresponding joint torques are given by

$$\overline{\mathbf{\Gamma}}_{sens} = \mathbf{J}^T\overline{\mathbf{F}}_{sens}. \tag{3}$$

Equation 3 gives sensed torques in the subspace of force control. In essence, we are estimating torques using the end-effector forces, but only in the constrained directions. These estimates can be used in conjunction with a set of desired torques, $\mathbf{\Gamma}_d$, to control joint torques in the constrained subspace. This results in a vector of applied actuator torques, $\mathbf{\Gamma}_c$. In the motion-control subspace, desired torques are fed directly to the actuators with no feedback.

Because the individual controllers may have different bandwidths and DC gains, $\mathbf{\Gamma}_c$ may contain components that are not in the constrained subspace. These components must be eliminated to avoid coupling with motion-control directions. As a result, when end-effector-based torque sensing is used for compliant motion, an additional projection must be performed. This projection guarantees that no coupling is introduced by $\mathbf{\Gamma}_c$ between the position and force-control directions, and is performed using the equation

$$\mathbf{\Gamma}_{c,p} = \mathbf{J}^T\overline{\mathbf{\Omega}}\mathbf{J}^{-T}\mathbf{\Gamma}_c. \tag{4}$$

The projection of $\mathbf{\Gamma}_c$, guarantees that force-control torques only operate in valid force-control directions. The equation for multiple-arm motions is similar, and is described in [7].

3. Transmission Dynamics

To model the actuator/transmission system, we will use the equations developed by Pfeffer et al. in [5]. However, since our experimental results [7] indicate a significant amount of damping at the motor, we will retain the viscous damping term at the motor to account for its effect. With this change, the actuator/transmission model for each joint, shown in Figure 3, is described by the equations:

$$I_m\ddot{q}_m + \beta_m\dot{q}_m + k_t(q_m - Nq_l) \;=\; \tau_m - \tau_f \tag{5}$$

Joint	1	2	3	4	5	6
ω $\frac{\text{rad}}{\text{s}}$	113.1	207.3	138.2	78.5	78.5	69.1
ζ	0.24	0.35	0.12	0.22	0.25	0.35

Table 1. Fixed-Link Natural Frequencies and Damping Ratios for the PUMA 560

$$\begin{aligned} I_l\ddot{q}_l + \beta_l\dot{q}_l + Nk_t(Nq_l - q_m) &= 0 \\ -Nk_t(Nq_l - q_m) &= \tau_l \end{aligned}$$

In this model, I_m is the inertia of the motor, and I_l is the inertia of link i, all outboard links, and the load, about the joint's axis of rotation. Current to the motor causes a torque of τ_m which is transmitted through a shaft of stiffness k_t and a gear reduction with gear ratio N to the link. The final torque applied at the link has a value of τ_l. During operation, the motor and link are subjected to viscous damping with coefficients β_m and β_l, respectively. The position of the motor is measured by q_m, and that of the link by q_l. Finally, a non-linear friction term, $\tau_f(q_m, \dot{q}_m, q_l, \dot{q}_l)$, arises due to coulomb friction and other non-linear effects in the transmission. Since the dominant friction torque acts at the motor, τ_f is modeled at that point. In reality, τ_f also accounts for the effects of backlash, link friction, and other unmodeled behavior.

If we neglect the non-linear friction term, we can use Equations 5 to determine the link torque transfer function:

$$\frac{\tau_l}{\tau_d} = \frac{\frac{k_t}{I_m}\left(s + \frac{\beta_l}{I_l}\right)}{s^3 + \left(\frac{\beta_l}{I_l} + \frac{\beta_m}{I_m}\right)s^2 + \left(\frac{\beta_m\beta_l}{I_m I_l} + \frac{k_t(I_l + N^2 I_m)}{I_m I_l}\right)s + \frac{k_t(\beta_l + N^2\beta_m)}{I_m I_l}} \tag{6}$$

where $\tau_d = N\tau_m$ is the desired torque output of the transmission. When the arm is rigidly connected to the environment, I_l approaches infinity, and Equation 6 reduces to

$$\frac{\tau_l}{\tau_d} = \frac{\omega_o^2}{s^2 + 2\zeta\omega_o s + \omega_o^2} \tag{7}$$

where

$$\omega_o^2 = \frac{k_t}{I_m}, \quad \text{and} \quad \zeta = \frac{\beta_m}{2\omega_o I_m}$$

Table 1 presents experimentally obtained values of ω_o and ζ for each link. Equation 6 shows that these frequencies represent a lower bound on the natural frequency of the link over all configurations. Therefore, they are conservative, and can be used with the transfer function in Equation 7 to develop and analyze torque controller designs for the PUMA 560.

3.1. Disturbances

By definition, disturbances are any unmodeled effects in the physical system. The primary purpose of torque control is to reject these disturbances, enabling the motor/transmission system to act as an ideal torque source within a certain bandwidth. In Figure 4, we see the feedback model for a single joint-torque controller, along with some of the common disturbance inputs. In this figure, r is the reference input to the system and y is its output. P represents the actuator/transmission system, which is being controlled with a dynamic feedback compensator, K. R is a pre-filter for the

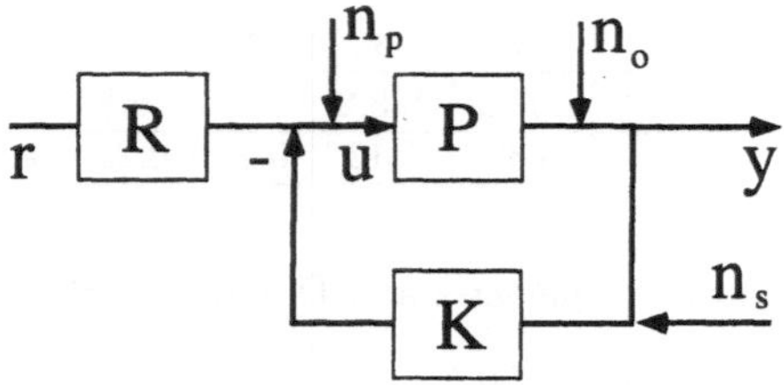

Figure 4. The Disturbance Model for Torque Control

reference input, whose output is applied to the plant after subtracting the feedback signal from K. This input to the motor, u, is also affected by process noise, n_p, which results from brush friction in the motors, backlash, cogging, amplifier noise, quantization, and other unmodeled effects. The corresponding transmission output is affected by output noise, n_o, which is primarily due to contact torque variations from motion of the environment and torque disturbances due to unmodeled dynamic coupling between links. Finally, the torque applied to the link, y, is sensed and fed to the controller, K. This signal is affected by sensor noise, n_s, due to quantization effects, electrical noise, and sensor calibration errors.

From a control standpoint, each of these disturbances acts like an auxiliary input. To provide adequate disturbance rejection and good tracking performance, the dynamic behavior of all inputs and disturbances on the system output, y, and control effort, u, must be considered. If we treat the feedback model of Figure 4 as a multi-input, multi-output system, we can formulate transfer functions in the frequency domain relating each input to the system output and the actuator effort. This allows us to accurately analyze the performance of different controllers with respect to each type of disturbance in addition to its input/output behavior.

For our model of the PUMA 560 transmission, the dominant disturbances are due to brush friction in n_p and output disturbances in n_o. Sensor disturbances are much smaller because sensor readings are digitized at the point of origin, nearly eliminating the effect of electrical noise. As a result, we will consider only the effect of n_p and n_o on system performance, neglecting n_s. With this assumption, we can represent our system of transfer functions in matrix form as

$$\begin{bmatrix} y(s) \\ u(s) \end{bmatrix} = \mathbf{H}(s) \begin{bmatrix} r(s) \\ n_p(s) \\ n_o(s) \end{bmatrix}. \tag{8}$$

where

$$\mathbf{H}(s) = \begin{bmatrix} \frac{R(s)P(s)}{1+P(s)K(s)} & \frac{P(s)}{1+P(s)K(s)} & \frac{1}{1+P(s)K(s)} \\ \frac{R(s)}{1+P(s)K(s)} & \frac{-P(s)K(s)}{1+P(s)K(s)} & \frac{-K(s)}{1+P(s)K(s)} \end{bmatrix} \tag{9}$$

All six of these transfer functions must have certain characteristics in order to provide acceptable performance. Those corresponding to disturbance inputs should be "small," to provide good disturbance rejection. The transfer function from reference input to output should have a high bandwidth, and the corresponding actuator effort should not be excessive.

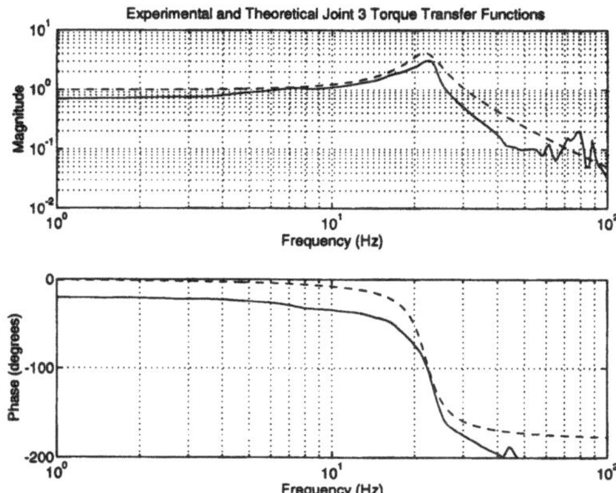

Figure 5. Theoretical and Experimental Transmission Flexibility for Joint Three

Another important reason to examine all six relationships in Equation 9 is that one of them has a special property. The transfer function relating n_o to y is called a disturbance sensitivity transfer function. Bode's integral theorem [4] states that, in order for the closed-loop system to be stable, this transfer function must satisfy the property

$$\int_{\omega=0}^{\infty} ln \left| \frac{1}{P(\omega)K(\omega)} \right| d\omega = 0. \tag{10}$$

Equation 10 implies that, in order to reject disturbances at some frequencies, they *must* be amplified at other frequencies such that the total area under the magnitude plot of the disturbance sensitivity transfer function is zero. For discrete systems, the problem is even worse. The limits of integration are reduced to $[0, \pi]$. In controller design, therefore, the idea is to find a controller that amplifies frequencies in a range where little or no disturbance energy is present. Equation 10 also shows the need for much higher servo rates than those implied by I/O bandwidth alone.

4. Controller Design

With a dynamic model of the transmission and a method of sensing joint torques, we are now in a position to design and implement torque control for the PUMA 560. We will discuss this design process using the transmission model for joint three. Figure 5 shows the experimentally obtained transfer function for this joint, along with a second-order approximation based on our model.

The large amount of extra phase in the experimental results is due to static and coulomb friction acting at the motor. This friction is a function of joint position, and can exhibit very high spatial frequencies [1]. As a consequence, large controller disturbances due to friction will occur over a broad range of frequencies.

To compensate for the large steady-state component of this friction, we need a torque controller with very high DC gain. The simplest choice is a PID controller:

$$\tau_c(s) = \tau_d(s) - [\frac{K_l}{s} + K_p + \frac{K_v s}{s+\omega}](\tau(s) - \tau_d(s)); \tag{11}$$

where τ_c is the control torque, τ_d is the reference torque, and τ is the measured

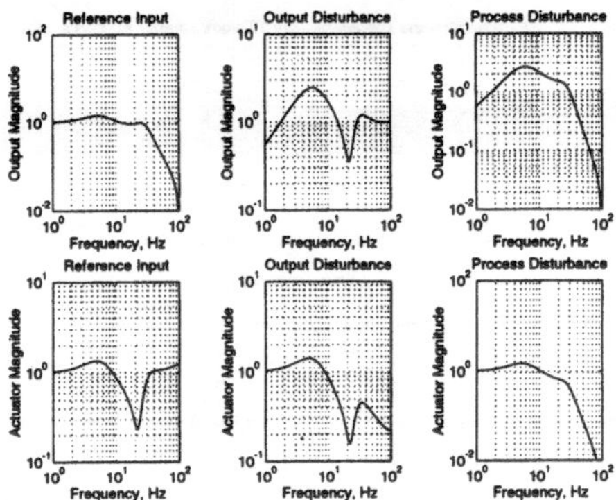

Figure 6. Closed-Loop Transfer Functions

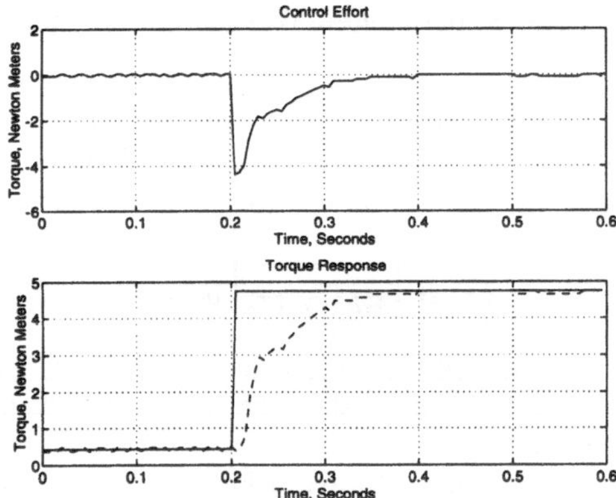

Figure 7. Experimental Fixed-Link Step Response

torque at the joint. K_p, K_v, and K_l are the proportional, derivative, and integral gains, respectively and ω is the velocity filter cutoff frequency. The feedforward term is included to ensure a DC gain of unity for designs where K_l is zero, although some studies suggest that this term should be omitted when an integral term is used [6]. To partially compensate for computational delays, we will also modify this controller to be *predictive*, as described in [2]. This significantly reduces the effect of delays on system performance.

Figure 6 shows the magnitudes of the six closed-loop transfer function discussed earlier for this PID design. Those in the top row relate reference input, process noise, and output noise, respectively, to the torque output. The magnitude plots in the bottom row relate these inputs to actuator effort. They are important to ensure that the actuators do not saturate over the desired range of inputs and expected disturbances. Experimentally, this controller performs well. A sample step response is shown in Figure 7. Response time is reasonable, and steady-state errors are small.

4.1. Disturbance Rejection

The results in Figure 7 show that our controller has good low-frequency rejection of process disturbances. We will now examine the effect of output disturbances on

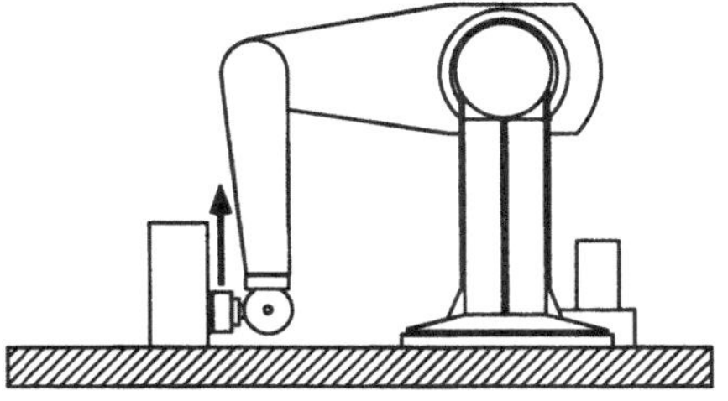

Figure 8. Experimental Setup for Puma 560 Disturbance Sensitivity Tests

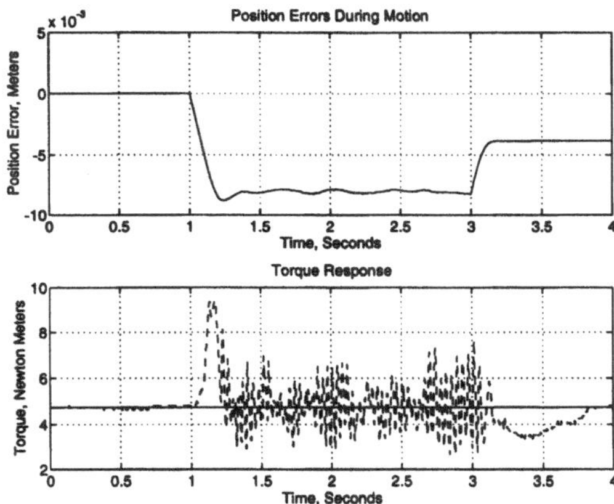

Figure 9. Closed-Loop Sliding Motion

our PID torque controller. By positioning the manipulator properly, as shown in Figure 8, we can isolate the effect of environmental contact forces during unified position/force control to a torque on joint three. To introduce output disturbances in the torque control loop, we command the arm to maintain a horizontal force of 11N on the rigid aluminum post while sliding up at 0.05 meters per second. Motion will be controlled with a simple, operational-space PD controller.

Figure 9 shows the PID controller's performance during compliant motion. The top plot represents the position error in the vertical direction, while the bottom figure shows the desired and sensed torques for joint three. Motion of the arm causes output disturbances to the torque controller, resulting in torque variations during motion. The disturbance sensitivity transfer functions discussed earlier represent the degree to which these output disturbances are amplified or attenuated at each frequency. Figure 10 compares this transfer function to the experimentally obtained ratio of closed-loop to open-loop disturbance energy at each frequency. The agreement with theory is very good, especially since little disturbance energy was present below 20Hz, increasing the experimental variance in this part of the curve.

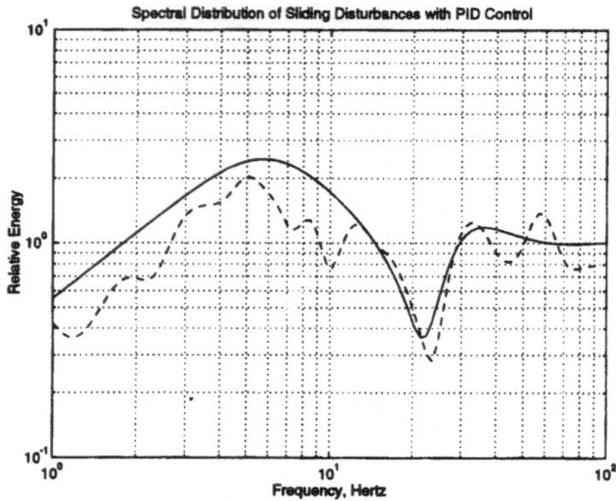

Figure 10. Disturbance Energy Ratio During Closed-Loop Sliding Motion

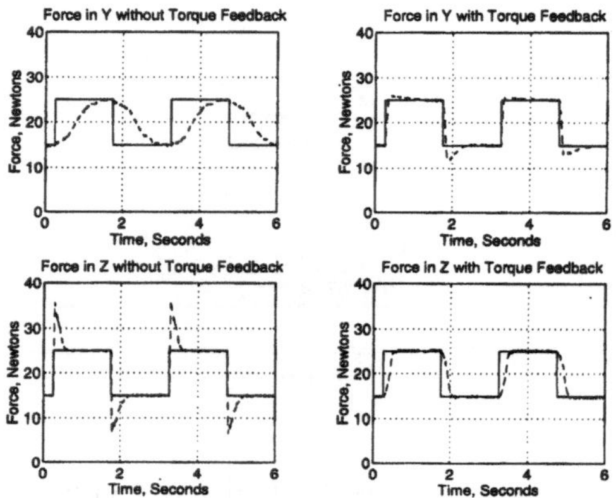

Figure 11. Experimental PID Performance With and Without Torque Feedback

5. Results

Once an adequate controller design has been implemented for each joint, high-performance force control is possible. To illustrate the improvements that are possible using torque feedback, Figure 11 shows a comparison of end-effector force control and torque-control at two robot configurations having very different inertial characteristics. In the two left hand plots, a PUMA 560 was controlled to track a square-wave force input at two different configurations without torque feedback, using only PID control in operational-space. In the right-hand plots, the same tasks are performed using a PID torque feedback loop instead. Performance is significantly improved, and varies much less with configuration. This robustness is especially important for compliant motions and cooperative manipulation, where an object may be moved through large distances.

6. Conclusions

In this paper, we have presented a technique that allows joint torque control on conventional robots, using a wrist force sensor. The result is a significant improvement in performance, and a more uniform behavior throughout the workspace. This is largely due to the fact that design of controllers at the level of joint torques takes full advantage of the actuator/transmission model. This allows an accurate analysis of disturbances, and the use of more advanced techniques to reduce their affect.

To evaluate the performance of the proposed method, we have conducted extensive experimentation on a PUMA 560 manipulator. These results have shown the proposed approach to be an effective method of control that is quite robust with respect to changes in configuration. Our analysis of disturbances has provided experimental verification that output disturbance sensitivity is a significant factor in controller design for compliant motion tasks. It has also illustrated the importance of high servo rates for adequate disturbance rejection. These results show that only through a combination of improved mechanical design to reduce disturbances, accurate dynamic modeling, and high-bandwidth torque control, can high-performance manipulation be achieved.

References

[1] Brian Armstrong-Helouvry. *Control of Machines with Friction.* Kluwer Academic Publishers, Boston, Massachusetts, 1991.

[2] Gene F. Franklin, J. David Powell, and Michael L. Workman. *Digital Control of Dynamic Systems*, 2 ed. Addison-Wesley, Reading, Massachusetts, 1990.

[3] Oussama Khatib. "A unified approach to motion and force control of robot manipulators: The operational space formulation". *IEEE Journal on Robotics and Automation*, 3(1), 1987.

[4] J.M. Maciejowski. *Multivariable Feedback Design.* Addison-Wesley, Workingham, England, 1989.

[5] Lawrence E. Pfeffer, Oussama Khatib, and John Hake. "Joint torque sensory feedback in the control of a puma manipulator". *IEEE Transactions on Robotics and Automation*, 5(4), pp. 418–424, 1989.

[6] Richard Volpe and Pradeep Khosla. "An analysis of manipulator force control strategies applied to an experimentally derived model". In *Proceedings of the 1992 IEEE/RSJ International Conference on Intelligent Robots and Systems*, pp. 1989–1997, July 1992.

[7] J. David Williams. *Characterization and Control of Multiple-Grasp Robotic Systems.* PhD thesis, Stanford University, Department of Mechanical Engineering, June 1995.

Chapter 12

Autonomous Robots

One form of autonomy sought in robotics is the ability to locomote a vehicle through an environment. Maintenance of vehicle stability, perception and navigation are central to such capability. The papers in this chapter report on several novel approaches to solving these collective problems.

Ha and Yuta describe a 2-wheeled inverted pendulum robot that can simultaneously balance and navigate. The system utilizes adaptive stabilization control to maintain balance even when collisions with the environment occur.

Lacroix and Chatila discuss an approach to autonomous robot navigation in an unknown environment. In particular they deal with long range navigation in which the planner must select sub-goals and appropriate sensor queries to accomplish the goal.

Pai, Barman, and Ralph describe a new class of spherically symmetric legged robots. They discuss locomotion planning issues and present results obtained with a tetrahedral, 4-legged robot.

Tsujita and Tsuchiya address the problem of stabilizing biped locomotion using reaction wheels mounted on the robot. Accelerating a reaction wheel provides a pure torque on the robot that can be used for postural stabilization. The authors' experiments validated the utility of the reaction wheel concept.

Pissard-Gibollet, Kapellos, Rives, and Borrelly present a robotic system which achieves operator defined goals by a sequence of visual servoing tasks. The results are demonstrated on a manipulator-equipped mobile robot.

Indoor Navigation of an Inverse Pendulum Type Autonomous Mobile Robot with Adaptive Stabilization Control System

Yun-Su Ha and Shin'ichi Yuta

Intelligent Robot Laboratory
Institute of Information Sciences and Electronics
University of Tsukuba
1-1-1 Tennodai, Tsukuba, 305 JAPAN
e-mail: hys,yuta@roboken.is.tsukuba.ac.jp

Abstract

We are developing a wheeled inverse pendulum type autonomous mobile robot which can navigate autonomously while keeping its own balance. In this paper, we propose a robust locomotion control system for the inverse pendulum robot, which changes its control mode automatically according to the changes of operating conditions or its dynamics. This paper also reports the results of the navigation of the experimental inverse pendulum robot on which the proposed algorithm is implemented in the real indoor environment.

1. Introduction

We developed a wheeled inverse pendulum type self-contained autonomous mobile robot which can navigate in a real world environment while keeping its balance. Some previous research on wheeled inverse pendulum type robots have been reported [1][2][3][4]. However, we could not find a report on sensor based navigation which was successful by such a robot. We have already reported on the trajectory control algorithm [5] for a navigation of such a robot and succeeded in an ultrasonic sensor based indoor navigation experiment using our experimental inverse pendulum type autonomous mobile robot named "Yamabico Kurara".

However, we found through experiments that the algorithm [5] was not adaptive enough to changes of its operating conditions or dynamics. In navigation experiments using a small type mobile robot, when the robot drives off course due to an error, the operator sometimes may have to pick up the robot and place it upon the desired path. Also, the robot may collide with obstacles because of the change of environment and the shortage of external sensor capability. In these cases, the inverse pendulum robot with the control algorithm [5] failed in keeping its balance because of the change of its dynamics. To adapt such situation, the robot should observe its own dynamics and the controller should select a suitable control law.

In this paper, we propose a new adaptive locomotion control algorithm for an inverse pendulum robot, which changes its control mode autonomously according to the observed operating conditions or dynamics. The most important point of this algorithm is "to discriminate its model among the possible models in real time" and "switching its control mode autonomously". The proposed algorithm was implemented on the experimental robot Yamabico Kurara, and navigational experiments have been performed in the real indoor environment.

2. Inverse Pendulum Robot Platform Yamabico Kurara

Figure 1. The experimental inverse pendulum type self-contained mobile robot Yamabico Kurara

Yamabico [6] is a series of self-contained autonomous mobile robot platforms for experimental research which were developed by the authors' group. The inverse pendulum robot used for the experiment in this research is "Yamabico Kurara". It is a standard type Yamabico robot from which the front and rear casters had been removed. Figure 1 shows Yamabico Kurara which is now controlling itself to keep its own balance. We treated Yamabico Kurara as a target robot for the analysis and controller design in this research.

This robot has rotary encoders (resolution:2000) and a vibration type gyro sensor (TOKIMEC Co. TFG-160) to measure wheel rotation and the body's inclination angular velocity. Both wheels are driven by DC motors of 10 Watts. Ultrasonic range sensors to recognize the environment are attached on the front, left and right sides of body.

3. The problems in real world navigation

We have reported on the trajectory control algorithm [5] for navigation of inverse pendulum robot and succeeded in an ultrasonic sensor based indoor navigation experiment using Yamabico Kurara. As a result, the robot showed very stable traveling if the robot could navigate successfully in the given environment.

However, we found through the experiments that inverse pendulum robot with a fixed controller is not adaptive enough to changes of operating conditions or its dynamics. For example, when the robot drove off course due to an error in its program or map, the operator sometimes had to pick up the robot and place it upon the desired path. Also, the robot might collide with obstacles because of the environment change and the shortage of external sensor capability. In these cases, the robot failed in keeping its balance because of the dynamics change caused by collisions or being lifted-up. Hence, a locomotion control algorithm which can cope with the change of operating conditions or dynamics was required for realizing more robust navigation.

4. Adaptive stabilization control using on-line model discrimination

4.1. Basic method

The proposed approach is to make the robot select a appropriate controller by itself to correspond with the operating conditions or current dynamics, by monitoring its dynamics continuously. Figure 2 illustrates the proposed control system. Several controllers to cope with each situation which may take place frequently through the ordinary operation, are prepared previously to realize a robust control system.

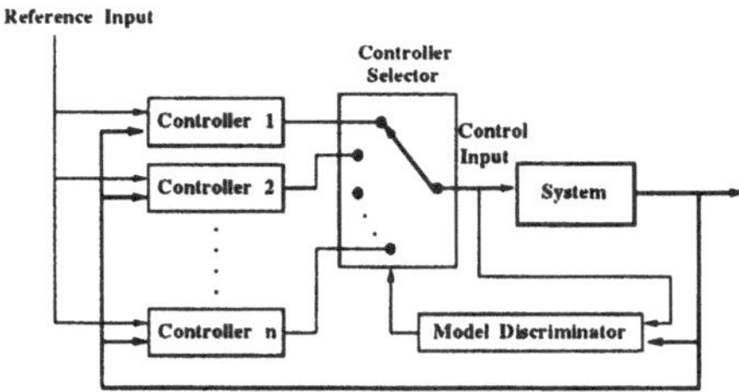

Figure 2. Illustration of the proposed adaptive control system

4.2. On-line model discrimination

The important thing to realize in the control system above is "how to discriminate robot's current situation in real time". We considered a way comparing the output of simulators and the output of a real system to discriminate current situation.

4.2.1. Design of simulator

At first, we considered three situations which may take place frequently through the ordinary navigation experiment : moving successfully while keeping balance on the ground, being lifted-up from the ground, and engaging in a collision state.

(a) Design of a simulator for the situation moving while keeping balance on the ground(Simulator 1)

One dimensional model [1] for inverse pendulum robot on a plane is given as

$$\dot{X} = AX + Bu \tag{1}$$

where,

$$A = \begin{pmatrix} 0 & 1 & 0 \\ a_1 & a_3 & a_5 \\ a_2 & a_4 & a_6 \end{pmatrix}, B = \begin{pmatrix} 0 \\ b_1 \\ b_2 \end{pmatrix}, X = [x_1\ x_2\ x_3]^T.$$

State variable x_1 and x_2 are the inclination angle and its angular velocity of the body respectively, and x_3 is the wheel's rotation angular velocity. The equation (1) represents an unstable system, and it could not be used for simulator directly, since the calculated value diverges due to initial condition or error. To solve such problem, we considered using a state observer. In the real system, using a gyro sensor to measure posture, the drift error of gyro sensor can not be neglected. Hence we augmented the system by regarding the drift as a new state variable

$$x_4 = constant. \tag{2}$$

Considering that the inclination angular velocity of the body and the wheel's rotation angular velocity are measured with a gyro sensor and rotary encoder respectively, the state observer used for simulator is designed as

$$\dot{\hat{X}}_a = A_a\hat{X}_a + B_a u + G_a(y - C_a\hat{X}_a) \tag{3}$$

where,

$$A_a = \begin{pmatrix} A & 0 \\ 0 & 0 \end{pmatrix}, B_a = \begin{pmatrix} B \\ 0 \end{pmatrix}, C_a = \begin{pmatrix} 0 & 1 & 0 & 1 \\ 0 & 0 & 1 & 0 \end{pmatrix}, G_a = \begin{pmatrix} g_{11} & g_{12} \\ g_{21} & g_{22} \\ g_{31} & g_{32} \\ g_{41} & g_{42} \end{pmatrix}$$

, $\hat{X}_a = [\hat{X}\ \hat{x_4}]^T$
$\hat{X}_a$: estimated state variables vector
G_a: gain matrix
C_a: output matrix.
G_a is decided appropriately to have suitable response.

(b) Design of a simulator for the situation being lifted off the ground(Simulator 2)

When the robot is lifted off the ground, the input current of DC motor does not affect the posture of the robot. Hence, a model of robot can be represented as

$$\dot{\omega} = -\frac{1}{T}\omega + \frac{k_m}{T}u \tag{4}$$

where, T and k_m are the time constant and the steady state gain of this system, and ω is the wheel's rotation angular velocity. Equation (4) can be used directly as a simulator because it is stable and its time constant is small.

(c) Design of a simulator for the situation engaged in a collision state(Simulator 3)

When the inverse pendulum robot collides with some object in the environment, its dynamics depend on the material of the object or the colliding state. Hence, we considered a method using the kinematics of the robot while colliding to detect collision. When the robot is controlled to go straight and collides with the front wall, the robot falls down slowly while keeping balance, ie. the wheel rotates backward slowly, probably because of the control algorithm for posture keeping. It is known through the experiments. Figure 3 shows the kinematic model of inverse pendulum robot after collision with front wall. Let the inclination angle of the body be ϕ, the rotational angle of wheel be θ, the distance from the center of wheel shaft to contact point with obstacle be d and the radius of wheel be r respectively.

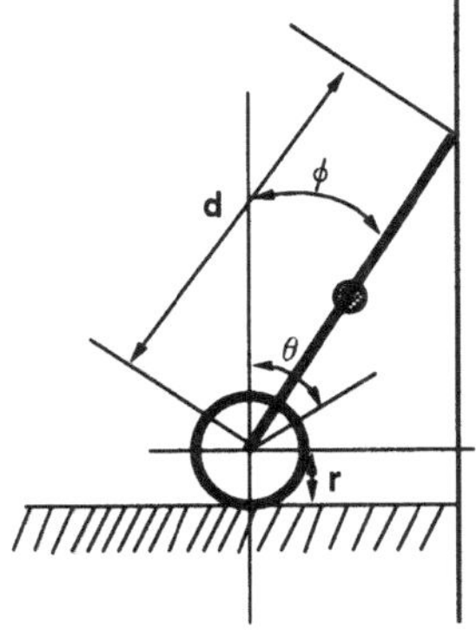

Figure 3. A model of collided inverse pendulum robot

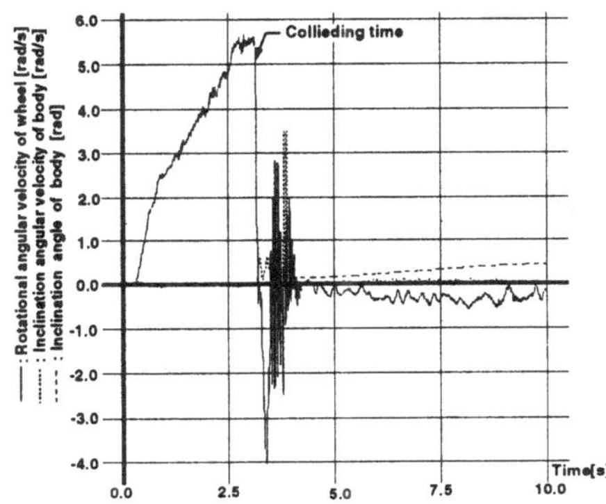

Figure 4. Motion of inverse pendulum robot when collided with wall

Figure 4 shows a variation of posture and velocity of robot which collided with front obstacle actually. We can obtain the kinematic equation (5) from Figure 3, 4.

$$\theta = -\frac{d}{r}\sin\phi \tag{5}$$

Differentiating equation (5), $\dot{\theta}$ is given as

$$\dot{\theta} = -\frac{d}{r}\cos\phi\dot{\phi}. \tag{6}$$

By observing the calculated value from equation (6) and the measured value of the encoder sensor, we can know the colliding state of the robot.

4.2.2. Discrimination of three states

To discriminate each state of the robot, the output of encoder and gyro sensor are compared with the simulated values by the three simulators. However, it is not easy because of the noise in sensory data. So the moving average of the absolute value of the difference of measured value and simulated data are taken. The delay caused by the moving average duration is decided not to seriously affect robot control.

5. Implementation and experimental results

5.1. Determination of parameter

The proposed algorithm was implemented on the Yamabico Kurara to experiment. Figure 5 shows the proposed adaptive locomotion control system. In the experiment, the coefficient matrices A, B are calculated by using the parameters of Yamabico Kurara. The gain matrix G_a of state observer was determined through trial and error by simulation. Finally A, B, G_a were given as:

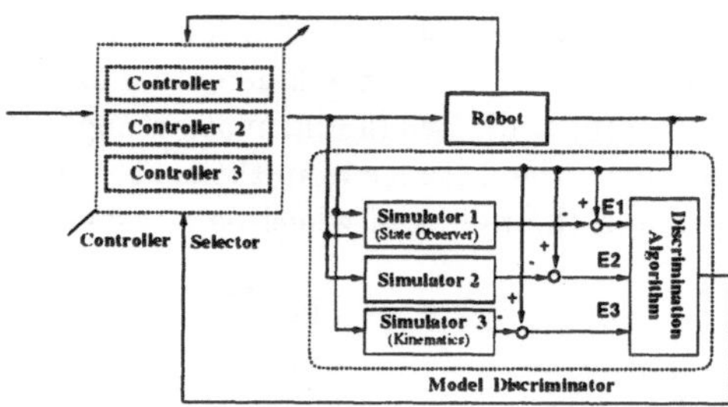

Figure 5. Locomotion control system for IPTMR

$$A = \begin{pmatrix} 0 & 1 & 0 \\ 38.23 & -0.085 & 0.121 \\ -57.76 & 0.347 & -0.56 \end{pmatrix}, \; B = \begin{pmatrix} 0 \\ -13.7 \\ 56.06 \end{pmatrix}, \; G_a = \begin{pmatrix} 50.0 & 0.0 \\ 100.0 & 0.0 \\ 0.7738 & 30.0 \\ -0.099 & 0.0 \end{pmatrix}$$

In equation (4), T and k_m were given 0.025, 23.3 by step response experiments, respectively.

By using the above parameters, we examined and compared both measured and estimated wheel rotational angular velocity, when the robot keeping its balance on the ground was lifted-up. The results are shown in Figure 6, 7 where, M_1 and M_2 represent the moving averaged values of absolute difference for inverse posture controlling and lifted-up status respectively. We also observed M_1 and M_2 when the robot was put back on the ground from the lifted up state. The result is shown in Figure 8. From these experiments, we could easily choose a threshold to discriminate these states.

In the collision experiment, the real angular velocity of the wheels and the calculated value from equation (6) when the robot collide with an obstacle, is shown in Figure 9. Thresholds were determined from these experiments.

5.2. Implementation of adaptive controller

The locomotion control system is designed to stop revolution of its wheel and to save dead reckoning data when robot is lifted up. Oppositely, when the robot is put upon the ground from being lifted up, it is made to wait for instruction from the master controller while keeping balancing control mode after reinitializing the parameters for balance control. In the case of collision, we made the robot detecting collision and part from obstecle while keeping its balance. Also, the control system is made to inform master controller of the change of situations above.

5.3. Experiments of adaptive control

We made experiments of posture control with changing operating conditions. Figure 10 shows the results of the experiment when the operator picked up the robot and put it on the ground. A dashed line shows the result of model discrimination, and 0, 1 means "on the ground state" and "lifted up state"of the robot, respectively. Figure 11 shows the experimental results of collision. We can see that the robot discriminate collision at time A, rising the body to up-right posture at interval B and part from obstacle at interval C.

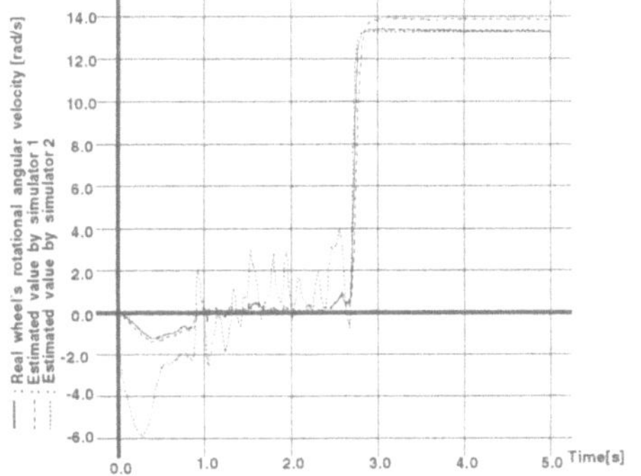

Figure 6. Measured and estimated the wheel's rotational angular velocity when lift-up after inverse pendulum control

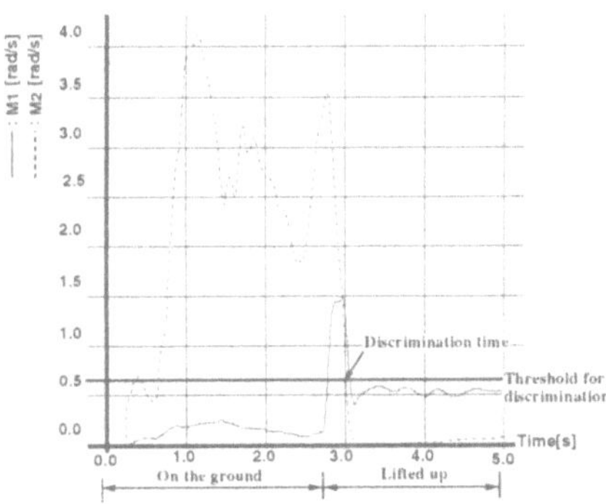

Figure 7. Moving average of the estimation errors when lift-up

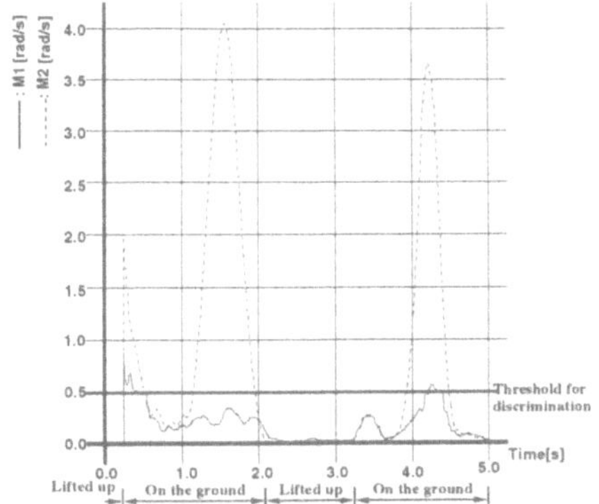

Figure 8. Moving average of the estimation errors when put down

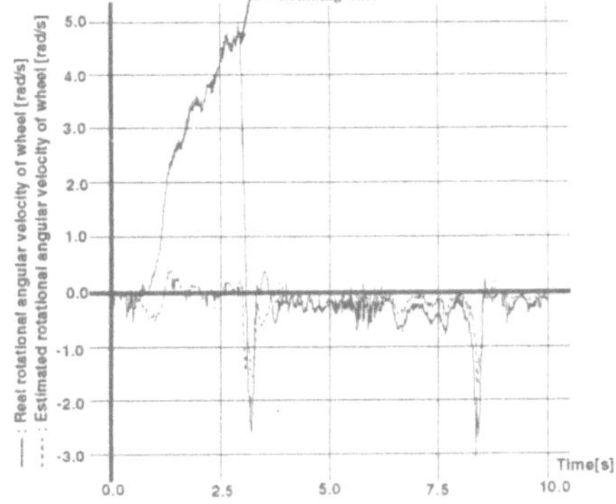

Figure 9. Measured and estimated values of wheel's rotational angular velocity in collision state

5.4. Integrated indoor navigation of Yamabico Kurara

We integrated a robust indoor navigation system which include the proposed algorithm to cope with the changes of operating condition. The intelligent Robot laboratory at University of Tsukuba was used as the experimental environment. In the experiments, even when an unexpected thin obstacle was put on the planed path, which is difficulte to detect by ultrasonic sensors, the robot detected the collision by itself and could cope with it. So the robot succeeded to navigate itself to the goal point while keeping its balance in spite of collision. An experimental scene is shown in Figure 12.

6. Conclusion

In this paper, we proposed a robust control system of an inverse pendulum robot with the ability to select autonomously its control mode to correspond with operating

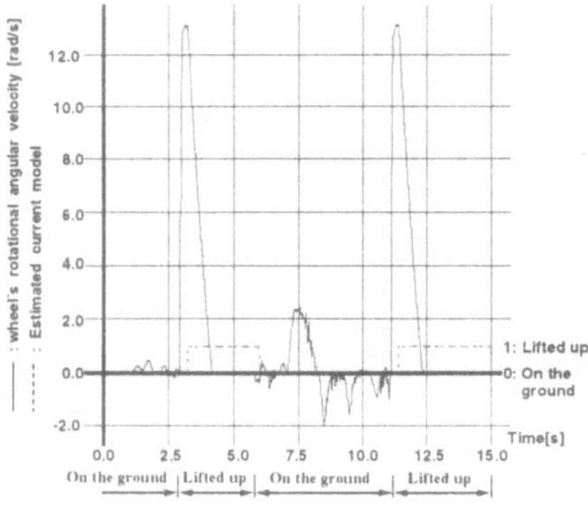

Figure 10. A discrimination results of lift up

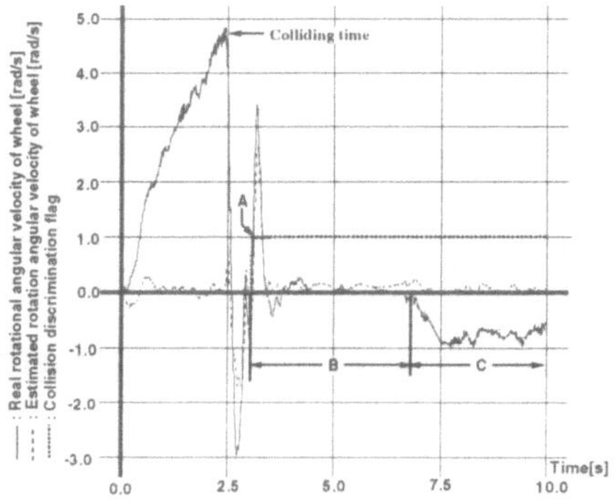

Figure 11. A discrimination result of collision

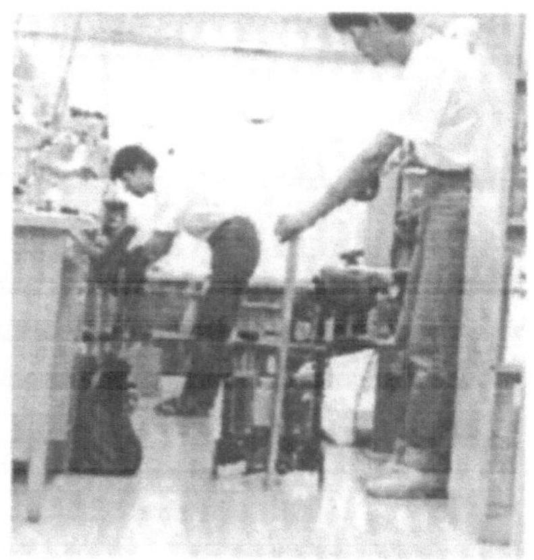

Figure 12. An experimental scene of indoor navigation

conditions and dynamics. It makes a big improvement in the robustness of the navigation system.

References

[1] K. Yamafuji and T. Kawamura, "Postural control of a monoaxial bicycle", Journal of the Robotics Society of Japan, Vol.7 No.4 pp.74-79 , 1989 (in Japanese).

[2] K. Yamafuji and A. Koshiyama, "Postural and Driving Control of a Variable-Configuration-Type Parallel Bicycle", Journal of Robotics and Mechatronics, Vol.3, No.5, pp.365-372, 1991.

[3] O. Matsumoto, S. Kajita and K. Tani, "Attitude estimation of the the wheeled inverted pendulum using adaptive observer", Proc. of 9th Academic Conf. of the Robotics Society of Japan, pp.909-910, 1991 (in Japanese).

[4] E. Koyanagi, S. Iida, K. Kimoto and S. Yuta, "A wheeled inverse pendulum type self-contained mobile robot and its two-dimensional trajectory control", Proc.of ISMCR'92, pp.891-898, 1992.

[5] Y. Ha and S. Yuta, "Trajectory Tracking Control for Navigation of Self-Contained

Mobile Inverse Pendulum", Proc. of the 1994 IEEE/RSJ Int. Conf. on Intelligent Robots and Systems, pp.1875-1882, 1994.

[6] S. Yuta, S. Suzuki, S. Iida,"Implementation of a small size experimental self-contained autonomous robot - sensors, vehicle control, and description of sensor based behavior", Proc. of the 1991 ISER (Int. Symposium on Experimental Robots), pp.344-359, published by Springer-Verlag,1993.

Motion and Perception Strategies for Outdoor Mobile Robot Navigation in Unknown Environments

Simon Lacroix* and Raja Chatila
LAAS - CNRS
7, ave du Colonel Roche 31077 Toulouse Cedex - France
E-mail: {simon, raja}@laas.fr

Abstract

This paper presents an experimented approach to autonomous robot navigation in an unknown natural environment. The approach involves several levels of reasoning, several environment representations, and three different motion modes. We focus on the "navigation level" of the whole system, which is in charge of reaching a distant goal by selecting sub-goals to reach, motion modes to apply, and perception tasks to execute for this purpose. We present how a terrain model dedicated to the navigation process is built on the 3D data acquired by the robot, and we describe an approach to tackle the difficult problem of planning perception and motion tasks. Experimental results on a realistic test site are presented and discussed.

1. Introduction

The problem of long range navigation in unknown outdoors environments is not very frequently addressed. Systems that were actually experimented were demonstrated for road following [1], or motion in rather limited environment conditions [2, 3]. Two important achievements in totally unstructured environments were Ambler [4] and the navigation of the UGV [5].

A canonical task a robot should be able to achieve in such an environment is the `''Go To [goal]''` task (or *navigation task*), where `goal` is a distant point to reach autonomously : any more complex robotic mission (exploration, sample collecting...) will include one or more instances of this task. Our approach to achieve this navigation task in an unknown natural environment involves several levels of reasoning, several environment representations, and three different motion modes. It raises a need for a specific decisional level (the *navigation* level), that is in charge of deciding which environment representation to update, which sub-goal to reach, and which motion mode to apply. The paper is especially devoted to this level, which is a key component of the system, since it controls the perception and motion activities of the robot.

The paper is organised along the following outline : the following section presents more precisely our general adaptive and hierarchical approach to autonomous navigation in unknown outdoor environments, pointing out the importance of the navigation level. Section 3 then presents how the terrain representation required by this

*Currently on a post-doc funded by INRIA at the Centre for Intelligent Machines, McGill University, Montréal (simon@cim.mcgill.ca)

decisional level is incrementally built on the basis of 3D data, produced either by a Laser Range Finder (LRF) or by stereo-vision. The algorithms that perform the selection of sub-goals and perception tasks (*i.e.* that compose the navigation level) are described in section 4. Experimental results conclude the paper.

2. A General Strategy for Navigation in Outdoor Unknown Environments

For any robotic task in general, and for the navigation task in particular, we favour an approach where the robot explicitly reasons on environment representations and its capabilities to decide the actions to perform [6, 7]. As opposed to a behaviour-based approach, we are convinced that such an approach is mandatory to develop machines that can be *controlled* and *programmed* by a remote human operator.

2.1. An adaptive approach

According to a general "economy of means" principle due to limitations of on-board processing capacities, memory and energy that we put on the system (realistic constraints for several applications were the robot has to be operational on a remote site, *e.g.* for planetary exploration of scientific missions in hostile places such as the Antarctic), and to achieve a time-efficient behaviour, we choose an *adaptive* approach in which the robot adapts its behaviour to the nature of the terrain [8, 7]. Hence, three motion modes are considered :

- A **reflex** motion mode : on large flat and lightly cluttered zones, the robot locomotion commands are determined on the basis of a goal (heading or position) and informations provided by "obstacle detector" sensors. The terrain representation required by this mode is just the description of the borders of the region within which it can be applied ;

- A **2D planned** motion mode : when the environment is mainly flat, but cluttered with obstacles, the robots locomotion commands are determined by a trajectory plan. The trajectory planner reasons on a binary description of the environment, which is described in terms of *Crossable/Non-Crossable* areas.

- A **3D planned** motion mode : when the environment is highly constrained (uneven terrain), collision and stability constraints have to be checked to determine the robot locomotion commands. This is done thanks to a 3D trajectory planner [9], that reasons on a fine 3D description of the terrain (a numerical terrain model - NTM [10]) ;

Choosing to endow the robot with three different motion modes enables "smart" and efficient behaviours, but complicates quite a lot the system : it must be able to deal with several different terrain representations and planning processes. It must especially have the ability to determine which motion mode to apply, and therefore which terrain representation to update : this is performed thanks to a specific planning level, the *navigation planner.*

2.2. The navigation planner

We assume that the terrain on which the robot must fulfill a navigation task is initially unknown, or mapped with a very low resolution. It is then only possible for an operator to specify a graph of *routes*, *i.e.* large corridors within which the robot has to move autonomously. In this context, the navigation task ''Go To'' is achieved thanks to three layers of planning (figure 1) :

- The *route planner* chooses long-term paths to the goal on the basis of the initial informations (the route graph that covers the whole area in which the mission takes place). The route planner selects a sub-goal for the navigation planning level ;

- The *navigation planner* (or *path planner*) reasons on a global qualitative representation of the terrain (the *region map*), built from the data acquired by the robot's sensors. It selects *(i)* the next perception task to perform, *(ii)* the sub-goal to reach and *(iii)* the motion mode to apply (which comes to select and control the trajectory planners) ;

- Finally, the *trajectory planner* determines the trajectory to execute (in one of the above-mentioned three motion modes) to reach the goal defined by the navigation planning level.

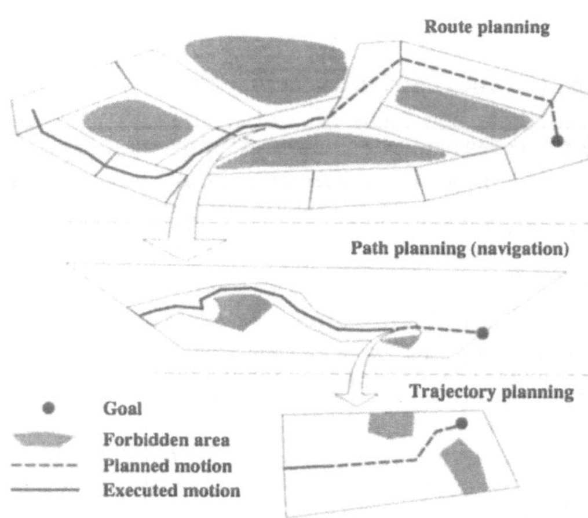

Figure 1. *Hierarchical organisation of the the three levels of planning involved in our approach, and corresponding terrain representations : the route planner reasons on the initial coarse environment model, that can cover several kilometres. The navigation planner reasons on a qualitative terrain representation built from the robot sensor's data (the* region map, *that can cover several hundreds of meters) ; and finally the trajectory planners reason on a local and precise description of the environment.*

Splitting the decisional processes into three layer of planning has the advantage to structure the problem : each planning layer controls the one that is directly below, by specifying its goal and its working domain. It has also the great advantage of helping to analyse and solve failing cases : when a planner fails to reach its goal, it means that the environment representation of the upper layer is erroneous and therefore that it has to be revised.

The navigation planner is *systematically* activated at each step of the incremental execution of the task : each time 3D data are acquired, they are analysed to provide a description of the perceived zone in terms of navigation classes, and this description is fused to update the *region map.* The introduction of the navigation planning layer defines a particular instance of the usual "perception-decision-action" loop, where the "decision" part is split into two distinct processes : navigation and trajectory planning.

2.3. Several terrain representations

Each of the three different motion modes requires a particular terrain representation. The navigation planner also requires a specific terrain representation, and during navigation, an exteroceptive localisation process has to be activated frequently, which requires an other terrain representation. Aiming at building a "universal" terrain model that contains all the necessary informations for these various processes is extremely difficult, inefficient, and moreover not really useful. It is more direct and easier to build different representations adapted to their use : the environment model is *multi-layered* and *heterogeneous*. Several perception processes coexist then in the system, each dedicated to the extraction of specific representations : perception is *multi-purpose*.

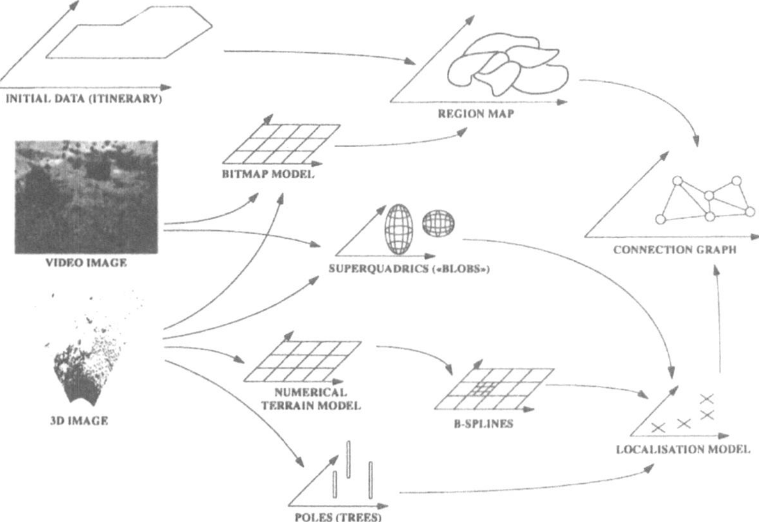

Figure 2. *The various representations used in the system : one can distinguish the numerical terrain model [10] necessary to the 3D trajectory planner, the region map (and connection graph) dedicated to the navigation planner, and three different ways to build a localisation model :*(i) *by modelling objects (rocks) with superquadrics [11],* (ii) *by detecting interesting zones in the NTM thanks to a B-spline based model [12], or* (iii) *by detecting poles in the 3D raw data*

Figure 2 presents the various terrain representations required during navigation : arrows represent the constructive dependencies between them. Coherence relationships between these various representations are to be maintained when necessary, which remains an open issue.

3. Building the region map

3.1. 3D data classification

Applied each time 3D data are acquired (either by a laser range finder or a correlation stereo-vision algorithm), the classification process produces a description of the locally perceived area in term in *terrain classes* [13]. It relies on a specific discretization of the perceived area that respects the sensor resolution (figure 3), that defines "cells" on which different characteristics (attributes) are determined : density (number of points contained in a cell compared with a nominal density defined by the discretization rates), mean altitude, variance on the altitude, mean normal vector and corresponding variances...

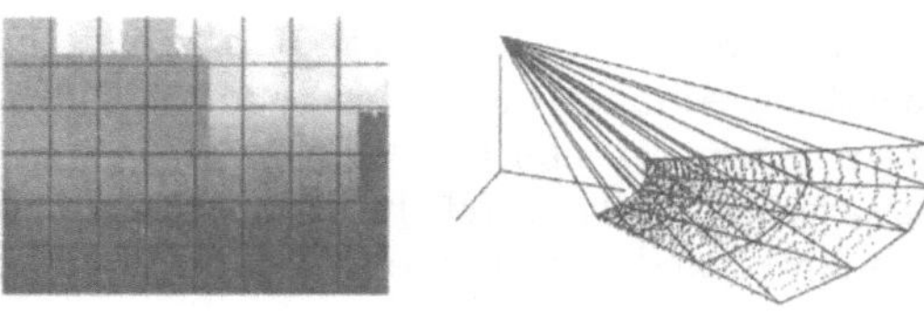

Figure 3. *Discretization used for the classification procedure : a regular discretization in the sensor frame (left : a 3D image is represented as a video image, where the gray levels corresponds to the points depth) defines a discretization of the perceived zone that respects the sensor resolution (right)*

A non-parametric Bayesian classification procedure is used to label each cell : a learning phase based on prototypes classified by a human lead to the determination of probability density functions, and the classical Bayesian approach is applied, which provides an estimate of the partial probability for each possible label. A decision function that privileges false alarms (*i.e.* labelling a flat area as obstacle or uneven) instead of the non-detections (*i.e.* the opposite: labelling an obstacle as a flat area) is used (figure 4). A simpler but faster technique based on thresholds on the cell attributes has also been implemented.

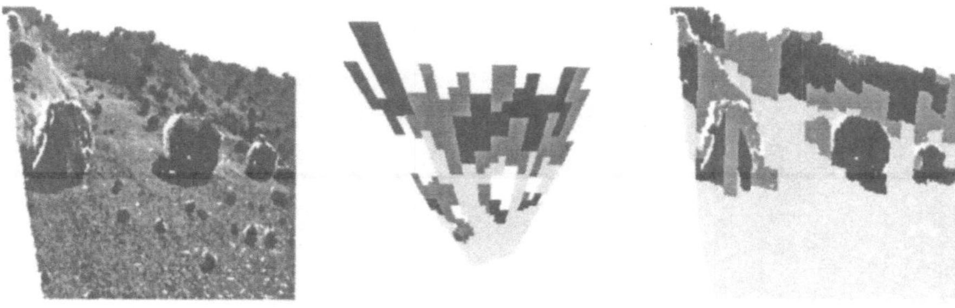

Figure 4. *Classification of a correlated stereo image : correlated pixels (left) ; classification of the perceived zone (center), and reprojection of the result in the camera frame (right). From clear to dark : unknown, flat, uneven and obstacle)*

This technique proved its efficiency and robustness on several hundreds of 3D images. Its main interest is that it provides an estimate of the confidence of its results : this information is given by the *entropy* of a cell. Moreover, a statistical analysis of the cell labelling confidence as a function of its label and distance to the sensor defines a predictive model of the classification process.

3.2. Incremental fusion

The partial probabilities of a cell to belong to a terrain class and the variance on their elevation allow to perform a fusion procedure of several classified images, provided the robot position is known (figure 5). The fusion procedure is performed on a bitmap, in the pixels of which are encoded cell attributes determined by the classification procedure (label, label confidence, elevation and variance on the elevation).

3.3. Model structuration and management

For the purpose of navigation planning, the global bitmap model is structured into a region map, that defines a connection graph. Planning a *path* (as opposed to planning a *trajectory*) does not require a precise evaluation of the static and kinematic

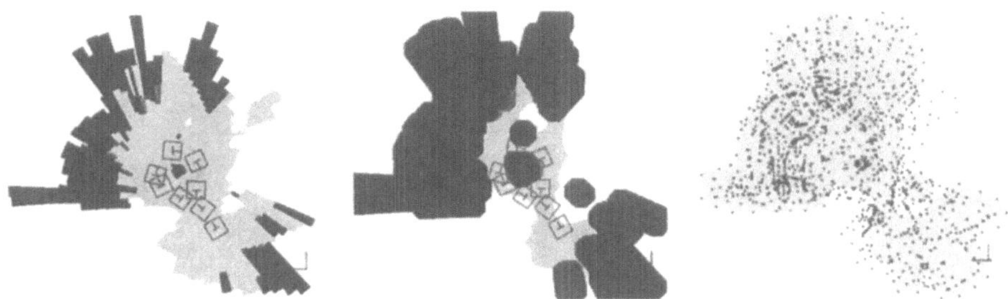

Figure 5. *The model structuration procedure. From left to right : a terrain model resulting from the fusion of 8 classified laser images ; the same model after constrained zones growing, and the nodes of the connection graph defined by the region segmentation*

constraints on the robot : we simply consider a robot point model, and therefore perform a constrained zones growing in the bitmap before segmenting it into regions. The regions define a connection graph, whose nodes are on their borders, and whose arcs correspond to a region crossing (figure 5). The terrain is described as a bitmap only in the surroundings of the robot's current position, whereas the region model (much more compact) is kept in memory during the whole mission.

4. Navigation planning

Each time 3D data are acquired, classified and fused in the global model, the robot has to answer autonomously the following questions :

- Where to go ? (sub-goal selection)
- How to go there ? (motion mode selection)
- Where to perceive ? (data acquisition control)
- What to do with the acquired data ? (perception task selection)

For that purpose, the navigation planner reasons on the robot capabilities (action models for perception and motion tasks) and the connection graph defined by the region map.

4.1. Planning motion versus planning perception

A straightforward fact is that motion and perception tasks are strongly interdependent : executing a motion requires to have formerly modelled the environment, and to acquire some specific data, a motion is often necessary to go the adequate observation position. Planning motion tasks in an environment modelled as a connection graph is easily solved by classical search techniques that find optimal paths minimising some criteria (in our case time and energy, that are respectively related to the terrain classes and elevation variations).

To plan perception tasks, one must be able to predict the results of such tasks (which requires a model of the perception processes), and the *utility* of these results to the mission to achieve. If the utility of a localisation task can be modelled by a simple function that expresses the gain on the the robot position precision as a function of the number and distance of perceivable landmarks, estimating the utility of a modelling task is a much more difficult issue (figure 6).

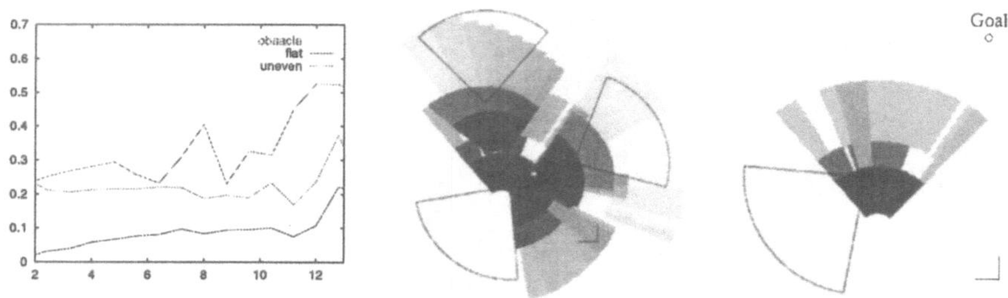

Figure 6. *The confidence model of the classification procedure (left : mean cell entropy as a function of the perceived distance) and the labelling confidence of the terrain model (represented as grey levels in the images) allow to determine the classification task that maximises the gain in confidence from any given position in the model (center). But the result of such a task may be of a poor reach interest to reach a specified goal (right). In such a case, one should be able to determine* **where** *a better information is required* **in order to** *reach the goal*

4.2. Approach

A direct and brute force approach to answer the former questions would be to perform a search in the connection graph, in which *all* the possible perception tasks would be predicted and evaluated at *each* node encountered during the search. Besides its drastic algorithmic complexity, this approach appeared unrealistic because of the difficulty to express the utility of a predicted classification task to reach the goal.

We therefore choose a different approach to tackle the problem : the perception task selection is *subordinated* to the motion task. A search algorithm provides an *optimal* path, that is analysed afterwards to deduce the perceptions tasks to perform. The "optimality" criterion takes here a crucial importance : it is a linear combination of time and energy consumed, weighted by the terrain class to cross *and* the confidence of the terrain labelling (figure 7).

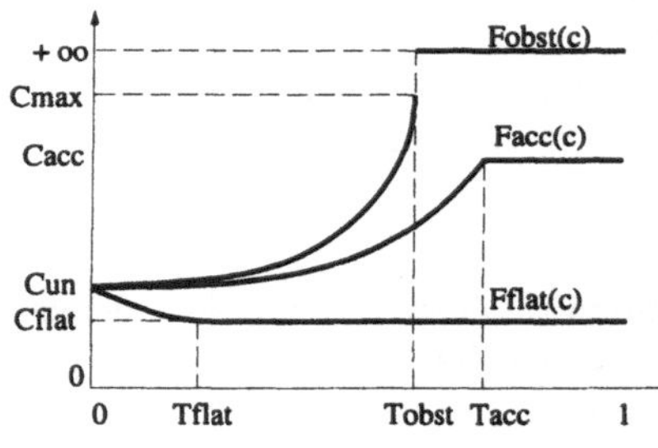

Figure 7. *Weighting functions of an arc cost, as a function of the region label confidence. An experimental statistical analysis helped to define thresholds on the confidence labelling over which the classification result can be considered faithful (thresholds T_{flat}, T_{acc} and T_{obst} respectively for regions labelled flat, uneven and obtacle). When the region confidence is over these thresholds, their crossing cost is equal to a* nominal *cost (C_{flat}, C_{acc} and $C_{obst} = \infty$) ; when the region confidence is under these thresholds, the costs decreases to reach the unknown region cost C_{un} when $c = 0$.*

The influence of these cost weighting functions is illustrated on figure 8. They come to consider *implicitly* the modelling capabilities of the robot : tolerating to cross obstacle areas labelled with a low confidence means that the robot is able to acquire easily informations on this area.

Off course, the returned path is not executed directly, it is analysed according the following procedure :

1. The sub-goal to reach is the last node of the path that lies in a crossable area ;

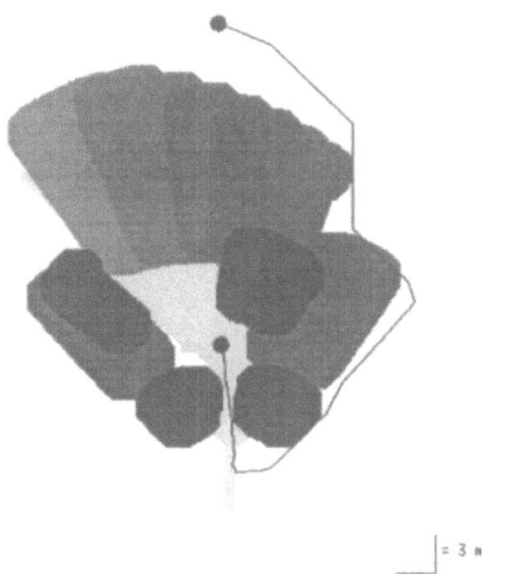

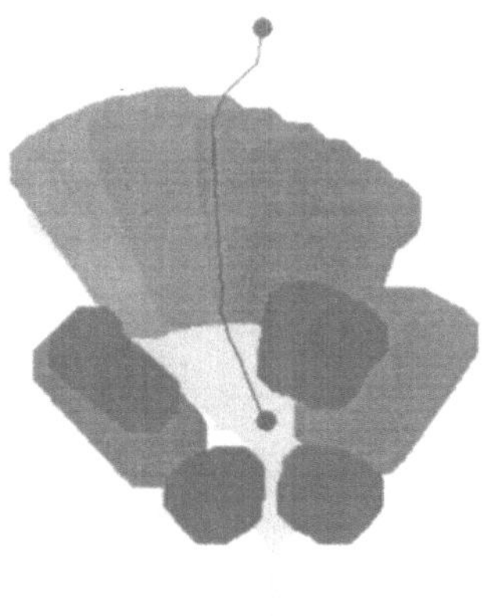

Figure 8. *Influence of the cost weighting functions (the gray levels represent the confidence of the region labelled "obstacle"). When not considering this confidence, the optimal path returns backward to reach the goal (left). When the cost weighting functions are applied, the optimal path crosses an obstacle area labelled with a low confidence (right)*

2. The label of the regions crossed to reach this sub-goal determines the motion mode to apply ;
3. And finally the rest of the path that reaches the global goal determines the aiming angle of the sensor.

One can interpret the result of this analysis as a way to answer the question asked in figure 6 : the cost weighting functions help to find the interesting regions to model in order to reach the goal.

Figure 9 presents a way to determine paths along which are planned localisation tasks. This approach has only been tested in simulation experiments, but seems to be helpful to plan safe paths.

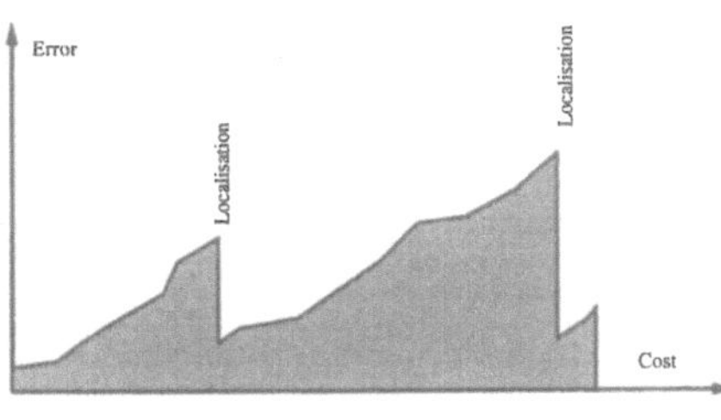

Figure 9. *Introduction of the uncertainty on the robot position in the cost to control localisation tasks. Provided a model of the evolution of the robot position uncertainty as it moves and a model of the localisation task, minimising the integral of the robot position precision as a function of the cost expressed in term of time and energy determines paths that gets closer to landmarks. The choice of minimising an integral (as opposed to a linear combination for instance) comes to defines motions with a big uncertainty on the position as more "risky" than motions with a low uncertainty*

5. Results and discussion

The terrain modelling procedures and navigation planning algorithm have been intensively tested with the mobile robot Adam[1]. We performed experiments on the

[1]Advanced Demonstrator for Autonomy and Mobility, is property of Framatome and Matra Marconi Space and is currently lent to LAAS

Figure 10. *ADAM in the Geroms test site*

Geroms test site in the French space agency CNES[2], where Adam achieved several ''Go To [goal]'' missions, travelling over 80 meters, avoiding obstacles and getting out of dead-ends (for more details concerning Adam and the experimental setup, refer to [8]).

Figure 11 presents two typical behaviours of the navigation algorithm in a dead-end, and figure 12 shows the trajectory followed by the robot to avoid this dead-end, on the elevation model built after 10 data acquisitions.

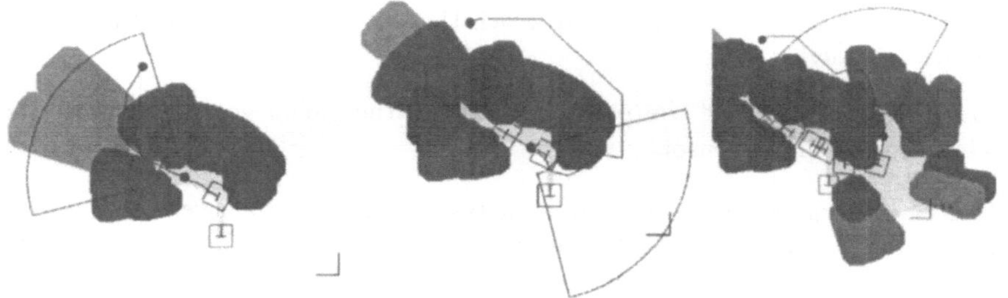

Figure 11. *The navigation planner explores a dead-end : it first tries to go through the bottom of the dead-end, which is modelled as an obstacle region, but with a low confidence level (left) ; after having perceived this region and confirmed that is must be labelled as obstacle, the planner decides to go back (center), and finally finds a way that reaches the goal(right)*

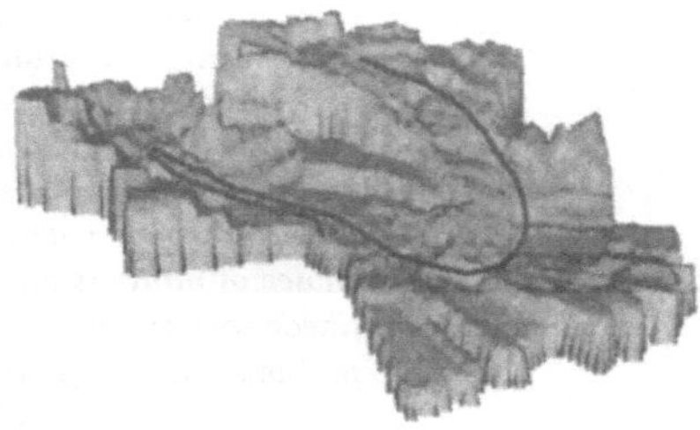

Figure 12. *Elevations encoded in the terrain model built during the dead-end exploration, and trajectory executed to reach the global goal. 80 meters have been travelled, and 10 perceptions have been activated.*

These experiment have proved the possibility to build and maintain quickly

[2]Centre National d'Études Spatiales

a global qualitative terrain model, on which smart decisions can be taken. The navigation algorithm proposed proved its efficiency on most of our experiments. It includes the sub-goal and next perception task determination, two issues that should never be considered indenpedently.

The problem of navigation as we understand it nevertheless requires some more attention. In particular, we are considering the framework of the decision theory to express the utility of a classification task. By considering more explicitly the sensor model (detectability and uncertainty models), we hope to have more optimal decisions.

References

[1] C. Thorpe, M. Hebert, T. Kanade, and S. Shafer. Toward autonomous driving : the cmu navlab. part i : Perception. *IEEE Expert*, 6(4), August 1991.

[2] C.R. Weisbin, M. Montenerlo, and W. Whittaker. Evolving directions in nasa's planetary rover requirements end technology. In *Missions, Technologies and Design of Planetary Mobile Vehicules. Centre National d'Etudes Spatiales, France*, Sept 1992.

[3] B. Wilcox and D. Gennery. A mars rover for the 1990's. *Journal of the British Interplanetary Society*, 40:484–488, 1987.

[4] E. Krotkov, M. Hebert, M. Buffa, F. Cozman, and L. Robert. Stereo friving and position estimation for autonomous planetary rovers. In *IARP 2nd Workshop on Robotics in Space, Montreal, Canada*, 1994.

[5] M. Hebert. Pixel-based range processing for autonomous driving. In *IEEE International Conference on Robotics and Automation, San Diego, California*, 1994.

[6] G. Giralt, R. Chatila, and R. Alami. Remote intervention, robot autonomy and teleprogramming : generic concepts and real-world application cases. In *IEEE International Conference on Intelligent Robots and Systems, Yokohama (Japan)*, July 1993.

[7] S. Lacroix, R. Chatila, S. Fleury, M. Herrb, and T. Simeon. Autonomous navigation in outdoor environment : Adaptative approach and experiment. In *IEEE International Conference on Robotics and Automation, San Diego, California*, 1994.

[8] R. Chatila, S. Fleury, M. Herrb, S. Lacroix, and C. Proust. Autonmous navigation in natural environment. In *Third International Symposium on Experimental Robotics, Kyoto (Japan)*, October 1993.

[9] T. Simeon and B. Dacre-Wright. A practical motion planner for all-terrain mobile robots. In *IEEE International Conference on Intelligent Robots and Systems, Yokohama (Japan)*, July 1993.

[10] F. Nashashibi, P. Fillatreau, B. Dacre-Wright, and T. Simeon. 3d autonomous navigation in a natural environment. In *IEEE International Conference on Robotics and Automation, San Diego (USA)*, May 1994.

[11] S. Betge-Brezetz, R. Chatila, and M. Devy. Natural scene understanding for mobile robot navigation. In *IEEE International Conference on Robotics and Automation, San Diego (USA)*, May 1994.

[12] P. Fillatreau, M. Devy, and R. Prajoux. Modelling of unstructured terrain and feature extraction using b-spline surfaces. In *International Conference on Advanced Robotics, Tokyo (Japan)*, November 1993.

[13] S. Lacroix, P. Phillatreau, F. Nashashibi, R. Chatila, and M. Devy. Perception for autonomous navigation in a natural environment. In *Workshop on Computer Vision for Space Applications, Antibes, France*, Sept. 1993.

Programming Symmetric Platonic Beast Robots *

Dinesh K. Pai   Roderick A. Barman   Scott K. Ralph
Department of Computer Science
University of British Columbia
Vancouver, Canada
{pai|rodb|ralph}@cs.ubc.ca

Abstract

We describe a new class of high degree of freedom, spherically symmetric legged robots called "Platonic Beasts", and programming alternatives for coordinated locomotion in these robots. The project is motivated by applications of robotics in hazardous, uncertain environments, theoretical issues in legged locomotion, and issues in programming high degree of freedom robots.

1. Introduction

We have developed a new family of symmetric, multilimbed robots called Platonic Beasts. A robot in this family is kinematically equivalent to a symmetric polyhedron, such as one of the five Platonic solids (hence the name), with identical multi-purpose limbs attached to its vertices. We have built a prototype with four limbs in a tetrahedral arrangement (12 actuated degrees of freedom) controlled by 4 embedded microcontrollers (see [7] for details of the design).

A part of our motivation is to explore fault tolerance with respect to falls and other failures with a high degree of freedom robot that has no preferred orientation. Rough terrain mobility is generally believed to be the *raison d'être* of legged robots. However such terrain is not merely uneven but also highly uncertain, for instance, due to difficulties in sensing the terrain geometry, integrity, and material properties. Therefore the ability to recover from falls and other failures is critical, especially in hazardous or remote environments. This was demonstrated in August 1994 the NASA/CMU Dante II mission to Mount Spurr. The robot successfully accomplished all of its scientific goals, but on return the robot tipped over and could not recover even though it was physically unharmed, and required an expensive and dangerous helicopter rescue.

The spherical symmetry of the robot also enables new gaits such as the "rolling" gait in which the robot approximates a rolling motion by synthesizing a sphere with its limbs. This is a new example of a circulating gait [1] and minimizes the number of footfalls. To our knowledge, the only other legged robot with such a gait is the CMU Ambler [1] which uses symmetry in the horizontal plane.

*This work was supported in part by NSERC, the Institute for Robotics and Intelligent Systems, and the BC Advanced Systems Institute.

Figure 1. Platonic beast robot prototype

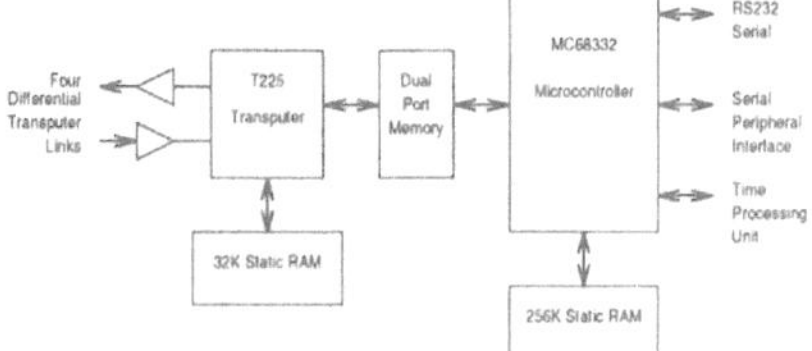

Figure 2. Embedded Module for Distributed Control

Perhaps the greatest difficulty encountered with effective use of such high degree of freedom robots is the lack of good programming models and software tools to support these models. In this paper we describe several programming tools for programming these robots. These include traditional joint trajectory control and cartesian trajectory control of individual limbs; however these are not sufficiently easy to use for specifying locomotion and other complex tasks.

In §4, we describe robot programming in terms of body trajectories, with the limbs coordinated to provide support. A higher level programming model that is useful for specifying locomotion is described in §5 and §6; we specify a sequence of limb permutations corresponding to "tumble" steps which can be derived from a canonical tumble step (see Figure 4) by appropriate transformations.

2. Platonic Beast Robot

We have built a prototype of the simplest member of the family, the 4-beast, based kinematically on the tetrahedron. Structurally, the limbs are attached to the centroids of alternate faces of an octahedron; the limb computer and control electronics for a limb are mounted on an adjacent face. The robot is small – it weighs less than 5kg, the body is an octahedron of side 17cm, and each limb has a reach of 25cm from the surface of the body. Figure 1 shows the prototype robot.

The prototype was constructed using the principles of modular design at two levels. First, all limbs of the robot are made of identical UBC-Zebra link modules which can be rapidly assembled to construct kinematic chains (see, e.g., [2] for other work on modular robot links). Second, each limb is a module with identical kinematic configuration, control and computing resources. The links are designed to allow the limbs to be assembled in different kinematic configurations. Further details on the mechanical design of the prototype can be found in [7].

In the first prototype (Mark I), each limb was controlled by a separate limb computer, based on the Motorola MC68332 32-bit microcontroller. The limb computer processes all sensing associated with the limb and communicates with other limb computers and the development host over serial lines. The communication with the host is via a serial line shared by the limb computers. Communication with the host computer uses a simple communication protocol that time multiplexes over the single serial line. The MC68332 processor core is a variant of the common Motorola

M68K series of processors; we can therefore use standard GNU software – compilers, linkers, assemblers, debuggers and libraries – for development of the target software on a variety of host platforms.

The Mark II prototype, currently under development, has a more scalable multiprocessor architecture for supporting intensive on-board processing for vision and locomotion planning. The architecture is based on the Embedded Module for Distributed Control (EMDC), a small self-contained dual-processor computer developed in our lab by Barman, Kingdon, Mackworth and Pai. The EMDC has a MC68332 32-bit microcontroller with 256K of static RAM coupled to an INMOS T225 16-bit transputer through a small dual-port memory (see Figure 2). The MC68332 handles the control and servoing tasks while the T225 is responsible for loading code into the system and commmunicating with other EMDCs and transputer modules. The T225 has four on-chip full-duplex 20 megabit/second communication links used for sharing data between nodes and for loading code into the system from the host.

To date, we have demonstrated coordinated control of the Platonic Beast prototype and locomotion using the rolling gait. In the current kinematic configuration the robot can locomote with this gait on terrain of slope up to 20 degrees, and on terrain with sparse, but known, footholds. Details of the locomotion are found in §4, §5 and §6, where we discuss the programming of these very high degree of freedom machines. Figure 4 shows the robot climbing terrain with a 20 degree slope.

3. Limb Level Programming

The link modules can be configured to form limbs in several ways, and the kinematics of a limb depends on the configuration. The motion of each link module is controlled independently, using a PID controller running at 100Hz. User programs can provide setpoints to the joint level servos.

The limbs are numbered 1,...,4 as shown in Figure 3. Two different assignments which are mirror images of each other are possible. We have picked the "right-handed" assignment.

We use the following notation below. A coordinate frame with index $A \in \mathbb{Z}$, is denoted ${}_A\mathsf{E}$. It consists of a reference point ${}_A\mathsf{E}_o$ called the "origin" of the coordinate frame, and set of orthonormal basis vectors ${}_A\mathsf{E}_x$, ${}_A\mathsf{E}_y$, ${}_A\mathsf{E}_z$. The matrix of coordinates of a vector[1] p in coordinate frame ${}_A\mathsf{E}$ is denoted ${}^A\mathsf{p}$. The homogeneous transformation of coordinate vectors from frame A to frame B is written as ${}^B_A\mathsf{E}$ and is given by the 4×4 matrix

$$ {}^B_A\mathsf{E} = \left({}^B_A\mathsf{E}_x \; {}^B_A\mathsf{E}_y \; {}^B_A\mathsf{E}_z \; {}^B_A\mathsf{E}_o \right) $$

where, for instance, a column ${}^B_A\mathsf{E}_x$ is the vector ${}_A\mathsf{E}_x$ expressed in homogeneous coordinates with respect to the frame B.

A reference frame ${}_l\mathsf{E}$ for the base of each limb $l = 1, \ldots, 4$ is attached to a limb face of the platonic beast. Specifically, the origin of the ${}_l\mathsf{E}$ frame is located at the center of the face. The z-axis unit vector is along the outward normal to the face. The orientation of the x-axis is also shown in Figure 3. The body reference ${}_0\mathsf{E}$ is located at the centroid of the body as shown in Figure 3.

In the prototype, each limb forms a 3R positioning manipulator; the forward and inverse kinematics for positioning the end point of the limb relative to the base

[1]Or other contravariant quantity.

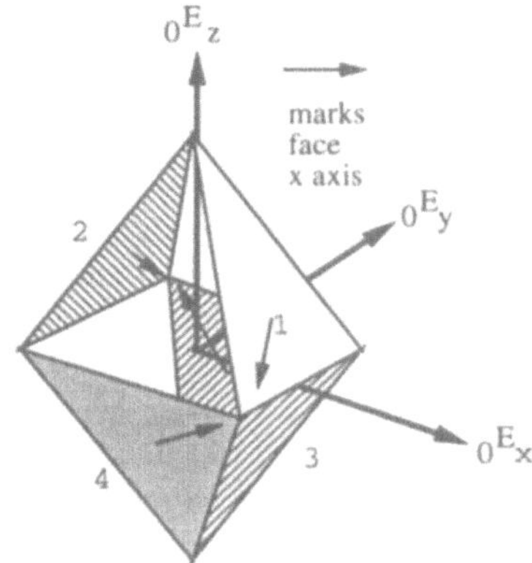

Figure 3. Assignment of body and limb coordinate frames

of the limb is simple (e.g., see [3]). Thus motion can be conveniently specified with respect to the body frame ${}_0\mathsf{E}$; cubic spline trajectory interpolation is also provided.

4. Body Level Programming

The body frame ${}_0\mathsf{E}$ is convenient for dealing with the kinematics of limbs but is not convenient for specifying motions of the robot, especially for locomotion. To simplify programming, we define some new reference frames that depend on which feet are in contact with the ground as well as the intended direction of motion. Complications arise due to the coordination of limbs and the need to specify the motion of the robot relative to the terrain; note that the body can be in any orientation with respect to the terrain.

As discussed in §6, we can specify the intended direction of motion by specifying an even permutation of the limb numbers (1234). Assume, without loss of generality, that the current permutation of the limbs is (1234) and by convention limbs 1, 2 and 3 are on the ground, while limb 4 is possibly free. Let us denote the location of the foot of limb i as $\mathsf{p}(i)$.

We define the *ground frame*, ${}_g\mathsf{E}$, as a frame fixed with respect to the ground with origin at $\mathsf{p}(1)$, the location of the foot of limb 1. The y-axis of the frame points towards foot 2; specifically, it is

$$ {}_g\mathsf{E}_y \stackrel{def}{=} \frac{(\mathsf{p}(2) - \mathsf{p}(1))}{|\mathsf{p}(2) - \mathsf{p}(1)|}. $$

The z-axis is orthogonal to the plane defined by the feet 1, 2 and 3; it is taken to be the unit vector along ${}_g\mathsf{E}_y \times (\mathsf{p}(3) - \mathsf{p}(1))$. This determines the ground frame and the transformation ${}^0_g\mathsf{E}$ completely.

For specifying motions relative to the body, it is often convenient to define a second frame called the *heading frame*, ${}_h\mathsf{E}$, parallel to the ground frame but located at the centroid of the body at the start of the specified motion. The origin of ${}_h\mathsf{E}$ coincides with that of ${}_0\mathsf{E}$. We also define the *instantaneous heading frame* at a time t, ${}_{h'}\mathsf{E}$, parallel to the heading frame, but located at the centroid of the body at time t during the motion.

4.1. Kinematics

The total motion of the body relative to the heading coordinate frame is given by

$$ {}^{h}_{h}\mathsf{M} \stackrel{def}{=} \mathsf{Tx}\,\mathsf{Ry}. \tag{1} $$

Here, Tx and Ry are general translations and rotations, respectively.

We can compute the joint angles at the end of the motion as follows. The foot position of limb l relative to base frame ${}_{l}\mathsf{E}$ (denoted ${}^{l}\mathsf{p}(l)$) is easily computed from the forward kinematics of the three link chain. The foot position relative to the ground frame is then given by

$$ {}^{g}\mathsf{p}(l) = {}^{g}_{0}\mathsf{E}\, {}^{0}_{l}\mathsf{E}\, {}^{l}\mathsf{p}(l), \tag{2} $$

where ${}^{g}_{0}\mathsf{E} = ({}^{0}_{g}\mathsf{E})^{-1}$. The new position of the body frame after the motion can be computed relative to the ground frame as

$$ {}^{g}_{0'}\mathsf{E} = {}^{g}_{h}\mathsf{E}\; \mathsf{Tx}\; \mathsf{Ry}\; {}^{h}_{0}\mathsf{E}. \tag{3} $$

Therefore, inverting and expanding, we have

$$ {}^{0'}_{g}\mathsf{E} = {}^{0}_{g}\mathsf{E}\, {}^{g}_{h}\mathsf{E}\; \mathsf{Ry}^{-1}\; \mathsf{Tx}^{-1}\; {}^{h}_{g}\mathsf{E}. \tag{4} $$

Hence the new location of the foot of limb l relative to its limb reference frame is given by

$$ {}^{l}\mathsf{p}(l)' = {}^{l}_{0}\mathsf{E}\, {}^{0'}_{g}\mathsf{E}\, {}^{g}\mathsf{p}(l). \tag{5} $$

From this the new joint angles can be obtained from the inverse kinematics of the limb (see, e.g., [4]).

4.2. Interpolation

For large displacements we interpolate the motion of the robot to achieve a uniform motion of the body as follows. We will focus on the limbs in contact with the ground; the rest are treated using standard trajectory interpolation techniques [4].

The twist ${}^{h}\mu$, corresponding to ${}^{h}_{h}\mathsf{M}$ can be computed as follows (see, e.g, [5]). We use the following notation [6]: if

$$ {}^{h}\mu = \begin{pmatrix} w \\ v \end{pmatrix}, $$

then

$$ [{}^{h}\mu] \stackrel{def}{=} \begin{pmatrix} 0 & -w_z & w_y & v_x \\ w_z & 0 & -w_x & v_y \\ -w_y & w_x & 0 & v_z \\ 0 & 0 & 0 & 0 \end{pmatrix}, $$

and by definition

$$ {}^{h}_{h}\mathsf{M} = \exp\,[{}^{h}\mu]. $$

One possible interpolation strategy is to move the body with a constant spatial velocity $\frac{1}{T}({}^{h}\mu)$ where T is the desired duration of motion. While this is uniform motion in $SE(3)$, the centroid of the body need not move in a straight line and is therefore difficult for programmers to visualize.

Instead we will move the body with constant linear velocity relative to the heading frame ${}_h\mathsf{E}$ with uniform rotation about the *instantaneous* heading frame ${}_{h'}\mathsf{E}$ located at the instantaneous centroid of the body. Specifically, define ${}^h\xi$ and ${}^h\eta$ from

$$\exp([{}^h\xi]T) = \mathsf{Tx}, \exp([{}^h\eta]T) = \mathsf{Ry}. \tag{6}$$

Clearly, these are of the form

$${}^h\eta = \begin{pmatrix} \omega \\ 0 \end{pmatrix}, {}^h\xi = \begin{pmatrix} 0 \\ \nu \end{pmatrix}. \tag{7}$$

where ω is the angular velocity and ν is the translational velocity. We define the motion at time t from the start of the motion segment as $\mathsf{Ry}' = \mathsf{Rot}(\omega t)$, $\mathsf{Tx}' = \mathsf{Trans}(\nu t)$

From equation 5,

$$\begin{aligned} {}^l\mathsf{p}(l)' &= {}^l_0\mathsf{E}\ {}^{0'}_g\mathsf{E}\ {}^g\mathsf{p}(l) & (8)\\ &= {}^l_0\mathsf{E}\ {}^0_g\mathsf{E}\ {}^g_h\mathsf{E}\ \mathsf{Ry}'^{-1}\ \mathsf{Tx}'^{-1}\ {}^h_g\mathsf{E}\ {}^g\mathsf{p}(l) & (9)\\ &= {}^l_h\mathsf{E}\ \mathsf{Ry}'^{-1}\ \mathsf{Tx}'^{-1}\ {}^h\mathsf{p}(l). & (10) \end{aligned}$$

Therefore,

$$\begin{aligned} \frac{d}{dt}{}^l\mathsf{p}(l)' &= {}^l_h\mathsf{E}\ \Big(\mathsf{Ry}'^{-1}\ [-{}^h\eta]\ \mathsf{Tx}'^{-1} \\ &\qquad +\mathsf{Ry}'^{-1}\ [-{}^h\xi]\ \mathsf{Tx}'^{-1}\Big)\ {}^h\mathsf{p}(l) & (11)\\ &= {}^l_h\mathsf{E}\ \mathsf{Ry}'^{-1}\ (-[{}^h\eta + {}^h\xi])\ \mathsf{Tx}'^{-1}\ {}^h\mathsf{p}(l), \end{aligned}$$

where

$$[{}^h\eta + {}^h\xi] = \begin{pmatrix} 0 & -\omega_z & \omega_y & \nu_x \\ \omega_z & 0 & -\omega_x & \nu_y \\ -\omega_y & \omega_x & 0 & \nu_z \\ 0 & 0 & 0 & 0 \end{pmatrix}.$$

In equation 11 we have used the fact that $\frac{d}{dt}\exp([\alpha]t) = [\alpha]\exp([\alpha]t) = \exp([\alpha]t)[\alpha]$.

Note that ω, ν, ${}^l_h\mathsf{E} = {}^l_0\mathsf{E}\ {}^0_g\mathsf{E}\ {}^g_h\mathsf{E}$, and ${}^h\mathsf{p}(l) = {}^h_l\mathsf{E}\ {}^l\mathsf{p}(l)$ can all be computed at the beginning of a motion segment. Hence, during the motion, we can simultaneously compute both the position and velocity of the foot relative to the limb frame as

$$\begin{aligned} \Gamma &= {}^l_h\mathsf{E}\ \mathsf{Ry}'^{-1}, & (12)\\ P &= \mathsf{Tx}'^{-1}\ {}^h\mathsf{p}(l), & (13)\\ {}^l\mathsf{p}(l)' &= \Gamma P, & (14)\\ \frac{d}{dt}{}^l\mathsf{p}(l)' &= \Gamma(-[{}^h\eta + {}^h\xi])P. & (15) \end{aligned}$$

A piecewise cubic joint trajectory which produces this body trajectory approximately is then computed. At knot points[2] we can compute the joint angles θ_l as corresponding to ${}^l\mathsf{p}(l)'$ from the inverse kinematics of the limb. The joint velocities $\dot{\theta}_l$ are computed using the limb Jacobian matrix J_l, by solving the linear system $J_l\dot{\theta}_l = \frac{d}{dt}{}^l\mathsf{p}(l)'$.

[2]We currently use constant knot spacing; it is not difficult to modify this to adaptive knot spacing schemes.

5. Programming a Tumble Step

A very convenient method of generating gait is by a sequence of isomorphic steps we call *tumbles*. Suppose the limbs are numbered 1 to 4, using a right hand screw rule. A canonical tumble step starts with limbs 1, 2, and 3 in contact with the terrain and limb 4 free, on top of the robot. The ending configuration of the tumble has the robot with limbs 1, 4 and 2 on the ground and limb 3 in the air.

We first describe the tumble step on horizontal terrain, and then the extension to arbitrary terrain. A tumble step can be achieved by translation of the body along the x-axis of the heading frame while rotating the body about an axis through the centroid, parallel to the y-axis ${}_h\mathsf{E}_y$. During the motion, the location of the feet in contact with the terrain will remain constant relative to the ground frame.

In the general case there is a brief dynamic "tip" episode during which only two limbs are in contact with the ground. For the choice of kinematic parameters of our prototype, all four limbs can simultaneously contact the ground, and thus the duration of the tip episode can be reduced to zero. Such a zero tip gait will be slower, but will minimize the impact loads on the limb. However, we typically plan for a brief tip episode, in part to compensate for uncertainty in terrain modeling and sensing.

The tumble step is thus partitioned into three phases: *approach, transfer, depart.* The contact point of limb i with the terrain (the "foot") is labeled p_i and the projection of the center of gravity of the robot on the horizontal plane is labeled p_0. The points p_1, p_2, and p_3 define the support polygon for the robot. During the approach phase, p_0 is moved near to the (p_1, p_2) edge of the support triangle; at the same time the free limb is moved on to a circle orthogonal to the edge (p_1, p_2), with its center on the edge and passing through the desired p_4 location. The location on the circle is not critical; at the end of the approach phase, the foot is at a small elevation over the terrain. During the transfer phase, p_0 is moved across the edge (p_1, p_2), causing the p_3 contact to break. The exact configuration at which this happens need not be known exactly; locating the free limb on the circle ensures that the configuration after the tip episode will have limb 4 correctly positioned. Finally the depart phase is the inverse of the approach phase in which the p_0 is moved to the center of the triangle formed by p_1, p_2, and p_4.

The configuration of the robot at the boundary of each phase is computed as follows. To translate the body a distance d while rotating the body by an angle ϕ, we can define the motions

$$\mathsf{Ry} = \begin{pmatrix} \cos\phi & 0 & \sin\phi & 0 \\ 0 & 1 & 0 & 0 \\ -\sin\phi & 0 & \cos\phi & 0 \\ 0 & 0 & 0 & 1 \end{pmatrix}, \quad \mathsf{Tx} = \begin{pmatrix} 1 & 0 & 0 & d \\ 0 & 1 & 0 & 0 \\ 0 & 0 & 1 & 0 \\ 0 & 0 & 0 & 1 \end{pmatrix}$$

We can now use the trajectory generator of §4 to generate a smooth trajectory to achieve the total body motion ${}_h^h\mathsf{M} = \mathsf{TxRy}$.

The tumble step can similarly be performed over arbitrary terrain, which is assumed to be given by a terrain elevation function $h : \mathbb{R}^2 \rightarrow \mathbb{R}$. This could be in the form of a digital elevation map. Only the local terrain needs to be known for the tumble step, so this information could even be obtained by detecting contact with the foot.

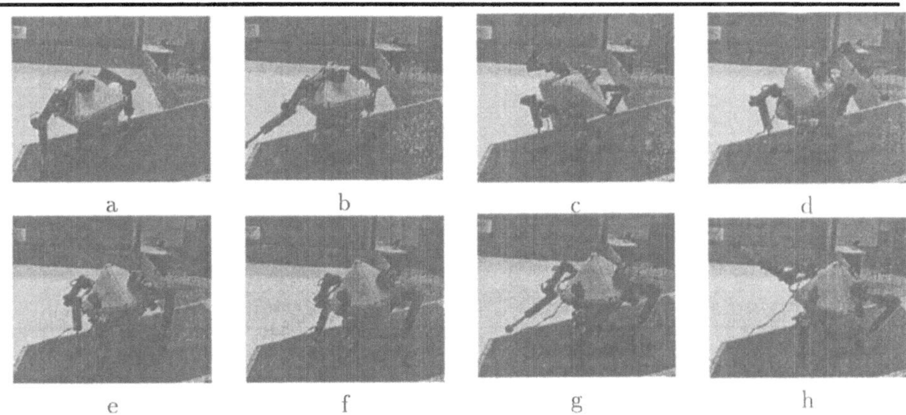

Figure 4. Canonical tumble with 4-beast prototype up a 20 degree slope

The basic idea is simple – we interpret the permutation sequence as specifying a sequence of support triangles in the *horizontal* plane. The actual foot placements and the position of the body is raised by height h above the plane; the orientation of the body is also chosen to conform to the local terrain. This approach has the advantage that since static stability depends only on the projection of the center of gravity on the horizontal plane, the stability margin of a tumble sequence is independent of terrain. Figure 4 shows a tumble on terrain with a slope of 20 degrees in the direction of motion.

The terrain data is currently only used for planning statically stable motion during a tumble step assuming that the contacts do not slip; we do not currently model the finite friction of real terrain. It is therefore possible to specify an infeasible tumble sequence on rough or steep terrain. In the near future we plan to use the multiresolution motion planner of [8] to find safe paths and tumble sequences.

6. Locomotion Programming

The spherical symmetry of the platonic beasts allows the robot to locomote with a *rolling gait* (see Figure 4). The gait is generated by a sequence of isomorphic *tumbles* described in §5. We therefore denote a tumble step by a permutation $\tau = (l_1 l_2 l_3 l_4)$. By convention, the tumble occurs with the limbs l_1 and l_2 in contact, limb l_3 starts the tumble in contact and ends free, while limb l_4 starts free and ends in contact. The identity permutation, (1234), is called the canonical tumble. Different tumbles can be uniquely labeled by the even permutations.

Figure 5 shows the successive support polygons for the rolling gait along a straight path. Each triangle in the figure is a support polygon; the numbers at the vertices indicate the supporting limb and the number at the center of the polygon indicates the limb that is free. The locomotion in Figure 5 is therefore specified at a high level by the sequence of permutations:

```
(1 2 3 4)
(1 4 2 3)
(3 4 1 2)
(3 2 4 1)
```

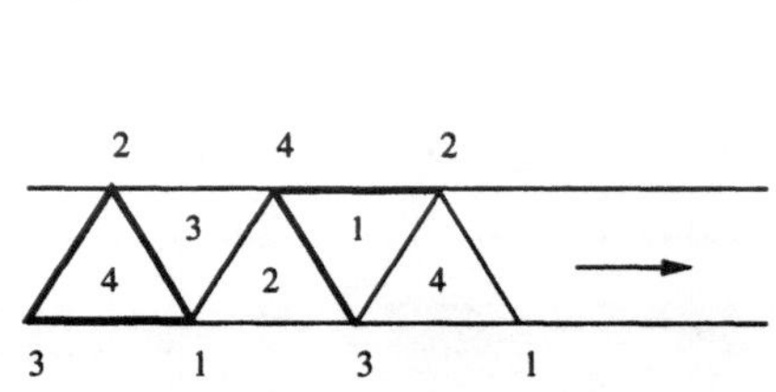

Figure 5. Straight path.

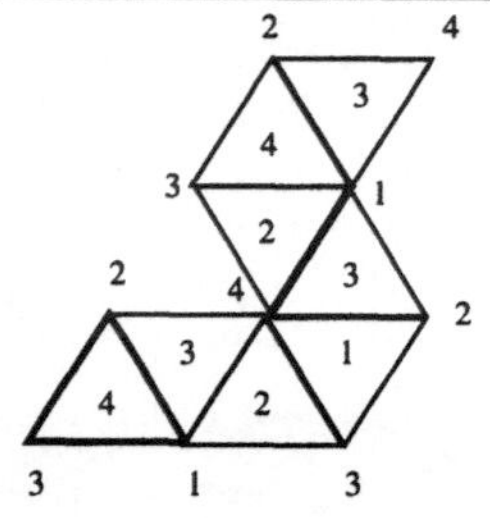

Figure 6. General path.

Similarly, the locomotion shown in Figure 6 is given by:

```
(1 2 3 4)
(1 4 2 3)
(3 4 1 2)
(2 4 3 1)
(1 4 2 3)
(1 3 4 2)
(1 2 3 4)
```

The shape of the generating support triangle for the canonical tumble can be separately specified. By default, we assume that this is an equilateral triangle in the horizontal plane, since this maximizes the minimum stability margin. The size of the triangle can be obtained from the specified stability margin.

7. Conclusions and future work

We have developed a new family of multi-limbed robots called Platonic Beasts and outlined the design and programming of an experimental protype. The spherical symmetry of these robots provides several advantages including the ability to recover from any orientation, a novel rolling gait, and more generally, greater flexibility in limb utilization. The robot supports a number of programming models for specifying locomotion and other tasks. We believe this class of robots is well suited for some application areas, particularly locomotion in hazardous environments with uncertain terrain.

We plan to extend the programming models to support locomotion on rough terrain. We are implementing a constraint programming model called "Least Constraint" [9] in the Mark II robot, which has a better computing and communications architecture. We have developed a multiresolution motion planner for rough terrain [8] which has been tested in simulation with real terrain data; we plan to test it with the Platonic Beast. Other issues being currently investigated include fault detection and selecting optimal configurations of the robot for various tasks [10].

References

[1] J. E. Bares and W. L. Whittaker, "Walking robot with a circulating gait," in *Proceedings of the IEEE International Conference on Intelligent Robots and Systems*, pp. 809–818, 1990.

[2] D. Schmitz, P. Khosla, and T. Kanade, "The CMU reconfigurable modular manipulator system." CMU-RI-TR-88-7, The Robotics Institute, Carnegie Mellon University, 1988.

[3] R. P. Paul, *Robot Manipulators: Mathematics, Programming, and Control.* The MIT Press, 1981.

[4] J. J. Craig, *Introduction to robotics : mechanics and control.* Addison-Wesley, 1989.

[5] R. Murray, Z. Li, and S. S. Sastry, *A mathematical introduction to robotic manipulation.* CRC Press, 1994.

[6] R. W. Brockett, "Robotic manipulators and the product of exponentials formula," *Mathematical Theory of Networks and Systems*, pp. 120–129, 1984.

[7] D. K. Pai, R. Barman, and S. Ralph, "Platonic Beasts: Spherically Symmetric Multilimbed Robots," to appear in *Autonomous Robots*, 2:4, December 1995.

[8] D. K. Pai and L.-M. Reissell, "Multiresolution rough terrain motion planning." Dept. of Computer Science Technical Report 94-33, University of British Columbia, November 1994.

[9] D. K. Pai, "Least constraint: A framework for the control of complex mechanical systems," in *Proceedings of the American Control Conference*, pp. 1615 – 1621, 1991.

[10] K. van den Doel and D. K. Pai, "Performance measures for robot manipulators: A unified approach." Accepted for publication International Journal of Robotics Research, October 1994.

An Experimental Study on Motion Control of a Biped Locomotion Machine using Reaction Wheels

Katsuyoshi Tsujita and Kazuo Tsuchiya
Osaka University,
2-1 Yamadaoka, Suita, Osaka 565, Japan
tsujita@dyna.ccm.eng.osaka-u.ac.jp,
tsuchiya@duna.ccm.eng.osaka-u.ac.jp

Abstract

This paper deals with motion control of a biped locomotion machine. A biped locomotion machine is statically unstable during a single supporting phase because of the gravitational force. Motion control of a biped locomotion machine is to control this unstable attitude and to establish a steady walk. In this paper, reaction wheels are used as torquers in order to control an attitude of the system. A reaction wheel can generate a reaction torque without any interactions with the gravitational field and this, in turn, simplifies the calculation of the motion of the wheels to generate the attitude control torques. In this paper, based on the computed torque method, a motion control is accomplished: Three types of controller are proposed and verified the efficiencies by hardware experiments.

1. Introduction

Research of locomotion machine is important because they can move on rough terrains. From the point of view of dynamical stability, locomotion machines that have been proposed so far are divided into two groups; The one is statically stable and the other is dynamically stable. The statically stable type keeps a statically stable state during a whole walking cycle, for example, quadruped locomotion machines whose gait patterns are called WALK, and hexapod locomotion machines. On the other hand, the dynamically stable type has a statically unstable attitude in a walking cycle, for example, quadruped locomotion machines whose gait patterns are called TROT, and biped locomotion machines. This paper deals with motion control of a biped locomotion machine. A biped locomotion machine is statically unstable during a single supporting phase because of the gravitational force. Motion control of a biped locomotion machine is to control this unstable attitude and to establish a steady walk. In order to control this unstable attitude in single supporting phase, two methods can be considered; The one uses special actuators (torquers) that generate the attitude control torques, and the other does not use such actuators. The method which does not use any actuators for attitude control, controls the unstable attitude of the main body by selecting the landing time of the swinging leg. This method makes good use of an inverse-pendulum motion of a locomotion machine in the gravitational field, but it is difficult to control the attitude of the main body precisely. On the other hand, the method which uses actuators usually employs two types of actuators, an external torquer equipped on the end of the leg to generate a control torque acting on the locomotion machine through an interaction with the ground, and an internal torquer equipped on the system to generate a control torque through an internal motion of the actuator. The method using external torquers is efficient one, but has some limitations of roughness of the terrain. In this paper, reaction wheels are used as torquers in order to control the attitude of the system. A reaction wheel can generate a reaction torque without any interactions with the gravitational field. Based on this dynamical feature, it can be simplified to calculate the motions of the wheels to generate required control torques. In this paper, motion control is accomplished using the computed torque method: Three types of controller are proposed and verified the efficiencies by hardware experiments.

2. Formulations

We consider a biped locomotion machine composed of the main body and two legs indicated in Fig. 1. Each leg is composed of two links connected to each other through one DOF(degree of freedom) rotational joint, and connected to the main body with one DOF rotational joints. On the main body, two wheels (pitch and roll wheel) are equipped. The main body and the legs are labeled as body 0, 1 and 2, and the roll and pitch wheels are as body 3 and 4. The links of the legs are labeled as link 1 and 2 from the link attached to the main body to the end link of the leg, and the joints of the legs as joint 1 and 2. We define unit vectors fixed to an inertial space as $\boldsymbol{a}_{-11}$, $\boldsymbol{a}_{-12}$, $\boldsymbol{a}_{-13}$, and unit vectors fixed to the main body whose origins are on the center of mass of the main body as $\boldsymbol{a}_{01}$, $\boldsymbol{a}_{02}$, $\boldsymbol{a}_{03}$, where the direction of $\boldsymbol{a}_{01}$ is corresponded to the sagittal direction of the main body and the direction of $\boldsymbol{a}_{03}$ is vertically up. Similarly, unit vectors whose origins are on the joint j of the leg i and which are fixed to the link j are defined as $\boldsymbol{a}_{j1}^{(i)}$, $\boldsymbol{a}_{j2}^{(i)}$, $\boldsymbol{a}_{j3}^{(i)}$, $(i = 1,2)$, the directions of $\boldsymbol{a}_{j1}^{(i)}$ $\boldsymbol{a}_{j3}^{(i)}$ coincide with the axis of link and the axis of rotation of the joint, and unit vectors $\boldsymbol{a}_{j1}^{(i)}$, $\boldsymbol{a}_{j2}^{(i)}$, $\boldsymbol{a}_{j3}^{(i)}$, $(i = 3,4)$ are set on the roll and pitch wheels. The directions of $\boldsymbol{a}_{j3}^{(3)}$, $\boldsymbol{a}_{j3}^{(4)}$ are oriented to the directions of $\boldsymbol{a}_{01}$ and $\boldsymbol{a}_{02}$, respectively. Using these unit vectors, we define the following unit column-matrices.

$$[\boldsymbol{a}_{-1}]^T = [\boldsymbol{a}_{-11}\ \boldsymbol{a}_{-12}\ \boldsymbol{a}_{-13}]\ [\boldsymbol{a}_0]^T = [\boldsymbol{a}_{01}\ \boldsymbol{a}_{02}\ \boldsymbol{a}_{03}]\ [\boldsymbol{a}_j^{(i)}]^T = [\boldsymbol{a}_{j1}^{(i)}\ \boldsymbol{a}_{j2}^{(i)}\ \boldsymbol{a}_{j3}^{(i)}]$$

We also define the following vectors.

$\boldsymbol{\omega}_{0-1} = [\boldsymbol{a}_0]^T\omega_{0-1}$: Angular velocity vector of $[\boldsymbol{a}_0]$ to $[\boldsymbol{a}_{-1}]$.

$\boldsymbol{\omega}_{jj-1}^{(i)} = [\boldsymbol{a}_j^{(i)}]^T\omega_{jj-1}^{(i)}$: Angular velocity vector of $[\boldsymbol{a}_j^{(i)}]$ to $[\boldsymbol{a}_{j-1}^{(i)}]$.

$\boldsymbol{r}_0 = [\boldsymbol{a}_{-1}]^T r_0$: Position vector from the origin of $[\boldsymbol{a}_{-1}]$ to the origin of $[\boldsymbol{a}_0]$.

$\boldsymbol{r}_j^{(i)} = [\boldsymbol{a}_{j-1}^{(i)}]^T r_j^{(i)}$: Position vector from the origin of $[\boldsymbol{a}_{j-1}^{(i)}]$ to the origin of $[\boldsymbol{a}_j^{(i)}]$.

$\boldsymbol{\rho}_0 = [\boldsymbol{a}_0]^T\rho_0$: Position vector from the origin of $[\boldsymbol{a}_0]$ to each point in the main body.

$\boldsymbol{R}_j^{(i)} = [\boldsymbol{a}_j^{(i)}]^T R_j^{(i)}$: Position vector from the origin of $[\boldsymbol{a}_j^{(i)}]$ to center of mass of link j.

$\boldsymbol{\rho}_j^{(i)} = [\boldsymbol{a}_j^{(i)}]^T\rho_j^{(i)}$: Position vector from the center of mass of link j to each point of link j.

$\boldsymbol{g} = [\boldsymbol{a}_{-1}]^T g$: Acceleration vector of gravity

We express cross product of a vector $\boldsymbol{x} = [\boldsymbol{a}]^T x$ as $\tilde{x}$ in $[\boldsymbol{a}]$

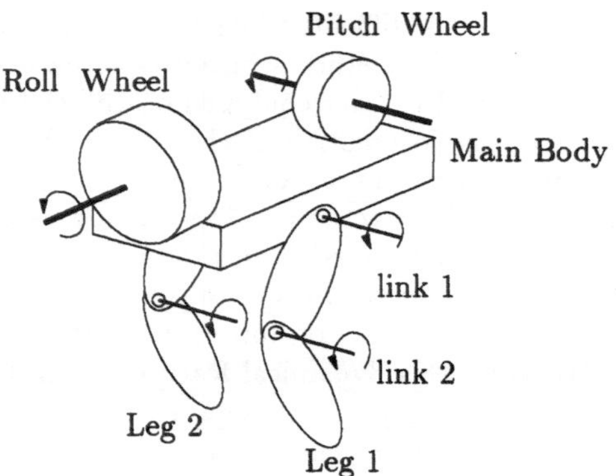

Fig. 1. Schematic Model of a Biped Locomotion Machine

We use the vector z as the state variable.

$$z^T = [\dot{r}_0^T \ \omega_{0-1}^T \ \cdots \ \omega_{jj-1}^{(i)T} \ \cdots] \tag{1}$$

Generalized momenta concerning to state variable z are given as follows;

$$\begin{bmatrix} p_0^T & l_0^T & l_j^{(i)T} \end{bmatrix}^T = H\{(\mathcal{L}_r + \mathcal{L}_R)M(\mathcal{L}_r^T + \mathcal{L}_R^T) + J\}H^T z \tag{2}$$

where p_0, l_0 are the linear and the angular momentum of the whole system, and $l_j^{(i)}$ is the angular momentum of sub-system composed of the links from link j to the end link of leg i.

$$\mathcal{L}_r = \begin{bmatrix} I & I & I & I & I & I \\ O & \tilde{r}_0^{(1)T} & \tilde{r}_0^{(1)T} & \tilde{r}_0^{(2)T} & \cdots & \tilde{r}_0^{(4)T} \\ O & O & \tilde{r}_0^{(1)T} & \cdots & & O \\ \vdots & & \ddots & & \cdots & O \\ \vdots & & & \ddots & \ddots & \vdots \\ O & \cdots & \cdots & \cdots & O & O \end{bmatrix} \quad \mathcal{L}_R = \begin{bmatrix} O & & & \cdots & & O \\ O & & & \cdots & & O \\ O & \tilde{R}_1^{(1)T} & O & \cdots & & O \\ \vdots & O & \tilde{R}_2^{(1)T} & O & \cdots & O \\ \vdots & & \ddots & \ddots & \ddots & \vdots \\ \vdots & & & \ddots & \ddots & O \\ O & \cdots & \cdots & \cdots & O & O \end{bmatrix}$$

$$H = \begin{bmatrix} I & O & \cdots & \cdots & \cdots & O \\ O & I & A_{10}^{(1)T} & A_{20}^{(1)T} & \cdots & A_{10}^{(4)T} \\ \vdots & O & I & A_{21}^{(1)T} & \cdots & O \\ \vdots & & \ddots & I & \cdots & \vdots \\ \vdots & & & \ddots & \cdots & \vdots \\ \vdots & & & & \ddots & \vdots \\ O & \cdots & \cdots & \cdots & & I \end{bmatrix} \quad \begin{array}{l} M = diag\begin{bmatrix} M_0^{(0)} & M_1^{(1)} & \cdots & M_1^{(4)} \end{bmatrix} \\ J = diag\begin{bmatrix} O & J_0 & J_1^{(1)} & \cdots & J_1^{(4)} \end{bmatrix} \end{array}$$

where, J_0 is an inertia matrix of the main body about the center of mass and $J_j^{(i)}$ is an inertia matrix of each link about the center of mass, respectively. M_0 and $M_j^{(i)}$ are the mass matrices of the main body and each link, respectively.

Equations of motion concerning to generalized momenta are derived as follows;

$$\dot{p}_0 = f_0 + D_0\lambda \tag{3}$$

$$\dot{l}_0 + \tilde{\omega}_{0-1}^T l_0 + \tilde{v}_{0r}^{(0)T} p_0 = t_0 + E_0\lambda \tag{4}$$

$$\dot{l}_j^{(i)} + \tilde{\hat{\omega}}_j^{(i)T} l_j^{(i)} + \tilde{v}_{jr}^{(i)T} \hat{p}_j^{(i)} = t_j^{(i)} + E_j^{(i)}\lambda + \tau_j^{(i)} \tag{5}$$

$$(i = 1,\ 2)\ (j = 1,\ 2)$$

$$\dot{l}_1^{(3)} + \tilde{\hat{\omega}}_1^{(3)T} l_1^{(3)} = \tau_1^{(3)} + E_1^{(3)}\lambda \tag{6}$$

$$\dot{l}_1^{(4)} + \tilde{\hat{\omega}}_1^{(4)T} l_1^{(4)} = \tau_1^{(4)} + E_1^{(4)}\lambda \tag{7}$$

where

$$\hat{\omega} = \begin{bmatrix} \dot{r}_0 \\ \hat{\omega}_0 \\ \hat{\omega}_1^{(1)} \\ \hat{\omega}_2^{(1)} \\ \hat{\omega}_1^{(2)} \\ \vdots \\ \hat{\omega}_1^{(4)} \end{bmatrix} = \begin{bmatrix} \dot{r}_0 \\ \omega_{0-1} \\ A_{10}^{(1)}\hat{\omega}_0 + \omega_{10}^{(1)} \\ A_{21}^{(1)}\hat{\omega}_1^{(1)} + \omega_{21}^{(1)} \\ A_{10}^{(2)}\hat{\omega}_0 + \omega_{10}^{(2)} \\ \vdots \\ A_{10}^{(4)}\hat{\omega}_0 + \omega_{10}^{(4)} \end{bmatrix} \qquad v_r = \begin{bmatrix} v_{0r} \\ v_{1r}^{(1)} \\ v_{2r}^{(1)} \\ \vdots \\ v_{1r}^{(4)} \end{bmatrix} = \begin{bmatrix} A_{0-1}\dot{r}_0 \\ A_{10}^{(1)}(v_{0r}^{(0)} + \tilde{r}_0^{(1)}\hat{\omega}_0) \\ A_{21}^{(1)}(v_{1r}^{(1)} + \tilde{r}_1^{(1)}\hat{\omega}_1^{(1)}) \\ \vdots \\ A_{10}^{(4)}(v_{0r}^{(0)} + \tilde{r}_0^{(4)}\hat{\omega}_0) \end{bmatrix}$$

$$v_R = \begin{bmatrix} v_{0R} \\ v_{1R}^{(1)} \\ v_{2R}^{(1)} \\ \vdots \\ v_{1R}^{(4)} \end{bmatrix} = \begin{bmatrix} O \\ \tilde{R}_1^{(1)}\hat{\omega}_1^{(1)} \\ \tilde{R}_2^{(1)}\hat{\omega}_2^{(1)} \\ \vdots \\ O \end{bmatrix} \qquad p = \begin{bmatrix} p_0^{(0)} \\ p_1^{(1)} \\ p_2^{(1)} \\ \vdots \\ p_1^{(4)} \end{bmatrix} = \begin{bmatrix} M_0 v_{0r}^{(0)} \\ M_1^{(1)}(v_{1r}^{(1)} + v_{1R}^{(1)}) \\ M_2^{(1)}(v_{2r}^{(1)} + v_{2R}^{(1)}) \\ \vdots \\ M_1^{(4)} v_{1r}^{(4)} \end{bmatrix}$$

$$\hat{p} = \begin{bmatrix} p_0 \\ \hat{p}_1^{(1)} \\ \hat{p}_2^{(1)} \\ \hat{p}_1^{(2)} \\ \vdots \\ \hat{p}_1^{(4)} \end{bmatrix} = \begin{bmatrix} p_0^{(0)} + A_{10}^{(1)T}\hat{p}_1^{(1)} + \cdots + A_{10}^{(4)T}\hat{p}_1^{(4)} \\ p_1^{(1)} + A_{21}^{(1)T}\hat{p}_2^{(1)} \\ p_2^{(1)} \\ p_1^{(2)} + A_{21}^{(2)T}\hat{p}_2^{(2)} \\ \vdots \\ p_1^{(4)} \end{bmatrix} \qquad n = \begin{bmatrix} f_0 \\ t_0 \\ t_j^{(i)} \end{bmatrix} = H(\mathcal{L}_r + \mathcal{L}_R)Mg$$

and kinematical constraints of each joint and geometrical constraints of the end of the supporting leg are expressed as follows;

$$\begin{bmatrix} \dot{r}_0^T & \omega_{0-1}^T & \omega_{jj-1}^{(i)T} & \cdots \end{bmatrix} \begin{bmatrix} D_0^{(0)T} & E_0^{(0)T} & E_j^{(i)T} & \cdots \end{bmatrix}^T = 0 \tag{8}$$

The equation of attitude motion of the whole system about the supporting point is given as the equation of generalized momentum $l_0^{(s)}$ corresponding to the angular velocity $\omega_0^{(s)}$, the angular velocity of the link which touches the ground.
where,

$$l_0^{(s)} = Q_{11}p_0 + A_{0-1}^T l_0 \tag{9}$$

$$\omega_0^{(s)} = A_{0-1}^T \omega_{0-1} + A_{1-1}^{(s)T}\omega_{10}^{(s)} + A_{2-1}^{(s)T}\omega_{21}^{(s)} \tag{10}$$

The equation of motion of generalized momentum $l_0^{(s)}$ is expressed in terms of angular velocity ω_0 as

$$K_1\dot{\omega}_0 + \dot{K}_1\omega_0 + h_1 = G - \sum_{k=3}^{4} A_{1-1}^{(k)T}(I - W^{(k)})\tau_1^{(k)} \tag{11}$$

where, K_1 is an inertia matrix corresponding to the rotational motion of the total system about the supporting point, G is a gravity term, and $W^{(k)}$ is a projection matrix.

It may be noted that Eq. (11) indicates that driving torque $\tau_1^{(3)}$ and $\tau_1^{(4)}$ of the wheels can directly control the attitude of the main body.

3. Motion Control

We propose three kinds of motion controller, a linear feedback controller, a quasi-linear feedback controller and non-linear feedback controller. These controllers are composed of two controllers, a leg controller and an attitude controller; A leg controller controls the motions of the legs by the motors mounted on the joints rapidly and an attitude controller controls the attitude of the system by the wheels slowly. The architectures of the system proposed here are indicated by Figs. 2 to 4.

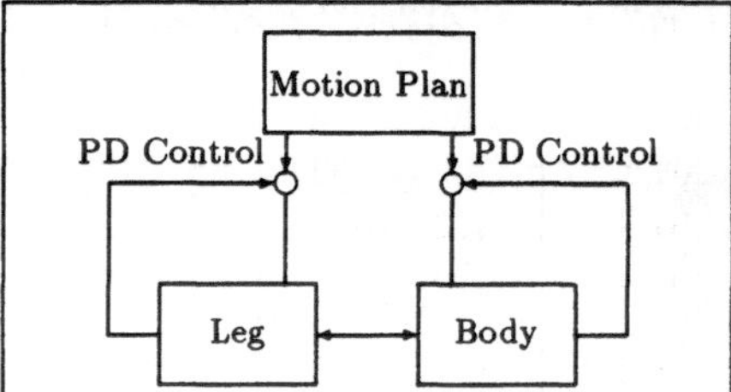

Fig. 2. Linear-feedback controller

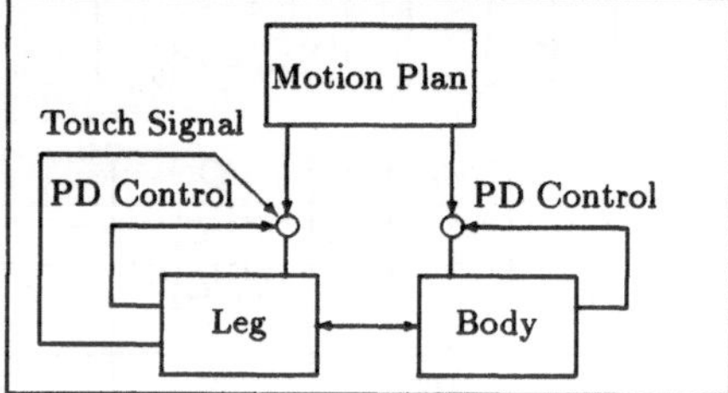

Fig. 3. Quasi-linear feedback controller

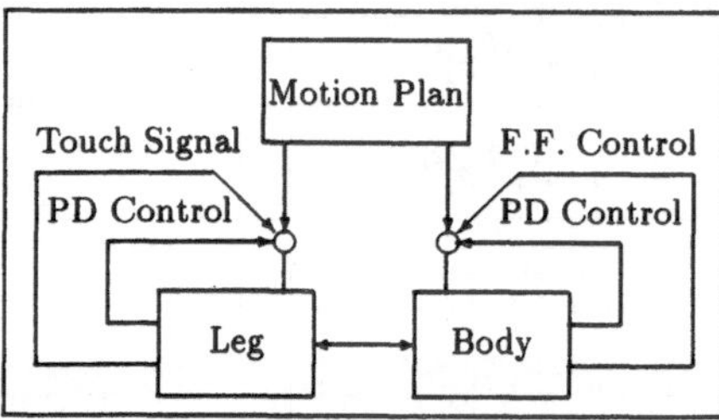

Fig. 4. Non-linear feedback controller

3.1 Design of Leg Controller

The motion of each leg is controlled by motion controller with local high-gain PD control. Commanded values of the angle and the angular velocity of the joint j of the body i, $\theta_j^{(i)c}$, $\omega_j^{(i)c}$ are calculated as follows;

(i) Linear Feedback Controller

$$\theta_j^{(i)c} = \theta_j^{(i)d}, \ \omega_j^{(i)c} = \omega_j^{(i)d} \tag{12}$$

(ii) Quasi-Linear and Non-Linear Feedback Controller

$$\theta_j^{(i)c} = \theta_j^{(i)c}(\theta_j^{(i)d}, \theta_0^{(s)m}, t^{(s)}), \quad \omega_j^{(i)c} = \omega_j^{(i)c}(\omega_j^{(i)d}, \omega_0^{m}, t^{(s)}) \tag{13}$$

where, $\theta_0^{(s)m}$ is measured angle of the link which touches the ground, ω_0^m is measured value of angular velocity of the main body, and $t^{(s)}$ is output signal of the touch sensor on the leg which touches the ground.
Quasi-linear and Non-linear feedback controllers utilize the desired values modified by the output signal of the touch sensor and angle sensors which equipped on the end of the legs and angular velocity sensor equipped on the main body as commanded values, while the linear feedback controller uses desired values as commanded values.
Input torque of the joints of the legs are designed using the desired trajectories as the commanded values as follows;

$$\tau_j^{(i)} = K_{pj}(\theta_j^{(i)c} - \theta_j^{(i)}) + K_{dj}(\omega_j^{(i)c} - \omega_j^{(i)}) \tag{14}$$

3.2 Design of Attitude Controller

The attitude of the system is controlled by the wheel controllers that utilize the reaction torque of the wheels as the control torque.
Here, since feedback gains of each joint-controller are designed so high for each link to follow the desired trajectories rapidly, on the assumption that the trajectory of each leg coincides with the desired trajectory, we can rewrite the equation of motion of the total system in terms of the state variable of the main body only as

$$K_1\dot{\omega}_0 + \dot{K}_1\omega_0 = F(\theta_0, \omega_0) \tag{15}$$

Input torque of the wheels are designed as follows;
(i) Linear and Quasi-Linear Feedback Controller

$$\tau_1^{(w)} = K_p^{(w)}(\theta_0^d - \theta_0) + K_d^{(w)}(\omega_0^d - \omega_0) \tag{16}$$

(ii) Non-Linear Feedback Controller

$$\tau_1^{(w)} = K_1(\theta)\hat{\tau}^{(w)} + h(\theta_0^{(s)m}, \omega_0^m, \theta_j^{(i)m}, \omega_j^{(i)m}, \dot{\omega}_j^{(i)}) \tag{17}$$

$$\hat{\tau}^{(w)} = \dot{\omega}_0^d + K_p^{(w)}(\theta_0^d - \theta_0) + K_d^{(w)}(\omega_0^d - \omega_0) \tag{18}$$

where, θ_0^d and ω_0^d are the desired values of the angle of attitude and the angular velocity of the main body, and $\theta_j^{(i)m}$ and $\omega_j^{(i)m}$ are the measured values of the angle, and angular velocity of the link j of leg i. In the case of linear and quasi-linear feedback controllers, the torques of the wheels are consist of a linear feedback term with desired values θ_0^d and ω_0^d. On the other hand, in the case of non-linear feedback controller, the torques of the wheels are composed of the linear feedback terms and non-linear feedback terms calculated by measured values $\theta_j^{(i)m}$, $\omega_j^{(i)m}$ and $\dot{\omega}_j^{(i)m}$ via the inverse dynamics. The values of feedback gains $K_p^{(w)}$ and $K_d^{(w)}$ are determined as follows; Denoting the state variables as

$$\dot{\omega}_0 = \dot{\omega}_0^* + \Delta\dot{\omega}_0, \quad \omega_0 = \omega_0^* + \Delta\omega_0, \quad \theta_0 = \theta_0^* + \Delta\theta_0$$

and substituting them into Eq. (15), and linearizing it, we get the following equation.

$$\Delta\dot{\omega}_0 + 2Z\Omega\Delta\omega_0 + \Omega^2\Delta\theta_0 = 0 \tag{19}$$

Z and Ω are damping ratio and natural frequency, respectively.
Using these equations, parameters of the wheel controllers are determined for the attitude motion of the main body to become stable.

4. Experiments

The performances of the controller proposed are verified by hardware experiments. Figure 5 shows the architecture of the hardware equipment. The desired trajectory of each leg is generated by personal–computer(CPU:Intel Pentium 90MHz). Angles and angular velocities of joints of the legs are measured by optical encoders equipped to the joints of the legs. And an angle of the leg which touched the ground and a touch signal are measured by an angle sensor equipped to the end of the leg and a touch-sensor, respectively. Angular velocity of the main body is measured by a gyro-sensor equipped on the main body. Motions of the legs are controlled by motor controllers on DSP board where reference signals are generated by the personal–computer. Attitudes of the main body are controlled by wheel controllers on the personal–computer. Figures 6. to 8. show the results of hardware experiments, in case of linear, quasi-linear and non-linear feedback controller, respectively.

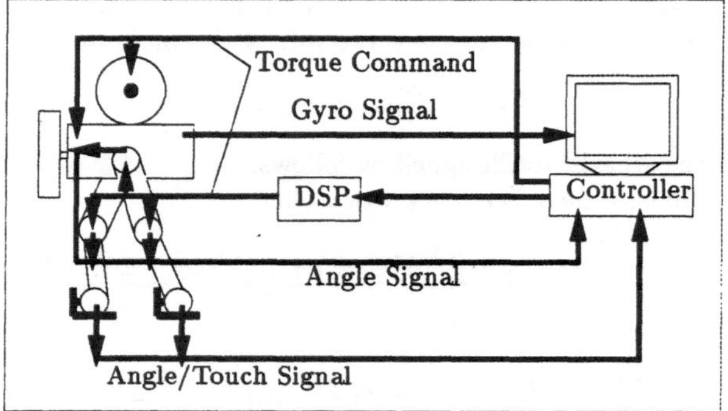

Fig. 5. The architecture of the hardware equipment

All the controllers have established steadily stable walk of the system. Among of them, the non-linear feedback controller realizes a precise control of the attitude of the main body.

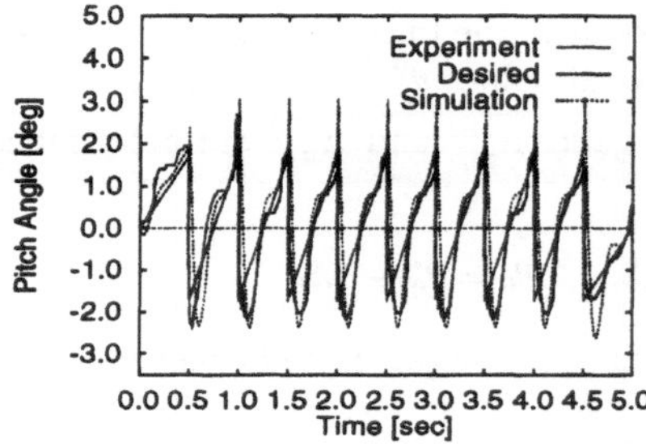

Fig. 6. Time history of pitch angle (linear feedback)

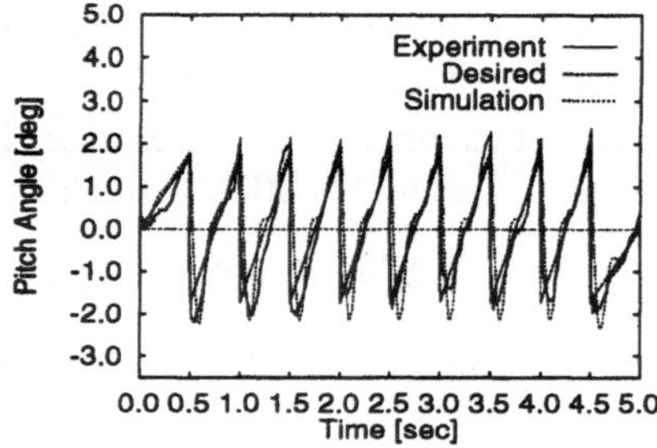

Fig. 7. Time history of pitch angle (quasi-linear feedback)

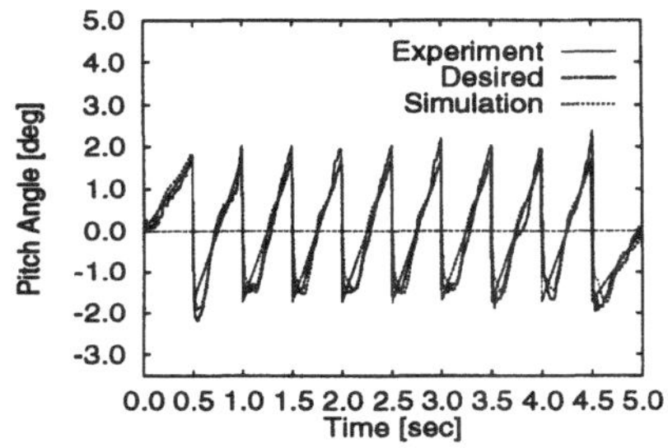

Fig. 8. Time history of pitch angle (non-linear feedback)

5. Conclusion

In this paper, three types of motion controller of a biped locomotion machine are proposed. They utilize reaction wheels as internal torquers. A reaction wheel can generate control torque without any interactions with the gravitational field. Therefore, it becomes easy to calculate the motion of the wheel to generate required torque.

The controllers proposed here are composed of leg controller and attitude controller, and have a hierarchical architecture: Leg controller uses high gain PD feedback control for the motions of the legs to follow the desired trajectories rapidly, and on the assumption that the motions of the legs coincide with the desired trajectories, attitude controller controls the attitude of the system.

We have developed the hardware model, and verified the performances experimentally. The controllers have established steadily stable walk of the system.

References

[1] A. Sano and J. Furusho, Analysis of Dynamic Quadruped Locomotion Based on Quasi-Angular-Momentum, Trans. Society of Instrument and Control Engineers, **24**-12 (1988), 1299-1305 (in Japanese)

[2] S. Arimoto and F. Miyazaki, A Hierarchical Control Scheme for Biped Robots, Journal of the Robotics Society of Japan, **1**-3 (1983), 167-175 (in Japanese)

[3] Q. Li, A. Takanishi and I. Kato, Learning Control of Compensative Trunk Motion for Biped Walking Robot based on ZMP Stability Criterion, Journal of the Robotics Society of Japan, **11**-4 (1993), 557-563 (in Japanese)

[4] S. Kajita, T. Yamaura, Dynamic Walking Control of a Biped Robot Along a Potential Energy Conserving Orbit, IEEE Trans. Robotics and Automation, **8**-4 (1992), 431-438

[5] K. Tsujita and K. Tsuchiya, Motion Control of a Biped Locomotion Machine, Proceedings of Second International Conference on Motion and Vibration Control, (1994), 180-185

[6] S. Kajita, K. Tani, Experimental Study of Biped Dynamic Walking in the Linear Inverted Pendulum Mode, Proceedings of IEEE Conference on Robotics and Automation, (1995), 2885-2891

Real-Time Programming of Mobile Robot Actions Using Advanced Control Techniques

R. Pissard-Gibollet, K. Kapellos, P. Rives, J.J. Borrelly

INRIA-centre de Sophia Antipolis,
2004 Route des Lucioles, 06565 Valbonne, France
e-mail : {pissard,rives,kapellos,borrelly}@sophia.inria.fr

Abstract

In this paper, we present a robotic application performed by sequencing visual servoing tasks. The theoritical framework used is the sensor-based control for the continuous parts of the application and discrete events system theory for its logical aspects. The design and analysis of the whole system are coherently handled using ORCCAD concepts. We focus our attention on the programming aspects of these theories and concepts, from application level specification up to real-time implementation and results analysis. An effective experimentation on our mobile hand-eye robot validating and illustrating this approach is fully detailed.

1. Introduction

Programming mobile robots in order to achieve a desired objective in a reliable way, operating in a structured and partially known environment, requires to solve many appealing problems. They range from task planning, reactive behavior synthesis up to control law choice and design. Even though, at the high level, problems of path planning and off line programming have been broadly addressed, the state is not so clear at the control level, where the interaction of the mobile robot with its environment must be considered. In most cases, the trajectory provided by the path planner is directly played at the servo level without taking into account perturbations due to real interactions between the robot and its local environment. As a consequence, the control law works in open loop with regard to this environment. A manner to perform a robotic task in a more robust and reliable way, is to control explicitely the local interactions between the robot and its environment. It can be efficiently done by using a sensor-based control approach. Furthermore, we claim that a complex robotic task can be successfully performed by sequencing elementary sensor-based control tasks.

The paper is organized as follows. In the first part we briefly recall the theoritical framework used to model elementary tasks and their composition. The second part describes the real experiment on a robotic task of *reaching a target* with our experimental testbed constitued by a mobile robot carrying a hand-eye system using dedicated vision hardware. A special attention will be put on the programming of the application ranging from specification to real-time programming and results analysis.

2. Continuous Control Laws Design

We assume that, in order to correctly complete a task, we need a low level of task specification which explicitely integrates the interaction between the robots and its local environment. When the task can be defined as a servoing on a local part of the environment, we use a previously developped framework in vision-based control. It allows us to perform elementary visual servoing tasks by means of robust closed loop control laws using data vision. Let us give a brief overview of the approach, while a more detailed presentation can be found in [1, 2, 3].

The basic assumption about the sensors is that the vector signal $\underline{s}$ provided by the sensor is a function of the relative position and orientation $\bar{r}$ between the sensor, associated to a frame F_S, and the target, associated to a frame F_T. We may thus write :

$$\underline{s}(\bar{r},t) = \underline{s}(F_S, F_T)$$

A Jacobian matrix L^T of the vision feature $\underline{s}$ with regard to the relative displacement (velocities screw T_{ST}) between the camera and the environment can be computed to get : $\dot{\underline{s}}(\bar{r},t) = L^T.T_{ST}$

Of course, the formal expression of L^T depends on the type of geometrical primitives (point, line, ellipse...) and their parametrization [1].

To specify a task and to select the visual signal, we use the notion of *virtual linkage* between the robot sensor and its environment. It is characterized by the velocity screw T^* which leave $\underline{s}$ invariant during the motion :

$$\dot{\underline{s}}(\bar{r},t) = L^T.T^* = 0$$

Using the general formalism of task function [3], we can express the goal of a task in term of the regulation of an output function $\underline{e}(\underline{r},t)$. Applying this formalism to the sensor based task, this function can be written like : $\underline{e}(\bar{r},t) = \underline{s}(\bar{r},t) - \underline{s}_d(t)$ where $\underline{s}_d$ is the desired visual signal. It has been shown in [1] that a very simple *gradient based approach* is sufficient to ensure an exponential regulation of $\underline{e}$ by using the following desired velocity screw T_d as control input. For a positioning task we have :

$$T_d = -\lambda.L^{T+}_{\underline{s}=\underline{s}_d}.(\underline{s}(\bar{r}) - \underline{s}_d) \tag{1}$$

The transposition of such a scheme to nonholonomic mobile robots is not straightforward. One way to overpass this problem consists in adding some degrees of mobility to the camera by mounting it on a motorized device like a manipulator or a "head". Considering the whole mechanical system (cart + manipulator) like a single kinematic chain, it becomes possible to fully control the motion of the camera without beeing limited by the cart nonholonomic constraints [4, 5].

On the other hand when we do not, or can not, take into account sensory data from the environment to design the control laws, it is possible to directly design it in the cartesian space. But it is now well established, that it does not exist a smooth continuous state-space feedback able to stabilize a nonholonomic mobile robot around a fixed point. So we used recent results in mobile robot control [6] and we have implemented a control law using a *time-varying* feedback technique. Using the state error vector $(x\, y\, \tilde{\theta})^T$ the control can be written :

$$\begin{cases} v = y^{\frac{1}{3}}\, sin(\omega t) + g_1\, x\, cos(\tilde{\theta}) \\ \dot{\theta} = \frac{2}{3}\, g_2\, y^{\frac{1}{3}}\, v \frac{sin(\tilde{\theta})}{\tilde{\theta}} + g_3\, \tilde{\theta} \end{cases} \tag{2}$$

3. Robotic Actions Concepts

3.1. The Robot-task (RT)

Control laws as (1),(2) without ambiguity characterize, in continuous time, the physical motion of the robot during the $[0, T]$ interval of their validity. Nevertheless, when we want to execute this motion in a realistic environment we need to take into account and react in time to various situations, at least for ensuring the integrity of the robot. These two tightly coupled aspects of a robotic action are coherently captured by the *Robot-task* (RT) definition as proposed by ORCCAD concepts [7]. Let's remind that a RT models an elementary robotic action. It is formaly defined as the entire parametrized specification of a control law, and a logical behavior associated with a set of events which may occur just before, during and just after the task execution.

The behavior of the system is handled by the framework of reactive systems theory: it consists of the legal sequences of input/output signals received/emitted by the system. Its specification is methodic; events are typed in pre-conditions, exceptions with three types of reaction and post-conditions. The control-law activation starts at the instant that all pre-conditions are satisfied. During its execution exceptions are monitored; they are handled either localy changing in-line a control parameter or globaly asking from the application to interrupt the current RT and activate a recovery program or imposing the total application interruption driving the robot in a safe position. Finaly the action stops when the set of post-conditions is satisfied.

3.2. The Robot-Procedure (RPr)

A more complex robotic application is therefore seen as the composition of the RTs necessary to accomplish the desired objective. Composition is obtained using dedicated operators as the sequence, the conditional, the iteration, and different levels of prehemption. In ORCCAD system the Robot-procedure (RPr) formalism is proposed to methodicaly specify, verify and implement RTs controllers in order to design complex robotic actions. It clearly separates the composition of the actions driving to a nominal execution from those required to recover from an exception not handled localy by the RTs. To the whole can be associated pre and post-conditions driving to the specification of an entity which can be used to compose other ones; structural programing is therefore obtained.

RPr formalism is translated to adequate languages (ESTEREL and TIMEDARGOS) providing the behavior controller with nice semantics. These languages may be compiled into a wide class of models, usually labeled transition systems. This allows to methodicaly verify a large set of behavioral and quantitative temporal properties, including crucial properties of liveliness and safety, as well as the conformity with applications requirements. From a practical point of view we use the ESTEREL compiler to obtain the program that schedules the evolution of the systems behavior.

4. The Robotic System

We have developped a versatile testbed in order to validate sensor-based control approaches in real experiments. This robotic system ([8]) uses a nonholonomic wheeled cart carrying a two d.o.f head with a CCD camera and recently equiped with a belt of eight sounders (see figure 1a.

The on board computer architecture is built around a VME backplane. The robot control is assumed by a Motorola MVME 167 board and a 6 axes custom-made servocontrol board. The management of the sounder belt is done by an other Motorola MVME 162 board.

Difficulties in image processing are induced by strong real time constraints and the processing of large dataflow from the sequence of images. To overpass them we have developped a vision system [9] which is characterized by its modularity and its real time capabilities. Its architecture is based both on VLSI chips for low level processing and on multi DSP's processors for more elaborated processings. For facilities of development, the machine vision has been implemented in an independant VME rack outside of the robot. However, for experiments which require high autonomy, these boards can be plugged into the on board rack.

During the development step, an external link is used both for power and communication (ethernet thin link, three video channels and two serial links).

Figure 1. The application

a. View of the Robotic System

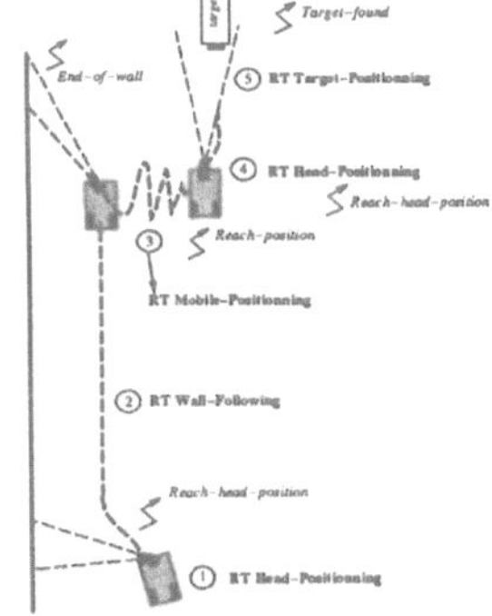

b. Description of the Application

5. A Robotic Application Specification

5.1. Description of our Application

Our robotic system is used to validate and illustrate the theoritical aspects previously presented. In this application, our long term objective is to explore an indoor environment.

The application, presented in this paper, consists in *reaching a target* and it is performed by our robotic system. Informally it is specified as follows: the robot must cross the room by following a wall, reach a region of interest, find a target in this region and go in front of it. To perform this application, sophisticated control laws must be sequenced; their activations are conditionned by the occurence of various types of events determining the end of the wall, the presence of the target, some failures. A nominal execution of this robotic application is illustrated in figure 1b.

5.2. Decomposition in Robot-task

Once the application is described, the next step towards its realization consists in identifying all the elementary tasks (RTs) needed to perform it. The basic idea is to associate a control law to a sub-objective and to identify the set of events related to its execution. Therefore a RT is constructed by the characterization of these elements. In this experiment, using only vision and odometry, we have defined four

RTs.

Head Positioning

In a priori known environment, the robot must be able to move its head (two axis on pitch and yaw) in order to look at an object of interest. For example, using *MoveHead*, the robot can look up to see the final target or look toward the floor to find a skirting board. This elementary action is finished when the head reaches the desired position (post-condition event *Reach-head-Position*).

Visual Wall Following

To cross the room the robot can follow a wall. The parallel lines corresponding to a skirting board at the bottom of a wall (see the two first images on figure 2) are used for its control. This visual-servoing task is handled by the vision-based control reminded in the theoritical part. For this task *WallFollowing*, thanks to the vision system, line parameters are extracted for the visual servoing control loop and events are detected as the *End-of-wall* (post-condition) or the exception *Target-lost.*

Visual Positioning in front of a Target

To position the robot in front of the target we have implemented the visual-servoing task *Pos-Target* (control law (1)). The selected $3D$ target is a cone which projects onto the image frame in two ellipses corresponding to the two circles bounding the cone (see the two last images on figure 2). The positioning task consists in aligning the cone axis with the optical axis of the camera and to tune the distance between the camera and the target along with this axis [4, 5]. Like in the previous visual servoing Robot-task, an event *Target-lost* monitors a possible failure. The correct end of this action is managed by an event indicating the end of task duration.

Cartesian Mobile Robot Positioning

To drive the robot in an area and to stabilize it around a fixed point, we use the time-varying feedback expressed in control law (2). Despite that the cartesian position of the mobile robot is reconstructed using only odometry, the precision of robot positioning is sufficient. When the robot reaches the desired position, the post-condition *Reach-position* is satisfied inducing the end of this Robot-task.

Figure 2. Images during the two visual-servoing RT

a. Wall following

b. End wall

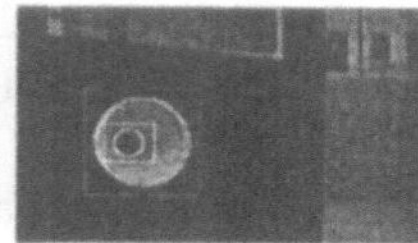
c. Initial target

d. Final target

5.3. Application Synthesis with RPrs

The RPr *ReachTarget* (see below) is designed to specify the evolution of the application as the composition of the previously defined RTs. It simply states that the application will start after user confirmation, pre-condition *start*; its nominal execution consists on the sequencing of five actions. Initialy the RT *MoveHead* fixes the head to the indicated position. It is followed by the *SafeWallFollow* (see bellow) which, after detecting the visual motif drives the mobile accross the room until the end of the wall avoiding eventualy the obstacles. Afterwards, *MoveCart* drives the mobile to reach the target location and *MoveHead* directs the robot look toward the target. Finally, *PosTarget* brings the robot in front of it. During the execution *Target-Lost, Robot-Fail...* events are handled as failure exceptions asking for application interruption.

Safe WallFollow (see below) is active continuously, since the end of the wall detection, the *One WallFollow* which, using *WallFollow* and *ObstacleAvoidance* displace the mobile along the wall and handles the case of obstacle presence.

The translation of *ReachTarget* RPr to a dedicated language, ESTEREL in the particular case, gives us both, the model describing its evolution and the program that controls it. Figure 3a gives an abstract view of the controller considering only the activation and the end of the RTs and the most significant events. On the other hand the control program uses the services of a real-time software that we developped to control RTs.

Figure 3. Design of the application

a. RPr design

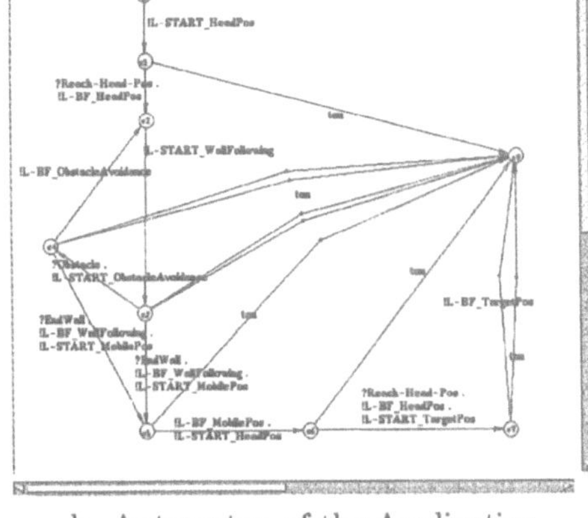

b. Automaton of the Application

6. Real-Time Programming

A real-time program must be logically correct to produce the correct outputs but also it must be temporally correct to produce them at the correct time ; control laws performance together with robot safety highly depends on it.

The real-time software used to realize this experiment has two main parts : the first one assumes the robot control ; the second concerns the vision processing. Their communication is based on a client/server architecture ; the robot controller manages the application and sees the vision program as an *intelligent sensor* (see 6.2). Let us briefly describe this software before presenting the experiment results.

6.1. Robot Control Software

The aim of our robot control software is to support robotic applications constructed as a composition of RTs . It is naturally divided in three parts [10] concerning :

• The computation of the continuous part of the system (control law and observers) : the tasks are usually periodics, communications and message passing between them are asynchronous of producer/consummer type.

• The discret event controller : synchronous approach is used to compose the behaviors of the RTs as well the coordination of RTs switching. This permits to enhance the reliability of our system by formally analyzing/proving interesting properties of the controller.

• The interface between synchronous and asynchronous parts of the application : save communications, preserving the order of messages, are established in a systematic way between the asynchronous observers and the synchronous controller.

The use of adapted programming and debugging tools for each part of the application is of prime necessity : object oriented language (C++) for programming facilities, synchronous language (ESTEREL) use for critical controllers and efficient real-time operating system with large environment capabilities (WINDRIVER tools : VxWorks 5.1.1, WindView).

6.2. Vision Software

To perform the vision processing, we use a dedicated machine vision of which the hardware and software architecture are described in [9]. It implements the concept of active window. An active window is attached to a particular region of interest in the image and has in charge to extract a desired feature in this region. To each active window is associated a temporal filter able to perform the tracking of the feature along the sequence. Several active windows can be defined, at the same time, in the image and different processings can be associated to each window.

In our application, the two visual servoing RTs use services coming from the vision software application, allowing to launch/stop active windows in images. These windows (see figure 2) return parameters for the visual servoing (on lines or ellipses) but also events (*target-lost, End-of-wall*). Services of target recognition in image and automatic windows positioning initialization are not currently developped but one way is investigated in [11].

7. Experimental Results

Let us present the effective experiment which consists of the RPR *ReachTarget* nominal execution by our robotic system. Actually, a set of tools allows the user to analyze the real execution of the robotic application from a continuous and discrete time point of view. They are used off-line after the execution and allows its analysis at different levels.

Firstly robot trajectory can be visualized using a 3D animation tool to estimate roughly the robot evolution (see figure 4a). For a finest estimation from the automatic control point of view, the evolution of continuous variables, such as robot state or sensors data, can be plotted. In addition occurences of events and activation of RTs can be associated to these plots. For example, in figure 4b, we verify that, during the RT1 *MoveHead* execution, the robot wheels velocities are null until the head is positioned which is signaled by the occurence of *Reach-head-Pos* event. Its presence activates the controller which transits from state e_1 to e_2 activating the RT2 *WallFollowing* which induces the wheels velocities evolution. Finally, all the real-time mecanisms involved in the application execution can be analyzed; in figure 5, the switch between the RTs *WallFollowing* and *MoveCart* is visualized.

8. Conclusion

Using a generic framework given by ORCCAD lying on continuous and discrete approaches, we have shown experimentaly how a "complex robotic task" can be successfully performed by sequencing elementary sensor-based control tasks. A special attention has been given to the programming/debugging/analyzing tools and real-

Figure 4. Mobile Robot Data Views

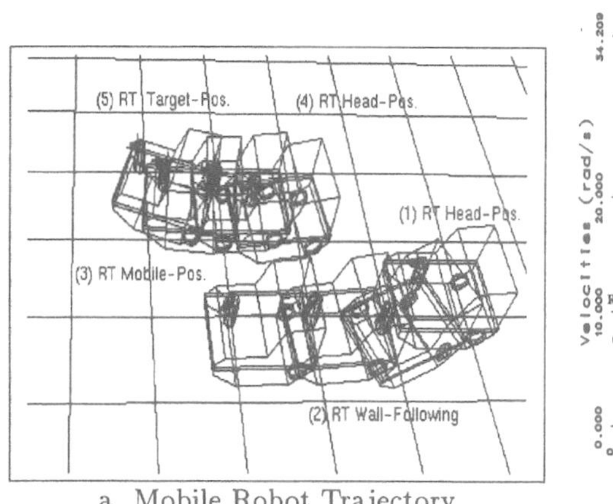

a. Mobile Robot Trajectory

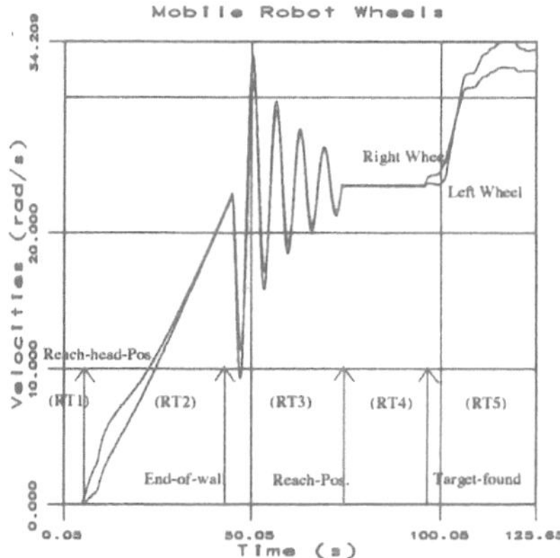

b. Velocities on Robot Wheels

Figure 5. Timing of a Robot-Task Switching

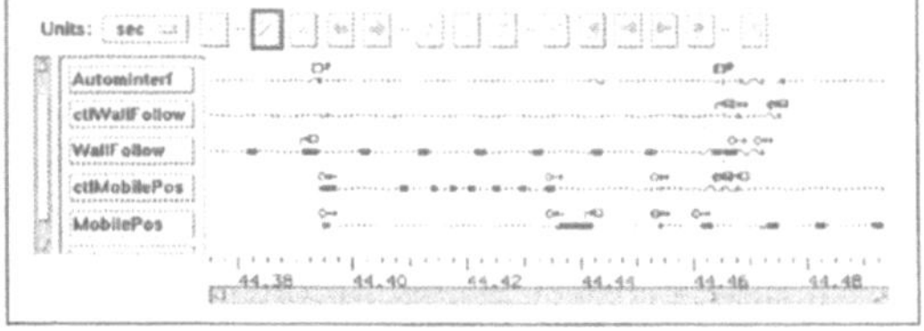

time performances of the system. The results show that work remains to be done for robustifying the experimental aspects. For example, more capabilities must be taken into account with regard to events failures or obstacles. Concerning the sensor aspects, we have to implement more elaborated vision algorithms and to integrate sounder-based tasks. In the future, we expect to achieve more complex application concerning automatic exploration and cartography of partial known structured environment.

References

[1] F. Chaumette, *La commande référencée vision: une approche aux problèmes d'asservissements visuels en robotique.* PhD thesis, Université de Rennes, Juillet 1990.

[2] B. Espiau, F. Chaumette, and P. Rives, "A new approach to visual servoing in robotics," *IEEE Transactions on Robotics and Automation*, vol. 8, pp. 313–326, June 1992.

[3] C. Samson, B. Espiau, and M. L. Borgne, *Robot control: the task function appproach.* Oxford University Press, 1990.

[4] R. Pissard-Gibollet and P. Rives, "Applying visual servoing techniques to control a mobile hand-eye system," in *IEEE Int. Conf. on robotics and automation*, (Nagoya, Japan), May 1995.

[5] R. Pissard-Gibollet, *Conception et Commande par Asservissement Visuel d'un Robot Mobile.* PhD thesis, Ecole des Mines de Paris, December 1993.

[6] C. Samson, "Time varying feedback stablisation of a car like wheeled mobile robot,"

The International Journal of Robotics Research, vol. 12, pp. 55–64, February 1993.

[7] D. Simon, B. Espiau, E. Castillo, and K. Kapellos, "Computer-aided design of generic robot controller handling reactivity and real-time control issues," *IEEE Transactions on Control Systems and Technology*, Fall 1993.

[8] P. Rives, R. Pissard-Gibollet, and K. Kapellos, "Development of a reactive mobile robot using real time vision," in *ISER Third International Symposium on Experimental Robotics*, (Kyoto, Japan), October 1993.

[9] P. Rives, J. Borrelly, J. Gallice, and P. Martinet, "A versatile parallel architecture for vision based control applications," in *Workshop on Computer Architecture for Machine Perception*, (New-Orleans USA), December 1993.

[10] E. Coste-Manière, "A synchronous/asynchronous approach to robot programming," in EUROMICRO *Workshop on Real-Time Systems*, (Oulu Finland), pp. 268–273, June 1993.

[11] D. Djian, P. Probert, and P. Rives, "Active sensing using bayes nets," in *to appear ICAR Int. Conf. on Advanced Robotics*, (Spain), September 1995.

Lecture Notes in Control and Information Sciences

Edited by M. Thoma

1993–1996 Published Titles:

Vol. 186: Sreenath, N.
Systems Representation of Global Climate Change Models. Foundation for a Systems Science Approach.
288 pp. 1993 [3-540-19824-5]

Vol. 187: Morecki, A.; Bianchi, G.; Jaworeck, K. (Eds)
RoManSy 9: Proceedings of the Ninth CISM-IFToMM Symposium on Theory and Practice of Robots and Manipulators.
476 pp. 1993 [3-540-19834-2]

Vol. 188: Naidu, D. Subbaram
Aeroassisted Orbital Transfer: Guidai. and Control Strategies
192 pp. 1993 [3-540-19819-9]

Vol. 189: Ilchmann, A.
Non-Identifier-Based High-Gain Adaptive Control
220 pp. 1993 [3-540-19845-8]

Vol. 190: Chatila, R.; Hirzinger, G. (Eds)
Experimental Robotics II: The 2nd International Symposium, Toulouse, France, June 25-27 1991
580 pp. 1993 [3-540-19851-2]

Vol. 191: Blondel, V.
Simultaneous Stabilization of Linear Systems
212 pp. 1993 [3-540-19862-8]

Vol. 192: Smith, R.S.; Dahleh, M. (Eds)
The Modeling of Uncertainty in Control Systems
412 pp. 1993 [3-540-19870-9]

Vol. 193: Zinober, A.S.I. (Ed.)
Variable Structure and Lyapunov Control
428 pp. 1993 [3-540-19869-5]

Vol. 194: Cao, Xi-Ren
Realization Probabilities: The Dynamics of Queuing Systems
336 pp. 1993 [3-540-19872-5]

Vol. 195: Liu, D.; Michel, A.N.
Dynamical Systems with Saturation Nonlinearities: Analysis and Design
212 pp. 1994 [3-540-19888-1]

Vol. 196: Battilotti, S.
Noninteracting Control with Stability for Nonlinear Systems
196 pp. 1994 [3-540-19891-1]

Vol. 197: Henry, J.; Yvon, J.P. (Eds)
System Modelling and Optimization
975 pp approx. 1994 [3-540-19893-8]

Vol. 198: Winter, H.; Nüßer, H.-G. (Eds)
Advanced Technologies for Air Traffic Flow Management
225 pp approx. 1994 [3-540-19895-4]

Vol. 199: Cohen, G.; Quadrat, J.-P. (Eds)
11th International Conference on Analysis and Optimization of Systems – Discrete Event Systems: Sophia-Antipolis, June 15–16–17, 1994
648 pp. 1994 [3-540-19896-2]

Vol. 200: Yoshikawa, T.; Miyazaki, F. (Eds)
Experimental Robotics III: The 3rd International Symposium, Kyoto, Japan, October 28-30, 1993
624 pp. 1994 [3-540-19905-5]

Vol. 201: Kogan, J.
Robust Stability and Convexity
192 pp. 1994 [3-540-19919-5]

Vol. 202: Francis, B.A.; Tannenbaum, A.R. (Eds)
Feedback Control, Nonlinear Systems, and Complexity
288 pp. 1995 [3-540-19943-8]

Vol. 203: Popkov, Y.S.
Macrosystems Theory and its Applications: Equilibrium Models
344 pp. 1995 [3-540-19955-1]

Vol. 204: Takahashi, S.; Takahara, Y.
Logical Approach to Systems Theory
192 pp. 1995 [3-540-19956-X]

Vol. 205: Kotta, U.
Inversion Method in the Discrete-time Nonlinear Control Systems Synthesis Problems
168 pp. 1995 [3-540-19966-7]

Vol. 206: Aganovic, Z.;.Gajic, Z.
Linear Optimal Control of Bilinear Systems with Applications to Singular Perturbations and Weak Coupling
133 pp. 1995 [3-540-19976-4]

Vol. 207: Gabasov, R.; Kirillova, F.M.; Prischepova, S.V.
Optimal Feedback Control
224 pp. 1995 [3-540-19991-8]

Vol. 208: Khalil, H.K.; Chow, J.H.; Ioannou, P.A. (Eds)
Proceedings of Workshop on Advances in Control and its Applications
300 pp. 1995 [3-540-19993-4]

Vol. 209: Foias, C.; Özbay, H.; Tannenbaum, A.
Robust Control of Infinite Dimensional Systems: Frequency Domain Methods
230 pp. 1995 [3-540-19994-2]

Vol. 210: De Wilde, P.
Neural Network Models: An Analysis
164 pp. 1996 [3-540-19995-0]

Vol. 211: Gawronski, W.
Balanced Control of Flexible Structures
280 pp. 1996 [3-540-76017-2]

Vol. 212: Sanchez, A.
Formal Specification and Synthesis of Procedural Controllers for Process Systems
248 pp. 1996 [3-540-76021-0]

Vol. 213: Patra, A.; Rao, G.P.
General Hybrid Orthogonal Functions and their Applications in Systems and Control
144 pp. 1996 [3-540-76039-3]

Vol. 214: Yin, G.; Zhang, Q. (Eds)
Recent Advances in Control and Optimization of Manufacturing Systems
240 pp. 1996 [3-540-76055-5]

Vol. 215: Bonivento, C.; Marro, G.; Zanasi, R. (Eds)
Colloquium on Automatic Control
240 pp. 1996 [3-540-76060-1]

Vol. 216: Kulhavý, R.
Recursive Nonlinear Estimation: A Geometric Approach
244 pp. 1996 [3-540-76063-6]

Vol. 217: Garofalo, F.; Glielmo, L. (Eds)
Robust Control via Variable Structure and Lyapunov Techniques
336 pp. 1996 [3-540-76067-9]

Vol. 218: van der Schaft, A.
L_2 Gain and Passivity Techniques in Nonlinear Control
176 pp. 1996 [3-540-76074-1]

Vol. 219: Berger, M.-O.; Deriche, R.; Herlin, I.; Jaffré, J.; Morel, J.-M. (Eds)
ICAOS '96: 12th International Conference on Analysis and Optimization of Systems - Images, Wavelets and PDEs:
Paris, June 26-28 1996
378 pp. 1996 [3-540-76076-8]

Vol. 220: Brogliato, B.
Nonsmooth Impact Mechanics: Models, Dynamics and Control
420 pp. 1996 [3-540-76079-2]

Vol. 221: Kelkar, A.; Joshi, S.
Control of Nonlinear Multibody Flexible Space Structures
160 pp. 1996 [3-540-76093-8]

Vol. 222: Morse, A.S.
Control Using Logic-Based Switching
288 pp. 1997 [3-540-76097-0]